3264 AND ALL THAT
A Second Course in Algebraic Geometry

The enumeration of solutions to systems of polynomial equations in several variables has been an active area of mathematics since the early work of Leibniz. In the 19th century, Chasles calculated that there are 3264 smooth conic plane curves tangent to five given general conics – a landmark in the field and perhaps the first important "excess intersection" problem.

Such computations in intersection theory were part of the motivation of Poincaré's development of topology, and also figured in Hilbert's Problems from 1900. Since then, intersection theory has become a topic of central importance in mathematics, with applications to topology, number theory and mathematical physics.

This book can form the basis of a second course in algebraic geometry. As motivation, it takes concrete problems from enumerative geometry and intersection theory. Its aim is to provide intuition and technique so that the student develops the ability to solve geometric problems.

The authors explain and illustrate key ideas such as rational equivalence, Chow rings, Grassmanians, Schubert calculus and Chern classes, excess intersection theory and the Grothendieck Riemann–Roch theorem. The geometric applications range from the 27 lines on a cubic surface through the existence of special divisors on Riemann surfaces.

Readers will appreciate the abundance of examples, many provided as exercises with solutions available online.

3264 AND ALL THAT

A Second Course in Algebraic Geometry

DAVID EISENBUD

University of California, Berkeley

JOE HARRIS

Harvard University

CAMBRIDGE
UNIVERSITY PRESS

CAMBRIDGE
UNIVERSITY PRESS

University Printing House, Cambridge CB2 8BS, United Kingdom

One Liberty Plaza, 20th Floor, New York, NY 10006, USA

477 Williamstown Road, Port Melbourne, VIC 3207, Australia

4843/24, 2nd Floor, Ansari Road, Daryaganj, Delhi - 110002, India

79 Anson Road, #06-04/06, Singapore 079906

Cambridge University Press is part of the University of Cambridge.

It furthers the University's mission by disseminating knowledge in the pursuit of education, learning and research at the highest international levels of excellence.

www.cambridge.org
Information on this title: www.cambridge.org/9781107017085

© David Eisenbud and Joe Harris 2016

First published 2016

A catalogue record for this publication is available from the British Library

ISBN 978-1-107-01708-5 Hardback
ISBN 978-1-107-60272-4 Paperback

Contents

Chapter 3   Introduction to Grassmannians and lines in $\mathbb{P}^3$

Chapter 13 Excess intersections and the Chow ring of a blow-up

Chapter 14 The Grothendieck Riemann–Roch theorem

Preface

We have been working on this project for over ten years, and at times we have felt that we have only brought on ourselves a plague of locus. However, our spirits have been lightened, and the project made far easier and more successful than it would have been, by the interest and help of many people.

First of all, we thank Bill Fulton, who created much of the modern approach to intersection theory, and who directly informed our view of the subject from the beginning.

Many people have helped us by reading early versions of the text and providing comments and corrections. Foremost among these is Paolo Aluffi, who gave extensive and detailed comments; we also benefited greatly from the advice of Francesco Cavazzani and Izzet Coşkun. We would also thank Mike Roth and Stephanie Yang, who provided notes on the early iterations of a course on which much of this text is based, as well as students who contributed corrections, including Sitan Chen, Jun Hou Fung, Changho Han, Chi-Yun Hsu, Hannah Larson, Ravi Jagadeesan, Aaron Landesman, Yogesh More, Arpon Raksit, Ashvin Swaminathan, Arnav Tripathy, Isabel Vogt and Lynnelle Ye.

Silvio Levy made many of the many illustrations in this book (and occasionally corrected our mathematical errors too!). Devlin Mallory then took over as copyeditor, and completed the rest of the figures. We are grateful to both of them for their many improvements to this text (and to Cambridge University Press for hiring Devlin!).

We are all familiar with the after-the-fact tone — weary, self-justificatory, aggrieved, apologetic — shared by ship captains appearing before boards of inquiry to explain how they came to run their vessels aground, and by authors composing forewords.

–John Lanchester

Chapter 0

Introduction

Why you want to read this book

Algebraic geometry is one of the central subjects of mathematics. All but the most analytic of number theorists speak our language, as do mathematical physicists, complex analysts, homotopy theorists, symplectic geometers, representation theorists.... How else could you get between such apparently disparate fields as topology and number theory in one hop, except via algebraic geometry?

And intersection theory is at the heart of algebraic geometry. From the very beginnings of the subject, the fact that the number of solutions to a system of polynomial equations is, in many circumstances, constant as we vary the coefficients of those polynomials has fascinated algebraic geometers. The distant extensions of this idea still drive the field forward.

At the outset of the 19th century, it was to extend this "preservation of number" that algebraic geometers made two important choices: to work over the complex numbers rather than the real numbers, and to work in projective space rather than affine space. (With these choices the two points of intersection of a line and an ellipse have somewhere to go as the ellipse moves away from the real points of the line, and the same for the point of intersection of two lines as the lines become parallel.) Over the course of the century, geometers refined the art of counting solutions to geometric problems — introducing the central notion of a *parameter space*, proposing the notions of an equivalence relation on cycles and a product on the equivalence classes and using these in many subtle calculations. These constructions were fundamental to the developing study of algebraic curves and surfaces.

In a different field, it was the search for a mathematically precise way of describing intersections that underlay Poincaré's study of what became algebraic topology. We owe Poincaré duality and a great deal more in algebraic topology directly to this search. The difficulties Poincaré encountered in working with continuous spaces (now called manifolds) led him to develop the idea of a simplicial complex as well.

Despite a lack of precise foundations, 19th century enumerative geometry rose to impressive heights: for example, Schubert, whose *Kalkül der abzählenden Geometrie* (originally published in 1879, and reprinted 100 years later in [1979]) represents the summit of intersection theory in the late 19th century, calculated the number of twisted cubics tangent to 12 quadrics — and got the right answer (5,819,539,783,680). Imagine landing a jumbo jet blindfolded!

At the outset of the 20th century, Hilbert made finding rigorous foundations for Schubert calculus one of his celebrated problems, and the quest to put intersection theory on a sound footing drove much of algebraic geometry for the following century; the search for a definition of multiplicity fueled the subject of commutative algebra in work of van der Waerden, Zariski, Samuel, Weil and Serre. This progress culminated, towards the end of the century, in the work of Fulton and MacPherson and then in Fulton's landmark book *Intersection theory* [1984], which both greatly extended the range of intersection theory and for the first time put the subject on a precise and rigorous foundation.

The development of intersection theory is far from finished. Today the focus includes virtual fundamental cycles, quantum intersection rings, Gromov–Witten theory and the extension of intersection theory from schemes to stacks. In a different direction, there are computer systems that can do many of the computations in this book and many more; see for example the package *Schubert2* in *Macaulay2* (Grayson and Stillman [2015]) and the library *Schubert* in SINGULAR (Decker et al. [2015]).

A central part of a central subject of mathematics — of course you would want to read this book!

Why we wrote this book

Given the centrality of the subject, it is not surprising how much of algebraic geometry one encounters in learning enumerative geometry. And that is how this book came to be written, and why: Like van der Waerden, we found that intersection theory makes for a great "second course" in algebraic geometry, weaving together threads from all over the subject. Moreover, the new ideas encountered in this setting are not merely more abstract definitions for the student to memorize, but tools that help answer concrete questions.

This is reflected in the organization of the contents. A good example of this is Chapter 6 ("Lines on hypersurfaces"). The stated goal of the chapter is to describe the class, in the Grassmannian $\mathbb{G}(1, n)$ of lines in $\mathbb{P}^n$, of the scheme $F_1(X) \subset \mathbb{G}(1, n)$ of lines lying on a given hypersurface $X \subset \mathbb{P}^n$, as an application of the new technique of Chern classes. But this raises a question: how can we characterize the scheme structure on $F_1(X)$, and what can we say about the geometry of this scheme? In short, this is an ideal time to introduce the notion of a *Hilbert scheme*, which gives a general framework for these questions; in the present setting, we can explicitly write down the equations defining $F_1(X)$, and prove theorems about its local geometry. In the end, a large part of the chapter is devoted to this discussion, which is as it should be: A reader may or may not have any use for the knowledge that a general quintic hypersurface $X \subset \mathbb{P}^4$ contains exactly 2875 lines, but a functional understanding of Hilbert schemes is a fundamental tool in algebraic geometry.

What's with the title?

The number in the title of this book is a reference to the solution of a classic problem in enumerative geometry: the determination, by Chasles, of the number of smooth conic plane curves tangent to five given general conics. The problem is emblematic of the dual nature of the subject. On the one hand, the number itself is of little significance: life would not be materially different if there were more or fewer. But the fact that the problem is well-posed — that there is a Zariski open subset of the space of 5-tuples $(C_1, \ldots, C_5)$ of conics for which the number of conics tangent to all five is constant, and that we can in fact determine that number — is at the heart of algebraic geometry. And the insights developed in the pursuit of a rigorous derivation of the number — the recognition of the need for, and the introduction of, a new parameter space for plane conics, and the understanding of why intersection products are well-defined for this space — are landmarks in the development of algebraic geometry.

The rest of the title is from "1066 & All That" by W. C. Sellar and R. J. Yeatman, a parody of English history textbooks; in many ways the number 3264 of conics tangent to five general conics is as emblematic of enumerative geometry as the date 1066 of the Battle of Hastings is of English history.

What is in this book

We are dealing here with a fundamental and almost paradoxical difficulty. Stated briefly, it is that learning is sequential but knowledge is not. A branch of mathematics [. . .] consists of an intricate network of interrelated facts, each of which contributes to the understanding of those around it. When confronted with this network for the first time, we are forced to follow a particular path, which involves a somewhat arbitrary ordering of the facts.

–Robert Osserman.

Where to begin? To start with the technical underpinnings of a subject risks losing the reader before the point of all the preliminary work is made clear, but to defer the logical foundations carries its own dangers — as the unproved assertions mount up, the reader may well feel adrift.

Intersection theory poses a particular challenge in this regard, since the development of its foundations is so demanding. It is possible, however, to state fairly simply and precisely the main foundational results of the subject, at least in the limited context of intersections on smooth projective varieties. The reader who is willing to take these results on faith for a little while, and accept this restriction, can then be shown what the subject is good for, in the form of examples and applications. This is the path we have chosen in this book, as we will now describe.

Overture

The first two chapters may be thought of as an overture to the subject, introducing the central themes that will play out in the remainder of the book. In the first chapter, we introduce rational equivalence, the Chow ring, the pullback and pushforward maps — the "dogma" of the subject. (In regard to the existence of an intersection product and pullback maps, we do not give proofs; instead, we refer the reader to Fulton [1984].) We follow this in the second chapter with a range of simple examples to give the reader a sense of the themes to come: the computation of Chow rings of affine and projective spaces, their products and (some) blow-ups. To illustrate how intersection theory is used in algebraic geometry, we examine loci of various types of singular cubic plane curves, thought of as subvarieties of the projective space $\mathbb{P}^9$ parametrizing plane cubics. Finally, we briefly discuss intersection products of curves on surfaces, an important early example of the subject.

Grassmannians

The intersection rings of the Grassmannians are archetypal examples of intersection theory. Chapters 3 and 4 are devoted to them and their underlying geometry. Here we introduce *Schubert cycles*, whose classes form a basis for the Chow ring, and use them to solve a number of geometric problems, illustrating again how intersection theory is used to solve enumerative problems.

Chern classes

We then come to a watershed in the subject. Chapter 5 takes up in earnest a notion at the center of modern intersection theory, and indeed of modern algebraic geometry: Chern classes. As with the development of intersection theory, we focus on the classical characterization of Chern classes as degeneracy loci of collections of sections. This interpretation provides useful intuition and is basic to many applications of the theory.

Applications, I: Using the tools

We illustrate the use of Chern classes by taking up two classical problems: Chapter 6 deals with the question of how many lines lie on a hypersurface (for example, the fact that there are exactly 27 lines on each smooth cubic surface and 2875 lines on a general quintic threefold), and Chapter 7 looks at the singular hypersurfaces in a one-dimensional family (for example, the fact that a general pencil of plane curves of degree d has $3(d-1)^2$ singular elements). Using the basic technique of *linearization*, these problems can be translated into problems of computing Chern classes. These and the next few chapters are organized around geometric problems involving constructions of useful vector bundles and the calculation of their Chern classes.

Parameter spaces

Chapter 8 concerns an area in which intersection theory has had a profound influence on modern algebraic geometry: *parameter spaces* and their compactifications. This is illustrated with the five conic problem; there is also a discussion of the modern example of Kontsevich spaces, and an application of these.

Applications, II: Further developments

The remainder of the book introduces a series of increasingly advanced topics. Chapters 9, 10 and 11 deal with a situation ubiquitous in the subject, the intersection theory of projective bundles, and its applications to subjects such as projective duality and the enumerative geometry of contact conditions.

Chern classes are defined in terms of the loci where collections of sections of a vector bundle become dependent. These can be interpreted as loci where maps from trivial vector bundles drop rank. The Porteous formula, proved and applied in Chapter 12, generalizes this, expressing the classes of the loci where a map between two general vector bundles has a given rank or less in terms of the Chern classes of the two bundles involved.

Advanced topics

Next, we come to some of the developments of the modern theory of intersections. In Chapter 13, we introduce the notion of "excess" intersections and the *excess intersection formula*, one of the subjects that was particularly mysterious in the 19th century but elucidated by Fulton and MacPherson. This theory makes it possible to describe the intersection class of two cycles, even if the dimension of their intersection is "too large." Central to this development is the idea of *specialization to the normal cone*, a construction fundamental to the work of Fulton and MacPherson; we use this to prove the famous "key formula" comparing intersections of cycles in a subvariety $Z \subset X$ to the intersections of those cycles in X, and use this in turn to give a description of the Chow ring of a blow-up.

Chapter 14 contains an account of Riemann–Roch formulas, leading up to a description of Grothendieck's version. The chapter concludes with a number of examples and applications showing how Grothendieck's formula can be used.

Appendices

The moving lemma

The literature contains a number of papers proving various parts of the moving lemma (see below for a statement). We give a careful proof of the first half of the lemma in Appendix A.

Cohomology and base change

Many results in this book will be proved by constructing an appropriate vector bundle and computing its Chern classes. The theorem on cohomology and base change (Theorem B.5) is a key tool in these constructions: We use it to show that, under appropriate hypotheses, the direct image of a sheaf is a vector bundle. We present a complete discussion of this important result in Appendix B.

Topology of algebraic varieties

When we treat algebraic varieties over an arbitrary field we use the Zariski topology, where an open set is defined as the locus where a polynomial function takes nonzero values. But if the ground field is the complex numbers, we can also use the "classical" topology: With this topology, a smooth projective variety over $\mathbb{C}$ is a compact, complex manifold, and tools like singular homology can help us study its geometry. Appendix C explains some of what is known in this direction, and also compares some of the possible substitutes for the Chow ring.

The Brill–Noether theorem

Appendix D explains an application of enumerative geometry to a problem that is central in the study of algebraic curves and their moduli spaces: the existence of special linear series on curves. We give the Kempf/Kleiman–Laksov proof of this theorem, which draws upon many of the ideas and techniques of the book, plus a new one: the use of topological cohomology in the context of intersection theory. This is also a wonderful illustration of the way in which enumerative geometry can be the essential ingredient in the proof of a purely qualitative result.

Relation of this book to *Intersection theory*

Fulton's book *Intersection theory* [1984] is a great work. It sets up for the first time a rigorous framework for intersection theory, and does so in a generality significantly extending and refining what was known before and laying out an enormous number of applications. It stands as an encyclopedic reference for the subject.

By contrast, the present volume is intended as a textbook in algebraic geometry, a second course, in which the classical side of intersection theory is a starting point for exploring many topics in geometry. We describe the intersection product at the outset, but do not attempt to give a rigorous proof of its existence, focusing instead on basic examples. We use concrete problems to motivate the introduction of new tools from all over algebraic geometry. Our book is not a substitute for Fulton's; it has a different aim. We do hope that it will provide the reader with intuition and motivation that will make reading Fulton's book easier.

Existence of the intersection product

The *moving lemma* was for most of a century the foundation on which intersection theory was supposed to rest. It has two parts:

(a) Given classes $\alpha, \beta \in A(X)$ in the Chow group of a smooth, projective variety X, we can find representative cycles A and B intersecting generically transversely.

(b) The class of the intersection of these cycles is independent of the choice of A and B.

Using these assertions it is easy to define the intersection product on the Chow groups of a smooth variety: $\alpha\beta$ is defined to be the class of $A \cap B$, where A and B are cycles representing the classes α and β and intersecting generically transversely, and this is how intersection products were defined. The problem is that, while the first part can be and was proved rigorously, as far as we know there was prior to the publication of Fulton's book in 1984 no complete proof of the second part. Of course, part (b) is an immediate consequence of the existence of a well-defined intersection product (Fulton [1984, Section 8.3]), and so we refer the reader to Fulton's book for this key existence result.

Nonetheless, we feel that part (a) of the moving lemma is useful in shaping one's intuition about intersection products. Moreover, given the existence statement, part (a) of the moving lemma allows simpler and more intuitive proofs of a number of the basic assertions of the theory, and we will use it in that way. We therefore give a proof of part (a) in Appendix A, following Severi's ideas.

Keynote problems

To highlight the sort of problems we will learn to solve, and to motivate the material we present, we will begin each chapter with some *keynote questions*.

Exercises

One of the wonderful things about the subject of enumerative geometry is the abundance of illuminating examples that are accessible to explicit computation. We have included many of these as exercises. We have been greatly aided by Francesco Cavazzani; in particular, he has prepared solutions, which appear on a web site associated to this book.

Prerequisites, notation and conventions

What you need to know before starting

When it comes to prerequisites, there are two distinct questions: what you should know to start reading this book; and what you should be prepared to learn along the way.

Of these, the second is by far the more important. In the course of developing and applying intersection theory, we introduce many key techniques of algebraic geometry, such as deformation theory, specialization methods, characteristic classes, Hilbert schemes, commutative and homological algebra and topological methods. That is not to say that you need to know these things going in. Just the opposite, in fact: Reading this book is an occasion to learn them.

So what do you need before starting?

(a) An undergraduate course in classical algebraic geometry or its equivalent, comprising the elementary theory of affine and projective varieties. *An invitation to algebraic geometry* (Smith et al. [2000]) contains almost everything required. Other books that cover this material include *Undergraduate algebraic geometry* (Reid [1988]), *Introduction to algebraic geometry* (Hassett [2007]), *Elementary algebraic geometry* (Hulek [2012]) and, at a somewhat more advanced level, *Algebraic geometry, I: Complex projective varieties* (Mumford [1976]), *Basic algebraic geometry, I* (Shafarevich [1994]) and *Algebraic geometry: a first course* (Harris [1995]). The last three include much more than we will use here.

(b) An acquaintance with the language of schemes. This would be amply covered by the first three chapters of *The geometry of schemes* (Eisenbud and Harris [2000]).

(c) An acquaintance with coherent sheaves and their cohomology. For this, *Faisceax algébriques cohérents* (Serre [1955]) remains an excellent source (it is written in the language of varieties, but applies nearly word-for-word to projective schemes over a field, the context in which this book is written).

In particular, *Algebraic geometry* (Hartshorne [1977]) contains much more than you need to know to get started.

Language

Throughout this book, a *scheme* X will be a separated scheme of finite type over an algebraically closed field $\Bbbk$ of characteristic 0. (We will occasionally point out the ways in which the characteristic p situation differs from that of characteristic 0, and how we might modify our statements and proofs in that setting.) In practice, all the schemes considered will be quasi-projective. We use the term *integral* to mean reduced and irreducible; by a *variety* we will mean an integral scheme. (The terms "curve" and "surface," however, refer to one-dimensional and two-dimensional schemes; in particular, they are not presumed to be integral.) A subvariety $Y \subset X$ will be presumed closed unless otherwise specified. If X is a variety we write $\Bbbk(X)$ for the field of rational functions on X. A *sheaf* on X will be a coherent sheaf unless otherwise noted.

By a *point* we mean a closed point. Recall that a *locally closed* subscheme U of a scheme X is a scheme that is an open subset of a closed subscheme of X. We use the term "subscheme" (without any modifier) to mean a closed subscheme, and similarly for "subvariety."

A consequence of the finite-type hypothesis is that any subscheme Y of X has a *primary decomposition*: locally, we can write the ideal of Y as an irredundant intersection of primary ideals with distinct associated primes. We can correspondingly write Y globally as an irredundant union of closed subschemes Y_i whose supports are distinct subvarieties of X. In this expression, the subschemes Y_i whose supports are maximal — corresponding to the minimal primes in the primary decomposition — are uniquely determined by Y; they are called the *irreducible components* of Y. The remaining subschemes are called *embedded components*; they are not determined by Y, though their supports are.

If a family of objects is parametrized by a scheme B, we will say that a "general" member of the family has a given property P if the set $U(P) \subset B$ of members of the family with that property contains an open dense subset of B. When we say that a "very general" member has this property we will mean that $U(P)$ contains the complement of a countable union of proper subvarieties of B.

By the *projectivization* of a vector space V, denoted $\mathbb{P}V$, we will mean the scheme $\mathrm{Proj}(\mathrm{Sym}\, V^*)$ (where by $\mathrm{Sym}\, V$ we mean the symmetric algebra of V); this is the space whose closed points correspond to one-dimensional subspaces of V. This is opposite to the usage in, for example, Grothendieck and Hartshorne, where the points of $\mathbb{P}V$ correspond to one-dimensional quotients of V (that is, their $\mathbb{P}V$ is our $\mathbb{P}V^*$), but is in agreement with Fulton.

If X and $Y \subset \mathbb{P}^n$ are subvarieties of projective space, we define the *join* of X and Y, denoted $\overline{X, Y}$, to be the closure of the union of lines meeting X and Y at distinct points. If $X = \Gamma \subset \mathbb{P}^n$ is a linear space, this is just the cone over Y with vertex Γ; if X and Y are both linear subspaces, this is simply their span.

There is a one-to-one correspondence between vector bundles on a scheme X and locally free sheaves on X. We will use the terms interchangeably, generally preferring "line bundle" and "vector bundle" to "invertible sheaf" and "locally free sheaf." When we speak of the *fiber* of a vector bundle $\mathcal{F}$ on X at a point $p \in X$, we will mean the (finite-dimensional) vector space $\mathcal{F} \otimes \kappa(p)$, where $\kappa(p)$ is the residue field at p.

By a *linear system*, or *linear series*, on a scheme X, we will mean a pair $\mathcal{D} = (\mathcal{L}, V)$, where $\mathcal{L}$ is a line bundle on X and $V \subset H^0(\mathcal{L})$ a vector space of sections. Associating to a section $\sigma \in V \subset H^0(\mathcal{L})$ its zero locus $V(\sigma)$, we can also think of a linear system as a family $\{V(\sigma) \,|\, \sigma \in V\}$ of divisors $D \subset X$ parametrized by the projective space $\mathbb{P}V$; in this setting, we will sometimes abuse notation slightly and write $D \in \mathcal{D}$. By the *dimension* of the linear series we mean the dimension of the projective space $\mathbb{P}V$ parametrizing it, that is, $\dim V - 1$. Specifically, a one-dimensional linear system is called a *pencil*, a two-dimensional system is called a *net* and a three-dimensional linear system is called a *web*.

We write $\mathcal{O}_{X,Y}$ for the local ring of X along Y, and, more generally, if $\mathcal{F}$ is a sheaf of $\mathcal{O}_X$-modules we write $\mathcal{F}_Y$ for the corresponding $\mathcal{O}_{X,Y}$-module.

We can identify the Zariski tangent space to the affine space $\mathbb{A}^n$ with $\mathbb{A}^n$ itself. If $X \subset \mathbb{A}^n$ is a subscheme, by the *affine tangent space* to X at a point p we will mean the affine linear subspace $p + T_p X \subset \mathbb{A}^n$. If $X \subset \mathbb{P}^n$ is a subscheme, by the *projective tangent space* to X at $p \in X$, denoted $\mathbb{T}_p X \subset \mathbb{P}^n$, we will mean the closure in $\mathbb{P}^n$ of the affine tangent space to $X \cap \mathbb{A}^n$ for any open subset $\mathbb{A}^n \subset \mathbb{P}^n$ containing p. Concretely, if X is the zero locus of polynomials F_α (that is, $X = V(I) \subset \mathbb{P}^n$ is the subscheme defined by the ideal $I = (\{F_\alpha\}) \subset \Bbbk[Z_0, \ldots, Z_n]$), the projective tangent space is the common zero locus of the linear forms

$$L_\alpha(Z) = \frac{\partial F_\alpha}{\partial Z_0}(p)Z_0 + \cdots + \frac{\partial F_\alpha}{\partial Z_n}(p)Z_n.$$

By a *one-parameter family* we will always mean a family $\mathcal{X} \to B$ with B smooth and one-dimensional (an open subset of a smooth curve, or spec of a DVR or power series ring in one variable), with marked point $0 \in B$. In this context, "with parameter t" means t is a local coordinate on the curve, or a generator of the maximal ideal of the DVR or power series ring.

Basic results on dimension and smoothness

There are a number of theorems in algebraic geometry that we will use repeatedly; we give the statements and references here. When X is a scheme, by the *dimension* of X we mean the Krull dimension, denoted $\dim X$. If X is an irreducible variety and $Y \subset X$ is a subvariety, then the *codimension* of Y in X, written $\mathrm{codim}_X Y$ (or simply $\mathrm{codim}\, Y$ when X is clear from context), is $\dim X - \dim Y$; more generally, if X is any scheme and Y is a subvariety, then $\mathrm{codim}_X Y$ denotes the minimum of

$$\{\mathrm{codim}_{X'} Y \mid X' \text{ is a reduced irreducible component of } X\}.$$

More on dimension and codimension can be found in Eisenbud [1995].

We will often use the following basic result of commutative algebra:

Theorem 0.1 (Krull's principal ideal theorem). *An ideal generated by n elements in a Noetherian ring has codimension $\leq n$.*

See Eisenbud [1995, Theorem 10.2] for a discussion and proof. We will also use the following important extension of the principal ideal theorem:

Theorem 0.2 (Generalized principal ideal theorem). *If $f : Y \to X$ is a morphism of varieties and X is smooth, then, for any subvariety $A \subset X$,*

$$\mathrm{codim}\, f^{-1} A \leq \mathrm{codim}\, A.$$

In particular, if A, B are subvarieties of X, and C is an irreducible component of $A \cap B$, then $\mathrm{codim}\, C \leq \mathrm{codim}\, A + \mathrm{codim}\, B$.

The proof of this result can be reduced to the case of an intersection of two subvarieties, one of which is locally a complete intersection, by expressing the inverse image $f^{-1} A$ as an intersection with the graph $\Gamma_f \subset X \times Y$ of f. In this form it follows from Krull's theorem. The result holds in greater generality; see Serre [2000, Theorem V.3]. Smoothness is necessary for this (Example 2.22).

A module M is said to be of finite length if it has a finite maximal sequence of submodules. Such a sequence is called a *composition series*, and we will call the length of the sequence the *length* of the module. The following theorem shows this length is well-defined:

Theorem 0.3 (Jordan–Hölder theorem). *A module M of finite length over a commutative local ring R has a maximal sequence of submodules $M \supsetneq M_1 \supsetneq \cdots \supsetneq M_k = 0$ Moreover, any two such maximal sequences are isomorphic; that is, they have the same length and composition factors (up to isomorphism).*

Theorem 0.4 (Chinese remainder theorem). *A module of finite length over a commutative ring is the direct sum of its localizations at finitely many maximal ideals.*

For discussion and proof see Eisenbud [1995, Chapter 2], especially Theorem 2.13.

Theorem 0.5 (Bertini). *If $\mathcal{D}$ is a linear system on a variety X in characteristic 0, the general member of $\mathcal{D}$ is smooth outside the base locus of $\mathcal{D}$ and the singular locus of X.*

Note that applying Bertini repeatedly, we see as well that if $D_1, \ldots, D_k$ are general members of the linear system $\mathcal{D}$ then the intersection $\bigcap D_i$ is smooth of dimension $\dim X - k$ away from the base locus of $\mathcal{D}$ and the singular locus of X.

    This is the form in which we will usually apply Bertini. But there is another version that is equivalent in characteristic 0 but allows for an extension to positive characteristic:

Theorem 0.6 (Bertini). *If $f : X \to \mathbb{P}^n$ is any generically separated morphism from a smooth, quasi-projective variety X to projective space, then the preimage $f^{-1}(H)$ of a general hyperplane $H \subset \mathbb{P}^n$ is smooth.*

Chapter 1

Introducing the Chow ring

Keynote Questions

As we indicated in the introduction, we will preface each chapter of this book with a series of "keynote questions:" examples of the sort of concrete problems that can be solved using the ideas and techniques introduced in that chapter. In general, the answers to these questions will be found in the same chapter. In the present case, we will not develop our roster of examples sufficiently to answer the keynote questions below until the second chapter; we include them here so that the reader can have some idea of "what the subject is good for" in advance.

(1) Let F_0, F_1 and $F_2 \in \Bbbk[X, Y, Z]$ be three general homogeneous cubic polynomials in three variables. Up to scalars, how many linear combinations $t_0 F_0 + t_1 F_1 + t_2 F_2$ factor as a product of a linear and a quadratic polynomial? (Answer on page 65.)

(2) Let F_0, F_1, F_2 and $F_3 \in \Bbbk[X, Y, Z]$ be four general homogeneous cubic polynomials in three variables. How many linear combinations $t_0 F_0 + t_1 F_1 + t_2 F_2 + t_3 F_3$ factor as a product of three linear polynomials? (Answer on page 65.)

(3) If A, B, C are general homogeneous quadratic polynomials in three variables, for how many triples $t = (t_0, t_1, t_2)$ do we have

$$(A(t), B(t), C(t)) = (t_0, t_1, t_2)?$$

(Answer on page 55.)

(4) Let $S \subset \mathbb{P}^3$ be a smooth cubic surface and $L \subset \mathbb{P}^3$ a general line. How many planes containing L are tangent to S? (Answer on page 50.)

(5) Let $L \subset \mathbb{P}^3$ be a line, and let S and $T \subset \mathbb{P}^3$ be surfaces of degrees s and t containing L. Suppose that the intersection $S \cap T$ is the union of L and a smooth curve C. What are the degree and genus of C? (Answer on page 71.)

1.1 The goal of intersection theory

Though intersection theory has many and surprising applications, in its most basic form it gives information about the intersection of two subvarieties of a given variety. An early incarnation, and in some sense the model for all of intersection theory, is the theorem of Bézout: If plane curves $A, B \subset \mathbb{P}^2$ intersect transversely, then they intersect in $(\deg A)(\deg B)$ points (see Figure 1.3 on page 18).

If A is a line, this is a special case of Gauss' fundamental theorem of algebra: A polynomial $f(x)$ in one complex variable has $\deg f$ roots, if the roots are counted with multiplicity. Late in the 19th century it was understood how to attribute multiplicities to the intersections of any two plane curves without common components (we shall describe this in Section 1.3.7 below), so Bézout's theorem could be extended: The intersection of two plane curves without common components consists a collection of points with multiplicities adding up to $(\deg A)(\deg B)$.

In modern geometry we need to understand intersections of subvarieties in much greater generality. In this book we will mostly consider intersections of arbitrary subvarieties in a smooth ambient variety X. The goal of this chapter is to introduce a ring $A(X)$, called the *Chow ring* of X, and to associate to every subscheme $A \subset X$ a class $[A]$ in $A(X)$ generalizing the degree of a curve in $\mathbb{P}^2$. In Section 1.3.7 we will explain a far-reaching extension of Bézout's theorem:

Theorem 1.1 (Bézout's theorem for dimensionally transverse intersections). *If $A, B \subset X$ are subvarieties of a smooth variety X and $\operatorname{codim}(A \cap B) = \operatorname{codim} A + \operatorname{codim} B$, then we can associate to each irreducible component C_i of $A \cap B$ a positive integer $m_{C_i}(A, B)$ in such a way that*

$$[A][B] = \sum m_{C_i}(A, B) \cdot [C_i].$$

The integer $m_{C_i}(A, B)$ is called the *intersection multiplicity of A and B along C_i*; giving a correct definition in this generality occupied algebraic geometers for most of the first half of the 20th century.

Though Theorem 1.1 is restricted to the case where the subvarieties A, B meet only in codimension $\operatorname{codim} A + \operatorname{codim} B$ (the case of *dimensionally proper intersection*), there is a very useful extension to the case where the codimensions of the components of the intersection are arbitrary; this will be discussed in Chapter 13.

Many important applications involve subvarieties defined as zero loci of sections of a vector bundle $\mathcal{E}$ on a variety X, and this idea has potent generalizations. It turns out that there is a way of defining classes $c_i(\mathcal{E}) \in A(X)$, called the *Chern classes* of $\mathcal{E}$, and the theory of Chern classes is a pillar of intersection theory. The third and final section of this chapter takes up a special case of the general theory that is of particular importance and relatively easy to describe: the first Chern class of a line bundle. This allows us to introduce the *canonical class*, a distinguished element of the Chow ring of any smooth variety, and show how to calculate it in simple cases. The general theory of Chern classes will be taken up in Chapter 5.

1.2 The Chow ring

We now turn to the definition and basic properties of the Chow ring. Then we introduce excision and Mayer–Vietoris theorems that allow us to calculate the Chow rings of many varieties. Most importantly we describe the functoriality of the Chow ring: the existence, under suitable circumstances, of pushforward and pullback maps.

Chow groups form a sort of homology theory for quasi-projective varieties; that is, they are abelian groups associated to a geometric object that are described as a group of cycles modulo an equivalence relation. In the case of a smooth variety, the intersection product makes the Chow groups into a graded ring, the Chow ring. This is analogous to the ring structure on the homology of smooth compact manifolds that can be imported, using Poincaré duality, from the natural ring structure on cohomology.

Throughout this book we will work over an algebraically closed ground field $\Bbbk$ of characteristic 0. Virtually everything we do could be formulated over arbitrary fields (though not every statement remains true in characteristic p), and occasionally we comment on how one would do this.

1.2.1 Cycles

Let X be any algebraic variety (or, more generally, scheme). The *group of cycles* on X, denoted $Z(X)$, is the free abelian group generated by the set of subvarieties (reduced irreducible subschemes) of X. The group $Z(X)$ is graded by dimension: we write $Z_k(X)$ for the group of cycles that are formal linear combinations of subvarieties of dimension k (these are called *k-cycles*), so that $Z(X) = \bigoplus_k Z_k(X)$. A cycle $Z = \sum n_i Y_i$, where the Y_i are subvarieties, is *effective* if the coefficients n_i are all nonnegative. A *divisor* (sometimes called a *Weil divisor*) is an $(n-1)$-cycle on a pure n-dimensional scheme. It follows from the definition that $Z(X) = Z(X_{\mathrm{red}})$; that is, $Z(X)$ is insensitive to whatever nonreduced structure X may have.

To any closed subscheme $Y \subset X$ we associate an effective cycle $\langle Y \rangle$: If $Y \subset X$ is a subscheme, and $Y_1, \ldots, Y_s$ are the irreducible components of the reduced scheme Y_{red}, then, because our schemes are Noetherian, each local ring $\mathcal{O}_{Y,Y_i}$ has a finite composition series. Writing l_i for its length, which is well-defined by the Jordan–Hölder theorem (Theorem 0.3), we define the cycle $\langle Y \rangle$ to be the formal combination $\sum l_i Y_i$. (The coefficient l_i is called the *multiplicity* of the scheme Y along the irreducible component Y_i, and written $\mathrm{mult}_{Y_i}(Y)$; we will discuss this notion, and its relation to the notion of intersection multiplicity, in Section 1.3.8.)

In this sense cycles may be viewed as coarse approximations to subschemes.

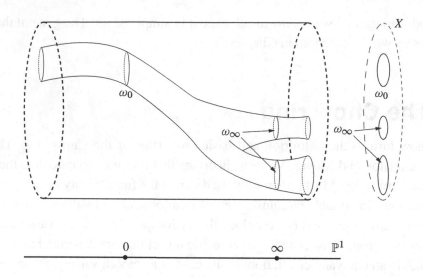

Figure 1.1 Rational equivalence between two cycles ω_0 and ω_∞ on X.

1.2.2 Rational equivalence and the Chow group

The *Chow group* of X is the group of cycles of X modulo *rational equivalence*. Informally, two cycles $A_0, A_1 \in Z(X)$ are rationally equivalent if there is a rationally parametrized family of cycles interpolating between them — that is, a cycle on $\mathbb{P}^1 \times X$ whose restrictions to two fibers $\{t_0\} \times X$ and $\{t_1\} \times X$ are A_0 and A_1. Here is the formal definition:

Definition 1.2. Let $\mathrm{Rat}(X) \subset Z(X)$ be the subgroup generated by differences of the form

$$\langle \Phi \cap (\{t_0\} \times X) \rangle - \langle \Phi \cap (\{t_1\} \times X) \rangle,$$

where $t_0, t_1 \in \mathbb{P}^1$ and Φ is a subvariety of $\mathbb{P}^1 \times X$ not contained in any fiber $\{t\} \times X$. We say that two cycles are *rationally equivalent* if their difference is in $\mathrm{Rat}(X)$, and we say that two subschemes are rationally equivalent if their associated cycles are rationally equivalent — see Figures 1.1 and 1.2.

Definition 1.3. The *Chow group* of X is the quotient

$$A(X) = Z(X)/\mathrm{Rat}(X),$$

the *group of rational equivalence classes of cycles on X*. If $Y \in Z(X)$ is a cycle, we write $[Y] \in A(X)$ for its equivalence class; if $Y \subset X$ is a subscheme, we abuse notation slightly and denote simply by $[Y]$ the class of the cycle $\langle Y \rangle$ associated to Y.

It follows from the principal ideal theorem (Theorem 0.1) that the Chow group is graded by dimension:

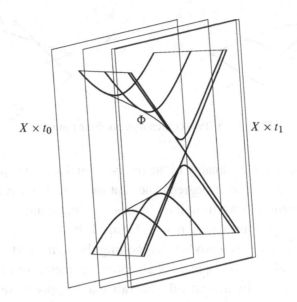

Figure 1.2 Rational equivalence between a hyperbola and the union of two lines in $\mathbb{P}^2$.

Proposition 1.4. *If X is a scheme then the Chow group of X is graded by dimension; that is,*

$$A(X) = \bigoplus A_k(X),$$

with $A_k(X)$ the group of rational equivalence classes of k-cycles.

Proof: If $\Phi \subset \mathbb{P}^1 \times X$ is an irreducible variety not contained in a fiber over X then, in an appropriate affine open set $\Phi \cap (\mathbb{A}^1 \times X) \subset \Phi$, the scheme $\Phi \cap (\{t_0\} \times X)$ is defined by the vanishing of the single nonzerodivisor $t - t_0$. It follows that the components of this intersection are all of codimension exactly 1 in Φ, and similarly for $\Phi \cap (\{t_1\} \times X)$. Thus all the varieties involved in the rational equivalence defined by Φ have the same dimension. $\square$

When X is equidimensional we may define the *codimension* of a subvariety $Y \subset X$ as $\dim X - \dim Y$, and it follows that we may also grade the Chow group by codimension. When X is also smooth, we will write $A^c(X)$ for the group $A_{\dim X - c}$, and think of it as the group of codimension-c cycles, modulo rational equivalence. (It would occasionally be convenient to adopt the same notation when X is singular, but this would conflict with established convention — see the discussion in Section 2.5 below.)

1.2.3 Transversality and the Chow ring

We said at the outset that much of what we hope to do in intersection theory is modeled on the classical Bézout theorem: that if plane curves $A, B \subset \mathbb{P}^2$ of degrees d and e intersect transversely then they intersect in de points. Two things about this

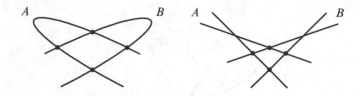

Figure 1.3 Two conics meet in four points.

result are striking. First, the cardinality of the intersection does not depend on the choice of curves, beyond knowing their degrees and that they meet transversely. Given this invariance, the theorem follows from the obvious fact that a union of d general lines meets a union of e general lines in de points (Figure 1.3).

Second, the answer, de, is a product, suggesting that some sort of ring structure is present. A great deal of the development of algebraic geometry over the past 200 years is bound up in the attempt to understand, generalize and apply these ideas, leading to precise notions of the sense in which intersection of subvarieties resembles multiplication. What makes the Chow groups useful is that, under good circumstances, the rational equivalence class of the intersection of two subvarieties A, B depends only on the rational equivalence classes of A and B, and this gives a product structure on the Chow groups of a smooth variety.

To make this statement precise we need some definitions. We say that subvarieties A, B of a variety X intersect *transversely* at a point p if A, B and X are all smooth at p and the tangent spaces to A and B at p together span the tangent space to X; that is,

$$T_p A + T_p B = T_p X,$$

or equivalently

$$\mathrm{codim}(T_p A \cap T_p B) = \mathrm{codim}\, T_p A + \mathrm{codim}\, T_p B.$$

We will say that subvarieties $A, B \subset X$ are *generically transverse*, or that they intersect *generically transversely*, if they meet transversely at a general point of each component C of $A \cap B$. The terminology is justified by the fact that the set of points of $A \cap B$ at which A and B are transverse is open. We extend the terminology to cycles by saying that two cycles $A = \sum n_i A_i$ and $B = \sum m_j B_j$ are generically transverse if each A_i is generically transverse to each B_j.

More generally, we will say subvarieties $A_i \subset X$ intersect transversely at a smooth point $p \in X$ if p is a smooth point on each A_i and $\mathrm{codim}\big(\bigcap T_p A_i\big) = \sum \mathrm{codim}\, T_p A_i$, and we say that they intersect generically transversely if there is a dense set of points in the intersection at which they are transverse.

As an example, if A and B have complementary dimensions in X (that is, if $\dim A + \dim B = \dim X$), then A and B are generically transverse if and only if they are transverse everywhere; that is, their intersection consists of finitely many points and they intersect transversely at each of them. (In this case we will accordingly drop the modifier "generically.") If codim A + codim B > dim X, then A and B are generically transverse if and only if they are disjoint.

Theorem–Definition 1.5. *If X is a smooth quasi-projective variety, then there is a unique product structure on $A(X)$ satisfying the condition:*

(∗) If two subvarieties A, B of X are generically transverse, then

$$[A][B] = [A \cap B].$$

This structure makes

$$A(X) = \bigoplus_{c=0}^{\dim X} A^c(X)$$

into an associative, commutative ring, graded by codimension, called the Chow ring *of X.*

Fulton [1984] gave a direct construction of the product of cycles on any smooth variety over any field, and proved that the products of rationally equivalent cycles are rationally equivalent. In a setting where the first half of the moving lemma (Theorem 1.6 below) holds, such as a smooth, quasi-projective variety over an algebraically closed field, this product is characterized by the condition (∗) of Theorem–Definition 1.5.

Even if X is smooth and A, B are subvarieties such that every component of $A \cap B$ has the expected codimension codim A + codim B, we cannot define $[A][B] \in A(X)$ to be $[A \cap B]$, because the class $[A \cap B]$ depends on more than the rational equivalence classes of A and B. This problem can be solved by assigning intersection multiplicities to the components; see Section 1.3.7.

1.2.4 The moving lemma

Historically, the proof of Theorem–Definition 1.5 was based on the *moving lemma*. This has two parts:

Theorem 1.6 (Moving lemma). *Let X be a smooth quasi-projective variety.*

(a) *For every $\alpha, \beta \in A(X)$ there are generically transverse cycles $A, B \in Z(X)$ with $[A] = \alpha$ and $[B] = \beta$.*
(b) *The class $[A \cap B]$ is independent of the choice of such cycles A and B.*

A proof of the first part is given in Appendix A; this is sufficient to establish the uniqueness of a ring structure on $A(X)$ satisfying the condition (∗) of Theorem–Definition 1.5.

The second part, which historically was used to prove the existence portion of Theorem–Definition 1.5, is more problematic; as far as we know, no complete proof existed prior to the publication of Fulton [1984].

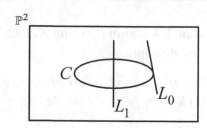

Figure 1.4 The cycle L_0 can be "moved" to the rationally equivalent cycle L_1, which is transverse to the given subvariety C.

The first half of the moving lemma is useful in shaping our understanding of intersection products and occasionally as a tool in the proof of assertions about them, and we will refer to it when relevant.

On a singular variety the moving lemma may fail: For example, if $X \subset \mathbb{P}^3$ is a quadric cone then any two cycles representing the class of a line of X meet at the origin, a singular point of X, and thus cannot be generically transverse (see Exercise 1.36). Further, the hypothesis of smoothness in Theorem 1.5 cannot be avoided: We will also see in Section 2.5 examples of varieties X where no intersection product satisfying the basic condition $(*)$ of Theorem 1.5 can be defined. The news is not uniformly negative: Intersection products *can* be defined on singular varieties if we impose some restrictions on the classes involved, as we will see in Proposition 1.31.

Kleiman's transversality theorem

There is one circumstance in which the first half of the moving lemma is relatively easy: when a sufficiently large group of automorphisms acts on X, we can use automorphisms to move cycles to make them transverse. Here is a special case of a result of Kleiman:

Theorem 1.7 (Kleiman's theorem in characteristic 0). *Suppose that an algebraic group G acts transitively on a variety X over an algebraically closed field of characteristic 0, and that $A \subset X$ is a subvariety.*

(a) *If $B \subset X$ is another subvariety, then there is an open dense set of $g \in G$ such that gA is generically transverse to B.*

(b) *More generally, if $\varphi : Y \to X$ is a morphism of varieties, then for general $g \in G$ the preimage $\varphi^{-1}(gA)$ is generically reduced and of the same codimension as A.*

(c) *If G is affine, then $[gA] = [A] \in A(X)$ for any $g \in G$.*

Proof: (a) This is the special case $Y = B$ of (b).

(b) Let the dimensions of X, A, Y and G be n, a, b and m respectively. If $x \in X$, then the map $G \to X : g \mapsto gx$ is surjective and its fibers are the cosets of the stabilizer of x in G. Since all these fibers have the same dimension, this dimension must be $m - n$. Set

$$\Gamma = \{(x, y, g) \in A \times Y \times G \mid gx = \varphi(y)\}.$$

Because G acts transitively on X, the projection $\pi : \Gamma \to A \times Y$ is surjective. Its fibers are the cosets of stabilizers of points in X, and hence have dimension $m - n$. It follows that Γ has dimension

$$\dim \Gamma = a + b + m - n.$$

On the other hand, the fiber over g of the projection $\Gamma \to G$ is isomorphic to $\varphi^{-1}(gA)$. Thus either this intersection is empty for general g, or else it has dimension $a + b - n$, as required.

Since X is a variety it is smooth at a general point. Since G acts transitively, all points of X look alike, so X is smooth. Since any algebraic group in characteristic 0 is smooth (see for example Lecture 25 of Mumford [1966]), the fibers of the projection to $A \times Y$ are also smooth, so Γ itself is smooth over $A_{sm} \times Y_{sm}$. Since field extensions in characteristic 0 are separable, the projection $(\Gamma \setminus \Gamma_{sing}) \to G$ is smooth over a nonempty open set of G, where Γ_{sing} is the singular locus of Γ. That is, the general fiber of the projection of Γ to G is smooth outside Γ_{sing}. If the projection of Γ_{sing} to G is not dominant, then $\varphi^{-1}(gA)$ is smooth for general g.

To complete the proof of generic transversality, we may assume that the projection $\Gamma_{sing} \to G$ is dominant. Since G is smooth, the principal ideal theorem shows that every component of every fiber of $\Gamma \to G$ has codimension $\leq \dim G$, and thus every component of the general fiber has codimension exactly $\dim G$ in Γ. Since $\Gamma_{sing} \to G$ is dominant, its general fiber has dimension $\dim \Gamma_{sing} - \dim G < \dim \Gamma - \dim G$, so no component of a general fiber can be contained in Γ_{sing}. Thus $\varphi^{-1}(gA)$ is generically reduced for general $g \in G$.

(c) We will prove this part only for the case where G is a product of copies of GL_n, as this is the only case we will use. For the general result, see Theorem 18.2 of Borel [1991].

In this case G is an open set in a product M of vector spaces of matrices. Let L be the line joining 1 to g in M. The subvariety

$$Z = \{(g, x) \in (G \cap L) \times X \mid g^{-1}x \in A\}$$

gives a rational equivalence between A and gA.   □

The conclusion fails in positive characteristic, even for Grassmannians; examples can be found in Kleiman [1974] and Roberts [1972b]. However, Kleiman showed that the conclusion holds in general under the stronger hypothesis that G acts transitively on nonzero tangent vectors to X (each tangent space to the Grassmannian is naturally identified with a space of homomorphisms — see Section 3.2.4 — and the automorphisms preserve the ranks of these homomorphisms, so they do not act transitively on tangent vectors).

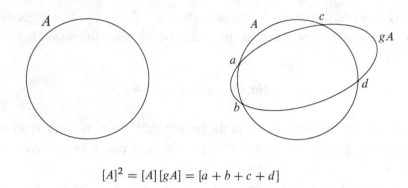

$$[A]^2 = [A][gA] = [a+b+c+d]$$

Figure 1.5 The cycle A meets a general translate of itself generically transversely.

1.3 Some techniques for computing the Chow ring

1.3.1 The fundamental class

If X is a scheme, then the *fundamental class* of X is $[X] \in A(X)$. It is always nonzero. We can immediately prove this and a little more, and these first results suffice to compute the Chow ring of a zero-dimensional scheme:

Proposition 1.8. *Let X be a scheme.*

(a) $A(X) = A(X_{\mathrm{red}})$.

(b) *If X is irreducible of dimension k, then $A_k(X) \cong \mathbb{Z}$ and is generated by the fundamental class of X. More generally, if the irreducible components of X are $X_1, \ldots, X_m$, then the classes $[X_i]$ generate a free abelian subgroup of rank m in $A(X)$.*

Proof: (a) Since both cycles and rational equivalences are generated by varieties we have $Z(X) = Z(X_{\mathrm{red}})$ and $\mathrm{Rat}(X) = \mathrm{Rat}(X_{\mathrm{red}})$.

(b) By definition the $[X_i]$ are among the generators of $A(X)$. Further, $\mathrm{Rat}(X)$ is generated by varieties in $\mathbb{P}^1 \times X$, each of which is contained in some $\mathbb{P}^1 \times X_i$. $\qquad \square$

Example 1.9 (Zero-dimensional schemes). From Proposition 1.8 it follows that the Chow group of a zero-dimensional scheme is the free abelian group on the components.

1.3.2 Rational equivalence via divisors

The next simplest case is that of curves, and it is not hard to see that the Chow group of 0-cycles on a curve is the divisor class group.

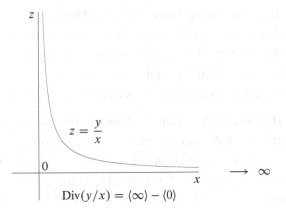

Figure 1.6 Graph of the rational function $z = y/x$ on the open set $y = 1$ in $\mathbb{P}^1$, showing that $[V(y)] - [V(x)] = 0$ in $A(\mathbb{P}^1)$.

More generally, for any variety X we can express the group $\mathrm{Rat}(X)$ of cycles rationally equivalent to 0 in terms of divisor classes: First, suppose that X is an affine variety. If $f \in \mathcal{O}_X$ is a function on X other than 0, then by Krull's principal ideal theorem (Theorem 0.1) the irreducible components of the subscheme defined by f are all of codimension 1, so the cycle defined by this subscheme is a divisor; we call it the *divisor of f*, denoted $\mathrm{Div}(f)$. If Y is any irreducible codimension-1 subscheme of X, we write $\mathrm{ord}_Y(f)$ for the order of vanishing of f along Y, so we have

$$\mathrm{Div}(f) = \sum_{Y \subset X \text{ irreducible}} \mathrm{ord}_Y(f)\langle Y \rangle.$$

If f, g are functions on X and $\alpha = f/g$, then we define the divisor $\mathrm{Div}(\alpha) = \mathrm{Div}(f/g)$ to be $\mathrm{Div}(f) - \mathrm{Div}(g)$; see Figure 1.6. This is well-defined because $\mathrm{ord}_Y(ab) = \mathrm{ord}_Y(a) + \mathrm{ord}_Y(b)$ for any functions defined on an open set. We denote by $\mathrm{Div}_0(\alpha)$ and $\mathrm{Div}_\infty(\alpha)$ the positive and negative parts of $\mathrm{Div}(\alpha)$ — in other words, the divisor of zeros of α and the divisor of poles of α, respectively.

We extend the definition of the divisor associated to a rational function to varieties X that are not affine as follows. The field of rational functions on X is the same as the field of rational functions on any open affine subset U of X, so if α is a rational function on X then we get a divisor $\mathrm{Div}(\alpha|_U)$ on each open subset $U \subset X$ by restricting α. These agree on overlaps, and thus define a divisor $\mathrm{Div}(\alpha)$ on X itself. We will see that the association $\alpha \mapsto \mathrm{Div}(\alpha)$ is a homomorphism from the multiplicative group of nonzero rational functions to the additive group of divisors on X.

Proposition 1.10. *If X is any scheme, then the group $\mathrm{Rat}(X) \subset Z(X)$ is generated by all divisors of rational functions on all subvarieties of X. In particular, if X is irreducible of dimension n, then $A_{n-1}(X)$ is equal to the divisor class group of X.*

See Fulton [1984, Proposition 1.6] for the proof.

Example 1.11. It follows from Proposition 1.10 that two 0-cycles on a curve C (by which we mean here a one-dimensional variety) are rationally equivalent if and only if they differ by the divisor of a rational function. In particular, the cycles associated to two points on C are rationally equivalent if and only if C is birational to $\mathbb{P}^1$, the isomorphism being given by a rational function that defines the rational equivalence.

Example 1.12. If X is an affine variety whose coordinate ring R does not have unique factorization, then there may not be a "best" way of choosing an expression of a rational function α on X as a fraction, and $\mathrm{Div}_0(\alpha)$ need not be the same thing as $\mathrm{Div}(f)$ for any one representation $\alpha = f/g$ of α. For example, on the cone $Q = V(XZ - Y^2) \subset \mathbb{A}^3$, the rational function $\alpha = X/Y$ has divisor $L - M$, where L is the line $X = Y = 0$ and M the line $Y = Z = 0$; but, as the reader can check, α cannot be written in any neighborhood of the vertex $(0, 0, 0)$ of Q as a ratio $\alpha = f/g$ with $\mathrm{Div}(f) = L$ and $\mathrm{Div}(g) = M$.

1.3.3 Affine space

Affine spaces are basic building blocks for many rational varieties, such as projective spaces and Grassmannians, and it is easy to compute their Chow groups directly:

Proposition 1.13. $A(\mathbb{A}^n) = \mathbb{Z} \cdot [\mathbb{A}^n]$.

Proof: Let $Y \subset \mathbb{A}^n$ be a proper subvariety, and choose coordinates $z = z_1, \ldots, z_n$ on $\mathbb{A}^n$ so that the origin does not lie in Y. We let

$$W^\circ = \{(t, tz) \subset (\mathbb{A}^1 \setminus \{0\}) \times \mathbb{A}^n \mid z \in Y\} = V(\{f(z/t) \mid f(z) \text{ vanishes on } Y\}).$$

The fiber of W° over a point $t \in \mathbb{A}^1 \setminus \{0\}$ is tY, that is, the image of Y under the automorphism of $\mathbb{A}^n$ given by multiplication by t. Let $W \subset \mathbb{P}^1 \times \mathbb{A}^n$ be the closure of W° in $\mathbb{P}^1 \times \mathbb{A}^n$. Note that W°, being the image of $(\mathbb{A}^1 \setminus 0) \times Y$, is irreducible, and hence so is W.

The fiber of W over the point $t = 1$ is just Y. On the other hand, since the origin in $\mathbb{A}^n$ does not lie in Y there is some polynomial $g(z)$ that vanishes on Y and has a nonzero constant term c. The function $G(t, z) = g(z/t)$ on $(\mathbb{A}^1 \setminus 0) \times \mathbb{A}^n$ then extends to a regular function on $(\mathbb{P}^1 \setminus 0) \times \mathbb{A}^n$ with constant value c on the fiber $\infty \times \mathbb{A}^n$. Thus the fiber of W over $t = \infty \in \mathbb{P}^1$ is empty, establishing the equivalence $Y \sim 0$ (see Figure 1.7). $\qquad \square$

See Section 3.5.2 for a more systematic treatment of this idea. If you are curious about the fiber of W over $t = 0$, see Exercise 1.34.

1.3.4 Mayer–Vietoris and excision

We will use the next proposition in conjunction with Proposition 1.13 to find generators for the Chow groups of projective spaces and Grassmannians.

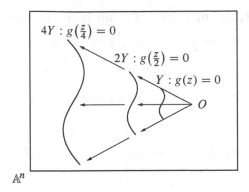

Figure 1.7 Scalar multiplication gives a rational equivalence between an affine variety not containing the origin and the empty set.

Proposition 1.10 makes it obvious that, if $Y \subset X$ is a closed subscheme, then the identification of the cycles on $\mathbb{P}^1 \times Y$ as cycles on $\mathbb{P}^1 \times X$ induces a map $\mathrm{Rat}(Y) \to \mathrm{Rat}(X)$, and thus a map $A(Y) \to A(X)$ (this is a special case of "proper pushforward;" see Section 1.3.6). Further, the intersection of a subvariety of X with the open set $U = X \setminus Y$ is a subvariety of U (possibly empty), so there is a restriction homomorphism $Z(X) \to Z(U)$. The rational equivalences restrict too, so we get a homomorphism of Chow groups $A(X) \to A(U)$ (this is a special case of "flat pullback;" see Section 1.3.6.)

Proposition 1.14. *Let X be a scheme.*

(a) *(Mayer–Vietoris) If X_1, X_2 are closed subschemes of X, then there is a right exact sequence*

$$A(X_1 \cap X_2) \longrightarrow A(X_1) \oplus A(X_2) \longrightarrow A(X_1 \cup X_2) \longrightarrow 0.$$

(b) *(Excision) If $Y \subset X$ is a closed subscheme and $U = X \setminus Y$ is its complement, then the inclusion and restriction maps of cycles give a right exact sequence*

$$A(Y) \longrightarrow A(X) \longrightarrow A(U) \longrightarrow 0.$$

If X is smooth, then the map $A(X) \to A(U)$ is a ring homomorphism.

Before starting the proof, we note that we can restate the definition of the Chow group by saying that there is a right exact sequence

$$Z(\mathbb{P}^1 \times X) \longrightarrow Z(X) \longrightarrow A(X) \longrightarrow 0,$$

where the left-hand map takes the any subvariety $\Phi \subset \mathbb{P}^1 \times X$ to 0 if Φ is contained in a fiber, and otherwise to

$$\langle \Phi \cap (\{t_0\} \times X) \rangle - \langle \Phi \cap (\{t_1\} \times X) \rangle.$$

Proof of Proposition 1.14: (b) There is a commutative diagram

$$
\begin{array}{ccccccccc}
0 & \longrightarrow & Z(Y \times \mathbb{P}^1) & \longrightarrow & Z(X \times \mathbb{P}^1) & \longrightarrow & Z(U \times \mathbb{P}^1) & \longrightarrow & 0 \\
 & & \downarrow \partial_Y & & \downarrow \partial_X & & \downarrow \partial_U & & \\
0 & \longrightarrow & Z(Y) & \longrightarrow & Z(X) & \longrightarrow & Z(U) & \longrightarrow & 0 \\
 & & \downarrow & & \downarrow & & \downarrow & & \\
 & & A(Y) & \longrightarrow & A(X) & \longrightarrow & A(U) & & \\
 & & \downarrow & & \downarrow & & \downarrow & & \\
 & & 0 & & 0 & & 0 & &
\end{array}
$$

where the map $Z(Y) \to Z(X)$ takes the class $[A] \in Z(Y)$, where A is a subvariety of Y, to $[A]$ itself, considered as a class in X, and similarly for $Z(Y \times \mathbb{P}^1) \to Z(X \times \mathbb{P}^1)$. The map $Z(X) \to Z(U)$ takes each free generator $[A]$ to the generator $[A \cap U]$, and similarly for $Z(X \times \mathbb{P}^1) \to Z(U \times \mathbb{P}^1)$. The two middle rows and all three columns are evidently exact. A diagram chase shows that the map $A(X) \to A(U)$ is surjective, and the bottom row of the diagram above is right exact, yielding part (b).

(a) Let $Y = X_1 \cap X_2$. We may assume $X = X_1 \cup X_2$. We may argue exactly as in part (b) from the diagram

$$
\begin{array}{ccccccccc}
0 & \longrightarrow & Z(Y \times \mathbb{P}^1) & \longrightarrow & Z(X_1 \times \mathbb{P}^1) \oplus Z(X_2 \times \mathbb{P}^1) & \longrightarrow & Z(X \times \mathbb{P}^1) & \longrightarrow & 0 \\
 & & \downarrow \partial & & \downarrow \partial \oplus \partial & & \downarrow \partial & & \\
0 & \longrightarrow & Z(Y) & \longrightarrow & Z(X_1) \oplus Z(X_2) & \longrightarrow & Z(X) & \longrightarrow & 0 \\
 & & \downarrow & & \downarrow & & \downarrow & & \\
 & & A(Y) & \longrightarrow & A(X_1) \oplus A(X_2) & \longrightarrow & A(X) & \longrightarrow & 0
\end{array}
$$

where, for example, the map $Z(Y) \to Z(X_1) \oplus Z(X_2)$ takes a generator $[A] \in Z(Y)$ to $([A], -[A]) \in Z(X_1) \oplus Z(X_2)$ and the map $Z(X_1) \oplus Z(X_2) \to Z(X)$ is addition. $\square$

The map $A(Y) \to A(X)$ of part (b) sends the class $[Z] \in A(Y)$ of a subvariety Z of Y to the class $[Z] \in A(X)$ of the same subvariety, now viewed as a subvariety of X. As we will see in Section 1.3.6, this is a special case of the pushforward map $f_* : A(Y) \to A(X)$ associated to any proper map $f : Y \to X$. The map $A(X) \to A(U)$, sending the class $[Z] \in A(X)$ of a subvariety of X to the class $[Z \cap U] \in A(U)$ of its intersection with U, is a special case of a pullback map, also described in Section 1.3.6.

Corollary 1.15. *If $U \subset \mathbb{A}^n$ is a nonempty open set, then $A(U) = A_n(U) = \mathbb{Z} \cdot [U]$.*

1.3.5 Affine stratifications

In general we will work with very partial knowledge of the Chow groups of a variety, but when X admits an *affine stratification* — a special kind of decomposition into a union of affine spaces — we can know them completely. This will help us compute the Chow groups of projective space, Grassmannians, and many other interesting rational varieties.

We say that a scheme X is *stratified* by a finite collection of irreducible, locally closed subschemes U_i if X is a disjoint union of the U_i and, in addition, the closure of any U_i is a union of U_j — in other words, if $\overline{U_i}$ meets U_j, then $\overline{U_i}$ contains U_j. The sets U_i are called the *strata* of the stratification, while the closures $Y_i := \overline{U_i}$ are called the *closed strata*. (If we want to emphasize the distinction, we will sometimes refer to the strata U_i as the *open strata* of the stratification, even though they are not open in X.) The stratification can be recovered from the closed strata Y_i: we have

$$U_i = Y_i \setminus \bigcup_{Y_j \subsetneq Y_i} Y_j.$$

Definition 1.16. We say that a stratification of X with strata U_i is:

- *affine* if each open stratum is isomorphic to some $\mathbb{A}^k$; and
- *quasi-affine* if each U_i is isomorphic to an open subset of some $\mathbb{A}^k$.

For example, a complete flag of subspaces $\mathbb{P}^0 \subset \mathbb{P}^1 \subset \cdots \subset \mathbb{P}^n$ gives an affine stratification of projective space; the closed strata are just the $\mathbb{P}^i$ and the open strata are affine spaces $U_i = \mathbb{P}^i \setminus \mathbb{P}^{i-1} \cong \mathbb{A}^i$.

Proposition 1.17. *If a scheme X has a quasi-affine stratification, then $A(X)$ is generated by the classes of the closed strata.*

Proof of Proposition 1.17: We will induct on the number of strata U_i. If this number is 1 then the assertion is Corollary 1.15.

Let U_0 be a minimal stratum. Since the closure of U_0 is a union of strata, U_0 must already be closed. It follows that $U := X \setminus U_0$ is stratified by the strata other than U_0. By induction, $A(U)$ is generated by the classes of the closures of these strata, and, by Corollary 1.15, $A(U_0)$ is generated by $[U_0]$. By excision (part (b) of Proposition 1.14) the sequence

$$\mathbb{Z} \cdot [U_0] = A(U_0) \longrightarrow A(X) \longrightarrow A(X \setminus U_0) \longrightarrow 0$$

is right exact. Since the classes in $A(U)$ of the closed strata in U come from the classes of (the same) closed strata in X, it follows that $A(X)$ is generated by the classes of the closed strata. $\qquad\square$

In general, the classes of the strata in a quasi-affine stratification of a scheme X may be zero in $A(X)$; for example, the affine line, with $A(\mathbb{A}^1) = \mathbb{Z}$, also has a quasi-affine stratification consisting of a single point and its complement, and we have already seen that the class of a point is 0. But in the case of an affine stratification, the classes are not only nonzero, they are independent:

Theorem 1.18 (Totaro [2014]). *The classes of the strata in an affine stratification of a scheme X form a basis of $A(X)$.*

We will often use results that are consequences of this deep theorem, although in our cases much more elementary proofs are available, as we shall see.

1.3.6 Functoriality

A key to working with Chow groups is to understand how they behave with respect to morphisms between varieties. To know what to expect, think of the analogous situation with homology and cohomology. A smooth complex projective variety of (complex) dimension n is a compact oriented $2n$-manifold, so $H_{2m}(X, \mathbb{Z})$ can be identified canonically with $H^{2n-2m}(X, \mathbb{Z})$ (singular homology and cohomology). If we think of $A(X)$ as being analogous to $H_*(X, \mathbb{Z})$, then we should expect $A_m(X)$ to be a covariant functor from smooth projective varieties to groups, via some sort of pushforward maps preserving dimension. If we think of $A(X)$ as analogous to $H^*(X, \mathbb{Z})$, then we should expect $A(X)$ to be a contravariant functor from smooth projective varieties to rings, via some sort of pullback maps preserving codimension. Both these expectations are realized.

Proper pushforward

If $f : Y \to X$ is a proper map of schemes, then the image of a subvariety $A \subset Y$ is a subvariety $f(A) \subset X$. One might at first guess that the pushforward could be defined by sending the class of A to the class of $f(A)$, and this would not be far off the mark. But this would not preserve rational equivalence (an example is pictured in Figure 1.8). Rather, we must take multiplicities into account.

If $A \subset Y$ is a subvariety and $\dim A = \dim f(A)$, then $f|_A : A \to f(A)$ is *generically finite*, in the sense that the field of rational functions $\Bbbk(A)$ is a finite extension of the field $\Bbbk(f(A))$ (this follows because they are both finitely generated fields, of the same transcendence degree $\dim A$ over the ground field). Geometrically the condition can be expressed by saying that, for a general point $x \in f(A)$, the preimage $y := f|_A^{-1}(x)$ in A is a finite scheme. In this case the degree $n := [\Bbbk(A) : \Bbbk(f(A))]$ of the extension of rational function fields is equal to the degree of y over x for a dense open subset of $x \in f(A)$, and this common value n is called the *degree* of the covering of $f(A)$ by A. We must count $f(A)$ with multiplicity n in the pushforward cycle:

Definition 1.19 (Pushforward for cycles). Let $f : Y \to X$ be a proper map of schemes, and let $A \subset Y$ be a subvariety.

(a) If $f(A)$ has strictly lower dimension than A, then we set $f_*\langle A \rangle = 0$.

(b) If $\dim f(A) = \dim A$ and $f|_A$ has degree n, then we set $f_*\langle A \rangle = n \cdot \langle f(A) \rangle$.

(c) We extend f_* to all cycles on Y by linearity; that is, for any collection of subvarieties $A_i \subset Y$, we set $f_*(\sum m_i \langle A_i \rangle) = \sum m_i f_*\langle A_i \rangle$.

$$a + b + c \sim d + e + f \sim 2g + h$$

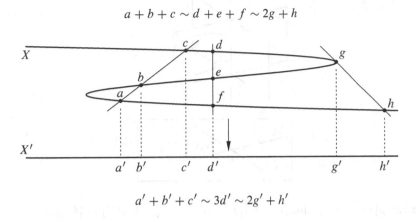

$$a' + b' + c' \sim 3d' \sim 2g' + h'$$

Figure 1.8 Pushforwards of equivalent cycles are equivalent.

With this definition, the pushforward of cycles preserves rational equivalence:

Theorem 1.20. *If $f : Y \to X$ is a proper map of schemes, then the map $f_* : Z(Y) \to Z(X)$ defined above induces a map of groups $f_* : A_k(Y) \to A_k(X)$ for each k.*

For a proof see Fulton [1984, Section 1.4].

It is often hard to prove that a given class in $A(X)$ is nonzero, but the fact that the pushforward map is well-defined gives us a start:

Proposition 1.21. *If X is proper over $\operatorname{Spec} \Bbbk$, then there is a unique map $\deg : A(X) \to \mathbb{Z}$ taking the class $[p]$ of each closed point $p \in X$ to 1 and vanishing on the class of any cycle of pure dimension > 0.*

As stated, Proposition 1.21 uses our standing hypothesis that the ground field is algebraically closed. Without this restriction we would have to count each (closed) point by the degree of its residue field extension over the ground field.

We will typically use this proposition together with the intersection product: If A is a k-dimensional subvariety of a smooth projective variety X and B is a k-codimensional subvariety of X such that $A \cap B$ is finite and nonempty, then the map

$$A_k(X) \to \mathbb{Z} : [Z] \mapsto \deg([Z][B])$$

sends $[A]$ to a nonzero integer. Thus no integer multiple $m[A]$ of the class A could be 0.

Pullback

We next turn to the pullback. Let $f : Y \to X$ be a morphism and $A \subset X$ a subvariety of codimension c. A good pullback map $f^* : Z(X) \to Z(Y)$ on cycles should preserve rational equivalence, and, in the nicest case, for example when $f^{-1}(A)$ is generically reduced of codimension c, it should be geometric, in the sense that

$$[P] = f^*[L_1]$$

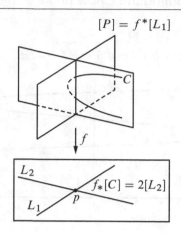

Figure 1.9 $2[p] = f_*([P][C]) = f_*([f^*L_1][C]) = [L_1]f_*[C] = [L_1][2L_2]$.

$f^*\langle A \rangle = \langle f^{-1}(A) \rangle$. This equality does not hold for all cycles, but does hold when A is a Cohen–Macaulay variety. (Recall that a scheme is said to be Cohen–Macaulay if all its local rings are Cohen–Macaulay. For a treatment of Cohen–Macaulay rings see Eisenbud [1995, Chapter 18].)

We start with a definition:

Definition 1.22. Let $f : Y \to X$ be a morphism of smooth varieties. We say a subvariety $A \subset X$ is *generically transverse* to f if the preimage $f^{-1}(A)$ is generically reduced and $\mathrm{codim}_Y(f^{-1}(A)) = \mathrm{codim}_X(A)$.

With that said, we have the following fundamental theorem:

Theorem 1.23. *Let $f : Y \to X$ be a map of smooth quasi-projective varieties.*

(a) *There is a unique map of groups $f^* : A^c(X) \to A^c(Y)$ such that whenever $A \subset X$ is a subvariety generically transverse to f we have*

$$f^*([A]) = [f^{-1}(A)].$$

This equality is also true without the hypothesis of generic transversality as long as $\mathrm{codim}_Y(f^{-1}(A)) = \mathrm{codim}_X(A)$ and A is Cohen–Macaulay. The map f^* is a ring homomorphism, and makes A into a contravariant functor from the category of smooth projective varieties to the category of graded rings.

(b) *(Push-pull formula) The map $f_* : A(Y) \to A(X)$ is a map of graded modules over the graded ring $A(X)$. More explicitly, if $\alpha \in A^k(X)$ and $\beta \in A_l(Y)$, then*

$$f_*(f^*\alpha \cdot \beta) = \alpha \cdot f_*\beta \in A_{l-k}(X).$$

The last statement of this theorem is the result of applying appropriate multiplicities to the set-theoretic equality $f(f^{-1}(A) \cap B) = A \cap f(B)$; see Figure 1.9.

One simple case of a projective morphism is the inclusion map from a closed subvariety $i : Y \subset X$. When X and Y are smooth, our definitions of intersections and pullbacks make it clear that, if A is any subvariety of X, then $[A][Y]$ is represented by the same cycle as $i^*([A])$ — except that these are considered as classes in different varieties. More precisely, we can write

$$[A][Y] = i_*(i^*[A]).$$

In this case the extra content of Theorem 1.23 is that this cycle is well-defined as a cycle on Y, not only as a cycle on X. Fulton [1984, Section 8.1] showed that it is even well-defined as a class on $X \cap Y$, and, more generally, he proved the existence of such a refined version of the pullback under a proper, locally complete intersection morphism (of which a map of smooth projective varieties is an example).

The uniqueness statement in Theorem 1.23 follows at once upon combining the moving lemma with the following:

Theorem 1.24. *If $f : Y \to X$ is a morphism of smooth quasi-projective varieties, then there is a finite collection of subvarieties $X_i \subset X$ such that if a subvariety $A \subset X$ is generically transverse to each X_i then A is generically transverse to f.*

(See Theorem A.6.) Note that this result depends on characteristic 0; it fails when f is not generically separable.

Pullback in the flat case

The flat case is simpler than the projective case for two reasons: first, the preimage of a subvariety of codimension k is always of codimension k; second, rational functions on the target pull back to rational functions on the source. We will use the flat case to analyze maps of affine space bundles.

Theorem 1.25. *Let $f : Y \to X$ be a flat map of schemes. The map $f^* : A(X) \to A(Y)$ defined on cycles by*

$$f^*(\langle A \rangle) := \langle f^{-1}(A) \rangle \quad \text{for every subvariety } A \subset X$$

preserves rational equivalence, and thus induces a map of Chow groups preserving the grading by codimension.

When X and Y are smooth and f is flat, the two pullback maps agree, as one sees at once from the uniqueness statement in Theorem 1.23.

1.3.7 Dimensional transversality and multiplicities

When two subvarieties A, B of a smooth variety X meet generically transversely, then we have

$$[A][B] = [A \cap B] \in A(X). \tag{$*$}$$

Does this formula hold more generally? Clearly it cannot hold unless the intersection $A \cap B$ has the expected dimension.

Theorem 1.26. *Let $A, B \subset X$ be subvarieties of a smooth variety X such that every irreducible component C of the intersection $A \cap B$ has codimension* $\operatorname{codim} C = \operatorname{codim} A + \operatorname{codim} B$. *For each such component C there is a positive integer $m_C(A, B)$, called the* intersection multiplicity *of A and B along C, such that:*

(a) $[A][B] = \sum m_C(A, B)[C] \in A(X)$.

(b) $m_C(A, B) = 1$ *if and only if A and B intersect transversely at a general point of C.*

(c) *In case A and B are Cohen–Macaulay at a general point of C, then $m_C(A, B)$ is the multiplicity of the component of the scheme $A \cap B$ supported on C. In particular, if A and B are everywhere Cohen–Macaulay, then*

$$[A][B] = [A \cap B].$$

(d) $m_C(A, B)$ *depends only on the local structure of A and B at a general point of C.*

For further discussion of this result see Hartshorne [1977, Appendix A], and for a full treatment see Fulton [1984, Chapter 7]. In view of Theorem 1.26, we make a definition:

Definition 1.27. Two subschemes A and B of a variety X are *dimensionally transverse* if for every irreducible component C of $A \cap B$ we have $\operatorname{codim} C = \operatorname{codim} A + \operatorname{codim} B$.

The reader should be aware that what we call "dimensionally transverse" is often called "proper" in the literature. We prefer "dimensionally transverse" since it suggests the meaning (and "proper" means so many different things!).

Recall that if X is smooth and C is a component of $A \cap B$, then by Theorem 0.2 we have $\operatorname{codim} C \leq \operatorname{codim} A + \operatorname{codim} B$, so in this case the condition of dimensional transversality is that A and B intersect in the smallest possible dimension. (But note that $A \cap B$ may also be empty. In this case too, A and B are transverse.)

The Cohen–Macaulay hypothesis in part (c) is necessary: in Example 2.6 we will see a case where the intersection multiplicity is not given by the multiplicities of the components of the intersection scheme.

Given that we sometimes have $[A \cap B] \neq [A][B]$, it is natural to look for a correction term. This was found by Jean-Pierre Serre; we will describe it in Theorem 2.7, following Example 2.6.

Remarkably, it is often possible to describe the intersection product $[A][B]$ of the classes of subvarieties $A, B \subset X$ geometrically even when they are not dimensionally transverse. See Chapter 13.

Just as we say that cycles $A = \sum m_i A_i$ and $B = \sum n_j Bj$ are generically transverse if A_i and B_j are generically transverse for all i, j, we say that A and B are dimensionally transverse if A_i and B_j are dimensionally transverse for every i, j.

The following explains the amount by which generic transversality is stronger than dimensional transversality.

Proposition 1.28. *Subschemes A and B of a variety X are generically transverse if and only if they are dimensionally transverse and each irreducible component of A ∩ B contains a point where X is smooth and A ∩ B is reduced.*

In particular, the proposition shows that, if X is smooth and A, B are dimensionally transverse subschemes that meet in a subvariety C, then A and B are generically transverse along C. The hypothesis that X is smooth cannot be dropped: For example, in the coordinate ring $\Bbbk[s^2, st, t^2]$ the ideal (s^2) defines a double line through the vertex that meets the reduced line defined by (st, t^2) in a reduced point.

Proof: If A and B are generically transverse, then each irreducible component C of $A \cap B$ contains a smooth point $p \in X$ such that A and B are smooth and transverse at p. It follows that C is smooth at p, and thus, in particular, C is reduced at p.

To prove the converse, let C be an irreducible component of $A \cap B$. Since the set of smooth points of X is open, and since by hypothesis C contains one, the smooth points of X that are contained in C form an open dense subset of C. Since $A \cap B$ is generically reduced, the open set where C is reduced is also dense, and it follows that the same is true for the smooth locus of C. Thus there is a point $p \in C$ that is smooth on both C and X. We must show that A and B are smooth at p.

The Zariski tangent space to C at p is the intersection of the Zariski tangent spaces $T_p A$ and $T_p B$ in $T_p X$. Since C and X are smooth at p,

$$\dim C = \dim T_p C = \dim T_p A + \dim T_p B - \dim T_p X$$
$$= \dim T_p A + \dim T_p B - \dim X.$$

By hypothesis,

$$\dim C = \dim A + \dim B - \dim X.$$

Since $\dim T_p A \geq \dim A$ and $\dim T_p B \geq \dim B$, we must have $\dim T_p A = \dim A$ and $\dim T_p B = \dim B$, proving that A and B are smooth at p as well. Since the tangent spaces of A, B, X at p are equal to the corresponding Zariski tangent spaces, the equality

$$\dim T_p C = \dim T_p A + \dim T_p B - \dim T_p X$$

above completes the proof. □

1.3.8 The multiplicity of a scheme at a point

In connection with the discussion of intersection multiplicities above, we collect here the basic facts about the multiplicity of a scheme at a point; for details, see Eisenbud [1995, Chapter 12]. We will also indicate, at least in some cases, how intersection multiplicities are related to multiplicities of schemes.

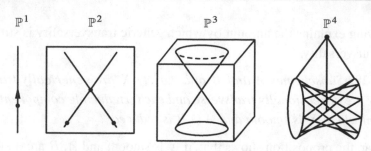

Figure 1.10 Ordinary double points of hypersurfaces of dimension 0, 1, 2 and 3.

Any discussion of the multiplicity of a scheme at a point begins with the case of a hypersurface in a smooth n-dimensional variety Z. In this case, we can be very explicit: If $p \in Z$ and $X \subset Z$ is a hypersurface given locally around p as the zero locus of a regular function f, we can choose local coordinates $z = (z_1, \ldots, z_n)$ on Z in a neighborhood of p and expand f around p, writing

$$f(z) = f_0 + f_1(z) + f_2(z) + \cdots$$

with $f_k(z)$ homogeneous of degree k. The hypersurface X contains p if $f_0 = f(p) = 0$, and is then singular at p if $f_1 = 0$. In general, we say that X has *multiplicity m at p* if $f_0 = \cdots = f_{m-1} = 0$ and $f_m \neq 0$; we write $\text{mult}_p(X)$ for the multiplicity of X at p. (If $m = 2$ we say that p is a *double point* of X; if $m = 3$ we say that p is a *triple point*, and so on.) We define the *tangent cone TC_pX* of X at p to be the zero locus of f_m in the affine space $\mathbb{A}^n$ with coordinates $(z_1, \ldots, z_n)$, and similarly we define the *projectivized tangent cone $\mathbb{T}C_pX$* of X at p to be the scheme in $\mathbb{P}^{n-1}$ defined by f_m.

We can say this purely in terms of the local ring $\mathcal{O}_{Z,p}$, without the need to invoke local coordinates: If $\mathfrak{m} \subset \mathcal{O}_{Z,p}$ is the maximal ideal, the multiplicity of X at p is the largest m such that $f \in \mathfrak{m}^m$. We can then take f_m to be the image of f in the quotient $\mathfrak{m}^m/\mathfrak{m}^{m+1}$. Note that since

$$\mathfrak{m}^m/\mathfrak{m}^{m+1} = \text{Sym}^m(\mathfrak{m}/\mathfrak{m}^2) = \text{Sym}^m T_p^* Z,$$

the vector space of homogeneous polynomials of degree m on the Zariski tangent space $T_p Z$, we can view the projectivized tangent cone as a subscheme of $\mathbb{P}T_p Z$. (Note that the projectivized tangent cone may be nonreduced even though X itself is reduced at p, as in the case of a cusp, given locally as the zero locus of $y^2 - x^3$.) The multiplicity can also be characterized in these terms simply as the degree of the projectivized tangent cone.

For example, the simplest possible singularity of a hypersurface X, generalizing the case of a node of a plane curve, is called an *ordinary double point*. This is a point $p \in X$ such that the equation of X can be written in local coordinates with $p = 0$ as above with $f_0 = f_1 = 0$ and where f_2 is a *nondegenerate quadratic form* — that is, the projectivized tangent cone to X at p is a smooth quadric. Indeed, examples are the cones over smooth quadrics — see Figure 1.10. (Here it is important that the characteristic is not 2: A quadric in $\mathbb{P}^{n-1}$ is smooth if the generator f_2 of its ideal, together with the derivatives of f_2, is an irrelevant ideal; when the characteristic is not 2, Euler's formula $2f_2 = \sum z_i \partial f_2 / \partial z_i$ shows that it is equivalent to assume that the partial derivatives of f_2 are linearly independent, and this is the property we will often use. In characteristic 2 — where a symmetric bilinear form is also skew-symmetric — *no* quadratic form in an odd number of variables has this property.)

How do we extend this definition to arbitrary schemes X? The answer is to start by defining the tangent cones. We can do this explicitly in terms of local coordinates $z = (z_1, \ldots, z_n)$ on a smooth ambient variety Z containing X: We define the tangent cone to be the subscheme of $\mathbb{A}^n$ defined by the leading terms of *all* elements of the ideal $I \subset \mathcal{O}_{Z,p}$ of X at p, and the projectivized tangent cone to be the corresponding subscheme of $\mathbb{P}^{n-1}$.

As before, this can be said without recourse to local coordinates (or, for that matter, any ambient variety Z). To start, we filter the local ring $\mathcal{O}_{X,p}$ by powers of its maximal ideal $\mathfrak{m}$:

$$\mathcal{O}_{X,p} \supset \mathfrak{m} \supset \mathfrak{m}^2 \supset \mathfrak{m}^3 \supset \cdots .$$

We then form the associated graded ring

$$A = \Bbbk \oplus \mathfrak{m}/\mathfrak{m}^2 \oplus \mathfrak{m}^2/\mathfrak{m}^3 \oplus \cdots ,$$

and define the tangent cone and projectivized tangent cone to be Spec A and Proj A respectively. Note that since the ring A is generated in degree 1, we have a surjection

$$\mathrm{Sym}(\mathfrak{m}/\mathfrak{m}^2) = \mathrm{Sym}(T_p^* X) \to A,$$

so that these can be viewed naturally as subschemes of the Zariski tangent space $T_p X$ and its projectivization, respectively. As we will see shortly, one important feature of these constructions is that we always have

$$\dim TC_p X = \dim X \quad \text{and} \quad \dim \mathbb{T}C_p X = \dim X - 1,$$

even though the dimension of the Zariski tangent space may be larger.

We now define the multiplicity $\mathrm{mult}_p(X)$ of X at p to be the degree of the projectivized tangent cone $\mathbb{T}C_p X$, viewed as a subscheme of the projective space $\mathbb{P}T_p X$. In purely algebraic terms, we can express this directly in terms of the Hilbert polynomial of the graded ring A: If we set

$$h_A(m) = \dim_{\Bbbk} A_m,$$

then for $m \gg 0$ the function h_A will be equal to a polynomial $p_A(m)$ of degree $\dim X - 1$, called the *Hilbert polynomial* of A. The multiplicity $\mathrm{mult}_p(X)$ is then equal to $(\dim X - 1)!$ times the leading coefficient of the Hilbert polynomial $p_A(m)$.

It follows from the theory that the multiplicity $\text{mult}_{Y_i}(Y)$ of a scheme Y along an irreducible component Y_i of Y, as introduced in Section 1.2.1 in connection with the definition of the cycle associated to a scheme, is equal to the multiplicity of Y at a general point of Y_i.

Tangent cones and blow-ups

There is another characterization of the projectivized tangent cone that will be very useful to us in what follows.

We start by recalling some basic facts about blow-ups. Blowing-up is an operation that associates to any scheme Z and subscheme Y a morphism $\pi : B = \text{Bl}_Y(Z) \to Z$. The general operation is described and characterized in Chapter 4 of Eisenbud and Harris [2000]; in the present circumstances, we will be concerned with the case where Y is a smooth point $p \in Z$.

The *exceptional divisor* $E \subset B$ is defined to be $\pi^{-1}(Y)$, the preimage of Y in B. If $X \subset Z$ is any subscheme, we define its *strict transform* $\widetilde{X} \subset B$ to be the closure in B of the preimage $\pi^{-1}(X \setminus Y \cap X)$ of X away from Y.

Suppose that X is embedded in a smooth ambient variety Z of dimension n, and consider the blow-up of Z at p. In this case the exceptional divisor E is isomorphic to $\mathbb{P}^{n-1}$. Unwinding the definitions, we can see that *the projectivized tangent cone $\mathbb{T}C_pX$ to X at p is the intersection of $\widetilde{X}$ with $E \cong \mathbb{P}^{n-1}$*. This gives us immediately that $\dim \mathbb{T}C_pX = \dim X - 1$.

Again, we can say this without having to choose an embedding of X in a smooth Z: Since blow-ups behave well with respect to pullbacks (see Proposition IV-21 of Eisenbud and Harris [2000]), we could simply say that $\mathbb{T}C_pX$ is the exceptional divisor in the blow-up $\text{Bl}_p(X)$.

Multiplicities and intersection multiplicities

The notions of multiplicity (of a scheme at a point) and intersection multiplicities (of two subschemes meeting dimensionally transversely in a smooth ambient variety) are closely linked: If $p \in X \subset \mathbb{P}^n$ is a point on a subscheme of pure dimension k and $\Lambda \cong \mathbb{P}^{n-k} \subset \mathbb{P}^n$ is a general $(n - k)$-plane containing p, then the intersection multiplicity $m_p(X, \Lambda)$ is equal to $\text{mult}_p(X)$.

This statement can be generalized substantially:

Proposition 1.29. *Let X and Y be two subschemes of complementary dimension intersecting dimensionally properly in a smooth variety Z, and $p \in X \cap Y$ any point of intersection. If the projectivized tangent cones $\mathbb{T}C_pX$ and $\mathbb{T}C_pY$ are disjoint in $\mathbb{P}T_pZ$, then*

$$m_p(X, Y) = \text{mult}_p(X) \cdot \text{mult}_p(Y).$$

This proposition is proved in Section 2.1.10. In general, there is only the inequality $m_p(X, Y) \geq \text{mult}_p(X) \cdot \text{mult}_p(Y)$; see Fulton [1984, Chapter 12].

1.4 The first Chern class of a line bundle

Many of the most interesting and useful classes in the Chow groups come from vector bundles via the theory of Chern classes. The simplest case is that of the first Chern class of a line bundle, which we will now describe. We will introduce the theory in more generality in Chapter 5.

If $\mathcal{L}$ is a line bundle on a variety X and σ is a rational section, then on an open affine set U of a covering of X we may write σ in the form f_U/g_U and define $\mathrm{Div}(\sigma)|_U = \mathrm{Div}(f) - \mathrm{Div}(g)$. This definition agrees where two affine open sets overlap, and thus defines a divisor on X, which is a *Cartier divisor* (see Hartshorne [1977, Section II.6]). Moreover, if τ is another rational section of $\mathcal{L}$ then $\alpha = \sigma/\tau$ is a well-defined rational function, so

$$\mathrm{Div}(\sigma) - \mathrm{Div}(\tau) = \mathrm{Div}(\alpha) \equiv 0 \bmod \mathrm{Rat}(X).$$

Thus for any line bundle $\mathcal{L}$ on a quasi-projective scheme X we may define the *first Chern class*

$$c_1(\mathcal{L}) \in A(X)$$

to be the rational equivalence class of the divisor σ for any nonzero rational section σ. (If we were working over an arbitrary scheme, we would have to insist that the numerator and denominator of our section were locally nonzerodivisors.) Note that there is no distinguished cycle in the equivalence class. As a first example, we see that $c_1(\mathcal{O}_{\mathbb{P}^n}(d))$ is the class of any hypersurface of degree d; in the notation of Section 2.1 it is $d\zeta$, where ζ is the class of a hyperplane.

Recall that the Picard group $\mathrm{Pic}(X)$ is by definition the group of isomorphism classes of line bundles $\mathcal{L}$ on X, with addition law $[\mathcal{L}] + [\mathcal{L}'] = [\mathcal{L} \otimes \mathcal{L}']$.

Proposition 1.30. *If X is a variety of dimension n, then c_1 is a group homomorphism*

$$c_1 : \mathrm{Pic}(X) \to A_{n-1}(X).$$

If X is smooth, then c_1 is an isomorphism.

If $Y \subset X$ is a divisor in a smooth variety X, then the ideal sheaf of Y is a line bundle denoted $\mathcal{O}_X(-Y)$, and its inverse in the Picard group is denoted $\mathcal{O}_X(Y)$. The inverse of the map c_1 above takes $[Y]$ to $\mathcal{O}_X(Y)$.

Proof of Proposition 1.30: To see that c_1 is a group homomorphism, suppose that $\mathcal{L}$ and $\mathcal{L}'$ are line bundles on X. If σ and σ' are rational sections of $\mathcal{L}$ and $\mathcal{L}'$ respectively, then $\sigma \otimes \sigma'$ is a rational section of $\mathcal{L} \otimes \mathcal{L}'$ whose divisor is $\mathrm{Div}(\sigma) + \mathrm{Div}(\sigma')$.

Now assume that X is smooth and projective. Since the local rings of X are unique factorization domains, every codimension-1 subvariety is a Cartier divisor, so to any divisor we can associate a unique line bundle and a rational section. Forgetting the section, we get a line bundle, and thus a map from the group of divisors to $\mathrm{Pic}(X)$. By

Proposition 1.10, rationally equivalent divisors differ by the divisor of a rational function, and thus correspond to different rational sections of the same bundle. It follows that the map on divisors induces a map on $A_{n-1}(X)$, inverse to the map c_1. □

If X is singular, the map $c_1 : \text{Pic}(X) \to A_{n-1}(X)$ is in general neither injective or surjective. For example, if X is an irreducible plane cubic with a node, then $c_1 : \text{Pic}(X) \to A_1(X)$ is not a monomorphism (Exercise 1.35). On the other hand, if $X \subset \mathbb{P}^3$ is a quadric cone with vertex p, then $A_1(X) = \mathbb{Z}$ and is generated by the class of a line, and the image of $c_1 : \text{Pic}(X) \to A_1(X)$ is $2\mathbb{Z}$ (Exercise 1.36).

Another case when the moving lemma is easy is when the class of the cycle to be moved has the form $c_1(\mathcal{L})$ for some line bundle $\mathcal{L}$. We also get a useful formula for the product of any class with $c_1(\mathcal{L})$:

Proposition 1.31. *Suppose that X is a smooth quasi-projective variety and $\mathcal{L}$ is a line bundle on X. If $Y_1, \ldots, Y_n$ are any subvarieties of X, then there is a cycle in the class of $c_1(\mathcal{L})$ that is generically transverse to each Y_i. If X is smooth and $Y \subset X$ is any subvariety, then*

$$c_1(\mathcal{L}) \cdot [Y] = c_1(\mathcal{L}|_Y).$$

The class $c_1(\mathcal{L}|_Y)$ on the right-hand side of the formula is actually a class in $A(Y)$, so to be precise we should have written $i_*(c_1(\mathcal{L}|_Y))$, where $i : Y \hookrightarrow X$ is the inclusion and i_* the pushforward map, first encountered in Proposition 1.14 and defined in general in Section 1.3.6. This imprecision points to an important theoretical fact: Even on a singular variety (or scheme) X one can form the intersection product of any class with the first Chern class of a line bundle, defined (when the class is the class of a subscheme) via the prescription $c_1(\mathcal{L}) \cdot [Y] = c_1(\mathcal{L}|_Y)$ above.

This intersection is actually defined by the formula as a class on Y, not just a class on X. This is the beginning of the theory of "refined intersection products" defined in Fulton [1984]. When we define other Chern classes of vector bundles we shall see that the same construction works in that more general case.

We imposed the hypothesis of smoothness in Proposition 1.31 because we have only discussed products in this context. In fact, the formula could be used to define an action of a class of the form $c_1(\mathcal{L})$ on $A(X)$ much more generally. This is the point of view taken by Fulton.

Sketch of proof of Proposition 1.31: Since X is quasi-projective, there is an ample bundle $\mathcal{L}'$ on X. For a sufficiently large integer n both the line bundles $\mathcal{L}'^{\otimes n}$ and $\mathcal{L}'^{\otimes n} \otimes \mathcal{L}$ are very ample, so by Bertini's theorem there are sections $\sigma \in H^0(\mathcal{L}'^{\otimes n})$ and $\tau \in H^0(\mathcal{L}'^{\otimes n} \otimes \mathcal{L})$ whose zero loci $\text{Div}(\sigma)$ and $\text{Div}(\tau)$ are generically transverse to each Y_i. The class $c_1(\mathcal{L})$ is rationally equivalent to the cycle $\text{Div}(\sigma) - \text{Div}(\tau)$, proving the first assertion. Moreover, $c_1(\mathcal{L})[Y_i] = [\text{Div}(\sigma) \cap Y_i] - [\text{Div}(\tau) \cap Y_i]$ by Theorem–Definition 1.5. Since $\text{Div}(\sigma) \cap Y_i = \text{Div}(\sigma|_{Y_i})$, and similarly for τ, we are done. □

genus	$\deg(K_X)$	topology	curvature	$\dim \mathrm{Aut}(X)$	cover	points
0	< 0	$\chi > 0$	> 0	3	$\mathbb{CP}^1$	infinite
1	0	$\chi = 0$	0	1	$\mathbb{C}$	infinite
≥ 2	> 0	$\chi < 0$	< 0	finite	Δ	finite

Table 1.1 Behavior of curves for $\deg(K_X) < 0$, $\deg(K_X) = 0$ and $\deg(K_X) > 0$.

1.4.1 The canonical class

Perhaps the most fundamental example of the first Chern class of a line bundle is the *canonical class*, which we will define here; in the following section, we will describe the *adjunction formula*, which gives us a way to calculate it in many cases.

Let X be a smooth n-dimensional variety. By the *canonical bundle* ω_X of X we mean the top exterior power $\wedge^n \Omega_X$ of the cotangent bundle Ω_X of X; this is the line bundle whose sections are regular n-forms. By the *canonical class* we mean the first Chern class $c_1(\omega_X) \in A^1(X)$ of this line bundle. Perhaps reflecting the German language history of the subject, this class is commonly denoted by K_X.

The canonical class is probably the single most important indicator of the behavior of X, geometrically, topologically and arithmetically. For example, the only topological invariant of a smooth projective curve X over the complex field $\mathbb{C}$ is its genus $g = g(X)$, and we have

$$\deg(K_X) = 2g - 2.$$

Virtually every aspect of the geometry over $\mathbb{C}$ and the arithmetic over $\mathbb{Q}$ of X are fundamentally different depending on whether $\deg K_X$ is negative, zero or positive, corresponding to $g = 0, 1$ or $g \geq 2$, as can be seen in Table 1.1. (Here the topology is represented by the topological Euler characteristic, the differential geometry by the curvature of a metric with constant curvature, the complex analysis by the isomorphism class as a complex manifold of the universal cover and the arithmetic by the number of rational points over a suitably large finite extension of $\mathbb{Q}$.)

Example 1.32 (Projective space). We can easily determine the canonical class of a projective space. To do this, we have only to write down a rational n-form ω on $\mathbb{P}^n$ and determine its divisors of zeros and poles. For example, if $X_0, \ldots, X_n$ are homogeneous coordinates on $\mathbb{P}^n$ and

$$x_i = \frac{X_i}{X_0}, \quad i = 1, \ldots, n,$$

are affine coordinates on the open set $U \cong \mathbb{A}^n \subset \mathbb{P}^n$ given by $X_0 \neq 0$, we may take ω to be the rational n-form given in U by

$$\omega = dx_1 \wedge \cdots \wedge dx_n.$$

The form ω is regular and nonzero in U, so we have only to determine its order of zero or pole along the hyperplane $H = V(X_0)$ at infinity. To this end, let $U' \subset \mathbb{P}^n$ be the open set $X_n \neq 0$, and take affine coordinates $y_0, \ldots, y_{n-1}$ on U' with $y_i = X_i / X_n$. We have

$$x_i = \begin{cases} y_i / y_0 & \text{for } i = 1, \ldots, n-1, \\ 1/y_0 & \text{for } i = n, \end{cases}$$

so that

$$dx_i = \begin{cases} (1/y_0)dy_i - (y_i/y_0^2)dy_0 & \text{for } i = 1, \ldots, n-1, \\ -(1/y_0^2)dy_0 & \text{for } i = n. \end{cases}$$

Taking wedge products, we see that

$$\omega = dx_1 \wedge \cdots \wedge dx_n = \frac{(-1)^n}{y_0^{n+1}} dy_0 \wedge \cdots \wedge dy_{n-1},$$

whence

$$\mathrm{Div}(\omega) = -(n+1)H,$$

so

$$K_{\mathbb{P}^n} = -(n+1)\zeta,$$

where $\zeta \in A^1(\mathbb{P}^n)$ is the class of a hyperplane.

1.4.2 The adjunction formula

Let X again be a smooth variety of dimension n, and suppose that $Y \subset X$ is a smooth $(n-1)$-dimensional subvariety. There is a natural way to relate the canonical class of Y to that of X: If we compare the tangent bundle $\mathcal{T}_Y$ of Y with the restriction $\mathcal{T}_X|_Y$ to Y of the tangent bundle $\mathcal{T}_X$ of X, we get an exact sequence

$$0 \longrightarrow \mathcal{T}_Y \longrightarrow \mathcal{T}_X|_Y \longrightarrow \mathcal{N}_{Y/X} \longrightarrow 0,$$

where the right-hand term $\mathcal{N}_{Y/X}$ is called the *normal bundle* of Y in X. Taking exterior powers, this gives an equality of line bundles

$$(\wedge^n \mathcal{T}_X)|_Y \cong \wedge^{n-1} \mathcal{T}_Y \otimes \mathcal{N}_{Y/X},$$

so that

$$\wedge^{n-1} \mathcal{T}_Y \cong (\wedge^n \mathcal{T}_X)|_Y \otimes \mathcal{N}_{Y/X}^*,$$

and dualizing we have

$$\omega_Y \cong \omega_X|_Y \otimes \mathcal{N}_{Y/X}.$$

Moreover, we can compute $\mathcal{N}_{Y/X}$ in another way. There is an exact sequence

$$0 \longrightarrow \mathcal{I}_{Y/X}/\mathcal{I}_{Y/X}^2 \overset{\delta}{\longrightarrow} \Omega_X|_Y \longrightarrow \Omega_Y \longrightarrow 0,$$

where the map δ sends the germ of a function to the germ of its differential (see, for example, Eisenbud [1995, Proposition 16.3]). This identifies $\mathcal{N}_{Y/X}^*$ with the locally free sheaf $\mathcal{I}_{Y/X}/\mathcal{I}_{Y/X}^2$. When Y is a Cartier divisor in X, the case of primary interest for us, the ideal sheaf $I_{Y/X}$ of Y in X is the line bundle $\mathcal{O}_X(-Y)$, and the sheaf $I_{Y/X}/I_{Y/X}^2 = \mathcal{O}_Y \otimes I_{Y/X}$ is its restriction to Y, denoted $\mathcal{O}_Y(-Y)$; thus

$$\mathcal{N}_{Y/X} \cong \mathcal{O}_X(Y)|_Y.$$

Combining this with the previous expression, we have what is commonly called the *adjunction formula*:

Proposition 1.33 (Adjunction formula). *If $Y \subset X$ is a smooth $(n-1)$-dimensional subvariety of a smooth n-dimensional variety, then*

$$\omega_Y = \omega_X|_Y \otimes \mathcal{O}_X(Y)|_Y,$$

which we usually write as $\omega_X(Y)|_Y$. In particular, if Y is a smooth curve in a smooth complete surface X, then the degree of K_Y is given by an intersection product:

$$\deg K_Y = \deg\big((K_X + [Y])[Y]\big).$$

1.4.3 Canonical classes of hypersurfaces and complete intersections

We can combine the adjunction formula with the calculation in Example 1.32 to calculate the canonical classes of hypersurfaces, and more generally of complete intersections, in projective space. To start, let $X \subset \mathbb{P}^n$ be a smooth hypersurface of degree d. We have

$$\omega_X = \omega_{\mathbb{P}^n}(X)|_X = \mathcal{O}_X(d - n - 1).$$

Thus

$$K_X = (d - n - 1)\zeta,$$

where $\zeta = c_1(\mathcal{O}_X(1)) \in A^1(X)$ is the class of a hyperplane section of X.

More generally, suppose

$$X = Z_1 \cap \cdots \cap Z_k$$

is a smooth complete intersection of hypersurfaces $Z_1, \ldots, Z_k$ of degrees $d_1, \ldots, d_k$. Applying adjunction repeatedly to the partial intersections $Z_1 \cap \cdots \cap Z_i$, we see that

$$\omega_X = \mathcal{O}_X\left(-n - 1 + \sum d_i\right)$$

and so

$$K_X = \left(-n - 1 + \sum d_i\right)\zeta.$$

This argument is not complete, because even though X is assumed smooth the partial intersections $Z_1 \cap \cdots \cap Z_i$ may not be. One way to complete it is to extend the definition of the canonical bundle to possibly singular complete intersections — the adjunction formula is true in this greater generality. Alternatively, if we order the hypersurfaces $Z_i = V(F_i)$ so that $d_1 \geq \cdots \geq d_k$ and replace F_i by a linear combination

$$F_i' = F_i + \sum_{j=i+1}^{k} G_j F_j,$$

with G_j general of degree $d_i - d_j$, the hypersurfaces $Z_i' = V(F_i')$ will have intersection X, and by Bertini's theorem the partial intersections will be smooth.

1.5 Exercises

Exercise 1.34. Let $Y \subset \mathbb{A}^n$ be a subvariety not containing the origin, and let $W \subset \mathbb{P}^1 \times \mathbb{A}^n$ be the closure of the locus

$$W^\circ = \{(t, z) \mid z \in t \cdot Y\},$$

as in the proof of Proposition 1.13. Show that the fiber of W over $t = 0$ is the cone with vertex the origin $0 \in \mathbb{A}^n$ over the intersection $\overline{Y} \cap H_\infty$, where $\overline{Y} \subset \mathbb{P}^n$ is the closure of Y in $\mathbb{P}^n$ and $H_\infty = \mathbb{P}^n \setminus \mathbb{A}^n$ is the hyperplane at infinity.

Exercise 1.35. Show that if X is an irreducible plane cubic with a node, then $c_1 : \operatorname{Pic}(X) \to A_1(X)$ is not a monomorphism, as follows: Show that there is no biregular map from X to $\mathbb{P}^1$. Use this to show that if $p \neq q \in X$ are smooth points, then the line bundles $\mathcal{O}_X(p)$ and $\mathcal{O}_X(q)$ are nonisomorphic. Show, however, that the zero loci of their unique sections, the points p and q, are rationally equivalent.

Exercise 1.36. Show that if $X \subset \mathbb{P}^3$ is a quadric cone with vertex p then $A_1(X) = \mathbb{Z}$ and is generated by the class of a line, and show that the image of $c_1 : \operatorname{Pic}(X) \to A_1(X)$ is $2\mathbb{Z}$ by showing that the image consists of the subgroup of classes of curves lying on X that have even degree as curves in $\mathbb{P}^3$. In particular, the class of a line on X is not in the image.

Hint: Do this by showing that no curve $C \subset X$ of odd degree can be a Cartier divisor on X: If such a curve meets the general line of the ruling of X at δ points away from p and has multiplicity m at p, then intersecting C with a general plane through p we see that $\deg(C) = 2\delta + m$; it follows that m is odd, and hence that C cannot be Cartier at p. Thus, the class $[M]$ of a line of the ruling cannot be $c_1(\mathcal{L})$ for any line bundle $\mathcal{L}$.

Chapter 2

First examples

Keynote Questions

(a) Let $F_0, F_1, F_2 \in \Bbbk[X, Y, Z]$ be three general homogeneous cubic polynomials in three variables. Up to scalars, how many linear combinations $t_0 F_0 + t_1 F_1 + t_2 F_2$ factor as a product of a linear and a quadratic polynomial? (Answer on page 65.)

(b) Let $F_0, F_1, F_2, F_3 \in \Bbbk[X, Y, Z]$ be four general homogeneous cubic polynomials in three variables. How many linear combinations $t_0 F_0 + t_1 F_1 + t_2 F_2 + t_3 F_3$ factor as a product of three linear polynomials? (Answer on page 65.)

(c) Let A, B, C be general homogeneous polynomials of degree d in three variables. Up to scalars, for how many triples $t = (t_0, t_1, t_2) \neq (0, 0, 0)$ is $(A(t), B(t), C(t))$ a scalar multiple of (t_0, t_1, t_2)? (Answer on page 55.)

(d) Let S_d denote the space of homogeneous polynomials of degree d in two variables. If $V \subset S_d$ and $W \subset S_e$ are general linear spaces of dimensions a and b with $a + b = d + 2$, how many pairs $(f, g) \in V \times W$ are there (up to multiplication of each of f and g by scalars) such that $f \mid g$? (Answer on page 56.)

(e) Let $S \subset \mathbb{P}^3$ be a smooth cubic surface and $L \subset \mathbb{P}^3$ a general line. How many planes containing L are tangent to S? (Answer on page 50.)

(f) Let $L \subset \mathbb{P}^3$ be a line, and let S and $T \subset \mathbb{P}^3$ be surfaces of degrees s and t containing L. Suppose that the intersection $S \cap T$ is the union of L and a smooth curve C. What are the degree and genus of C? (Answer on page 71.)

In this chapter we illustrate the general theory introduced in the preceding chapter with a series of examples and applications.

The first section is a series of progressively more interesting computations of Chow rings of familiar varieties, with easy applications. Following this, in Section 2.2 we see an example of a different kind: We use facts about the Chow ring to describe some geometrically interesting loci in the projective space of cubic plane curves.

Finally, in Section 2.4 we briefly describe intersection theory on surfaces, a setting in which the theory takes a particularly simple and useful form. As one application, we describe in Section 2.4.3 the notion of *linkage*, a tool used classically to understand the geometry of curves in $\mathbb{P}^3$.

2.1 The Chow rings of $\mathbb{P}^n$ and some related varieties

So far we have not seen any concrete examples of the intersection product or pullback. The first interesting case where this occurs is projective space.

Theorem 2.1. *The Chow ring of $\mathbb{P}^n$ is*

$$A(\mathbb{P}^n) = \mathbb{Z}[\zeta]/(\zeta^{n+1}),$$

where $\zeta \in A^1(\mathbb{P}^n)$ is the rational equivalence class of a hyperplane; more generally, the class of a variety of codimension k and degree d is $d\zeta^k$.

In particular, the theorem implies that $A^m(\mathbb{P}^n) \cong \mathbb{Z}$ for $0 \leq m \leq n$, generated by the class of an $(n-m)$-plane. The natural proof, given below, uses the intersection product.

Proof: Let $\{p\} \subset \mathbb{P}^1 \subset \cdots \subset \mathbb{P}^n$ be a complete flag of subspaces. Applying Proposition 1.17 to the affine stratification with strata $U_i = \mathbb{P}^i \setminus \mathbb{P}^{i-1}$, we see that $A^k(\mathbb{P}^n)$ is generated by the class of $\mathbb{P}^{n-k}$, and thus by the class of any $(n-k)$-plane $L \subset \mathbb{P}^n$. Using Proposition 1.21, we get $A^n(\mathbb{P}^n) = \mathbb{Z}$. Since a general $(n-k)$-plane L intersects a general k-plane M transversely in one point, multiplication by $[M]$ induces a surjective map $A^k(\mathbb{P}^n) \to A^n(\mathbb{P}^n) = \mathbb{Z}$, so $A^k(\mathbb{P}^n) = \mathbb{Z}$ for all k.

An $(n-k)$-plane $L \subset \mathbb{P}^n$ is the transverse intersection of k hyperplanes, so

$$[L] = \zeta^k,$$

where $\zeta \in A^1(\mathbb{P}^n)$ is the class of a hyperplane. Finally, since a subvariety $X \subset \mathbb{P}^n$ of dimension $n-k$ and degree d intersects a general k-plane transversely in d points, we have $\deg([X]\zeta^{n-k}) = d$. Since $\deg(\zeta^n) = 1$, we conclude that $[X] = d\zeta^k$. $\qquad\square$

Here are two interesting qualitative results that follow from Theorem 2.1:

Corollary 2.2. *A morphism from $\mathbb{P}^n$ to a quasi-projective variety of dimension strictly less than n is constant.*

Proof: Let $\varphi : \mathbb{P}^n \to X \subset \mathbb{P}^m$ be the map, which we may assume is surjective onto X. The preimage of a general hyperplane section of X is disjoint from the preimage of a general point of X. But if $0 < \dim X < n$ then the preimage of a hyperplane section of X has dimension $n-1$ and the preimage of a point has dimension > 0. Since any two such subvarieties of $\mathbb{P}^n$ must meet, this is a contradiction. $\qquad\square$

Corollary 2.3. *If $X \subset \mathbb{P}^n$ is a variety of dimension m and degree d then $A_m(\mathbb{P}^n \setminus X) \cong \mathbb{Z}/(d)$, while if $m < m' \leq n$ then $A_{m'}(\mathbb{P}^n \setminus X) = \mathbb{Z}$. In particular, m and d are determined by the isomorphism class of $\mathbb{P}^n \setminus X$.*

Proof: Part (b) of Proposition 1.14 shows that there are exact sequences $A_i(X) \to A_i(\mathbb{P}^n) \to A_i(\mathbb{P}^n \setminus X) \to 0$. Furthermore $A_m(X) = \mathbb{Z}$ by part (b) of Proposition 1.8, while $A_{m'}(X) = 0$ for $m < m' \leq n$. By Theorem 2.1, we have $A_i(\mathbb{P}^n) = \mathbb{Z}$ for $0 \leq i \leq n$, and the image of the generator of $A_m(X)$ in $A_m(\mathbb{P}^n)$ is d times the generator of $A_i(\mathbb{P}^n)$. The results of the corollary follow. □

Theorem 2.1 implies the analog of Poincaré duality for $A(\mathbb{P}^n)$: $A_k(\mathbb{P}^n)$ is dual to $A^k(\mathbb{P}^n)$ via the intersection product. The reader should be aware that in cases where the Chow groups and the homology groups are different, Poincaré duality generally does *not* hold for the Chow ring; for example, when X is a variety, $A_{\dim X}(X) \cong \mathbb{Z}$, but $A_0(X)$ need not even be finitely generated.

One aspect of Theorem 2.1 may, upon reflection, seem strange: why is it that only the dimension and degree of a variety $X \subset \mathbb{P}^n$ are preserved under rational equivalence, and not other quantities such as (in the case of X a curve) the arithmetic genus?

First of all, to understand why this may appear curious, we recall from Eisenbud and Harris [2000, Proposition III-56] (see also Corollary B.12) that, if B is reduced and connected, then a closed subscheme $\mathcal{Y} \subset B \times \mathbb{P}^n$ is flat over B if and only if the fibers all have the same Hilbert polynomial. Thus, for example, if $Z \subset \mathbb{P}^1 \times \mathbb{P}^n$ is an irreducible surface dominating $\mathbb{P}^1$, then the fibers Z_0 and Z_∞ will be one-dimensional subschemes of $\mathbb{P}^n$ having not only the same degree, but also the same arithmetic genus. Why does this not contradict the assertion of Theorem 2.1 that curves C and $C' \subset \mathbb{P}^3$ of the same degree d but different genera *are* rationally equivalent?

The explanation is that both can be deformed, in families parametrized by $\mathbb{P}^1$, to schemes C_0, C_0' supported on a line $L \subset \mathbb{P}^3$ and having multiplicity d, so that $\langle C \rangle \sim \langle C_0 \rangle = d\langle L \rangle$ as cycles, and likewise for C'. The difference in the genera of C and C' will be reflected in two things: the scheme structure along the line in the flat limits C_0 and C_0', and the presence and multiplicity of embedded points in these limits.

For an example of the former, note that the schemes $C_0 = V((x,y)^2)$ and $C_0' = V(x, y^3)$ are both supported on the line $L = V(x, y)$, and both have multiplicity 3, but the arithmetic genus of C_0 is 0, while that of C_0' is 1 (after all, it is a plane cubic!). But the mechanism by which we associate a cycle to a scheme does not see the difference in the scheme structure; we have $\langle C_0 \rangle = \langle C_0' \rangle = 3\langle L \rangle$. Similarly, a twisted cubic curve $C \subset \mathbb{P}^3$ can be deformed to a scheme generically isomorphic to either C_0 or C_0'; the difference in the arithmetic genus is accounted for by the fact that in the latter case the limiting scheme will necessarily have an embedded point. But again, rational equivalence does not "see" the embedded point; we have $[C] = 3[L]$ regardless.

2.1.1 Bézout's theorem

As an immediate consequence of Theorem 2.1, we get a general form of Bézout's theorem:

Corollary 2.4 (Bézout's theorem). *If $X_1, \ldots, X_k \subset \mathbb{P}^n$ are subvarieties of codimensions $c_1, \ldots, c_k$, with $\sum c_i \leq n$, and the X_i intersect generically transversely, then*

$$\deg(X_1 \cap \cdots \cap X_k) = \prod \deg(X_i).$$

In particular, two subvarieties $X, Y \subset \mathbb{P}^n$ having complementary dimension and intersecting transversely will intersect in exactly $\deg(X) \cdot \deg(Y)$ points.

Using multiplicities we can extend this formula to the more general case where we assume only that the varieties intersect dimensionally transversely (that is, all components of the intersection $Z = \bigcap X_i$ have codimension equal to $\sum c_i$), as long as the X_i are generically Cohen–Macaulay along each component of their intersection. In this case, the intersection multiplicity $m_{Z_\alpha}(X_1, \ldots, X_k)$ of the X_i along a component Z_α of their intersection, as described in Section 1.3.7, is equal to the multiplicity of the scheme Z at a general point of Z_α.

Corollary 2.5. *Suppose $X_1, \ldots, X_k \subset \mathbb{P}^n$ are subvarieties of codimensions $c_1, \ldots, c_k$ whose intersection is a scheme Z of pure dimension $n - \sum c_i$, with irreducible components $Z_1, \ldots, Z_t$. If the X_i are Cohen–Macaulay at a general point of each Z_α, then*

$$[Z] = \sum [Z_j] = \prod [X_i];$$

equivalently,

$$\deg Z = \sum \deg Z_j = \prod \deg X_i.$$

Note that by the degree of a subscheme $Z \subset \mathbb{P}^n$ of dimension m we mean $m!$ times the leading coefficient of the Hilbert polynomial; in case Z is irreducible this will be equal to the degree of the reduced scheme Z_{red} times the multiplicity of the scheme, and more generally it will be given by

$$\deg(Z) = \sum \text{mult}_{Z_i}(Z) \deg(Z_{\text{red}}),$$

where the Z_i are the irreducible components of Z of maximal dimension m.

The Cohen–Macaulay hypothesis is satisfied if, for example, the X_i are all hypersurfaces; thus the classical case of two curves intersecting in $\mathbb{P}^2$ is covered.

There is a standard example that shows that the Cohen–Macaulay hypothesis is necessary:

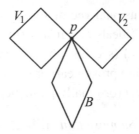

Figure 2.1 Let $A = V_1 \cup V_2 \subset \mathbb{P}^4$, where the V_i are general 2-planes, and let B be a 2-plane passing through the point $V_1 \cap V_2$. The degree of the product $[B][A]$ in $A(\mathbb{P}^4)$ is 2, as one sees by moving B to a plane B' transverse to A, but the length of the local ring of $B \cap A$ is 3.

Example 2.6. Let $X = \mathbb{P}^4$, let $V_1, V_2 \subset \mathbb{P}^4$ be general 2-planes and let $A = V_1 \cup V_2$.

Since V_1 and V_2 are general, they meet in a single point p. Let B be a 2-plane that passes through p and does not meet A anywhere else, and let B' be a 2-plane that does not pass through p and meets each of V_1, V_2 in a single (necessarily reduced) point. The cycles $\langle B \rangle$ and $\langle B' \rangle$ are rationally equivalent in $\mathbb{P}^4$. The intersection $B' \cap A$ consists of two reduced points, so $\deg(B' \cap A) = 2$ (see Figure 2.1).

However, the degree of the scheme $B \cap A$ is strictly greater than 2: Since the Zariski tangent space to the scheme $A = V_1 \cup V_2$ at the point p is all of $T_p(\mathbb{P}^4)$, the tangent space to the intersection $B \cap A$ at p must be all of $T_p(B)$. In other words, $B \cap A$ must contain the "fat point" at p in the plane B (that is, the scheme defined by the square of the ideal of p in B), and so must have degree at least 3.

In fact, we can see that the degree of the scheme $B \cap A$ is equal to 3 by a local calculation, as follows. Since B meets A only at the point p, we have to show that the length of the Artinian ring $\mathcal{O}_{\mathbb{P}^4,p}/(\mathcal{I}(B) + \mathcal{I}(A))\mathcal{O}_{\mathbb{P}^4,p}$ is 3. Let $S = \Bbbk[x_0, \ldots, x_4]$ be the homogeneous coordinate ring of $\mathbb{P}^4$. We may choose V_1, V_2 and B to have homogeneous ideals

$$I(A) = (x_0, x_1) \cap (x_2, x_3) = (x_0 x_2, x_0 x_3, x_1 x_2, x_1 x_3),$$
$$I(B) = (x_0 - x_2, x_1 - x_3).$$

Modulo $I(B)$, we can eliminate the variables x_2 and x_3 and the ideal $I(A)$ becomes $(x_0^2, x_0 x_1, x_1^2)$. Passing to the affine open subset where $x_4 \neq 0$, this is the square of the maximal ideal corresponding to the origin in B. Therefore $\mathcal{O}_{\mathbb{P}^4,p}/(\mathcal{I}(B) + \mathcal{I}(A))\mathcal{O}_{\mathbb{P}^4,p}$ has basis $\{1, x_0/x_4, x_1/x_4\}$, and hence its length is 3.

Given that we sometimes have $[A \cap B] \neq [A][B]$, it is natural to look for a correction term. In the example above, the set-theoretic intersection is a point, so this comes down to looking for a formula that will predict the difference in multiplicities $3 - 2 = 1$. Of course the correction term should reflect nontransversality, and one measure of nontransversality is the quotient $I(A) \cap I(B)/(I(A) \cdot I(B))$. In the case above one can compute this, finding that the quotient is a finite-dimensional vector space of length 1 — just the correction term we need. Now for any pair of ideals I, J in any ring R, the quotient $(I \cap J)/(I \cdot J)$ is isomorphic to $\mathrm{Tor}_1^R(R/I, R/J)$ (see Eisenbud [1995, Exercise A3.17]). With this information, knowing a special case proven earlier by Auslander and Buchsbaum, Serre [2000] produced a general formula (originally published in 1957):

Theorem 2.7 (Serre's formula). *Suppose that $A, B \subset X$ are dimensionally transverse subschemes of a smooth scheme X and Z is an irreducible component of $A \cap B$. The intersection multiplicity of A and B along Z is*

$$m_Z(A, B) = \sum_{i=0}^{\dim X} (-1)^i \, \mathrm{length}_{\mathcal{O}_{A \cap B, Z}} \left(\mathrm{Tor}_i^{\mathcal{O}_{X,Z}} (\mathcal{O}_{A,Z}, \mathcal{O}_{B,Z}) \right).$$

The first term of the alternating sum in Serre's formula is

$$\mathrm{length}_{\mathcal{O}_{A \cap B, Z}} \, \mathrm{Tor}_0^{\mathcal{O}_{X,Z}} (\mathcal{O}_{A,Z}, \mathcal{O}_{B,Z}) = \mathrm{length}_{\mathcal{O}_{A \cap B, Z}} \mathcal{O}_{X,Z} / (\mathcal{I}_A + \mathcal{I}_B),$$

which is precisely the multiplicity of Z in the subscheme $A \cap B$; the remaining terms, involving higher Tors, are zero in the Cohen–Macaulay case and may be viewed as correction terms. We note that this formula is used relatively rarely in practice, since there are many alternatives, such as the one given by Fulton [1984, Chapter 7].

2.1.2 Degrees of Veronese varieties

Let

$$\nu = \nu_{n,d} : \mathbb{P}^n \to \mathbb{P}^N, \quad \text{with } N = \binom{n+d}{n} - 1,$$

be the *Veronese map*

$$[Z_0, \dots, Z_n] \mapsto [\dots, Z^I, \dots],$$

where Z^I ranges over all monomials of degree d in $n + 1$ variables. The image $\Phi = \Phi_{n,d} \subset \mathbb{P}^N$ of the Veronese map $\nu = \nu_{n,d}$ is called the *d-th Veronese variety of* $\mathbb{P}^n$, as is any subvariety of $\mathbb{P}^N$ projectively equivalent to it. This variety may be characterized, up to automorphisms of the target $\mathbb{P}^N$, as the image of the map associated to the complete linear system $|\mathcal{O}_{\mathbb{P}^n}(d)|$; in other words, by the property that the preimages $\nu^{-1}(H) \subset \mathbb{P}^n$ of hyperplanes $H \subset \mathbb{P}^N$ comprise all hypersurfaces of degree d in $\mathbb{P}^n$.

There is another attractive description, at least in characteristic 0: writing $\mathbb{P}^n = \mathbb{P}V$, where V is an $(n + 1)$-dimensional vector space, $\nu_{n,d}$ is projectively equivalent to the map taking $\mathbb{P}V \to \mathbb{P}\,\mathrm{Sym}^d V$ by $[v] \mapsto [v^d]$; for if the coordinates of v are $v_0, \dots, v_n$ then the coordinates of v^d are

$$\frac{d!}{\prod_i d_i!} (v_0^{d_0} \cdots v_n^{d_n}).$$

If the characteristic is 0 then the coefficients are nonzero, so we may rescale by an automorphism of $\mathbb{P}^N$ to get the standard Veronese map above.

We can use Corollary 2.4 to compute the degrees of Veronese varieties:

Proposition 2.8. *The degree of* $\Phi_{n,d}$ *is* d^n.

Proof: The degree of Φ is the cardinality of its intersection with n general hyperplanes $H_1, \ldots, H_n \subset \mathbb{P}^N$; since the map ν is one-to-one, this is in turn the cardinality of the intersection $f^{-1}(H_1) \cap \cdots \cap f^{-1}(H_n) \subset \mathbb{P}^n$. The preimages of the hyperplanes H_i are n general hypersurfaces of degree d in $\mathbb{P}^n$. By Bézout's theorem, the cardinality of their intersection is d^n. $\qquad\square$

2.1.3 Degree of the dual of a hypersurface

The same idea allows us to compute the degree of the *dual variety* of a smooth hypersurface $X \subset \mathbb{P}^n$ of degree d, that is, the set of points $X^* \subset \mathbb{P}^{n*}$ corresponding to hyperplanes of $\mathbb{P}^n$ that are tangent to X. (In Chapter 10 we will generalize this notion substantially, discussing the duals of varieties of higher codimension and singular varieties as well.)

The set X^* is a variety because it is the image of X under the *Gauss map* $\mathcal{G}_X : X \to \mathbb{P}^{n*}$, a morphism that sends a point $p \in X$ to its tangent hyperplane $\mathbb{T}_p X$; in coordinates, if X is the zero locus of the homogeneous polynomial $F(Z_0, \ldots, Z_n)$, then $\mathcal{G}_X$ is given by the formula

$$\mathcal{G}_X : p \mapsto \left[\frac{\partial F}{\partial Z_0}(p), \ldots, \frac{\partial F}{\partial Z_n}(p) \right].$$

To see that this map is well-defined, note first that, since X is smooth, the partials of F have no common zeros on X (and this implies, by Euler's relation, that they do not have any common zeros in $\mathbb{P}^n$). Thus $\mathcal{G}_X$ defines a morphism $\mathbb{P}^n \to \mathbb{P}^{n*}$. When $p \in X$, Euler's relation shows that the vector $\mathcal{G}_X(p)$ is orthogonal to the vector $\tilde{p}$ representing the point p; thus the linear functional represented by $\mathcal{G}_X(p)$ induces a functional on the tangent space to $\mathbb{P}^n$, and the zero locus of this functional is the tangent space to X at p.

If $d = 1$, the map $\mathcal{G}_X$ is constant and X^* is a point. But if $d > 1$, then the fact that the partials of F have no common zeros says that the map $\mathcal{G}_X$ is finite: If $\mathcal{G}_X$ were constant on a complete curve $C \subset X$, the restrictions to C of the partials of F would be scalar multiples of each other, and so would have a common zero.

In particular, if $X \subset \mathbb{P}^n$ is a smooth hypersurface of degree $d \geq 2$, the dual variety $X^* \subset \mathbb{P}^{n*}$ is again a hypersurface, though not usually smooth. The smoothness hypothesis is necessary here; for example, the dual Q^* of the quadric cone $Q = V(XZ - Y^2) \subset \mathbb{P}^3$ is a conic curve in $\mathbb{P}^{3*}$.

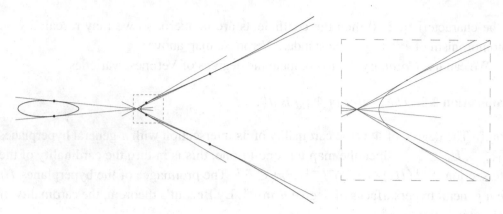

Figure 2.2 Six of the lines through a general point are tangent to a smooth plane cubic (but often not all the lines are defined over $\mathbb{R}$).

We will see in Corollary 10.21 that when X is a smooth hypersurface the map $\mathcal{G}_X$ is birational onto its image as well as finite. (This requires the hypothesis of characteristic 0; strangely enough, it may be false in characteristic p, where for example a general tangent line to a smooth plane curve may be bitangent!) We will use this now to deduce the degree of the dual hypersurface:

Proposition 2.9. *If $X \subset \mathbb{P}^n$ is a smooth hypersurface of degree $d > 1$, then the dual of X is a hypersurface of degree $d(d-1)^{n-1}$.*

Proof: The degree of the dual variety $X^* \subset \mathbb{P}^{n*}$ is the number of points of intersection of X^* and $n-1$ general hyperplanes $H_i \subset \mathbb{P}^{n*}$. Since by Corollary 10.21 the map $\mathcal{G}_X : X \to X^* \subset \mathbb{P}^{n*}$ is birational, this is the same as the number of points of intersection of the preimages $\mathcal{G}_X^{-1}(H_i)$. Since $\mathcal{G}_X$ is given by the partial derivatives of the defining equation F of X, the preimages of these hyperplanes are the intersections of X with the hypersurfaces $Z_i \subset \mathbb{P}^n$ of degree $d-1$ in $\mathbb{P}^n$ given by general linear combinations of these partial derivatives. Inasmuch as the partials of F have no common zeros, Bertini's theorem (Theorem 0.5) tells us that the hypersurfaces given by $n-1$ general linear combinations will intersect transversely with X. By Bézout's theorem the number of these points of intersection is the product of the degrees of the hypersurfaces, that is, $d(d-1)^{n-1}$. $\qquad\square$

For example, suppose that X is a smooth cubic curve in $\mathbb{P}^2$. By the above formula, the degree of X^* is 6. Since a general line in $\mathbb{P}^{2*}$ corresponds to the set of lines through a general point $p \in \mathbb{P}^2$, there will be exactly six lines in $\mathbb{P}^2$ through p tangent to X, as shown in Figure 2.2.

Proposition 2.9 gives us the answer to Keynote Question (e): Since the planes containing the line L form a general line in the dual projective space $\mathbb{P}^{3*}$, the number of such planes tangent to a smooth cubic surface $S \subset \mathbb{P}^3$ is $3 \cdot 2^2 = 12$.

2.1.4 Products of projective spaces

Though the Chow ring of a smooth variety behaves like cohomology in many ways, there are important differences. For example the cohomology ring of the product of two spaces is given modulo torsion by the Künneth formula $H^*(X \times Y) = H^*(X) \otimes H^*(Y)$, but in general there is no analogous Künneth formula for the Chow rings of products of varieties. Even for a product of two smooth curves C and D of genera $g, h \geq 1$ we have no algorithm for calculating $A^1(C \times D)$, and no idea at all what $A^2(C \times D)$ looks like, beyond the fact that it cannot be in any sense finite-dimensional (Mumford [1962]).

However the Chow ring of the product of a variety with a projective space does obey the Künneth formula, as we will prove in a more general context in Theorem 9.6 (Totaro [2014] proved it for products of any two varieties with affine stratifications). For the moment we will content ourselves with the special case where both factors are projective spaces:

Theorem 2.10. *The Chow ring of $\mathbb{P}^r \times \mathbb{P}^s$ is given by the formula*

$$A(\mathbb{P}^r \times \mathbb{P}^s) \cong A(\mathbb{P}^r) \otimes A(\mathbb{P}^s).$$

Equivalently, if $\alpha, \beta \in A^1(\mathbb{P}^r \times \mathbb{P}^s)$ denote the pullbacks, via the projection maps, of the hyperplane classes on $\mathbb{P}^r$ and $\mathbb{P}^s$ respectively, then

$$A(\mathbb{P}^r \times \mathbb{P}^s) \cong \mathbb{Z}[\alpha, \beta]/(\alpha^{r+1}, \beta^{s+1}).$$

Moreover, the class of the hypersurface defined by a bihomogeneous form of bidegree (d, e) on $\mathbb{P}^r \times \mathbb{P}^s$ is $d\alpha + e\beta$.

Proof: We proceed exactly as in Theorem 2.1. We may construct an affine stratification of $\mathbb{P}^r \times \mathbb{P}^s$ by choosing flags of subspaces

$$\Lambda_0 \subset \Lambda_1 \subset \cdots \subset \Lambda_{r-1} \subset \Lambda_r = \mathbb{P}^r \quad \text{and} \quad \Gamma_0 \subset \Gamma_1 \subset \cdots \subset \Gamma_{s-1} \subset \Gamma_s = \mathbb{P}^s,$$

with $\dim \Lambda_i = i = \dim \Gamma_i$, and taking the closed strata to be

$$\Xi_{a,b} = \Lambda_{r-a} \times \Gamma_{s-b} \subset \mathbb{P}^r \times \mathbb{P}^s.$$

The open strata

$$\widetilde{\Xi}_{a,b} := \Xi_{a,b} \setminus (\Xi_{a-1,b} \cup \Xi_{a,b-1})$$

of this stratification are affine spaces. Invoking Proposition 1.17, we conclude that the Chow groups of $\mathbb{P}^r \times \mathbb{P}^s$ are generated by the classes $\varphi_{a,b} = [\Xi_{a,b}] \in A^{a+b}(\mathbb{P}^r \times \mathbb{P}^s)$. Since $\Xi_{a,b}$ is the transverse intersection of the pullbacks of a hyperplanes in $\mathbb{P}^r$ and b hyperplanes in $\mathbb{P}^s$, we have

$$\varphi_{a,b} = \alpha^a \beta^b,$$

and in particular $\alpha^{r+1} = \beta^{s+1} = 0$. This shows that $A(\mathbb{P}^r \times \mathbb{P}^s)$ is a homomorphic image of

$$\mathbb{Z}[\alpha, \beta]/(\alpha^{r+1}, \beta^{s+1}) = \mathbb{Z}[\alpha]/(\alpha^{r+1}) \otimes_{\mathbb{Z}} \mathbb{Z}[\beta]/(\beta^{s+1}).$$

On the other hand, $\Xi_{r,s}$ is a single point, so $\deg \varphi_{r,s} = 1$. The pairing

$$A^{p+q}(\mathbb{P}^r \times \mathbb{P}^s) \times A^{r+s-p-q}(\mathbb{P}^r \times \mathbb{P}^s) \to \mathbb{Z}, \quad ([X], [Y]) \to \deg([X][Y])$$

sends $(\alpha^p \beta^q, \alpha^m \beta^n)$ to 1 if $p + m = r$ and $q + n = s$, because in this case the intersection is transverse and consists of one point, and to 0 otherwise, since then the intersection is empty. This shows that the monomials of bidegree (p, q), for $0 \le p \le r$ and $0 \le q \le s$, are linearly independent over $\mathbb{Z}$, proving the first statement.

If $F(X, Y)$ is a bihomogeneous polynomial with bidegree (d, e), then, because $F(X, Y)/X_0^d Y_0^e$ is a rational function on $\mathbb{P}^r \times \mathbb{P}^s$, the class of the hypersurface X defined by $F = 0$ is d times the class of the hypersurface $X_0 = 0$ plus e times the class of the hypersurface $Y_0 = 0$; that is, $[X] = d\alpha + e\beta$. □

2.1.5 Degrees of Segre varieties

The Segre variety $\Sigma_{r,s}$ is by definition the image of $\mathbb{P}^r \times \mathbb{P}^s$ in $\mathbb{P}^{(r+1)(s+1)-1}$ under the map

$$\sigma_{r,s} : ([X_0, \ldots, X_r], [Y_0, \ldots, Y_s]) \mapsto [\ldots, X_i Y_j, \ldots].$$

The map $\sigma_{r,s}$ is an embedding because on each open set where one of the X_i and one of the Y_j are nonzero the rest of the coordinates can be recovered from the products.

If V and W are vector spaces of dimensions $r + 1$ and $s + 1$, we may express $\sigma_{r,s}$ without choosing bases by the formula

$$\sigma_{r,s} : \mathbb{P}V \times \mathbb{P}W \to \mathbb{P}(V \otimes W),$$
$$(v, w) \mapsto v \otimes w.$$

For example, the map $\sigma_{1,1}$ is defined by the four forms $a = X_0 Y_0$, $b = X_0 Y_1$, $c = X_1 Y_0$, $d = X_1 Y_1$, and these satisfy the equation $ac - bd = 0$; thus the Segre variety $\Sigma_{1,1}$ is the nonsingular quadric in $\mathbb{P}^3$.

Proposition 2.11. *The degree of the Segre embedding of* $\mathbb{P}^r \times \mathbb{P}^s$ *is*

$$\deg \Sigma_{r,s} = \binom{r+s}{r}.$$

Proof: The degree of $\Sigma_{r,s}$ is the number of points in which it meets the intersection of $r + s$ hypersurfaces in $\mathbb{P}^{(r+1)(s+1)-1}$. Since $\sigma_{r,s}$ is an embedding, we may compute this number by pulling back these hypersurfaces to $\mathbb{P}^r \times \mathbb{P}^s$ and computing in the Chow ring of $\mathbb{P}^r \times \mathbb{P}^s$. Thus $\deg \Sigma_{r,s} = \deg(\alpha + \beta)^{r+s}$, which gives the desired formula because $(\alpha + \beta)^{r+s} = \binom{r+s}{r} \alpha^r \beta^s$. □

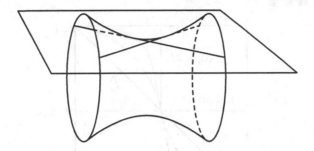

Figure 2.3 A tangent plane to a quadric in $\mathbb{P}^3$ meets the quadric in two lines, one from each ruling.

For instance, the Segre variety $\mathbb{P}^1 \times \mathbb{P}^r \subset \mathbb{P}^{2r+1}$ has degree $r + 1$. These varieties are among those called *rational normal scrolls* (see Section 9.1.1). The simplest of these is the smooth quadric surface $Q \subset \mathbb{P}^3$, which is the Segre image of $\mathbb{P}^1 \times \mathbb{P}^1$; the pullbacks α and β of the point classes via the two projections are the classes of the lines of the two rulings of Q, and we have $\zeta = \alpha + \beta$, where ζ is the hyperplane class on $\mathbb{P}^3$ restricted to Q — a fact that is apparent if we look at the intersection of Q with any tangent plane, as in Figure 2.3.

This discussion can be generalized to arbitrary products of projective spaces (see Exercise 2.30).

2.1.6 The class of the diagonal

Next we will find the class δ of the diagonal $\Delta \subset \mathbb{P}^r \times \mathbb{P}^r$ in the Chow group $A^r(\mathbb{P}^r \times \mathbb{P}^r)$, and more generally the class γ_f of the graph of a map $f : \mathbb{P}^r \to \mathbb{P}^s$. Apart from the applications of such a formula, this will introduce the *method of undetermined coefficients*, which we will use many times in the course of this book. (Another approach to this problem, via specialization, is given in Exercise 2.31.)

By Theorem 2.10, we have

$$A(\mathbb{P}^r \times \mathbb{P}^r) = \mathbb{Z}[\alpha, \beta]/(\alpha^{r+1}, \beta^{r+1}),$$

where $\alpha, \beta \in A^1(\mathbb{P}^r \times \mathbb{P}^r)$ are the pullbacks, via the two projection maps, of the hyperplane class in $A^1(\mathbb{P}^r)$. The class $\delta = [\Delta]$ of the diagonal is expressible as a linear combination

$$\delta = c_0 \alpha^r + c_1 \alpha^{r-1}\beta + c_2 \alpha^{r-2}\beta^2 + \cdots + c_r \beta^r$$

for some $c_0, \ldots, c_r \in \mathbb{Z}$. We can determine the coefficients c_i by taking the product of both sides of this expression with various classes of complementary codimension: Specifically, if we intersect both sides with the class $\alpha^i \beta^{r-i}$ and take degrees, we have

$$c_i = \deg(\delta \cdot \alpha^i \beta^{r-i}).$$

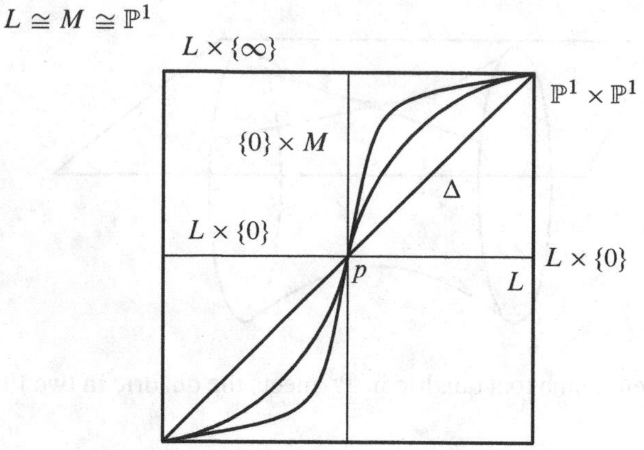

$$L \cong M \cong \mathbb{P}^1$$

$L \times \{\infty\}$

$\{0\} \times M$

$\mathbb{P}^1 \times \mathbb{P}^1$

$L \times \{0\}$

Δ

p

L

$L \times \{0\}$

Figure 2.4 $[\Delta][L \times \{0\}] = 1 = [\Delta][\{0\} \times M]$, so $[\Delta] = [\{0\} \times M] + [L \times \{0\}]$, as one also sees from the degeneration in the figure.

We can evaluate the product $\delta \cdot \alpha^i \beta^{r-i}$ directly: If Λ and Γ are general linear subspaces of codimension i and $r - i$, respectively, then $[\Lambda \times \Gamma] = \alpha^i \beta^{r-i}$. Moreover,

$$(\Lambda \times \Gamma) \cap \Delta \cong \Lambda \cap \Gamma$$

is a reduced point, so

$$c_i = \deg(\delta \cdot \alpha^i \beta^{r-i}) = \#(\Delta \cap (\Lambda \times \Gamma)) = \#(\Lambda \cap \Gamma) = 1.$$

Thus

$$\delta = \alpha^r + \alpha^{r-1}\beta + \cdots + \alpha\beta^{r-1} + \beta^r.$$

See Figure 2.4. (This formula and its derivation will be familiar to anyone who has had a course in algebraic topology. As partisans we cannot resist pointing out that algebraic geometry had it first!)

2.1.7 The class of a graph

Let $f : \mathbb{P}^r \to \mathbb{P}^s$ be the morphism given by $(s + 1)$ homogeneous polynomials F_i of degree d that have no common zeros:

$$f : [X_0, \ldots, X_r] \mapsto [F_0(X), F_1(X), \ldots, F_s(X)].$$

By Corollary 2.2, we must have $s \geq r$. Let $\Gamma_f \subset \mathbb{P}^r \times \mathbb{P}^s$ be the graph of f. What is its class $\gamma_f = [\Gamma_f] \in A^s(\mathbb{P}^r \times \mathbb{P}^s)$?

As before, we can write

$$\gamma_f = c_0\alpha^r\beta^{s-r} + c_1\alpha^{r-1}\beta^{s-r+1} + c_2\alpha^{r-2}\beta^{s-r+2} + \cdots + c_r\beta^s$$

for some $c_0, \ldots, c_r \in \mathbb{Z}$, and as before we can determine the coefficients c_i in this expression by intersecting both sides with a cycle of complementary dimension:

$$c_i = \deg(\gamma_f \cdot \alpha^i \beta^{r-i}) = \#(\Gamma_f \cap (\Lambda \times \Phi))$$

for general linear subspaces $\Lambda \cong \mathbb{P}^{r-i}$ and $\Phi \cong \mathbb{P}^{s-r+i} \subset \mathbb{P}^s$. By Theorem 1.7 the intersection $\Gamma_f \cap (\Lambda \times \Phi)$ is generically transverse.

Finally, $\Gamma_f \cap (\Lambda \times \Phi)$ is the zero locus in Λ of $r - i$ general linear combinations of the polynomials $F_0, \ldots, F_s$. By Bertini's theorem, the corresponding hypersurfaces will intersect transversely, and by Bézout's theorem the intersection will consist of d^{r-i} points. Thus we arrive at the formula:

Proposition 2.12. *If $f : \mathbb{P}^r \to \mathbb{P}^s$ is a regular map given by polynomials of degree d on $\mathbb{P}^r$, the class γ_f of the graph of f is given by*

$$\gamma_f = \sum_{i=0}^{r} d^i \alpha^i \beta^{s-i} \in A^s(\mathbb{P}^r \times \mathbb{P}^s).$$

Using this formula, we can answer a general form of Keynote Question (c). A sequence $F_0, \ldots, F_r$ of general homogeneous polynomials of degree d in $r + 1$ variables defines a map $f : \mathbb{P}^r \to \mathbb{P}^r$, and we can count the fixed points

$$\{t = [t_0, \ldots, t_r] \in \mathbb{P}^r \mid f(t) = t\}.$$

Since the F_i are general, we can take them to be general translates under $\mathrm{GL}_{r+1} \times \mathrm{GL}_{r+1}$ of arbitrary polynomials, so the cardinality of this set is the degree of the intersection of the graph γ_f of f with the diagonal $\Delta \subset \mathbb{P}^r \times \mathbb{P}^r$. This is

$$\deg(\delta \cdot \gamma_f) = \deg((\alpha^r + \alpha^{r-1}\beta + \cdots + \beta^r) \cdot (d^r \alpha^r + d^{r-1}\alpha^{r-1}\beta + \cdots + \beta^r))$$
$$= d^r + d^{r-1} + \cdots + d + 1;$$

in particular, if A, B, C are general forms of degree d in three variables then there are exactly $d^2 + d + 1$ points $t = [t_0, t_1, t_2] \in \mathbb{P}^2$ such that $[A(t), B(t), C(t)] = [t_0, t_1, t_2]$, and this is the answer to Keynote Question (c).

Note that in the case $d = 1$ and $s = r$, Proposition 2.12 implies that a general $(r + 1) \times (r + 1)$ matrix has $r + 1$ eigenvalues. It also follows that an arbitrary matrix has at least one eigenvalue.

2.1.8 Nested pairs of divisors on $\mathbb{P}^1$

We consider here one more example of an intersection theory problem involving products of projective spaces; this one will allow us to answer Keynote Question (d). To set this up, let $\mathbb{P}^d = \mathbb{P} H^0(\mathcal{O}_{\mathbb{P}^1}(d))$ be the projectivization of the space of homogeneous polynomials of degree d on $\mathbb{P}^1$ (equivalently, the space of effective divisors of degree d on $\mathbb{P}^1$). For any pair of natural numbers d and e with $e \geq d$, we consider the locus

$$\Phi = \{(f, g) \in \mathbb{P}^d \times \mathbb{P}^e \mid f \mid g\}.$$

Alternatively, if we think of $\mathbb{P}^d$ as parametrizing divisors of degree d on $\mathbb{P}^1$, we can write this as

$$\Phi = \{(D, E) \in \mathbb{P}^d \times \mathbb{P}^e \mid E \geq D\}.$$

Since the projection map $\pi : \Phi \to \mathbb{P}^d$ has fibers isomorphic to $\mathbb{P}^{e-d}$, we see that Φ is irreducible of dimension e, or codimension d, in $\mathbb{P}^d \times \mathbb{P}^e$. We ask: What is the class of Φ in $A^d(\mathbb{P}^d \times \mathbb{P}^e)$?

Let $\sigma, \tau \in A^1(\mathbb{P}^d \times \mathbb{P}^e)$ be the pullbacks of the hyperplane classes in $\mathbb{P}^d$ and $\mathbb{P}^e$, respectively. A priori, we can write

$$[\Phi] = \sum c_i \sigma^i \tau^{d-i},$$

where each coefficient c_i is given by the degree of the product $[\Phi] \cdot \sigma^{d-i} \tau^{e-d+i}$; that is, the number of points of intersection of Φ with the product $\Lambda \times \Gamma$ of general linear spaces $\Lambda \cong \mathbb{P}^i \subset \mathbb{P}^d$ and $\Gamma \cong \mathbb{P}^{d-i} \subset \mathbb{P}^e$. This is exactly the number asked for in Keynote Question (d), but it may not be clear at first glance how to evaluate it.

The key to doing this is the observation is that, abstractly, the variety Φ is isomorphic to a product $\mathbb{P}^d \times \mathbb{P}^{e-d}$: Specifically, it is the image of $\mathbb{P}^d \times \mathbb{P}^{e-d}$ under the map

$$\alpha : \mathbb{P}^d \times \mathbb{P}^{e-d} \to \mathbb{P}^d \times \mathbb{P}^e,$$
$$(D, D') \mapsto (D, D + D').$$

Furthermore, the pullback map $\alpha^* : A(\mathbb{P}^d \times \mathbb{P}^e) \to A(\mathbb{P}^d \times \mathbb{P}^{e-d})$ is readily described. Let $\sigma, \mu \in A^1(\mathbb{P}^d \times \mathbb{P}^{e-d})$ be the pullbacks of the hyperplane classes from $\mathbb{P}^d$ and $\mathbb{P}^{e-d}$, respectively. Since α commutes with the projection on the first factor, we see that $\alpha^*(\sigma) = \sigma$; since the composition $\mathbb{P}^d \times \mathbb{P}^{e-d} \to \mathbb{P}^d \times \mathbb{P}^e \to \mathbb{P}^e$ is given by bilinear forms on $\mathbb{P}^d \times \mathbb{P}^{e-d}$, we have $\alpha^*(\tau) = \sigma + \mu$. To evaluate the coefficient c_i, we write

$$\deg([\Phi] \cdot \sigma^{d-i} \tau^{e-d+i}) = \deg(\alpha^*(\sigma^{d-i} \tau^{e-d+i}))$$
$$= \deg(\sigma^{d-i}(\sigma + \mu)^{e-d+i})$$
$$= \binom{e-d+i}{i};$$

thus

$$[\Phi] = \sum \binom{e-d+i}{i} \sigma^i \tau^{d-i},$$

and correspondingly the answer to Keynote Question (d) is $\binom{e-d+a-1}{a-1}$.

2.1.9 The blow-up of $\mathbb{P}^n$ at a point

We will see in Chapter 13 how to describe the Chow ring of a blow-up in general. In this chapter, both to illustrate some of the techniques introduced so far and because the formulas derived will be useful in the interim, we will discuss two special cases: here the

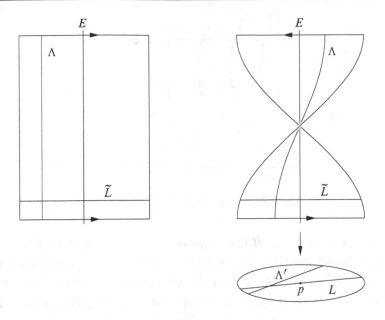

Figure 2.5 Blow-up of $\mathbb{P}^2$.

blow-up of $\mathbb{P}^n$ at a point for any $n \geq 2$ and in Section 2.4.4 the blow-up of any smooth surface at a point.

Recall that the *blow-up* of $\mathbb{P}^n$ at a point p is the morphism $\pi : B \to \mathbb{P}^n$, where $B \subset \mathbb{P}^n \times \mathbb{P}^{n-1}$ is the closure of the graph of the projection $\pi_p : \mathbb{P}^n \setminus \{p\} \to \mathbb{P}^{n-1}$ from p, and π is the projection on the first factor:

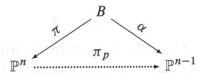

Since the graph of the projection is isomorphic to the source $\mathbb{P}^n \setminus \{p\}$, B is irreducible. It is not hard to write explicit equations for B and to show that it is smooth as well; see, for example, Section IV.2 of Eisenbud and Harris [2000].

The *exceptional divisor* $E \subset B$ is defined to be $\pi^{-1}(p)$, the preimage of p in B, which, as a subset of $\mathbb{P}^n \times \mathbb{P}^{n-1}$, is $\{p\} \times \mathbb{P}^{n-1}$. Some other obvious divisors on B are the preimages of the hyperplanes of $\mathbb{P}^n$. If the hyperplane $H \subset \mathbb{P}^n$ contains p, then its preimage is the sum of two irreducible divisors, E and $\widetilde{H}$; the latter is called the *strict transform*, or *proper transform*, of H. More generally, if $Z \subset \mathbb{P}^n$ is any subvariety, we define the strict transform of Z to be the closure in B of the preimage $\pi^{-1}(Z \setminus \{p\})$. See Figure 2.5.

To compute the Chow ring of B, we start from a stratification of B, using the geometry of the projection map $\alpha : B \to \mathbb{P}^{n-1}$ to the second factor. We do this by first choosing a stratification of the target $\mathbb{P}^{n-1}$, and taking the preimages in B of these strata.

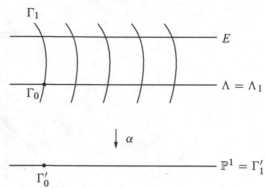

Figure 2.6 Blow-up of $\mathbb{P}^2$ as $\mathbb{P}^1$-bundle.

Then we choose a divisor $\Lambda \subset B$ that maps isomorphically by α to $\mathbb{P}^{n-1}$ — a *section* of α — and take, as additional strata, the intersections of these preimages with Λ.

We will choose as our section the preimage $\Lambda = \pi^{-1}(\Lambda')$ of a hyperplane $\Lambda' \cong \mathbb{P}^{n-1} \subset \mathbb{P}^n$ not containing the point p. (There are other possible choices of a section, such as the exceptional divisor $E \subset B$; see Exercise 2.37.)

To carry this out, let

$$\Gamma_0' \subset \Gamma_1' \subset \cdots \subset \Gamma_{n-2}' \subset \Gamma_{n-1}' = \mathbb{P}^{n-1}$$

be a flag of linear subspaces and, for $k = 1, 2, \ldots, n$, let

$$\Gamma_k = \alpha^{-1}(\Gamma_{k-1}') \subset B.$$

Since the fibers of $\alpha : B \to \mathbb{P}^{n-1}$ are projective lines, the dimension of Γ_k is k. Next, for $k = 0, 1, \ldots, n-1$, we set

$$\Lambda_k = \Gamma_{k+1} \cap \Lambda,$$

so that Λ_k is the preimage of Γ_k' under the isomorphism $\alpha|_\Lambda : \Lambda \to \mathbb{P}^{n-1}$.

The subvarieties $\Gamma_1, \ldots, \Gamma_n, \Lambda_0, \ldots, \Lambda_{n-1}$ are the closed strata of a stratification of B, with inclusion relations

$$
\begin{array}{ccccccccc}
\Lambda_0 & \hookrightarrow & \Lambda_1 & \hookrightarrow & \cdots & \hookrightarrow & \Lambda_{n-2} & \hookrightarrow & \Lambda_{n-1} \\
& \searrow & & \searrow & & & & \searrow & & \searrow \\
& & \Gamma_1 & \hookrightarrow & \Gamma_2 & \hookrightarrow & \cdots & \hookrightarrow & \Gamma_{n-1} & \hookrightarrow & \Gamma_n = B
\end{array}
$$

As we will soon see, this is an affine stratification, so that the classes of the closed strata generate the Chow group $A(B)$. (In fact, the open strata are isomorphic to affine spaces, and it follows from Totaro [2014] that they generate $A(B)$ freely; we will verify this independently when we determine the intersection products.)

To visualize this, we think of the blow-up B as the total space of a $\mathbb{P}^1$-bundle over $\mathbb{P}^{n-1}$ via the projection map α; for example, this is the picture that arises if we take the standard picture of the blow-up of $\mathbb{P}^2$ at a point (shown in Figure 2.5) and "unwind" it as in Figure 2.6.

Proposition 2.13. *Let B be the blow-up of $\mathbb{P}^n$ at a point, with $n \geq 2$. With notation as above, the Chow ring $A(B)$ is the free abelian group on the generators $[\Lambda_k] = [\Lambda_{n-1}]^{n-k}$ for $k = 0, \ldots, n-1$ and $[\Gamma_k] = [\Gamma_{n-1}]^{n-k}$ for $k = 1, \ldots, n$. The class of the exceptional divisor E is $[\Lambda_{n-1}] - [\Gamma_{n-1}]$. If we set $\lambda = [\Lambda_{n-1}]$ and $e = [E]$, then*

$$A(B) \cong \frac{\mathbb{Z}[\lambda, e]}{(\lambda e, \lambda^n + (-1)^n e^n)}$$

as rings.

Proof: We start by verifying that the open strata $\Gamma_1^\circ, \ldots, \Gamma_n^\circ, \Lambda_0^\circ, \ldots, \Lambda_{n-1}^\circ$ of the stratification of B with closed strata Γ_k, Λ_k are isomorphic to affine spaces. This is immediate for the strata Λ_k°. For the strata Γ_k°, we choose coordinates $(x_0, \ldots, x_n)$ on $\mathbb{P}^n$ so that $p = (1, 0, \ldots, 0)$ and $\Lambda' \subset \mathbb{P}^n$ is the hyperplane $x_0 = 0$. By definition,

$$B = \{((x_0, \ldots, x_n), (y_1, \ldots, y_n)) \in \mathbb{P}^n \times \mathbb{P}^{n-1} \mid x_i y_j = x_j y_i \text{ for all } i, j \geq 1\}.$$

Say the $(k-1)$-plane $\Gamma'_{k-1} \subset \mathbb{P}^{n-1}$ is given by $y_1 = \cdots = y_{n-k} = 0$. We can write the open stratum $\Gamma_k^\circ = \alpha^{-1}(\Gamma'_{k-1} \setminus \Gamma'_{k-2}) \cap (B \setminus \Lambda)$ as

$$\Gamma_k^\circ = \{((1, 0, \ldots, 0, \lambda, \lambda y_{n-k+2}, \ldots, \lambda y_n), (0, \ldots, 0, 1, y_{n-k+2}, \ldots, y_n))\}.$$

The functions $\lambda, y_{n-k+2}, \ldots, y_n$ give an isomorphism of Γ_k° with $\mathbb{A}^k$.

It follows that the classes

$$\lambda_k = [\Lambda_k] \quad \text{and} \quad \gamma_k = [\Gamma_k] \quad \text{in } A_k(B)$$

generate the Chow groups of B.

We next compute the intersection products. Since Λ_k is the preimage of a k-plane in $\mathbb{P}^n$ not containing p, and any two such planes are linearly equivalent in $\mathbb{P}^n$, the classes of their pullbacks are all equal to λ_k. Similarly, the class of the proper transform of any k-plane in $\mathbb{P}^n$ containing p is γ_k. Having these representative cycles for the classes λ_k and γ_k makes it easy to determine their intersection products.

For example, a general k-plane in $\mathbb{P}^n$ intersects a general l-plane transversely in a general $(k + l - n)$-plane; thus

$$\lambda_k \lambda_l = \lambda_{k+l-n} \quad \text{for all } k + l \geq n.$$

Similarly, the intersection of a general k-plane in $\mathbb{P}^n$ containing p with a general l-plane not containing p is a general $(k + l - n)$-plane not containing p, so that

$$\gamma_k \lambda_l = \lambda_{k+l-n} \quad \text{for all } k + l \geq n,$$

and likewise

$$\gamma_k \gamma_l = \gamma_{k+l-n} \quad \text{for all } k + l \geq n + 1.$$

Note the restriction $k + l \geq n + 1$ on the last set of products: In the case $k + l = n$, the proper transforms of a general k-plane through p and a general l-plane through p are disjoint.

This determines the Chow ring of B. The pairing $A_k(B) \times A_{n-k}(B) \to A_0(B) \cong \mathbb{Z}$ is given by

$$\lambda_k \lambda_{n-k} = \lambda_k \gamma_{n-k} = \gamma_k \lambda_{n-k} = 1 \quad \text{and} \quad \gamma_k \gamma_{n-k} = 0.$$

This is nondegenerate, so the classes $\lambda_0, \ldots, \lambda_{n-1}$ and $\gamma_1, \ldots, \gamma_n$ freely generate $A(B)$.

It follows that we can express the class of the exceptional divisor E in terms of the generators Λ_{n-1} and Γ_{n-1} of $A_{n-1}(B)$. The most geometric way to do this is to observe that Λ'_{n-1} is linearly equivalent in $\mathbb{P}^n$ to a hyperplane $\Sigma \subset \mathbb{P}^n$ containing p, so the pullback of Σ is linearly equivalent to the union of the exceptional divisor E and a divisor D. Since D projects to a hyperplane of $\mathbb{P}^{n-1}$, it is contained in the preimage Γ of such a hyperplane. Since Γ is a $\mathbb{P}^1$-bundle over its image, it is irreducible. We see upon comparing dimensions that $D = \Gamma$. Since any two hyperplanes in $\mathbb{P}^{n-1}$ are rationally equivalent, so are their preimages in B; thus $\Lambda_{n-1} \sim D + E \sim \Gamma_{n-1} + E$, or $[E] = \lambda_{n-1} - \gamma_{n-1}$.

We now turn to the ring structure of $A(B)$. Let $\lambda = [\Lambda_{n-1}]$ and $e = [E] = \lambda - \gamma_{n-1}$. Since $\Lambda_{n-1} \cap E = \varnothing$, we have

$$\lambda e = 0.$$

Also,

$$\lambda_k = \lambda^{n-k} \quad \text{for } k = 0, \ldots, n-1,$$

and, since $\gamma_{n-1} = \lambda - e$,

$$\gamma_k = \gamma_{n-1}^{n-k} = (\lambda - e)^{n-k} = \lambda^{n-k} + (-1)^{n-k} e^{n-k} \quad \text{for } k = 1, \ldots, n.$$

It follows that λ and e generate $A(B)$ as a ring. In addition to the relation $\lambda e = 0$, they satisfy the relation

$$0 = \gamma_{n-1}^n = (\lambda - e)^n = \lambda^n + (-1)^n e^n.$$

Thus the Chow ring is a homomorphic image of the ring

$$A' := \mathbb{Z}[\lambda, e]/(\lambda e, \lambda^n + (-1)^n e^n).$$

For $m = 1, \ldots, n - 1$, it is clear that every homogeneous element of degree m in A' is a $\mathbb{Z}$-linear combination of e^m and λ^m. Since for $0 < m < n$ the group $A^m(B)$ is a free $\mathbb{Z}$-module of rank 2, this implies that the map $A' \twoheadrightarrow A$ is an isomorphism. $\qquad \square$

We have computed the intersection products of the Λ_k and Γ_k by taking representatives that meet transversely (indeed, the possibility of doing so motivated our choice of Λ as a cross section of α above). Since E is the only irreducible variety in the class $[E]$ we cannot give a representative for e^2 quite as easily. But as we have seen,

$[E] = [\Lambda_{n-1}] - [\Gamma_{n-1}]$ and both Λ_{n-1} and Γ_{n-1} are transverse to E (this illustrates the conclusion of the moving lemma!). It follows that

$$e^2 = [E \cap (\Lambda - \Gamma)] = -[E \cap \Gamma_{n-1}].$$

Since E projects isomorphically to $\mathbb{P}^{n-1}$ and Γ projects to a hyperplane in $\mathbb{P}^{n-1}$, we see that $E \cap \Gamma_{n-1}$ is a hyperplane in E; that is, $[E]^2$ is the *negative* of the class of a hyperplane in E.

The Chow ring of the blow-up of $\mathbb{P}^3$ along a line is worked out in Exercises 2.38–2.40. More generally, we will see how to describe the Chow ring of a general projective bundle in Chapter 9, and the Chow ring of a more general blow-up in Chapter 13.

2.1.10 Intersection multiplicities via blow-ups

We can use the description of the Chow ring of the blow-up B of $\mathbb{P}^n$ at a point to prove Proposition 1.29, relating the intersection multiplicity of two subvarieties $X, Y \subset \mathbb{P}^n$ of complementary dimension at a point to the multiplicities of X and Y at p. (The same argument will apply to subvarieties of an arbitrary smooth variety once we have described the Chow ring of a general blow-up in Section 13.6.) The idea is to compare the intersection $X \cap Y \subset \mathbb{P}^n$ of X and Y in $\mathbb{P}^n$ with the intersection $\widetilde{X} \cap \widetilde{Y} \subset B$ of their proper transforms in the blow-up.

We start by finding the class of the proper transforms:

Proposition 2.14. *Let $X \subset \mathbb{P}^n$ be a k-dimensional variety and $\widetilde{X} \subset B$ its proper transform in the blow-up B of $\mathbb{P}^n$ at a point p. If X has degree d and multiplicity $m = \text{mult}_p(X)$ at p, then the class of the proper transform is*

$$[\widetilde{X}] = (d - m)\lambda_k + m\gamma_k \in A(B).$$

Proof: This follows from two things: the definition of the multiplicity of X at p as the degree of the projectivized tangent cone $\mathbb{T}C_p X$ (Section 1.3.8), and the identification of the projectivized tangent cone $\mathbb{T}C_p X$ to X at p with the intersection of the proper transform $\widetilde{X}$ with the exceptional divisor $E \cong \mathbb{P}^{n-1} \subset B$ (on page 36).

Given these, the proposition follows from the observation that if $i : E \hookrightarrow B$ is the inclusion, then $i^*(\lambda_k) = 0$ (λ_k is represented by the cycle Λ_k, which is disjoint from E) and $i^*(\gamma_k)$ is the class of a $(k-1)$-plane in $E \cong \mathbb{P}^{n-1}$ (γ_k is represented by the cycle Γ_k, which intersects E transversely in a $(k-1)$-plane). This says that the coefficient of γ_k in the expression above for $[\widetilde{X}]$ must be the multiplicity $m = \text{mult}_p(X)$; the coefficient of λ_k similarly follows by restricting to a hyperplane not containing p. $\square$

Now suppose we are in the setting of Proposition 1.29: $X, Y \subset \mathbb{P}^n$ are dimensionally transverse subvarieties of complementary dimensions k and $n - k$, having multiplicities m and m' respectively at p. If, as we supposed in the statement of the proposition, the projectivized tangent cones to X and Y at p are disjoint (that is, $\widetilde{X} \cap \widetilde{Y} \cap E = \varnothing$), then the intersection multiplicity $m_p(X, Y)$ of X and Y at p is simply the difference between the intersection number $\deg([X][Y])$ of X and Y in $\mathbb{P}^n$ and the intersection number $\deg([\widetilde{X}][\widetilde{Y}])$ of their proper transforms in B; by Proposition 2.14 and our description of the Chow ring $A(B)$, this is just $m m'$.

2.2 Loci of singular plane cubics

This section represents an important shift in viewpoint, from studying the Chow rings of common and useful algebraic varieties to studying Chow rings of *parameter spaces*. It is a hallmark of algebraic geometry that the set of varieties (and more generally, schemes, morphisms, bundles and other geometric objects) with specified numerical invariants may often be given the structure of a scheme itself, sometimes called a parameter space. Applying intersection theory to the study of such a parameter space, we learn something about the geometry of the objects parametrized, and about geometrically characterized classes of these objects. This gets us into the subject of *enumerative geometry*, and was one of the principal motivations for the development of intersection theory in the 19th century.

By way of illustration, we will focus on the family of curves of degree 3 in $\mathbb{P}^2$: plane cubics. Plane cubics are parametrized by the set of homogeneous cubic polynomials $F(X, Y, Z)$ in three variables, modulo scalars, that is, by $\mathbb{P}^9$.

There is a continuous family of isomorphism classes of smooth plane cubics, parametrized naturally by the affine line (see Hartshorne [1977]), but there are only a finite number of isomorphism classes of singular plane cubics:

- irreducible plane cubics with a node;
- irreducible plane cubics with a cusp;
- plane cubics consisting of a smooth conic and a line meeting it transversely;
- plane cubics consisting of a smooth conic and a line tangent to it;
- plane cubics consisting of three nonconcurrent lines ("triangles");
- plane cubics consisting of three concurrent lines ("asterisks");
- cubics consisting of a double line and a line; and finally
- cubics consisting of a triple line.

These are illustrated in Figures 2.7–2.9, where the arrows represent specialization, as explained below.

The locus in $\mathbb{P}^9$ of points corresponding to singular curves of each type is an orbit of PGL_3 and a locally closed subset of $\mathbb{P}^9$. These loci, together with the open subset $U \subset \mathbb{P}^9$ of smooth cubics, give a stratification of $\mathbb{P}^9$. We may ask: What are the closed strata of this stratification like? What are their dimensions? What containment relations hold among them? Where is each one smooth and singular? What are their tangent spaces and tangent cones? What are their degrees as subvarieties of $\mathbb{P}^9$?

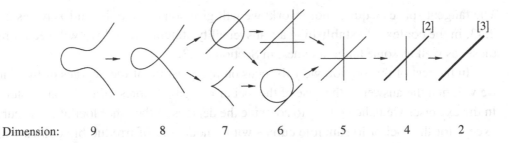

Dimension: 9   8   7   6   5   4   2

Figure 2.7 Hierarchy of singular plane cubic curves.

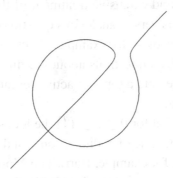

Figure 2.8 Nodal cubic about to become the union of a conic and a transverse line:
$y^2 - x^2(x + 1) + 100(x - y)\left(\left(x - \frac{1}{2}\right)^2 + \left(y - \frac{1}{2}\right)^2 - \frac{1}{2}\right)$.

Some of these questions are easy to answer. For example, the dimensions are given in Figure 2.7, and the reader can verify them as an exercise. The specialization relationships (when one orbit is contained in the closure of another, as indicated by arrows in the chart) are also easy, because to establish that one orbit lies in the closure of another it suffices to exhibit a one-parameter family $\{C_t \subset \mathbb{P}^2\}$ of plane cubics with an open set of parameter values t corresponding to one type and a point corresponding to the other. The noninclusion relations are subtler — why, for example, is a triangle not a specialization of a cuspidal cubic? — but can also be proven by focusing on the singularities of the curves.

Figure 2.9 Cuspidal cubic about to become the union of a conic and a tangent line:
$y^2 - x^3 + 7y(x^2 + (y - 1)^2 - 1)$.

The tangent spaces require more work; we will give some examples in Exercises 2.42–2.43, in the context of establishing a transversality statement, and we will see more of these, as well as some tangent cones, in Section 7.7.3.

In the rest of this section we will focus on the question of the degrees of these loci; we will find the answer in the case of the loci of reducible cubics, triangles and asterisks. In the exercises we indicate how to compute the degrees of the other loci of plane cubics, except for the loci of irreducible cubics with a node and of irreducible cubics with a cusp; these will be computed in Section 7.3.2 and Section 11.4 respectively.

The calculations here barely scratch the surface of the subject; see for example Aluffi [1990; 1991] for a beautiful and extensive treatment of the enumerative geometry of plane cubics. Moreover, the answers to analogous questions for higher-degree curves or hypersurfaces of higher dimension — for example, about the stratification by singularity type — remain mysterious. Even questions about the dimension and irreducibility of these loci are mostly open; they are a topic of active research. See Greuel et al. [2007] for an introduction to this area.

For example, it is known that for $0 \le \delta \le \binom{d}{2}$ the locus of plane curves of degree d having exactly δ nodes is irreducible of codimension δ in the projective space $\mathbb{P}^N$ of all plane curves of degree d (see, for example, Harris and Morrison [1998]), and its degree has also been determined (Caporaso and Harris [1998]). But we do not know the answers to the analogous questions for plane curves with δ nodes and κ cusps, and when it comes to more complicated singularities even existence questions are open. For example, for $d > 6$ it is not known whether there exists a rational plane curve $C \subset \mathbb{P}^2$ of degree d whose singularities consist of just one double point.

2.2.1 Reducible cubics

Let $\Gamma \subset \mathbb{P}^9$ be the closure of the locus of cubics consisting of a conic and a transverse line (equivalently, the locus of reducible and/or nonreduced cubics). We can describe Γ as the image of the map

$$\tau : \mathbb{P}^2 \times \mathbb{P}^5 \to \mathbb{P}^9$$

from the product of the space $\mathbb{P}^2$ of homogeneous linear forms and the space $\mathbb{P}^5$ of homogeneous quadratic polynomials to $\mathbb{P}^9$, given simply by multiplication: $(F, G) \mapsto FG$. Inasmuch as the coefficients of the product FG are bilinear in the coefficients of F and G, the pullback $\tau^*(\zeta)$ of the hyperplane class $\zeta \in A^1(\mathbb{P}^9)$ is the sum

$$\tau^*(\zeta) = \alpha + \beta,$$

where α and β are the pullbacks to $\mathbb{P}^2 \times \mathbb{P}^5$ of the hyperplane classes on $\mathbb{P}^2$ and $\mathbb{P}^5$.

By unique factorization of polynomials, the map τ is birational onto its image; it follows that the degree of Γ is given by

$$\deg(\Gamma) = \deg(\tau^*(\zeta)^7) = \deg((\alpha + \beta)^7) = 21,$$

and this is the answer to Keynote Question (a).

Another way to calculate the degree of Γ is described in Exercises 2.42–2.44.

2.2.2 Triangles

A similar analysis gives the answer to Keynote Question (b) — how many cubics in a three-dimensional linear system factor completely, as a product of three linear forms. Here, the key object is the closure $\Sigma \subset \mathbb{P}^9$ of the locus of such totally reducible cubics, which we may call *triangles*; the keynote question asks us for the number of points of intersection of Σ with a general 3-plane. By Bertini's theorem this is the degree of Σ.

Since Σ is the image of the map

$$\mu : \mathbb{P}^2 \times \mathbb{P}^2 \times \mathbb{P}^2 \to \mathbb{P}^9,$$

$$([L_1], [L_2], [L_3]) \mapsto [L_1 L_2 L_3],$$

we can proceed as before, with the one difference that the map is now no longer birational, but rather is generically six-to-one. Thus if $\alpha_1, \alpha_2, \alpha_3 \in A^1(\mathbb{P}^2 \times \mathbb{P}^2 \times \mathbb{P}^2)$ are the pullbacks of the hyperplane classes in the factors $\mathbb{P}^2$ via the three projections, so that

$$\mu^*(\zeta) = \alpha_1 + \alpha_2 + \alpha_3,$$

we get

$$\deg(\Sigma) = \tfrac{1}{6} \deg (\alpha_1 + \alpha_2 + \alpha_3)^6 = \tfrac{1}{6} \binom{6}{2,2,2} = 15.$$

This is the answer to Keynote Question (b): In a general three-dimensional linear system of cubics, there will be exactly 15 triangles.

2.2.3 Asterisks

By an *asterisk*, we mean a cubic consisting of the sum of three concurrent lines. To see that the closure of this locus is indeed a subvariety of $\mathbb{P}^9$ and to calculate its degree, let

$$\mu : \mathbb{P}^2 \times \mathbb{P}^2 \times \mathbb{P}^2 \to \mathbb{P}^9$$

be as in Section 2.2.2, and consider the subset

$$\Phi = \{(L_1, L_2, L_3) \in \mathbb{P}^2 \times \mathbb{P}^2 \times \mathbb{P}^2 \mid L_1 \cap L_2 \cap L_3 \neq \varnothing\};$$

the locus $A \subset \mathbb{P}^9$ of asterisks is then the image $\mu(\Phi)$ of Φ under the map μ. If we write the line L_i as the zero locus of the linear form

$$a_{i,1} X + a_{i,2} Y + a_{i,3} Z,$$

then the condition that $L_1 \cap L_2 \cap L_3 \neq \varnothing$ is equivalent to the equality

$$\begin{vmatrix} a_{1,1} & a_{1,2} & a_{1,3} \\ a_{2,1} & a_{2,2} & a_{2,3} \\ a_{3,1} & a_{3,2} & a_{3,3} \end{vmatrix} = 0.$$

The left-hand side of this equation is a homogeneous trilinear form on $\mathbb{P}^2 \times \mathbb{P}^2 \times \mathbb{P}^2$, from which we see that Φ is a closed subset of $\mathbb{P}^2 \times \mathbb{P}^2 \times \mathbb{P}^2$ and A is a closed subset of $\mathbb{P}^9$. Moreover, we see that the class of Φ is

$$[\Phi] = \alpha_1 + \alpha_2 + \alpha_3 \in A^1(\mathbb{P}^2 \times \mathbb{P}^2 \times \mathbb{P}^2),$$

so that the pullback via μ of five general hyperplanes in $\mathbb{P}^9$ will intersect Φ in

$$\deg([\Phi](\alpha_1 + \alpha_2 + \alpha_3)^5) = \deg(\alpha_1 + \alpha_2 + \alpha_3)^6 = \binom{6}{2,2,2} = 90$$

points. Since the map $\mu|_\Phi : \Phi \to A$ has degree 6, it follows that the degree of the locus $A \subset \mathbb{P}^9$ of asterisks is 15.

2.3 The circles of Apollonius

Apollonius posed the problem of constructing the circles tangent to three given circles. Using Bézout's theorem we can count them.

Theorem 2.15. *If D_1, D_2 and D_3 are three general circles, there are exactly eight circles tangent to all three.*

2.3.1 What is a circle?

We first need to say what we mean by a circle in complex projective space. While circles are usually characterized in terms of a metric, in fact they have a purely algebro-geometric definition. Starting from the affine equation

$$(x - a)^2 + (y - b)^2 = r^2 \tag{2.1}$$

of a circle of radius r centered at a point (a, b) in $\mathbb{A}^2$, and homogenizing with respect to a new variable z, we get $(x - az)^2 + (y - bz)^2 = r^2 z^2$. We think of the line $z = 0$ as the "line at infinity," and we see that the circle passes through the two points

$$\circ_+ = (1, i, 0) \quad \text{and} \quad \circ_- := (1, -i, 0)$$

on the line at infinity; these are called the *circular points*. Conversely, it is an easy exercise to see that the equation of any smooth conic passing through the two circular points can be put into put into the form (2.1).

We thus define a *circle* to be a conic in $\mathbb{P}^2$ with coordinates x, y, z passing through the two circular points on the line at infinity $z = 0$; equivalently, a circle is a conic $C = V(f) \subset \mathbb{P}^2$ whose defining equation f lies in the ideal $(z, x^2 + y^2)$. (This formulation makes sense over any field of characteristic $\neq 2$.) We see from this that the set of circles is a three-dimensional linear subspace in the space $\mathbb{P}^5$ of all conics in $\mathbb{P}^2$.

Much geometry can be done in this context. For example, a direct calculation shows that the *center* of the circle is the point of intersection of the tangent lines to the circle at the circular points; in particular, the coordinates of the center are rational functions of the coefficients of its defining equation.

Note that when we characterize circles as conics containing the circular points p, q at infinity, we are including singular conics that pass through these points, and we see that there are two kinds of singular circles: unions of the line at infinity $\overline{\circ_+, \circ_-}$ with another line in $\mathbb{P}^2$, and unions $L \cup M$ of lines with $\circ_+ \in L$ and $\circ_- \in M$. It is easy to see from the equations that these are the limits of smooth circles of radius r as $r \to \infty$ and $r \to 0$, respectively. (When the radius of a circle goes to 0, we may think the circle shrinks to a point, but that is because we are seeing only points in $\mathbb{R}^2$: over $\mathbb{C}$, the conic $x^2 + y^2 = 0$ consists of the two lines $x = \pm iy$.)

2.3.2 Circles tangent to a given circle

Next, we have to define what we mean when we say two circles are tangent. Let $D \subset \mathbb{P}^2$ be a smooth circle. If C is any other circle, we can write the intersection $C \cap D$, viewed as a divisor on D, as the sum

$$C \cap D = \circ_+ + \circ_- + p + q.$$

In these terms, we make the following definition:

Definition 2.16. We say that the circle C is tangent to the circle D if $p = q$.

In other words, C and D are tangent if they have two coincident intersections in addition to their intersection at the circular points; this includes the case where C, D have intersection multiplicity 3 at p or q. Let Z_D be the variety of circles tangent to a given smooth circle D. We will show that Z_D is a quadric cone in the $\mathbb{P}^3$ of circles.

It is visually obvious that the family of circles in $\mathbb{R}^2$ tangent to a given circle is two-dimensional. To prove this algebraically we consider the incidence correspondence

$$\Phi = \{(r, C) \in D \times \mathbb{P}^3 \mid C \text{ is tangent to } D \text{ at } r\},$$

where when r is a circular point the condition should be interpreted as saying the intersection multiplicity $m_r(C, D)$ is ≥ 3. The condition that a curve $f = 0$ meet a curve D with multiplicity m at a smooth point $r \in D$ means that the function $f|_D$ vanishes to order m at r; it is thus m linear conditions on the coefficients of the equation f. This shows that, for each point $r \in D$, the fiber of Φ over r is a $\mathbb{P}^1$, cut out by two linear equations in the space of circles. It follows that Φ is irreducible of dimension 2. Since almost all circles tangent to D are tangent at a single point, the map $\Phi \to \mathbb{P}^3$ sending (r, C) to C is birational. Thus the image Z_D of Φ in $\mathbb{P}^3$ also two-dimensional.

To show that $Z_D \subset \mathbb{P}^3$ is a quadric, let $L \subset \mathbb{P}^3$ be a general line, corresponding to a pencil of circles $\{C_t\}_{t \in \mathbb{P}^1}$. If f and g are the defining equations of C_0 and C_∞, the rational function f/g has two zeros (where C_0 meets D, aside from the circular points) and two poles (where C_∞ meets D, aside from the circular points), so f/g gives a map $D \to \mathbb{P}^1$ of degree 2.

The circles C_t tangent to D correspond to the branch points of this map; by the classical Riemann–Hurwitz formula, there will be two such points. Thus the degree of $L \cap Z_D$ is 2, and we see that Z_D is a quadric surface. On the other hand, if $C \neq D$ is tangent to D at $r \in D$, then every member of the linear space of circles jointing C to D satisfies the linear condition for tangency at r, so $Z_D \subset \mathbb{P}^3$ is a cone with vertex corresponding to D, as claimed.

2.3.3 Conclusion of the argument

Now let D_1, D_2, D_3 be three circles. If the intersection $A := Z_{D_1} \cap Z_{D_2} \cap Z_{D_3}$ is finite, then Bézout's theorem implies that $\deg A = 2^3 = 8$. To prove that the intersection is finite for nearly all triples of circles D_i, we consider the incidence correspondence

$$\Psi := \{(D_1, D_2, D_3, C) \in (\mathbb{P}^3)^4 \mid C \text{ is tangent to each of the } D_i\}.$$

If we project onto the last factor, the fiber is Z_C^3, and thus has dimension 6, so $\dim \Psi = 9$. Thus the projection to the nine-dimensional space consisting of all triples (D_1, D_2, D_3) cannot have generic fiber of positive dimension.

We have now shown that, counting with multiplicity, there are eight circles tangent to three general circles D_1, D_2, D_3. To prove that there are really eight distinct circles, we would need to prove that the intersection $Z_{D_1} \cap Z_{D_2} \cap Z_{D_3}$ is transverse. In Section 8.2.3 we will see how to do this directly, by identifying explicitly the tangent spaces to the loci Z_D. For now we will be content to give an example of the situation where the eight circles are distinct: it is shown on the cover of this book!

Another approach to the circles of Apollonius, via the notion of *theta-characteristics*, is given in Harris [1982]. There is also an analogous notion of a *sphere* in $\mathbb{P}^3$; see for example Exercise 13.32.

2.4 Curves on surfaces

Aside from enumerative problems, intersection products appeared in algebraic geometry as a central tool in the theory of surfaces, developed mostly by the Italians in the late 19th and early 20th centuries. In this section we describe some of the basic ideas. This will serve to illustrate the use of intersection products in a simple setting, and also provide us with formulas that will be useful throughout the book. A different treatment of some of this material is in the last chapter of Hartshorne [1977]; and much more can be found, for example, in Beauville's beautiful book on algebraic surfaces [1996].

Throughout this section we will use some classical notation: If S is a smooth projective surface and $\alpha, \beta \in A^1(S)$, we will write $\alpha \cdot \beta$ for the degree $\deg(\alpha\beta)$ of their product $\alpha\beta \in A^2(S)$, and we refer to this as the *intersection number* of the two classes. Further, if $C \subset S$ is a curve we will abuse notation and write C for the class $[C] \in A^1(S)$. Thus, for example, if $C, D \subset S$ are two curves, we will write $C \cdot D$ in place of $\deg([C] \cdot [D])$ and we will write C^2 for $\deg([C]^2)$. The reader should not be misled by this notation into thinking that $A^2(S) = \mathbb{Z}$ — as we have already remarked, the group $A^2(S)$ need not even be finite-dimensional in any reasonable sense.

2.4.1 The genus formula

One of the first formulas in which intersection products appeared was the *genus formula*, a straightforward rearrangement of the adjunction formula that describes the genus of a smooth curve on a smooth projective surface (we will generalize it to some singular curves in Section 2.4.6). If $C \subset S$ is a smooth curve of genus g on a smooth surface, then

$$K_C = (K_S + C)|_C;$$

since the degree of the canonical class of C is $2g - 2$, this yields

$$g = \frac{C^2 + K_S \cdot C}{2} + 1. \tag{2.2}$$

Example 2.17 (Plane curves). By way of examples, consider first a smooth curve $C \subset \mathbb{P}^2$ of degree d. If we let $\zeta \in A^1(\mathbb{P}^2)$ be the class of a line, we have $[C] = d\zeta$ and $K_{\mathbb{P}^2} = -3\zeta$, so the genus of C is

$$g = \frac{-3d + d^2}{2} + 1 = \frac{(d-1)(d-2)}{2}.$$

Thus we recover, for example, the well-known fact that lines and smooth conics have genus 0 while smooth cubics have genus 1.

Example 2.18 (Curves on a quadric). Now suppose that $Q \subset \mathbb{P}^3$ is a smooth quadric surface, and that $C \subset Q$ is a smooth curve of bidegree (d, e) — that is, a curve linearly equivalent to d times a line of one ruling plus e times a line of the other (equivalently, in terms of the isomorphism $Q \cong \mathbb{P}^1 \times \mathbb{P}^1$, the zero locus of a bihomogeneous polynomial of bidegree (d, e)). Let α and $\beta \in A^1(Q)$ be the classes of the lines of the two rulings of Q, as in the discussion in Section 2.1.5 above, and let $\zeta = \alpha + \beta$ be the class of a plane section of Q. Applying adjunction to $Q \subset \mathbb{P}^3$, we have

$$K_Q = (K_{\mathbb{P}^3} + Q)|_Q = -2\zeta = -2\alpha - 2\beta.$$

Thus, by the genus formula,

$$g = \frac{(d\alpha + e\beta)^2 - 2(\alpha + \beta)(d\alpha + e\beta)}{2} + 1$$

$$= \frac{2de - 2d - 2e}{2} + 1$$

$$= (d - 1)(e - 1).$$

2.4.2 The self-intersection of a curve on a surface

We can sometimes use the genus formula to determine the self-intersection of a curve on a surface. For example, suppose that $S \subset \mathbb{P}^3$ is a smooth surface of degree d and $L \subset S$ is a line. Letting $\zeta \in A^1(S)$ denote the plane class and applying adjunction to $S \subset \mathbb{P}^3$, we have $K_S = (d - 4)\zeta$, so that $L \cdot K_S = d - 4$; since the genus of L is 0, the genus formula yields

$$0 = \frac{L^2 + d - 4}{2} + 1,$$

or

$$L^2 = 2 - d.$$

The cases $d = 1$ (a line on a plane) and $d = 2$ are probably familiar already; in the case $d \geq 3$, the formula implies the qualitative statement that *a smooth surface $S \subset \mathbb{P}^3$ of degree 3 or more can contain only finitely many lines.* (See Exercise 2.60 below for a sketch of a proof, and Exercise 2.59 for an alternative derivation of $L^2 = 2 - d$.)

We note in passing that we could similarly ask for the degree of the self-intersection of a 2-plane $\Lambda \cong \mathbb{P}^2 \subset X$ on a smooth hypersurface $X \subset \mathbb{P}^5$. This is far harder (as the reader may verify, neither of the techniques suggested in this chapter for calculating the self-intersection of a line on a smooth surface $S \subset \mathbb{P}^3$ will work); the answer is given in Exercise 13.22.

2.4.3 Linked curves in $\mathbb{P}^3$

Another application of the genus formula yields a classical relation between what are called *linked curves* in $\mathbb{P}^3$.

Let $S, T \subset \mathbb{P}^3$ be smooth surfaces of degrees s and t, and suppose that the scheme-theoretic intersection $S \cap T$ consists of the union of two smooth curves C and D with no common components. Let the degrees of C and D be c and d, and let their genera be g and h respectively. By Bézout's theorem, we have

$$c + d = st,$$

so that the degree of C determines the degree of D. What is much less obvious is that the degree and genus of C determine the degree and genus of D. Here is one way to derive the formula.

To start, we use the genus formula (2.2) to determine the self-intersection of C on S: Since $K_S = (s - 4)\zeta$, we have

$$g = \frac{C^2 + K_S \cdot C}{2} + 1 = \frac{C^2 + (s - 4)c}{2} + 1,$$

and hence

$$C^2 = 2g - 2 - (s - 4)c$$

(generalizing our formula in Section 2.4.2 for the self-intersection of a line). Next, since $[C] + [D] = t\zeta \in A^1(S)$, we can write the intersection number of C and D on S as

$$C \cdot D = C(t\zeta - C) = tc - (2g - 2 - (s - 4)c) = (s + t - 4)c - (2g - 2).$$

This in turn allows us to determine the self-intersection of D on S:

$$D^2 = D(t\zeta - C) = td - ((s + t - 4)c - (2g - 2)).$$

Applying the genus formula to D, we obtain

$$h = \frac{D^2 + K_S \cdot D}{2} + 1$$
$$= \frac{td - ((s + t - 4)c - (2g - 2)) + (s - 4)d}{2} + 1.$$

Simplifying, we get

$$h - g = \frac{s + t - 4}{2}(d - c); \tag{2.3}$$

in English, *the difference in the genera of C and D is proportional to the difference in their degrees, with ratio* $(s + t - 4)/2$.

The answer to Keynote Question (f) is a special case of this: If $L \subset \mathbb{P}^3$ is a line, and S and T general surfaces of degrees s and t containing L, then, writing $S \cap T = L \cup C$, we see that C is a curve of degree $st - 1$ and genus

$$h = \frac{(s + t - 4)(st - 2)}{2}.$$

As is often the case with enumerative formulas, this is just the beginning of a much larger picture. The theory of *liaison* describes the relationship between the geometry of linked curves such as C and D above. The theory in general is far more broadly applicable (the curves C and D need only be Cohen–Macaulay, and we need no hypotheses at all on the surfaces S and T), and ultimately provides a complete answer to the question of when two given curves $C, D \subset \mathbb{P}^3$ can be connected by a series of curves $C = C_0, C_1, \ldots, C_{n-1}, C_n = D$ with C_i and C_{i+1} linked as above. We will see a typical application of the notion of linkage in Exercise 2.62 below; for the general theory, see Peskine and Szpiro [1974].

2.4.4 The blow-up of a surface

The blow-up of a point on a surface plays an important role in the theory of surfaces, and we will now explain a little of this theory. Locally (in the analytic or étale topology), such blow-ups look like the blow-up of $\mathbb{P}^2$ at a point, which was treated in Section 2.1.9.

To fix notation, we let $p \in S$ be a point in a smooth projective surface and write $\pi : \widetilde{S} \to S$ for the blow-up of S at p. We write $E = \pi^{-1}(p) \subset \widetilde{S}$ for the preimage of p, called the *exceptional divisor*, and $e \in A^1(\widetilde{S})$ for its class. We will use the following definitions and facts:

- $\pi : \widetilde{S} \to S$ is birational, and if $q \in E \subset \widetilde{S}$ is any point of the exceptional divisor, then there are generators z, w for the maximal ideal of $\mathcal{O}_{\widetilde{S},q}$ and generators x, y for the maximal ideal of $\mathcal{O}_{S,p}$ such that $\pi^*x = zw$, $\pi^*y = w$, and E is defined locally by the equation $w = 0$. In particular, $\widetilde{S}$ is smooth and E is a Cartier divisor.
- If C is a smooth curve through p, then the *proper transform* $\widetilde{C}$ of C, which is by definition the closure in $\widetilde{S}$ of $\pi^{-1}(C \setminus \{p\})$, meets E transversely in one point.
- More generally, if C has an *ordinary m-fold point at* p, then $\widetilde{C}$ meets E transversely in m distinct points. Here we say that C has an ordinary m-fold point at p if the completion of the local ring of C at p has the form

$$\widehat{\mathcal{O}}_{C,p} \cong \Bbbk[\![x, y]\!] \Big/ \left(\prod_{i=1}^{m} (x - \lambda_i y) \right)$$

for some distinct $\lambda_1, \ldots, \lambda_m \in \Bbbk$; geometrically, this says that, near p, C consists of the union of m smooth branches meeting pairwise transversely at p.

We can completely describe $A(\widetilde{S})$ in terms of $A(S)$:

Proposition 2.19. *Let S be a smooth projective surface and $\pi : \widetilde{S} \to S$ the blow-up of S at a point p; let $e \in A^1(\widetilde{S})$ be the class of the exceptional divisor.*

(a) $A(\widetilde{S}) = A(S) \oplus \mathbb{Z}e$ as abelian groups.

(b) $\pi^*\alpha \cdot \pi^*\beta = \pi^*(\alpha\beta)$ for any $\alpha, \beta \in A^1(S)$.

(c) $e \cdot \pi^*\alpha = 0$ for any $\alpha \in A^1(S)$.

(d) $e^2 = -[q]$ for any point $q \in E$ (in particular, $\deg(e^2) = -1$).

Proof: We first show that π_* and π^* are inverse isomorphisms between $A^2(S)$ and $A^2(\widetilde{S})$. By the moving lemma, if $\alpha \in A_0(S)$ is any class, we can write $\alpha = [A]$ for some $A \in Z_0(S)$ with support disjoint from p; thus $\pi_*\pi^*\alpha = \alpha$. Likewise, if $\alpha \in A_0(\widetilde{S})$ is any class, we can write $\alpha = [A]$ for some $A \in Z_0(\widetilde{S})$ with support disjoint from E; thus $\pi^*\pi_*\alpha = \alpha$.

We next turn to A^1. If $\alpha \in A^1(S)$ is any class, we can write $\alpha = [A]$ for some $A \in Z_1(S)$ with support disjoint from p; thus $\pi_*\pi^*\alpha = \alpha$. On the other hand, the kernel of the pushforward map $\pi_* : Z_1(\tilde{S}) \to Z_1(S)$ is just the subgroup generated by e, the class of E. Thus we have an exact sequence

$$0 \longrightarrow \langle e \rangle \longrightarrow A^1(\tilde{S}) \longrightarrow A^1(S) \longrightarrow 0,$$

with $\pi^* : A^1(S) \to A^1(\tilde{S})$ splitting the sequence.

It remains to show that the class e is not torsion in $A^1(\tilde{S})$. This follows from the formula $\deg e^2 = -1$, which we will prove independently below.

Part (b) of the proposition simply recalls the fact that π^* is a ring homomorphism. For part (c) we use the push-pull formula:

$$\pi_*(e \cdot \pi^*\alpha) = \pi_* e \cdot \alpha = 0.$$

For part (d), let $C \subset S$ be any curve smooth at p, so that the proper transform $\tilde{C} \subset \tilde{S}$ of C will intersect E transversely at one point q. We have then

$$\pi^*[C] = [\tilde{C}] + e,$$

and intersecting both sides with the class e yields

$$0 = [q] + e^2,$$

so the self-intersection number of e is $\deg e^2 = -1$. $\qquad\square$

2.4.5 Canonical class of a blow-up

We can express the canonical class of $\tilde{S}$ in terms of the canonical class of S as follows:

Proposition 2.20. *With notation as above,*

$$K_{\tilde{S}} = \pi^* K_S + e.$$

Proof: We must show that if ω is a rational 2-form on S, regular and nonzero at p, then the pullback $\pi^*\omega$ vanishes simply along E. Let $q \in E \subset \tilde{S}$, and let (z, w) be generators of the maximal ideal of $\mathcal{O}_{\tilde{S},q}$ such that there are generators (x, y) for the maximal ideal of $\mathcal{O}_{S,p}$ with

$$\pi^* x = zw \quad \text{and} \quad \pi^* y = w.$$

It follows that

$$\pi^* dx = z\,dw + w\,dz \quad \text{and} \quad \pi^* dy = dw.$$

Thus

$$\pi^*(dx \wedge dy) = w(dz \wedge dw).$$

Since the local equation of E at q is $w = 0$, this shows that $\pi^* dx$ vanishes simply along E, as required.   □

2.4.6 The genus formula with singularities

It will be useful in a number of situations to have a version of the genus formula (2.2) that gives the geometric genus of a possibly singular curve $C \subset S$. (The *geometric genus* of a reduced curve is the genus of its normalization.) To start with the simplest case, suppose that $C \subset S$ is a curve smooth away from a point $p \in C$ of multiplicity m. Assume moreover that p is an ordinary m-fold point, so that in particular the proper transform $\tilde{C}$ is smooth. We can invoke the genus formula on $\tilde{S}$ to give a formula for the genus g of $\tilde{C}$ in terms of intersection numbers on S.

As divisors,

$$\pi^* C = \tilde{C} + m E,$$

so that

$$[\tilde{C}] = \pi^*[C] - me.$$

From Proposition 2.20, we have

$$K_{\tilde{S}} = \pi^* K_S + e,$$

and, putting this together with the genus formula for $\tilde{C} \subset \tilde{S}$ and Proposition 2.19, we have

$$
\begin{aligned}
g &= \frac{\tilde{C}^2 + K_{\tilde{S}} \cdot \tilde{C}}{2} + 1 \\
&= \frac{(\pi^* C - me)^2 + (\pi^* K_S + e)(\pi^* C - me)}{2} + 1 \\
&= \frac{C^2 + K_S \cdot C}{2} + 1 - \binom{m}{2}.
\end{aligned}
$$

More generally, if $C \subset S$ has singular points $p_1, \ldots, p_\delta$ of multiplicity $m_1, \ldots, m_\delta$, and the proper transform $\tilde{C}$ of C in the blow-up $\mathrm{Bl}_{\{p_1,\ldots,p_\delta\}}$ of S at the points p_i is smooth (as will, in particular, be the case if the p_i are ordinary m_i-fold points of C), we have

$$g = \frac{C^2 + K_S \cdot C}{2} + 1 - \sum \binom{m_i}{2}.$$

One can extend this further, to general singular curves $C \subset S$, by using iterated blow-ups, or by generalizing the adjunction formula, using the fact that any curve on a smooth surface has a canonical bundle (see for example Hartshorne [1977, Theorem III.7.11]).

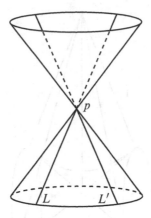

Figure 2.10 The degree of intersection of two lines on a quadric cone is $\frac{1}{2}$.

2.5 Intersections on singular varieties

In this section we discuss the problems of defining intersection products on singular varieties. To begin with, the moving lemma may fail if X is even mildly singular:

Example 2.21 (Figure 2.10). Let $C \subset \mathbb{P}^2 \subset \mathbb{P}^3$ be a smooth conic and $X = \overline{p, C} \subset \mathbb{P}^3$ the cone with vertex $p \notin \mathbb{P}^2$. Let $L \subset X$ be a line (which necessarily contains p). We claim that every cycle on X that is rationally equivalent to L has support containing p, and thus the conclusion of part (a) of the moving lemma does not hold for X.

To show this, we first remark that the degrees of any two rationally equivalent curves on X are the same; that is, there is a function $\deg : A_1(X) \to \mathbb{Z}$ taking each irreducible curve to its degree. For, if $i : X \to \mathbb{P}^3$ is the inclusion, then for any curve D on X we have

$$\deg D = \deg(\zeta \cdot i_*([D])),$$

where ζ is the class of a hyperplane in $\mathbb{P}^3$. In particular $\deg L = 1$ is odd.

Now let $D \subset X$ be any curve not containing p. We claim that the degree of D must be even. To see this, observe that the projection map $\pi_p : D \to C$ is a finite map whose fibers are the intersections of D with the lines of X; it follows that a general line in X will intersect D transversely in $\deg(\pi_p)$ points. Now let $H \subset \mathbb{P}^3$ be a general plane through p. H intersects X in the union of two general lines $L, L' \subset X$, and so meets D transversely in $2 \deg(\pi_p)$ points, so $\deg D$ is even. It follows that any cycle of dimension 1 on X, effective or not, whose support does not contain p has even degree, and hence cannot be rationally equivalent to L.

Retaining the notation of Example 2.21, one might hope to define an intersection product on $A(X)$ even without the moving lemma. It seems natural to think that since two distinct lines $L, L' \subset X$ through p meet in the reduced point p, we would have $[L][L'] = [p]$. However, if ζ is the class of a general plane section $H \cap X$ of X through p,

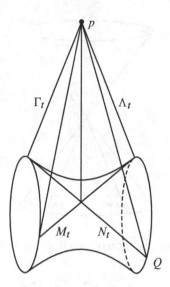

Figure 2.11 The intersection product of $[\Lambda_t]$ and the class of a line cannot be defined.

then (since such a hyperplane meets each L transversely in one point) we might also expect $\zeta[L] = [p]$. But ζ is rationally equivalent to the union of two lines through p. Thus, if both expectations were satisfied, we would have

$$[p] = \zeta[L'] = 2[L][L'] = 2[p].$$

Applying the degree map, we would get the contradiction $1 = 2$.

There is a way around the difficulty, if we work in the ring $A(X) \otimes \mathbb{Q}$: We can take the product of the classes of two lines to be one-half the class of the point p, and our contradiction is resolved. As Mumford has pointed out, something similar can be done for all normal surfaces (see Example 8.3.11 of Fulton [1984]). But in higher dimensions there are more difficult problems, as the following example shows:

Example 2.22. Let $Q \subset \mathbb{P}^3 \subset \mathbb{P}^4$ be a smooth quadric surface, and let $X = \overline{p, Q}$ be the cone in $\mathbb{P}^4$ with vertex $p \notin \mathbb{P}^3$. The quadric Q contains two families of lines $\{M_t\}$ and $\{N_t\}$, and the cone X is correspondingly swept out by the two families of 2-planes $\{\Lambda_t = \overline{p, M_t}\}$ and $\{\Gamma_t = \overline{p, N_t}\}$; see Figure 2.11.

Now, any line $L \subset X$ not passing through the vertex p maps, under projection from p, to a line of Q; that is, it must lie either in a plane Λ_t or in a plane Γ_t; lines on X that do pass through p lie on one plane of each type. Note that since lines $M_t, M_{t'} \subset Q$ of the same ruling are disjoint for $t \neq t'$, while lines M_t and $N_{t'}$ of opposite rulings meet in a point, a general line $M \subset X$ lying in a plane Λ_t is disjoint from $\Lambda_{t'}$ for $t \neq t'$ and meets each plane Γ_s transversely in a point. Thus, if there were any intersection product on $A(X)$ satisfying the fundamental condition $(*)$ of Theorem 1.5, we would have

$$[M][\Lambda_t] = 0 \quad \text{and} \quad [M][\Gamma_t] = [q]$$

for some point $q \in X$. Likewise, for a general line $N \subset X$ lying in a plane Γ_t, the opposite would be true; that is, we would have

$$[N][\Lambda_t] = [r] \quad \text{and} \quad [N][\Gamma_t] = 0.$$

But the lines M and N — indeed, any two lines on X — are rationally equivalent! Since any two lines in Λ_t are rationally equivalent, the line M is rationally equivalent to the line of intersection $\Lambda_t \cap \Gamma_s$. Since any two lines in Γ_s are rationally equivalent, the line of intersection (and thus also M) is rationally equivalent to an arbitrary line in Γ_s. Since a point cannot be rationally equivalent to 0 on X, we have a contradiction. Thus products such as $[M][\Lambda_t]$ cannot be defined in $A(X)$.

Despite this trouble, one can still define $f_M^*[\Lambda_t]$ and $f_N^*[\Lambda_t]$ using methods of Fulton [1984]. In fact, one can define the pullback f^* for an inclusion morphism $f : B \hookrightarrow X$ that is a "regular embedding" (which means that B is locally a complete intersection in X), or for the composition of such a morphism with a flat map.

Example 2.22 also shows that, even though f_{M*} is well-defined, pullbacks cannot be defined, at least in a way that makes the push-pull formula valid. If X were smooth, then by the push-pull formula $[M][\Lambda_t]$ would be equal to $f_{M*}([M]f_M^*[\Lambda_t])$, where the product $[M]f_M^*[\Lambda_t]$ should be interpreted as being in $A(M)$. This product is well-defined, as are the pullback and pushforward. But they do not allow us to compute the product $[M][\Lambda_t]$; since $[M] = [N]$ in $A(X)$, we would arrive at the contradiction

$$0 = f_{M*}(f_M^*[\Lambda_t]) = [M][\Lambda_t] = [N][\Lambda_t] = f_{N*}(f_N^*[\Lambda_t]) = [r].$$

There are, however, certain cycles (such as those represented by Chern classes of bundles) with which one can intersect, and this leads to a notion of "Chow cohomology" groups $A^*(X)$, which play a role relative to the Chow groups analogous to the role of cohomology relative to homology in the topological context: we have intersection products

$$A^c(X) \otimes A^d(X) \to A^{c+d}(X)$$

and

$$A^c(X) \otimes A_k(X) \to A_{k-c}(X)$$

analogous to cup and cap products in topology. In the present volume we will avoid all of this by sticking for the most part to the case of intersections on smooth varieties, where we can simply equate $A^c(X) = A_{\dim X - c}(X)$; for the full treatment, see Fulton [1984, Chapters 6, 8 and 17], and, for a visionary account of what might be possible, Srinivas [2010].

2.6 Exercises

Exercise 2.23. Let $\nu = \nu_{2,2} : \mathbb{P}^2 \to \mathbb{P}^5$ be the quadratic Veronese map. If $C \subset \mathbb{P}^2$ is a plane curve of degree d, show that the image $\nu(C)$ has degree $2d$. (In particular, this means that the Veronese surface $S \subset \mathbb{P}^5$ contains only curves of even degree!) More generally, if $\nu = \nu_{n,d} : \mathbb{P}^n \to \mathbb{P}^N$ is the degree-d Veronese map and $X \subset \mathbb{P}^n$ is a variety of dimension k and degree e, show that the image $\nu(X)$ has degree $d^k e$.

Exercise 2.24. Let $\sigma = \sigma_{r,s} : \mathbb{P}^r \times \mathbb{P}^s \to \mathbb{P}^{(r+1)(s+1)-1}$ be the Segre map, and let $X \subset \mathbb{P}^r \times \mathbb{P}^s$ be a subvariety of codimension k. Let the class $[X] \in A^k(\mathbb{P}^r \times \mathbb{P}^s)$ be given by

$$[X] = c_0 \alpha^k + c_1 \alpha^{k-1}\beta + \cdots + c_k \beta^k$$

(where $\alpha, \beta \in A^1(\mathbb{P}^r \times \mathbb{P}^s)$ are the pullbacks of the hyperplane classes, and we take $c_i = 0$ if $i > s$ or $k - i > r$).

(a) Show that all $c_i \geq 0$.
(b) Calculate the degree of the image $\sigma(X) \subset \mathbb{P}^{(r+1)(s+1)-1}$.
(c) Using (a) and (b), show that any linear space $\Lambda \subset \Sigma_{r,s} \subset \mathbb{P}^{(r+1)(s+1)-1}$ contained in the Segre variety lies in a fiber of either the map $\Sigma_{r,s} \cong \mathbb{P}^r \times \mathbb{P}^s \to \mathbb{P}^r$ or the corresponding map to $\mathbb{P}^s$.

Exercise 2.25. Let $\varphi : \mathbb{P}^2 \to \mathbb{P}^2$ be the rational map given by

$$\varphi : (x_0, x_1, x_2) \dashrightarrow \left(\frac{1}{x_0}, \frac{1}{x_1}, \frac{1}{x_2} \right),$$

or, equivalently,

$$\varphi : (x_0, x_1, x_2) \mapsto (x_1 x_2, x_0 x_2, x_0 x_1),$$

and let $\Gamma_\varphi \subset \mathbb{P}^2 \times \mathbb{P}^2$ be the graph of φ. Find the class

$$[\Gamma_\varphi] \in A^2(\mathbb{P}^2 \times \mathbb{P}^2).$$

Exercise 2.26. Let $\sigma : \mathbb{P}^2 \times \mathbb{P}^2 \to \mathbb{P}^8$ be the Segre map. Find the class of the graph of σ in $A(\mathbb{P}^2 \times \mathbb{P}^2 \times \mathbb{P}^8)$.

Exercise 2.27. Let $s : \mathbb{P}^2 \times \mathbb{P}^2 \dashrightarrow \mathbb{P}^{2*}$ be the rational map sending $(p, q) \in \mathbb{P}^2 \times \mathbb{P}^2$ to the line $\overline{p, q}$. Find the class of the graph of s in $A(\mathbb{P}^2 \times \mathbb{P}^2 \times \mathbb{P}^{2*})$.

Exercise 2.28. Let $X \subset \mathbb{P}^n$ be a hypersurface of degree d. Suppose that X has an ordinary double point (that is, a point $p \in X$ such that the projective tangent cone $\mathbb{T}C_p X$ is a smooth quadric), and is otherwise smooth. What is the degree of the dual hypersurface $X^* \subset \mathbb{P}^{n*}$?

Exercise 2.29. Let $p \in X \subset \mathbb{P}^n$ be a variety of degree d and dimension k, and suppose that $p \in X$ is a point of multiplicity m (see Section 1.3.8 for the definition). Assuming that the projection map $\pi_p : X \to \mathbb{P}^{n-1}$ is birational onto its image, what is the degree of $\pi_p(X)$?

Hint: Use Proposition 2.14.

Exercise 2.30. Show that the Chow ring of a product of projective spaces $\mathbb{P}^{r_1} \times \cdots \times \mathbb{P}^{r_k}$ is

$$A(\mathbb{P}^{r_1} \times \cdots \times \mathbb{P}^{r_k}) = \bigotimes A(\mathbb{P}^{r_i})$$
$$= \mathbb{Z}[\alpha_1, \ldots, \alpha_k]/(\alpha_1^{r_1+1}, \ldots, \alpha_k^{r_k+1}),$$

where $\alpha_1, \ldots, \alpha_k$ are the pullbacks of the hyperplane classes from the factors. Use this to calculate the degree of the image of the Segre embedding

$$\sigma : \mathbb{P}^{r_1} \times \cdots \times \mathbb{P}^{r_k} \hookrightarrow \mathbb{P}^{(r_1+1)\cdots(r_k+1)-1}$$

corresponding to the multilinear map $V_1 \times \cdots \times V_k \to V_1 \otimes \cdots \otimes V_k$.

Exercise 2.31. For $t \neq 0$, let $A_t : \mathbb{P}^r \to \mathbb{P}^r$ be the automorphism

$$[X_0, X_1, X_2, \ldots, X_r] \mapsto [X_0, tX_1, t^2 X_2, \ldots, t^r X_r].$$

Let $\Phi \subset \mathbb{A}^1 \times \mathbb{P}^r \times \mathbb{P}^r$ be the closure of the locus

$$\Phi^\circ = \{(t, p, q) \mid t \neq 0 \text{ and } q = A_t(p)\}.$$

Describe the fiber of Φ over the point $t = 0$, and deduce once again the formula of Section 2.1.6 for the class of the diagonal in $\mathbb{P}^r \times \mathbb{P}^r$.

In the simplest case, this construction is a rational equivalence between a smooth plane section of a quadric $Q \cong \mathbb{P}^1 \times \mathbb{P}^1 \subset \mathbb{P}^3$ (the diagonal, in terms of suitable identifications of the factors with $\mathbb{P}^1$) and a singular one (the sum of a line from each ruling), as in Figure 2.12.

Exercise 2.32. Let

$$\Psi = \{(p, q, r) \in \mathbb{P}^n \times \mathbb{P}^n \times \mathbb{P}^n \mid p, q \text{ and } r \text{ are collinear in } \mathbb{P}^n\}.$$

(Note that this includes all diagonals.)

(a) Show that this is a closed subvariety of codimension $n - 1$ in $\mathbb{P}^n \times \mathbb{P}^n \times \mathbb{P}^n$.

(b) Use the method of undetermined coefficients to find the class

$$\psi = [\Psi] \in A^{n-1}(\mathbb{P}^n \times \mathbb{P}^n \times \mathbb{P}^n).$$

(We will see a way to calculate the class $[\psi]$ using Porteous' formula in Exercise 12.9.)

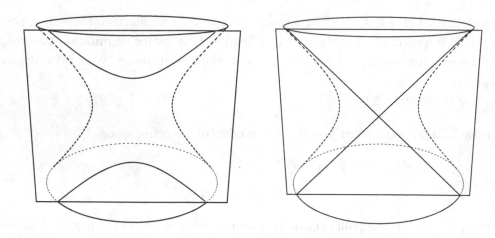

Figure 2.12 The diagonal in $\mathbb{P}^1 \times \mathbb{P}^1$ is equivalent to a sum of fibers.

Exercise 2.33. Suppose that $(F_0, \ldots, F_r)$ and $(G_0, \ldots, G_r)$ are general $(r + 1)$-tuples of homogeneous polynomials in $r + 1$ variables, of degrees d and e respectively, so that in particular the maps $f : \mathbb{P}^r \to \mathbb{P}^r$ and $g : \mathbb{P}^r \to \mathbb{P}^r$ sending x to $(F_0(x), \ldots, F_r(x))$ and $(G_0(x), \ldots, G_r(x))$ are regular. For how many points $x = (x_0, \ldots, x_r) \in \mathbb{P}^r$ do we have $f(x) = g(x)$?

The next two exercises set up Exercise 2.36, which considers when a point $p \in \mathbb{P}^2$ will be collinear with its images under several maps:

Exercise 2.34. Consider the locus $\Phi \subset (\mathbb{P}^2)^4$ of 4-tuples of collinear points. Find the class $\varphi = [\Phi] \in A^2((\mathbb{P}^2)^4)$ of Φ by the method of undetermined coefficients, that is, by intersecting with cycles of complementary dimension.

Exercise 2.35. With $\Phi \subset (\mathbb{P}^2)^4$ as in the preceding problem, calculate the class $\varphi = [\Phi]$ by using the result of Exercise 2.32 on the locus $\Psi \subset (\mathbb{P}^2)^3$ of triples of collinear points and considering the intersection of the loci $\Psi_{1,2,3}$ and $\Psi_{1,2,4}$ of 4-tuples (p_1, p_2, p_3, p_4) with (p_1, p_2, p_3) and (p_1, p_2, p_4) each collinear.

Exercise 2.36. Let A, B and $C : \mathbb{P}^2 \to \mathbb{P}^2$ be three general automorphisms. For how many points $p \in \mathbb{P}^2$ are the points $p, A(p), B(p)$ and $C(p)$ collinear?

Exercise 2.37. Let B be the blow-up of $\mathbb{P}^n$ at a point p, with exceptional divisor E as in Section 2.1.9. With notation as in that section, show that there is an affine stratification with closed strata Γ_k for $k = 1, \ldots, n$ and $E_k := \Gamma_k \cap E$ for $k = 0, \ldots, n - 1$. Let e_k be the class of E_k. Show that $e_{n-1} = \lambda_{n-1} - \gamma_{n-1}$ to describe the classes γ_k in terms of λ_k and e_k and vice versa. Conclude that the classes $\gamma_k = [\Gamma_k]$ and e_k form a basis for the Chow group $A(B)$.

Exercises 2.38–2.40 deal with the blow-up of $\mathbb{P}^3$ along a line. To fix notation, let $\pi : X \to \mathbb{P}^3$ be the blow-up of $\mathbb{P}^3$ along a line $L \subset \mathbb{P}^3$, that is, the graph $X \subset \mathbb{P}^3 \times \mathbb{P}^1$ of the rational map $\pi_L : \mathbb{P}^3 \dashrightarrow \mathbb{P}^1$ given by projection from L. Let $\alpha : X \to \mathbb{P}^1$ be projection on the second factor.

Exercise 2.38. Let $H \subset \mathbb{P}^3$ be a plane containing L and $\tilde{H} \subset X$ its proper transform. Let $J \subset \mathbb{P}^3$ be a plane transverse to L, $\tilde{J} \subset X$ its proper transform (which is equal to its preimage in X) and $M \subset J$ a line not meeting L. Show that the subvarieties

$$X, \ \tilde{H}, \ \tilde{J}, \ \tilde{J} \cap \tilde{H}, \ M, \ M \cap \tilde{H}$$

are the closed strata of an affine stratification of X, with open strata isomorphic to affine spaces. In particular, since only one (the subvariety $M \cap \tilde{H}$) is a point, deduce that $A^3(X) \cong \mathbb{Z}$.

Exercise 2.39. Let $h = [\tilde{H}]$, $j = [\tilde{J}] \in A^1(X)$ and $m = [M] \in A^2(X)$ be the classes of the corresponding strata. Show that

$$h^2 = 0, \ \ j^2 = m \ \ \text{and} \ \ \deg(jm) = \deg(hm) = 1.$$

Conclude that

$$A(X) = \mathbb{Z}[h, j]/(h^2, j^3 - hj^2).$$

Exercise 2.40. Now let $E \subset X$ be the exceptional divisor of the blow-up, and $e = [E] \in A^1(X)$ its class. What is the class e^2?

Exercise 2.41. Let $\mathbb{P}^5$ be the space of conic curves in $\mathbb{P}^2$.

(a) Find the dimension and degree of the locus of double lines (in characteristic $\neq 2$).
(b) Find the dimension and degree of the locus $\Delta \subset \mathbb{P}^5$ of singular conics (that is, line pairs and double lines).

Exercises 2.42–2.54 deal with some of the loci in the space $\mathbb{P}^9$ of plane cubics described in Section 2.2.

Exercise 2.42. Let $\mathbb{P}^9$ be the space of plane cubics and $\Gamma \subset \mathbb{P}^9$ the locus of reducible cubics. Let $L, C \subset \mathbb{P}^2$ be a line and a smooth conic intersecting transversely at two points $p, q \in \mathbb{P}^2$; let $L + C$ be the corresponding point of Γ. Show that Γ is smooth at $L + C$, with tangent space

$$\mathbb{T}_{L+C}\Gamma = \mathbb{P}\{\text{homogeneous cubic polynomials } F \mid F(p) = F(q) = 0\}.$$

Exercise 2.43. Using the preceding exercise, show that, if $p_1, \ldots, p_7 \in \mathbb{P}^2$ are general points and $H_i \subset \mathbb{P}^9$ is the hyperplane of cubics containing p_i, then the hyperplanes $H_1, \ldots, H_7$ intersect Γ transversely — that is, the degree of Γ is the number of reducible cubics through $p_1, \ldots, p_7$.

Exercise 2.44. Calculate the number of reducible plane cubics passing through seven general points $p_1, \ldots, p_7 \in \mathbb{P}^2$, and hence, by the preceding exercise, the degree of Γ.

Exercise 2.45. We can also calculate the degree of the locus $\Sigma \subset \mathbb{P}^9$ of triangles (that is, totally reducible cubics) directly, as in Exercises 2.42–2.44. To start, show that if $C = L_1 + L_2 + L_3$ is a triangle with three distinct vertices — that is, points $p_{i,j} = L_i \cap L_j$ of pairwise intersection — then Σ is smooth at C with tangent space

$$\mathbb{T}_{L+C}\Sigma = \mathbb{P}\{\text{homogeneous cubic polynomials } F \mid F(p_{i,j}) = 0 \text{ for all } i, j\}.$$

Exercise 2.46. Using the preceding exercise,

(a) show that if $p_1, \ldots, p_6 \in \mathbb{P}^2$ are general points, then the degree of Σ is the number of triangles containing $p_1, \ldots, p_6$; and
(b) calculate this number directly.

Exercise 2.47. Consider a general asterisk — that is, the sum $C = L_1 + L_2 + L_3$ of three distinct lines all passing through a point p. Show that the variety $\Sigma \subset \mathbb{P}^9$ of triangles is smooth at C, with tangent space the space of cubics double at p. Deduce that the space $A \subset \mathbb{P}^9$ of asterisks is also smooth at C.

Exercise 2.48. Let $p_1, \ldots, p_5 \in \mathbb{P}^2$ be general points. Show that any asterisk containing $\{p_1, \ldots, p_5\}$ consists, possibly after relabeling the points, of the sum of the line $L_1 = \overline{p_1, p_2}$, the line $L_2 = \overline{p_3, p_4}$ and the line $L_3 = \overline{p_5, (L_1 \cap L_2)}$.

Exercise 2.49. Using the preceding two exercises, show that, if $p_1, \ldots, p_5 \in \mathbb{P}^2$ are general points, then the hyperplanes H_{p_i} intersect the locus $A \subset \mathbb{P}^9$ of asterisks transversely, and calculate the degree of A accordingly.

Exercise 2.50. Show that (in characteristic $\neq 3$) the locus $Z \subset \mathbb{P}^9$ of triple lines is a cubic Veronese surface, and deduce that its degree is 9.

Exercise 2.51. Let $X \subset \mathbb{P}^9$ be the locus of cubics of the form $2L + M$ for L and M lines in $\mathbb{P}^2$.

(a) Show that X is the image of $\mathbb{P}^2 \times \mathbb{P}^2$ under a regular map such that the pullback of a general hyperplane in $\mathbb{P}^9$ is a hypersurface of bidegree $(2, 1)$.
(b) Use this to find the degree of X.

Exercise 2.52. If you try to find the degree of the locus X of the preceding problem by intersecting X with hyperplanes $H_{p_1}, \ldots, H_{p_4}$, where

$$H_p = \{C \in \mathbb{P}^9 \mid p \in C\},$$

you get the wrong answer (according to the preceding problem). Why? Can you account for the discrepancy?

Exercise 2.53. Let $\mathbb{P}^2$ denote the space of lines in the plane and $\mathbb{P}^5$ the space of plane conics. Let $\Phi \subset \mathbb{P}^2 \times \mathbb{P}^5$ be the closure of the locus of pairs

$$\{(L, C) \mid C \text{ is smooth and } L \text{ is tangent to } C\}.$$

Show that Φ is a hypersurface, and, assuming characteristic 0, find its class $[\Phi] \in A^1(\mathbb{P}^2 \times \mathbb{P}^5)$.

Exercise 2.54. Let $Y \subset \mathbb{P}^9$ be the closure of the locus of reducible cubics consisting of a smooth conic and a tangent line. Use the result of Exercise 2.53 to determine the degree of Y.

Exercise 2.55. Let $\mathbb{P}^{14}$ be the space of quartic curves in $\mathbb{P}^2$, and let $\Sigma \subset \mathbb{P}^{14}$ be the closure of the space of reducible quartics. What are the irreducible components of Σ, and what are their dimensions and degrees?

Exercise 2.56. Find the dimension and degree of the locus $\Omega \subset \mathbb{P}^{14}$ of totally reducible quartics (that is, quartic polynomials that factor as a product of four linear forms).

Exercise 2.57. Again let $\mathbb{P}^{14}$ be the space of plane quartic curves, and let $\Theta \subset \mathbb{P}^{14}$ be the locus of sums of four concurrent lines. Using the result of Exercise 2.34, find the degree of Θ.

Exercise 2.58. Find the degree of the locus $A \subset \mathbb{P}^{14}$ of the preceding problem, this time by calculating the number of sums of four concurrent lines containing six general points $p_1, \ldots, p_6 \in \mathbb{P}^2$, assuming transversality.

A natural generalization of the locus of asterisks, or of sums of four concurrent lines, would be the locus, in the space $\mathbb{P}^N$ of hypersurfaces of degree d in $\mathbb{P}^n$, of *cones*. We will indeed be able to calculate the degree of this locus in general, but it will require more advanced techniques than we have at our disposal here; see Section 7.3.4 for the answer.

Exercise 2.59. Let $S \subset \mathbb{P}^3$ be a smooth surface of degree d and $L \subset S$ a line. Calculate the degree of the self-intersection of the class $\lambda = [L] \in A^1(S)$ by considering the intersection of S with a general plane $H \subset \mathbb{P}^3$ containing L.

Exercise 2.60. Let S be a smooth surface. Show that if $C \subset S$ is any irreducible curve such that the corresponding point in the Hilbert scheme $\mathcal{H}$ of curves on S (see Section 6.3) lies on a positive-dimensional irreducible component of $\mathcal{H}$, then the degree $\deg(\gamma^2)$ of the self-intersection of the class $\gamma = [C] \in A^1(S)$ is nonnegative. Using this and the preceding exercise, prove the statement made in Section 2.4.2 that *a smooth surface $S \subset \mathbb{P}^3$ of degree 3 or more can contain only finitely many lines*.

Exercise 2.61. Let $C \subset \mathbb{P}^3$ be a smooth quintic curve. Show that

(a) if C has genus 2, it must lie on a quadric surface;
(b) if C has genus 1, it cannot lie on a quadric surface; and

(c) if C has genus 0, it may or may not lie on a quadric surface (that is, some rational quintic curves do lie on quadrics and some do not).

Exercise 2.62. Let $C \subset \mathbb{P}^3$ be a smooth quintic curve of genus 2. Show that C lies on a quadric surface Q and a cubic surface S with intersection $Q \cap S$ consisting of the union of C and a line.

Exercise 2.63. Use the result of Exercise 2.62 — showing that a smooth quintic curve of genus 2 is linked to a line in the complete intersection of a quadric and a cubic — to find the dimension of the subset of the Hilbert scheme corresponding to smooth curves of degree 5 and genus 2.

Chapter 3

Introduction to Grassmannians and lines in $\mathbb{P}^3$

Keynote Questions

(a) Given four general lines $L_1, \ldots, L_4 \subset \mathbb{P}^3$, how many lines $L \subset \mathbb{P}^3$ will meet all four? (Answer on page 110.)

(b) Given four curves $C_1, \ldots, C_4 \subset \mathbb{P}^3$ of degrees $d_1, \ldots, d_4$, how many lines will meet general translates of all four? (Answer on page 112.)

(c) If $C, C' \subset \mathbb{P}^3$ are two general twisted cubic curves, how many chords do they have in common? That is, how many lines will meet each twice? (Answer on page 115.)

(d) If $Q_1, \ldots, Q_4 \subset \mathbb{P}^3$ are four general quadric surfaces, how many lines are tangent to all four? (Answer on page 125.)

3.1 Enumerative formulas

In this chapter we introduce Grassmannian varieties through enumerative problems, of which the keynote questions above are examples. To clarify this context we begin by discussing enumerative problems in general and their relation to the intersection theory described in the preceding chapters.

In Section 3.2 we lay out the basic facts about Grassmannians in general. (Sections 3.2.5 and 3.2.6 may be omitted on the first reading, but will be important in later chapters.)

Starting in Section 3.3 we focus on the Grassmannian of lines in $\mathbb{P}^3$. We calculate the Chow ring and then, in Sections 3.4 and 3.6, use this to solve some enumerative problems involving lines, curves and surfaces in $\mathbb{P}^3$. In Section 3.5 we introduce the key technique of *specialization*, using it to re-derive some of these formulas.

3.1.1 What are enumerative problems, and how do we solve them?

Enumerative problems in algebraic geometry ask us to describe the set Φ of objects of a certain type satisfying a number of conditions — for example, the set of lines in $\mathbb{P}^3$ meeting each of four given lines, as in Keynote Question (a), or meeting each of four given curves $C_i \subset \mathbb{P}^3$, as in Keynote Question (b). In the most common situation, we expect Φ to be finite and we ask for its cardinality, whence the name enumerative geometry. Enumerative problems are interesting in their own right, but — as van der Waerden is quoted as saying in the introduction — they are also a wonderful way to learn some of the more advanced ideas and techniques of algebraic geometry, which is why they play such a central role in this text.

There are a number of steps common to most enumerative problems, all of which will be illustrated in the examples of this chapter. If we are asked to describe the set Φ of objects of a certain type that satisfy a number of conditions, we typically carry out the following five steps:

■ *Find or construct a suitable parameter space $\mathcal{H}$ for the objects we seek.* Suitable, for us, will mean that $\mathcal{H}$ should be projective and smooth, so that we can carry out calculations in the Chow ring $A(\mathcal{H})$. Most importantly, though, for each condition imposed, the locus $Z_i \subset \mathcal{H}$ of objects satisfying that condition should be a closed subscheme (which means in turn that the set $\Phi = \bigcap Z_i$ of solutions to our geometric problem will likewise have the structure of a subscheme of $\mathcal{H}$).

In our examples, the natural choice of parameter space $\mathcal{H}$ is the Grassmannian $G = \mathbb{G}(1, 3)$ parametrizing lines in $\mathbb{P}^3$, which we will construct and describe in Sections 3.2.1 and 3.2.2 below; as we will see, it is indeed smooth and projective of dimension 4. As we will see in Sections 3.3.1 and 3.4.2, moreover, the locus $\Sigma_C \subset G$ of lines $\Lambda \subset \mathbb{P}^3$ meeting a given curve $C \subset \mathbb{P}^3$ will indeed be a closed subscheme of codimension 1.

■ *Describe the Chow ring $A(\mathcal{H})$ of $\mathcal{H}$.* This is what we will undertake in Section 3.3 below; in the case of the Grassmannian $\mathbb{G}(1, 3)$, we will be able to give a complete description of its Chow ring. (In some circumstances, we may have to work with the cohomology ring rather than the Chow ring, as in Appendix D, or with a subring of $A(\mathcal{H})$ including the classes of the subschemes Z_i, as in Chapter 8.)

■ *Find the classes $[Z_i] \in A(\mathcal{H})$ of the loci of objects satisfying the conditions imposed.* Thus, in the case of Keynote Question (b), we have to determine the class in $A(G)$ of the locus $Z_i \subset G$ of lines meeting the curve C_i; the answer is given in Section 3.4.2.

■ *Calculate the product of the classes found in the preceding step.* If we have done everything correctly up to this point, this should be a straightforward combination of the two preceding steps.

At this point, we have what is known as an *enumerative formula*: It describes the class, in $A(\mathcal{H})$, of the scheme $\Phi \subset \mathcal{H}$ of solutions to our geometric problem, *under the assumption that this locus has the expected dimension and is generically reduced*—that is, the cycles $Z_i \subset \mathcal{H}$ intersect generically transversely. (If the cycles Z_i are all locally Cohen–Macaulay, then by Section 1.3.7 the enumerative formula describes the class of the subscheme $\Phi \subset \mathcal{H}$ under the weaker hypothesis that Φ has the expected dimension; that is, the cycles Z_i are dimensionally transverse.)

■ *Verify that the set of solutions, viewed as a subscheme of $\mathcal{H}$, indeed has the expected dimension, and investigate its geometry.* We will discuss, in the following section, what exactly we have proven if we simply stop at the conclusion of the last step. But ideally we would like to complete the analysis and say when the cycles $Z_i \subset \mathcal{H}$ do in fact meet generically transversely or dimensionally transversely. In particular, if the geometric problem posed depends on choices—the number of lines meeting each of four curves C_i, for example, depends on the C_i—we would like to be able to say that for general choices the corresponding scheme Φ is indeed generically reduced.

Thus, for example, in the case of Keynote Question (b), the analysis described above and carried out in Section 3.4.2 will tell us that if the subscheme $\Phi \subset G$ of lines meeting each of four curves $C_i \subset \mathbb{P}^3$ is zero-dimensional then it has degree $2 \prod \deg(C_i)$. But it does *not* tell us that the actual number of lines meeting each of the four curves is in fact $2 \prod \deg(C_i)$ for general C_i, or for that matter for any. That is addressed in Section 3.4.2 in characteristic 0; we will also see another approach to this question in Exercises 3.30–3.33 that also works in positive characteristic.

One reason this last step is sometimes given short shrift is that it is often the hardest. For example, it typically involves knowledge of the local geometry of the subschemes $Z_i \subset \mathcal{H}$—their smoothness or singularity, and their tangent spaces or tangent cones accordingly—and this is usually finer information than their dimensions and classes. But it is necessary, if the result of the first four steps is to give a description of the actual set of solutions, and it is also a great occasion to learn some of the relevant geometry.

3.1.2 The content of an enumerative formula

Because the last step in the process described above is sometimes beyond our reach, it is worth saying exactly what has been proved when we carry out just the first four steps in the process.

In general, the computation of the product $\alpha = \prod[Z_i] \in A(\mathcal{H})$ of the classes of some effective cycles Z_i in a space $\mathcal{H}$ tells us the following:

(a) If $\alpha \neq 0$ (for example, if $\alpha \in A_0(\mathcal{H})$ and $\deg(\alpha) \neq 0$), *we can conclude that the intersection $\bigcap Z_i$ is nonempty.* This is the source of many applications of enumerative geometry; for example, it is the basis of the Kempf/Kleiman–Laksov proof of the existence half of the Brill–Noether theorem, described in Appendix D.

(b) If the cycles Z_i intersect in the expected dimension, then the class α is a positive linear combination of the classes of the components of the intersection $\bigcap Z_i$. In particular, if $\alpha \in A_0(\mathcal{H})$ has dimension 0, then *the number of points of $\bigcap Z_i$ is at most $\deg(\alpha)$.* This in turn implies:

(i) If $\alpha \in A_0(\mathcal{H})$ and $\deg(\alpha) < 0$, we may conclude that the intersection $\bigcap Z_i$ is infinite rather than finite. More generally, if α is not the class of an effective cycle, *we can conclude that $\bigcap Z_i$ has dimension greater than the expected dimension.*

(ii) If $\alpha \in A_0(\mathcal{H})$ and $\deg(\alpha) = 0$, then the intersection $\bigcap Z_i$ must either be empty or infinite. (In general, if $\alpha = 0$ we can conclude that either $\bigcap Z_i = \varnothing$ or $\bigcap Z_i$ has dimension greater than the expected dimension.)

So, suppose we have carried out the first four steps in the process of the preceding section in the case of Keynote Question (a): We have described the Grassmannian $G = \mathbb{G}(1, 3)$ and its Chow ring, found the class $\sigma_1 = [Z]$ of the cycle Z of lines meeting a given line $L \subset \mathbb{P}^3$, and calculated that $\deg(\sigma_1^4) = 2$. What does this tell us?

Without a verification of transversality, the formula $\deg \sigma_1^4 = 2$ really only tells us that the number of intersections is either infinite or 1 or 2. Beyond this, it says that *if* the number of "solutions to the problem" — in this case, lines in $\mathbb{P}^3$ that meet the four given lines — is finite, then there are two *counted with multiplicity* — that is, either two solutions with multiplicity 1, or one solution with multiplicity 2. In order to say more, we need to be able to say when the intersection $\bigcap Z_i$ has the expected dimension; we need to be able to detect transversality and, ideally, to calculate the multiplicity of a given solution. (The third of these is often the hardest. For example, in the calculation of the number of lines meeting four given curves $C_i \subset \mathbb{P}^3$, we see in Exercises 3.30–3.33 how to check the condition of transversality, but there is no simple formula for the multiplicity when the intersection is not transverse.)

A common aspect of enumerative problems is that they themselves may vary with parameters: If we ask how many lines meet each of four curves C_i, the problem varies with the choice of curves C_i. In these situations, a good benchmark of our understanding is whether we can count the actual number of solutions for a *general* such problem: for example, whether we can prove that if $C_1, \ldots, C_4$ are general conics, then there are exactly 32 lines meeting all four. Thus, in most of the examples of enumerative geometry we will encounter in this book, there are two aspects to the problem. The first is to find the "expected" number of solutions by carrying out the first four steps of the preceding section to arrive at an enumerative formula. The second is to verify transversality — in other words, that the actual cardinality of the set of solutions is indeed this expected number — when the problem is suitably general.

3.2 Introduction to Grassmannians

A *Grassmann variety*, or *Grassmannian*, is a projective variety whose closed points correspond to the vector subspaces of a certain dimension in a given vector space. Projective spaces, which parametrize one-dimensional subspaces, are the most familiar examples. In this chapter we will begin the study of Grassmannians in general, and then focus on the geometry and Chow ring of the Grassmannian of lines in $\mathbb{P}^3$, the first and most intuitively accessible example beyond projective spaces.

Our goal in doing this is to introduce the reader to some ideas that will be developed in much greater generality (and complexity) in later chapters: the Grassmannian (as an example of parameter spaces), the methods of *undetermined coefficients* and *specialization* for computing intersection products more complicated than those mentioned in Chapter 2, and questions of transversality, treated via the tangent spaces to parameter spaces. For more information about Grassmannians, the reader may consult the books of Harris [1995] for basic geometry of the Grassmannian, Griffiths and Harris [1994] for the basics of the Schubert calculus and Fulton [1997] for combinatorial formulas, as well as the classic treatment in the second volume of Hodge and Pedoe [1952].

As a set, we take the Grassmannian $G = G(k, V)$ to be the set of k-dimensional vector subspaces of the vector space V. We give this set the structure of a projective variety by giving an inclusion in a projective space, called the *Plücker embedding*, and showing that the image is the zero locus of a certain collection of homogeneous polynomials.

A k-dimensional vector subspace of an n-dimensional vector space V is the same as a $(k - 1)$-dimensional linear subspace of $\mathbb{P}V \cong \mathbb{P}^{n-1}$, so the Grassmannian $G(k, V)$ could also be thought of as parametrizing $(k - 1)$-dimensional subspaces of $\mathbb{P}V$. We will write the Grassmannian $G(k, V)$ as $\mathbb{G}(k - 1, \mathbb{P}V)$ when we wish to think of it this way. When there is no need to specify the vector space V but only its dimension, say n, we will write simply $G(k, n)$ or $\mathbb{G}(k - 1, n - 1)$. Note also that there is a natural identification

$$G(k, V) = G(n - k, V^*)$$

sending a k-dimensional subspace $\Lambda \subset V$ to its annihilator $\Lambda^\perp \subset V^*$.

There are two points of potential confusion in the notation. First, if $\Lambda \subset V$ is a k-dimensional vector subspace of an n-dimensional vector space V, we will often use the same symbol Λ to denote the corresponding point in $G = G(k, V)$. When we need to make the distinction explicit, we will write $[\Lambda] \in G$ for the point corresponding to the plane $\Lambda \subset V$. Second, when we consider the Grassmannian $G = \mathbb{G}(k, \mathbb{P}V)$ we will sometimes need to work with the corresponding vector subspaces of V. In these circumstances, if $\Lambda \subset \mathbb{P}V$ is a k-plane, we will write $\widetilde{\Lambda}$ for the corresponding $(k + 1)$-dimensional vector subspace of V.

3.2.1 The Plücker embedding

To embed the set of k-dimensional vector subspaces of a given vector space V in a projective space, we associate to a k-dimensional subspace $\Lambda \subset V$ the one-dimensional subspace

$$\textstyle\bigwedge^k \Lambda \subset \bigwedge^k V;$$

that is, if Λ has basis $v_1, \ldots, v_k$, we associate to it the point of $\mathbb{P}(\bigwedge^k V)$ corresponding to the line spanned by $v_1 \wedge \cdots \wedge v_k$. This gives us a map of sets

$$G(k, V) \to \mathbb{P}(\textstyle\bigwedge^k V) \cong \mathbb{P}^{\binom{n}{k}-1},$$

called the *Plücker embedding*. To see that this map is one-to-one, observe that if $v_1, \ldots, v_k$ are a basis of $\Lambda \subset V$, then a vector v annihilates $\eta = v_1 \wedge \cdots \wedge v_k$ in the exterior algebra if and only if v is in the span Λ of $v_1, \ldots, v_k$; thus η determines Λ.

Concretely, if we choose a basis $\{e_1, \ldots, e_n\}$ for V, and so identify V with $\Bbbk^n$, we may represent Λ as the row space of a $k \times n$ matrix

$$A = \begin{pmatrix} a_{1,1} & a_{1,2} & \cdots & a_{1,n} \\ a_{2,1} & a_{2,2} & \cdots & a_{2,n} \\ \vdots & \vdots & \ddots & \vdots \\ a_{k,1} & a_{k,2} & \cdots & a_{k,n} \end{pmatrix}.$$

In these terms, a basis for $\bigwedge^k V$ is given by the set of products

$$\{e_{i_1} \wedge \cdots \wedge e_{i_k}\}_{1 \leq i_1 < \cdots < i_k \leq n},$$

and if $v_1, \ldots, v_k$ is a basis for Λ then we may write a nonzero element of $\bigwedge^k \Lambda$ in the form

$$v_1 \wedge \cdots \wedge v_k = \sum_{1 \leq i_1 < \cdots < i_k \leq n} p_{i_1, \ldots, i_k} e_{i_1} \wedge \cdots \wedge e_{i_k}.$$

Here the scalar $p_{i_1, \ldots, i_k}$ is the determinant of the submatrix (that is, *minor*) of A made from the columns $i_1, \ldots, i_k$. These $p_{i_1, \ldots, i_k}$ are called the *Plücker coordinates* of Λ.

The matrix A is not unique, since we can multiply on the left by any invertible $k \times k$ matrix Ω without changing the row space, but the collection of $k \times k$ minors of A, viewed as a vector in $\Bbbk^{\binom{n}{k}}$, is well-defined up to scalars: Multiplying by Ω multiplies each such minor by $\det(\Omega)$. (Conversely, if another matrix A' has the same row space, then we can write $A' = \Omega A$ for some invertible $k \times k$ matrix Ω.)

A quick-and-dirty way to see that the image $G \hookrightarrow \mathbb{P}(\bigwedge^k V)$ of the Plücker embedding — the locus of vectors $\eta \in \bigwedge^k V$ that are expressible as a wedge product $v_1 \wedge \cdots \wedge v_k$ of k vectors $v_i \in V$ — is a closed algebraic set is to use the ring structure of the exterior algebra $\bigwedge V$. Writing out an element $\eta \in \bigwedge^k V$ in coordinates as above, we see that $e_i \wedge \eta = 0$ if and only if η can be written as $e_i \wedge \eta'$ for some $\eta' \in \bigwedge^{k-1} V$. Since there is nothing special about the vector $e_i \in V$ we could replace it by any nonzero element $v \in V$. Repeating this idea, we see that a nonzero element $\eta \in \bigwedge^k V$ can be written in the form $v_1 \wedge \cdots \wedge v_k$ for some (necessarily independent) $v_1, \ldots, v_k \in V$ if and only if the kernel of the multiplication map

$$V \xrightarrow{\wedge \eta} \bigwedge^{k+1} V$$

has dimension at least k. That is, the image of the Plücker embedding is

$$G = \{\eta \in \textstyle\bigwedge^k V \mid \mathrm{rank}(V \xrightarrow{\wedge \eta} \textstyle\bigwedge^{k+1} V) \leq n - k\},$$

and this is the zero locus of the homogeneous polynomials of degree $n-k+1$ on $\bigwedge^k V$ that are the $(n-k+1)$-st-order minors of the map $\wedge \eta : V \to \bigwedge^{k+1} V$ written out as a matrix.

Once we know that G is an algebraic set, it follows that G is a variety: Its ideal is the kernel of the map of polynomial rings

$$\Bbbk[p_{i_1,\ldots,i_k}]_{1 \leq i_1 < \cdots < i_k \leq n} \to \Bbbk[x_{i,j}]_{\substack{1 \leq i \leq k \\ 1 \leq j \leq n}}$$

sending $p_{i_1,\ldots,i_k}$ to the corresponding Plücker coordinate of the generic matrix $(x_{i,j})$, and is thus prime.

Though the equations for G just given have degree $n - k + 1$, the generators of the ideal of homogeneous forms vanishing on $G \subset \mathbb{P}(\bigwedge^k V)$ are actually quadratic polynomials, known as the *Plücker relations*. We will be able to describe these quadratic polynomials explicitly following Proposition 3.2; for fuller accounts (including a proof that they do indeed generate the homogeneous ideal of $G \subset \mathbb{P}(\bigwedge^k V)$) and some of their beautiful combinatorial structure, we refer the reader to De Concini et al. [1980, Section 2] or Fulton [1997, Section 9.1].

From here on, we will view $G(k, V)$ as being endowed with the structure of a projective variety via the Plücker embedding. As will follow from the description of its covering by affine spaces in the following subsection, it is a smooth variety of dimension $k(n - k)$. The smoothness statement follows in any case from the fact that $\mathrm{GL}(V)$ acts transitively on it by linear transformations of the projective space $\mathbb{P}(\bigwedge^k V)$.

Example 3.1. The first example of a Grassmannian other than projective space is the Grassmannian $G(2, 4) = \mathbb{G}(1, 3)$. Let V be a four-dimensional vector space, and consider the Plücker embedding of $G(2, V) = \mathbb{G}(1, \mathbb{P}V)$ in $\mathbb{P}(\bigwedge^2 V) \cong \mathbb{P}^5$. Since (as we will see shortly) $\dim G(2, 4) = 4$, this will be a hypersurface. From the discussion above, we know that the equation of $G(2, 4)$ in this embedding is a polynomial relation among the minors $p_{i,j}$ of a generic 2×4 matrix

$$\begin{pmatrix} a_{1,1} & a_{1,2} & a_{1,3} & a_{1,4} \\ a_{2,1} & a_{2,2} & a_{2,3} & a_{2,4} \end{pmatrix}.$$

One way to obtain this relation is to note that the determinant of the 4×4 matrix with repeated rows

$$\begin{pmatrix} a_{1,1} & a_{1,2} & a_{1,3} & a_{1,4} \\ a_{2,1} & a_{2,2} & a_{2,3} & a_{2,4} \\ a_{1,1} & a_{1,2} & a_{1,3} & a_{1,4} \\ a_{2,1} & a_{2,2} & a_{2,3} & a_{2,4} \end{pmatrix}$$

must be 0. Expanding this determinant as a sum of products of minors of the first two rows and of the last two rows, all of which are Plücker coordinates, we obtain

$$p_{1,2}p_{3,4} - p_{1,3}p_{2,4} + p_{1,4}p_{2,3} = 0. \tag{3.1}$$

As this is an irreducible polynomial (and $\dim G(2,4) = 4$), it generates the homogeneous ideal of $G(2,4) \subset \mathbb{P}^5$, which is thus a smooth quadric.

In fact, for any n, the ideal of the Grassmannian of 2-planes $G(2,n)$ is cut out by quadratic polynomials in the Plücker coordinates similar to the polynomial (3.1) above. More precisely, if $e_1, \ldots, e_n$ is a basis of V and $\eta = \sum p_{a,b} e_a \wedge e_b \in \wedge^2 V$, then the polynomials

$$\{g_{a,b,c,d} := p_{a,b}p_{c,d} - p_{a,c}p_{b,d} + p_{a,d}p_{b,c} = 0 \mid 1 \le a < b < c < d \le n\}$$

minimally generate the ideal of the Grassmannian. These are the *Plücker relations* in the special case of the Grassmannian $G(2,n)$. We will describe the Plücker relations in general following Proposition 3.2.

Another way to characterize the collection of polynomials $\{g_{a,b,c,d}\}$ defining $\mathbb{G}(1,n)$, in characteristic not equal to 2, is that they are the coefficients of the element $\eta^2 = 0 \in \wedge^4 V$ — in other words, an element $\eta \in \wedge^2 V$ is decomposable if and only if $\eta \wedge \eta = 0$. These coefficients may be characterized (up to a factor of 2) as the *Pfaffians* of a skew-symmetric matrix.

Exercises 3.17–3.22 describe a number of aspects of the projective geometry of the Grassmannian in the Plücker embedding.

3.2.2 Covering by affine spaces; local coordinates

Like a projective space, a Grassmannian $G = G(k,V)$ can be covered by Zariski open subsets isomorphic to affine space. To see this, fix an $(n-k)$-dimensional subspace $\Gamma \subset V$, and let U_Γ be the subset of k-planes that do not meet Γ:

$$U_\Gamma = \{\Lambda \in G \mid \Lambda \cap \Gamma = 0\}.$$

This is a Zariski open subset of G: In fact, if we take $w_1, \ldots, w_{n-k}$ to be any basis for Γ and set $\eta = w_1 \wedge \cdots \wedge w_{n-k}$, then we have

$$U_\Gamma = \{[\omega] \in G \subset \mathbb{P}(\wedge^k V) \mid \omega \wedge \eta \ne 0\},$$

from which we see that U_Γ is the complement of the hyperplane section of G corresponding to the vanishing of a Plücker coordinate (though not all hyperplane sections of G in the Plücker embedding have this form).

We claim now that the open set $U_\Gamma \subset G(k,n)$ is isomorphic to affine space $\mathbb{A}^{k(n-k)}$. To see this, we first choose an arbitrary point $[\Omega] \in U_\Gamma$ that will play the role of the origin; that is, fix a k-plane $\Omega \subset V$ complementary to Γ, so that we have a direct-sum decomposition $V = \Omega \oplus \Gamma$. Any k-dimensional subspace $\Lambda \subset V$ complementary to Γ projects to Γ modulo Ω — call this map π_Γ — and projects isomorphically to Ω modulo Γ — call this map π_Ω. Thus Λ is the graph of the linear map

$$\varphi : \Omega \xrightarrow{\pi_\Omega^{-1}} \Lambda \subset V = \Omega \oplus \Gamma \xrightarrow{\pi_\Gamma} \Gamma.$$

Conversely, the graph of any map $\varphi : \Omega \to \Gamma$ is a subspace $\Lambda \subset \Omega \oplus \Gamma = V$ complementary to Γ. These two correspondences establish a bijection

$$U_\Gamma \cong \mathrm{Hom}(\Omega, \Gamma) \cong \mathbb{A}^{k(n-k)}.$$

To make this explicit, suppose we choose a basis for V consisting of a basis $e_1, \ldots, e_k$ for Ω followed by a basis $e_{k+1}, \ldots, e_n$ for Γ. If $\Lambda \in U_\Gamma$ is a k-plane then the preimages $\pi_\Omega^{-1} e_1, \ldots, \pi_\Omega^{-1} e_k \in \Lambda$ form a basis for Λ. Thus Λ is the row space of the matrix

$$B = \begin{pmatrix} 1 & 0 & \cdots & 0 & a_{1,1} & a_{1,2} & \cdots & a_{1,n-k} \\ 0 & 1 & \cdots & 0 & a_{2,1} & a_{2,2} & \cdots & a_{2,n-k} \\ \vdots & \vdots & \ddots & 0 & \vdots & \vdots & \ddots & \vdots \\ 0 & 0 & \cdots & 1 & a_{k,1} & a_{k,2} & \cdots & a_{k,n-k} \end{pmatrix},$$

where $A = (a_{i,j})$ is the matrix representing the linear transformation $\varphi : \Omega \to \Gamma$ in the given bases. Since there is a unique vector in Λ projecting (mod Γ) to each $e_i \in \Omega$, this matrix representation is unique. The bijection defined above sends $\Lambda \in U_\Gamma$ to the linear transformation $\Omega \to \Gamma$ given by the transpose of the matrix $A = (a_{i,j})$.

If we start with any representation of Λ as the span of the rows of a $k \times n$ matrix B' with respect to the given basis of V, then the Plücker coordinate $p_{1,2,\ldots,k}$, which is the determinant of the submatrix consisting of the first k columns of B', is nonzero. Multiplying B' on the left by the inverse of this submatrix gives us back the matrix B above, and thus the $k \times k$ minors of B are the $k \times k$ minors of B' multiplied by the inverse of the determinant of the first $k \times k$ minor of B'.

On the other hand, we can realize the entry $a_{i,j}$ of A, up to sign, as a $k \times k$ minor of B: It is (up to sign) the determinant of the $k \times k$ submatrix in which we take all the first k columns except for the i-th, and put in instead the $(k+j)$-th column. Thus we may write

$$\pm a_{i,j} = \frac{p_{1,\ldots,i-1,\hat{i},i+1,\ldots,k,k+j}(\Lambda)}{p_{1,\ldots,k}(\Lambda)},$$

and this expression shows that $a_{i,j}$ is a regular function on U_Γ. Thus the bijection $U_\Gamma \cong \mathbb{A}^{k(n-k)}$ is a biregular isomorphism.

More generally, it turns out that the ratios $p_{a_1,...,a_k}(\Lambda)/p_{1,2,...,k}(\Lambda)$ of Plücker coordinates are, up to sign, precisely the determinants of submatrices (of all sizes) of A. To express the result, suppose that I and J are sets of indices. Write A_I^J for the minor of the matrix A involving rows with indices in I and columns with indices in J. Write I' for the complement (in the set of row indices $\{1,...,k\}$) of I, and, if $J = \{j_1,...,j_t\}$, write $J + k$ for the "translated" set of indices $\{j_1 + k, ..., j_t + k\}$. With this notation, the $t \times t$ minor A_I^J of A is equal, up to sign, to the $k \times k$ minor of B involving columns $I' \cup J$. To see this as a regular function on U_Γ we need only divide by the minor involving columns $1,...,k$:

Proposition 3.2. *With notation as above, suppose that* $I = \{i_1,...,i_{k-t}\}$ *are row indices and* $J = \{j_1,...,j_t\}$ *are column indices with each* $j_i > k$. *We have*

$$\pm \det A_I^J = \frac{p_{I' \cup (J+k)}(\Lambda)}{p_{1,...,k}(\Lambda)}.$$

For example, the 3×3 minor of the matrix

$$B = \begin{matrix} \mathbf{1} & 2 & 3 & 4 & \mathbf{5} & \mathbf{6} & 7 \\ \begin{pmatrix} 1 & 0 & 0 & a_{1,1} & a_{1,2} & a_{1,3} & a_{1,4} \\ 0 & 1 & 0 & a_{2,1} & \boldsymbol{a_{2,2}} & \boldsymbol{a_{2,3}} & a_{2,4} \\ 0 & 0 & 1 & a_{3,1} & \boldsymbol{a_{3,2}} & \boldsymbol{a_{3,3}} & a_{3,4,} \end{pmatrix} \end{matrix}$$

involving columns 1, 5 and 6, is, up to sign, the 2×2 minor

$$\det \begin{pmatrix} a_{2,2} & a_{2,3} \\ a_{3,2} & a_{3,3} \end{pmatrix}$$

of the matrix $(a_{i,j})$.

Proof: The expression in Plücker coordinates on the right is independent of the matrix representation chosen for Λ, so we may compute the two Plücker coordinates in terms of the matrix B in the form given above, so $p_{1,...,k}(\Lambda) = 1$ and $p_{I' \cup (J+k)}(\Lambda)$ is the minor of B involving the columns $I' \cup (J + k)$. Expanding this minor in terms of the $(k-t) \times (k-t)$ minors involving the rows of I', we see that all but the term $\pm 1 \cdot A_I^J$ are zero. $\square$

Having established Proposition 3.2, it is easy to describe the *Plücker relations*, the quadratic polynomials in the Plücker coordinates that generate the homogeneous ideal of $G(k, V) \subset \mathbb{P}(\wedge^k V)$. With notation as above, consider the expansion of any $t \times t$ minor of A along one of its rows or columns. Replacing each factor of each term that appears by the ratio of two Plücker coordinates, with denominator $p_{1,...,k}(\Lambda)$, and multiplying through by $p_{1,...,k}(\Lambda)^2$, we get a homogeneous quadratic polynomial in the p_I satisfied identically in U_Γ and hence in all of $G(k, V)$. For more information we refer the reader to De Concini et al. [1980, Section 2] or Fulton [1997, Section 9.1].

3.2.3 Universal sub and quotient bundles

In this section and the following, we will introduce the *universal bundles* on the Grassmannian $G(k, n)$ and show how to describe the tangent bundle to $G(k, n)$ in terms of them. These constructions are of fundamental importance in understanding the geometry of Grassmannians.

Let V be an n-dimensional vector space, $G = G(k, V)$ the Grassmannian of k-planes in V, and let $\mathcal{V} := G \times V$ be the trivial vector bundle of rank n on G whose fiber at every point is the vector space V (here we are thinking of a vector bundle as a variety, rather than as a locally free sheaf). We write $\mathcal{S}$ for the rank-k subbundle of $\mathcal{V}$ whose fiber at a point $[\Lambda] \in G$ is the subspace Λ itself; that is,

$$\mathcal{S}_{[\Lambda]} = \Lambda \subset V = \mathcal{V}_{[\Lambda]}.$$

$\mathcal{S}$ is called the *universal subbundle* on G; the quotient $\mathcal{Q} = \mathcal{V}/\mathcal{S}$ is called the *universal quotient bundle*. In the case $k = 1$—that is, $G = \mathbb{P}V \cong \mathbb{P}^{n-1}$—the universal subbundle $\mathcal{S}$ is the line bundle $\mathcal{O}_{\mathbb{P}V}(-1)$; similarly, in the case $k = n - 1$ (so $G = \mathbb{P}V^*$) the universal quotient bundle $\mathcal{Q}$ is the line bundle $\mathcal{O}_{\mathbb{P}V^*}(1)$.

We have said "the rank-k subbundle of $\mathcal{V}$ whose fiber at a point $[\Lambda] \in G$ is the subspace Λ itself," and this certainly describes *at most* one bundle, since we have unambiguously defined a subset of $\mathcal{V} = G \times V$. Who would doubt that it is an algebraic subbundle of $\mathcal{V}$? To prove this, however, something more is necessary. Most primitively, we must check that it is trivial on an affine open cover, and that the transition functions are regular on the overlap of any two open sets of the cover. Alternatively, and equivalently, we may show that the subset $\mathcal{S}$ is an algebraic subset, and that over an open cover it is isomorphic, as an algebraic variety, to a trivial bundle. Here is a proof:

Proposition 3.3. *The subset $\mathcal{S}$ of $\mathcal{V}$ whose fiber over a point $[\Lambda] \in G = G(k, V)$ is the subspace $\Lambda \subset V$ is a vector bundle over G.*

Of course, it follows that $\mathcal{Q} = \mathcal{V}/\mathcal{S}$ is also a vector bundle.

Proof: Let S be the incidence correspondence

$$S = \{(\Lambda, v) \in G \times V \mid v \in \Lambda\}.$$

The set S is an algebraic subset of $G \times V$, since if we represent Λ by a vector $\eta \in \bigwedge^k \Lambda \subset \bigwedge^k V$, it is given by the equation $\eta \wedge v = 0 \in \bigwedge^{k+1} V$. Explicitly, if Λ is the row space of the matrix A, as in Section 3.2.2, then the condition $v \in \Lambda$ is equivalent to the vanishing of the $(k + 1)$-st-order minors of the matrix obtained from A' by adjoining v as the $(k + 1)$-st row. These minors can be expressed (by expanding along the new row of A') as bilinear functions in the coordinates of v and the Plücker coordinates, proving that S is an algebraic subset.

Now pick a subspace $\Gamma \subset V$ of dimension $n - k$ and consider the preimage of $U_\Gamma \subset G$. Choosing a complement Ω to Γ as before, we can identify U_Γ with $\mathrm{Hom}(\Omega, \Gamma)$. Moreover, if $\Lambda \in U_\Gamma$ then the projection $\beta_{\Omega, \Gamma} : V \to \Omega$ with kernel Γ takes $S_{[\Lambda]} = \Lambda \subset V$ isomorphically to Ω. In other words, this projection gives an isomorphism S_{U_Γ} to the trivial bundle $\Omega \times U_\Gamma$. This proves that S is actually a vector bundle, which we identify as $\mathcal{S}$. □

The following result is the reason that we refer to $\mathcal{S}$ as the universal subbundle. A proof may be found in Eisenbud and Harris [2000].

Theorem 3.4. *If X is any scheme then the morphisms $\varphi : X \to G$ are in a one-to-one correspondence with rank-k subbundles $\mathcal{F} \subset V \otimes \mathcal{O}_X$ such that φ corresponds to the bundle $\mathcal{F} = \varphi^* \mathcal{S}$.*

There is also a projective analog of the vector bundle $\mathcal{S}$. Viewing G as $\mathbb{G}(k-1, \mathbb{P}V)$ (that is, as parametrizing $(k-1)$-planes in $\mathbb{P}V$), we set

$$\Phi = \{(\Lambda, p) \in G \times \mathbb{P}V \mid p \in \Lambda\}.$$

The space Φ can also be realized as the projectivization of the universal subbundle $\mathcal{S}$, where by the *projectivization* of a vector bundle $\mathcal{E}$ on a scheme X we mean $\mathbb{P}\mathcal{E} :=$ $\mathrm{Proj}(\mathrm{Sym}\,\mathcal{E}^*)$ — a locally trivial fiber bundle over X whose fiber over a point $p \in X$ is $\mathbb{P}(\mathcal{E}_p)$. (We will see more of the space Φ in Section 4.8.1, where we will discuss flag manifolds in general and Φ in particular; we will deal with projective bundles in general in Chapter 9.) $\Phi = \mathbb{P}\mathcal{S}$ is called the *universal k-plane* over G.

Theorem 3.4 may be interpreted as saying that *the Grassmannian represents the functor of families of k-dimensional subspaces of V*, in the sense that the contravariant functor from schemes to sets given on objects by $X \mapsto \mathrm{Mor}(X, G(k, V))$ is naturally isomorphic to the functor given by $X \mapsto \{\text{rank-}k \text{ subbundles of } V \otimes \mathcal{O}_X\}$. Again, in the language that we will develop in Section 6.3, this says that *the Grassmannian $\mathbb{G}(k-1, \mathbb{P}V)$ is the Hilbert scheme of $(k-1)$-planes in $\mathbb{P}V$.* See Eisenbud and Harris [2000, Chapter 6] for an introduction to these ideas and a proof of this statement.

3.2.4 The tangent bundle of the Grassmannian

Knowledge of the tangent bundle of the Grassmannian is the key to its geometry. It turns out that the tangent bundle can be expressed in terms of the universal bundles $\mathcal{S}$ and $\mathcal{Q}$:

Theorem 3.5. *The tangent bundle $\mathcal{T}_G$ to the Grassmannian $G = G(k, V)$ is isomorphic to $\mathcal{H}om_G(\mathcal{S}, \mathcal{Q})$, where $\mathcal{S}$ and $\mathcal{Q}$ are the universal sub and quotient bundles.*

Proof: Consider the open affine set

$$U_\Gamma = \{\Lambda \in G \mid \Lambda \cap \Gamma = 0\}$$

described in Section 3.2.2, where Γ is a subspace of V of dimension $n - k$. Fixing a point $[\Omega] \in U_\Gamma$ and decomposing V as $\Omega \oplus \Gamma$, we get an identification of U_Γ with the vector space $\mathrm{Hom}(\Omega, \Gamma)$ under which the point $[\Omega]$ goes to the linear transformation 0. In particular, the tangent bundle $\mathcal{T}_G$ restricted to U_Γ is the trivial bundle and the fiber over $[\Omega]$ is $\mathrm{Hom}(\Omega, \Gamma)$.

The bundle $\mathcal{S}|_{U_\Gamma}$ is isomorphic to the trivial bundle $\Omega \times U_\Gamma$ by the composite map

$$\mathcal{S}|_{U_\Gamma} \to V \times U_\Gamma \to V/\Gamma \times U_\Gamma = \Omega \times U_\Gamma,$$

and the bundle $\mathcal{Q}|_{U_\Gamma}$ is isomorphic to the trivial bundle $\Gamma \times U_\Gamma$ via the tautological projection $V \otimes \mathcal{O}_G \to \mathcal{Q}$. This gives an identification of fibers, depending on Γ:

$$(\mathcal{T}_G)_\Omega = \mathrm{Hom}(\Omega, \Gamma) = \mathrm{Hom}(\mathcal{S}_\Omega, \mathcal{Q}_\Omega).$$

To prove that these identifications extend to an isomorphism $\mathcal{T}_G \cong \mathcal{H}om_G(\mathcal{S}, \mathcal{Q})$, we must check that the gluing map for $\mathcal{T}_G$ and that for $\mathcal{H}om_G(\mathcal{S}, \mathcal{Q})$ on an intersection $U = U_\Gamma \cap U_{\Gamma'}$ containing the point $[\Omega]$ agree on the fiber over $[\Omega]$ (and thus agree as maps of bundles). We may regard $U \subset U_\Gamma = \mathrm{Hom}(\Omega, \Gamma)$ as the set of linear transformations whose graphs do not meet Γ', and this representation is related to the representation of $U \subset U_{\Gamma'}$ by the isomorphisms $\Gamma \xrightarrow{\ \alpha\ } V/\Omega \xleftarrow{\ \beta\ } \Gamma'$. The gluing

$$d\varphi : (\mathcal{T}_G|_{U_\Gamma})|_{U_{\Gamma'}} \xrightarrow{\ \sim\ } (\mathcal{T}_G|_{U_{\Gamma'}})|_{U_\Gamma}$$

along this set is by the differential of the composite linear transformation

$$\varphi : \mathrm{Hom}(\Omega, \Gamma) \xrightarrow{\ \alpha\ } \mathrm{Hom}(\Omega, V/\Omega) \xrightarrow{\ \beta^{-1}\ } \mathrm{Hom}(\Omega, \Gamma')$$

induced by these isomorphisms. Of course, the differential of a linear transformation is the same linear transformation. The same isomorphisms give the gluing of the bundle $\mathcal{H}om_G(\mathcal{S}, \mathcal{Q})$.   □

From the identification of tangent vectors to $G = G(k, V)$ at Λ with the space $\mathrm{Hom}(\Lambda, V/\Lambda)$, we can see that not all tangent vectors at a given point are alike: We can associate to any tangent vector its *rank*, and this will be preserved under automorphisms of G (see Exercise 3.24 and, for a nice application, Exercise 3.23). In particular, this means that when $1 < k < \dim V - 1$ the automorphism group of $G(k, V)$ does *not* act transitively on nonzero tangent vectors, and hence Kleiman's theorem (Theorem 1.7) does not apply in positive characteristic. Nevertheless, the conclusions it gives for intersections of Schubert cycles are correct in all characteristics (and may be proven by a different method).

The Euler sequence on $\mathbb{P}^n$

The isomorphism of Theorem 3.5 is already useful in the case of projective space $\mathbb{P}^n = \mathbb{G}(0, n)$. In this setting Theorem 3.5 gives rise to the *Euler sequence*.

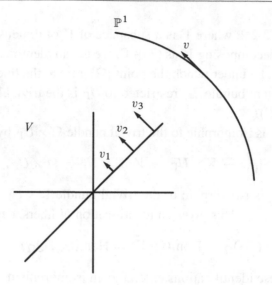

Figure 3.1 The tangent vectors v_1, v_2, v_3 to $\mathbb{A}^2$ all map to the tangent vector v to $\mathbb{P}^1$.

Let V be an $(n + 1)$-dimensional vector space and $\mathbb{P}^n = \mathbb{P}V$ its projectivization. We consider the quotient map

$$q : U = V \setminus \{0\} \to \mathbb{P}^n$$

sending a nonzero vector $v \in V$ to the corresponding point $p = [v] \in \mathbb{P}^n$. The tangent space to U at v is the same as the tangent space to V at v, which is to say the vector space V itself, and the kernel of the differential

$$dq_v : T_v U \to T_p \mathbb{P}^n$$

is the one-dimensional subspace $\tilde{p} = \langle v \rangle \subset V$ spanned by v. Thus dq_v induces an isomorphism

$$V/\tilde{p} \xrightarrow{\sim} T_p \mathbb{P}^n,$$

as illustrated in Figure 3.1.

This isomorphism does not, however, give a natural identification of the vector spaces $V/\tilde{p}$ and $T_p \mathbb{P}^n$. Even though both these vector spaces depend only on the point $p \in \mathbb{P}^n$, *the isomorphism dq_v between them depends on the choice of the vector v.* Indeed, if λ is any nonzero scalar, the differential $dq_{\lambda v}$ is equal to dq_v divided by λ. But, by the same token, if $l : \langle v \rangle \to \Bbbk$ is any linear functional, then the map $l(v) \cdot dq_v$ *is* independent of the choice of v, and so we have a natural identification

$$\langle v \rangle^* \otimes V/\langle v \rangle \xrightarrow{\sim} T_{[v]}\mathbb{P}^n.$$

This is the identification

$$\mathcal{T}_{\mathbb{P}^n} \cong \mathcal{O}_{\mathbb{P}^n}(1) \otimes \mathcal{Q} = \mathcal{H}om(\mathcal{O}_{\mathbb{P}^n}(-1), \mathcal{Q}) = \mathcal{H}om(\mathcal{S}, \mathcal{Q})$$

asserted (more generally) in Theorem 3.5.

To put it another way, in terms of coordinates $x_0, \ldots, x_n$ on V, a constant vector field $\partial/\partial x_i$ on V does not give rise to a vector field on $\mathbb{P}V$, but the vector field

$$w(x) = x_j \frac{\partial}{\partial x_i}$$

on V does. This gives us a map

$$\mathcal{O}_{\mathbb{P}^n}(1) \otimes V \to \mathcal{T}_{\mathbb{P}V},$$

whose kernel is the Euler vector field

$$e(x) = \sum x_i \frac{\partial}{\partial x_i}.$$

The resulting exact sequence

$$0 \longrightarrow \mathcal{O}_{\mathbb{P}V} \xrightarrow{1 \mapsto e} \mathcal{O}_{\mathbb{P}^n}(1) \otimes V \longrightarrow \mathcal{T}_{\mathbb{P}V} \longrightarrow 0$$

is called the *Euler sequence*. To relate this to the identification of the tangent bundle above, start with the universal sequence on $\mathbb{P}V$:

$$0 \longrightarrow \mathcal{S} \longrightarrow \mathcal{O}_{\mathbb{P}V} \otimes V \longrightarrow \mathcal{Q} \longrightarrow 0.$$

Now tensor with the line bundle $\mathcal{S}^* = \mathcal{O}_{\mathbb{P}V}(1)$; since $\mathcal{S} \otimes \mathcal{S}^* \cong \mathcal{O}_{\mathbb{P}V}$, we arrive at the sequence

$$0 \longrightarrow \mathcal{O}_{\mathbb{P}V} \longrightarrow \mathcal{O}_{\mathbb{P}^n}(1) \otimes V \longrightarrow \mathcal{S}^* \otimes \mathcal{Q} \longrightarrow 0.$$

By Theorem 3.5 the term on the right is $\mathcal{T}_{\mathbb{P}^n}$, and we obtain the Euler sequence again.

3.2.5 The differential of a morphism to the Grassmannian

Suppose that the morphism $\varphi : X \to G(k, n)$ corresponds via the universal property to a subbundle $\mathcal{E} \subset \mathcal{O}_X^n$, so that $\mathcal{E}$ is the pullback of the universal subbundle $\mathcal{S}$ on $G(k, n)$. Set $\mathcal{F} = \mathcal{O}^n/\mathcal{E}$, so that $\mathcal{F}$ is the pullback of the universal quotient bundle $\mathcal{Q}$ on $G(k, n)$. The differential of φ is by definition a homomorphism of vector bundles

$$d\varphi : \mathcal{T}_X \to \varphi^* \mathcal{T}_{G(k,n)} = \varphi^* \mathcal{H}om_{G(k,n)}(\mathcal{S}, \mathcal{Q}) = \mathcal{H}om_X(\mathcal{E}, \mathcal{F}).$$

The local description of the Grassmannian above makes it easy to identify this homomorphism locally.

A global section of the $\mathcal{T}_X$ is called a *vector field*. Recall that a vector field may be identified with a derivation $\partial : \mathcal{O}_X \to \mathcal{O}_X$ (Eisenbud [1995, Chapter 16]). This works even if X is singular: In that case we *define* $\mathcal{T}_X$ to be the dual of the sheaf of differential forms; of course then $\mathcal{T}_X$ is a coherent sheaf, not necessarily a vector bundle. (A famous question posed by Zariski (see Lipman [1965]) asks whether, with this definition — and always assuming that the characteristic is 0 — $\mathcal{T}_X$ is a vector bundle if and only if X is smooth. See Hochster [1977] for some partial results.)

Proposition 3.6. *Let X be a variety and $\varphi : X \to G(k, n)$ the morphism corresponding to a subbundle $i : \mathcal{E} \to \mathcal{O}_X^n$; set $\mathcal{F} = \mathcal{O}_X^n / \mathcal{E}$ and let $\pi : \mathcal{O}_X^n \to \mathcal{F}$ be the projection. Let $U \subset X$ be an open subset over which $\mathcal{E}$ is trivial and $\psi : \mathcal{O}_U^k \cong \mathcal{E}_U$ a trivialization of $\mathcal{E}$ over U. If ∂ is a vector field on U, then in U the homomorphism $(d\varphi)(\partial) \in \mathcal{H}om_U(\mathcal{E}, \mathcal{F})$ is the composition $\pi \circ \partial(\alpha)$, where $\partial(\alpha)$ is the derivative with respect to ∂ of the composite map*

$$\alpha = i \circ \varphi : \mathcal{O}_U^k \xrightarrow{\psi} \mathcal{E}_U \xrightarrow{i} \mathcal{O}_U^n;$$

that is, $(d\varphi)(\partial)$ is the map obtained by applying ∂ to each entry of a matrix representing α and composing the result with the projection $\pi : \mathcal{O}_U^n \to \mathcal{F}_U$.

Note that the map $(d\varphi)(\partial)$ described above depends only on the subbundle $\mathcal{E}$, and not on the trivialization $\mathcal{O}_X^k \cong \mathcal{E}$ chosen: If $\beta : \mathcal{O}_X^k \to \mathcal{O}_X^k$ is an invertible matrix over $\mathcal{O}_X$, then

$$\partial(\alpha\beta) = (\partial\alpha)\beta + \alpha(\partial\beta),$$

and the second term vanishes when we project to $\mathcal{F}$.

Proof: The desired result follows at once from the description of the affine spaces covering the Grassmannian: We can change bases in $\mathcal{O}_U^n$ so that α is given by the $k \times n$ matrix

$$\begin{pmatrix} 1 & 0 & \cdots & 0 & \alpha_{1,1} & \alpha_{1,2} & \cdots & \alpha_{1,n-k} \\ 0 & 1 & \cdots & 0 & \alpha_{2,1} & \alpha_{2,2} & \cdots & \alpha_{2,n-k} \\ \vdots & \vdots & \ddots & \vdots & \vdots & \vdots & \ddots & \vdots \\ 0 & 0 & 0 & 1 & \alpha_{k,1} & \alpha_{k,2} & \cdots & \alpha_{k,n-k} \end{pmatrix},$$

where the $\alpha_{i,j}$ are functions on U, and give the morphism φ in local coordinates. The derivative of φ, applied to ∂, is then by definition obtained by applying ∂ to each of the coordinate functions $\alpha_{i,j}$. $\square$

3.2.6 Tangent spaces via the universal property

There is another way to approach the tangent space, which depends on a pretty and well-known bit of algebra. Let $\mathcal{O}$ be a local ring with maximal ideal $\mathfrak{m}$, and suppose for simplicity that $\mathcal{O}$ contains a copy of its residue field $\Bbbk$.

Proposition–Definition 3.7. *There are natural one-to-one correspondences between the following sets:*

(a) $\mathrm{Hom}_{\Bbbk}(\mathfrak{m}/\mathfrak{m}^2, \Bbbk)$ *(homomorphisms of $\Bbbk$-vector spaces).*
(b) $\mathrm{Der}_{\Bbbk}(\mathcal{O}, \Bbbk)$ *($\Bbbk$-linear derivations; that is, $\Bbbk$-vector space homomorphisms $d : \mathcal{O} \to \Bbbk$ that satisfy Leibniz' rule $d(fg) = f d(g) + g d(f)$).*
(c) $\mathrm{Hom}_{\Bbbk\text{-algebras}}(\mathcal{O}, \Bbbk[\epsilon]/(\epsilon^2))$ *(homomorphisms of $\Bbbk$-algebras).*
(d) $\mathrm{Mor}(D, \mathrm{Spec}\, \mathcal{O})$, *where $D = \mathrm{Spec}\, \Bbbk[\epsilon]/(\epsilon^2)$ (morphisms of schemes).*

The first two of these are naturally $\Bbbk$-*vector spaces, and the correspondence preserves this structure; we regard the other two as equipped with this structure as well. Any of these spaces, with its vector space structure, is called the* Zariski tangent space *of* Spec $\mathcal{O}$ *at its closed point.*

When $\mathcal{O}$ is the local ring $\mathcal{O}_{X,x}$ of a variety (or scheme) at a closed point x, we think of the Zariski tangent space of $\mathcal{O}$ as the Zariski tangent space of X at x, and of course the set in (d) is the same as the set of morphisms of $\Bbbk$-schemes carrying the (unique) point of D — which we will call 0 — to x.

Proof: The sets in (c) and (d) are the same by definition. If φ is as in (c), then $\varphi|_{\mathfrak{m}}$ annihilates $\mathfrak{m}^2$ and induces a vector space homomorphism $\overline{\varphi} : \mathfrak{m}/\mathfrak{m}^2 \to (\epsilon)/(\epsilon^2) \cong \Bbbk$ as in (a). Similarly, a derivation d as in (b) induces a $\Bbbk$-linear map $\overline{d|_{\mathfrak{m}}} : \mathfrak{m}/\mathfrak{m}^2 \to \Bbbk$. We leave to the reader the construction of the inverse correspondences. $\qquad\square$

Consider the case $X = G(k, V)$. The tangent space at $x = [\Lambda]$ is, by the argument above, the collection of maps $D \to G(k, V)$ sending 0 to $[\Lambda]$. By Theorem 3.4, giving a map $D \to G(k, V)$ is the same as the giving a rank-k subbundle $\mathcal{W}$ of $V \times D$; the map takes $0 \in D$ to $[\Lambda] \in \mathbb{G}(k, V)$ if and only if the fiber $\mathcal{W}_0$ is equal to Λ.

We can understand the identification of the tangent space $T_\Lambda G(k, V)$ to the Grassmannian with the space $\operatorname{Hom}(\Lambda, V/\Lambda)$ using this description together with the universal property of the Grassmannian described in Theorem 3.4.

Since D is affine, a vector bundle over D is the same as a locally free module over $\Bbbk[\epsilon]/(\epsilon^2)$. Since this ring is local, Nakayama's lemma shows that such a module is free (see for example Eisenbud [1995, Exercise 4.11]). Thus only the inclusion $\Lambda \times D \to V \times D$ varies.

Putting this together, we get a new way to look at the identification of the tangent spaces to the Grassmannian:

Proposition 3.8. *Let* $\Lambda \subset V$ *be a* k-*dimensional subspace, and let* $\varphi : \Lambda \to V/\Lambda$ *be a homomorphism. As an element of the tangent space to the Grassmannian* $G(k, V)$ *at the point* $[\Lambda]$, φ *corresponds to the free submodule*

$$\Lambda \otimes \Bbbk[\epsilon]/(\epsilon^2) \to V \otimes \Bbbk[\epsilon]/(\epsilon^2), \quad v \otimes 1 \mapsto v \otimes 1 + \varphi'(v) \otimes \epsilon,$$

where $\varphi' : \Lambda \to V$ *is any map that when composed with the projection* $V \to V/\Lambda$ *gives* φ.

Proof: Any map $\Lambda \times D \to V \times D$ that reduces to the inclusion modulo ϵ has the form $v \otimes 1 \mapsto v \otimes 1 + \varphi'(v) \otimes \epsilon$ for some φ'. If we work in the affine coordinates corresponding to a subspace Γ complementary to Λ and use the splitting $V = \Lambda \oplus \Omega$, then the point $\Lambda \subset V$ corresponds to the matrix

$$
B = \begin{pmatrix}
1 & 0 & \cdots & 0 & 0 & \cdots & 0 \\
0 & 1 & \cdots & 0 & 0 & \cdots & 0 \\
\vdots & \vdots & \ddots & 0 & 0 & \cdots & 0 \\
0 & 0 & \cdots & 1 & 0 & \cdots & 0
\end{pmatrix}.
$$

In this matrix representation φ is represented by the last $n - k$ columns of φ', and taking a different lifting of φ corresponds to making a different choice of the first $k \times k$ block of φ'.

We can do row operations to clear all the ϵ terms from the first $k \times k$ block, adding a multiple of ϵ times certain rows to other rows. This corresponds to composing with an automorphism of $\Lambda \times D$, and thus does not change the image of $\Lambda \times D \to V \times D$. Since we add after multiplying by ϵ, this does not change the block representing φ. Thus we may assume that the first $k \times k$ block of φ' is 0; equivalently, the first $k \times k$ block corresponding to the map $\Lambda \times D \to V \times D$ is the identity. □

3.3 The Chow ring of $\mathbb{G}(1, 3)$

Before launching into the geometry of general Grassmannians in the next chapter, we will spend the remainder of this chapter studying the geometry of $\mathbb{G}(1, 3)$, the Grassmannian of lines in $\mathbb{P}^3$. This is the simplest case beyond the projective spaces. The general results are in many ways similar, but more combinatorics is involved, and in the case of lines in $\mathbb{P}^3$ it is possible to visualize more of what is going on. Once the reader has absorbed the case of $\mathbb{G}(1, 3)$ the general results will seem more natural.

3.3.1 Schubert cycles in $\mathbb{G}(1, 3)$

To start, we fix a *complete flag* $\mathcal{V}$ on $\mathbb{P}^3$; that is, a choice of a point $p \in \mathbb{P}^3$, a line $L \subset \mathbb{P}^3$ containing p, and a plane $H \subset \mathbb{P}^3$ containing L (Figure 3.2).

We can give a stratification of $\mathbb{G}(1, 3)$ by considering the loci of lines $\Lambda \in \mathbb{G}(1, 3)$ having specified dimension of intersection with each of the subspaces p, L and H. These are called *Schubert cells* and their closures, which are irreducible subvarieties, are called *Schubert cycles* (or sometimes Schubert varieties); the classes of these cycles are the *Schubert classes*. As we shall see, the Schubert cells form an affine stratification of $\mathbb{G}(1, 3)$, and it will follow from Proposition 1.17 that the Schubert classes generate the Chow group $A(\mathbb{G}(1, 3))$. Using intersection theory, we will be able to show that in fact $A(\mathbb{G}(1, 3))$ is a free $\mathbb{Z}$-module having the Schubert classes as free generators. In the next chapter, we will see that the same situation is repeated for all Grassmannians.

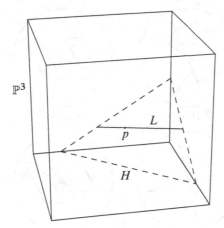

Figure 3.2 A complete flag $p \subset L \subset H$ in $\mathbb{P}^3$.

More formally, we begin not with the Schubert cells but with the Schubert cycles:

$$\Sigma_{0,0} = \mathbb{G}(1,3),$$
$$\Sigma_{1,0} = \{\Lambda \mid \Lambda \cap L \neq \varnothing\},$$
$$\Sigma_{2,0} = \{\Lambda \mid p \in \Lambda\},$$
$$\Sigma_{1,1} = \{\Lambda \mid \Lambda \subset H\},$$
$$\Sigma_{2,1} = \{\Lambda \mid p \in \Lambda \subset H\},$$
$$\Sigma_{2,2} = \{\Lambda \mid \Lambda = L\}.$$

The four nontrivial ones are illustrated in Figure 3.3. In each case we take the reduced scheme structure. To show, for example, that $\Sigma_{1,0}$ is an irreducible variety, we note first that $\Sigma_{1,0}$ is the image of the incidence correspondence

$$\Gamma = \{(L', p) \in \mathbb{G}(1,3) \times L \mid p \in L'\}.$$

The fiber of Γ under projection to L, the set of lines through a given point $p \in L$, is isomorphic to $\mathbb{P}^2$; since all fibers are irreducible and of the same dimension and the projection is proper, it follows that Γ, and with it $\Sigma_{1,0}$, are irreducible. The proof that the other Schubert cycles are irreducible follows in exactly the same way.

Thus $\Sigma_{a,b}$ denotes the set of lines meeting the $(2-a)$-dimensional plane of V in a point and the $(3-b)$-dimensional plane of V in a line. This system of indexing may seem peculiar at first; the reasons for it will be clearer when we discuss Schubert cycles in general in the following chapter. For now, we will mention that the codimension of $\Sigma_{a,b}$ is $a+b$, as will be clear in examples and as we will prove in general in Theorem 4.1.

We often drop the second index when it is 0, writing for example Σ_1 instead of $\Sigma_{1,0}$. When the choice of flag is relevant, we will sometimes indicate the dependence by writing $\Sigma(V)$, or simply note the dependence on the relevant flag elements by writing, for example, $\Sigma_1(L)$ for the cycle of lines Λ meeting L.

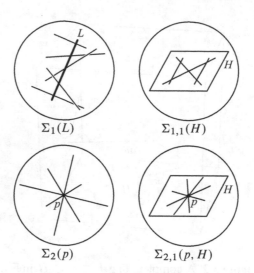

$\Sigma_1(L)$ $\Sigma_{1,1}(H)$

$\Sigma_2(p)$ $\Sigma_{2,1}(p, H)$

Figure 3.3 Schubert cycles in $\mathbb{G}(1,3)$.

It is easy to see that there are inclusions

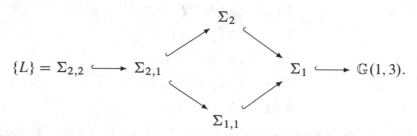

For each index (a, b) we define the Schubert cell $\Sigma_{a,b}^\circ$ to be the complement in $\Sigma_{a,b}$ of the union of all the other Schubert cycles properly contained in $\Sigma_{a,b}$. To show that the $\Sigma_{a,b}$ form an affine stratification, it suffices to show that each $\Sigma_{a,b}^\circ$ is isomorphic to an affine space. We will do the most complicated case, leaving the others for the reader (the general case of a Schubert cell in $G(k, n)$ is done in Theorem 4.1).

Example 3.9. We will show that the set

$$\Sigma_1^\circ = \Sigma_1 \setminus (\Sigma_2 \cup \Sigma_{1,1}) = \{\Lambda \mid \Lambda \cap L \neq \varnothing \text{ but } p \notin \Lambda \text{ and } \Lambda \not\subset H\}$$

is isomorphic to $\mathbb{A}^3$. Let H' be a general plane containing the point p but not containing the line L. Any line meeting L but not passing through p and not contained in H meets H' in a unique point contained in $H' \setminus (H' \cap H)$ (Figure 3.4). Thus we have maps

$$\Sigma_1^\circ \to (L \setminus \{p\}) \cong \mathbb{A}^1 \quad \text{and} \quad \Sigma_1^\circ \to (H' \setminus (H \cap H')) \cong \mathbb{A}^2$$

sending Λ to $\Lambda \cap L$ and $\Lambda \cap H'$ respectively. The product of these maps gives us an isomorphism

$$\Sigma_1^\circ \cong \mathbb{A}^1 \times \mathbb{A}^2 = \mathbb{A}^3.$$

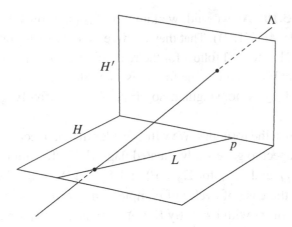

Figure 3.4 The map $\Lambda \mapsto (\Lambda \cap L, \Lambda \cap H')$ defines an isomorphism $\Sigma_1^\circ \to (L \setminus \{p\}) \times (H \setminus H \cap H') \cong \mathbb{A}^1 \times \mathbb{A}^2 \cong \mathbb{A}^3$.

By Theorem 1.7, the class $[\Sigma_{a,b}] \in A^{a+b}(\mathbb{G}(1,3))$ does not depend on the choice of flag, since any two flags differ by a transformation in GL_4; we will denote the class of $\Sigma_{a,b}$ by

$$\sigma_{a,b} = [\Sigma_{a,b}] \in A^{a+b}(\mathbb{G}(1,3)).$$

By Proposition 1.8, the group $A^0(\mathbb{G}(1,3))$ is isomorphic to $\mathbb{Z}$ and is generated by the fundamental class $\sigma_{0,0} = [\mathbb{G}(1,3)]$; by Theorem 1.7, the group $A^4(\mathbb{G}(1,3))$ is also isomorphic to $\mathbb{Z}$ and is generated by the class $\sigma_{2,2}$ of a point in $\mathbb{G}(1,3)$. (In particular any two points in $\mathbb{G}(1,3)$ are rationally equivalent.)

3.3.2 Ring structure

We can now determine the structure of the Chow ring of $\mathbb{G}(1,3)$ completely:

Theorem 3.10. *The six Schubert classes* $\sigma_{a,b} \in A^{a+b}(\mathbb{G}(1,3))$, $0 \le b \le a \le 2$, *freely generate* $A(\mathbb{G}(1,3))$ *as a graded abelian group, and satisfy the multiplicative relations*

$$\sigma_1^2 = \sigma_{1,1} + \sigma_2 \qquad\qquad (A^1 \times A^1 \to A^2);$$
$$\sigma_1\sigma_{1,1} = \sigma_1\sigma_2 = \sigma_{2,1} \qquad (A^1 \times A^2 \to A^3);$$
$$\sigma_1\sigma_{2,1} = \sigma_{2,2} \qquad\qquad (A^1 \times A^3 \to A^4);$$
$$\sigma_{1,1}^2 = \sigma_2^2 = \sigma_{2,2}, \quad \sigma_{1,1}\sigma_2 = 0 \quad (A^2 \times A^2 \to A^4).$$

From these formulas we deduce that $\sigma_1^3 = 2\sigma_{2,1}, \sigma_1^4 = 2\sigma_{2,2}$, and $\sigma_1^2\sigma_{1,1} = \sigma_1^2\sigma_2 = \sigma_{2,2}$. Since $\dim(\mathbb{G}(1,3)) = 4$, any product that would have degree > 4, such as $\sigma_2\sigma_{2,1}$, is 0.

Proof of Theorem 3.10: As we said, we know by Proposition 1.17 that the Schubert classes $\sigma_{a,b}$ generate $A(\mathbb{G}(1,3))$. That they are free generators follows for $A^4(\mathbb{G}(1,3))$ from Proposition 1.21, and will follow for the remaining Chow groups from the intersections products above: For example, the formulas show that the matrix of the intersection pairing on $\sigma_{1,1}$ and σ_2 is nonsingular, so $A^2(\mathbb{G}(1,3))$ is freely generated by these two classes.

It remains to prove the formulas. We will consider the intersections of pairs of cycles, taking these with respect to generically situated flags $\mathcal{V}, \mathcal{V}'$. To simplify notation we will henceforth write $\Sigma_{a,b}$ and $\Sigma'_{a,b}$ for $\Sigma_{a,b}(\mathcal{V})$ and $\Sigma_{a,b}(\mathcal{V}')$, respectively.

We begin with the case of cycles of complementary dimension, starting with the intersection number of σ_2 with itself. By Kleiman transversality we have

$$\sigma_2^2 = \#(\Sigma_2 \cap \Sigma'_2) \cdot \sigma_{2,2},$$

and since the intersection

$$\Sigma_2 \cap \Sigma'_2 = \{\Lambda \mid p \in \Lambda \text{ and } p' \in \Lambda\}$$

consists of one point (corresponding to the unique line $\Lambda = \overline{p, p'}$ through p and p'), we conclude that

$$\sigma_2^2 = \sigma_{2,2}.$$

Similarly,

$$\sigma_{1,1}^2 = \#(\Sigma_{1,1} \cap \Sigma'_{1,1}) \cdot \sigma_{2,2};$$

since

$$\Sigma_{1,1} \cap \Sigma'_{1,1} = \{\Lambda \mid \Lambda \subset H \text{ and } \Lambda \subset H'\}$$

consists of the unique line $\Lambda = H \cap H'$, we conclude that

$$\sigma_{1,1}^2 = \sigma_{2,2}$$

as well. On the other hand, $\Sigma_2 = \{\Lambda \mid p \in \Lambda\}$ and $\Sigma'_{1,1} = \{\Lambda \mid \Lambda \subset H'\}$ are disjoint, since $p \notin H'$, so that

$$\sigma_2 \sigma_{1,1} = 0.$$

Finally,

$$\Sigma_1 \cap \Sigma'_{2,1} = \{\Lambda \mid \Lambda \cap L \neq \varnothing \text{ and } p' \in \Lambda \subset H'\}.$$

Since L will intersect H' in one point q, and any line Λ satisfying all the above conditions can only be the line $\overline{p', q}$ (Figure 3.5), this intersection is again a single point. Thus

$$\sigma_1 \sigma_{2,1} = \sigma_{2,2}$$

as well.

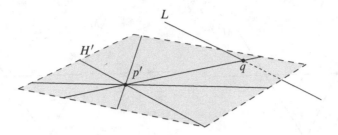

Figure 3.5 $\Sigma_1(L) \cap \Sigma'_{2,1}(p', H') = \{\overline{p', q}\}$.

We now turn to the intersections of cycles whose codimensions sum to less than 4. First, the intersection $\Sigma_1 \cap \Sigma'_2$ is the locus of lines Λ meeting L and containing the point p', which is to say the Schubert cycle $\Sigma_{2,1}$ with respect to a flag containing the point p' and the plane $\overline{p', L}$, so we have

$$\sigma_1 \sigma_2 = \sigma_{2,1}.$$

In a similar fashion, the intersection of Σ_1 and $\Sigma'_{1,1}$ is a cycle of the form $\Sigma_{2,1}$ with respect to a certain flag; specifically, it is the locus of lines containing the point $L \cap H'$ and lying in H', so that

$$\sigma_1 \sigma_{1,1} = \sigma_{2,1}.$$

The last and most interesting computation to be made is the product σ_1^2. (This is such a crucial case that we will prove it twice: here and in Section 3.5.1!) The difference between this case and the preceding ones is that the locus $\Sigma_1 \cap \Sigma'_1$ of lines meeting each of the two general lines L and L' is not a Schubert cycle.

We will use the method of undetermined coefficients, first introduced in Section 2.1.6. We have by now established that $A^2(\mathbb{G}(1,3))$ is freely generated by the classes $\sigma_{1,1}$ and σ_2, so that we may write

$$\sigma_1^2 = \alpha \sigma_2 + \beta \sigma_{1,1} \tag{3.2}$$

for some (unique) α and $\beta \in \mathbb{Z}$. We can then determine the coefficients α and β by taking the product of both sides of (3.2) with classes of complementary dimension.

One way to do this is by invoking the associativity of $A(\mathbb{G}(1,3))$ and the previous calculations: We have

$$(\alpha \sigma_2 + \beta \sigma_{1,1})\sigma_2 = \sigma_1^2 \cdot \sigma_2 = \sigma_1(\sigma_1 \sigma_2) = \sigma_1 \sigma_{2,1} = \sigma_{2,2},$$

and since $\sigma_2^2 = \sigma_{2,2}$ and $\sigma_{1,1}\sigma_2 = 0$ we get $\alpha = 1$. Similarly, from

$$(\alpha \sigma_2 + \beta \sigma_{1,1})\sigma_{1,1} = \sigma_1^2 \cdot \sigma_{1,1} = \sigma_1(\sigma_1 \sigma_{1,1}) = \sigma_1 \sigma_{2,1} = \sigma_{2,2}$$

and $\sigma_{1,1}^2 = \sigma_{2,2}$ we see that $\beta = 1$. In sum, we have

$$\sigma_1^2 = \sigma_2 + \sigma_{1,1},$$

and this completes our description of the Chow ring $A(\mathbb{G}(1,3))$. $\square$

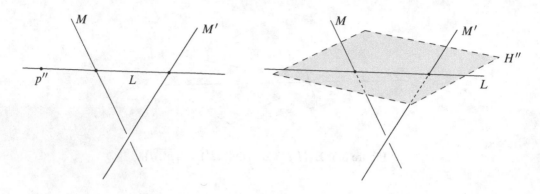

Figure 3.6 $\Sigma_2(p'')\cap\Sigma_1(M)\cap\Sigma_1(M') = \{L\}$; $\Sigma_{1,1}(H'')\cap\Sigma_1(M)\cap\Sigma_1(M') = \{L\}$.

It is instructive to compute $\sigma_1^2\sigma_2$ and $\sigma_1^2\sigma_{1,1}$ geometrically, without invoking associativity as in the proof above. To determine α, we used

$$\sigma_1^2 \cdot \sigma_2 = (\alpha\sigma_2 + \beta\sigma_{1,1}) \cdot \sigma_2 = \alpha\sigma_{2,2}.$$

By Kleiman transversality, we have

$$\alpha = \#\left\{ \Lambda \ \middle| \ \begin{array}{l} \Lambda \cap L \neq \varnothing, \\ \Lambda \cap L' \neq \varnothing \text{ and} \\ p'' \in \Lambda \end{array} \right\}$$

for L and L' general lines and p'' a general point in $\mathbb{P}^3$. Any line Λ satisfying the three conditions must lie in each of the planes $\overline{p'', L}$ and $\overline{p'', L'}$, and so must be their intersection; thus $\alpha = 1$.

Similarly, to determine β we used

$$\sigma_1^2 \cdot \sigma_{1,1} = (\alpha\sigma_2 + \beta\sigma_{1,1}) \cdot \sigma_{1,1} = \beta\sigma_{2,2}.$$

Again, by generic transversality, we get:

$$\beta = \#\left\{ \Lambda \ \middle| \ \begin{array}{l} \Lambda \cap L \neq \varnothing, \\ \Lambda \cap L' \neq \varnothing \text{ and} \\ \Lambda \subset H'' \end{array} \right\}$$

for L and L' general lines and H'' a general plane in $\mathbb{P}^3$. The only line Λ satisfying these conditions is the line joining the points $L \cap H''$ and $L' \cap H''$, so again $\beta = 1$ (Figure 3.6).

Tangent spaces to Schubert cycles

The generic transversality of the cycles $\Sigma_{a,b}$ and $\Sigma_{a',b'}$, guaranteed by Kleiman's theorem in characteristic 0, played an essential role in the computation above. By describing the tangent spaces to the Schubert cycles, we can prove this transversality directly and hence extend the results to characteristic p.

We will carry this out here for the intersection $\Sigma_2 \cap \Sigma_2'$. Tangent spaces to other Schubert cycles in $\mathbb{G}(1,3)$ are described in Exercises 3.26 and 3.27; they will be treated in general in Theorem 4.1. The key identification is given in the following result:

Proposition 3.11. *Let* $\Sigma = \Sigma_2(p)$ *be the Schubert cycle of lines in* $\mathbb{P}^3 = \mathbb{P}V$ *that contain* p, *and suppose that* $L \in \Sigma_2(p)$. *Writing* $\tilde{L} \subset V$ *for the two-dimensional subspace corresponding to* L, *and identifying* $T_L\, \mathbb{G}(1,3)$ *with* $\mathrm{Hom}(\tilde{L}, V/\tilde{L})$, *we have*

$$T_L \Sigma = \{\varphi \mid \varphi(\tilde{p}) = 0\}.$$

Given Proposition 3.11, it follows immediately that for general $p, p' \in \mathbb{P}^3$ the cycles $\Sigma_2(p)$ and $\Sigma_2(p')$ meet transversely: If $p \neq p'$, then at the unique point $L = \overline{p, p'}$ of intersection of the Schubert cycles $\Sigma = \Sigma_2(p)$ and $\Sigma' = \Sigma_2(p')$, we have

$$T_{[L]}\Sigma \cap T_{[L]}\Sigma' = \{\varphi \mid \varphi(\tilde{p}) = \varphi(\tilde{p}') = 0\} = \{0\},$$

since $\tilde{p}$ and $\tilde{p}'$ span $\tilde{L}$.

Proof of Proposition 3.11: We choose a subspace $\Gamma \subset V$ complementary to $\tilde{L}$ and identify the open subset

$$U_\Gamma = \{\Lambda \in \mathbb{G}(1,3) \mid \Lambda \cap \Gamma = 0\}$$

with the vector space $\mathrm{Hom}(\tilde{L}, \Gamma)$ by thinking of a 2-plane $\Lambda \in U_\Gamma$ as the graph of a linear map from $\tilde{L}$ to Γ, just as in the beginning of the proof of Theorem 3.5. It is immediate from the identification that $U_\Gamma \cap \Sigma_2$ is the linear space in $\mathrm{Hom}(\tilde{L}, \Gamma)$ consisting of maps φ such that $\tilde{p} \subset \mathrm{Ker}(\varphi)$. Thus its tangent space at a point $[L]$ has the same description. $\qquad\square$

As a consequence of Theorem 3.10, we have the following description of the Chow ring of $\mathbb{G}(1,3)$:

Corollary 3.12. $$A(\mathbb{G}(1,3)) = \frac{\mathbb{Z}[\sigma_1, \sigma_2]}{(\sigma_1^3 - 2\sigma_1\sigma_2,\ \sigma_1^2\sigma_2 - \sigma_2^2)}.$$

We will generalize this to the Chow ring of any Grassmannian, and prove it by applying the theory of Chern classes, in Chapter 5. A point to note is that the given presentation of the Chow ring has the same number of generators as relations — that is, given that the Chow ring $A(\mathbb{G}(1,3))$ has Krull dimension 0, it is a *complete intersection*. The analogous statement is true for all Grassmannians.

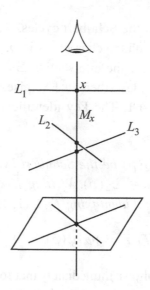

Figure 3.7 An apparent double point: when viewed from x, the lines L_2 and L_3 appear to cross at a point in the direction M_x, and therefore there is a unique line through x meeting L_2 and L_3.

3.4 Lines and curves in $\mathbb{P}^3$

In this section and the next we present several applications of the computations above.

3.4.1 How many lines meet four general lines?

This is Keynote Question (a). Since σ_1 is the class of the locus $\Sigma_1(L)$ of lines meeting a given line L, and generic translates of $\Sigma_1(L)$ are generically transverse, the number is

$$\deg \sigma_1^4 = 2.$$

We can see the geometry behind this computation—and answer more refined questions about the situation—as follows. Suppose that the lines are $L_1, \ldots, L_4$, and consider first the lines that meet just the first three. To begin with, we claim that *if $x \in L_1$ is any point, there is a unique line $M_x \subset \mathbb{P}^3$ passing through x and meeting L_2 and L_3*, as in Figure 3.7. To see this, note that if we project L_2 and L_3 from x to a plane H, we get two general lines in H, and these lines meet in a unique point y. The line $M_x := \overline{x, y}$ is then the unique line in $\mathbb{P}^3$ containing x and meeting L_2 and L_3. (Informally: If we look at L_2 and L_3, sighting from the point x, we see an "apparent crossing" in the direction of the line M_x—see Figure 3.7.) Moreover, if $x \neq x'$ then the lines $M_x, M_{x'}$ are disjoint: If they had a common point, they would lie in a plane, and all three of L_1, L_2, L_3 would be coplanar, contradicting our hypothesis of generality.

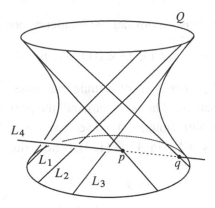

Figure 3.8 Two lines that meet each of $L_1, \ldots, L_4$.

The union of the lines M_x is a surface that we can easily identify. There is a three-dimensional family of quadratic polynomials on each $L_i \cong \mathbb{P}^1$. Each restriction map $H^0(\mathcal{O}_{\mathbb{P}^3}(2)) \to H^0(\mathcal{O}_{L_i}(2))$ is linear, so its kernel has codimension ≤ 3. Since there is a 10-dimensional vector space of quadratic polynomials on $\mathbb{P}^3$, there is thus at least one quadric surface Q containing L_1, L_2 and L_3. By Bézout's theorem, any line meeting each of L_1, L_2, L_3, and thus meeting Q at least three times, must be contained in Q. Since the union of the lines M_x is a nondegenerate surface (the lines are pairwise disjoint, and so cannot lie in a plane), it follows that Q is unique, and is equal to the disjoint union

$$\coprod_{x \in L_1} M_x = Q.$$

Since the degree of Q is 2, and L_4 is general, L_4 meets Q in two distinct points p and q; the two lines M_p and M_q passing through p and q are the unique lines meeting all of $L_1, \ldots, L_4$ (Figure 3.8). Thus we see again that the answer to our question is 2.

For which sets of four lines are there more or fewer than two distinct lines meeting all four? The geometric construction above will enable the reader to answer this question; see Exercise 3.29.

3.4.2 Lines meeting a curve of degree d

We do not know a geometric argument such as the above one for four lines that would enable us to answer the corresponding question for four curves, Keynote Question (b). In this case, intersection theory is essential. The basic computation is the following:

Proposition 3.13. *Let $C \subset \mathbb{P}^3$ be a curve of degree d. If*

$$\Gamma_C := \{L \in \mathbb{G}(1,3) \mid L \cap C \neq \varnothing\}$$

is the locus of lines meeting C, then the class of Γ_C is

$$[\Gamma_C] = d \cdot \sigma_1 \in A^1(\mathbb{G}(1,3)).$$

Proof: To see that Γ_C is a divisor, consider the incidence correspondence

$$\Sigma = \{(p, L) \in C \times \mathbb{G}(1,3) \mid p \in L\},$$

whose image in $\mathbb{G}(1,3)$ is Γ_C. The fibers of Σ under the projection to C are all projective planes, so Σ has pure dimension 3. On the other hand, the projection to Γ_C is generically one-to-one, so Γ_C also has pure dimension 3. (See Exercise 3.20 for a generalization.)

Now let γ_C be the class of Γ_C in $A^1(\mathbb{G}(1,3))$, and write

$$\gamma_C = \alpha \cdot \sigma_1$$

for some $\alpha \in \mathbb{Z}$. To determine α, we intersect both sides with the class $\sigma_{2,1}$ and get

$$\deg \gamma_C \cdot \sigma_{2,1} = \alpha \deg(\sigma_1 \cdot \sigma_{2,1}) = \alpha.$$

If (p, H) is a general pair consisting of a point $p \in \mathbb{P}^3$ and a plane $H \subset \mathbb{P}^3$ containing p, the Schubert cycle

$$\Sigma_{2,1}(p, H) = \{L \mid p \in L \subset H\}$$

will intersect the cycle Γ_C transversely. (This follows from Kleiman's theorem in characteristic 0, and can be proven in all characteristics by using the description of the tangent spaces to $\Sigma_1(p, H)$ in Exercise 3.27 and of the tangent spaces to Γ_C in Exercise 3.30.) Therefore

$$\alpha = \#(\Gamma_C \cap \Sigma_{2,1}(p, H)) = \#\{L \mid p \in L \subset H \text{ and } L \cap C \neq \varnothing\}.$$

To evaluate this number, note that H (being general) will intersect C transversely in d points $\{q_1, \ldots, q_d\}$; since $p \in H$ is general, no two of the points q_i will be collinear with p. Thus the intersection $\Gamma_C \cap \Sigma_{2,1}(p, H)$ will consist of the d lines $\overline{p, q_i}$, as in Figure 3.9. It follows that $\alpha = d$, so

$$\gamma_C = d \cdot \sigma_1. \hspace{3cm} \square$$

We will revisit Proposition 3.13 in Section 3.5.3, where we will see how to calculate γ_C by the method of *specialization*.

Proposition 3.13 makes it easy to answer Keynote Question (b): If $C_1, \ldots, C_4 \subset \mathbb{P}^3$ are general translates of curves of degrees $d_1, \ldots, d_4$, then the cycles Γ_{C_i} are generically transverse by Kleiman transversality, so the number of lines meeting all four is

$$\deg \prod_{i=1}^{4} [\Gamma_{C_i}] = \deg \prod_{i=1}^{4} (d_i \sigma_1) = 2 \prod_{i=1}^{4} d_i.$$

One can verify the necessary transversality by using our description of the tangent spaces, too; as a bonus, we can see exactly when transversality fails. This is the content of Exercises 3.30–3.33.

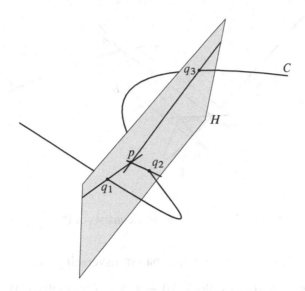

Figure 3.9 The intersection of Γ_C with $\Sigma_{2,1}(p, H)$.

3.4.3 Chords to a space curve

Consider now a smooth, nondegenerate space curve $C \subset \mathbb{P}^n$ of degree d and genus g. We define the locus $\Psi_2(C) \subset \mathbb{G}(1,n)$ of *chords*, or *secant lines* to C, to be the closure in $\mathbb{G}(1,n)$ of the locus of lines of the form $\overline{p,q}$ with p and q distinct points of C. Inasmuch as $\Psi_2(C)$ is the image of the rational map $\tau : C \times C \dashrightarrow \mathbb{G}(1,n)$ sending (p,q) to $\overline{p,q}$, we see that $\Psi_2(C)$ will have dimension 2.

Note that we could also characterize $\Psi_2(C)$ as the locus of lines $L \subset \mathbb{P}^n$ such that the scheme-theoretic intersection $L \cap C$ has degree at least 2. As we will see in Exercise 3.38, this characterization differs from the definition given when we consider singular curves, or (as we will see in Exercise 3.39) higher-dimensional secant planes to curves; but for smooth curves in $\mathbb{P}^n$ we will show in Exercise 3.37 they agree, and we can adopt either one. (For much more about secant planes to curves in general, see the discussion in Section 10.3.)

Let us now restrict ourselves to the case $n = 3$ of smooth, nondegenerate space curves $C \subset \mathbb{P}^3$, and ask: What is the class, in $A^2(\mathbb{G}(1,3))$, of the locus $\Psi_2(C)$ of secant lines to C? We can answer this question by intersecting with Schubert cycles of complementary codimension (in this case, codimension 2). We know that

$$[\Psi_2(C)] = \alpha\sigma_2 + \beta\sigma_{1,1}$$

for some integers α and β. To find the coefficient β we take a general plane $H \subset \mathbb{P}^3$, and consider the Schubert cycle

$$\Sigma_{1,1}(H) = \{L \in \mathbb{G}(1,3) \mid L \subset H\}.$$

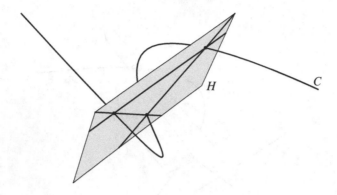

Figure 3.10 $\Sigma_{1,1}(H) \cap \Psi_2(C)$ consists of $\binom{\deg C}{2}$ lines.

By our calculation of $A(\mathbb{G}(1,3))$ and Kleiman transversality, we have

$$\beta = \deg(\sigma_{1,1} \cdot [\Psi_2(C)]) = \#(\Sigma_{1,1}(H) \cap \Psi_2(C)).$$

The cardinality of this intersection is easy to determine: The plane H will intersect C in d points $p_1, \ldots, p_d$, no three of which will be collinear (Arbarello et al. [1985, Section 3.1]), so that there will be exactly $\binom{d}{2}$ lines $\overline{p_i, p_j}$ joining these points pairwise; thus

$$\beta = \binom{d}{2}$$

(see Figure 3.10).

Similarly, to find α we let $p \in \mathbb{P}^3$ be a general point and

$$\Sigma_2(p) = \{L \in \mathbb{G}(1,3) \mid p \in L\};$$

we have as before

$$\alpha = \deg(\sigma_2 \cdot [\Psi_2(C)]) = \#(\Sigma_2(p) \cap \Psi_2(C)).$$

To count this intersection — that is, the number of chords to C through the point p — consider the projection $\pi_p : C \to \mathbb{P}^2$. This map is birational onto its image $\overline{C} \subset \mathbb{P}^2$, which will be a curve having only nodes as singularities (see Exercise 3.34), and these nodes correspond exactly to the chords to C through p. (These chords were classically called the *apparent nodes* of C (Figure 3.11): If you were looking at C with your eye at the point p, and had no depth perception, they are the nodes you would see.) By the genus formula for singular curves (Section 2.4.6), this number is

$$\alpha = \binom{d-1}{2} - g.$$

Thus we have proven:

Proposition 3.14. *If $C \subset \mathbb{P}^3_{\mathbb{C}}$ is a smooth nondegenerate curve of degree d and genus g, then the class of the locus of chords to C is*

$$[\Psi_2(C)] = \left(\binom{d-1}{2} - g\right)\sigma_2 + \binom{d}{2}\sigma_{1,1} \in A^2(\mathbb{G}(1,3)).$$

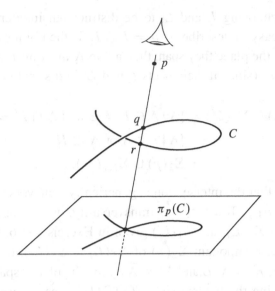

Figure 3.11 Another apparent node.

We can use this to answer the third of the keynote questions of this chapter: If C and C' are general twisted cubic curves, by Kleiman's theorem the cycles $S = \Psi_2(C)$ and $S' = \Psi_2(C')$ intersect transversely; since the class of each is $\sigma_2 + 3\sigma_{1,1}$, we have

$$\#(S \cap S') = \deg(\sigma_2 + 3\sigma_{1,1})^2 = 10.$$

Exercises 3.40 and 3.41 explain how to use the tangent space to the Grassmannian to prove generic transversality, and thus verify this result, in all characteristics.

3.5 Specialization

There is a another powerful approach to evaluating the intersection products of interesting subvarieties: *specialization*. In this section we will discuss some of its variations.

3.5.1 Schubert calculus by static specialization

As a first illustration we show how to compute the class $\sigma_1^2 \in A(\mathbb{G}(1,3))$ by specialization. The reader will find a far-reaching generalization to the Chow rings of Grassmannians and even to more general flag varieties in the algorithms of Vakil [2006a] and Coşkun [2009].

The idea is that instead of intersecting two general cycles $\Sigma_1(L)$ and $\Sigma_1(L')$ representing σ_1, we choose a *special* pair of lines L, L'. The goal is to choose L and L' special enough that the class of the intersection $\Sigma_1(L) \cap \Sigma_1(L')$ is readily identifiable, but at the same time not so special that the intersection fails to be generically transverse.

We do this by choosing L and L' to be distinct but incident. The intersection $\Sigma_1(L) \cap \Sigma_1(L')$ is easy to describe: If $p = L \cap L'$ is the point of intersection of the lines and $H = \overline{L, L'}$ the plane they span, then a line Λ meeting L and L' either passes through p or lies in H (since it then meets L and L' in distinct points). Thus, as sets, we have

$$\Sigma_1(L) \cap \Sigma_1(L') = \{\Lambda \mid \Lambda \cap L \neq \emptyset \text{ and } \Lambda \cap L' \neq \emptyset\}$$
$$= \{\Lambda \mid p \in \Lambda \text{ or } \Lambda \subset H\}$$
$$= \Sigma_2(p) \cup \Sigma_{1,1}(H).$$

If we now show that the intersection is generically transverse, we get the desired formula $\sigma_1^2 = \sigma_2 + \sigma_{1,1}$. To check this transversality, we can use the description of the tangent spaces to $\Sigma_1(L)$ and $\Sigma_1(L')$ given in Exercise 3.26. First, suppose Λ is a general point of the component $\Sigma_2(p)$ of $\Sigma_1(L) \cap \Sigma_1(L')$, that is, a general line through p; we will let $K = \overline{\Lambda, L}$ and $K' = \overline{\Lambda, L'}$ be the planes spanned by Λ together with L and L'. Viewing the tangent space $T_\Lambda(\mathbb{G}(1,3))$ as the vector space of linear maps $\varphi : \tilde{\Lambda} \to V/\tilde{\Lambda}$, we have

$$T_\Lambda(\Sigma_1(L)) = \{\varphi \mid \varphi(\tilde{p}) \subset \tilde{K}/\tilde{\Lambda}\} \quad \text{and} \quad T_\Lambda(\Sigma_1(L')) = \{\varphi \mid \varphi(\tilde{p}) \subset \tilde{K'}/\tilde{\Lambda}\}.$$

Since K and K' are distinct, they intersect in Λ, so that the intersection is

$$T_\Lambda(\Sigma_1(L)) \cap T_\Lambda(\Sigma_1(L')) = \{\varphi \mid \varphi(\tilde{p}) = 0\}$$

Since this is two-dimensional, the intersection $\Sigma_1(L) \cap \Sigma_1(L')$ is transverse at $[\Lambda]$.

Similarly, if Λ is a general point of the component $\Sigma_{1,1}(H)$ of $\Sigma_1(L) \cap \Sigma_1(L')$, so that Λ meets L and L' in distinct points q and q', we have

$$T_\Lambda(\Sigma_1(L)) = \{\varphi \mid \varphi(\tilde{q}) \subset \tilde{H}/\tilde{\Lambda}\} \quad \text{and} \quad T_\Lambda(\Sigma_1(L')) = \{\varphi \mid \varphi(\tilde{q'}) \subset \tilde{H}/\tilde{\Lambda}\},$$

so

$$T_\Lambda(\Sigma_1(L)) \cap T_\Lambda(\Sigma_1(L')) = \{\varphi \mid \varphi(\tilde{\Lambda}) \subset \tilde{H}\}.$$

Again this is two-dimensional and we conclude that $\Sigma_1(L) \cap \Sigma_1(L')$ is transverse at $[\Lambda]$.

Before going on, we mention that the computation of σ_1^2 given here is an example of the simplest kind of specialization argument, what we may call *static specialization*: We are able to find cycles representing the two given classes that are special enough that the class of the intersection is readily identifiable, but general enough that they still intersect properly.

In general, we may not be able to find such cycles. Such situations call for a more powerful and broadly applicable technique, called *dynamic specialization*. There, we consider a one-parameter family of pairs of cycles (A_t, B_t) specializing from a "general" pair to a special pair (A_0, B_0), which may not intersect dimensionally transversely at all! The key idea is to ask not for the intersection $A_0 \cap B_0$ of the limiting cycles, but rather for the limit $\lim_{t \to 0}(A_t \cap B_t)$ of their intersections. For an example of dynamic specialization, see Section 4.4 of the following chapter, where we consider in the Grassmannian $\mathbb{G}(1, 4)$ of lines in $\mathbb{P}^4$ the self-intersection of the cycle of lines meeting a given line in $\mathbb{P}^4$.

3.5.2 Dynamic projection

Problems situated in projective space tend to be especially amenable to specialization techniques: We can use the large automorphism group of $\mathbb{P}^n$ to morph the objects we are dealing with into potentially simpler, more tractable ones. One fundamental example of this is the technique of *dynamic projection*, which we will describe here and use in the following section to re-derive the formulas for the class of the locus of lines incident to a curve.

Fix two disjoint planes A (the "attractor") and R (the "repellor") that span $\mathbb{P}^n$. Choose coordinates $x_0, \ldots, x_r, y_0, \ldots, y_a$ on $\mathbb{P}^n$ so that the equations of A are $\{x_i = 0\}$ and the equations of R are $\{y_i = 0\}$, and consider the action Ψ of the multiplicative group G_m on $\mathbb{P}^n$ given by

$$\psi_t : (x_0, \ldots, x_r, y_0, \ldots, y_a) \mapsto (tx_0, \ldots, tx_r, y_0, \ldots, y_a)$$
$$= (x_0, \ldots, x_r, t^{-1}y_0, \ldots, t^{-1}y_a).$$

(In what follows, we will abbreviate $(x_0, \ldots, x_r, y_0, \ldots, y_a)$ to (x, y); for example, we will write $\psi_t(x, y) = (tx, y)$.) It is clear that the points of A and R remain fixed under the action of G_m. On the other hand, we can say intuitively that a point not in A or R will "flow toward A" as t approaches zero, and will "flow toward R" as t approaches ∞. More precisely, note that any point $p \notin A \cup R$ lies on a unique line that meets both A and R. (The span $\overline{p, A}$, being an $(a + 1)$-plane, must meet R. Since A and R are disjoint, $\overline{p, A}$ can meet R only in a point $q \in R$; the line $\overline{p, q}$ is then the unique line containing p and meeting A and R.) This line is the closure of the orbit of p under the given action of G_m. In particular, any point in $\mathbb{P}^n \setminus R$ has a well-defined limit in A as t approaches zero.

Now suppose $X \subset \mathbb{P}^n$ is any variety. We consider the images of X under the automorphisms ψ_t, and in particular their flat limit as $t \to 0$. In other words, we set

$$Z^\circ = \{(t, p) \in \mathbb{A}^1 \times \mathbb{P}^n \mid t \neq 0 \text{ and } p \in \psi_t(X)\};$$

we let $Z \subset \mathbb{A}^1 \times \mathbb{P}^n$ be the closure of Z°, and look at the fiber X_0 of Z over $t = 0$. (Note that even if X is a variety, X_0 may well be nonreduced.) We think of X_0 as the limit of the varieties $\psi_t(X)$ as t approaches 0 (see Figure 3.12 for an illustration; see also Eisenbud and Harris [2000, Chapter 2] for a discussion of flat limits in general).

The following properties of the limit X_0 make it easy to analyze some interesting cases:

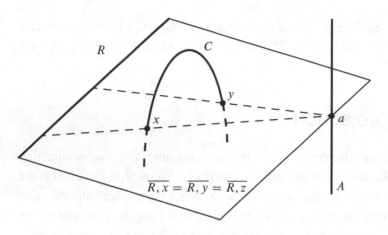

Figure 3.12 Dynamic projection of a conic in $\mathbb{P}^3$ from a line R to a line A.

Proposition 3.15. *With notation as above:*

(a) $X_0 \subset \mathbb{P}^n$ *is stable under the action of G_m.*

(b) $X_0 \cap R = X \cap R$.

(c) $(X_0)_{\text{red}}$ *is contained in the cone over $X_0 \cap R$ with base A (in case $X_0 \cap R = \varnothing$, we take this to mean $(X_0)_{\text{red}} \subset A$).*

In addition, we know (as for any rational equivalence) that X_0 is equidimensional and $\dim X_0 = \dim X$, and, by Eisenbud [1995, Exercise 6.11], that the Hilbert polynomial of X_0 is the same as the Hilbert polynomial of X.

Proof: For the first part, consider the action of G_m on the product $\mathbb{A}^1 \times \mathbb{P}^n$ given as the product of the standard action of G_m on $\mathbb{A}^1$ and the action Ψ above of G_m on $\mathbb{P}^n$; that is,

$$\varphi_t : (s, p) \to (ts, \psi_t(p)).$$

This carries Z to itself and the fiber $\{0\} \times \mathbb{P}^n$ to itself, so it carries X_0 to itself. But it acts on the fiber $\{0\} \times \mathbb{P}^n$ via the action Ψ above; thus X_0 is invariant under Ψ.

The second point is more subtle. (In particular, it is asymmetric: The same statement, with R replaced by A, would be false.) It is not, however, intuitively unreasonable: Since points in $\mathbb{P}^n \setminus R$ flow away from R as $t \to 0$, the only way a point $p \in R$ can be a limit of points $\psi_t(p_t)$ is if it is there all along, that is, if $p \in X \cap R$.

In any case, note first that one inclusion is immediate: Since R is fixed pointwise by the automorphisms ψ_t, we have

$$\mathbb{P}^1 \times (X \cap R) \subset Z$$

and hence $X \cap R \subset X_0 \cap R$. To see the other inclusion, we want to show that the ideal $I(X \cap R)$ is contained in $I(X_0 \cap R)$. Let $f(x) \in I(X \cap R)$. We can then write

$$f(x) = g(x, 0) \quad \text{for some } g(x, y) \in I(X).$$

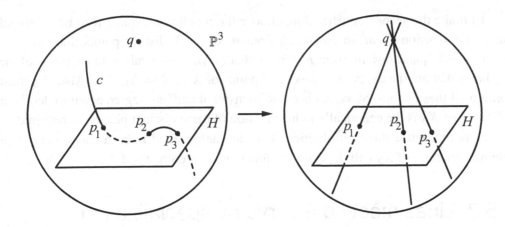

Figure 3.13 A space curve C specializes to a union of lines.

Now observe that

$$I(Z) \supset \{f(x, ty) \mid f \in I(X)\},$$

so $h(t, x, y) = g(x, ty) \in I(Z)$. Setting $t = 0$, we see that $g(x, 0) \in I(X_0)$, and hence $f \in I(X_0 \cap R)$.

To prove the third assertion, note that the G_m-orbit of any point not contained in $R \cup A$ is a straight line joining a point of R to a point of A. Since X_0 is stable under G_m, it is the union of such lines, together with any points of $A \cup R$ it contains. $\square$

We sometimes call this construction *dynamic projection*: We are realizing the projection map

$$\pi_R : \mathbb{P}^n \setminus R \to A,$$
$$(x, y) \mapsto (0, y),$$

as the limit of a family of automorphisms ψ_t of $\mathbb{P}^n$. As we will see, though, considering the limit of the images $\psi_t(X)$ yields more information than simply taking the projection $\pi_R(X)$. See Figures 3.12 and 3.13 for examples.

Example 3.16. Let $X \subset \mathbb{P}^n$ be a subvariety of dimension m and degree d. We will exhibit a dynamic projection of X whose limit is a d-fold m-plane (that is, a scheme whose support is an m-plane and that has multiplicity d at the general point of that plane), and another whose limit is the generically reduced union of d distinct m-planes containing a fixed $(m-1)$-plane.

To make the first construction, let $A \subset \mathbb{P}^n$ be any m-dimensional subspace, and choose R to be an $(n - m - 1)$-plane $R \subset \mathbb{P}^n$ disjoint from X and from A. Since $X \cap R = \varnothing$, we see that $X_0 \cap R = \varnothing$ as well, and it follows that $(X_0)_{\mathrm{red}} \subset A$. Since $\dim X_0 = \dim X = \dim A$, we see that the support of X_0 is exactly A, and computing the degree we have $\langle X_0 \rangle = d \langle A \rangle$, as claimed.

To make the second construction, choose the repellor subspace R to be a general plane of dimension $n - m$, so that $R \cap X$ consists of $\deg X$ distinct points, and take A to be an $(m - 1)$-plane disjoint from R. We see that $(X_0)_{\mathrm{red}}$ is contained in the union of the m-planes that are the cones over the $\deg X$ points of $X \cap R = X_0 \cap R$. Also, X_0 must contain all these points. Since X_0 is equidimensional and has degree equal to $\deg X$, it follows that X_0 is the generically reduced union of these distinct planes, as required.

Note that while the multiplicities of X_0 are determined in both cases, the actual scheme structure of X_0 will depend very much on the geometry of X.

3.5.3 Lines meeting a curve by specialization

As an example of how dynamic projection can be used in specialization arguments, we revisit the computation of the class γ_C of the locus $\Gamma_C \subset \mathbb{G}(1,3)$ of lines meeting a curve $C \subset \mathbb{P}^3$ from Proposition 3.13.

Let $C \subset \mathbb{P}^3$ be a curve of degree d. Choose a plane $H \subset \mathbb{P}^3$ intersecting C transversely in points $p_1, \ldots, p_d$, and $q \in \mathbb{P}^3$ any point not lying on H. Consider the one-parameter group $\{A_t\} \subset \mathrm{PGL}_4$ with repellor plane H and attractor q; that is, choose coordinates $[Z_0, \ldots, Z_3]$ on $\mathbb{P}^3$ such that $q = [1, 0, 0, 0]$ and H is given by $Z_0 = 0$, and consider for $t \neq 0$ the automorphisms of $\mathbb{P}^3$ given by

$$A_t = \begin{pmatrix} 1 & 0 & 0 & 0 \\ 0 & t & 0 & 0 \\ 0 & 0 & t & 0 \\ 0 & 0 & 0 & t \end{pmatrix}.$$

Let $C_t = A_t(C)$, and let $\Phi \subset \mathbb{A}^1 \times \mathbb{P}^3$ be the closure of the locus

$$\Phi^\circ = \{(t, p) \in \mathbb{A}^1 \times \mathbb{P}^3 \mid t \neq 0 \text{ and } p \in C_t\}.$$

As we saw in the preceding section, the limit of the curves C_t as $t \to 0$ (that is, the fiber of Φ over $t = 0$) is supported on the union of the d lines $\overline{p_i, q}$, and has multiplicity 1 at a general point of each, as shown in Figure 3.13.

We can use this construction to give a rational equivalence between the cycle Γ_C and the sum of the Schubert cycles $\Sigma_1(\overline{p_i, q})$ in $\mathbb{G}(1,3)$. Explicitly, take $\Psi \subset \mathbb{A}^1 \times \mathbb{G}(1,3)$ to be the closure of the locus

$$\{(t, \Lambda) \in \mathbb{A}^1 \times \mathbb{G}(1,3) \mid t \neq 0 \text{ and } \Lambda \cap C_t \neq \varnothing\}.$$

As we will verify in Exercises 3.35 and 3.36, the fiber Ψ_0 of Ψ over $t = 0$ is supported on the union of the Schubert cycles $\Sigma_1(\overline{p_i, q})$ and has multiplicity 1 along each, establishing the rational equivalence $\gamma_C = d \cdot \sigma_1$.

The fiber of Φ over $t = 0$ (that is, the flat limit $\lim_{t\to 0} C_t$ of the curves C_t) is *not* necessarily equal to the union of the d lines $\overline{p_i, q}$: it may have an embedded point at the point q. Nonetheless, the fiber Ψ_0, being a divisor in $\mathbb{G}(1, 3)$, will not have embedded components.

3.5.4 Chords via specialization: multiplicity problems

One of the main difficulties in using specialization is the appearance of multiplicities. We will now illustrate this problem by trying to compute, via specialization, the class of the chords to a smooth curve in $\mathbb{P}^3$.

Consider again the family of curves $C_t := A_t(C)$ described in the previous section. What is the limit as $t \to 0$ of the cycles $\Psi_2(C_t) \subset \mathbb{G}(1, 3)$ of chords to C_t? To interpret this question, let $\Pi \subset \mathbb{A}^1 \times \mathbb{G}(1, 3)$ be the closure of the locus

$$\Pi^\circ = \{(t, \Lambda) \in \mathbb{A}^1 \times \mathbb{G}(1, 3) \,|\, t \neq 0 \text{ and } \Lambda \in \Psi_2(C_t)\}.$$

What is the fiber Π_0 of this family?

The support of Π_0 is easy to identify. It is contained in the locus of lines whose intersection with the flat limit $C_0 = \lim_{t\to 0} C_t$ contains a scheme of degree at least 2, which is to say the union of the Schubert cycles $\Sigma_{1,1}(\overline{p_i, p_j, q})$ of lines lying in a plane spanned by a pair of the lines $\overline{p_i, q}$, and the Schubert cycle $\Sigma_2(q)$ of lines containing the point q. Moreover one can show that the Schubert cycles $\Sigma_{1,1}(\overline{p_i, p_j, q})$ all appear with multiplicity 1 in the limiting cycle Π_0, from which we can deduce that the coefficient of $\sigma_{1,1}$ in the class of $\Psi_2(C)$ is $\binom{d}{2}$.

The hard part is determining the multiplicity with which the cycle $\Sigma_2(q)$ appears in Π_0: This will depend in part on the multiplicity of the embedded point of C_0 at q, which will in turn depend on the genus g of C (see for example Exercises 3.43 and 3.44). Note the contrast with the calculation in Section 3.5.3 of the class of the locus Γ_C of incident lines via specialization: There, the embedded component of the limit scheme $\lim_{t\to 0} \Gamma_{C_t}$ also depended on the genus of C, but did not affect the limiting cycle.

An alternative approach to this problem would be to use a different specialization to capture the coefficient of σ_2: Specifically, we could take the one-parameter subgroup with repellor a general point q and attractor a general plane $H \subset \mathbb{P}^3$. The limiting scheme $C_0 = \lim_{t\to 0} C_t$ will be a plane curve of degree d with $\delta = \binom{d-1}{2} - g$ nodes $r_1, \ldots, r_\delta$, with a spatial embedded point of multiplicity 1 at each node. The limit of the corresponding cycles $\Psi_2(C_t) \subset \mathbb{G}(1, 3)$ will correspondingly be supported on the union of the Schubert cycle $\Sigma_{1,1}(H)$ and the δ Schubert cycles $\Sigma_2(r_i)$. In this case the coefficient of the Schubert cycle $\Sigma_{1,1}(H)$ is the mysterious one (though calculable: given that a general line $\Lambda \subset H$ meets C_0 in d points, we can show that it is the limit of $\binom{d}{2}$ chords to C_t as $t \to 0$). On the other hand, one can show that the Schubert cycles $\Sigma_2(r_i)$ all appear with multiplicity 1 in the limit of the cycles $\Psi_2(C_t)$, from which we can read off the coefficient δ of σ_2 in the class of $\Psi_2(C)$.

We will fill in some of the details involved in this calculation in Exercise 3.45.

3.5.5 Common chords to twisted cubics via specialization

To illustrate the artfulness possible in specialization arguments, we give a different specialization approach to counting the common chords of two twisted cubics: We will not degenerate the twisted cubics; we will just specialize them to a general pair of twisted cubic curves C, C' lying on the same smooth quadric surface Q, of types $(1, 2)$ and $(2, 1)$ respectively.

The point is, no line of either ruling of Q will be a chord of both C and C' (the lines of one ruling are chords of C but not of C', and vice versa for lines of the other ruling). But since $C \cup C' \subset Q$, any line meeting $C \cup C'$ in three or more distinct points must lie in Q. It follows that *the only common chords to C and C' will be the lines joining the points of intersection $C \cap C'$ pairwise*; since the number of such points is $\#(C \cap C') = \deg([C][C']) = 5$, the number of common chords will be $\binom{5}{2} = 10$. Of course, to deduce the general formula from this analysis, we have to check that the intersection $\Psi_2(C) \cap \Psi_2(C')$ is transverse; we will leave this as Exercise 3.46.

What would happen if we specialized C and C' to twisted cubics lying on Q, both having type $(1, 2)$? Now there would only be four points of $C \cap C'$, giving rise to $\binom{4}{2} = 6$ common chords. But now the lines of one ruling of Q would *all* be common chords to both. Thus $\Psi_2(C) \cap \Psi_2(C')$ would have a positive-dimensional component: Explicitly, $\Psi_2(C) \cap \Psi_2(C')$ would consist of six isolated points and one copy of $\mathbb{P}^1$. It might seem that in these circumstances we could not deduce anything about the intersection number $\deg([\Psi_2(C)] \cdot [\Psi_2(C')])$ from the actual intersection, but in fact the excess intersection formula of Chapter 13 can be used in this case to determine $\deg([\Psi_2(C)] \cdot [\Psi_2(C')])$; see Exercise 13.35.

3.6 Lines and surfaces in $\mathbb{P}^3$

3.6.1 Lines lying on a quadric

Let $Q \subset \mathbb{P}^3$ be a smooth quadric surface and $F = F_1(Q) \subset \mathbb{G}(1, 3)$ the locus of lines contained in Q. In this section we will determine the class $[F] \in A(\mathbb{G}(1, 3))$. ($F$ is an example of a *Fano scheme*, a construction that will be treated extensively in Chapter 6.)

Via the isomorphism $Q \cong \mathbb{P}^1 \times \mathbb{P}^1$, the lines on Q are fibers of the two projections maps $Q \to \mathbb{P}^1$; in particular, we see that $\dim F = 1$. Since $A^3(\mathbb{G}(1, 3))$ is generated by $\sigma_{2,1}$, we must have

$$[F] = \alpha \cdot \sigma_{2,1}$$

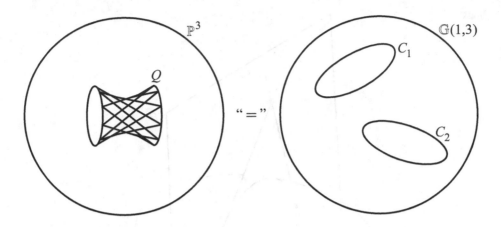

Figure 3.14 The rulings of a quadric surface $Q \subset \mathbb{P}^3$ correspond to conic curves $C_i \subset \mathbb{G}(1,3) \subset \mathbb{P}^5$; thus $[F_1(Q)] = 4\sigma_{2,1}$.

for some integer α. If $L \subset \mathbb{P}^3$ is a general line and $\Sigma_1(L) \subset \mathbb{G}(1,3)$ the Schubert cycle of lines meeting L, then by Kleiman transversality we have

$$\alpha = \deg([F] \cdot \sigma_1)$$
$$= \#(\Sigma_1(L) \cap F)$$
$$= \#\{M \in \mathbb{G}(1,3) \mid M \subset Q \text{ and } M \cap L \neq \varnothing\}.$$

Now L, being general, will intersect Q in two points, and through each of these points there will be two lines contained in Q; thus we have $\alpha = 4$ and

$$[F] = 4\sigma_{2,1}.$$

We will see how to calculate the class of the locus of linear spaces on a quadric hypersurface more generally in Section 4.6.

The variety F is actually the union $C_1 \cup C_2$ of two disjoint curves in the Grassmannian, corresponding to the two rulings of Q; each of these curves has class $2\sigma_{2,1}$, and thus has degree 2 as a curve in the Plücker embedding in $\mathbb{P}^5$ (see Figure 3.14). For details see Eisenbud and Harris [2000].

3.6.2 Tangent lines to a surface

Next, let $S \subset \mathbb{P}^3$ be any smooth surface of degree d, and consider the locus $T_1(S) \subset \mathbb{G}(1,3)$ of lines tangent to S. Let Φ be the incidence correspondence

$$\Phi = \{(q, L) \in S \times \mathbb{G}(1,3) \mid q \in L \subset \mathbb{T}_q S\},$$

where $\mathbb{T}_q S$ denotes the projective plane tangent to S at q. The projection $\Phi \to S$ on the first factor expresses Φ as a $\mathbb{P}^1$-bundle over S, from which we deduce that Φ, and hence its image $T_1(S)$ in $\mathbb{G}(1,3)$, is irreducible of dimension 3.

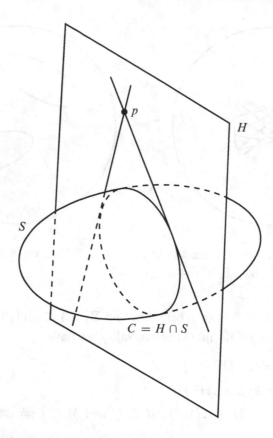

Figure 3.15 $\deg(\Sigma_{2,1}(p, H) \cap T_1(S)) = 2$.

To find the class of $T_1(S)$, we write

$$[T_1(S)] = \alpha \cdot \sigma_1,$$

and choose a general plane $H \subset \mathbb{P}^3$ and a general point $p \in H$. By Kleiman transversality,

$$
\begin{aligned}
\alpha &= [T_1(S)] \cdot \sigma_{2,1} \\
&= \#(\Sigma_{2,1}(p, H) \cap T_1(S)) \\
&= \#\{M \in \mathbb{G}(1,3) \mid q \in M \subset \mathbb{T}_q S \text{ for some } q \in S \text{ and } p \in M \subset H\}.
\end{aligned}
$$

Now H, being general, will intersect S in a smooth plane curve $C \subset H \cong \mathbb{P}^2$ of degree d, and, p being general in H, the line $p^* \subset \mathbb{P}^{2*}$ dual to p will intersect the dual curve $C^* \subset \mathbb{P}^{2*}$ transversely in $\deg(C^*)$ points. By Proposition 2.9, we have

$$\deg(C^*) = d(d - 1)$$

and hence

$$[T_1(S)] = d(d - 1)\sigma_1$$

(see Figure 3.15).

This gives the answer to the last keynote question of this chapter: How many lines are tangent to each of four general quadric surfaces Q_i? Once more, Kleiman's theorem assures us that the cycles $T_1(Q_i)$ intersect transversely, a fact we can verify in all characteristics by explicit calculation. The answer is thus

$$\deg \prod [T_1(Q_i)] = \deg(2\sigma_1)^4 = 32.$$

3.7 Exercises

Exercise 3.17. Let $\Lambda, \Gamma \in G$ be two points in the Grassmannian $G = G(k, V)$. Show that the line $\overline{\Lambda, \Gamma} \subset \mathbb{P}(\wedge^k V)$ is contained in G if and only if the intersection $\Lambda \cap \Gamma \subset V$ of the corresponding subspaces of V has dimension $k - 1$.

Exercise 3.18. Using the fact that the Grassmannian

$$G = G(k, V) \subset \mathbb{P}(\wedge^k V)$$

is cut out by quadratic equations, show that if $[\Lambda] \in G$ is the point corresponding to a k-plane Λ then the tangent plane $\mathbb{T}_{[\Lambda]} G \subset \mathbb{P}(\wedge^k V)$ intersects G in the locus

$$G \cap \mathbb{T}_{[\Lambda]} G = \{\Gamma \mid \dim(\Gamma \cap \Lambda) \geq k - 1\};$$

that is, the locus of k-planes meeting Λ in codimension 1.

Exercise 3.19. Let V be an $(n + 1)$-dimensional vector space, and consider the universal k-plane over $G = \mathbb{G}(k, \mathbb{P}V)$ introduced in Section 3.2.3:

$$\Phi = \{(\Lambda, p) \in G \times \mathbb{P}V \mid p \in \Lambda\}.$$

Show that this is a closed subvariety of $G \times \mathbb{P}V$ of dimension $k + (k + 1)(n - k)$, and that it is cut out on $G \times \mathbb{P}V$ by bilinear forms on $\mathbb{P}(\wedge^{k+1} V) \times \mathbb{P}V$.

Exercise 3.20. Use the preceding exercise to show that, if $X \subset \mathbb{P}^n$ is any subvariety of dimension $l < n - k$, then the locus

$$\Gamma_X = \{\Lambda \in \mathbb{G}(k, n) \mid X \cap \Lambda \neq \varnothing\}$$

of k-planes meeting X is a closed subvariety of $\mathbb{G}(k, n)$ of codimension $n - k - l$.

Exercise 3.21. Let $l < k < n$, and consider the locus of nested pairs of linear subspaces of $\mathbb{P}^n$ of dimensions l and k:

$$\mathbb{F}(l, k; n) = \{(\Gamma, \Lambda) \in \mathbb{G}(l, n) \times \mathbb{G}(k, n) \mid \Gamma \subset \Lambda\}.$$

Show that this is a closed subvariety of $\mathbb{G}(l, n) \times \mathbb{G}(k, n)$, and calculate its dimension. (These are examples of a further generalization of Grassmannians called *flag manifolds*, which we will explore further in Section 4.8.1.)

Exercise 3.22. Again let $l < k < n$, and for any $m \leq l$ consider the locus of pairs of linear subspaces of $\mathbb{P}^n$ of dimensions l and k intersecting in dimension at least m:

$$\mathbb{F}(l, k; m; n) = \{(\Gamma, \Lambda) \in \mathbb{G}(l, n) \times \mathbb{G}(k, n) \mid \dim(\Gamma \cap \Lambda) \geq m\}.$$

Show that this is a closed subvariety of $\mathbb{G}(l, n) \times \mathbb{G}(k, n)$ and calculate its dimension.

Exercise 3.23. Let $B \subset \mathbb{G}(1, n)$ be a curve in the Grassmannian of lines in $\mathbb{P}^n$, with the property that all nonzero tangent vectors to B have rank 1. Show that the lines in $\mathbb{P}^n$ parametrized by B either

(a) all lie in a fixed 2-plane;
(b) all pass through a fixed point; or
(c) are all tangent to a fixed curve $C \subset \mathbb{P}^n$.

(Note that the last possibility actually subsumes the first.)

Exercise 3.24. Show that an automorphism of $G(k, n)$ carries tangent vectors to tangent vectors of the same rank (in the sense of Section 3.2.4), and hence for $1 < k < n$ the group of automorphisms of $G(k, n)$ cannot act transitively on nonzero tangent vectors. Show, on the other hand, that the group of automorphisms of $G(k, n)$ *does* act transitively on tangent vectors of a given rank.

Exercise 3.25. In Example 3.9, we demonstrated that the open Schubert cell $\Sigma_1^\circ = \Sigma_1 \setminus (\Sigma_2 \cup \Sigma_{1,1})$ is isomorphic to the affine space $\mathbb{A}^3$. For each of the remaining Schubert indices a, b, show that the Schubert cell $\Sigma_{a,b}^\circ \subset \mathbb{G}(1, 3)$ is isomorphic to the affine space of dimension $4 - a - b$.

Exercise 3.26. Consider the Schubert cycle

$$\Sigma_1 = \{\Lambda \in \mathbb{G}(1, 3) \mid \Lambda \cap L \neq \emptyset\}.$$

Suppose that $\Lambda \in \Sigma_1$ and $\Lambda \neq L$, so that $\Lambda \cap L$ is a point q and the span $\overline{\Lambda, L}$ a plane K. Show that Λ is a smooth point of Σ_1, and that its tangent space is

$$T_\Lambda(\Sigma_1) = \{\varphi \in \operatorname{Hom}(\tilde{\Lambda}, V/\tilde{\Lambda}) \mid \varphi(\tilde{q}) \subset \tilde{K}/\tilde{\Lambda}\}.$$

Exercise 3.27. Consider the Schubert cycle

$$\Sigma_{2,1} = \Sigma_{2,1}(p, H) = \{\Lambda \in \mathbb{G}(1, 3) \mid p \in \Lambda \subset H\}.$$

Show that $\Sigma_{2,1}$ is smooth, and that its tangent space at a point Λ is

$$T_\Lambda(\Sigma_{2.1}) = \{\varphi \in \operatorname{Hom}(\tilde{\Lambda}, V/\tilde{\Lambda}) \mid \varphi(\tilde{p}) = 0 \text{ and } \operatorname{Im}(\varphi) \subset \tilde{H}/\tilde{\Lambda}\}.$$

Exercise 3.28. Use the preceding two exercises to show in arbitrary characteristic that general Schubert cycles $\Sigma_1, \Sigma_{2,1} \subset \mathbb{G}(1, 3)$ intersect transversely, and deduce the equality $\deg(\sigma_1 \cdot \sigma_{2,1}) = 1$.

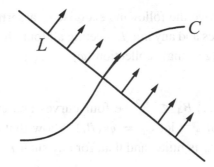

Figure 3.16 Deformation of a line L preserving incidence with a curve C.

Exercise 3.29. Let $L_1, \ldots, L_4 \subset \mathbb{P}^3$ be four pairwise skew lines and $\Lambda \subset \mathbb{P}^3$ a line meeting all four; set

$$p_i = \Lambda \cap L_i \quad \text{and} \quad H_i = \overline{\Lambda, L_i}.$$

Show that $[\Lambda] \in G$ fails to be a transverse point of intersection of the Schubert cycles $\Sigma_1(L_i)$ exactly when the cross-ratio of the four points $p_1, \ldots, p_4 \in \Lambda$ equals the cross-ratio of the four planes $H_1, \ldots, H_4$ in the pencil of planes containing Λ.

Exercises 3.30–3.33 deal with a question raised in Section 3.4.2: If $C_1, \ldots, C_4 \subset \mathbb{P}^3$ are general translates of four curves in $\mathbb{P}^3$, do the corresponding cycles $\Gamma_{C_i} \subset \mathbb{G}(1, 3)$ of lines meeting the C_i intersect transversely?

To start with, we have to identify the smooth locus of the cycle $\Gamma_C \subset \mathbb{G}(1, 3)$ of lines meeting a given curve C, and its tangent spaces at these points; this is the content of the next exercise, which is a direct generalization of Exercise 3.26 above.

Exercise 3.30. Let $C \subset \mathbb{P}^3$ be any curve, and $L \subset \mathbb{P}^3$ a line meeting C at one smooth point p of C and not tangent to C. Show that the cycle $\Gamma_C \subset \mathbb{G}(1, 3)$ of lines meeting C is smooth at the point $[L]$, and that its tangent space at $[L]$ is the space of linear maps $\tilde{L} \to \Bbbk^4/\tilde{L}$ carrying the one-dimensional subspace $\tilde{p} \subset \tilde{L}$ to the one-dimensional subspace $(\tilde{\mathbb{T}}_p C + \tilde{L})/\tilde{L}$ of $\Bbbk^4/\tilde{L}$ (see Figure 3.16).

Next, we have to verify that, for general translates C_i of any four curves, the corresponding cycles Γ_{C_i} are smooth at each of the points of their intersection. A key fact will be the irreducibility of the relevant incidence correspondence:

Exercise 3.31. Let $B_1, \ldots, B_4 \subset \mathbb{P}^3$ be four irreducible curves and let $\varphi_1, \ldots, \varphi_4 \in \mathrm{PGL}_4$ be four general automorphisms of $\mathbb{P}^3$; let $C_i = \varphi_i(B_i)$. Show that the incidence correspondence

$$\Phi = \{(\varphi_1, \ldots, \varphi_4, L) \in (\mathrm{PGL}_4)^4 \times \mathbb{G}(1, 3) \mid L \cap \varphi_i(B_i) \neq \varnothing \text{ for all } i\}$$

is irreducible.

Using this, we can prove the following exercise — asserting that for general translates C_i of four given curves and any line L meeting all four, the cycles Γ_{C_i} are smooth at $[L]$ — simply by exhibiting a single collection $(\varphi_1, \ldots, \varphi_4, L)$ satisfying the conditions in question:

Exercise 3.32. Let $B_1, \ldots, B_4 \subset \mathbb{P}^3$ be four curves and $\varphi_1, \ldots, \varphi_4 \subset \mathrm{PGL}_4$ four general automorphisms of $\mathbb{P}^3$; let $C_i = \varphi_i(B_i)$. Show that the set of lines $L \subset \mathbb{P}^3$ meeting C_1, C_2, C_3 and C_4 is finite, and that, for any such L,

(a) L meets each C_i at only one point p_i;
(b) p_i is a smooth point of C_i; and
(c) L is not tangent to C_i for any i.

Exercise 3.33. Let $C_1, \ldots, C_4 \subset \mathbb{P}^3$ be any four curves, and $L \subset \mathbb{P}^3$ a line meeting all four and satisfying the conclusions of Exercise 3.32. Use the result of Exercise 3.30 to give a necessary and sufficient condition for the four cycles $\Gamma_{C_i} \subset \mathbb{G}(1,3)$ to intersect transversely at $[L]$, and show directly that this condition is satisfied for all lines meeting $C_1, \ldots, C_4$ when the C_i are general translates of given curves.

Exercise 3.34. Let $C \subset \mathbb{P}^3$ be a smooth curve and $p \in \mathbb{P}^3$ a general point. Show that

(a) p does not lie on any tangent line to C;
(b) p does not lie on any trisecant line to C; and
(c) p does not lie on any *stationary secant* to C (that is, a secant line $\overline{q,r}$ to C such that the tangent lines $\mathbb{T}_q C$ and $\mathbb{T}_r C$ meet).

Deduce from these facts that the projection $\pi_p : C \to \mathbb{P}^2$ is birational onto a plane curve $C_0 \subset \mathbb{P}^2$ having only nodes as singularities. (Note that as a consequence the same is true for the projection of a smooth curve $C \subset \mathbb{P}^n$ from a general $(n-3)$-plane to $\mathbb{P}^2$.)

Exercises 3.35 and 3.36 deal with the approach, described in Section 3.5.3, to calculating the class of the variety $\Sigma_C \subset \mathbb{G}(1,3)$ of lines incident to a space curve $C \subset \mathbb{P}^3$ by specialization. Recall from that section that we choose a general plane $H \subset \mathbb{P}^3$ meeting C at d points p_i and a general point $q \in \mathbb{P}^3$, and let $\{A_t\}$ be the one-parameter subgroup of PGL_4 with attractor q and repellor H; we let $C_t = A_t(C)$ and take $\Psi \subset \mathbb{A}^1 \times \mathbb{G}(1,3)$ to be the closure of the locus

$$\Psi^\circ = \{(t, \Lambda) \in \mathbb{A}^1 \times \mathbb{G}(1,3) \mid t \neq 0 \text{ and } \Lambda \cap C_t \neq \varnothing\}.$$

Exercise 3.35. Show that the support of the fiber Ψ_0 is exactly the union of the Schubert cycles $\Sigma_1(\overline{p_i, q})$.

Exercise 3.36. Show that Ψ_0 has multiplicity 1 at a general point of each Schubert cycle $\Sigma_1(\overline{p_i, q})$.

Exercise 3.37. Let $C \subset \mathbb{P}^r$ be a smooth curve. Show that the rational map $\varphi :$ $C^2 \dashrightarrow \mathbb{G}(1, r)$ sending (p, q) to the line $\overline{p, q}$ when $p \neq q$ actually extends to a regular map on all of C^2 sending (p, p) to the projective tangent line $\mathbb{T}_p C$. Use this to show that the image of φ coincides with the locus of lines $L \subset \mathbb{P}^r$ such that the scheme-theoretic intersection $L \cap C$ has degree at least 2.

Exercise 3.38. Show by example that the conclusion of the preceding exercise is false in general if we do not assume $C \subset \mathbb{P}^r$ to be smooth. Is it still true if we allow C to have mild singularities, such as nodes?

Exercise 3.39. Similarly, show by example that the conclusion of Exercise 3.37 is false if we consider higher-dimensional secant planes: For example, the image of the rational map

$$\varphi : C^3 \dashrightarrow \mathbb{G}(2, r),$$
$$(p, q, r) \mapsto \overline{p, q, r},$$

need not coincide with the locus of 2-planes $\Lambda \subset \mathbb{P}^r$ whose scheme-theoretic intersection with C has degree at least 3.

Exercise 3.40. Show that the smooth locus of $S = \Psi_2(C)$ contains the locus of lines $L \subset \mathbb{P}^3$ such that the scheme-theoretic intersection $L \cap C$ consists of two reduced points, and for such a line L identify the tangent plane $T_L S$ as a subspace of $T_L \mathbb{G}$. (When is a tangent line to C a smooth point of $\Psi_2(C)$?)

Exercise 3.41. Use the result of the preceding exercise to show that if $C, C' \subset \mathbb{P}^3$ are two general twisted cubic curves, then the varieties $\Psi_2(C), \Psi_2(C') \subset \mathbb{G}(1, 3)$ of chords to C and C' intersect transversely.

Exercise 3.42. Let $C \subset \mathbb{P}^3$ be a smooth, nondegenerate curve of degree d and genus g, and let $L, M \subset \mathbb{P}^3$ be general lines.

(a) Find the number of chords to C meeting both L and M by applying Proposition 3.14.
(b) Verify this count by considering the product morphism

$$\pi_L \times \pi_M : C \to \mathbb{P}^1 \times \mathbb{P}^1$$

(where $\pi_L, \pi_M : C \to \mathbb{P}^1$ are the projections from L and M) and comparing the arithmetic and geometric genera of the image curve.

Exercise 3.43. Let $C \subset \mathbb{P}^3$ be a smooth, irreducible nondegenerate curve of degree d, and let $\Phi \subset \mathbb{A}^1 \times \mathbb{P}^3$ be the family of curves specializing C to a scheme supported on the union of lines joining a point $p \in \mathbb{P}^3$ to the points of a plane section of C, as constructed in Section 3.5.3. Show that C_0 may have an embedded point at p, and that the multiplicity of this embedded point may depend on the genus of the curve C, by considering the examples of curves of degrees 4 and 5.

Exercise 3.44. In the situation of the preceding problem, let $\Psi_2(C_t) \subset \mathbb{G}(1,3)$ be the locus of chords to C_t for $t \neq 0$. Suppose that the degree of C is 4. Show that the component $\Sigma_2(p)$ will be in the flat limit with multiplicity depending on the genus of C.

Exercise 3.45. Again, suppose $C \subset \mathbb{P}^3$ is any curve of degree d; choose a general plane $H \subset \mathbb{P}^3$ and point $p \in \mathbb{P}^3$, and consider the one-parameter group $\{A_t\} \subset \mathrm{PGL}_4$ with repellor point p and attractor plane H — that is, choose coordinates $[Z_0, \ldots, Z_3]$ on $\mathbb{P}^3$ such that $p = [0, 0, 0, 1]$ and H is given by $Z_3 = 0$, and consider for $t \neq 0$ the automorphisms of $\mathbb{P}^3$ given by

$$A_t = \begin{pmatrix} 1 & 0 & 0 & 0 \\ 0 & 1 & 0 & 0 \\ 0 & 0 & 1 & 0 \\ 0 & 0 & 0 & t \end{pmatrix}.$$

Let $C_t = A_t(C)$, and for $t \neq 0$ let $\Psi_2(C_t) \subset \mathbb{G}(1,3)$ be the locus of chords to C_t. Show that the Schubert cycle $\Sigma_{1,1}(H)$ appears as a component of multiplicity $\binom{d}{2}$ in the limiting scheme $\lim_{t \to 0} \Psi_2(C_t)$.

Hint: Let $\Psi \subset \mathbb{A}^1 \times \mathbb{G}(1,3)$ be the closure of the family

$$\Psi^\circ = \{(t, L) \mid t \neq 0 \text{ and } L \in \Psi_2(C_t)\},$$

and show that if $L \subset H$ is a general line then in a neighborhood of the point $(0, L) \in \mathbb{A}^1 \times G$ the family Ψ consists of the union of $\binom{d}{2}$ smooth sheets, each intersecting the fiber $\{0\} \times \mathbb{G}(1,3)$ transversely in the Schubert cycle $\Sigma_{1,1}(H)$.

Exercise 3.46. Let $C, C' \subset Q \subset \mathbb{P}^3$ be general twisted cubic curves lying on a smooth quadric surface Q, of types $(1, 2)$ and $(2, 1)$ respectively. Show that the intersection $\Psi_2(C) \cap \Psi_2(C')$ of the corresponding cycles of chords is transverse.

Exercise 3.47. Let $C \subset \mathbb{P}^3$ be a smooth curve of degree d and genus g, and let $T(C) \subset \mathbb{G}(1,3)$ be the locus of its tangent lines. Find the class $[T(C)] \in A^3(\mathbb{G}(1,3))$ of $T(C)$ in the Grassmannian $\mathbb{G}(1,3)$.

Exercise 3.48. Let $C \subset \mathbb{P}^3$ be a smooth curve of degree d and genus g, and let $S \subset \mathbb{P}^3$ be a general surface of degree e. How many tangent lines to C are tangent to S?

Chapter 4

Grassmannians in general

Keynote Questions

(a) If $V_1, \ldots, V_4 \cong \mathbb{P}^n \subset \mathbb{P}^{2n+1}$ are four general n-planes, how many lines $L \subset \mathbb{P}^{2n+1}$ meet all four? (Answer on page 150.)

(b) Let $C \subset \mathbb{G}(1,3) \subset \mathbb{P}^5$ be a twisted cubic curve contained in the Grassmannian $\mathbb{G}(1,3) \subset \mathbb{P}^5$ of lines in $\mathbb{P}^3$, and let

$$S = \bigcup_{[\Lambda] \in C} \Lambda \subset \mathbb{P}^3$$

be the surface swept out by the lines corresponding to points of C. What is the degree of S? How can we describe the geometry of S? (Answer on page 145.)

(c) If $Q_1, Q_2 \subset \mathbb{P}^4$ are general quadric hypersurfaces and $S = Q_1 \cap Q_2$ their surface of intersection, how many lines does S contain? More generally, if Q_1 and Q_2 are general quadric hypersurfaces in $\mathbb{P}^{2n}$ and $X = Q_1 \cap Q_2$, how many $(n-1)$-planes does X contain? (Answer on page 157.)

(d) What is the degree of the Grassmannian $\mathbb{G}(1,n)$ of lines in $\mathbb{P}^n$, embedded in projective space via the Plücker embedding? (Answer on page 150.)

In this chapter, we will extend the ideas developed in Chapter 3 by introducing Schubert cycles and classes on $G(k,n)$, the Grassmannian of k-dimensional subspaces in an n-dimensional vector space V, and analyzing their intersections, a subject that goes by the name of the *Schubert calculus*. Of course we may also consider $G(k,n)$ in its projective guise as $\mathbb{G}(k-1, n-1)$, the Grassmannian of projective $(k-1)$-planes in $\mathbb{P}^{n-1}$, and in places where projective geometry is more natural (such as Sections 4.2.3 and 4.4) we will switch to the projective notation.

4.1 Schubert cells and Schubert cycles

Let $G = G(k, V)$ be the Grassmannian of k-dimensional subspaces of an n-dimensional vector space V. Generalizing the example of $\mathbb{G}(1, 3) = G(2, 4)$, the center of our study will be a collection of subvarieties of $G(k, n)$ called *Schubert varieties* or *Schubert cycles*, defined in terms of a chosen *complete flag* $\mathcal{V}$ in V, that is, a nested sequence of subspaces

$$0 \subset V_1 \subset \cdots \subset V_{n-1} \subset V_n = V$$

with $\dim V_i = i$.

The Schubert cycles are indexed, in a way that will be motivated below, by sequences $a = (a_1, \ldots, a_k)$ of integers with

$$n - k \geq a_1 \geq a_2 \geq \cdots \geq a_k \geq 0.$$

(Such sequences are often described by *Young diagrams* — see Section 4.5.) For such a sequence a, we define the *Schubert cycle* $\Sigma_a(\mathcal{V}) \subset G$ to be the closed subset

$$\Sigma_a(\mathcal{V}) = \{\Lambda \in G \mid \dim(V_{n-k+i-a_i} \cap \Lambda) \geq i \text{ for all } i\}.$$

Theorem 1.7 shows that the class $[\Sigma_a(\mathcal{V})] \in A(G)$ does not depend on the choice of flag, since any two flags differ by an element of GL_n. In general, when dealing with a property independent of the choice of $\mathcal{V}$, we will shorten the name to Σ_a, and we define

$$\sigma_a := [\Sigma_a] \in A(G);$$

these, naturally, are called *Schubert classes*. We shall see (in Corollary 4.7) that $A(G)$ is a free abelian group and that the classes σ_a form a basis.

To simplify notation, we generally suppress trailing zeros in the indices, writing $\Sigma_{a_1, \ldots, a_s}$ in place of $\Sigma_{a_1, \ldots, a_s, 0, \ldots, 0}$. Also, we use the shorthand Σ_{p^r} to denote $\Sigma_{p, \ldots, p}$, with r indices equal to p.

To elucidate the rather awkward-looking definition of $\Sigma_a(\mathcal{V})$, suppose that $\Lambda \subset V$ is a k-plane. If Λ is general, then $V_i \cap \Lambda = 0$ for $i \leq n - k$, while $\dim V_{n-k+i} \cap \Lambda = i$ for $i > n - k$. Thus we may describe Σ_a as the set of Λ such that $\dim V_j \cap \Lambda \geq i$ occurs for a value of j that is a_i steps sooner than expected.

Equivalently, we may consider the sequence of subspaces of Λ

$$0 \subset (V_1 \cap \Lambda) \subset (V_2 \cap \Lambda) \subset \cdots \subset (V_{n-1} \cap \Lambda) \subset (V_n \cap \Lambda) = \Lambda. \tag{4.1}$$

Each subspace in this sequence is either equal to the one before it or of dimension one greater, and the latter phenomenon occurs exactly k times. The Schubert cycle $\Sigma_a(\mathcal{V})$ is the locus of planes Λ for which "the i-th jump in the sequence (4.1) occurs at least a_i steps early."

Here are two common special cases to bear in mind:

- The cycle of k-subspaces Λ meeting a given space of dimension l nontrivially is the Schubert cycle

$$\Sigma_{n-k+1-l}(\mathcal{V}) = \{\Lambda \mid \Lambda \cap V_l \neq 0\}.$$

In particular, the Schubert cycle of k-dimensional subspaces meeting a given $(n-k)$-dimensional subspace nontrivially is

$$\Sigma_1(\mathcal{V}) = \{\Lambda \mid \Lambda \cap V_{n-k} \neq 0\}.$$

This is a hyperplane section of G in the Plücker embedding. (But not every hyperplane section of G is of this form. This follows by a dimension count: the family of $(n-k)$-planes — the Grassmannian $G(n-k, n)$ — has dimension $k(n-k)$, whereas the space of linear forms in the Plücker coordinates has dimension $\binom{n}{k} - 1$.)

- The sub-Grassmannian of k-subspaces contained in a given l-subspace is the Schubert cycle

$$\Sigma_{(n-l)^k}(\mathcal{V}) = \{\Lambda \mid \Lambda \subset V_l\}.$$

Similarly, the sub-Grassmannian of planes containing a given r-plane is the Schubert cycle

$$\Sigma_{(n-k)^r}(\mathcal{V}) = \{\Lambda \mid V_r \subset \Lambda\}.$$

The cycles Σ_i, defined for $0 \leq i \leq n-k$, and the cycles Σ_{1^i}, defined for $0 \leq i \leq k$, are called *special* Schubert cycles. As we shall see in Section 5.8, their classes are intimately connected with the tautological sub and quotient bundles on G, and each of the corresponding sequences of classes forms a minimal generating set for the algebra $A(G)$.

Our indexing of the Schubert cycles is by no means the only one in use, but it has several good properties:

- It reflects the partial order of the Schubert cycles defined with respect to a given flag $\mathcal{V}$ by inclusion: if we order the indices termwise, that is, $(a_1, \ldots, a_k) \leq (a'_1, \ldots, a'_k)$ if and only if $a_i \leq a'_i$ for $1 \leq i \leq k$ (writing $a < a'$ when $a \leq a'$ and $a \neq a'$; that is, $a_i < a'_i$ for some i), then

$$\Sigma_a \subset \Sigma_b \iff a \geq b.$$

This follows immediately from the definition.

- It makes the codimension of a Schubert cycle apparent: By Theorem 4.1 below,

$$\mathrm{codim}(\Sigma_a \subset G) = \sum a_i,$$

so that $|a| := \sum a_i$ is the degree of σ_a in $A(G)$.

- It is preserved under pullback via the natural inclusions

$$i : G(k, n) \hookrightarrow G(k+1, n+1)$$

(whose image is the set of $(k + 1)$-subspaces containing V_1) and

$$j : G(k, n) \hookrightarrow G(k, n + 1)$$

(whose image is the set of k-subspaces contained in V_n); that is,

$$i^*(\sigma_a) = \sigma_a \quad \text{and} \quad j^*(\sigma_a) = \sigma_a.$$

Here we adopt the convention that when $a_1 > n - k$, or when $a_{k+1} > 0$, we take $\sigma_a = 0$ as a class in $A(G(k, n))$. (This convention is consistent with the restriction to sub-Grassmannians. For example, $\Sigma_{n-k+1} \subset G(k, n + 1)$ is the subset of the k-planes containing a fixed general one-dimensional subspace, and thus the intersection of Σ_{n-k+1} with the $G(k, n)$ of subspaces contained in a fixed codimension-1 subspace is empty, so that $j^*\sigma_{n-k+1} = 0 \in A(G(k, n))$.) It follows that if we establish a formula

$$\sigma_a \sigma_b = \sum \gamma_{a,b;c} \, \sigma_c$$

in the Chow ring of $G(k, n)$, the same formula holds true in all $G(k', n')$ with $k' \leq k$ and $n' - k' \leq n - k$. Whenever it happens that i^* or j^* is an isomorphism on $A^{|a|+|b|}$, the formula will also hold in $A(G(k, n + 1))$ or $A(G(k + 1, n + 1))$, respectively. Conditions for this are given in Exercise 4.32.

There is a natural isomorphism $G(k, V) \cong G(n - k, V^*)$ obtained by associating to a k-dimensional subspace $\Lambda \subset V$ the $(n - k)$-dimensional subspace $\Lambda^\perp \subset V^*$ consisting of all those linear functionals on V that annihilate Λ. This duality carries each Schubert cycle to another Schubert cycle. For example, one checks immediately that $\Sigma_i(W)$, which is the set of k-planes Λ meeting a fixed $(n - k + 1 - i)$-plane W nontrivially, is carried into the Schubert cycle Σ_{1^i} of $(n - k)$-planes Λ' such that $\dim(\Lambda' \cap W^\perp) \geq i$, that is, such that $\Lambda' + W^\perp \subsetneq V$. See Section 4.5 for the general case.

4.1.1 Schubert classes and Chern classes

Schubert classes provide fundamental invariants of vector bundles. Recall from Theorem 3.4 that, if $\mathcal{E}$ is a vector bundle of rank r generated by a space $W \cong \Bbbk^n$ of global sections on a variety X, then there is a map $X \to G(n - r, W)$ sending a point $x \in X$ to the subspace in W consisting of the sections vanishing at x. The pullbacks of the Schubert classes σ_a give a fundamental set of invariants of $\mathcal{E}$ called the *Chern classes* of $\mathcal{E}$ — see Section 5.6.2. We will see that every Schubert class is a polynomial in the special Schubert classes (see Section 4.7).

4.1.2 The affine stratification by Schubert cells

As in the case of $\mathbb{G}(1,3) = G(2,4)$, the Grassmannian $G(k,n)$ has an affine stratification. To see this, set

$$\Sigma_a^\circ = \Sigma_a \setminus \left(\bigcup_{b>a} \Sigma_b \right).$$

The Σ_a° are called *Schubert cells*.

Theorem 4.1. *The locally closed subset* $\Sigma_a^\circ \subset G$ *is isomorphic to the affine space* $\mathbb{A}^{k(n-k)-|a|}$; *in particular* Σ_a° *is smooth and irreducible, and the Schubert variety* Σ_a *is irreducible and of codimension* $|a|$ *in* $G(k,n)$. *The tangent space to* Σ_a° *at a point* $[\Lambda]$ *is the subspace of* $T_{[\Lambda]}G = \mathrm{Hom}(\Lambda, V/\Lambda)$ *consisting of those elements* φ *that send*

$$V_{n-k+i-a_i} \cap \Lambda \subset \Lambda$$

into

$$\frac{V_{n-k+i-a_i} + \Lambda}{\Lambda} \subset V/\Lambda$$

for $i = 1, \ldots, k$.

Proof: Choose a basis $(e_1, \ldots, e_n)$ for V so that

$$V_i = \langle e_1, \ldots, e_i \rangle.$$

Suppose $[\Lambda] \in \Sigma_a$, and consider the sequence (4.1) of subspaces of Λ. By definition, the first nonzero subspace in the sequence will be $V_{n-k+1-a_1} \cap \Lambda$, the first of dimension 2 will be $V_{n-k+2-a_2} \cap \Lambda$, and so on. Thus we may choose a basis $(v_1, \ldots, v_k)$ for Λ with $v_1 \in V_{n-k+1-a_1}$, $v_2 \in V_{n-k+2-a_2}$, and so on. In terms of this basis, and the basis $(e_1, \ldots, e_n)$ for V, the matrix representative of Λ has the form

$$\begin{pmatrix} * & * & * & 0 & 0 & 0 & 0 & 0 & 0 \\ * & * & * & * & * & 0 & 0 & 0 & 0 \\ * & * & * & * & * & * & 0 & 0 & 0 \\ * & * & * & * & * & * & * & * & 0 \end{pmatrix}.$$

(This particular matrix corresponds to the case $k = 4$, $n = 9$ and $a = (3,2,2,1)$.) If Λ were general in G, and we chose a basis for Λ in this way, the corresponding matrix would look like

$$\begin{pmatrix} * & * & * & * & * & * & 0 & 0 & 0 \\ * & * & * & * & * & * & * & 0 & 0 \\ * & * & * & * & * & * & * & * & 0 \\ * & * & * & * & * & * & * & * & * \end{pmatrix}.$$

Thus the Schubert index a_i is the number of "extra zeros" in row i.

Now suppose that $\Lambda \in \Sigma_a^\circ$; that is, $\Lambda \in \Sigma_a$ but not in any of the smaller varieties $\Sigma_{a'}$ for $a' > a$. In this case $v_i \notin V_{n-k+i-a_i-1}$, so, for each i, the coefficient of $e_{n-k+i-a_i}$ in the expression of v_i as a linear combination of the e_α is nonzero, and this condition characterizes elements of Σ_a° among elements of Σ_a. (It follows in particular that Σ_a is the closure of Σ_a°.) Given that the coefficient of $e_{n-k+i-a_i}$ in v_i is nonzero, we can multiply v_i by a scalar to make the coefficient 1, obtaining a basis for Λ represented by the rows of a matrix of the form

$$
\begin{pmatrix}
* & * & 1 & 0 & 0 & 0 & 0 & 0 & 0 \\
* & * & * & * & 1 & 0 & 0 & 0 & 0 \\
* & * & * & * & * & 1 & 0 & 0 & 0 \\
* & * & * & * & * & * & * & 1 & 0
\end{pmatrix},
$$

where the 1 in the i-th row appears in the $(n - k + i - a_i)$-th column for $i = 1, \ldots, k$.

Finally, we can subtract a linear combination of $v_1, \ldots, v_{i-1}$ from v_i to kill the coefficients of $e_{n-k+j-a_j}$ in the expression of v_i as a linear combination of the e_α for $j < i$, to arrive at a basis of Λ given by the row vectors of the matrix

$$
A = \begin{pmatrix}
* & * & 1 & 0 & 0 & 0 & 0 & 0 & 0 \\
* & * & 0 & * & 1 & 0 & 0 & 0 & 0 \\
* & * & 0 & * & 0 & 1 & 0 & 0 & 0 \\
* & * & 0 & * & 0 & 0 & * & 1 & 0
\end{pmatrix}.
$$

Setting $b = \{n - k + 1 - a_1, \ldots, n - a_k\}$, we may describe this by saying that the b-th submatrix A_b of A (that is, the submatrix involving columns from b) is the identity matrix. We claim that Λ has a unique basis of this form. Indeed, any other basis of Λ has a matrix obtained from this one by left multiplication by a unique invertible $k \times k$ matrix g, and thus has submatrix $A_b = g$.

It follows that Σ_a° is contained in the open subset $U \subset G$ consisting of planes Λ complementary to the span of the $n - k$ basis vectors whose indices are not in b. By the same argument, any element of $U = \Sigma_0^\circ$ has a unique basis given by the rows of a $k \times n$ matrix with submatrix $A_b = I$, that is, of the form

$$
A = \begin{pmatrix}
* & * & 1 & * & 0 & 0 & * & 0 & * \\
* & * & 0 & * & 1 & 0 & * & 0 & * \\
* & * & 0 & * & 0 & 1 & * & 0 & * \\
* & * & 0 & * & 0 & 0 & * & 1 & *
\end{pmatrix}.
$$

Thus Σ_a° is a coordinate subspace of $U \cong \mathbb{A}^{k(n-k)}$ defined by the vanishing of $|a|$ coordinates, and it follows that Σ_a° is smooth and irreducible, and of codimension $|a|$, as claimed. Since Σ_a is the closure of Σ_a°, it is also irreducible and of codimension $|a|$ in $G(k, n)$ (but it may be singular; for example, one sees from the Plücker relation (Example 3.1) that $\Sigma_1 \subset G(2, 4)$ is the cone in $\mathbb{P}^4$ over a smooth quadric in $\mathbb{P}^3$).

The statement about tangent spaces follows from the explicit coordinate description of Σ_a above. We identify the open set $U \cong \mathbb{A}^{k(n-k)}$ with the set of $k \times n$ matrices having an identity matrix in positions from b. Since the tangent space to an affine space may be identified with the corresponding vector space, the tangent space $\mathrm{Hom}(\Lambda, V/\Lambda)$ to $G(k, n)$ at Λ is given by the set of matrices in the positions from b' complementary to those in b, or more properly by the transposes of these matrices. Given such a tangent vector, we may complete it uniquely to a $k \times n$ matrix with submatrix $A_b = I$, and this (or rather its transpose) corresponds to the lifting $\Lambda \to V = V/\Lambda \oplus \Lambda$ inducing the identity map $\Lambda \to \Lambda$. Thus the set of tangent directions at Λ to the affine subspace Σ_a° is identified with the set of matrices in that subspace, and this corresponds precisely to the set of maps in $\mathrm{Hom}(\Lambda, V/\Lambda)$ whose lifting as above sends $V_{n-k+i-a_i}$ into $V_{n-k+i-a_i} + \Lambda$, as claimed. $\qquad\square$

From Proposition 1.17 we see that $A(G)$ is at least generated as an abelian group by the classes σ_a, and the existence of the degree homomorphism $\deg : A^{k(n-k)} \to \mathbb{Z}$ that counts points shows that $A^{k(n-k)}(G)$ is actually free on the class of a point, which is the generator $\sigma_{(n-k)^k}$. In Corollary 4.7 we will prove that all the $A^i(G)$ are free, by intersection theory and results on transversality.

The description of the tangent spaces in Theorem 4.1 can be used to prove this transversality. Here is an example:

Corollary 4.2. *Let $G = G(k, n)$. Then*

$$(\sigma_{n-k})^k = (\sigma_{1^k})^{n-k} = \sigma_{(n-k)^k} \in A^{k(n-k)}(G);$$

that is, $(\sigma_{n-k})^k$ and $(\sigma_{1^k})^{n-k}$ are both equal to the class of a point in the Chow ring of G.

Proof: We know that $A_0(G)$ is generated by the class $\sigma_{(n-k)^k}$ of a point, so it suffices to show that both $(\sigma_{n-k})^k$ and $(\sigma_{1^k})^{(n-k)}$ are of degree 1.

We regard G as the variety of k-dimensional subspaces Λ of the n-dimensional vector space V. If $H \subset V$ is a codimension-1 subspace, then

$$\Sigma_{1^k}(H) = \{\Lambda \subset V \mid \Lambda \subset H\},$$

and the tangent space to $\Sigma_{1^k}(H)$ at the point corresponding to Λ is

$$T_{[\Lambda]}(\Sigma_{1^k}(H)) = \{\varphi \in \mathrm{Hom}(\Lambda, V/\Lambda)\} \mid \varphi(\Lambda) \subset H\}.$$

If $H_1, \ldots, H_{n-k}$ are general codimension-1 subspaces, then there is a unique k-plane Λ in $\bigcap_{i=1}^k \Sigma_{1^k}(H_i)$, namely, the intersection $\Lambda = \bigcap_{i=1}^k H_i$. Further, the tangent spaces intersect only in the zero homomorphism, so the intersection is transverse. This proves that $(\sigma_{1^k})^{(n-k)}$ is the class of a point.

To prove the corresponding statement for $(\sigma_{n-k})^k$, we can make an analogous argument, or we can simply use duality: the isomorphism $G(k,n) \cong G(n-k,n)$ introduced above carries σ_{1^k} to σ_k, as we have already remarked, and preserves the degree of 0-cycles. $\square$

4.1.3 Equations of the Schubert cycles

It is a remarkable fact that under the Plücker embedding $G = G(k,n) \hookrightarrow \mathbb{P}^N$ every Schubert cycle $\Sigma_a(\mathcal{V}) \subset G$ defined relative to the standard flag $V_i = \langle e_1, \ldots, e_i \rangle$ is the intersection of G with a coordinate subspace of $\mathbb{P}^N$, that is, a subspace defined by the vanishing of an easily described subset of the Plücker coordinates. This is true even at the level of homogeneous ideals:

Theorem 4.3. *Let $\Sigma_a \subset G(k,n) \subset \mathbb{P}^N$ be a Schubert cycle, and let b be the strictly increasing k-tuple $b = (n-k+1-a_1, \ldots, n-k+2-a_2, \ldots)$. The homogeneous ideal of the Σ_a in $\mathbb{P}^N$ is generated by the homogeneous ideal of the Grassmannian (the Plücker relations, page 94) together with those Plücker coordinates $p_{b'}$ such that $b' \not\leq b$ in the termwise partial order.*

The equations of the Σ_a were studied in Hodge [1943], and this work led to the notions of a straightening law (Doubilet et al. [1974]) and Hodge algebra (De Concini et al. [1982]). A proof of Theorem 4.3 in terms of Hodge algebras may be found in the latter publication, along with a proof that the homogeneous coordinate ring of Σ_a is Cohen–Macaulay. The ideas have also been extended to homogeneous varieties for other reductive groups by Lakshmibai, Musili, Seshadri and their coauthors (see for example Seshadri [2007]). Avoiding this theory, we will prove Theorem 4.3 only in the easy case $G(2,4) = \mathbb{G}(1,3) \subset \mathbb{P}^5$. In Exercise 4.17 we invite the reader to give the easier proof of the set-theoretic version of Theorem 4.3.

Proof of Theorem 4.3 for $G(2,4)$: In $G(2,4)$, the Schubert cycle $\Sigma_{a,b}$ consists of those two-dimensional subspaces that meet V_{3-a} nontrivially and are contained in V_{4-b}. We must show that the homogeneous ideal of

$$\Sigma_{a,b} \subset G(2,4) \subset \mathbb{P}^5$$

is generated by the Plücker relation $g := p_{1,2}p_{3,4} - p_{1,3}p_{2,4} + p_{1,4}p_{2,3}$ together with the Plücker coordinates

$$\{p_{i,j} \mid (i,j) \not\leq (3-a, 4-b)\}$$

(note that the condition $(i,j) \not\leq (3-a, 4-b)$ means $i > 3-a$ or $j > 4-b$). Specifically:

- Σ_1 is the hyperplane section $p_{3,4} = 0 \subset G$; that is, it is the cone over the nonsingular quadric $\overline{g} = -p_{1,3}p_{2,4} + p_{14}p_{2,3}$ in $\mathbb{P}^3$.

- Σ_2 is the plane $p_{2,3} = p_{2,4} = p_{3,4} = 0$.
- $\Sigma_{1,1}$ is the plane $p_{1,4} = p_{2,4} = p_{3,4} = 0$.
- $\Sigma_{2,1}$ is the line $p_{1,4} = p_{2,3} = p_{2,4} = p_{3,4} = 0$.
- $\Sigma_{2,2}$ is the point $p_{1,3} = p_{1,4} = p_{2,3} = p_{2,4} = p_{3,4} = 0$.

A subspace $L \in \Sigma_{a,b}$ has a basis whose first vector is in V_{3-a}, and therefore has its last $a + 1$ coordinates equal to 0, and whose second vector is in V_{4-b}, and thus has its last b coordinates equal to 0. If B is the matrix whose rows are the coordinates of these two vectors, then $p_{i,j}$ is (up to sign) the determinant of the submatrix of B involving the columns i and j. It follows that if $i > 3 - a$ or $j > 4 - b$, then $p_{i,j}(L) = 0$, so the given subsets of Plücker coordinates do vanish on the Schubert cycles as claimed.

To show that the ideals of the Schubert cycles are generated by the relation g and the given subsets, observe that each of the subsets is the ideal of the irreducible subvariety described above, and these have the same dimensions as the Schubert cycles. For example, we know that $\dim \Sigma_{1,1} = 2$, and the ideal

$$(g, p_{1,4}, p_{2,4}, p_{3,4}) = (p_{1,4}, p_{2,4}, p_{3,4}) \subset \Bbbk[p_{1,2}, \ldots, p_{3,4}]$$

is the entire ideal of a plane. □

4.2 Intersection products

4.2.1 Transverse flags

Throughout, we let $G = G(k, V)$ be the Grassmannian of k-dimensional linear subspaces of an n-dimensional vector space V. We start with one useful definition. As we said, Kleiman's theorem assures us (in characteristic 0) that, for a general pair of flags $\mathcal{V}$ and $\mathcal{W}$ on V, the Schubert cycles $\Sigma_a(\mathcal{V})$, $\Sigma_b(\mathcal{W}) \subset G$ intersect generically transversely. In this case, we can actually say explicitly what "general" means:

Definition 4.4. We say that a pair of flags $\mathcal{V}$ and $\mathcal{W}$ on V are *transverse* if any of the following equivalent conditions hold:

(a) $V_i \cap W_{n-i} = 0$ for all i.
(b) $\dim(V_i \cap W_j) = \min(0, i + j - n)$ for all i, j.
(c) There exists a basis $e_1, \ldots, e_n$ for V in terms of which

$$V_i = \langle e_1, \ldots, e_i \rangle \quad \text{and} \quad W_j = \langle e_{n+1-j}, \ldots, e_n \rangle.$$

Note that any two transverse pairs can be carried into one another by a linear automorphism of V. Moreover, transverse pairs form a dense open subset in the space of all pairs of flags, so any statement proved for a general pair of flags (such as the generic transversality of the intersection $\Sigma_a(\mathcal{V}) \cap \Sigma_b(\mathcal{W}) \subset G$) holds for any transverse pair, and vice versa.

Here is a lemma that will prove useful in intersecting Schubert cycles, though we will not use its full strength until the proof of Pieri's formula (Proposition 4.9):

Lemma 4.5. *Let* $\Sigma_a(\mathcal{V}), \Sigma_b(\mathcal{W}) \subset G$ *be Schubert cycles defined relative to transverse flags* $\mathcal{V}$ *and* $\mathcal{W}$ *on* V. *If* $\Lambda \in \Sigma_a(\mathcal{V}) \cap \Sigma_b(\mathcal{W})$ *is a general point of their intersection, then:*

(a) Λ *does not lie in any strictly smaller Schubert cycle* $\Sigma_{a'}(\mathcal{V}) \subsetneq \Sigma_a(\mathcal{V})$.

(b) *The flags induced by* $\mathcal{V}$ *and* $\mathcal{W}$ *on* Λ *(that is, consisting of intersections with* Λ *with flag elements* V_α *and* W_β) *are transverse.*

Note that, by the first part, the flags $\Lambda^\mathcal{V}$ and $\Lambda^\mathcal{W}$ on Λ induced by $\mathcal{V}$ and $\mathcal{W}$ are, explicitly,

$$\Lambda_i^\mathcal{V} = \Lambda \cap V_{n-k+i-a_i} \quad \text{and} \quad \Lambda_i^\mathcal{W} = \Lambda \cap W_{n-k+i-b_i}, \quad i = 1, \ldots, k.$$

Proof of Lemma 4.5: The first part of the statement is immediate for dimension reasons: the flags $\mathcal{V}$ and $\mathcal{W}$ being transverse, the intersection $\Sigma_{a'}(\mathcal{V}) \cap \Sigma_b(\mathcal{W})$ will have dimension strictly less than $\Sigma_a(\mathcal{V}) \cap \Sigma_b(\mathcal{W})$.

As for the second part, we have to show that the subspaces $\Lambda_i^\mathcal{V}$ and $\Lambda_{k-i}^\mathcal{W}$ are complementary, that is, that

$$\Lambda \cap V_{n-k+i-a_i} \cap W_{n-i-b_{k-i}} = 0.$$

We do this by a dimension count: consider the incidence correspondence

$$\Phi = \left\{ (\Lambda, [v]) \in (\Sigma_a(\mathcal{V}) \cap \Sigma_b(\mathcal{W})) \times \mathbb{P}(V_{n-k+i-a_i} \cap W_{n-i-b_{k-i}}) \mid v \in \Lambda \right\}.$$

We will show that $\dim \Phi < \dim(\Sigma_a(\mathcal{V}) \cap \Sigma_b(\mathcal{W}))$, and thus the projection $\Phi \to \Sigma_a(\mathcal{V}) \cap \Sigma_b(\mathcal{W})$ cannot be dominant, proving the lemma. Note that by the first part of the lemma we can replace Φ by the preimage of the complement U of

$$\mathbb{P}(V_{n-k+i-a_i-1} \cap W_{n-i-b_{k-i}}) \quad \text{and} \quad \mathbb{P}(V_{n-k+i-a_i} \cap W_{n-i-b_{k-i}-1})$$

in $\mathbb{P}(V_{n-k+i-a_i} \cap W_{n-i-b_{k-i}})$.

Since the flags $\mathcal{V}$ and $\mathcal{W}$ are transverse, we have

$$\dim \mathbb{P}(V_{n-k+i-a_i} \cap W_{n-i-b_{k-i}}) = n - k - a_i - b_{k-i} - 1.$$

(If $a_i + b_{k-i} \geq n - k$, then the intersection $V_{n-k+i-a_i} \cap W_{n-i-b_{k-i}}$ is 0 and Φ is correspondingly empty, so we are done in that case.) Next, suppose that $[v] \in U \subset \mathbb{P}(V_{n-k+i-a_i} \cap W_{n-i-b_{k-i}})$. To describe the fiber of Φ over $[v]$, we consider the quotient space $V' = V/\langle v \rangle$, and the flags $\mathcal{V}'$ and $\mathcal{W}'$ on V' comprised of images of subspaces V_i and W_i under the projection $V \to V'$; that is,

$$V_j' = \begin{cases} (V_j + \langle v \rangle)/\langle v \rangle & \text{if } j < n - k + i - a_i, \\ V_{j+1}/\langle v \rangle & \text{if } j + 1 \geq n - k + i - a_i, \end{cases}$$

and similarly

$$W'_j = \begin{cases} (W_j + \langle v \rangle)/\langle v \rangle & \text{if } j < n - i - b_{k-i}, \\ W_{j+1}/\langle v \rangle & \text{if } j + 1 \geq n - i - b_{k-i}. \end{cases}$$

Now we just observe that, if $(\Lambda, [v]) \in \Phi$, then the plane $\Lambda' = \Lambda/\langle v \rangle \subset V'$ belongs to the Schubert cycles

$$\Sigma_{a_1, \ldots, \widehat{a_i}, \ldots, a_k}(V') \quad \text{and} \quad \Sigma_{b_1, \ldots, \widehat{b_{k-i}}, \ldots, b_k}(W') \subset G(k - 1, V').$$

Thus the fibers of Φ over $\mathbb{P}(V_{n-k+i-a_i} \cap W_{n-i-b_{k-i}})$ have dimension

$$(k-1)(n-k) - \sum_{j \neq i} a_j - \sum_{j \neq k-i} b_j = (k-1)(n-k) - (|a| - a_i) - (|b| - b_{k-i}),$$

and altogether we have

$$\begin{aligned} \dim \Phi &= (n - k - a_i - b_{k-i} - 1) + \big((k-1)(n-k) - (|a| - a_i) - (|b| - b_{k-i})\big) \\ &= k(n-k) - |a| - |b| - 1 \\ &< \dim(\Sigma_a(V) \cap \Sigma_b(W)), \end{aligned}$$

as desired. $\qquad\qquad\qquad\qquad\qquad\qquad\qquad\qquad\qquad\qquad\qquad\qquad\qquad\qquad\square$

4.2.2 Intersections in complementary dimension

As in the case of the Grassmannian $\mathbb{G}(1, 3)$, we start our description of the Chow ring of G by evaluating intersections of Schubert cycles in complementary codimension. Here as before we use the fact that Schubert cycles $\Sigma_a(V)$ and $\Sigma_b(W)$ defined in terms of general flags V, W always intersect generically transversely; this follows from Kleiman's theorem, or, in arbitrary characteristic, from Theorem 4.1.

Proposition 4.6. *If V and W are transverse flags in V and $\Sigma_a(V), \Sigma_b(W)$ are Schubert cycles with $|a| + |b| = k(n - k)$, then $\Sigma_a(V)$ and $\Sigma_b(W)$ intersect transversely in a unique point if $a_i + b_{k+1-i} = n - k$ for each $i = 1, \ldots, k$, and are disjoint otherwise. Thus*

$$\deg \sigma_a \sigma_b = \begin{cases} 1 & \text{if } a_i + b_{k-i+1} = n - k \text{ for all } i, \\ 0 & \text{otherwise.} \end{cases}$$

Proof: As observed, since the two flags V and W are transverse, the Schubert cycles will meet generically transversely, and hence (since the intersection is zero-dimensional) transversely. Thus

$$\deg \sigma_a \sigma_b = \#(\Sigma_a(V) \cap \Sigma_b(W))$$

$$= \#\left\{ \Lambda \ \middle| \ \begin{aligned} \dim(V_{n-k+i-a_i} \cap \Lambda) &\geq i, \\ \dim(W_{n-k+i-b_i} \cap \Lambda) &\geq i, \end{aligned} \ \text{for all } i \right\}.$$

To evaluate the cardinality of this set, consider the conditions in pairs; that is, for each i, consider the i-th condition associated to the Schubert cycle $\Sigma_a(\mathcal{V})$:

$$\dim(V_{n-k+i-a_i} \cap \Lambda) \geq i$$

in combination with the $(k - i + 1)$-st condition associated to $\Sigma_b(\mathcal{W})$:

$$\dim(W_{n-i+1-b_{k-i+1}} \cap \Lambda) \geq k - i + 1.$$

If these conditions are both satisfied, then the subspaces

$$V_{n-k+i-a_i} \cap \Lambda \quad \text{and} \quad W_{n-i+1-b_{k-i+1}} \cap \Lambda,$$

having greater than complementary dimension in Λ, must have nonzero intersection; in particular, we must have

$$V_{n-k+i-a_i} \cap W_{n-i+1-b_{k-i+1}} \neq 0,$$

and, since the flags $\mathcal{V}$ and $\mathcal{W}$ are general, this in turn says we must have

$$n - k + i - a_i + n - i + 1 - b_{k-i+1} \geq n + 1,$$

or, in other words,

$$a_i + b_{k-i+1} \leq n - k.$$

If equality holds in this last inequality, the subspaces $V_{n-k+i-a_i}$ and $W_{n-i+1-b_{k-i+1}}$ will meet in a one-dimensional vector space Γ_i, necessarily contained in Λ. (In the notation of Definition 4.4, $\Gamma_i = \langle e_{n-k+i-a_i} \rangle$.)

We have thus seen that $\Sigma_a(\mathcal{V})$ and $\Sigma_b(\mathcal{W})$ will be disjoint unless $a_i + b_{k-i+1} \leq n - k$ for all i. But from the equality

$$|a| + |b| = \sum_{i=1}^{k} (a_i + b_{k-i+1}) = k(n - k),$$

we see that if $a_i + b_{k-i+1} \leq n - k$ for all i, then we must have $a_i + b_{k-i+1} = n - k$ for all i. Moreover, in this case any Λ in the intersection $\Sigma_a(\mathcal{V}) \cap \Sigma_b(\mathcal{W})$ must contain each of the k subspaces Γ_i, so there is a unique such Λ, equal to the span of these one-dimensional spaces, as required. □

Corollary 4.7. *The Schubert classes form a free basis for $A(G)$, and the intersection forms $A^m(G) \times A^{\dim G - m}(G) \to \mathbb{Z}$ have the Schubert classes as dual bases.*

In view of the explicit duality between $A^m(G)$ and $A^{k(n-k)-m}(G)$ given by Proposition 4.6, it makes sense to introduce one more bit of notation: for any Schubert index $a = (a_1, \ldots, a_k)$, we will define the *dual index* to be the Schubert index $a^* = (n - k - a_k, \ldots, n - k - a_1)$. In these terms, Proposition 4.6 says that $\deg(\sigma_a \sigma_b) = 1$ if $b = a^*$ and is 0 otherwise.

Corollary 4.7 suggests a general approach to determining the coefficients in the expression of the class of a cycle as a linear combination of Schubert classes: If $\Gamma \subset G$ is any cycle of pure codimension m, we can write

$$[\Gamma] = \sum_{|a|=m} \gamma_a \sigma_a.$$

To find the coefficient γ_a, we intersect both sides with the Schubert cycle $\Sigma_{a*}(\mathcal{V}) = \Sigma_{n-k-a_k,\ldots,n-k-a_1}(\mathcal{V})$ for a general flag $\mathcal{V}$; we then have

$$\gamma_a = \deg([\Gamma] \cdot \sigma_{a*}) = \#(\Gamma \cap \Sigma_{a*}(\mathcal{V})).$$

We have used exactly this approach — called the method of *undetermined coefficients* — in calculating classes of various cycles in $\mathbb{G}(1,3)$ in the preceding chapter; Proposition 4.6 and Corollary 4.7 say that it is more generally applicable in any Grassmannian. Explicitly, we have:

Corollary 4.8. *If $\alpha \in A^m(G)$ is any class, then*

$$\alpha = \sum_{|a|=m} \deg(\alpha \sigma_{a*}) \cdot \sigma_a.$$

In particular, if σ_a and $\sigma_b \in A(G)$ are any Schubert classes on $G = G(k,n)$, then the product $\sigma_a \sigma_b$ is equal to

$$\sum_{|c|=|a|+|b|} \gamma_{a,b;c} \sigma_c,$$

where

$$\gamma_{a,b;c} = \deg(\sigma_a \sigma_b \sigma_{c*}).$$

Since for general flags $\mathcal{U}$, $\mathcal{V}$ and $\mathcal{W}$ the Schubert cycles $\Sigma_a(\mathcal{U})$, $\Sigma_b(\mathcal{V})$ and $\Sigma_{c*}(\mathcal{W})$ are generically transverse by Kleiman's theorem, the coefficients $\gamma_{a,b;c} = \deg(\sigma_a \sigma_b \sigma_{c*})$ are nonnegative integers. They are called *Littlewood–Richardson coefficients*, and they appear in many combinatorial and representation-theoretic contexts. If we adopt the convention that $\sigma_a = 0 \in A^{|a|}(G(k,n))$ if a fails to satisfy the conditions

$$n - k \geq a_1 \geq \cdots \geq a_k \geq 0 \quad \text{and} \quad a_l = 0 \text{ for all } l > k,$$

then the Littlewood–Richardson coefficients $\gamma_{a,b;c}$ depend only on the indices a, b and c, and not on k and n.

Corollary 4.8 shows that knowing the Littlewood–Richardson coefficients suffices to determine the products of all Schubert classes. In the case of the Grassmannians $G(2,n)$ they are either 0 or 1 (see Section 4.3), but a Littlewood–Richardson coefficient $\gamma_{a,b;c} > 1$ appears already in $G(3,6)$ (Exercise 4.35).

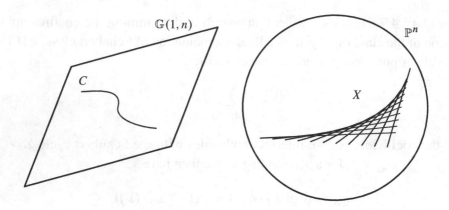

Figure 4.1 The surface $X \subset \mathbb{P}^n$ swept out by a one-parameter family $C \subset \mathbb{G}(1, n)$ of lines.

There exist beautiful algorithms for calculating the $\gamma_{a,b;c}$. We will give one in Section 4.7, and much more effective methods are given for example in Coşkun [2009] and Vakil [2006a]. But even simple questions such as, "when is $\gamma_{a,b;c} \neq 0$?" and "when is $\gamma_{a,b;c} > 1$?" do not seem to admit simple answers in general.

We will return to Schubert calculus shortly, but we take a moment here to use what we have already learned to answer Keynote Question (b).

4.2.3 Varieties swept out by linear spaces

Let $C \subset \mathbb{G}(k, n)$ be an irreducible curve, and consider the variety $X \subset \mathbb{P}^n$ swept out by the linear spaces corresponding to points of C; that is,

$$X = \bigcup_{[\Lambda] \in C} \Lambda \subset \mathbb{P}^n$$

(See Figure 4.1). We would like to relate the geometry of X to that of C; in particular, Keynote Question (b) asks us to find the degree of X when $C \subset \mathbb{G}(1, 3) \subset \mathbb{P}^5$ is a twisted cubic curve.

To begin with, observe that X is indeed a closed subvariety of $\mathbb{P}^n$: If

$$\Phi = \{(\Lambda, p) \in \mathbb{G}(k, n) \times \mathbb{P}^n \mid p \in \Lambda\}$$

is the universal k-plane over $\mathbb{G}(k, n)$, as described in Section 3.2.3, and $\alpha : \Phi \to \mathbb{G}(k, n)$ and $\beta : \Phi \to \mathbb{P}^n$ are the projections, then we can write

$$X = \beta(\alpha^{-1}(C)).$$

Now, suppose that a general point $x \in X$ lies on a unique k-plane $\Lambda \in C$ — that is, the map $\beta : \alpha^{-1}(C) \to X \subset \mathbb{P}^n$ is birational, so that in particular $\dim(X) = k + 1$. The degree of X is the number of points of intersection of X with a general $(n - k - 1)$-plane

$\Gamma \subset \mathbb{P}^n$; since each of these points is a general point of X, and so lies on a unique k-plane Λ, the number is the number of k-planes Λ that meet Γ. In other words, we have

$$
\begin{aligned}
\deg(X) &= \#(X \cap \Gamma) \\
&= \#(C \cap \Sigma_1(\Gamma)) \\
&= \deg([C] \cdot \sigma_1) \qquad \text{(by Kleiman's theorem)} \\
&= \deg(C),
\end{aligned}
$$

where by the degree of C we mean the degree under the Plücker embedding of $\mathbb{G}(k, n)$.

These ideas allow us to answer Keynote Question (b): The surface $X \subset \mathbb{P}^3$ swept out by the lines corresponding to a twisted cubic $C \subset \mathbb{G}(1, 3) \subset \mathbb{P}^5$, times the degree of the map β defined above, is equal to 3. Thus the surface X itself has degree 3 or 1. In the latter case, the curve C would be contained in a Schubert cycle $\Sigma_{1,1}$, and as we have seen in the description on page 138, this Schubert cycle is contained in the 2-plane in $\mathbb{P}^5$ defined by the vanishing of three Plücker coordinates. Since a twisted cubic is not contained in a 2-plane, this shows that the surface X has degree 3. More of the geometry of X is described in Exercises 4.23-4.25.

If $Z \subset \mathbb{G}(k, n)$ is a variety of any dimension m, we can form the variety $X \subset \mathbb{P}^n$ swept out by the planes of Z. Its degree — assuming it has the expected dimension $k + \dim Z$ and that a general point of X lies on only one plane $\Lambda \in Z$ — is expressible in terms of the Schubert coefficients of the class $[Z] \in A_m(\mathbb{G}(k, n))$, though it is not in general equal to the degree of Z. This is the content of Exercise 4.22; we will return to this question in Section 10.2, where we will see how to express the answer in terms of Chern and Segre classes.

4.2.4 Pieri's formula

One situation in which we can give a simple formula for the product of Schubert classes is when one of the classes has the special form $\sigma_b = \sigma_{b,0,\dots,0}$. Such classes are called *special Schubert classes*.

Proposition 4.9 (Pieri's formula). *For any Schubert class $\sigma_a \in A(G)$ and any integer b,*

$$
(\sigma_b \cdot \sigma_a) = \sum_{\substack{|c|=|a|+b \\ a_i \le c_i \le a_{i-1} \forall i}} \sigma_c
$$

Proof:[1] By Corollary 4.8, Pieri's formula is equivalent to the assertion that, for any Schubert index c with $|c| = |a| + b$,

$$
\deg(\sigma_a \sigma_b \sigma_{c^*}) = \begin{cases} 1 & \text{if } a_i \le c_i \le a_{i-1} \text{ for all } i, \\ 0 & \text{otherwise.} \end{cases}
$$

[1] This proof was shown to us by Izzet Coşkun

To prove this, we will look at the corresponding Schubert cycles $\Sigma_a(\mathcal{V})$, $\Sigma_b(\mathcal{U})$ and $\Sigma_{c*}(\mathcal{W})$, defined with respect to general flags $\mathcal{V}$, $\mathcal{U}$ and $\mathcal{W}$; we will show that their intersection is empty if c_i violates the condition $a_i \leq c_i \leq a_{i-1}$ for any i, and consists of a single point if these inequalities are all satisfied. By Kleiman's theorem, the intersection multiplicity will be 1 in the latter case.

By definition,

$$\Sigma_a(\mathcal{V}) = \{\Lambda \mid \dim(\Lambda \cap V_{n-k+i-a_i}) \geq i \text{ for all } i\}$$

and

$$\Sigma_{c*}(\mathcal{W}) = \{\Lambda \mid \dim(\Lambda \cap W_{i+c_{k+1-i}}) \geq i \text{ for all } i\}.$$

Set

$$A_i = V_{n-k+i-a_i} \cap W_{k+1-i+c_i},$$

so that either $A_i = 0$ or $\dim A_i = c_i - a_i + 1$. Combining the i-th condition in the first definition and the $(k + 1 - i)$-th condition in the second, we see that for any $\Lambda \in \Sigma_a(\mathcal{V}) \cap \Sigma_{c*}(\mathcal{W})$ we have

$$\Lambda \cap A_i \neq 0.$$

If $c_i < a_i$ for some i then $A_i = 0$, so that $\Sigma_a(\mathcal{V}) \cap \Sigma_{c*}(\mathcal{W}) = \varnothing$, and $\deg \sigma_a \sigma_b \sigma_{c*} = 0$, as required. Thus we may assume that $c_i \geq a_i$ for every i.

We claim that the A_i are linearly independent if and only if $c_i \leq a_{i-1}$ for all i. To see this, choose a basis e_i as in Section 4.2.2, so that $V_i = \langle e_1, \ldots, e_i \rangle$ and $W_j = \langle e_{n-j+1}, \ldots, e_n \rangle$. With this notation

$$A_i = \langle e_{n-k+i-c_i}, \ldots, e_{n-k+i-a_i} \rangle,$$

and the condition $c_i \leq a_{i-1}$ amounts to the condition that the two successive ranges of indices $n-k+i-1-c_{i-1}, \ldots, n-k+i-1-a_{i-1}$ and $n-k+i-c_i, \ldots, n-k+i-a_i$ do not overlap. In other words, if we let

$$A = \langle A_1, \ldots, A_k \rangle$$

be the span of the spaces A_i, then we have

$$\dim A \leq \sum c_i - a_i + 1 = k + b,$$

with equality holding if and only if $c_i \leq a_{i-1}$ for all i. Note that by Lemma 4.5 the plane Λ is spanned by its intersections with the A_i; that is, $\Lambda \subset A$.

Now we introduce the conditions associated with the special Schubert cycle $\Sigma_b(\mathcal{U})$. This is the set of k-planes that have nonzero intersection with a general linear subspace $U = U_{n-k+1-b} \subset V$ of dimension $n - k + 1 - b$. For there to be any $\Lambda \in \Sigma_a(\mathcal{V}) \cap \Sigma_{c*}(\mathcal{W})$ satisfying this condition requires that $A \cap U \neq 0$, and hence, since U is general, that $\dim A \geq k + b$. Thus, if $c_i > a_{i-1}$ for any i, then we will have

$\Sigma_a(\mathcal{V}) \cap \Sigma_{c^*}(\mathcal{W}) \cap \Sigma_b(\mathcal{U}) = \emptyset$. We can accordingly assume $c_i \le a_{i-1}$ for all i, and hence dim $A = k + b$.

Finally, since $U \subset V$ is a general subspace of codimension $k + b - 1$, it will meet A in a one-dimensional subspace. Choose v any nonzero vector in this intersection. Since $A = \bigoplus A_i$, we can write v uniquely as a sum

$$v = v_1 + \cdots + v_k \quad \text{with } v_i \in A_i.$$

Suppose now that $\Lambda \in \Sigma_a(\mathcal{V}) \cap \Sigma_b(\mathcal{U}) \cap \Sigma_{c^*}(\mathcal{W})$ satisfies all the Schubert conditions above. Since $\Lambda \subset A$ and $\Lambda \cap U \ne 0$, Λ must contain the vector v, and, since Λ is spanned by its intersections with the A_i, it follows that Λ must contain the vectors v_i as well. Thus, we see that the intersection $\Sigma_a(\mathcal{V}) \cap \Sigma_b(\mathcal{U}) \cap \Sigma_{c^*}(\mathcal{W})$ will consist of the single point corresponding to the plane $\Lambda = \langle v_1, \ldots, v_k \rangle$ spanned by the v_i, and we are done. $\square$

As a corollary of the Pieri formula, we can prove a relation among the special Schubert classes that is an important special case of a theorem of Whitney used for computing Chern classes (Theorem 5.3):

Corollary 4.10. *In* $A(G(k, n))$, *we have*

$$(1 + \sigma_1 + \sigma_2 + \cdots + \sigma_{n-k})(1 - \sigma_1 + \sigma_{1,1} - \sigma_{1,1,1} + \cdots + (-1)^k \sigma_{1^k}) = 1.$$

Proof: We can use Pieri to calculate the individual products appearing in this expression. To start, Pieri tells us that

$$\sigma_l \sigma_{1^m} = \sigma_{l,1^m} + \sigma_{l+1,1^{m-1}}.$$

When we write out the terms of degree d in the product on the left, then, the sum telescopes: For $d > 0$,

$$\sum_{i=0}^{d} (-1)^i \sigma_{d-i} \sigma_{1^i} = \sigma_d - (\sigma_d + \sigma_{d-1,1}) + (\sigma_{d-1,1} + \sigma_{d-2,1,1})$$
$$- \cdots + (-1)^{d-1}(\sigma_{2,1^{d-2}} + \sigma_{1^d}) + (-1)^d \sigma_{1^d}$$
$$= 0.   \qquad\square$$

4.3 Grassmannians of lines

Let $G = G(2, V)$ be the Grassmannian of two-dimensional subspaces of an $(n + 1)$-dimensional vector space V, or, equivalently, lines in the projective space $\mathbb{P}V \cong \mathbb{P}^n$. The Schubert cycles on G with respect to a flag $\mathcal{V}$ are of the form

$$\Sigma_{a_1, a_2}(\mathcal{V}) = \{\Lambda \mid \Lambda \cap V_{n-a_1} \ne 0 \text{ and } \Lambda \subset V_{n+1-a_2}\}.$$

In this case, Pieri's formula (Proposition 4.9) allows us to give a closed-form expression for the product of any two Schubert classes:

Proposition 4.11. *Assuming that* $a_1 - a_2 \geq b_1 - b_2$,

$$\sigma_{a_1,a_2}\sigma_{b_1,b_2} = \sigma_{a_1+b_1,a_2+b_2} + \sigma_{a_1+b_1-1,a_2+b_2+1} + \cdots + \sigma_{a_1+b_2,b_1+a_2}$$

$$= \sum_{\substack{|c|=|a|+|b| \\ a_1+b_1 \geq c_1 \geq a_1+b_2}} \sigma_{c_1,c_2}.$$

Proof: We will start with the simplest cases, where the intersection of general Schubert cycles is again a Schubert cycle: If $b_1 = b_2 = b$, then the Schubert cycle $\Sigma_{b,b}(\mathcal{W})$ is equal to

$$\{\Lambda \mid \Lambda \subset W_{n-b}\},$$

so that for any a_1, a_2 we have

$$\Sigma_{a_1,a_2}(\mathcal{V}) \cap \Sigma_{b,b}(\mathcal{W}) = \left\{ \Lambda \;\middle|\; \begin{array}{l} \Lambda \cap V_{n-1-a_1} \neq 0, \\ \qquad\qquad \Lambda \subset V_{n-a_2}, \\ \qquad\qquad \Lambda \subset W_{n-b} \end{array} \right\}$$

$$= \left\{ \Lambda \;\middle|\; \begin{array}{l} \Lambda \cap (V_{n-1-a_1} \cap W_{n-b}) \neq 0, \\ \qquad\qquad \Lambda \subset (V_{n-a_2} \cap W_{n-b}) \end{array} \right\}$$

$$= \Sigma_{a_1+b,a_2+b}(V_{n-1-a_1} \cap W_{n-b}, V_{n-a_2} \cap W_{n-b}).$$

Thus by Kleiman's theorem we have

$$\sigma_{a_1,a_2}\sigma_{b,b} = \sigma_{a_1+b,a_2+b}. \tag{4.2}$$

Now, suppose we want to intersect an arbitrary pair of Schubert classes σ_{a_1,a_2} and σ_{b_1,b_2}. We can write

$$\sigma_{a_1,a_2}\sigma_{b_1,b_2} = (\sigma_{a_1-a_2,0}\sigma_{a_2,a_2})(\sigma_{b_1-b_2,0}\sigma_{b_2,b_2})$$

$$= \sigma_{a_1-a_2,0}\,\sigma_{b_1-b_2,0}\,\sigma_{a_2+b_2,a_2+b_2},$$

and if we can evaluate the product of the first two terms in the last expression, we can use (4.2) to finish the calculation.

But this is exactly what Pieri gives us: if $a \geq b$, Pieri says that

$$\sigma_{a,0}\sigma_{b,0} = \sigma_{a+b,0} + \sigma_{a+b-1,1} + \cdots + \sigma_{a,b},$$

and the general statement follows. $\qquad\qquad\qquad\qquad\qquad\qquad\qquad\qquad\qquad\qquad\qquad\square$

We can use this description of the Chow ring of $G(1, n)$ (and a little combinatorics) to answer Keynote Question (d): What is the degree of the Grassmannian $G(1, n) = G(2, n + 1)$ under the Plücker embedding? We observe first that, since the hyperplane class on $\mathbb{P}(\wedge^2 \Bbbk^{n+1})$ pulls back to the class $\sigma_1 \in A^1(G(2, n + 1))$, we have

$$\deg(G(2, n + 1)) = \deg(\sigma_1^{2n-2}).$$

To evaluate this product, we make a directed graph with the Schubert classes σ_a in $G(2, n + 1)$ as vertices and with the inclusions among the corresponding Schubert cycles $\Sigma_a(\mathcal{V})$ indicated by arrows (the graph shown is the case $n = 5$):

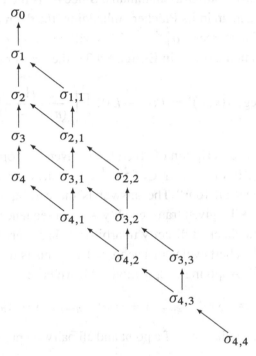

In terms of this graph, the rule expressed in Proposition 4.11 for multiplication by σ_1 is simple: The product of any Schubert class $\sigma_{a,b}$ with σ_1 is the sum of all immediate predecessors of $\sigma_{a,b}$ — that is, the Schubert classes in the row below $\sigma_{a,b}$ that are connected to $\sigma_{a,b}$ by an arrow. In particular, the degree $\deg((\sigma_1)^{2n-2})$ of the Grassmannian is the number of paths upward through this diagram starting with $\sigma_{n-1,n-1}$ and ending with $\sigma_{0,0}$. If we designate such a path by a sequence of $n - 1$ "1"s and $n - 1$ "2"s, corresponding to whether the first or second indices change (these are the vertical and diagonal arrows in the graph shown) reading from left to right, there are never more "2"s than "1"s. Equivalently, if we associate to a "1" a left parenthesis and to a "2" a right parenthesis, this is the number of ways in which $n - 1$ pairs of parentheses can appear in a grammatically correct sentence. This is called the $(n - 1)$-st *Catalan number*; a standard combinatorial argument (see, for example, Stanley [1999]) gives

$$c_{n-1} = \frac{(2n - 2)!}{n!(n - 1)!}.$$

In sum, we have:

Proposition 4.12. *The degree of the Grassmannian* $G(2, n + 1) \subset \mathbb{P}(\wedge^2 \Bbbk^{n+1})$ *is*

$$\deg G(2, n + 1) = \frac{(2n - 2)!}{n!(n - 1)!}.$$

This number also represents the answer to the enumerative problem of how many lines in $\mathbb{P}^n$ meet each of $2n - 2$ general $(n - 2)$-planes $V_1, \ldots, V_{2n-2} \subset \mathbb{P}^n$.

Pieri's formula (Proposition 4.9) gives us the means to answer the generalization of Keynote Question (d) to all Grassmannians: Since σ_1 is the class of the hyperplane section of the Grassmannian in its Plücker embedding, the degree of the Grassmannian in that embedding is the degree of $\sigma_1^{k(n-k)}$. This will be worked out (with the aid of the hook formula from combinatorics) in Exercise 4.38; the answer is that

$$\deg(G(k, n)) = (k(n - k))! \prod_{i=0}^{k-1} \frac{i!}{(n - k + i)!}.$$

We can also use the description of $A(\mathbb{G}(1, n))$ given in Proposition 4.11 to answer Keynote Question (a): If $V_1, \ldots, V_4 \subset \mathbb{P}^{2n+1}$ are four general n-planes, how many lines $L \subset \mathbb{P}^{2n+1}$ meet all four? The answer is the cardinality of the intersection $\bigcap \Sigma_n(V_i) \subset \mathbb{G}(1, 2n + 1)$; given transversality—a consequence of Kleiman's theorem in characteristic 0, and checked directly in arbitrary characteristic via the description of tangent spaces to Schubert cycles in Theorem 4.1—this is the degree of the product $\sigma_n^4 \in A(\mathbb{G}(1, 2n + 1))$. Applying Proposition 4.11, we have

$$\sigma_n^2 = \sigma_{2n} + \sigma_{2n-1,1} + \cdots + \sigma_{n+1,n-1} + \sigma_{n,n};$$

since each term squares to the class of a point and all pairwise products are zero, we have

$$\deg(\sigma_n^4) = n + 1,$$

and this is the answer to our question.

We will see in Exercise 4.26 another way to arrive at this number, in a manner analogous to the alternative solution to the four-line problem given in Section 3.4.1; Exercise 4.27 gives a nice geometric consequence.

4.4 Dynamic specialization

In Section 3.5.1, we started to discuss the method of *specialization*, and used it to determine the products of some Schubert classes. We can compute intersection numbers in other cases only by using a stronger and more broadly applicable version of this technique, called *dynamic specialization*.

Recall that in Section 3.5.1 we described an alternative approach to establishing the relation $\sigma_1^2 = \sigma_{11} + \sigma_2$ in the Chow ring of the Grassmannian $\mathbb{G}(1,3)$. Instead of taking two general translates of the Schubert cycle $\Sigma_1(L) \subset \mathbb{G}(1,3)$ — whose intersection was necessarily generically transverse, but whose intersection class required additional work to calculate — we considered the intersection $\Sigma_1(L) \cap \Sigma_1(L')$, where $L, L' \subset \mathbb{P}^3$ were not general, but incident lines. The benefit here is that now the intersection is visibly a union of Schubert cycles: Specifically, if $p = L \cap L'$ is their point of intersection and $H = \overline{L, L'}$ their span, we have

$$\Sigma_1(L) \cap \Sigma_1(L') = \Sigma_2(p) \cup \Sigma_{1,1}(H).$$

The trade-off is that we cannot just invoke Kleiman to see that the intersection is indeed generically transverse; this can however be established directly by using the description of the tangent spaces to the two cycles given in Theorem 4.1.

Suppose now we are dealing with the Grassmannian $G = \mathbb{G}(1,4)$ of lines in $\mathbb{P}^4$ and we try to use an analogous method to determine the product $\sigma_2^2 \in A^4(G)$ — that is, the class of the locus of lines meeting each of two given lines. We would try to find a pair of lines $L, M \subset \mathbb{P}^4$ such that the two cycles

$$\Sigma_2(L) = \{\Lambda \mid \Lambda \cap L \neq \varnothing\} \quad \text{and} \quad \Sigma_2(M) = \{\Lambda \mid \Lambda \cap M \neq \varnothing\}$$

representing the class σ_2 are special enough that the class of the intersection is clear, but still sufficiently general that they intersect generically transversely.

However, there are no such pairs of lines. If the lines L and M are disjoint, they are effectively a general pair, and the intersection is not a union of Schubert cycles. On the other hand, if L meets M at a point p, then the locus of lines through p forms a three-dimensional component of the intersection $\Sigma_2(L) \cap \Sigma_2(M)$, so the intersection is not even dimensionally transverse.

We can nevertheless consider a family of lines $\{M_t\}$ in $\mathbb{P}^4$, parametrized by $t \in \mathbb{A}^1$, with M_t disjoint from L for $t \neq 0$ and with M_0 meeting L at a point p. This gives a family of intersection cycles $\Sigma_2(L) \cap \Sigma_2(M_t)$. To make this precise, we consider the subvariety

$$\Phi^\circ = \{(t, \Lambda) \in \mathbb{A}^1 \times G \mid t \neq 0 \text{ and } \Lambda \in \Sigma_2(L) \cap \Sigma_2(M_t)\}$$

and its closure $\Phi \subset \mathbb{A}^1 \times G$. Since M_t is disjoint from L for $t \neq 0$, the fiber $\Phi_t = \Sigma_2(L) \cap \Sigma_2(M_t)$ of Φ over $t \neq 0$ represents the class σ_2^2, and it follows that Φ_0 does as well. The key point is that when we look at the fiber Φ_0 *we are looking not at the intersection $\Sigma_2(L) \cap \Sigma_2(M_0)$ of the limiting cycles, but rather at the flat limit of the intersection cycles $\Sigma_2(L) \cap \Sigma_2(M_t)$, which is necessarily of the expected dimension.*

The fiber Φ_0 is contained in the intersection $\Sigma_2(L) \cap \Sigma_2(M_0)$, but has smaller dimension. Thus a line Λ arising as the limit of lines Λ_t meeting both L and M_t must satisfy some additional condition beyond meeting both L and M_0, and to characterize Φ_0 we need to say what that condition is.

For $t \neq 0$, the lines L and M_t together span a hyperplane $H_t = \overline{L, M_t} \cong \mathbb{P}^3 \subset \mathbb{P}^4$. Let H_0 be the hyperplane that is the limit of the H_t as t goes to 0. If $\{\Lambda_t\}$ is a family of lines with Λ_t meeting both L and M_t for $t \neq 0$, then the limiting line Λ_0 must be contained in H_0.

Of course, if Λ_0 does not pass through the point $p = L \cap M_0$, then it must be contained in the 2-plane $P = \overline{L, M_0}$, so the new condition $\Lambda_0 \subset H_0$ is redundant. In sum, we conclude that the support of Φ_0 must be contained in the union of the two two-dimensional Schubert cycles

$$\Phi_0 \subset \{\Lambda \mid \Lambda \subset P\} \cup \{\Lambda \mid p_0 \in \Lambda \subset H_0\} = \Sigma_{2,2}(P) \cup \Sigma_{3,1}(p_0, H_0).$$

We will see in Exercise 4.28 that the support of Φ_0 is all of $\Sigma_{2,2}(P) \cup \Sigma_{3,1}(p_0, H_0)$, and in Exercise 4.29 that Φ_0 is generically reduced. Thus, the cycle associated to the scheme Φ_0 is exactly the sum $\Sigma_{2,2}(P) + \Sigma_{3,1}(p_0, H_0)$, and we can deduce the formula

$$\sigma_2^2 = \sigma_{3,1} + \sigma_{2,2} \in A^4(\mathbb{G}(1,4)).$$

This is a good example of the method of dynamic specialization, in which we consider not a special pair of cycles representing given Chow classes and intersecting generically transversely, but a family of representative pairs specializing from pairs that do intersect transversely to a pair that may not. We then must describe the limit of the intersections (not the intersection of the limits). This technique is a starting point for the general algorithms of Coşkun [2009] and Vakil [2006a]. For another example of its application, see Griffiths and Harris [1980].

Often, as in the examples cited above, to carry out the calculation of an intersection of Schubert cycles we may have to specialize in stages; see Exercise 4.30 for an example.

To see this idea carried out in a much broader context, see Fulton [1984, Chapter 11].

4.5 Young diagrams

For many purposes, it is convenient to represent the Schubert class $\sigma_{a_1,\dots,a_k}$ by a *Young diagram*; that is, as a collection of left-justified rows of boxes with the i-th row of length a_i. For example, $\sigma_{4,3,3,1,1}$ would be represented by

$$\sigma_{4,3,3,1,1} \quad \longleftrightarrow \quad \text{(Young diagram)} \quad .$$

(Warning: there are many different conventions in use for interpreting the correspondence between Schubert classes and Young diagrams!) The condition that $n-k \geq a_1 \geq a_k \geq 0$ means that the Young diagram fits into a box with k rows and $n-k$ columns, and the rows of the diagram are weakly decreasing in length from top to bottom. As another example, the relation between a Schubert class σ_a and the dual Schubert class σ_{a^*}, described in Proposition 4.6, could be described by saying that the Young diagrams of $\sigma = \sigma_a$ and $\tau = \sigma_{a^*}$, after rotating the latter $180°$, are complementary in the $k \times (n-k)$ box; if $\sigma = \sigma_{4,3,3,1,1} \in A(G(5,10))$, for example, then τ is as shown in the following:

σ	σ	σ	σ	τ
σ	σ	σ	τ	τ
σ	σ	σ	τ	τ
σ	τ	τ	τ	τ
σ	τ	τ	τ	τ

that is,

$$\tau \quad \longleftrightarrow \quad$$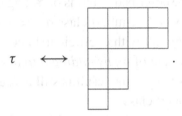

As a first application of this correspondence, we can count the Schubert classes as follows:

Corollary 4.13. $A(G(k,n)) \cong \mathbb{Z}^{\binom{n}{k}}$ *as abelian groups.*

Proof: The number of Schubert classes is the same as the number of Young diagrams that fit into a $k \times (n-k)$ box of squares B. To count these, we associate to each Young diagram Y in B its "right boundary" L: this is the path consisting of horizontal and vertical segments of unit length which starts from the upper-right corner of the $k \times (n-k)$ box and ends at the lower-left corner of the box, such that the squares in Y are those to the left of L. (For example, in the case of the Young diagram associated to $\sigma_{4,3,3,1,1} \subset G(5,10)$, illustrated above, we may describe L by the sequence $h, v, h, v, v, h, h, v, v$ where h and v denote horizontal and vertical segments, respectively, and we start from the upper-right corner.)

Of course the number of h terms in any such boundary must be $n-k$, the width of the box, and the number of v terms must be k, the height of the box. Thus the length of the boundary is n, and giving the boundary is equivalent to specifying which k steps will be vertical; that is, the number of Young diagrams in B is $\binom{n}{k}$, as required. $\square$

The correspondence between Schubert classes and Young diagrams behaves well with respect to many basic operations on Grassmannians. For example, under the duality $G(k,n) \cong G(n-k,n)$, the Schubert cycle corresponding to the Young diagram Y is taken to the Schubert cycle corresponding to the Young diagram Z that is the *transpose* of Y, that is, the diagram obtained by flipping Y around a 45° line running northwest-to-southeast. For example, if

$$\sigma_{3,2,1,1} \in A(G(4,7)) \longleftrightarrow$$

then the corresponding Schubert cycle in $G(3,7)$ is

$$\sigma_{4,2,1} \in A(G(3,7)) \longleftrightarrow$$

This is reasonably straightforward to verify, and is the subject of Exercise 4.31.

Pieri's formula can also be described in terms of Young diagrams: It says that for any Schubert class σ_b and any special Schubert class $\sigma_a = \sigma_{a,0,\dots,0}$, the Schubert classes appearing in the product $\sigma_a \sigma_b$ (all with coefficient 1) correspond to Young diagrams obtained from the Young diagram of σ_b *by adding a total of a boxes, with at most one box added to each column*, as long as the result is still a Young diagram: for example, if we want to multiply the Schubert class

$$\sigma_{4,2,1,1} \in A(G(4,8)) \longleftrightarrow$$

by the Schubert class σ_1, we can add a box in either the first, second, third or fifth row, to obtain the expression

$$\sigma_1 \sigma_{4,2,1,1} = \sigma_{5,2,1,1} + \sigma_{4,3,1,1} + \sigma_{4,2,2,1} + \sigma_{4,2,1,1,1}.$$

The combinatorics of Young diagrams is an extremely rich subject with many applications. For an introduction, see for example Fulton [1997].

4.5.1 Pieri's formula for the other special Schubert classes

Let V be an n-dimensional vector space and $\mathcal{V} = V_1 \subset \cdots \subset V_{n-1} \subset V_n = V$ a flag in V. As we observed in Section 4.1, for any integer a with $1 \leq a \leq n - k$ the isomorphism $G(k, V) \cong G(n - k, V^*)$ carries the special Schubert cycle

$$\Sigma_a = \{\Lambda \in G(k, V) \mid \Lambda \cap V_{n-k+1-a} \neq 0\}$$

defined relative to the flag $\mathcal{V}$ to the Schubert cycle

$$\Sigma_{1,\dots,1}(\mathcal{V}^{\perp}) = \{\Lambda \in G(n-k, V^*) \mid \dim(\Lambda \cap V_{n-k+a-1}) \geq a\}$$

defined relative to the flag $\mathcal{V}^{\perp}$ formed by the annihilators of the V_i. The Schubert classes $\sigma_{1,\dots,1}$ (often written σ_{1^a})

$$\sigma_{1^5} \quad \longleftrightarrow \quad$$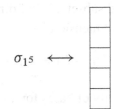

are also referred to as *special Schubert classes*. The correspondence between Schubert classes and Young diagrams makes it easy to translate Pieri's formula into a formula for multiplication by σ_{1^a}: The Schubert classes appearing in the product $\sigma_{1^a}\sigma_b$ (all with coefficient 1) correspond to Young diagrams obtained from the Young diagram of σ_b *by adding a total of a boxes, with at most one box added to each row*, as long as the result is still a Young diagram:

Theorem 4.14 (Pieri's formula, part II). *For any Schubert class $\sigma_b = \sigma_{b_1,\dots,b_k} \in A(G(k,n))$ and any integer a with $1 \leq a \leq n-k$,*

$$\sigma_{1^a}\sigma_b = \sum_{\substack{|c|=a+|b| \\ b_i \leq c_i \leq b_i+1 \, \forall i}} \sigma_c$$

4.6 Linear spaces on quadrics

We can generalize the calculation in Section 3.6 of the class of the locus of lines on a quadric surface to a description of the class of the locus of planes of any dimension on a smooth quadric hypersurface of any dimension.

To begin with, since our field $\Bbbk$ has characteristic $\neq 2$, a nonsingular form $Q(x)$ of degree 2 on $\mathbb{P}V$ can be written in the form $Q(x) = q(x, x)$, where $q(x, y)$ is a nonsingular symmetric bilinear form $V \times V \to \Bbbk$. A linear subspace $\mathbb{P}W \subset \mathbb{P}V$ lies on the quadric $Q(x) = 0$ if and only if W is *isotropic* for q, that is, $q(W, W) = 0$. Thus, we want to find the class of the locus $\Phi \subset G = G(k, V)$ of isotropic k-planes for q.

To start, we want to find the dimension of Φ. There are a number of ways to do this; probably the most elementary is to count bases for isotropic subspaces. To find a basis for an isotropic subspace, we can start with any vector v_1 with $q(v_1, v_1) = 0$, then choose $v_2 \in \langle v_1 \rangle^{\perp} \setminus \langle v_1 \rangle$ with $q(v_2, v_2) = 0$, $v_3 \in \langle v_1, v_2 \rangle^{\perp} \setminus \langle v_1, v_2 \rangle$ with $q(v_3, v_3) = 0$, and so on. Since $\langle v_1, \dots, v_i \rangle \subset \langle v_1, \dots, v_i \rangle^{\perp}$, this necessarily terminates when $i \geq n/2$; in other words, *a nondegenerate quadratic form will have no isotropic subspaces of dimension strictly greater than half the dimension of the ambient space.* (We could also see this by observing that q defines an isomorphism of V with its dual V^* carrying any isotropic subspace $\Lambda \subset V$ into its annihilator $\Lambda^\perp \subset V^*$.)

In this process, the allowable choices for v_1 correspond to points on the quadric $Q(x) = 0$; those for v_2 correspond to the points on the quadric $Q|_{v_1^\perp}$, and so forth. In general, the v_i form a locally closed subset of V of dimension $n - i$. Thus the space of all bases for isotropic k-planes has dimension

$$(n - 1) + \cdots + (n - k) = k(n - k) + \binom{k}{2}.$$

Since there is a k^2-dimensional family of bases for a given isotropic k-plane, the space of such planes has dimension

$$k(n - k) + \binom{k}{2} - k^2 = k(n - k) - \binom{k+1}{2},$$

or in other words the cycle Φ has codimension $\binom{k+1}{2}$ in $G(k, V)$ when $k \leq n/2$, and is empty otherwise.

Having determined the dimension of Φ, we ask now for its class in $A(G(k, V))$. Following the method of undetermined coefficients, we write

$$[\Phi] = \sum_{|a| = \binom{k+1}{2}} \gamma_a \sigma_a,$$

with

$$\gamma_a = \#(\Phi \cap \Sigma_{n-k-a_k, \ldots, n-k-a_1}(V))$$
$$= \#\{\Lambda \mid q|_\Lambda \equiv 0 \text{ and } \dim(\Lambda \cap V_{i+a_i}) \geq i \text{ for all } i\}.$$

To evaluate γ_a, suppose that $\Lambda \subset V$ is a k-plane in this intersection. The subspace $V_{a_i + i} \subset V$ being general, the restriction $q|_{V_{a_i+i}}$ of q to it will again be nondegenerate. Since $q|_{V_{a_i+i}}$ has an isotropic i-plane, we must have $a_i + i \geq 2i$, or in other words

$$a_i \geq i \quad \text{for all } i.$$

But by hypothesis $\sum a_i = \binom{k+1}{2}$, so we must have equality in each of these inequalities. In other words, $\gamma_a = 0$ *for all a except the index $a = (k, k - 1, \ldots, 2, 1)$.*

It remains to evaluate the coefficient

$$\gamma_{k,k-1,\ldots,2,1} = \#\{\Lambda \mid q(\Lambda, \Lambda) = 0 \text{ and } \dim(\Lambda \cap V_{2i}) \geq i \text{ for all } i\}, \quad (4.3)$$

where the equality holds by Kleiman's theorem. We claim that this number is 2^k.

We prove this inductively. To start, note that the restriction $q|_{V_2}$ of q to the two-dimensional space V_2 has two one-dimensional isotropic spaces, and Λ will necessarily contain exactly one of them: it cannot contain both, since Φ is disjoint from any Schubert cycle $\Sigma_b(V)$ with $|b| > k(n - k) - \binom{k+1}{2}$.

We may thus suppose that Λ contains the isotropic subspace $W \subset V_2$, so that Λ is contained in $W^\perp$. Now, since $q(W, W) \equiv 0$, q induces a nondegenerate quadratic form q' on the $(n-2)$-dimensional quotient $W' = W^\perp/W$, and the quotient space

$$\Lambda' = \Lambda/W \subset W^\perp/W$$

is a $(k-1)$-dimensional isotropic subspace for q'. Moreover, since the spaces V_{2i} are general subspaces of V containing V_2, the subspaces

$$V'_{2i-2} = (V_{2i} \cap W^\perp)/W \subset W^\perp/W$$

form a general flag in $W' = W^\perp/W$, and we have

$$\dim(\Lambda' \cap V'_{2i-2}) \geq i-1 \quad \text{for all } i.$$

Inductively, there are 2^{k-1} isotropic $(k-1)$-planes $\Lambda' \subset W'$ satisfying these conditions, and so there are 2^k planes $\Lambda \subset W$ satisfying the conditions of (4.3). We have proven:

Proposition 4.15. *Let q be a nondegenerate quadratic form on the n-dimensional vector space V and $\Phi \subset G(k, V)$ the variety of isotropic k-planes for q. Assuming $k \leq n/2$, the class of the cycle Φ is*

$$[\Phi] = 2^k \sigma_{k,k-1,\dots,2,1}.$$

As an immediate application of this result, we can answer Keynote Question (c). To begin with, we asked how many lines lie on the intersection of two quadrics in $\mathbb{P}^4$. To answer this, let $Q, Q' \subset \mathbb{P}^4$ be two general quadric hypersurfaces and $X = Q_1 \cap Q_2$. The set of lines on X is the intersection $\Phi \cap \Phi'$ of the cycles of lines lying on Q and Q'; by Kleiman's theorem these are transverse, and so we have

$$\#(\Phi \cap \Phi') = \deg(4\sigma_{2,1})^2 = 16.$$

More generally, if Q and $Q' \subset \mathbb{P}^{2n}$ are general quadrics, we ask how many $(n-1)$-planes are contained in their intersection; again, this is the intersection number

$$\#(\Phi \cap \Phi') = \deg(2^n \sigma_{n,n-1,\dots,1})^2 = 4^n.$$

4.7 Giambelli's formula

Pieri's formula tells us how to intersect an arbitrary Schubert class with one of the special Schubert classes $\sigma_b = \sigma_{b,0,\dots,0}$. Giambelli's formula is complementary, in that it tells us how to express an arbitrary Schubert class in terms of special ones; the two together give us (in principle) a way of calculating the product of two arbitrary Schubert classes.

We will state Giambelli's formula and indicate one method of proof; see Chapter 12 for some special cases and Fulton [1997] for a proof in general.

Proposition 4.16 (Giambelli's formula).

$$\sigma_{a_1,a_2,\ldots,a_k} = \begin{vmatrix} \sigma_{a_1} & \sigma_{a_1+1} & \sigma_{a_1+2} & \cdots & \sigma_{a_1+k-1} \\ \sigma_{a_2-1} & \sigma_{a_2} & \sigma_{a_2+1} & \cdots & \sigma_{a_2+k-2} \\ \sigma_{a_3-2} & \sigma_{a_3-1} & \sigma_{a_3} & \cdots & \sigma_{a_3+k-3} \\ \vdots & \vdots & \vdots & \ddots & \vdots \\ \sigma_{a_k-k+1} & \sigma_{a_k-k+2} & \sigma_{a_k-k+3} & \cdots & \sigma_{a_k} \end{vmatrix}.$$

Thus, for example, we have

$$\sigma_{2,1} = \begin{vmatrix} \sigma_2 & \sigma_3 \\ \sigma_0 & \sigma_1 \end{vmatrix} = \sigma_2\sigma_1 - \sigma_3,$$

which we can then use together with Pieri to evaluate $\sigma_{2,1}^2$, for example. Giambelli's formula also reproduces some formulas we have derived already by other means: For example, when $a_1 = a_2 = 1$ it gives

$$\sigma_{1,1} = \begin{vmatrix} \sigma_1 & \sigma_2 \\ \sigma_0 & \sigma_1 \end{vmatrix} = \sigma_1^2 - \sigma_2,$$

or in other words $\sigma_1^2 = \sigma_2 + \sigma_{1,1}$.

As the last two examples suggest, we could deduce Giambelli's formula from Pieri's formula. For example, in the 2×2 case, we can expand the determinant and apply Pieri's formula to obtain

$$\begin{vmatrix} \sigma_a & \sigma_{a+1} \\ \sigma_{b-1} & \sigma_b \end{vmatrix} = \sigma_a\sigma_b - \sigma_{a+1}\sigma_{b-1}$$

$$= (\sigma_{a,b} + \sigma_{a+1,b-1} + \cdots + \sigma_{a+b}) - (\sigma_{a+1,b-1} + \cdots + \sigma_{a+b})$$

$$= \sigma_{a,b}.$$

More generally, we could prove Giambelli's formula inductively by expanding the determinant in Proposition 4.16 by cofactors along the right-hand column; Exercise 4.39 asks the reader to do this in the 3×3 case.

Giambelli's formula implies that the Chow ring $A(G)$ of a Grassmannian G is generated as a ring by the special Schubert classes, and we can ask about the polynomial relations among these classes. There is a surprisingly simple and elegant description of these relations, which we will derive in Section 5.8, from the fact that the special Schubert classes are exactly the Chern classes of the universal bundles on the Grassmannian.

Giambelli's formula and Pieri's formula together give an algorithm for calculating the product of any two Schubert classes: Use Giambelli to express either as a polynomial in the special Schubert classes, and then use Pieri to evaluate the product of this polynomial with the other. But this is a terrible idea for computation except in low-dimensional examples: Because Giambelli's formula is determinantal, the number of products involved increases rapidly with k and n. Nor is it easy to use Giambelli's formula to prove qualitative results about products of Schubert classes; it is not even clear from this approach that such a product is necessarily a nonnegative linear combination of Schubert classes. The algorithms of Coşkun and Vakil referred to earlier (Coşkun [2009] and Vakil [2006a]) are far better in these regards.

4.8 Generalizations

Much of the analysis we have given here of the Chow rings of Grassmannians applies more generally to any compact homogeneous space for a semisimple algebraic group. In this section, we will describe some of these spaces, and indicate how the analysis goes in some of the simplest non-Grassmannian cases.

4.8.1 Flag manifolds

Let V be a vector space of dimension n and $(k_1, \ldots, k_m)$ any sequence of integers with $0 < k_1 < \cdots < k_m < n$. We define the *flag manifold* $F(k_1, \ldots, k_m; V)$ to be the space of nested sequences of subspaces of V of dimensions $k_1, \ldots, k_m$; that is,

$$F(k_1, \ldots, k_m; V) = \left\{ (\Lambda_1, \ldots, \Lambda_m) \in \prod G(k_i, V) \mid \Lambda_1 \subset \cdots \subset \Lambda_m \right\}.$$

As in the case of the Grassmannian, when only the dimension of V matters we also use the symbol $F(k_1, \ldots, k_m; n)$; also as in the case of Grassmannians, we will sometimes use the projective notation

$$\mathbb{F}(k_1, \ldots, k_m; \mathbb{P}V) = \left\{ (\Lambda_1, \ldots, \Lambda_m) \in \prod \mathbb{G}(k_i, \mathbb{P}V) \mid \Lambda_1 \subset \cdots \subset \Lambda_m \right\}$$

for $0 \le k_1 < \cdots < k_m < \dim \mathbb{P}V$. We leave as an exercise the verification that the condition $\Lambda_1 \subset \cdots \subset \Lambda_m$ defines a closed subscheme of the product $\prod G(k_i, V)$. (This follows immediately from the case $m = 2$, which is the content of Exercise 3.21.) In particular, $F(k_1, \ldots, k_m; V)$ is a projective variety.

At one extreme we have the case $m = n-1$, that is, $(k_1, \ldots, k_m) = (1, 2, \ldots, n-1)$; the variety

$$F(1, 2, \ldots, n - 1; V) \subset \prod_{k=1}^{n-1} G(k, V)$$

is called the *full flag manifold*, and maps to all the other flag manifolds $F(k_1, \ldots, k_m; V)$ via projections to subproducts of Grassmannians. At the other, the cases with $m = 1$ are just the ordinary Grassmannians, and the cases with $m = 2$ are called *two-step flag manifolds*. We have already encountered some of these: the variety

$$\mathbb{F}(0, k; V) = \{ (p, \Lambda) \in \mathbb{P}V \times \mathbb{G}(k, \mathbb{P}V) \mid p \in \Lambda \}$$

is often called the *universal k-plane in $\mathbb{P}V$.*

Many of the aspects of the geometry of Grassmannians we have explored in the last two chapters hold more generally for flag manifolds. In particular, a flag manifold $\mathbb{F}$ admits an affine stratification, the classes of whose closed strata (again called *Schubert classes*) freely generate the Chow ring $A(\mathbb{F})$ as a group. It is possible to describe the ring structure on $A(\mathbb{F})$ in terms of these generators (see Coşkun [2009]), but there is an alternative: as we will see in Chapter 9 it is possible (and easier in some settings) to determine the ring $A(\mathbb{F})$ by realizing the flag manifold as a series of projective bundles.

4.8.2 Lagrangian Grassmannians and beyond

There are important generalizations of flag manifolds that are homogeneous spaces for semisimple algebraic groups other than GL_n. For example:

(a) *Lagrangian Grassmannians*: If V is a vector space of dimension $2n$ with a nondegenerate skew-symmetric bilinear form $Q : V \times V \to \Bbbk$, the Lagrangian Grassmannians $LG(k, V)$ parametrize k-dimensional subspaces $\Lambda \subset V$ that are isotropic for Q; that is, such that $Q(\Lambda, \Lambda) = 0$. More generally, we have *Lagrangian flag manifolds*, parametrizing flags of such subspaces.

(b) *Orthogonal Grassmannians*: As in the previous case, we consider a vector space V with nondegenerate bilinear form Q, but now Q is symmetric. The orthogonal Grassmannian parametrizes isotropic subspaces, and likewise the orthogonal flag manifolds parametrize flags of isotropic subspaces. (In case dim V is even, we have to allow for the fact that the space of maximal isotropic planes has two connected components.)

The subgroup of GL_n that fixes a full flag in $\Bbbk^n$ is the group B of upper-triangular matrices. This is called a *Borel* subgroup of GL_n. Since GL_n acts transitively on flags, the set of all such flags is GL_n / B; it can be given the structure of an algebraic variety by taking the regular functions on the quotient to be B-invariant functions on GL_n. (With this structure, it is isomorphic to the flag manifold as we have defined it.) More generally, one could look at partial flags (for example, a single k-dimensional subspace); these are fixed by groups of block upper-triangular matrices, called *parabolic subgroups*. Thus for example the ordinary Grassmannian $G(k, n)$ has the form GL_n / P, where P is a parabolic subgroup.

It turns out that there is a natural way of defining Borel subgroups and parabolic subgroups in any semisimple group, and the Lagrangian and orthogonal Grassmannians may similarly be defined as quotients of the groups SO_n and Sp_n. The theory of general flag manifolds to which this leads is an extremely rich branch of mathematics. See for example Fulton and Harris [1991].

4.9 Exercises

Exercise 4.17. Use the description of the points of the Schubert cells given in Theorem 4.1 to show that Theorem 4.3 holds at least set-theoretically.

Exercise 4.18. Let $X \subset G(2,4)$ be an irreducible surface, and suppose that

$$[X] = \gamma_2 \sigma_2 + \gamma_{1,1} \sigma_{1,1} \in A^2(G(2,4)).$$

Show that γ_2 and $\gamma_{1,1}$ are nonnegative, and that if $\gamma_2 = 0$ then $\gamma_{1,1} = 1$. (In general, it is not known what pairs $(\gamma_2, \gamma_{1,1})$ occur!)

Exercise 4.19. Let $S \subset \mathbb{P}^4$ be a surface of degree d, and $\Gamma_S \subset \mathbb{G}(1,4)$ the variety of lines meeting S.

(a) Find the class $\gamma_S = [\Gamma_S] \in A^1(\mathbb{G}(1,4))$.
(b) Use this to answer the question: if $S_1, \ldots, S_6 \subset \mathbb{P}^4$ are general translates (under GL_5) of surfaces of degrees $d_1, \ldots, d_6$, how many lines in $\mathbb{P}^4$ will meet all six?

Exercise 4.20. Let $C \subset \mathbb{P}^4$ be a curve of degree d, and $\Gamma_C \subset \mathbb{G}(1,4)$ the variety of lines meeting C.

(a) Find the class $\gamma_C = [\Gamma_C] \in A^2(\mathbb{G}(1,4))$.
(b) Use this to answer the question: if C_1, C_2 and $C_3 \subset \mathbb{P}^4$ are general translates of curves of degrees d_1, d_2 and d_3, how many lines in $\mathbb{P}^4$ will meet all three?

The following exercise is the first of a series regarding the variety $T_1(S)$ of lines tangent to a surface S in $\mathbb{P}^n$. More will follow in Exercises 7.30, 10.39 and 12.20.

Exercise 4.21. Let $S \subset \mathbb{P}^n$ be a smooth surface of degree d whose general hyperplane section is a curve of genus g, and $T_1(S) \subset \mathbb{G}(1,n)$ the variety of lines tangent to S. To find the class of the cycle $T_1(S)$, we need the intersection numbers $[T_1(S)] \cdot \sigma_3$ and $[T_1(S)] \cdot \sigma_{2,1}$. Find the latter.

Exercise 4.22. Let $Z \subset \mathbb{G}(k,n)$ be a variety of dimension m, and consider the variety $X \subset \mathbb{P}^n$ swept out by the linear spaces corresponding to points of Z; that is,

$$X = \bigcup_{[\Lambda] \in Z} \Lambda \subset \mathbb{P}^n.$$

For simplicity, assume that a general point $x \in X$ lies on a unique k-plane $\Lambda \in Z$.

(a) Show that X has dimension $k + m$ and degree equal to the intersection number $\deg(\sigma_m \cdot [Z])$.
(b) Show that this is not in general the degree of Z.

Exercises 4.23-4.25 deal with the geometry of the surface described in Keynote Question (b): the surface $X \subset \mathbb{P}^3$ swept out by the lines corresponding to a general twisted cubic $C \subset \mathbb{G}(1,3)$, whose degree we worked out in Section 4.2.3. To make life easier, we will assume that C is general, and in particular that it lies in a general 3-plane section of $\mathbb{G}(1,3)$. See also Section 9.1.1.

Exercise 4.23. To start, use the fact that the dual of $\mathbb{G}(1,3) \subset \mathbb{P}^5$ has degree 2 to show that a general twisted cubic $C \subset \mathbb{G}(1,3)$ lies on the Schubert cycles $\Sigma_1(L)$ and $\Sigma_1(M)$ for some pair of skew lines $L, M \subset \mathbb{P}^3$.

Exercise 4.24. Show that for skew lines $L, M \subset \mathbb{P}^3$, the intersection $\Sigma_1(L) \cap \Sigma_1(M)$ is isomorphic to $L \times M$ via the map sending a point $[\Lambda] \in \Sigma_1(L) \cap \Sigma_1(M)$ to the pair $(\Lambda \cap L, \Lambda \cap M) \in L \times M$, and that it is the intersection of $\mathbb{G}(1,3)$ with the intersection of the hyperplanes spanned by $\Sigma_1(L)$ and $\Sigma_1(M)$.

Exercise 4.25. Using the fact that $C \subset \Sigma_1(L) \cap \Sigma_1(M)$ has bidegree $(2,1)$ in $\Sigma_1(L) \cap \Sigma_1(M) \cong L \times M \cong \mathbb{P}^1 \times \mathbb{P}^1$ (possibly after switching factors), show that for some degree-2 map $\varphi : L \to M$ the family of lines corresponding to C may be realized as the locus

$$C = \left\{ \overline{p, \varphi(p)} \mid p \in L \right\}.$$

Show correspondingly that the surface

$$X = \bigcup_{[\Lambda] \in C} \Lambda \subset \mathbb{P}^3$$

swept out by the lines of C is a cubic surface double along a line, and that it is the projection of a rational normal surface scroll $S(1,2) \subset \mathbb{P}^4$.

In Section 4.3 we calculated the number of lines meeting four general n-planes in $\mathbb{P}^{2n+1}$. In the following two exercises, we will see another way to do this (analogous to the alternative count of lines meeting four lines in $\mathbb{P}^3$ given in Section 3.4.1), and a nice geometric sidelight.

Exercise 4.26. Let $\Lambda_1, \ldots, \Lambda_4 \cong \mathbb{P}^n \subset \mathbb{P}^{2n+1}$ be four general n-planes. Calculate the number of lines meeting all four by showing that the union of the lines meeting Λ_1, Λ_2 and Λ_3 is a Segre variety $S_{1,n} = \mathbb{P}^1 \times \mathbb{P}^n \subset \mathbb{P}^{2n+1}$ and using the calculation in Section 2.1.5 of the degree of $S_{1,n}$.

Exercise 4.27. By the preceding exercise, we can associate to a general configuration $\Lambda_1, \ldots, \Lambda_4$ of k-planes in $\mathbb{P}^{2k+1}$ an unordered set of $k + 1$ cross-ratios. Show that two such configurations $\{\Lambda_i\}$ and $\{\Lambda_i'\}$ are projectively equivalent if and only if the corresponding sets of cross-ratios coincide.

The next two exercises deal with the example of dynamic specialization given in Section 4.4, and specifically with the family Φ of cycles described there.

Exercise 4.28. Show that the support of Φ_0 is all of $\Sigma_{2,2}(P) \cup \Sigma_{3,1}(p_0, H_0)$.

Exercise 4.29. Verify the last assertion made in the calculation of σ_2^2; that is, show that Φ_0 has multiplicity 1 along each component.
Hint: Argue that by applying a family of automorphisms of $\mathbb{P}^4$ we can assume that the plane H_t is constant, and use the calculation of the preceding chapter.

Exercise 4.30. A further wrinkle in the technique of dynamic specialization is that to carry out the calculation of an intersection of Schubert cycles we may have to specialize in stages. To see an example of this, use dynamic specialization to calculate the intersection σ_2^2 in the Grassmannian $\mathbb{G}(1,5)$.
Hint: You have to let the two 2-planes specialize first to a pair intersecting in a point, then to a pair intersecting in a line.

Exercise 4.31. Suppose that the Schubert class $\sigma_a \in A(G(k,n))$ corresponds to the Young diagram Y in a $k \times (n-k)$ box B. Show that under the duality $G(k,n) \cong G(n-k,n)$ the class σ_a is taken to the Schubert class σ_b corresponding to the Young diagram Z that is the *transpose* of Y, that is, the diagram obtained by flipping Y around a 45° line running northwest-to-southeast. For example, if

$$\sigma_{3,2,1,1} \in A(G(4,7)) \quad \longleftrightarrow \quad$$

then the corresponding Schubert class in $G(3,7)$ is

$$\sigma_{4,2,1} \in A(G(3,7)) \quad \longleftrightarrow \quad .$$

Exercise 4.32. Let $i : G(k,n) \to G(k+1,n+1)$ and $j : G(k,n) \to G(k,n+1)$ be the inclusions obtained by sending $\Lambda \subset \Bbbk^n$ to the span of Λ and e_{n+1} and to Λ respectively. Show that the map $i^* : A^d(G(k+1,n+1)) \to A^d(G(k,n))$ is a monomorphism if and only if $n - k \geq d$, and that $j^* : A^d(G(k,n+1)) \to A^d(G(k,n))$ is a monomorphism if and only if $k \geq d$. (Thus, for example, the formula

$$\sigma_1^2 = \sigma_2 + \sigma_{11},$$

which we established in $A(\mathbb{G}(1,3))$, *holds true in every Grassmannian.*)

Exercise 4.33. Let $C \subset \mathbb{P}^r$ be a smooth, irreducible, nondegenerate curve of degree d and genus g, and let $S_1(C) \subset \mathbb{G}(1,r)$ be the variety of chords to C, as defined in Section 3.4.3 above. Find the class $[S_1(C)] \in A_2(\mathbb{G}(1,r))$.

Exercise 4.34. Let $Q \subset \mathbb{P}^n$ be a smooth quadric hypersurface, and let $T_k(Q) \subset \mathbb{G}(k, n)$ be the locus of planes $\Lambda \subset \mathbb{P}^n$ such that $\Lambda \cap Q$ is singular. Show that

$$[T_k(Q)] = 2\sigma_1.$$

Exercise 4.35. Find the expression of $\sigma_{2,1}^2$ as a linear combination of Schubert classes in $A(G(3, 6))$. This is the first example of a product of two Schubert classes where another Schubert class appears with coefficient > 1.

Exercise 4.36. Using Pieri's formula, determine all products of Schubert classes in the Chow ring of the Grassmannian $\mathbb{G}(2, 5)$.

Exercise 4.37. Let Q, Q' and Q'' be three general quadrics in $\mathbb{P}^8$. How many 2-planes lie on all three? (Try first to do this without the tools introduced in Section 4.2.4.)

Exercise 4.38. Use Pieri to identify the degree of $\sigma_1^{k(n-k)}$ with the number of standard tableaux, that is, ways of filling in a $k \times (n-k)$ matrix with the integers $1, \ldots, k(n-k)$ in such a way that every row and column is strictly increasing. Then use the "hook formula" (see, for example, Fulton [1997]) to show that this number is

$$(k(n-k))! \prod_{i=0}^{k-1} \frac{i!}{(n-k+i)!}.$$

Exercise 4.39. Deduce Giambelli's formula in the 3×3 case (that is, the relation

$$\begin{vmatrix} \sigma_a & \sigma_{a+1} & \sigma_{a+2} \\ \sigma_{b-1} & \sigma_b & \sigma_{b+1} \\ \sigma_{c-2} & \sigma_{c-1} & \sigma_c \end{vmatrix} = \sigma_{a,b,c}$$

for any $a \geq b \geq c$) by assuming Giambelli in the 2×2 case, expanding the determinant by cofactors along the last column and applying Pieri.

Chapter 5

Chern classes

Keynote Questions

(a) Let $S \subset \mathbb{P}^3$ be a smooth cubic surface. How many lines $L \subset \mathbb{P}^3$ are contained in S? (Answer on page 253.)

(b) Let F and G be general homogeneous polynomials of degree 4 in four variables, and consider the corresponding family $\{S_t = V(t_0 F + t_1 G) \subset \mathbb{P}^3\}_{t \in \mathbb{P}^1}$ of quartic surfaces in $\mathbb{P}^3$. How many members S_t of the family contain a line? (Answer on page 233.)

(c) Let F and G be general homogeneous polynomials of degree d in three variables, and let $\{C_t = V(t_0 F + t_1 G) \subset \mathbb{P}^2\}_{t \in \mathbb{P}^1}$ be the corresponding family of plane curves of degree d. How many of the curves C_t will be singular? (Answer on page 268.)

In this chapter we will introduce the machinery for answering these questions; the answers themselves will be found in Chapters 6 and 7.

5.1 Introduction: Chern classes and the lines on a cubic surface

Cartier divisors — defined through the vanishing loci of sections of line bundles — are of enormous importance in algebraic geometry. More generally, it turns out that many interesting varieties of higher codimension may be described as the loci where sections of vector bundles vanish, or where collections of sections become dependent; this reduces some difficult problems about varieties to easier, linear problems.

Chern classes provide a systematic way of treating the classes of such loci, and are a central topic in intersection theory. They will play a major role in the rest of this book. We begin with an example of how they are used, and then proceed to a systematic discussion.

To illustrate, we explain the Chern class approach to a famous classical result:

Theorem 5.1. *Each smooth cubic surface in $\mathbb{P}^3$ contains exactly 27 distinct lines.*

Sketch: Given a smooth cubic surface $X \subset \mathbb{P}^3$ determined by the vanishing of a cubic form F in four variables, we wish to determine the degree of the locus in $\mathbb{G}(1, 3)$ of lines contained in X.

We *linearize* the problem using the observation that, if we fix a particular line L in $\mathbb{P}^3$, then the condition that L lie on X can be expressed as four linear conditions on the coefficients of F. To see this, note that the restriction map from the 20-dimensional vector space of cubic forms on $\mathbb{P}^3$ to the four-dimension vector space $V_L = H^0(\mathcal{O}_L(3))$ of cubic forms on a line $L \cong \mathbb{P}^1 \subset \mathbb{P}^3$ is a linear surjection, and the condition for the inclusion $L \subset X$ is that F maps to 0 in V_L.

As the line L varies over $\mathbb{G}(1, 3)$, the four-dimensional spaces V_L of cubic forms on the varying lines L fit together to form a vector bundle $\mathcal{V}$ of rank 4 on $\mathbb{G}(1, 3)$. A cubic form F on $\mathbb{P}^3$, through its restriction to each V_L, defines an algebraic global section σ_F of this vector bundle. Thus the locus of lines contained in the cubic surface X is the zero locus of the section σ_F. Assuming for the moment that this zero locus is zero-dimensional, we call its class in $A(\mathbb{G}(1, 3))$ the *fourth Chern class* of $\mathcal{V}$, denoted $c_4(\mathcal{V})$.

At this point all we have done is to give our ignorance a fancy name. But there are powerful tools for computing Chern classes of vector bundles, especially when those bundles can be built up from simpler bundles by linear-algebraic constructions. In the present situation, the spaces $H^0(\mathcal{O}_L(1))$ fit together to form the dual $\mathcal{S}^*$ of the tautological subbundle of rank 2 on $\mathbb{G}(1, 3)$, and the bundle $\mathcal{V}$ is the symmetric cube $\mathrm{Sym}^3 \mathcal{S}^*$ of $\mathcal{S}^*$, which allows us to express the Chern classes of $\mathcal{V}$ in terms of those of $\mathcal{S}^*$, as in Example 5.16. At the same time, it is not hard to calculate the Chern classes $c_i(\mathcal{S}^*)$ directly; we do this in Section 5.6.2. Putting these things together, we will show in Chapter 6 that

$$\deg c_4(\mathcal{V}) = 27.$$

Of course, to prove Theorem 5.1 one still has to show that the number of lines on any smooth cubic surface is finite, and that the zeros of σ_F all occur with multiplicity 1; this will also be carried out in Chapter 6. $\square$

There are proofs of Theorem 5.1 that do not involve vector bundles and Chern classes. For example, one can show that any smooth cubic surface X can be realized as the blow-up of $\mathbb{P}^2$ in six suitably general points, and using this one can analyze the geometry of X in detail (see for example Manin [1986] or Reid [1988]). But the Chern class approach applies equally to results where no such analysis is available.

For example, we will see in Chapter 6 how to use the Chern class method to show that a general quintic threefold in $\mathbb{P}^4$ contains exactly 2875 lines (a computation that played an important role in the discovery of mirror symmetry; see for example Morrison [1993]), and that a general hypersurface of degree 37 in $\mathbb{P}^{20}$ contains exactly

$$47984924096538345636727806051910707603936407618172699855515$$

lines, a fact of no larger significance whatsoever.

5.2 Characterizing Chern classes

Let $\mathcal{E}$ be a vector bundle on a variety X of dimension n. We will introduce Chern classes $c_i(\mathcal{E}) \in A_{n-i}(X)$, extending the definition of $c_1(\mathcal{L})$ for a line bundle $\mathcal{L}$ in Section 1.4. As with our treatment of the intersection product, we will give an appealingly intuitive characterization rather than a proof of existence.

Recall that we defined the first Chern class $c_1(\mathcal{L})$ of a line bundle $\mathcal{L}$ on a variety X to be

$$c_1(\mathcal{L}) = [\mathrm{Div}(\tau)] \in A_{n-1}(X)$$

for any rational section τ of $\mathcal{L}$. We define $c_i(\mathcal{L}) = 0$ for all $i \geq 2$. In this section we will characterize Chern classes $c_i(\mathcal{E})$ for any vector bundle $\mathcal{E}$ and any integer $i \geq 0$.

We first sketch the situation in the case of a bundle $\mathcal{E}$ generated by its global sections (this circumstance is in fact the case in most of our applications, and in particular in the example of the 27 lines given above). Let $r = \mathrm{rank}\,\mathcal{E}$.

In the case $r = 1$ already treated, the class $c_1(\mathcal{L})$ may be regarded as a measure of nontriviality: if $c_1(\mathcal{L}) = 0$, then $\mathcal{L}$ has a nowhere-vanishing section, whence $\mathcal{L} \cong \mathcal{O}_X$. We extend this idea of measuring nontriviality using the idea of the "degeneracy locus" of a collection of sections — roughly, this is the locus where the sections become linearly dependent in the fibers of $\mathcal{E}$. To make the meaning precise, we use multilinear algebra.

The bundle $\mathcal{E}$ is trivial if and only if it has r everywhere-independent global sections $\tau_0, \ldots, \tau_{r-1}$; in this case, any set of r general sections will do. Thus a first measure of nontriviality is the locus where r general sections $\tau_0, \ldots, \tau_{r-1}$ are dependent. If we write $\tau : \mathcal{O}_X^r \to \mathcal{E}$ for the map sending the i-th basis vector to τ_i, then this is the locus where τ fails to be a surjection, or, equivalently, where the determinant of τ is zero. We can interpret this as the vanishing of a special section of an exterior power of $\mathcal{E}$: It is the zero scheme of the section

$$\tau_0 \wedge \cdots \wedge \tau_{r-1} \in \textstyle\bigwedge^r \mathcal{E}.$$

Since $\mathrm{rank}\,\mathcal{E} = r$, the bundle $\bigwedge^r \mathcal{E}$ has rank 1 and the class of the zero locus is by definition $c_1(\bigwedge^r \mathcal{E})$; this is a class in $A_{\dim X - 1}(X)$ depending only on the isomorphism class of $\mathcal{E}$. We call it the *first Chern class of* $\mathcal{E}$, written $c_1(\mathcal{E})$.

More generally, we can consider for any i the scheme where $r - i$ general sections of $\mathcal{E}$ fail to be independent, defined by the vanishing of

$$\tau_0 \wedge \cdots \wedge \tau_{r-i} \in \textstyle\bigwedge^{r-i+1} \mathcal{E}.$$

This is called the *degeneracy locus* of the sections $\tau_0, \ldots, \tau_{r-i}$. Since these degeneracy loci are central to our understanding of (and applications of) Chern classes, we should first say what we expect them to look like.

To see how this should go, consider first the "degeneracy locus of one section." A section τ of $\mathcal{E}$ is locally given by r functions $f_1, \ldots, f_r$, so that by the principal ideal theorem the codimension of each component of $V(\tau)$ is at most r. Moreover, if $\mathcal{E}$ is generated by global sections and τ is a general section, then the function f_{i+1} will not vanish identically on any component of the locus where $f_1, \ldots, f_i$ vanish, and it follows that every component of $V(\tau)$ has codimension exactly r. Under our standing assumption of characteristic 0, a version of Bertini's theorem tells us that $V(\tau)$ is reduced as well. (This may fail in characteristic p, for example in the case of a line bundle whose complete linear system defines an inseparable morphism.) It turns out that this is typical.

Lemma 5.2. *Suppose that $\mathcal{E}$ is a vector bundle of rank r on a variety X, and let i be an integer with $1 \leq i \leq r$. Let $\tau_0, \ldots, \tau_{r-i}$ be global sections of $\mathcal{E}$, and let $D = V(\tau_0 \wedge \cdots \wedge \tau_{r-i})$ be the degeneracy locus where they are dependent.*

(a) No component of D has codimension $> i$.

(b) If the τ_i are general elements of a vector space $W \subset H^0(\mathcal{E})$ of global sections generating $\mathcal{E}$, then D is generically reduced and has codimension i in X.

Proof: (a) This is Macaulay's "generalized unmixedness theorem." He proved it for the case of polynomial rings, and the general case was proved by Eagon and Northcott — see for example Eisenbud [1995, Exercise 10.9].

(b) Let W be an m-dimensional vector space of global sections of $\mathcal{E}$ that generate $\mathcal{E}$, and let $\varphi : X \to G(m - r, W)$ be the associated morphism sending $p \in X$ to the kernel of the evaluation map $W \to \mathcal{E}_p$. If $U \subset W$ is a subspace of dimension $r - i + 1$ spanned by $\tau_0, \ldots, \tau_{r-i}$, then the locus $V(\tau_0 \wedge \cdots \wedge \tau_{r-i}) \subset X$ is the preimage $\varphi^{-1}(\Sigma)$ of the Schubert cycle

$$\Sigma_i(U) = \{\Lambda \in G(m - r, W) \mid \Lambda \cap U \neq 0\}$$

of $(m - r)$-planes in W meeting U nontrivially. By Kleiman's theorem (Theorem 1.7), if $U \subset W$ is general this locus is generically reduced of codimension i. $\qquad \square$

We can now characterize the Chern classes $c_i(\mathcal{E}) \in A^i(X)$ for vector bundles $\mathcal{E}$ on smooth varieties X and integers $i \geq 0$:

Theorem 5.3. *There is a unique way of assigning to each vector bundle $\mathcal{E}$ on a smooth quasi-projective variety X a class $c(\mathcal{E}) = 1 + c_1(\mathcal{E}) + c_2(\mathcal{E}) + \cdots \in A(X)$ in such a way that:*

(a) (Line bundles) If $\mathcal{L}$ is a line bundle on X then the Chern class of $\mathcal{L}$ is $1 + c_1(\mathcal{L})$, where $c_1(\mathcal{L}) \in A^1(X)$ is the class of the divisor of zeros minus the divisor of poles of any rational section of $\mathcal{L}$.

(b) (Bundles with enough sections) If $\tau_0, \ldots, \tau_{r-i}$ are global sections of $\mathcal{E}$, and the degeneracy locus D where they are dependent has codimension i, then $c_i(\mathcal{E}) = [D] \in A^i(X)$.

(c) (Whitney's formula) If

$$0 \longrightarrow \mathcal{E} \longrightarrow \mathcal{F} \longrightarrow \mathcal{G} \longrightarrow 0$$

is a short exact sequence of vector bundles on X then

$$c(\mathcal{F}) = c(\mathcal{E})c(\mathcal{G}) \in A(X).$$

(d) (Functoriality) If $\varphi : Y \to X$ is a morphism of smooth varieties, then

$$\varphi^*(c(\mathcal{E})) = c(\varphi^*(\mathcal{E})).$$

Although we will not prove Theorem 5.3 completely, we will explain some parts of the proof in Section 5.9 below. We will see below that these properties make many Chern class computations easy. Here are two tastes:

Corollary 5.4 (Sums of line bundles). *If $\mathcal{E}$ is the direct sum of line bundles $\mathcal{L}_i$, or more generally has a filtration whose quotients are line bundles $\mathcal{L}_i$, then*

$$c(\mathcal{E}) = \prod c(\mathcal{L}_i) = \prod (1 + c_1(\mathcal{L}_i));$$

that is, $c_i(\mathcal{E})$ is the result of applying the i-th elementary symmetric function to the classes $c_1(\mathcal{L}_i)$.

Proof: This follows from a repeated application of Whitney's formula.   □

Corollary 5.5. *If $\mathcal{E}$ is a vector bundle on X of rank $> \dim X$, and $\mathcal{E}$ is generated by its global sections, then $\mathcal{E}$ has a nowhere-vanishing global section.*

Proof: By part (b), the degeneracy locus (vanishing locus) of one generic section has codimension $> \dim X$.   □

The strong Bertini theorem

Corollary 5.5 has an interesting geometric consequence in the following strengthening of Bertini's theorem:

Proposition 5.6 (Strong Bertini). *Let X be a smooth, n-dimensional quasi-projective variety, and $\mathcal{D}$ the linear system of divisors on X corresponding to the subspace $W \subset H^0(\mathcal{L})$ of sections of a line bundle $\mathcal{L}$. If the base locus of $\mathcal{D}$—that is, the scheme-theoretic intersection*

$$Z = \bigcap_{D \in \mathcal{D}} D$$

— is a smooth k-dimensional subscheme of X, and $k < n/2$, then the general member of the linear system $\mathcal{D}$ is smooth everywhere.

The inequality $k < n/2$ is sharp. For example, take $X = \mathbb{P}^4$ and $\mathcal{D}$ the linear system of all hypersurfaces of degree $d \geq 2$ containing a fixed 2-plane Z. If $Y = V(F) \subset \mathbb{P}^4$ is any hypersurface of degree $d > 1$ containing Z, then the three partial derivatives of F corresponding to the coordinates on Z are identically zero and the two remaining partial derivatives of F must have a common zero somewhere along Z; thus Y is singular at some point of Z. For an extension of this example, see Exercise 5.45.

Proof of Proposition 5.6: To begin with, the classical Bertini theorem tells us that the general member D of the linear system $\mathcal{D}$ is smooth away from Z.

To see that it is also smooth along Z, suppose that D is the zero locus of a general section $\sigma \in W \subset H^0(\mathcal{L})$. Since σ vanishes on Z, it gives rise to a section $d\sigma$ of the tensor product $\mathcal{N}^*_{Z/X} \otimes \mathcal{L}$ of the conormal bundle $\mathcal{N}^*_{Z/X} = \mathcal{I}_{Z/X}/\mathcal{I}^2_{Z/X}$ with the line bundle $\mathcal{L}$; we can think of $d\sigma$ as the differential of σ along Z. The hypothesis that the sections $\sigma \in W$ generate the sheaf $\mathcal{I}_{Z/X} \otimes \mathcal{L}$, together with Lemma 5.2 and the fact that $\dim Z = k < n - k = \operatorname{rank}(\mathcal{N}^*_{Z/X})$, shows that $d\sigma$ is nowhere zero. □

5.3 Constructing Chern classes

A construction of Chern classes for a bundle of rank r that is generated by global sections is implicit in Theorem 5.3 (b): $c_i(\mathcal{E})$ is the degeneracy locus of $r - i + 1$ general global sections. An alternative way of stating the same thing is often useful. We have already proved this in Lemma 5.2, but it is worth stating it here explicitly:

Proposition 5.7. *Let $\mathcal{E}$ be a vector bundle of rank r on the smooth, quasi-projective variety X, and let $W \subset H^0(\mathcal{E})$ be an m-dimensional vector space of sections generating $\mathcal{E}$. If $\varphi : X \to G(m - r, W)$ denotes the associated morphism sending $p \in X$ to the kernel of the evaluation map $W \to \mathcal{E}_p$, then the i-th Chern class $c_i(\mathcal{E})$ is the pullback*

$$c_i(\mathcal{E}) = \varphi^*(\sigma_i)$$

of the Schubert class $\sigma_i \in A^i(G(m - r, W))$.

This allows us to construct Chern classes for globally generated bundles, and we will see in Section 5.9.1 how to prove basic facts about Chern classes, such as Whitney's formula, from this construction. To construct Chern classes for arbitrary bundles we use a different technique, the projectivization of a vector bundle. We will have much more to say about this construction in Chapter 9; for now we will simply state what is necessary to construct the Chern classes and to make use of a fundamental tool for computing with Chern classes, introduced in Section 5.4: the "splitting principle."

Definition 5.8. Let X be a scheme, and let $\mathcal{E}$ be a vector bundle of rank $r + 1$ on X. By the *projectivization* of E we will mean the natural morphism

$$\pi_{\mathcal{E}} : \mathbb{P}\mathcal{E} := \mathrm{Proj}(\mathrm{Sym}\,\mathcal{E}^*) \to X.$$

By a *projective bundle* over X we mean a morphism $\pi : Y \to X$ that can be realized as $\pi_{\mathcal{E}}$ for some vector bundle $\mathcal{E}$ over X.

Thus the closed points of $\mathbb{P}\mathcal{E}$ correspond to pairs (x, ξ) with $x \in X$ and ξ a one-dimensional subspace $\xi \subset \mathcal{E}_x$ of the fiber $\mathcal{E}_x$ of $\mathcal{E}$. Ordinary projective space is of course the special case in which X is a point and $\mathcal{E}$ is a vector space.

The bundle $\pi : \mathbb{P}\mathcal{E} \to X$ comes equipped with a tautological line bundle

$$\mathcal{S}_{\mathcal{E}} := \mathcal{O}_{\mathbb{P}\mathcal{E}}(-1) \subset \pi^*\mathcal{E},$$

constructed as the sheafification of the graded $\mathrm{Sym}\,\mathcal{E}^*$-module obtained by shifting the grading by -1, just as in the case of ordinary projective space.

Here is the result about projectivized vector bundles that serves to define the Chern classes in general:

Theorem 5.9. *Let $\mathcal{E}$ be a vector bundle of rank r on a smooth variety X, and let $\pi : \mathbb{P}\mathcal{E} \to X$ be the projectivized vector bundle. Let ζ be the first Chern class of the dual $\mathcal{S}_{\mathcal{E}}^*$ of the tautological bundle $\mathcal{S}_{\mathcal{E}}$ on $\mathbb{P}\mathcal{E}$.*

(a) The flat pullback map $\pi^* : A(X) \to A(\mathbb{P}\mathcal{E})$ is injective.

(b) The element $\zeta \in A(\mathbb{P}\mathcal{E})$ satisfies a unique monic polynomial $f(\zeta)$ of degree r with coefficients in $\pi^*(A(X))$.

Definition 5.10. Let $\mathcal{E}$ be a vector bundle of rank r on a smooth variety X. The Chern classes $c_i(\mathcal{E})$ are the unique elements of $A(X)$ such that

$$f(\zeta) = \zeta^r + \pi^*c_1(\mathcal{E})\zeta^{r-1} + \cdots + \pi^*c_r(\mathcal{E});$$

that is,

$$A(\mathbb{P}\mathcal{E}) = A(X)[\zeta]/(f(\zeta)).$$

In fact, this definition of Chern classes may be extended to singular varieties, as in Fulton [1984, Chapter 3], and this is a crucial element of the intersection theory of singular varieties: as we have seen (Example 2.22), it is simply not possible to define products of arbitrary classes on singular varieties in general, but it is possible to define products with Chern classes of a vector bundle by restricting the vector bundle. For a proof of Theorem 5.9 in the smooth case, see Theorem 9.6; for the proof in general, see Fulton [1984, Chapter 3].

5.4 The splitting principle

For more complicated examples, we will use Whitney's formula in conjunction with a result called the *splitting principle*, which may be stated as:

Theorem 5.11 (Splitting principle). *Any identity among Chern classes of bundles that is true for bundles that are direct sums of line bundles is true in general.*

This remarkable result is really a corollary of the construction of projectivized vector bundles, applied via the next result:

Lemma 5.12 (Splitting construction). *Let X be any smooth variety and $\mathcal{E}$ a vector bundle of rank r on X. There exists a smooth variety Y and a morphism $\varphi : Y \to X$ with the following two properties:*

(a) The pullback map $\varphi^* : A(X) \to A(Y)$ is injective.
(b) The pullback bundle $\varphi^*\mathcal{E}$ on Y admits a filtration

$$0 = \mathcal{E}_0 \subset \mathcal{E}_1 \subset \cdots \subset \mathcal{E}_{r-1} \subset \mathcal{E}_r = \varphi^*\mathcal{E}$$

by vector subbundles $\mathcal{E}_i \subset \varphi^*\mathcal{E}$ with successive quotients $\mathcal{E}_i/\mathcal{E}_{i-1}$ locally free of rank 1.

Proof: We may construct $\varphi : Y \to X$ by iterating the projectivized vector bundle construction: First, on $Y_1 := \mathbb{P}\mathcal{E}$ we have a tautological subbundle $\mathcal{S}_1 \subset \pi_{\mathcal{E}}^*(\mathcal{E})$. Writing $\mathcal{Q}_1$ for the quotient, we next construct $Y_2 := \mathbb{P}\mathcal{Q}_1$. On Y_2 we have exact sequences

$$0 \longrightarrow \pi_{\mathcal{Q}_1}^*(\mathcal{S}_1) \subset \pi_{\mathcal{Q}_1}^* \pi_{\mathcal{E}}^* \mathcal{E} \longrightarrow \pi_{\mathcal{Q}_1}^* \mathcal{Q}_1 \longrightarrow 0$$

and

$$0 \longrightarrow \mathcal{S}_2 \longrightarrow \pi_{\mathcal{Q}_1}^* \mathcal{Q}_1 \longrightarrow \mathcal{Q}_2 \longrightarrow 0.$$

Continuing this way for $r - 1$ steps we get a space $Y := Y_r$ such that the pullback of $\mathcal{E}$ to Y admits a filtration whose successive quotients are line bundles.

Finally, by Theorem 5.9, there is a class $\zeta \in A^1(\mathbb{P}\mathcal{E})$ in the Chow ring of any projective bundle $\pi : \mathbb{P}\mathcal{E} \to X$ that restricts to the hyperplane class on each fiber. By the push-pull formula, if $\mathcal{E}$ has rank r then for any class $\alpha \in A(X)$ we have

$$\pi_*(\zeta^{r-1}\pi^*\alpha) = \alpha,$$

from which we see that the pullback map $\pi^* : A(X) \to A(\mathbb{P}\mathcal{E})$ is injective. $\qquad\square$

We will study projective bundles much more extensively in Chapter 9; in particular, we give a more fleshed out version of this argument in the proof of Lemma 9.7.

Proof of Theorem 5.11: With notation as in the theorem, we can use Whitney's formula (part (c) of Theorem 5.3) and our a priori definition of the Chern class of a line bundle to describe the Chern class of the pullback:

$$c(\varphi^* \mathcal{E}) = \prod_{i=1}^{r} c(\mathcal{E}_i / \mathcal{E}_{i-1});$$

by the first part of the lemma, this determines the Chern classes of $\mathcal{E}$.　　□

5.5 Using Whitney's formula with the splitting principle

We will now illustrate the use of Whitney's formula with the splitting principle.

A first consequence is that the Chern classes of a bundle vanish above the rank, something we saw already in the case of bundles with enough sections.

Example 5.13 (Vanishing). If $\mathcal{E}$ is a vector bundle of rank r, then $c_i(\mathcal{E}) = 0$ for $i > r$. Reason: If $\mathcal{E}$ split as $\bigoplus_{i=1}^{r} \mathcal{L}_i$ for line bundles $\mathcal{L}_i$ then, since $c(\mathcal{L}_i) = 1 + c_1(\mathcal{L}_i)$, Whitney's formula would imply that

$$c(\mathcal{E}) = \prod_{i=1}^{r} (1 + c_1(\mathcal{L}_i)),$$

which has no terms of degree $> r$.

Example 5.14 (Duals). If $\mathcal{E} = \bigoplus \mathcal{L}_i$, then

$$c(\mathcal{E}^*) = \prod (1 + c_1(\mathcal{L}_i^*)) = \prod (1 - c_1(\mathcal{L}_i)),$$

since $c_1(\mathcal{L}^*) = -c_1(\mathcal{L})$ when $\mathcal{L}$ is a line bundle. Given this, Whitney's formula gives us the basic identity

$$c_i(\mathcal{E}^*) = (-1)^i c_i(\mathcal{E}).$$

By the splitting principle, this identity holds for any bundle.

Example 5.15 (Determinant of a bundle). By the *determinant* $\det \mathcal{E}$ of a bundle $\mathcal{E}$ we mean the line bundle that is the highest exterior power $\det \mathcal{E} := \bigwedge^{\text{rank} \mathcal{E}} \mathcal{E}$. We have already observed that if $\mathcal{E}$ is globally generated, then $c_1(\det \mathcal{E}) = c_1(\mathcal{E})$; this was one of our motivating examples. The splitting principle and Whitney's formula allow us to deduce this for arbitrary bundles: If we assume that $\mathcal{E} = \bigoplus \mathcal{L}_i$, then $\det \mathcal{E} = \bigotimes \mathcal{L}_i$ and hence

$$c_1(\det \mathcal{E}) = \sum c_1(\mathcal{L}_i) = c_1(\mathcal{E});$$

the splitting principle tells us this identity holds in general.

Example 5.16 (Symmetric squares). Suppose that $\mathcal{E}$ is a bundle of rank 2. If $\mathcal{E}$ splits as a direct sum $\mathcal{E} = \mathcal{L} \oplus \mathcal{M}$ of line bundles $\mathcal{L}$ and $\mathcal{M}$ with Chern classes $c_1(\mathcal{L}) = \alpha$ and $c_1(\mathcal{M}) = \beta$ then, by Whitney's formula, $c(\mathcal{E}) = (1 + \alpha)(1 + \beta)$, whence

$$c_1(\mathcal{E}) = \alpha + \beta \quad \text{and} \quad c_2(\mathcal{E}) = \alpha\beta.$$

Further, we would have

$$\mathrm{Sym}^2 \mathcal{E} = \mathcal{L}^{\otimes 2} \oplus (\mathcal{L} \otimes \mathcal{M}) \oplus \mathcal{M}^{\otimes 2},$$

from which we would deduce

$$\begin{aligned} c(\mathrm{Sym}^2 \mathcal{E}) &= (1 + 2\alpha)(1 + \alpha + \beta)(1 + 2\beta) \\ &= 1 + 2(\alpha + \beta) + (2\alpha^2 + 8\alpha\beta + 2\beta^2) + 4\alpha\beta(\alpha + \beta). \end{aligned}$$

This expression may be rewritten in a way that involves only the Chern classes of $\mathcal{E}$: As the reader may immediately check, it is equal to

$$1 + 2c_1(\mathcal{E}) + \left(2c_1(\mathcal{E})^2 + 4c_2(\mathcal{E})\right) + 4c_1(\mathcal{E})c_2(\mathcal{E}).$$

By the splitting principle, this is a valid expression for $c(\mathrm{Sym}^2 \mathcal{E})$ whether or not $\mathcal{E}$ actually splits.

We could use the same method to give formulas for the Chern classes of any symmetric or exterior power — or of any multilinear functor — applied to vector bundles whose Chern classes we know.

Together, the splitting principle and Whitney's formula give a powerful tool for calculating Chern classes, as we will see over and over in the remainder of this text; see Exercises 5.30–5.35 for more examples.

5.5.1 Tensor products with line bundles

As an application of the splitting principle, we will derive the relation between the Chern classes of a vector bundle $\mathcal{E}$ of rank r on a variety X and the Chern classes of the tensor product of $\mathcal{E}$ with a line bundle $\mathcal{L}$.

To do this, we start by assuming that $\mathcal{E}$ splits as a direct sum of line bundles

$$\mathcal{E} = \bigoplus_{i=1}^{r} \mathcal{M}_i;$$

let $\alpha_i = c_1(\mathcal{M}_i) \in A^1(X)$ be the first Chern class of $\mathcal{M}_i$, so that

$$c(\mathcal{E}) = \prod_{i=1}^{r}(1 + \alpha_i).$$

In other words, the elementary symmetric polynomials in the α_i are the Chern classes of $\mathcal{E}$:

$$\alpha_1 + \alpha_2 + \cdots + \alpha_r = c_1(\mathcal{E}),$$

$$\sum_{1 \le i < j \le r} \alpha_i \alpha_j = c_2(\mathcal{E}),$$

$$\vdots$$

$$\alpha_1 \alpha_2 \ldots \alpha_r = c_r(\mathcal{E}).$$

Now let $\beta = c_1(\mathcal{L})$ be the first Chern class of $\mathcal{L}$. Since

$$\mathcal{E} \otimes \mathcal{L} = \bigoplus_{i=1}^{r} \mathcal{M}_i \otimes \mathcal{L},$$

we have, by Whitney's formula,

$$c(\mathcal{E} \otimes \mathcal{L}) = \prod_{i=1}^{r}(1 + \alpha_i + \beta). \tag{5.1}$$

Now, we can express the product on the right as a polynomial in β and the elementary symmetric polynomials in the α_i: For example, we have

$$c_1(\mathcal{E} \otimes \mathcal{L}) = \sum_{i=1}^{r}(\alpha_i + \beta) = c_1(\mathcal{E}) + rc_1(\mathcal{L}),$$

and likewise

$$c_2(\mathcal{E} \otimes \mathcal{L}) = \sum_{1 \le i < j \le r}(\alpha_i + \beta)(\alpha_j + \beta)$$

$$= \sum_{1 \le i < j \le r} \alpha_i \alpha_j + (r-1)\beta \sum_{i=1}^{r} \alpha_i + \binom{r}{2}\beta^2$$

$$= c_2(\mathcal{E}) + (r-1)c_1(\mathcal{E})c_1(\mathcal{L}) + \binom{r}{2}c_1(\mathcal{L})^2,$$

and so on. In general, we have:

Proposition 5.17. *If $\mathcal{E}$ is a vector bundle of rank r and $\mathcal{L}$ is a line bundle, then*

$$c_k(\mathcal{E} \otimes \mathcal{L}) = \sum_{l=0}^{k}\binom{r-l}{k-l}c_1(\mathcal{L})^{k-l}c_l(\mathcal{E})$$

$$= \sum_{i=0}^{k}\binom{r-k+i}{i}c_1(\mathcal{L})^i c_{k-i}(\mathcal{E}).$$

Proof: This is just a matter of collecting the terms of degree l in the α_i and degree $k - l$ in β in the expression (5.1): we write

$$\prod_{i=1}^{r}(1 + \alpha_i + \beta) = \sum_{1 \le i_1 < \cdots < i_l \le r} (1 + \beta)^{r-l} \alpha_{i_1} \cdots \alpha_{i_l}$$

$$= \sum_{l} c_l(\mathcal{E})(1 + \beta)^{r-l},$$

and the proposition follows. $\qquad\square$

5.5.2 Tensor product of two bundles

Whitney's formula and the splitting principle yield a formula for the Chern class of the tensor product of two bundles of any rank. But, as we will see in Exercises 5.35–5.36, the formula in general is quite complicated. Special cases, however, are amenable to explicit calculation; for example, we can handle the case of the first Chern class $c_1(\mathcal{E} \otimes \mathcal{F})$:

Proposition 5.18. *If $\mathcal{E}, \mathcal{F}$ are vector bundles of ranks e and f respectively, then*

$$c_1(\mathcal{E} \otimes \mathcal{F}) = f \cdot c_1(\mathcal{E}) + e \cdot c_1(\mathcal{F}).$$

Proof: Suppose $\mathcal{E} = \bigoplus \mathcal{L}_i$ and $\mathcal{F} = \bigoplus \mathcal{M}_i$ are direct sums of line bundles, so that we can write

$$c(\mathcal{E}) = \prod_{i=1}^{e}(1 + \alpha_i) \quad \text{and} \quad c(\mathcal{F}) = \prod_{j=1}^{f}(1 + \beta_j)$$

with $c_1(\mathcal{L}_i) = \alpha_i$ and $c_1(\mathcal{M}_j) = \beta_j$; note that $c_1(\mathcal{E}) = \alpha_1 + \cdots + \alpha_e$ and $c_1(\mathcal{F}) = \beta_1 + \cdots + \beta_f$. We have then

$$\mathcal{E} \otimes \mathcal{F} = \bigoplus_{i,j=1,1}^{e,f} \mathcal{L}_i \otimes \mathcal{M}_j,$$

and correspondingly

$$c(\mathcal{E} \otimes \mathcal{F}) = \prod_{i,j=1,1}^{e,f} (1 + \alpha_i + \beta_j).$$

In particular, this gives

$$c_1(\mathcal{E} \otimes \mathcal{F}) = \sum_{i,j=1,1}^{e,f} (\alpha_i + \beta_j)$$

$$= f \sum_{i=1}^{e} \alpha_i + e \sum_{j=1}^{f} \beta_j$$

$$= f c_1(\mathcal{E}) + e c_1(\mathcal{F}). \qquad\square$$

There is one other case in which we can give a closed-form expression for a Chern class of a general tensor product: We will see, in Chapter 12, a formula for the top Chern class $c_{ef}(\mathcal{E} \otimes \mathcal{F})$ of a tensor product of bundles of ranks e and f.

There is also a different approach that allows us to express the characteristic classes of a general tensor product more comprehensibly: The *Chern character* $\mathrm{Ch}(\mathcal{E})$ of a vector bundle $\mathcal{E}$ is a certain formal power series in the Chern classes of $\mathcal{E}$, with rational coefficients, that satisfies the attractive formulas

$$\mathrm{Ch}(\mathcal{E} \oplus \mathcal{F}) = \mathrm{Ch}(\mathcal{E}) + \mathrm{Ch}(\mathcal{F}),$$
$$\mathrm{Ch}(\mathcal{E} \otimes \mathcal{F}) = \mathrm{Ch}(\mathcal{E}) \cdot \mathrm{Ch}(\mathcal{F}).$$

See Section 14.2.1 for more information.

5.6 Tautological bundles

We have seen how the splitting principle, in conjunction with Whitney's formula, allows us to express the Chern classes of bundles in terms of simpler ones. To apply this, of course, we need to have a roster of basic bundles whose Chern classes we know; in this section we will calculate the Chern classes of some of these.

5.6.1 Projective spaces

We start with the most basic of all bundles: the bundle $\mathcal{O}_{\mathbb{P}^r}(1)$ on projective space $\mathbb{P}^r$. We have

$$c_1(\mathcal{O}_{\mathbb{P}^r}(1)) = \zeta \in A^1(\mathbb{P}^r),$$

where ζ is the hyperplane class; similarly,

$$c_1(\mathcal{O}_{\mathbb{P}^r}(n)) = n \cdot \zeta \in A^1(\mathbb{P}^r)$$

for any $n \in \mathbb{Z}$.

This in turn allows us to compute the Chern class of the universal quotient bundle $\mathcal{Q}$ on $\mathbb{P}^r$: If $\mathbb{P}^r = \mathbb{P}V$, from the exact sequence

$$0 \longrightarrow \mathcal{S} = \mathcal{O}_{\mathbb{P}^r}(-1) \longrightarrow V \otimes \mathcal{O}_{\mathbb{P}^r} \longrightarrow \mathcal{Q} \longrightarrow 0,$$

we have

$$c(\mathcal{Q}) = \frac{1}{c(\mathcal{O}_{\mathbb{P}^r}(-1))} = \frac{1}{1 - \zeta} = 1 + \zeta + \zeta^2 + \cdots + \zeta^r.$$

Note that we could also arrive at this directly from the description of Chern classes as degeneracy loci of sections: An element $v \in V$ gives rise to a global section σ of the bundle $\mathcal{Q}$; given k elements $v_1, \ldots, v_k \in V$, the corresponding sections $\sigma_1, \ldots, \sigma_k$

of Q will be linearly dependent at a point $x \in \mathbb{P}^r$ exactly when x lies in the $\mathbb{P}^{k-1}$ corresponding to the subspace $W = \langle v_1, \ldots, v_k \rangle \subset V$ spanned by the v_i. Thus

$$c_{r-k+1}(Q) = [\mathbb{P}^{k-1}] = \zeta^{r-k+1} \in A^{r-k+1}(\mathbb{P}^r).$$

5.6.2 Grassmannians

Let us consider next the case of the Grassmannian $G = G(k, n)$ of k-planes in an n-dimensional vector space V, and its universal sub and quotient bundles S and Q.

We will start with Q, since this bundle is globally generated, so that we can determine its Chern classes directly as degeneracy loci. Specifically, elements $v \in V$ give rise to sections σ of Q simply by taking their images in each quotient of V; that is, for a k-plane $\Lambda \subset V$, we set

$$\sigma(\Lambda) = \bar{v} \in V/\Lambda.$$

Now, given a collection $v_1, \ldots, v_m \in V$, the corresponding sections will fail to be independent at a point $\Lambda \in G$ exactly when the corresponding $\bar{v}_i \in V/\Lambda$ are dependent, which is to say when Λ intersects the span $W = \langle v_1, \ldots, v_m \rangle \subset V$ in a nonzero subspace — that is, when

$$\mathbb{P}\Lambda \cap \mathbb{P}W \neq \varnothing.$$

We may recognize this locus as the Schubert cycle $\Sigma_{n-k-m+1}(W)$, from which we conclude that the Chern class of Q is the sum

$$c(Q) = 1 + \sigma_1 + \sigma_2 + \cdots + \sigma_{n-k}.$$

Unlike Q, the universal subbundle S does not have nonzero global sections, so we cannot use the characterization of Chern classes as degeneracy loci. But the dual bundle S^* does: If $l \in V^*$ is a linear form, we can define a section τ of S^* by restricting l to each k-plane $\Lambda \subset V$ in turn; in other words, we set

$$\tau(\Lambda) = l|_\Lambda.$$

Now, if we have m independent linear forms $l_1, \ldots, l_m \in V^*$, the corresponding sections of S^* will fail to be independent at the point $\Lambda \in G$ — that is, some linear combination of the l_i will vanish identically on Λ — exactly when Λ fails to intersect the common zero locus U of the l_i properly, that is, when

$$\dim(\mathbb{P}\Lambda \cap \mathbb{P}U) \geq k - m.$$

Again, this locus is a Schubert cycle in G, specifically the cycle $\Sigma_{1,1,\ldots,1}(U)$, and we conclude that

$$c(S^*) = 1 + \sigma_1 + \sigma_{1,1} + \cdots + \sigma_{1,1,\ldots,1};$$

from this we can deduce in turn that

$$c(\mathcal{S}) = 1 - \sigma_1 + \sigma_{1,1} + \cdots + (-1)^k \sigma_{1,1,\ldots,1}.$$

Note that this description of $c(\mathcal{S})$ can also be deduced from our knowledge of $c(\mathcal{Q})$ and Corollary 4.10.

5.7 Chern classes of varieties

The most important vector bundles on a smooth variety X are its tangent bundle $\mathcal{T}_X$ and its dual, the cotangent bundle Ω_X. Their Chern classes are so important in geometry that the Chern class of the tangent bundle is usually just called the *Chern class of X*.

For example, if X is a smooth curve then its tangent bundle is a line bundle, so its Chern class has the form $1 + c_1(\mathcal{T}_X)$. Here $c_1(\mathcal{T}_X) = -c_1(\Omega_X)$ is the anticanonical class, whose degree is $2 - 2g$, where g is the genus of X. In general, if X is a smooth complex projective manifold of dimension n then Theorem 5.21 below says that $\deg c_n(\mathcal{T}_X)$ is the topological Euler characteristic of X.

5.7.1 Tangent bundles of projective spaces

We start by calculating the Chern classes of the tangent bundle $\mathcal{T}_{\mathbb{P}^n}$ of projective space. This is straightforward, given the Euler sequence of Section 3.2.4: We have

$$0 \longrightarrow \mathcal{O}_{\mathbb{P}^n} \longrightarrow \mathcal{O}_{\mathbb{P}^n}(1)^{n+1} \longrightarrow \mathcal{T}_{\mathbb{P}^n} \longrightarrow 0$$

and hence

$$c(\mathcal{T}_{\mathbb{P}^n}) = (1 + \zeta)^{n+1},$$

where $\zeta \in A^1(\mathbb{P}^n)$ is the hyperplane class.

We could also derive this from the identification $\mathcal{T} = \mathrm{Hom}(\mathcal{S}, \mathcal{Q}) = \mathcal{S}^* \otimes \mathcal{Q}$, where $\mathcal{S} = \mathcal{O}_{\mathbb{P}^n}(-1)$ and $\mathcal{Q}$ are the universal sub and quotient bundles, by applying Proposition 5.17.

Note that this calculation implies the algebraic/projective version of the "hairy coconut" theorem: Since $c_n(\mathcal{T}_{\mathbb{P}^n}) = (n+1)\zeta^n \neq 0$, there does not exist a nowhere-zero vector field on $\mathbb{P}^n$.

5.7.2 Tangent bundles to hypersurfaces

We can combine the formula above for the Chern classes of the tangent bundle to projective space $\mathbb{P}^r$ and Whitney's formula to calculate the Chern classes of the tangent bundle to a smooth hypersurface $X \subset \mathbb{P}^n$ of degree d.

To do this, we use the standard normal bundle sequence

$$0 \longrightarrow \mathcal{T}_X \longrightarrow \mathcal{T}_{\mathbb{P}^n}|_X \longrightarrow \mathcal{N}_{X/\mathbb{P}^n} \longrightarrow 0$$

and the identification

$$\mathcal{N}_{X/\mathbb{P}^n} = \mathcal{O}_{\mathbb{P}^n}(X)|_X = \mathcal{O}_X(d)$$

established in Section 1.4.2. Letting ζ_X denote the restriction to X of the hyperplane class on $\mathbb{P}^n$, we can write

$$\begin{aligned} c(\mathcal{T}_X) &= \frac{c(\mathcal{T}_{\mathbb{P}^n}|_X)}{\mathcal{N}_{X/\mathbb{P}^n}} \\ &= \frac{(1 + \zeta_X)^{n+1}}{1 + d\zeta_X} \\ &= \left(1 + (n+1)\zeta_X + \binom{n+1}{2}\zeta_X^2 + \cdots\right)(1 - d\zeta_X + d^2\zeta_X^2 + \cdots). \end{aligned}$$

We can generalize this calculation to complete intersections:

Example 5.19 (Chern classes of complete intersections). Suppose that

$$X = Z_1 \cap \cdots \cap Z_k \subset \mathbb{P}^n$$

is the complete intersection of k hypersurfaces of degrees $d_1, \ldots, d_k$ defined by forms F_i of degrees d_i. The relations among the F_i are generated by the Koszul relations $F_j F_i - F_i F_j = 0$. This means that if we restrict to Y, where the F_i vanish, we get

$$\mathcal{I}_{Y/X}/\mathcal{I}_{Y/X}^2 = \mathcal{I}_{Y/X}|_Y = \bigoplus_i \mathcal{O}_Y(-d_i),$$

so the normal bundle $\mathcal{N} = \mathcal{N}_{X/\mathbb{P}^n}$ of X in $\mathbb{P}^n$ is a direct sum $\mathcal{N} = \bigoplus \mathcal{O}_X(d_i)$. Applying Whitney's formula, we get

$$c(\mathcal{T}_X) = \frac{(1 + \zeta_X)^{n+1}}{\prod(1 + d_i\zeta_X)}.$$

5.7.3 The topological Euler characteristic

Recall that the *topological Euler characteristic* of a manifold M is by definition $\chi_{\text{top}}(M) := \sum(-1)^i \dim_{\mathbb{Q}} H^i(M; \mathbb{Q})$, where $H^i(M; \mathbb{Q})$ is the singular cohomology group. When M is a smooth projective variety over $\mathbb{C}$, it may be regarded as a manifold with respect to the classical, or analytic, topology, so $\chi_{\text{top}}(M)$ makes sense in this case.

Theorem 5.20 (Poincaré–Hopf theorem). *If M is a smooth compact orientable manifold and σ is a vector field with isolated zeros, then*

$$\chi_{\text{top}}(M) = \sum_{\{x \mid \sigma(x)=0\}} \text{index}_x(\sigma).$$

A beautiful account of this classic result can be found in Milnor [1997]. Now suppose that M is a smooth complex projective variety. If the tangent bundle $\mathcal{T}_X$ is generated by global sections, then it has a section σ that vanishes at only finitely many points, and vanishes simply there. Since this section is represented locally by complex analytic functions, its index at each of its zeros will be 1, and we may replace the sum in the Poincaré–Hopf theorem by the number of its zeros — in other words, the degree of the top Chern class of $\mathcal{T}_X$. An elementary topological argument (see, for example, Chapter 3 of Griffiths and Harris [1994]) shows that this is true more generally:

Theorem 5.21. *If X is a smooth n-dimensional projective variety, then*

$$\chi_{\text{top}}(X) = \deg c_n(\mathcal{T}_X).$$

Example 5.22 (Euler characteristic of $\mathbb{P}^n$). Since $c(\mathcal{T}_{\mathbb{P}^n}) = (1 + \zeta)^{n+1}$, where $\zeta = c_1(\mathcal{O}_{\mathbb{P}^n}(1))$ is the class of a hyperplane, we deduce that

$$\chi_{\text{top}}(\mathbb{P}^n) = \deg(c_n(\mathcal{T}_{\mathbb{P}^n})) = n + 1.$$

Of course this is immediate from the fact that $H^{2i}(\mathbb{P}^n, \mathbb{Q}) = \mathbb{Q}$ for $i = 0, \ldots, n$ while $H^{2i+1}(\mathbb{P}^n, \mathbb{Q}) = 0$ for all i.

Example 5.23 (Blow-up of a surface). Sometimes one can use Theorem 5.21 to compute a Chern class. For example, the blow-up Y of a complex surface X at a point p can be described topologically as the union of $X \setminus D$ with a tubular neighborhood of the exceptional curve, which is a copy of $\mathbb{P}^1$. Thus

$$\chi_{\text{top}}(X) = \chi_{\text{top}}(X) - \chi_{\text{top}}(p) + \chi_{\text{top}}(\mathbb{P}^1) = \chi_{\text{top}}(X) - 1 + 2 = \chi_{\text{top}}(X) + 1,$$

and we deduce that $\deg c_2(\mathcal{T}_Y) = \deg c_2(\mathcal{T}_X) + 1$. (One can generalize this formula algebraically, and identify the class $c(\mathcal{T}_Y)$, by using the Chern classes of coherent sheaves that are not vector bundles; see for example Section 14.2.1, and, for the computation, Fulton [1984, Section 15.4].)

Example 5.24 (Euler characteristic of a hypersurface). Now let X be a smooth hypersurface of degree d in $\mathbb{P}^n$. From the normal bundle sequence

$$0 \longrightarrow \mathcal{T}_X \longrightarrow \mathcal{T}_{\mathbb{P}^n}|_X \longrightarrow \mathcal{N}_{X/\mathbb{P}^n} \longrightarrow 0$$

and the fact that $\mathcal{N}_{X/\mathbb{P}^n} \cong \mathcal{O}_X(d)$, we have

$$c(\mathcal{T}_X) = \frac{(1 + \zeta_X)^{n+1}}{(1 + d\zeta_X)} = ((1 + \zeta_X)^{n+1})(1 - d\zeta_X + d^2(\zeta_X)^2 + \cdots).$$

Taking the component of degree $\dim X = n - 1$, we get

$$c_{n-1}(\mathcal{T}_X) = \sum_{i=0}^{n-1} (-1)^i \binom{n+1}{n-1-i} d^i \zeta_X^{n-1}.$$

Since the degree of ζ_X^{n-1} is the number of points of intersection of $n - 1$ general hyperplanes on the $(n - 1)$-dimensional variety X, we have $\zeta_X^{n-1} = d$. Thus, finally,

$$\chi_{\mathrm{top}}(X) = \deg(c_{n-1}(\mathcal{T}_X)) = \sum_{i=0}^{n-1} (-1)^i \binom{n+1}{n-1-i} d^{i+1}.$$

We can get still more from this formula: The Lefschetz hyperplane theorem (see Section C.4) tells us that the integral cohomology groups of X are all equal to the corresponding cohomology groups of projective space, except for the middle one $H^{n-1}(X)$; that is, the Betti numbers $b_i = \dim_{\mathbb{Q}} H^i(M; \mathbb{Q})$ other than b_{n-1} are 1 in even dimensions and 0 in odd. (In fact, the analogous statement is true for any smooth complete intersection: All the cohomology groups except the middle are equal to those of projective space.) Thus the Euler characteristic determines the middle Betti number b_{n-1}. In Table 5.1, we give the results of this calculation in a few of the cases where it is most frequently used.

hypersurface	χ	b_{n-1}
quadric surface	4	2
cubic surface	9	7
quartic surface	24	22
quintic surface	55	53
quadric threefold	4	0
cubic threefold	−6	10
quartic threefold	−56	60
quintic threefold	−200	204
quadric fourfold	6	2
cubic fourfold	27	23

Table 5.1 Euler characteristics of favorite hypersurfaces.

It is interesting to compare this computation with what we already knew: A smooth quadric surface in $\mathbb{P}^3$ is isomorphic to $\mathbb{P}^1 \times \mathbb{P}^1$, from which we can see directly both the Euler characteristic and the second Betti number; a smooth cubic surface in $\mathbb{P}^3$ is the blow-up of $\mathbb{P}^2$ at six points, so the Euler characteristic is $3 + 6$; the quadric fourfold may also be viewed as the Plücker embedding of the Grassmannian $\mathbb{G}(1, 3)$, whose cohomology has as basis its six Schubert cycles, and whose middle cohomology in particular has basis given by the two Schubert cycles $\sigma_{1,1}$ and σ_2.

5.7.4 First Chern class of the Grassmannian

In theory, we should be able to use the identification of the Chern classes $c(\mathcal{S})$ and $c(\mathcal{Q})$ to derive the Chern class of the tangent bundle $\mathcal{T}_G$, which by Theorem 3.5 is isomorphic to $\operatorname{Hom}(\mathcal{S}, \mathcal{Q}) = \mathcal{S}^* \otimes \mathcal{Q}$. In general, unfortunately, this knowledge remains theoretical: As we indicated in Section 5.5.2, the formula for the Chern class of the tensor product of two bundles of higher rank is complicated. But we can at least use Proposition 5.18 to give the first Chern class $c_1(\mathcal{T}_G)$; since $c_1(\mathcal{S}^*) = c_1(\mathcal{Q}) = \sigma_1$, we have:

Proposition 5.25. *The first Chern class of the tangent bundle of the Grassmannian $G = G(k,n)$ is*

$$c_1(\mathcal{T}_G) = n \cdot \sigma_1.$$

We see from this also that the canonical class K_G of G is $-n\sigma_1$. Note that this agrees with our prior calculations in the case $k = 1$ of projective space $\mathbb{P}^{n-1}$, and in the case $k = 2$ and $n = 4$, where the Grassmannian $G(2,4)$ may be realized as a quadric hypersurface in $\mathbb{P}^5$ and we can apply the results of Section 5.7.2.

5.8 Generators and relations for $A(G(k,n))$

We have seen in Corollary 4.7 that the Chow ring of the Grassmannian is a free abelian group generated by the Schubert cycles. It follows moreover from Giambelli's formula (Proposition 4.16) that it is generated multiplicatively by just the *special Schubert cycles*, which are the Chern classes of the universal subbundle. We will now see that Whitney's formula and the fact that the Chern classes of a bundle vanish above the rank of the bundle provide a complete description of the relations among the special Schubert cycles, and that these form a complete intersection.

Theorem 5.26. *The Chow ring of the Grassmannian $G(k,n)$ has the form*

$$A(G(k,n)) = \mathbb{Z}[c_1, \dots, c_k]/I,$$

where $c_i \in A^i(G(k,n))$ is the i-th Chern class of the universal subbundle $\mathcal{S}$ and the ideal I is generated by the terms of total degree $n - k + 1, \dots, n$ in the power series expansion

$$\frac{1}{1 + c_1 + \dots + c_k} = 1 - (c_1 + \dots + c_k) + (c_1 + \dots + c_k)^2 - \dots \in \mathbb{Z}[\![c_1, \dots, c_k]\!].$$

Moreover, I is a complete intersection.

For example, the Chow ring of $G(3,7)$ is $\mathbb{Z}[c_1, c_2, c_3]/I$, where I is generated by the elements

$$c_1^5 + 4c_1^3 c_2 + 3c_1 c_2^2 + 3c_1^2 c_3 + 2c_2 c_3,$$
$$c_1^6 + 5c_1^4 c_2 + 6c_1^2 c_2^2 + c_2^3 + 4c_1^3 c_3 + 6c_1 c_2 c_3 + c_3^2,$$
$$c_1^7 + 6c_1^5 c_2 + 10c_1^3 c_2^2 + 4c_1 c_2^3 + 5c_1^4 c_3 + 12c_1^2 c_2 c_3 + 3c_2^2 c_3 + 3c_1 c_3^2,$$

and these elements form a regular sequence.

The proof of Theorem 5.26 uses two results from commutative algebra, Proposition 5.27 and Lemma 5.28, which are variations on some frequently used results; readers may wish to familiarize themselves with them before reading the proof of Theorem 5.26. Recall that the *socle* of a finite-dimensional graded algebra T is the submodule of elements annihilated by all elements of positive degree. In particular, if d is the largest degree such that $T_d \neq 0$, then the socle of T contains T_d. For a somewhat different proof, and the generalization to flag bundles of arbitrary vector bundles, see Grayson et al. [2012].

Proof: Set $A = A(G(k,n))$ and write t_i for the degree-i part of the power series expansion of $1/(1 + c_1 + \cdots + c_k)$, so that $t_0 = 1$, $t_1 = -c_1$, $t_2 = c_1^2 - c_2, \ldots$. Let $J = (t_{n-k+1}, \ldots, t_n)$, and let $R = \mathbb{Z}[c_1, \ldots, c_k]/J$.

Corollary 4.10, which is the special case of Whitney's formula (Theorem 5.3, part (c)) applied to the tautological sequence of vector bundles

$$0 \longrightarrow \mathcal{S} \longrightarrow \mathcal{O}_{G(k,n)}^n \longrightarrow \mathcal{Q} \longrightarrow 0$$

on $G(k,n)$, shows that $c(\mathcal{Q}) = 1/c(\mathcal{S})$. Since $\mathcal{Q}$ has rank $n - k$, the classes $c_i(\mathcal{Q})$ vanish for all $i > n - k$, and it follows that there is a ring homomorphism

$$\varphi : R \to A, \quad t_i \mapsto c_i(\mathcal{Q}).$$

Under this homomorphism, the class c_i goes to the Schubert cycle $c_i(\mathcal{S}) = (-1)^i \sigma_{1^i}$ (where the subscript denotes a sequence of 1 repeated i times). Recall from Corollary 4.2 that $\sigma_{1^k}^{n-k}$ is the class of a point.

We will show that for any field F the sequence $t_{n-k+1}, \ldots, t_n$ is a regular sequence in $R \otimes_{\mathbb{Z}} F$, and the induced map

$$R' := R \otimes_{\mathbb{Z}} F \xrightarrow{\varphi' := \varphi \otimes_{\mathbb{Z}} F} A' := A \otimes_{\mathbb{Z}} F$$

is an isomorphism. Since A is a finitely generated abelian group, the surjectivity of φ follows from this result using Nakayama's lemma and the two cases $F = \mathbb{Z}/(p)$ and $F = \mathbb{Q}$. On the other hand, by Corollary 4.7, A is a free abelian group so, as an abelian group, $\varphi(R)$ is free. Thus the kernel of φ is a summand of R, so the injectivity of φ follows from the injectivity, for every choice of F, of φ'. Using Lemma 5.28 inductively, this also follows that $t_{n-k+1}, \ldots, t_n$ is a regular sequence, proving the theorem.

To show that $t_{n-k+1}, \ldots, t_n$ is a regular sequence in R' it suffices, since the t_i have positive degree, to show that

$$F[c_1, \ldots, c_k]/J$$

has Krull dimension zero. Since F was arbitrary it suffices, by the Nullstellensatz, to show that, if $f_i \in F$ are substituted for the c_i in such a way that $t_{n-k+1} = \cdots = t_n = 0$, then all the f_i are zero.

Indeed, after such a substitution we see that $1/(1 + f_1 x + f_2 x^2 + \cdots + f_k x^k) = p(x) + q(x)$, where $p(x)$ is a polynomial of degree $\leq n - k$ and $q(x)$ is a rational function vanishing to order at least $n + 1$ at 0. We may rewrite this as

$$\frac{1 - p(x)(1 + f_1 x + f_2 x^2 + \cdots + f_k x^k)}{1 + f_1 x + \cdots + f_k x^k} = q(x).$$

However, the denominator of the left-hand side is nonzero at the origin, and the numerator has degree at most n. Since $q(x)$ vanishes to order at least $n + 1$ at the origin, both sides must be identically zero; that is $p(x) = 1, q(x) = 0$, and thus each $f_i = 0$, as required.

Combining this information with Proposition 5.27, we get:

- The dimension of R' (as a vector space over F) is $\binom{n}{k}$.
- The highest degree d such that $R'_d \neq 0$ is $k(n-k)$.
- Since a complete intersection is Gorenstein (Eisenbud [1995, Corollary 21.19]), every nonzero ideal of R' contains $R'_{k(n-k)}$.

We now return to the map φ'. By Corollary 4.13, the rank of $A(G(k,n))$ is also $\binom{n}{k}$; thus to show that φ' is an isomorphism, it suffices to show that its kernel is zero. We know that $(\sigma_{1^k})^{n-k}$ is in the image of φ', so $\mathrm{Ker}\, \varphi'$ does not contain $R_{k(n-k)}$. Since $R'_{k(n-k)}$ is the socle of R', the kernel of φ' must be zero. $\square$

We have used the following two results from commutative algebra:

Proposition 5.27. *Suppose that F is a field and that*

$$T = F[x_1, \ldots, x_k]/(g_1, \ldots, g_k)$$

is a zero-dimensional graded complete intersection with $\deg x_i = \delta_i > 0$ *and* $\deg g_i = \epsilon_i > 0$. *The Hilbert series of T is*

$$H_T(d) := \sum_{u=0}^{\infty} \dim_F T_u d^u = \frac{\prod_{i=1}^k (1 - d^{\epsilon_i})}{\prod_{i=1}^k (1 - d^{\delta_i})}.$$

The degree of the socle of T is $\sum_{i=0}^k \epsilon_i - \sum_{i=0}^k \delta_i$, and the dimension of T is

$$\dim_F T = \frac{\prod_{i=1}^k (\epsilon_i - 1)}{\prod_{i=1}^k (\delta_i - 1)}.$$

Proof: We begin with the Hilbert series. The polynomial ring $F[x_1, \ldots, x_k]$ is the tensor product of the one-variable polynomial rings $F[x_i]$, so

$$H_{F[x_1,\ldots,x_k]}(d) := \frac{1}{\prod_{i=1}^{k}(1 - d_i^{\delta})}.$$

We can put in the relations one-by-one using the exact sequences

$$0 \longrightarrow F[x_1, \ldots, x_k]/(g_1, \ldots, g_i)(-\epsilon_i) \xrightarrow{g_{i+1}} F[x_1, \ldots, x_k]/(g_1, \ldots, g_i)$$
$$\longrightarrow F[x_1, \ldots, x_k]/(g_1, \ldots, g_{i+1}) \longrightarrow 0,$$

and using induction we see that

$$H_T(d) = H_{F[x_1,\ldots,x_k]/(g_1,\ldots,g_k)}(d) = \frac{\prod_{i=1}^{k}(1 - d^{\epsilon_i})}{\prod_{i=1}^{k}(1 - d^{\delta_i})}.$$

A priori this is a rational function of degree $s := \sum_{i=1}^{k} \epsilon_i - \sum_{i=1}^{k} \delta_i$. Since we know from the computation above that T is a finite-dimensional vector space over F, the Hilbert series must be a polynomial. Thus it is a polynomial of degree s, so the largest degree in which T is nonzero is s.

The dimension of T is the value of $H_T(d)$ at $d = 1$. The product $(1-d)^k$ obviously divides both the numerator and the denominator of the expression for the Hilbert series above. After dividing, we get

$$H_T(d) = \frac{\prod_{i=1}^{k} \sum_{j=0}^{\epsilon_i-1} d^j}{\prod_{i=1}^{k} \sum_{j=0}^{\delta_i-1} d^j}.$$

Setting $d = 1$ in this expression gives us the desired result. $\qquad\square$

The other result from commutative algebra that we used is a version of the fact that regular sequences in a local ring can be permuted (Eisenbud [1995, Corollary 17.2]). The same result holds in the local case when every element of the regular sequence has positive degree, but the case we need is slightly different, since one element of the regular sequence is an integer. The result may also be viewed as a variation on the local criterion of flatness (Eisenbud [1995, Section 6.4]).

Lemma 5.28. *Suppose that R is a finitely generated graded algebra over $\mathbb{Z}$, with algebra generators in positive degrees, and that $f \in R$ is a homogeneous element. If R is free as a $\mathbb{Z}$-module and $f \otimes_{\mathbb{Z}} \mathbb{Z}/(p)$ is a monomorphism for every prime p, then f is a monomorphism and $R/(f)$ is free as a $\mathbb{Z}$-module as well.*

Proof: Since R is free, so is every submodule; in particular fR is free, and the kernel K of multiplication by f is a free summand of R. It follows that $K \otimes_{\mathbb{Z}} \mathbb{Z}/(p) \subset R \otimes_{\mathbb{Z}} \mathbb{Z}/(p)$. Since this ideal is obviously contained in the kernel of multiplication by f

on $R \otimes \mathbb{Z}/(p)$, we see that $K \otimes_{\mathbb{Z}} \mathbb{Z}/(p) = 0$. Since K is free, this implies that $K = 0$ as well; that is, f is a nonzerodivisor on R, and the diagram

$$
\begin{array}{ccccccccc}
0 & \longrightarrow & R(-1) & \overset{f}{\longrightarrow} & R & \longrightarrow & R/fR & \longrightarrow & 0 \\
& & \downarrow{\scriptstyle p} & & \downarrow{\scriptstyle p} & & \downarrow{\scriptstyle p} & & \\
0 & \longrightarrow & R(-1) & \overset{f}{\longrightarrow} & R & \longrightarrow & R/fR & \longrightarrow & 0
\end{array}
$$

has exact rows. A diagram chase (the *snake lemma*) shows that p is a nonzerodivisor on R/fR. Since p was an arbitrary prime, R/fR is a torsion-free abelian group. Since R is finitely generated and f is homogeneous, R/fR is a direct sum of finitely generated abelian groups, and torsion-freeness implies freeness. $\square$

5.9 Steps in the proofs of Theorem 5.3

Though the locus D in item (b) of Theorem 5.3 depends very much on the sections τ_i chosen, Theorem 5.3 asserts that the class $[D]$ does not, so long as it has the "expected" codimension. This point is worth understanding directly: We start with the case $k = r$ of the top Chern class. If τ and τ' are two sections of $\mathcal{E}$ whose zero loci are of codimension r, then we can interpolate between $V(\tau)$ and $V(\tau')$ with the family

$$
\Phi = \{([s,t], p) \in \mathbb{P}^1 \times X \mid s\tau(p) + t\tau'(p) = 0\}.
$$

This gives a rational equivalence between $V(\tau)$ and $V(\tau')$: Since Φ has codimension at most r everywhere, components of Φ intersecting the fibers over 0 or $\infty \in \mathbb{P}^1$ must dominate $\mathbb{P}^1$, and taking the union of these components we get a rational equivalence between the class of the zero locus of τ and that of τ'.

The same argument works in the general case: If both $\tau_0, \ldots, \tau_{r-i}$ and $\tau'_0, \ldots, \tau'_{r-i}$ are collections of sections with degeneracy loci of codimension i, we set

$$
\Phi = \left\{([s,t], p) \in \mathbb{P}^1 \times X \mid p \in V(s\tau_0 + t\tau'_0 \wedge \cdots \wedge s\tau_{r-i} + t\tau'_{r-i})\right\}.
$$

Using Lemma 5.2, one can show that the components of Φ dominating $\mathbb{P}^1$ give a rational equivalence between $V(\tau_0 \wedge \cdots \wedge \tau_{r-i})$ and $V(\tau'_0 \wedge \cdots \wedge \tau'_{r-i})$.

5.9.1 Whitney's formula for globally generated bundles

Though we will not prove the existence of Chern classes satisfying the properties of Theorem 5.3, it is instructive to see how Whitney's formula (property (c) in Theorem 5.3) follows in the case of a globally generated bundle from facts about the Grassmannian.

Suppose that $\mathcal{E}$ and $\mathcal{F}$ are globally generated bundles on a variety X. Denote the ranks of $\mathcal{E}$ and $\mathcal{F}$ by e and f respectively. We will show that

$$c(\mathcal{E} \oplus \mathcal{F}) = c(\mathcal{E})c(\mathcal{F}) \in A(X),$$

or equivalently

$$c_i(\mathcal{E} \oplus \mathcal{F}) = \sum_{i=j+k} c_j(\mathcal{E})c_k(\mathcal{F})$$

for $i \geq 0$.

In the extreme cases $i = 1$ and $i = e + f$ we can see this at once: In the first of these cases, Whitney's formula says that

$$c_1(\mathcal{E} \oplus \mathcal{F}) = c_1(\mathcal{E}) + c_1(\mathcal{F}).$$

If $\sigma_1, \ldots, \sigma_e \in H^0(\mathcal{E})$ and $\tau_1, \ldots, \tau_f \in H^0(\mathcal{F})$ are general sections, then the degeneracy locus of the $e + f$ sections

$$(\sigma_1, 0), \ldots, (\sigma_e, 0), (0, \tau_1), \ldots, (0, \tau_f) \in H^0(\mathcal{E} \oplus \mathcal{F})$$

is the sum, as divisors, of the degeneracy loci $V(\sigma_1 \wedge \cdots \wedge \sigma_e)$ and $V(\tau_1 \wedge \cdots \wedge \tau_f)$. Here we are using the identification

$$\wedge^{e+f}(\mathcal{E} \oplus \mathcal{F}) = \wedge^e \mathcal{E} \otimes \wedge^f \mathcal{F}.$$

In the second case, Whitney's formula says that

$$c_{e+f}(\mathcal{E} \oplus \mathcal{F}) = c_e(\mathcal{E})c_f(\mathcal{F}).$$

To see this, let σ and τ be general sections of $\mathcal{E}$ and $\mathcal{F}$ respectively. The zero locus $V((\sigma, \tau))$ of the section $(\sigma, \tau) \in H^0(\mathcal{E} \oplus \mathcal{F})$ is then the intersection of the zero loci $V(\sigma)$ and $V(\tau)$; by Lemma 5.2 applied to $\mathcal{F}|_{V(\sigma)}$, it will have the expected codimension $e + f$ and the equality above follows.

For the general case we adopt the alternative characterization of Chern classes of Proposition 5.7: If $V \subset H^0(\mathcal{E})$ is an n-dimensional subspace generating $\mathcal{E}$, we have a map $\varphi_V : X \to G(n - e, V)$ sending p to the subspace $V_p \subset V$ of sections vanishing at p; the k-th Chern class of $\mathcal{E}$ is then the pullback $\varphi_V^* \sigma_k$ of the Schubert class $\sigma_k \in A^k(G(n - e, V))$.

Let $V \subset H^0(\mathcal{E})$ and $W \subset H^0(\mathcal{F})$ be generating subspaces, of dimensions n and m; let φ_V and φ_W be the corresponding maps. The subspace $V \oplus W \subset H^0(\mathcal{E} \oplus \mathcal{F})$ is again generating, and gives a map

$$\varphi_{V \oplus W} : X \to G(n + m - e - f, V \oplus W).$$

Let

$$\varphi_V \times \varphi_W : X \to G(n - e, V) \times G(m - f, W)$$

be the product map. We have

$$\varphi_{V \oplus W} = \eta \circ (\varphi_V \times \varphi_W),$$

where $\eta : G(n - e, V) \times G(m - f, W) \to G(n + m - e - f, V \oplus W)$ is the map sending a pair of subspaces of V and W to their direct sum.

Lemma 5.29. *Let V and W be vector spaces of dimensions n and m. For any s and t, let*

$$\eta : G(s, V) \times G(t, W) \to G(s + t, V \oplus W)$$

be the map sending a pair (Λ, Γ) to $\Lambda \oplus \Gamma$. If α and β are the projection maps on $G(s, V) \times G(t, W)$, then, for any k,

$$\eta^*(\sigma_k) = \sum_{i+j=k} \alpha^* \sigma_i \cdot \beta^* \sigma_j.$$

Given Lemma 5.29, Whitney's formula (in our special case) follows: with φ_V, φ_W and $\varphi_{V \oplus W}$ as above, we have

$$c_k(\mathcal{E} \oplus \mathcal{F}) = \varphi_{V \oplus W}^*(\sigma_k) = \sum_{i+j=k} \varphi_V^*(\sigma_i) \varphi_V^*(\sigma_j) = \sum_{i+j=k} c_i(\mathcal{E}) c_j(\mathcal{F}).$$

Note that Lemma 5.29 is a direct (and substantial) generalization of the calculation in Section 2.1.4 of the class of the diagonal $\Delta \subset \mathbb{P}^r \times \mathbb{P}^r$. Specifically, if $V = W$, $m = n = r + 1$ and $s = t = 1$, then the diagonal $\Delta \subset \mathbb{P}^r \times \mathbb{P}^r$ is the preimage under the map $\eta : \mathbb{P}V \times \mathbb{P}V \to G(2, V \oplus V)$ of the Schubert cycle $\Sigma_n(V)$ of 2-planes intersecting the diagonal $V \subset V \oplus V$. Thus Lemma 5.29 in this case yields the formula of Section 2.1.4.

Proof: As in the earlier calculation of the class of the diagonal in $\mathbb{P}^r \times \mathbb{P}^r$, we will use the method of undetermined coefficients. Note that the product $G(s, V) \times G(t, W)$ can be stratified by products of Schubert cells; thus, by Proposition 1.17 the products $\alpha^* \sigma_a \cdot \beta^* \sigma_b$ span $A(G(s, V) \times G(t, W))$. (In particular, we have $A_0(G(s, V) \times G(t, W)) = \mathbb{Z}$.) Moreover, intersection products in complementary dimensions between classes of this type again have a simple form: We have

$$\deg\big((\alpha^* \sigma_a \beta^* \sigma_b)(\alpha^* \sigma_c \beta^* \sigma_d)\big) = \begin{cases} 1 & \text{if } a_i + c_{s-i+1} = n - s \text{ for all } i \text{ and} \\ & \quad b_j + d_{m-j+1} = m - t \text{ for all } j, \\ 0 & \text{otherwise.} \end{cases}$$

From this, we see that $A(G(s, V) \times G(t, W))$ is freely generated by the classes $\alpha^* \sigma_a \beta^* \sigma_b$, and that the intersection pairing in complementary dimensions is nondegenerate. Thus, to prove the equality of Lemma 5.29 it will be enough to show that both sides have the same product with any class $\alpha^* \sigma_a \cdot \beta^* \sigma_b$. Specifically, we need to show that for products $\alpha^* \sigma_a \cdot \beta^* \sigma_b$ of dimension k (that is, with $|a| + |b| = s(n - s) + t(m - t) - k$)

we have

$$\deg(\eta^*\sigma_k \cdot \alpha^*\sigma_a \cdot \beta^*\sigma_b) = \begin{cases} 1 & \text{if } a = (n-s,\dots,n-s,n-s-i) \text{ and} \\ & \quad b = (m-t,\dots,m-t,m-t-j) \\ & \quad \text{for some } i+j = k, \\ 0 & \text{otherwise.} \end{cases}$$

We start with the "otherwise" half. Note that, by the dimension condition $|a|+|b| = s(n-s)+t(m-t)-k$, the condition $a = (n-s,\dots,n-s,n-s-i)$ and $b = (m-t,\dots,m-t,m-t-j)$ for some $i+j = k$ is equivalent to saying that the sum of the last two indices a_s and b_t is $a_s + b_t = n-s+m-t-k$; in all other cases it will be strictly greater.

Start by choosing general flags $V_1 \subset \cdots \subset V_n = V$, $W_1 \subset \cdots \subset W_m = W$ and $U_1 \subset \cdots \subset U_{n+m} = V \oplus W$. Then

$$\Sigma_a(V) \subset \{\Lambda \subset V \mid \Lambda \subset V_{n-a_s}\}$$

and

$$\Sigma_b(W) \subset \{\Gamma \subset W \mid \Lambda \subset W_{m-a_t}\},$$

so

$$\eta(\alpha^{-1}\Sigma_a \cap \beta^{-1}\Sigma_b) \subset \{\Omega \subset V \oplus W \mid \Omega \subset V_{n-a_s} \oplus W_{m-a_t}\}.$$

But

$$\Sigma_k(U) = \{\Omega \subset V \oplus W \mid \Omega \cap U_{n-s+m-t-k+1} \neq 0\},$$

and, if $a_s + b_t > n-s+m-t-k$, then $(V_{n-a_s} \oplus W_{m-a_t}) \cap U_{n-s+m-t-k+1} = 0$; thus

$$\eta^{-1}\Sigma_k \cap \alpha^{-1}\Sigma_a \cap \beta^{-1}\Sigma_b = \varnothing$$

and the product of the corresponding classes is zero.

Similarly, in case $a = (n-s,\dots,n-s,n-s-i)$ and $b = (m-t,\dots,m-t,m-t-j)$ for some $i+j = k$, the intersection $U = (V_{n-a_s} \oplus W_{m-a_t}) \cap U_{n-s+m-t-k+1}$ will be one-dimensional. Since

$$\Sigma_a(V) = \left\{ \Lambda \subset V \ \middle| \ \begin{matrix} V_{s-1} \subset \Lambda \text{ and} \\ \Lambda \subset V_{n-a_s} \end{matrix} \right\}$$

and

$$\Sigma_b(W) = \left\{ \Gamma \subset W \ \middle| \ \begin{matrix} W_{t-1} \subset \Lambda \text{ and} \\ \Lambda \subset W_{m-a_t} \end{matrix} \right\},$$

we see that the intersection $\eta^{-1}\Sigma_k \cap \alpha^{-1}\Sigma_a \cap \beta^{-1}\Sigma_b$ will consist of the single point (Λ, Γ), where $\Lambda \subset V$ is the span of V_{s-1} and the projection $\pi_1(U)$ and likewise $\Gamma \subset W$ is the span of W_{t-1} and the image $\pi_2(U)$. That the intersection is transverse follows from Kleiman's theorem in characteristic 0, and from direct examination of the tangent spaces in general. $\qquad\square$

5.10 Exercises

Many of the following exercises give applications of Whitney's formula and the splitting principle. We will be assuming the basic facts that if

$$\mathcal{E} = \bigoplus_{i=1}^{e} \mathcal{L}_i \quad \text{and} \quad \mathcal{F} = \bigoplus_{i=1}^{f} \mathcal{M}_i$$

are direct sums of line bundles, then

$$\text{Sym}^k \mathcal{E} = \bigoplus_{1 \leq i_1 \leq \cdots \leq i_k \leq r} \mathcal{L}_{i_1} \otimes \cdots \otimes \mathcal{L}_{i_k},$$

$$\bigwedge^k \mathcal{E} = \bigoplus_{1 \leq i_1 < \cdots < i_k \leq r} \mathcal{L}_{i_1} \otimes \cdots \otimes \mathcal{L}_{i_k},$$

$$\mathcal{E} \otimes \mathcal{F} = \bigoplus_{i,j=1,1}^{e,f} \mathcal{L}_i \otimes \mathcal{M}_j.$$

Exercise 5.30. Let $\mathcal{E}$ be a vector bundle of rank 3. Express the Chern classes of $\bigwedge^2 \mathcal{E}$ in terms of those of $\mathcal{E}$ by invoking the splitting principle and Whitney's formula

Exercise 5.31. Verify your answer to the preceding exercise by observing that the wedge product map

$$\mathcal{E} \otimes \bigwedge^2 \mathcal{E} \to \bigwedge^3 \mathcal{E} = \det(\mathcal{E})$$

yields an identification $\bigwedge^2 \mathcal{E} = \mathcal{E}^* \otimes \det(\mathcal{E})$ and applying the formula for a tensor product with a line bundle.

Exercise 5.32. Let $\mathcal{E}$ be a vector bundle of rank 4. Express the Chern classes of $\bigwedge^2 \mathcal{E}$ in terms of those of $\mathcal{E}$.

Exercise 5.33. Let $\mathcal{E}$ be a vector bundle of rank 3. Express the Chern classes of $\text{Sym}^2 \mathcal{E}$ in terms of those of $\mathcal{E}$.

Exercise 5.34. Let $\mathcal{E}$ be a vector bundle of rank 2. Express the Chern classes of $\text{Sym}^3 \mathcal{E}$ in terms of those of $\mathcal{E}$.

Exercise 5.35. Let $\mathcal{E}$ and $\mathcal{F}$ be vector bundles of rank 2. Express the Chern classes of the tensor product $\mathcal{E} \otimes \mathcal{F}$ in terms of those of $\mathcal{E}$ and $\mathcal{F}$.

Exercise 5.36. Just to get a sense of how rapidly this gets complicated: Do the preceding exercise for a pair of vector bundles $\mathcal{E}$ and $\mathcal{F}$ of ranks 2 and 3.

Exercise 5.37. Apply Exercise 5.35 to find all the Chern classes of the tangent bundle $\mathcal{T}_G$ of the Grassmannian $G = G(2, 4)$.

Exercise 5.38. Find all the Chern classes of the tangent bundle $\mathcal{T}_Q$ of a quadric hypersurface $Q \subset \mathbb{P}^5$. Check that your answer agrees with your answer to the last exercise!

Exercise 5.39. Calculate the Chern classes of the tangent bundle of a product $\mathbb{P}^n \times \mathbb{P}^m$ of projective spaces

Exercise 5.40. Find the Euler characteristic of a smooth hypersurface of bidegree (a, b) in $\mathbb{P}^m \times \mathbb{P}^n$.

Exercise 5.41. Using Whitney's formula, show that for $n \geq 2$ the tangent bundle $\mathcal{T}_{\mathbb{P}^n}$ of projective space is not a direct sum of line bundles.

Exercise 5.42. Find the Betti numbers of the smooth intersection of a quadric and a cubic hypersurface in $\mathbb{P}^4$, and of the intersection of three quadrics in $\mathbb{P}^5$. (Both of these are examples of *K3 surfaces*, which are diffeomorphic to a smooth quartic surface in $\mathbb{P}^3$.)

Exercise 5.43. Find the Betti numbers of the smooth intersection of two quadrics in $\mathbb{P}^5$. This is the famous *quadric line complex*, about which you can read more in Griffiths and Harris [1994, Chapter 6].

Exercise 5.44. Show that the cohomology groups of a smooth quadric threefold $Q \subset \mathbb{P}^4$ are isomorphic to those of $\mathbb{P}^3$ ($\mathbb{Z}$ in even dimensions, 0 in odd), but its cohomology ring is different (the square of the generator of $H^2(Q, \mathbb{Z})$ is twice the generator of $H^4(Q, \mathbb{Z})$). (This is a useful example of the fact that two compact, oriented manifolds can have the same cohomology groups but different cohomology rings, if you are ever teaching a course in algebraic topology.)

Exercise 5.45. Let $S \subset \mathbb{P}^4$ be a smooth complete intersection of hypersurfaces of degrees d and e, and let $Y \subset \mathbb{P}^4$ be any hypersurface of degree f containing S. Show that if f is not equal to either d or e, then Y is necessarily singular.
Hint: Assume Y is smooth, and apply Whitney's formula to the sequence

$$0 \longrightarrow \mathcal{N}_{S/Y} \longrightarrow \mathcal{N}_{S/\mathbb{P}^4} \longrightarrow \mathcal{N}_{Y/\mathbb{P}^4}|_S \longrightarrow 0$$

to arrive at a contradiction.

Chapter 6

Lines on hypersurfaces

Keynote Questions

(a) Let $X \subset \mathbb{P}^4$ be a general quintic hypersurface. How many lines $L \subset \mathbb{P}^4$ does X contain? (Answer on page 228.)

(b) Let $\{X_t \subset \mathbb{P}^3\}_{t \in \mathbb{P}^1}$ be a general pencil of quartic surfaces. How many of the surfaces X_t contain a line? (Answer on page 233.)

(c) Let $\{X_t \subset \mathbb{P}^3\}_{t \in \mathbb{P}^1}$ be a general pencil of cubic surfaces, and consider the locus $C \subset \mathbb{G}(1,3)$ of all lines $L \subset \mathbb{P}^3$ that are contained in some member of this family. What is the genus of C? What is the degree of the surface $S \subset \mathbb{P}^3$ swept out by these lines? (Answers on pages 233 and 233.)

(d) Can a smooth quartic hypersurface in $\mathbb{P}^4$ contain a two-parameter family of lines? (Answer on page 238.)

In this chapter we will study the schemes parametrizing lines (and planes of higher dimension) on a hypersurface. These are called *Fano schemes*. There are two phases to the treatment. It turns out that the enumerative content of the keynote questions above, and many others, can be answered through a single type of Chern class computation. But there is another side of the story, involving beautiful and important techniques for working with the tangent spaces of Hilbert schemes, of which Fano schemes are examples. These ideas will allow us to verify that the "numbers" we compute really correspond to the geometry that they are meant to reflect. We will go even beyond these techniques and explore a little of the local structure of the Fano scheme. There are many open questions in this area, and the chapter ends with an exploration of one of them.

6.1 What to expect

For what n and d should we expect a general hypersurface $X \subset \mathbb{P}^n$ of degree d to contain lines? What is the dimension of the family of lines we would expect it to contain? When the dimension is zero, how many lines will there be?

To answer these questions, we introduce in this chapter a fundamental object, the *Fano scheme* $F_k(X) \subset \mathbb{G}(k,n)$ parametrizing k-planes on X, and then study its geometry.

We will defer for a moment a discussion of the scheme structure on $F_k(X)$, and start by answering the first two of the questions above, since these have to do only with the underlying set $F_k(X) \subset \mathbb{G}(k,n)$. Even so, the answers may not be apparent at first, since (as we shall see) the equations on the Grassmannian of lines that describe the locus of lines L contained in a given hypersurface $X \subset \mathbb{P}^n$ may be complicated. But if we reverse the roles of L and X — that is, ask for the locus, in the space $\mathbb{P}^N$ of all hypersurfaces of degree d in $\mathbb{P}^n$, of the hypersurfaces X that contain a given line L — the equations are much simpler; in fact, given L, the locus of X that contain L is simply a linear subspace of $\mathbb{P}^N$.

To capitalize on this, we use an incidence correspondence: We set $N = \binom{n+d}{d} - 1$ and let $\mathbb{P}^N$ be the projective space parametrizing all hypersurfaces of degree d, and consider the variety $\Phi = \Phi(n, d, k)$ given by the formula

$$\Phi = \{(X, L) \in \mathbb{P}^N \times \mathbb{G}(k,n) \mid L \subset X\}.$$

(As the title of this chapter suggests, our primary focus will be on the case $k = 1$ of lines, but many of the constructions we make can be carried out just as readily for arbitrary-dimensional linear spaces, as here.)

That $\Phi \subset \mathbb{P}^N \times \mathbb{G}(k,n)$ is a closed subset may be seen by a number of elementary arguments (see for example Harris [1995, Chapter 6]); in any event, we will give explicit equations of Φ in the next section. The variety Φ will be quite useful in many ways; we call it the *universal Fano scheme* of k-planes on hypersurfaces of degree d in $\mathbb{P}^n$, since the fiber over any point $X \in \mathbb{P}^N$ is the Fano scheme $F_k(X) \subset \mathbb{G}(k,n)$ of k-planes on X. To start, we have:

Proposition 6.1. *Let* $N = \binom{n+d}{d} - 1$. *The universal Fano scheme* $\Phi = \Phi(n, d, k) \subset \mathbb{P}^N \times \mathbb{G}(k, \mathbb{P}^n)$ *is a smooth irreducible variety of dimension*

$$\dim \Phi(n, d, k) = N + (k + 1)(n - k) - \binom{k+d}{k}.$$

Proof: As we said, the fibers of Φ over $\mathbb{G}(k,n)$ are readily described: For any plane $\Lambda \cong \mathbb{P}^k \subset \mathbb{P}^n$, the restriction map

$$H^0(\mathcal{O}_{\mathbb{P}^n}(d)) \to H^0(\mathcal{O}_\Lambda(d))$$

is a surjection, and the fiber of Φ over the point $\Lambda \in \mathbb{G}(k, n)$ is simply the projectivization of the kernel, that is, a projective space $\mathbb{P}^{N-\binom{k+d}{k}}$. $\square$

Proposition 6.1 gives us an "expected" answer to the questions raised at the beginning of this section, and also allows us to deduce the answer in some cases: We would expect the family of k-planes on a general hypersurface $X \subset \mathbb{P}^n$ of degree d — that is, the fiber of Φ over a general point $[X] \in \mathbb{P}^N$ — to have dimension $\varphi(n, d, k) := \dim \Phi - N$, and that in the case $\varphi < 0$ a general X will contain no k-planes. In fact, Proposition 6.1 immediately implies the second statement, and while it does not imply the first, it does imply a lower bound on the dimension of the family of such planes, should there be any. We collect these consequences in the following corollary:

Corollary 6.2. *(a)* *The dimension of any component of the family of k-planes on any hypersurface of degree d in $\mathbb{P}^n$ is at least*

$$\varphi(n, d, k) := (k + 1)(n - k) - \binom{k+d}{k}.$$

(b) *If $\varphi(n, d, k) < 0$, then the general hypersurface of degree d in $\mathbb{P}^n$ contains no k-planes.*

(c) *If $\varphi(n, d, k) \geq 0$ and the general hypersurface of degree d contains any k-planes, then every hypersurface of degree d contains k-planes, and every component of the family of k-planes on a general hypersurface of degree d has dimension exactly $\varphi(n, d, k)$.*

Proof: Part (b) is immediate: If $\dim \Phi < N$, then Φ cannot dominate $\mathbb{P}^N$. For part (a), we observe that since a fiber of $\Phi(n, d, k)$ over $\mathbb{P}^N$ is cut out by N equations, the principal ideal theorem (Theorem 0.1) gives the desired lower bound. As for part (c), if the general hypersurface of degree d contains a k-plane, then $\Phi(n, d, k)$ dominates $\mathbb{P}^N$, and, since $\Phi(n, d, k)$ is projective, the map to $\mathbb{P}^N$ is surjective with general fiber of dimension $\dim \Phi - N = \varphi$. $\square$

We shall eventually show (Corollary 6.32 and Theorem 6.28) that if $\varphi(n, d, k) \geq 0$ then, except in some cases where $k > 1$ and $d = 2$, the general hypersurface actually does contain k-planes, so the results above apply. For example, if $d \leq 2n - 3$, every hypersurface of degree d in $\mathbb{P}^n$ contains lines, and the family of lines on the general such hypersurface has dimension $\varphi(n, d, 1) = 2n - 3 - d$.

Corollary 6.2 shows that the general surface in $\mathbb{P}^3$ of degree $d \geq 4$ contains no lines. But we can say more, using the same sort of incidence correspondence argument made above. For example, we will see in Exercise 6.64 that a general surface $S \subset \mathbb{P}^3$ of degree $d \geq 4$ containing a line contains only one. This implies that the locus $\Sigma \subset \mathbb{P}^N$ of surfaces that do contain a line has codimension $d - 3$.

6.1.1 Definition of the Fano scheme

We begin by giving a direct definition of the Fano scheme $F_k(X)$ of k-planes on a scheme $X \subset \mathbb{P}^n$ that is local on $\mathbb{G}(k, n)$. We will return to the definition twice later in this chapter to give a global description and a universal property that justifies the idea that we are taking the "right" scheme structure. Note that it will suffice to give the definition of the Fano scheme $F_k(X)$ when X is a hypersurface in $\mathbb{P}^n$; for an arbitrary scheme $Y \subset \mathbb{P}^n$ we define

$$F_k(Y) = \bigcap_{\substack{Y \subset X \subset \mathbb{P}^n \\ X \text{ is a hypersurface}}} F_k(X).$$

To define $F_k(X)$ for a hypersurface X of degree d given by an equation $g = 0$, we use the idea that a plane L lies on X if and only if the restriction of g to L is zero. If we have a parametrization $\alpha : \mathbb{P}^k \to L$ of L, then we can pull back g via α; the condition $L \subset X$ is given by the vanishing of the coefficients of $\alpha^*(g)$.

In fact, we can give such parametrizations simultaneously for all planes $L \in \mathbb{G}(k, n)$ lying in an open set U of the open cover of $\mathbb{G}(k, \mathbb{P}^n)$ described in Section 3.2.2. Recall that such an open set U is defined as the set of all k-planes not meeting a fixed $(n-k-1)$-plane. If the latter is given by the vanishing of the first $k + 1$ coordinates, then U may be identified with the affine space of $(k + 1) \times (n - k)$ matrices: Any k-plane L belonging to U is the row space of a unique matrix of the form

$$A = \begin{pmatrix} 1 & 0 & \cdots & 0 & a_{0,k+1} & \cdots & a_{0,n+1} \\ 0 & 1 & \cdots & 0 & a_{1,k+1} & \cdots & a_{1,n+1} \\ \vdots & \vdots & \ddots & \vdots & \vdots & \ddots & \vdots \\ 0 & 0 & \cdots & 1 & a_{k,k+1} & \cdots & a_{k,n+1} \end{pmatrix}.$$

We can thus give a parametrization of L of the form

$$\mathbb{P}^k \ni (s_0, \dots, s_k) \xrightarrow{\alpha} (s_0 \cdots s_k)A = \left(s_0, \dots, s_k, \sum_i a_{i,k+1}s_i, \dots, \sum_i a_{i,n+1}s_i\right),$$

where the $a_{i,j}$ are coordinates on $U \cong \mathbb{A}^{(k+1)(n-k)}$ and the s_i are homogeneous coordinates on our fixed source $\mathbb{P}^k$.

Now suppose that $X \subset \mathbb{P}^n$ is the hypersurface $V(g)$ given by the polynomial

$$g(z_0, \dots, z_n) = \sum_{|\delta|=d} c_\delta z^\delta.$$

We substitute the $n + 1$ coordinates of the parametrization α for the variables $z_0, \dots, z_n$ of g, and arrive at a homogeneous polynomial of degree d in $s_0, \dots, s_k$:

$$\alpha^*(g) = \sum_{|\delta|=d} e_\delta s^\delta.$$

The coefficients $\{e_\delta\}$ of this polynomial are polynomials in the coordinates $\{a_{i,j}\}$ on $U \subset \mathbb{G}(k, n)$, which we take as defining equations for $F_k(X)$.

We should check, of course, that the scheme structure defined in this way agrees on the overlap $U \cap V$ of two such open sets. This is straightforward, but we will skip it: In what follows we will see a remarkable intrinsic characterization of the Fano scheme that will imply it.

Finally, we return to the universal Fano scheme $\Phi \subset \mathbb{P}^N \times \mathbb{G}(k, n)$ discussed in the preceding section. We promised to give equations for Φ, and we have: The coefficients e_δ of the polynomial $\alpha^*(g)$ above may be viewed as polynomials in both sets of variables $\{a_{i,j}\}$ and $\{c_\delta\}$, and we take $\Phi \subset \mathbb{P}^N \times \mathbb{G}(k, n)$ to be the subscheme defined locally by these polynomials. Note that for any hypersurface $X \subset \mathbb{P}^n$ the Fano scheme $F_k(X)$ is the scheme-theoretic fiber of Φ over the point $X \in \mathbb{P}^N$.

In fact, there is a simpler way to characterize the scheme structure of Φ: it is reduced. This follows from the fact that the scheme-theoretic fibers of the projection $\Phi \to \mathbb{G}(k, n)$ are projective spaces (the e_δ are homogeneous linear in the variables c_δ). For the same reason, Φ is smooth.

This is very much *not* to say that the Fano scheme $F_k(X)$ is either smooth or reduced for a given X. It does imply that $F_k(X)$ is smooth and reduced for a *general* hypersurface $X \subset \mathbb{P}^n$, but we will see many examples of nonreduced and/or singular Fano schemes; part of the challenge of the subject is to figure out under what circumstances this may happen. As a first example, you may wish to consider the Fano scheme $F_1(Q)$ of lines on a quadric surface $Q \subset \mathbb{P}^3$; as you can see from the equations, $F_1(Q)$ is smooth if Q is smooth, but everywhere nonreduced if Q is singular. Apart from this being the first nontrivial example of such phenomena, what makes this interesting is that we will also be able to see this from two other viewpoints, without coordinates: once when we describe the class $[F_1(Q)] \in A(\mathbb{G}(1, 3))$ in Section 6.2, and again at the end of Section 6.4.2 when we introduce the notion of first-order deformations.

Of course special hypersurfaces may well contain families of planes of dimension greater than $\varphi(n, d, k)$. We can easily give an upper bound on the possible dimension:

Proposition 6.3. *If $X \subset \mathbb{P}^n$ is an m-dimensional variety, then*

$$\dim F_k(X) \leq (m - k)(k + 1) = \dim \mathbb{G}(k, m),$$

with equality if and only if X is an m-plane.

Proof: We may assume without loss of generality that X is nondegenerate. Let $U \subset X^{k+1}$ be the open set consisting of $(k + 1)$-tuples of linearly independent points, and let

$$\Gamma = \{((p_0, \ldots, p_k), L) \in U \times F_k(X) \,|\, p_i \in L \text{ for all } i\}.$$

Via the projection $\Gamma \to U$, we see that $\dim \Gamma \leq m(k + 1)$. Since the fibers of the projection $\Gamma \to F_k(X)$ have dimension $k(k + 1)$, we conclude that $\dim F_k(X) \leq m(k + 1) - k(k + 1) = (m - k)(k + 1)$, as required.

Equality of dimensions can hold only if the projection $\Gamma \to U$ is dominant, that is, if X contains the plane spanned by any $k + 1$ general points of X, and this can happen only if X is a linear space.   □

In Section 6.8 we will discuss some open questions about these dimensions.

6.2 Fano schemes and Chern classes

To get global information about the Fano scheme of a hypersurface, we will express it as the zero locus of a section of a vector bundle on the Grassmannian. To understand the idea, suppose that $X \subset \mathbb{P}^n$ is the hypersurface $g = 0$, where g is a homogeneous form of degree d. As we have seen, the condition that X contain a particular k-dimensional linear space L is that g is sent to 0 by the restriction map

$$H^0(\mathcal{O}_{\mathbb{P}^n}(d)) \to H^0(\mathcal{O}_L(d)).$$

To describe the scheme $F_k(X) \subset \mathbb{G}(k, n)$ of k-planes on X, we will realize the family of vector spaces $H^0(\mathcal{O}_L(d))$ (with varying k-planes L) as the fibers of a vector bundle in such a way that the images of g in these vector spaces are the values of a section σ_g of the bundle:

Proposition 6.4. *Let V be an $(n + 1)$-dimensional vector space, and let $\mathcal{S} \subset V \otimes \mathcal{O}_G$ be the tautological rank-$(k + 1)$ subbundle on the Grassmannian $G = \mathbb{G}(k, \mathbb{P}V)$ of k-planes in $\mathbb{P}V \cong \mathbb{P}^n$. A form g of degree d on $\mathbb{P}V$ gives rise to a global section σ_g of $\mathrm{Sym}^d \mathcal{S}^*$ whose zero locus is $F_k(X)$, where X is the hypersurface $g = 0$.*

Thus, when $F_k(X)$ has expected codimension $\binom{k+d}{k} = \mathrm{rank}(\mathrm{Sym}^d \mathcal{S}^*)$ in G, we have

$$[F_k(X)] = c_{\binom{k+d}{k}}(\mathrm{Sym}^d \mathcal{S}^*) \in A(G).$$

Proof: The fiber of $\mathcal{S}$ over the point $[L] \in \mathbb{G}(k, \mathbb{P}V)$ representing the subspace $L \cong \mathbb{P}^k \subset \mathbb{P}V$ is the corresponding $(k + 1)$-dimensional subspace of V. The fiber of the dual bundle $\mathcal{S}^*$ at $[L]$ is thus the space of linear forms on L, that is to say $H^0(\mathcal{O}_{\mathbb{P}L}(1))$, and the dual map $V^* \otimes \mathcal{O}_G \to \mathcal{S}^*$ evaluated at a point $[L]$ takes a linear form $\varphi \in V^*$, thought of as a constant section of the trivial bundle $V^* \otimes \mathcal{O}_G$, to the restriction of φ to L. The vector space of forms of degree d on $\mathbb{P}V$ is $\mathrm{Sym}^d V^* = H^0(\mathcal{O}_{\mathbb{P}V}(d))$, and the induced map on symmetric powers

$$\mathrm{Sym}^d V^* \to \mathrm{Sym}^d \mathcal{S}^*$$

evaluated at L takes a form g of degree d to its restriction to L, as required.

Let $\sigma_g \in H^0(\mathrm{Sym}^d\, \mathcal{S}^*)$ be the global section of $\mathrm{Sym}^d\, \mathcal{S}^*$ that is the image of g. We claim that $F_k(X) \subset \mathbb{G}(k, \mathbb{P}V)$ is the zero locus of this section. It is enough to check this locally on an open covering of $\mathbb{G}(k, \mathbb{P}V)$, and we use the open covering by basic affine sets U described in Section 3.2.2. On such an open set, the bundle $\mathcal{S}$ is trivial, with the inclusion

$$\mathcal{S}|_U = \mathcal{O}_U^{k+1} \to V \otimes \mathcal{O}_U$$

given by the transpose of the matrix A. It follows that the dual map

$$A^* : H^0(\mathcal{O}_{\mathbb{P}^n}(1)) \otimes \mathcal{O}_U = V^* \otimes \mathcal{O}_U \to \mathcal{S}^*|_U$$

is the restriction of linear forms from $\mathbb{P}V$, and its d-th symmetric power is the restriction of forms of degree d. Thus the value of σ_g at the point of U corresponding to a plane L is the restriction of g to L, or in other words the result of the substitution given in Section 6.1.1, as required.   □

The fact that the class $[F_k(X)]$ depends only on d and n (assuming it has the expected dimension) has consequences by itself, even without calculating the actual class. For example, consider the lines on a quadric surface $Q \subset \mathbb{P}^3$. As we saw in Section 3.6.1, when Q is smooth the Fano scheme $F_1(Q)$ consists of two disjoint smooth conic curves in the Grassmannian $\mathbb{G}(1, 3) \subset \mathbb{P}^5$. But what happens when Q is a cone over a smooth conic curve? Here the support of $F_1(Q)$ is a single conic curve in $\mathbb{G}(1, 3)$, and we may deduce from this and Proposition 6.4 that $F_1(Q)$ *is everywhere nonreduced.*

6.2.1 Counting lines on cubics

We want to see how this works for the case of lines on a cubic surface $X \subset \mathbb{P}^3$. In the language above, we want to compute the class of the Fano scheme $F_1(X)$ in the Grassmannian $\mathbb{G} = \mathbb{G}(1, 3)$. We saw in Section 5.6.2 that the Chern class of $\mathcal{S}^*$ is

$$c(\mathcal{S}^*) = 1 + \sigma_1 + \sigma_{1,1}.$$

Since $\mathcal{S}$ has rank 2, the rank of $\mathrm{Sym}^3\, \mathcal{S}^*$ is 4, so we want to compute $c_4(\mathrm{Sym}^3\, \mathcal{S}^*)$. To do this, we will apply the splitting principle (Section 5.4), which implies that to compute the Chern class we may pretend that $\mathcal{S}^*$ splits into a direct sum of two line bundles $\mathcal{L}$ and $\mathcal{M}$. Suppose that

$$c(\mathcal{L}) = 1 + \alpha \quad \text{and} \quad c(\mathcal{M}) = 1 + \beta.$$

By the Whitney formula,

$$c(\mathcal{S}^*) = (1 + \alpha)(1 + \beta),$$

so that

$$\alpha + \beta = \sigma_1 \quad \text{and} \quad \alpha \cdot \beta = \sigma_{1,1}.$$

If $\mathcal{S}^*$ were to split as above, then the bundle $\mathrm{Sym}^3 \mathcal{S}^*$ would split as well:

$$\mathrm{Sym}^3 \mathcal{S}^* = \mathcal{L}^3 \oplus (\mathcal{L}^2 \otimes \mathcal{M}) \oplus (\mathcal{L} \otimes \mathcal{M}^2) \oplus \mathcal{M}^3,$$

so that we would have

$$c(\mathrm{Sym}^3 \mathcal{S}^*) = (1 + 3\alpha)(1 + 2\alpha + \beta)(1 + \alpha + 2\beta)(1 + 3\beta).$$

In particular, the top Chern class could be written

$$\begin{aligned}
c_4(\mathrm{Sym}^3 \mathcal{S}^*) &= 3\alpha(2\alpha + \beta)(\alpha + 2\beta)3\beta \\
&= 9\alpha\beta(2\alpha^2 + 5\alpha\beta + 2\beta^2) \\
&= 9\alpha\beta(2(\alpha + \beta)^2 + \alpha\beta).
\end{aligned}$$

Re-expressing this in terms of the Chern classes of $\mathcal{S}^*$ itself, we get

$$\begin{aligned}
c_4(\mathrm{Sym}^3 \mathcal{S}^*) &= 9\sigma_{1,1}(2\sigma_1^2 + \sigma_{1,1}) \\
&= 27\sigma_{2,2},
\end{aligned}$$

so

$$\deg(c_4((\mathrm{Sym}^3 \mathcal{S}^*))) = 27;$$

by the splitting principle, these formulas hold even though $\mathcal{S}$ does not in fact split.

The whole Chern class of $\mathrm{Sym}^3 \mathcal{S}^*$ can also be computed by hand in this way, or with the following commands in *Macaulay2*:

```
loadPackage "Schubert2"
G = flagBundle({2,2}, VariableNames=>{s,q})
-- sets G to be the Grassmannian of 2-planes in 4-space,
-- and gives the names $s_i$ and $q_i$ to the Chern classes
-- of the sub and quotient bundles, respectively.
(S,Q)=G.Bundles
-- names the sub and quotient bundles on G
chern symmetricPower(3,dual S)
```

which returns the output

```
             2                           2
o4 = 1 + 6q  + (21q  - 10q ) + 42q q  + 27q
           1       1     2       1 2      2

                 QQ[][s , s , q , q ]
                      1   2   1   2
o4 : ------------------------------------------
        (s  + q , s  + s q  + q , s q  + s q , s q )
          1    1   2    1 1    2   2 1    1 2   2 2
```

The answer on the first output line "o4" is written in terms of the Chern classes $q_i := c_i(Q)$, which generate the (rational) Chow ring of the Grassmannian, described on the second output line "o4."

Since the class $c_4(\text{Sym}^3 \, \mathcal{S}^*)$ is nonzero, we deduce that *every cubic surface must contain lines*, and thus that a general cubic surface contains only finitely many. Moreover, if a particular cubic surface $X \subset \mathbb{P}^3$ contains only finitely many lines, then the number of these lines, counted with the appropriate multiplicity (that is, the degree of the corresponding component of the zero scheme of σ_g), is 27. As we will soon see, the Fano scheme $F_1(X)$ of a smooth cubic surface X is necessarily of dimension zero and reduced, so the actual number of lines is always 27. In the next section we will develop a general technique that will allow us to prove this statement, and much more. We will also see, in Section 6.7, how to count lines in cases where X is singular.

6.3 Definition and existence of Hilbert schemes

It was Grothendieck's brilliant observation that the Grassmannian and the Fano scheme are special cases of a very general construction, the *Hilbert scheme*. Hilbert schemes are defined by a universal property that we will explain in this section, after making the property explicit for the Grassmannian and Fano schemes.

One of the useful properties of Hilbert schemes is a general formula for tangent spaces, which we will explain in the next section. For more remarks about Hilbert schemes in general, see Section 8.4.1.

6.3.1 A universal property of the Grassmannian

Recall from Theorem 3.4 that the Grassmannian $G = G(k + 1, V) = \mathbb{G}(k, \mathbb{P}V)$ of $(k + 1)$-planes in an $(n + 1)$-dimensional vector space V, with its tautological subbundle $\mathcal{S} \subset V \otimes \mathcal{O}_G$, has the following universal property: Given any scheme B and any rank-$(k + 1)$ subbundle $\mathcal{F}$ of the trivial bundle $V \otimes \mathcal{O}_B$, there is a unique morphism $\varphi : B \to G$ such that $\mathcal{F} = \varphi^* \mathcal{S}$. We could, of course, just as well express this in terms of a universal property of the quotient bundle $\mathcal{Q} = V \otimes \mathcal{O}_B / \mathcal{S}$, or of the $(n - k)$-subbundle $\mathcal{Q}^* \subset V^* \otimes \mathcal{O}_B$, which is the most convenient for what we will do in this section.

Similarly, the universal k-plane

$$\Phi = \{(\Lambda, p) \in G \times \mathbb{P}V \mid p \in \Lambda\}$$

is a universal family of k-planes in $\mathbb{P}V$ in the following sense: For any scheme B, we will say that a subscheme $\mathcal{L} \subset B \times \mathbb{P}V$ is a *flat family of k-planes in $\mathbb{P}V$* if the restriction $\pi : \mathcal{L} \to B$ of the projection $\pi_1 : B \times \mathbb{P}V \to B$ is flat, and the fibers over closed points of B are linearly embedded k-planes in $\mathbb{P}V$. We have then:

Proposition 6.5. *If $\pi : \mathcal{L} \subset B \times \mathbb{P}V \to B$ is a flat family of k-planes in $\mathbb{P}V$, then there is a unique map $\alpha : B \to G$ such that $\mathcal{L}$ is equal, as a subscheme of $B \times \mathbb{P}V$, to the pullback of the family $\mathcal{L}$ via α:*

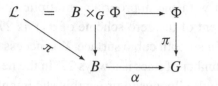

Proof: We will prove the proposition by showing that the desired property can be reduced to the universal property of Theorem 3.4. Though the reduction may appear technical, it is really just an application of Theorem B.5, together with the remark that the ideal of any k-plane in $\mathbb{P}V$ is generated by $n - k$ independent linear forms.

To simplify the notation we denote the ideal sheaves of $\Phi \subset G \times \mathbb{P}V$ and $\mathcal{L} \subset B \times \mathbb{P}V$ by $\mathcal{I}$ and $\mathcal{J}$ respectively, and we write $\mathcal{J}(1)$ for $\mathcal{J} \otimes \mu^* \mathcal{O}_{\mathbb{P}V}(1)$, where μ denotes the projection onto $\mathbb{P}V$. We define $\mathcal{I}(1)$ similarly. For any scheme B, we denote the trivial bundle with fiber V^* and base B by $\mathcal{O}_B \otimes V^*$.

The proof consists of the following steps: We will begin by showing that $\pi_* \mathcal{J}(1)$ is a subbundle of $\mathcal{O}_B \otimes V^*$. Since Φ satisfies the same hypotheses as $\mathcal{L}$, the same reasoning will show that the sheaf $\pi_* \mathcal{I}(1)$ is a subbundle of $\mathcal{O}_G \otimes V^*$. We will see that this subbundle is equal to the subbundle $\mathcal{Q}^*$. It follows that there is a unique map $\alpha : B \to G$ such that

$$\alpha^*(\pi_* \mathcal{I}(1)) = \pi_* \mathcal{J}(1).$$

Finally, we will show that this last equation is equivalent to the equality

$$\mathcal{L} = B \times_\alpha \Phi$$

as families of k-planes in $\mathbb{P}V$, where $B \times_\alpha \mathcal{L}$ denotes the pullback $B \times_G \mathcal{L}$ defined using the map α.

The fact that $\pi_*(\mathcal{J}(1))$ is a bundle follows from Theorem B.5 and the remark that the restriction of $\pi_*(\mathcal{J}(1))$ to a fiber b is the $(n - k)$-dimensional linear space of forms vanishing on the k-plane $\mathcal{L}'_b \subset \{b\} \times \mathbb{P}V \cong \mathbb{P}V$. The natural map $\pi_* \mathcal{J}(1) \to \pi_* \mathcal{O}_{B \times \mathbb{P}V}(1) = \mathcal{O}_B \otimes V^*$ is an inclusion on fibers, so $\pi_* \mathcal{J}(1)$ is a subbundle, as claimed.

To identify $\pi_* \mathcal{I}(1)$ with $\mathcal{Q}^*$, we remark that both are subbundles of $\mathcal{O}_B \otimes V^*$, and at each point $b \in B$ their fibers are the same subspace — namely, the space of linear forms vanishing on $\mathcal{L}_b$. It now follows from the universal property of Theorem 3.4 that there is a unique morphism $\alpha : B \to G$ such that $\alpha^* \pi_* \mathcal{I}(1) = \pi_* \mathcal{J}(1)$ as subbundles of $\mathcal{O}_B \otimes V^*$.

We claim that this property of α implies the equality $\mathcal{L} = B \times_\alpha \Phi$. To prove this, it suffices to show that $\mathcal{J} = (\alpha \times 1)^* \mathcal{I}$, or equivalently $\mathcal{J}(1) = (\alpha \times 1)^* \mathcal{I}(1)$. If we restrict $\mathcal{L}$ to the fiber over $b \in B$, we get a subspace of $\mathbb{P}V$ whose ideal is generated by the linear forms it contains. For $b \in B$, Theorem B.5 identifies this space of linear forms with the fiber of $\pi_* \mathcal{J}(1)$ at b. Thus there is a surjection

$$\pi^* \pi_* \mathcal{J}(1) \to \mathcal{J}(1).$$

Similar remarks hold for $\mathcal{I}$. Thus the commutative diagram

$$\pi^* \pi_* \mathcal{J}(1) = \pi^* \alpha^* \pi_* \mathcal{I}(1) = (\alpha \times 1)^* \pi^* \pi_* \mathcal{I}(1)$$

$$\mathcal{O}_{B \times \mathbb{P}V}$$

shows that the ideal sheaves of $\mathcal{L}$ and $B \times_\alpha \Phi$ are equal.

Finally, we prove the uniqueness of α. Suppose that $\mathcal{L} = B \times_{\alpha'} \Phi$ for some morphism α'. We will show that $\alpha' = \alpha$ by showing that $\pi_* \mathcal{J}(1) = \alpha'^* \pi_* \mathcal{I}(1)$. But the hypothesis implies that $\mathcal{J}(1) = (\alpha' \times 1)^* \mathcal{I}(1)$. From the definition of the pushforward, we get a natural map

$$\alpha'^* \pi_* \mathcal{I}(1) \to \pi_* (\alpha' \times 1)^* \mathcal{I}(1) = \pi_* \mathcal{J}(1)$$

that is an isomorphism fiber-by-fiber, so we are done.   □

6.3.2 A universal property of the Fano scheme

We realized the Fano scheme of a projective variety X as the subscheme of the Grassmannian consisting of planes lying in X, and as such it inherits a universal property:

Proposition 6.6. *If $X \subset \mathbb{P}^n$ is a subscheme, then the scheme $F_k(X)$ represents the functor of k-planes on X, in the sense that the correspondence above induces a one-to-one correspondence between morphisms of schemes $B \to F_k(X) \subset \mathbb{G}(k, n)$ and families of k-planes $\mathcal{L} \subset B \times X \subset B \times \mathbb{P}^n$ that are flat over B.*

Proof: This is a corollary of the statement for the Grassmannian: Suppose that $\mathbb{P}^n = \mathbb{P}V$ and X is defined by some homogeneous forms $g_i \in \mathrm{Sym}^{d_i} V^*$. Let $\mathcal{S}$ be the universal subbundle on $\mathbb{G}(k, n)$, so that the fiber of $\mathcal{S}^*$ at a point $[L] \in \mathbb{G}(k, n)$ is the space of linear forms on the corresponding k-plane $L \subset \mathbb{P}V$. Writing δ_{g_i} for the section of $\mathrm{Sym}^{d_i} \mathcal{S}^*$ that is the image of the form g_i, we see that g_i vanishes on L if and only if the sections σ_{g_i} vanish at the point $[L]$.   □

6.3.3 The Hilbert scheme and its universal property

Grothendieck's idea was to ask, more generally: given any projective scheme X and a subscheme Y, "How Y can move within X?" More precisely and ambitiously: Can we describe all flat families $B \times X \supset \mathcal{Y} \to B$ including Y as a fiber? Is there a universal such family?

When B is reduced, a family $\mathcal{Y}$ as above is flat if and only if the fibers all have the same Hilbert polynomial; in particular, any family over a reduced base whose fibers are all k-planes is automatically flat. (See for example Eisenbud and Harris [2000, Proposition III-56].) Grothendieck's idea was to define "the family of all subschemes" of X with Hilbert polynomial equal to $P_Y(d)$, the Hilbert polynomial of Y.

We might worry that this goes too far to be a generalization of the Fano scheme — could there be a subscheme of X that is not a k-plane but whose Hilbert polynomial is equal to that of a k-plane? The following result shows that all is well:

Proposition 6.7. *A subscheme* $Y \subset \mathbb{P}^n$ *is a linearly embedded k-plane* $\mathbb{P}^k$ *if and only if the Hilbert polynomial of Y is*

$$P(d) = \binom{d+k}{k} = \frac{(d+k)(d+k-1)\cdots(d+1)}{k(k-1)\cdots 1}.$$

Proof: Since the dimension of the d-th graded component of a polynomial ring on $k+1$ variables is $\binom{d+k}{k}$, the Hilbert polynomial of a linearly embedded k-plane is $P(d)$.

Conversely, suppose that Y has Hilbert polynomial P. From the degree and leading coefficient of P we see that Y is a scheme of dimension k and degree 1. Thus $L := Y_{\mathrm{red}} \subset Y$ is a linearly embedded k-plane. This inclusion induces a surjection of homogeneous coordinate rings $S_Y \to S_L$, and the equality of Hilbert polynomials shows that it is an isomorphism in high degrees. Since the inclusions $L \subset Y \subset \mathbb{P}^n$ can be recovered as $\mathrm{Proj}(S_L) \subset \mathrm{Proj}(S_Y) \subset \mathrm{Proj}(S)$, where S is the homogeneous coordinate ring of $\mathbb{P}^n$, and since $\mathrm{Proj}(S_Y)$ depends only on the high degree part of S_Y, this shows $L \subset Y$ is actually an equality. $\qquad\square$

Here is the general definition and existence theorem for Hilbert schemes, showing that there is a unique "most natural" scheme structure:

Proposition–Definition 6.8. *Let* $X \subset \mathbb{P}^n$ *be a closed subscheme, and let $P(d)$ be a polynomial. There exists a unique scheme* $\mathcal{H}_P(X)$*, called the* Hilbert scheme *of X for the Hilbert polynomial P, with a flat family*

$$\mathcal{H}_P(X) \times X \supset \mathcal{Y} \xrightarrow{\ \pi\ } \mathcal{H}_P(X)$$

of subschemes of X, called the universal family *of subschemes of X with Hilbert polynomial P, having the following properties:*

- *The fibers of π all have Hilbert polynomial equal to $P(d)$.*
- *For any flat family*

$$B \times X \supset \mathcal{Y}' \xrightarrow{\ \pi\ } B$$

whose fibers have Hilbert polynomial $P(d)$, there is a unique morphism $\alpha : B \to \mathcal{H}_P(X)$ such that $\mathcal{Y}'$ is equal to the pullback of $\mathcal{Y}$:

$$
\mathcal{Y}' \;=\; B \times_{\mathcal{H}_P(X)} \mathcal{Y} \longrightarrow \mathcal{Y}
$$

$$
\begin{array}{ccc}
 & & \downarrow \pi \\
B & \xrightarrow{\;\alpha\;} & \mathcal{H}_P(X)
\end{array}
$$

A compact way of stating the existence of the family $\mathcal{H}_P(X) \times X \supset \mathcal{Y} \to \mathcal{H}_P(X)$ and its universal property is to use the language of representable functors. Consider the contravariant functor from schemes to sets that is defined on objects by

$$
F_{X,P} : B \mapsto \big\{ \text{flat families } X \times B \supset \mathcal{Y}' \to B \text{ of subschemes of } X \subset \mathbb{P}^n
$$
$$
\text{whose fibers over closed points all have Hilbert polynomial } P \big\}
$$

and that takes a map $B' \to B$ to the map of sets taking a flat family over B to its pullback to a family over B'. In this language, the existence theorem says that $F_{X,P}$ is *representable* by the scheme $\mathcal{H}_P(X)$, in the sense that

$$
F_{X,P} \cong \mathrm{Mor}(-, \mathcal{H}_P(X))
$$

as functors. The universal family in $F_{X,P}(\mathcal{H}_P(X))$ then corresponds to the identity map in $\mathrm{Mor}(\mathcal{H}_P(X), \mathcal{H}_P(X))$. See for example Eisenbud and Harris [2000, Chapter VI] for more about this idea.

Proof of uniqueness in Proposition 6.8: As with any object with a universal property, the uniqueness of a map $\pi : \mathcal{Y} \to \mathcal{H}_P(X)$ with the given properties is easy: Given another such map $\pi' : \mathcal{Y}' \to B$, the universal properties of the two produce maps $B \to \mathcal{H}_P(X)$ and $\mathcal{H}_P(X) \to B$ whose composition $\mathcal{H}_P(X) \to B \to \mathcal{H}_P(X)$ is the unique map guaranteed by the definition that corresponds to the family $\pi : \mathcal{Y} \to \mathcal{H}_P(X)$ itself — that is, the identity map — and similarly for the composite $B \to \mathcal{H}_P(X) \to B$. □

6.3.4 Sketch of the construction of the Hilbert scheme

The construction of $\mathcal{H}_P(X)$ and the universal family is also relatively easy to describe, though the proofs of the necessary facts are deeper. There are several approaches, all along the lines of Grothendieck's original idea (see Grothendieck [1966b]), but the following (from Bayer [1982]) is perhaps the most explicit.

We first treat the case when $X = \mathbb{P}^n$, since (as in the case of the Fano schemes) we shall see that the general case reduces to this. Let $S = \mathbb{k}[x_0, \ldots, x_n]$ be the homogeneous coordinate ring of $\mathbb{P}^n$. The Hilbert scheme $\mathcal{H}_P(X)$ is constructed as a subscheme of the Grassmannian of $P(d)$-dimensional subspaces of S_d, the space of homogeneous forms of degree d, for suitably large d. The possibility of doing so is provided by the following basic result from commutative algebra, which combines ideas of Macaulay and Gotzmann (see Green [1989] for a coherent account).

Theorem 6.9. *With notation as above, there is an integer $d_0(P)$ (explicitly computable from the coefficients of P) such that if $d \geq d_0(P)$ then the saturated homogeneous ideal I of any subscheme of X with Hilbert polynomial P is generated in degrees $\leq d$ and $\dim(S_d/I_d) = P(d)$. Further, a subspace $U \subset S_d$ of dimension $P(d)$ generates an ideal with Hilbert polynomial $P(d)$ if and only if*

$$\dim(S_{d+1}/S_1 U) \geq P(d+1),$$

in which case

$$\dim(S_{d+1}/S_1 U) = P(d+1).$$

Example 6.10. If X is any hypersurface of degree s in $\mathbb{P}^n$, then the Hilbert function of S_X is

$$\dim(S_X)_d = \binom{n+d}{n} - \binom{n+d-s}{n},$$

which is equal to a polynomial $P(d)$ of degree $n-1$ for all d such that $d \geq s$, as one can check immediately. Conversely, given any scheme $X \subset \mathbb{P}^n$ with this Hilbert polynomial, we see that $\dim X = n - 1$, so X is a hypersurface, and the leading coefficient of the Hilbert polynomial tells us that $\deg X = s$. It follows that the saturated ideal of X is generated by a single form of degree s. In this case, every subspace $U \subset S_d$ generates an ideal with this Hilbert polynomial; the growth condition of the theorem is automatically satisfied.

Given Theorem 6.9, we choose $d \geq d_0(P)$, and take $\mathcal{H}_P(\mathbb{P}^n)$ to be the closed subscheme of the Grassmannian $G := G(\dim S_d - P(d), S_d)$ defined by determinantal equations saying that $\mathcal{H}_P(\mathbb{P}^n)$ consists of those $U \in G$ such that the vector space $S_1 U$ has the smallest possible dimension, which is $\dim S_{d+1} - P(d+1)$. Writing $\mathcal{S}$ for the universal subbundle of the trivial vector bundle $S_d \otimes \mathcal{O}_G$ on G, $\mathcal{H}_P(\mathbb{P}^n)$ is the subscheme defined by the condition that the composite map

$$S_1 \otimes \mathcal{S} \to S_1 \otimes S_d \otimes \mathcal{O}_G \to S_{d+1} \otimes \mathcal{O}_G$$

has corank $\geq P(d+1)$.

Further, we can construct the universal family $\mathcal{Y} \subset \mathbb{P}^n \times \mathcal{H}_P(\mathbb{P}^n)$ as follows. Let

$$\mathbb{P}^n \xleftarrow{\pi_1} \mathbb{P}^n \times G \xrightarrow{\pi_2} G$$

be the projection maps. There is a natural map $S_d \otimes \pi_1^* \mathcal{O}_{\mathbb{P}^n}(-d) \to \mathcal{O}_{\mathbb{P}^n \times G}$, and composing this with the inclusion we get a map of sheaves

$$\pi_2^* U \otimes \mathcal{O}_{\mathbb{P}^n}(-d) \to \mathcal{O}_{\mathbb{P}^n \times G}.$$

Let $\tilde{\mathcal{Y}}$ be the subscheme of $\mathbb{P}^n \times G$ defined by the image of this map, and let $\mathcal{Y} \to \mathcal{H}_P(\mathbb{P}^n)$ be the restriction to $\mathcal{H}_P(X) \subset G$ of the (non-)flat family given by the composite

$$\tilde{\mathcal{Y}} \subset \mathbb{P}^n \times G \xrightarrow{\pi_2} G.$$

The universal property (which we will not prove) shows that these construction are independent of the choice of $d \geq d_0$ (up to canonical isomorphism).

So far we have only defined $\mathcal{H}_P(\mathbb{P}^n)$, but we can use this to construct $\mathcal{H}_P(X)$ for any $X \subset \mathbb{P}^n$. Let $I = I(X) \subset S$ be the ideal corresponding to X, and suppose that I is generated in degrees $\leq e$. Given the Hilbert polynomial P, we choose $d \geq \max\{d_0(P), e\}$. Then to define $\mathcal{H}_P(X)$ we simply add equations to $\mathcal{H}_P(\mathbb{P}^n)$ implying that $I_d \otimes \pi_1^* \mathcal{O}_{\mathbb{P}^n}(-d)$ is contained in $U \otimes \pi_1^* \mathcal{O}_{\mathbb{P}^n}(-d)$. This can be translated into a rank condition on a map of vector bundles, as before.

Example 6.11 (Example 6.10, continued). The argument above shows that the Hilbert scheme of hypersurfaces of degree s in $\mathbb{P}^n$ is the projective space $\mathbb{P}^N$ of all homogeneous forms of degree s, and the universal family is the universal hypersurface

$$\mathcal{X} = \{(x, X) \in \mathbb{P}^n \times \mathbb{P}^N \mid x \in X\},$$

as one would hope.

Here is one way to understand the integer $d_0(P)$ that plays a central role in the construction. Recall that the set of monomials of given degree d can be ordered *lexicographically*, where

$$x^e := x_0^{e_0} \cdots x_n^{e_n} < x_0^{f_0} \cdots x_n^{f_n} =: x^f$$

if $e_i > f_i$ for the smallest i such that $e_i \neq f_i$ — informally put, if x^e involves more of the lowest-index variables than x^f. A monomial ideal $I \subset S$ is called *lexicographic* if, whenever $x^e < x^f$ are monomials of degree d and $x^f \in I$, then $x^e \in I$ too. It follows easily that the saturation of a lexicographic ideal is lexicographic.

Proposition 6.12. *Let $S = \Bbbk[x_0, \ldots, x_n]$.*

(a) *If I is any homogeneous ideal of S, then there is a lexicographic ideal J such that the Hilbert function of S/J is the same as that of S/I.*

(b) *If $P = P_I$ is the Hilbert polynomial S/I, then there is a unique saturated lexicographic ideal J_P with Hilbert polynomial P.*

The integer $d_0(P)$ may be taken to be the maximal degree of a generator of J_P.

For example, if I is the principal ideal generated by a form of degree s as in the example above, this proposition gives $d_0 = s$. See Green [1989] for further information.

6.4 Tangent spaces to Fano and Hilbert schemes

In order to use the Chern class calculation of Section 6.2.1 to count the number of distinct lines on a cubic surface, we need to know when the Fano scheme is reduced. In the zero-dimensional case, this is the same as being smooth, and the question can thus be approached through a computation of Zariski tangent spaces. Happily, we can give a simple description of the Zariski tangent spaces of any Hilbert scheme.

We first state the main assertions for Fano schemes. They will allow us to deduce the exact number of lines on a general hypersurface $X \subset \mathbb{P}^n$ of degree $d = 2n - 3$, along with other geometric facts. We will then compute the tangent spaces in the general setting of Hilbert schemes (Theorem 6.21). In Section 6.7 below, we will show how to calculate the multiplicity of $F_1(X)$ at L by writing down explicit local equations for $F_1(X) \subset \mathbb{G}(1, n)$.

6.4.1 Normal bundles and the smoothness of the Fano scheme

We will make use of the universal property of Fano schemes to give a geometric condition for the smoothness of $F_k(X)$ at a given point. Recall that if $Y \subset X$ is a smooth subvariety of the smooth variety X, then the *normal bundle* $\mathcal{N}_{Y/X}$ of Y in X is the cokernel of the map of tangent bundles $\mathcal{N}_{Y/X} = \operatorname{coker}(\mathcal{T}_Y \to \mathcal{T}_X|_Y)$ induced by the inclusion of $Y \subset X$. Recall also that the Zariski tangent space of a scheme F at a point p is by definition $\operatorname{Hom}_{\mathcal{O}_p}(\mathfrak{m}_p/\mathfrak{m}_p^2, \mathcal{O}_p/\mathfrak{m}_p)$, where $\mathfrak{m}_p$ is the maximal ideal of the local ring $\mathcal{O}_p$ of F at p.

The following theorem is a special case of a general result on Hilbert schemes, Theorem 6.21, which we will prove in the next section:

Theorem 6.13. *Suppose that $L \subset X$ is a k-plane in a smooth variety $X \subset \mathbb{P}^n$, and let $[L] \in F_k(X)$ be the corresponding point. The Zariski tangent space of $F_k(X)$ at $[L]$ is $H^0(\mathcal{N}_{L/X})$.*

The result is intuitively plausible if we think of a section of $\mathcal{N}_{L/X}$ as providing a normal vector at each point in X, with a corresponding infinitesimal motion of X.

For a case that is easy to understand, take $k = 0$. The Hilbert scheme of points on a variety X is X itself, as one checks from the definition. The tangent space at a point $x \in X$ is thus the Zariski tangent space to X at x, and this is — identifying sheaves on the space $\{x\}$ with vector spaces — equal to $\operatorname{Hom}_{\mathcal{O}_X}(\mathfrak{m}_{X,x}/\mathfrak{m}_{X,x}^2, \mathcal{O}_x) = \mathcal{N}_{x/X}$. Before introducing the general machinery of the proof, we explain how the result can be used.

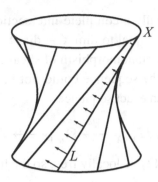

Figure 6.1 A tangent vector to the Fano scheme $F_1(X)$ at $[L]$ corresponds to a normal vector field along L in X.

Corollary 6.14. *Suppose that $L \subset X$ is a k-plane in a smooth variety $X \subset \mathbb{P}^n$, and let $[L] \in F_k(X)$ be the corresponding point. The dimension of $F_k(X)$ at $[L]$ is at most $\dim H^0(\mathcal{N}_{L/X})$. Moreover, $F_k(X)$ is smooth at $[L]$ if and only if equality holds.*

Proof of Corollary 6.14: By the principal ideal theorem, the dimension of the Zariski tangent space of a local ring is always at least the dimension of the ring, and equality holds if and only if the ring is regular. See Eisenbud [1995]. □

To apply Corollary 6.14, we need to be able to compute normal bundles, and this is often easy. For example, we have:

Proposition–Definition 6.15. *Suppose that $Y \subset X$ are schemes.*

(a) *If X and Y are smooth varieties then $\mathcal{N}_{Y/X} = \mathcal{H}om_{\mathcal{O}_Y}(\mathcal{I}_Y/\mathcal{I}_Y^2, \mathcal{O}_Y)$. For arbitrary schemes $Y \subset X$, we define $\mathcal{N}_{Y/X}$ by this formula.*

(b) *If $Y \subset X \subset W$ are schemes, and X is locally a complete intersection in W, then there is a left exact sequence of normal bundles*

$$0 \longrightarrow \mathcal{N}_{Y/X} \longrightarrow \mathcal{N}_{Y/W} \overset{\alpha}{\longrightarrow} \mathcal{N}_{X/W}|_Y.$$

If all three schemes are smooth, then α is an epimorphism.

(c) *If Y is a Cartier divisor on X then $\mathcal{N}_{Y/X} = \mathcal{O}_X(Y)$. More generally, if Y is the zero locus of a section of a bundle $\mathcal{E}$ of rank e on X, and Y has codimension e in X, then*

$$\mathcal{N}_{Y/X} = \mathcal{E}|_Y.$$

Proof: (a) For any inclusion of subschemes $Y \subset X$, there is a right exact sequence involving the cotangent sheaves of X and Y:

$$\mathcal{I}_{Y/X}/\mathcal{I}_{Y/X}^2 \overset{d}{\longrightarrow} \Omega_X|_Y \longrightarrow \Omega_Y \longrightarrow 0,$$

where d is the map taking the class of a (locally defined) function $f \in \mathcal{I}_{Y/X}$ to its differential $df \in \Omega_X|_Y$; see for example Eisenbud [1995, Proposition 16.12]. Since

X and Y are smooth, Y is locally a complete intersection in X, so $\mathcal{I}_{Y/X}/\mathcal{I}^2_{Y/X}$ is a locally free sheaf on Y of rank equal to $\dim X - \dim Y = \operatorname{rank} \Omega_X|_Y - \operatorname{rank} \Omega_Y$. If the left-hand map d were not a monomorphism of sheaves, then the image of d would have strictly smaller rank, so the sequence could not be exact at $\Omega_X|_Y$. Thus d is a monomorphism, and we have an exact sequence

$$0 \longrightarrow \mathcal{I}_{Y/X}/\mathcal{I}^2_{Y/X} \overset{d}{\longrightarrow} \Omega_X|_Y \longrightarrow \Omega_Y \longrightarrow 0$$

of bundles. Since Y is smooth, Ω_Y is locally free, so dualizing preserves exactness, and we get an exact sequence

$$0 \longleftarrow \mathcal{H}om_{\mathcal{O}_Y}(\mathcal{I}_{Y/X}/\mathcal{I}^2_{Y/X}, \mathcal{O}_Y) \longleftarrow \mathcal{T}_X|_Y \longleftarrow \mathcal{T}_Y \longleftarrow 0,$$

where the right-hand map is the differential of the inclusion $Y \subset X$, proving that $\mathcal{N}_{Y/X} = \mathcal{H}om_{\mathcal{O}_Y}(\mathcal{I}_Y/\mathcal{I}^2_Y, \mathcal{O}_Y)$.

(b) From the inclusions $Y \subset X \subset W$, we derive an exact sequence of ideal sheaves

$$0 \longrightarrow \mathcal{I}_{X/W} \longrightarrow \mathcal{I}_{Y/W} \longrightarrow \mathcal{I}_{Y/X} \longrightarrow 0.$$

Applying the functor $\mathcal{H}om_{\mathcal{O}_W}(-, \mathcal{O}_Y)$ gives a left exact sequence

$$0 \longrightarrow \mathcal{N}_{Y/X} \longrightarrow \mathcal{N}_{Y/W} \longrightarrow \mathcal{H}om(\mathcal{I}_{Y/X}, \mathcal{O}_Y).$$

Since $\mathcal{H}om(\mathcal{I}_{Y/X}, \mathcal{O}_Y) \cong \mathcal{H}om(\mathcal{I}_{Y/X} \otimes \mathcal{O}_Y, \mathcal{O}_Y)$ and $\mathcal{I}_{Y/X} \otimes \mathcal{O}_Y \cong \mathcal{I}_{Y/X}/\mathcal{I}^2_{Y/X}$, we get the desired sequence.

In the smooth case, we start with the exact sequence

$$0 \longrightarrow \mathcal{T}_X \longrightarrow \mathcal{T}_W|_X \longrightarrow \mathcal{N}_{X/W} \longrightarrow 0$$

that defines $\mathcal{N}_{X/W}$. We restrict to Y and factor out the subbundle $\mathcal{T}_Y$ from both $\mathcal{T}_X$ and $\mathcal{T}_W|_X$ to get the required exact sequence

$$0 \longrightarrow \mathcal{N}_{Y/X} \longrightarrow \mathcal{N}_{Y/W} \longrightarrow \mathcal{N}_{X/W}|_Y \longrightarrow 0.$$

(c) The first formula follows at once from part (a), since in that case $\mathcal{I}_{Y/X} = \mathcal{O}_X(-Y)$, and taking the dual of a bundle commutes with restriction.

For the second statement of part (c) we first give a geometric argument that works in the smooth case, and then a proof in general. Let Z be the total space of the bundle $\mathcal{E}$. The tangent bundle to Z restricted to the zero section $X \subset Z$ is $\mathcal{T}_X \oplus \mathcal{E}$.

Along the zero locus Y of σ, the derivative $D\sigma$ of σ is thus a map $\mathcal{T}_X|_Y \to \mathcal{T}_X|_Y \oplus \mathcal{E}_Y$. Since the component of $D\sigma$ that maps $\mathcal{T}_X|_Y$ to $\mathcal{E}_Y$ is zero along Y, the composite

$$\mathcal{T}_Y \longrightarrow \mathcal{T}_X|_Y \overset{D\sigma}{\longrightarrow} \mathcal{T}_X|_Y \oplus \mathcal{E}_Y \longrightarrow \mathcal{E}_Y$$

is zero. Locally at each point $y \in Y$, the image of $(\mathcal{T}_X)_y$ in $(\mathcal{T}_X)_y \oplus \mathcal{E}_y$ is the tangent space to $\sigma(X) \subset Z$. Since Y is smooth of codimension equal to the rank of $\mathcal{E}$, the

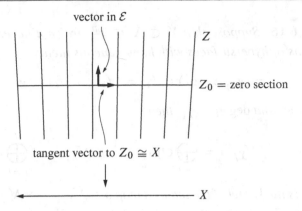

Figure 6.2 The tangent bundle to Z restricted to $Z_0 \cong X$ is $\mathcal{T}_X \oplus \mathcal{E}$.

manifold $\sigma(X)$ meets the zero locus $X \subset Z$ transversely. This means that $(\mathcal{T}_X)_y$ projects onto $\mathcal{E}_y$, and tells us that the composite map of bundles

$$\mathcal{T}_X|_Y \xrightarrow{\ D\sigma\ } \mathcal{T}_X|_Y \oplus \mathcal{E}_Y \longrightarrow \mathcal{E}_Y$$

is surjective. Considering the ranks, it follows that the sequence

$$0 \longrightarrow \mathcal{T}_Y \longrightarrow \mathcal{T}_X|_Y \longrightarrow \mathcal{E}_Y \longrightarrow 0$$

is exact; that is, $\mathcal{N}_{Y/X} = \mathcal{E}_Y$.

With a more algebraic approach, we can avoid the hypothesis that X or Y is smooth. We may think of σ as defining the map $\mathcal{O}_X \to \mathcal{E}$ that sends $1 \in \mathcal{O}_X$ to $\sigma \in \mathcal{E}$. Dualizing, the statement that Y is the zero locus of σ means that the ideal sheaf $\mathcal{I}_{Y/X}$ is the image of the map $\sigma^* : \mathcal{E}^* \to \mathcal{O}_X$. Since the codimension of Y is e, we see that Y is locally a complete intersection. Thus the kernel of σ^* is generated by the Koszul relations; that is, the sequence

$$\cdots \longrightarrow \wedge^2 \mathcal{E}^* \xrightarrow{\ \kappa\ } \mathcal{E}^* \longrightarrow \mathcal{I}_{Y/X} \longrightarrow 0$$

is exact, where $\kappa(e \wedge f) = \sigma^*(e)f - \sigma^*(f)e$. Because the coefficients in the map κ lie in $\mathcal{I}_{Y/X}$, they become zero on tensoring with $\mathcal{O}_Y = \mathcal{O}_X/\mathcal{I}_Y$, so we get the right exact sequence

$$\cdots \longrightarrow \wedge^2 \mathcal{E}^*|_Y \xrightarrow{\ 0\ } \mathcal{E}^*|_Y \longrightarrow \mathcal{I}_{Y/X}/\mathcal{I}_{Y/X}^2 \longrightarrow 0.$$

This shows that $\mathcal{E}^*|_Y \cong \mathcal{I}_{Y/X}/\mathcal{I}_{Y/X}^2$, whence $\mathcal{E}|_Y = \mathcal{E}^{**}|_Y = \mathcal{N}_{Y/X}$. $\square$

In the special case where Y is a complete intersection of X with divisors on $\mathbb{P}^n$ of degrees d_i, the normal bundle is $\mathcal{N}_{Y/X} = \bigoplus \mathcal{O}_X(d_i)$, so the last statement of Proposition 6.15 takes a particularly simple form. We can make it even more explicit when both X and Y are complete intersections:

Corollary 6.16. *Suppose that $Y \subset X \subset \mathbb{P}^n$ are (not necessarily smooth) complete intersections of hypersurfaces with homogeneous ideals*

$$I_X = (g_1, \ldots, g_s) \subset I_Y = (f_1, \ldots, f_t), \quad g_i = \sum_j a_{i,j} f_j.$$

If $\deg f_i = \varphi_i$ *and* $\deg g_i = \gamma_i$, *then*

$$\mathcal{N}_{Y/\mathbb{P}^n} = \bigoplus_{i=1}^{t} \mathcal{O}_Y(\varphi_i), \qquad \mathcal{N}_{X/\mathbb{P}^n} = \bigoplus_{i=1}^{s} \mathcal{O}_X(\gamma_i)$$

and $\mathcal{N}_{Y/X}$ *is the kernel of the induced map* $\alpha : \mathcal{N}_{Y/\mathbb{P}^n} \to \mathcal{N}_{X/\mathbb{P}^n}|_Y$ *given by the matrix* $(\bar{a}_{j,i})$, *where* $\bar{a}_{j,i}$ *denotes the restriction of* $a_{j,i}$ *to* Y.

Proof: The complete intersection X is the zero locus of the section $(g_1, \ldots, g_s)$ of the bundle $\mathcal{O}_{\mathbb{P}^n}(\gamma_1) \oplus \cdots \oplus \mathcal{O}_{\mathbb{P}^n}(\gamma_s)$, and similarly for Y. Using the formula of part (c), we see that

$$\mathcal{N}_{X/\mathbb{P}^n} = \bigoplus_{i=1}^{s} \mathcal{O}_X(\gamma_i),$$

and similarly for Y. The identification of α follows at once from part (a).   □

As an immediate application, we can finally show that there are exactly 27 distinct lines on every smooth cubic surface (pending, of course, the proof of Theorem 6.13):

Corollary 6.17. *Let* $X \subset \mathbb{P}^3$ *be a smooth surface of degree* $d \geq 3$. *If* $F_1(X) \neq \emptyset$, *then* $F_1(X)$ *is smooth and zero-dimensional. In particular, X contains at most finitely many lines, and if* $d = 3$ *then X contains exactly 27 distinct lines.*

See Corollary 6.27 for a strengthening.

Proof: Suppose $L \subset X$ is a line. As we saw in Section 2.4.2, the self-intersection number of L on X is negative, so the normal bundle $\mathcal{N}_{L/X}$ is a line bundle of negative degree. It follows that $\dim H^0(\mathcal{N}_{L/X}) = 0$, and Corollary 6.14 now implies that L is isolated and $F_1(X)$ is smooth at $[L]$.

In particular, in the case of the cubic surface the fact that the class of the Fano scheme is 27 points implies, with this result, that the Fano scheme actually consists of 27 reduced points.   □

We will also be able to see Corollary 6.17 geometrically once we have introduced the notion of first-order deformation in Section 6.4.2

6.4.2 First-order deformations as tangents to the Hilbert scheme

The proof of Theorem 6.13 and its generalization involves the idea of a *first-order deformation* of a subscheme, which is the main content of this section. Suppose that

Y is a closed subscheme of a scheme X, defined over the field $\Bbbk$. By a *deformation of $Y \subset X$ over a scheme T with distinguished point* $\operatorname{Spec} \Bbbk \in T$ we mean a subscheme $\mathcal{Y} \subset T \times X$, flat over T, whose fiber over the distinguished point $\operatorname{Spec} \Bbbk$ is equal to Y, that is, a diagram

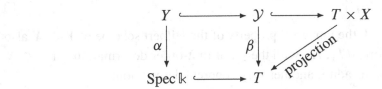

We think of the image of $\operatorname{Spec} \Bbbk \hookrightarrow T$ as a distinguished point of T, and we will denote it by $[Y]$.

A deformation is called *first-order* if its base T is the spectrum of a local ring of the form $R_m = \Bbbk[\epsilon_1, \ldots, \epsilon_m]/(\epsilon_1, \ldots, \epsilon_m)^2$ for some m. We set $T_m := \operatorname{Spec} R_m$. Note that this is a scheme with a unique closed point, which we shall denote by 0. We think of T_m as a first-order neighborhood of a point on a smooth m-dimensional variety.

It follows from the universal property of the Hilbert scheme that a first-order deformation of Y over T_m is the same thing as a morphism $T_m \to H$ sending 0 to $[Y]$.

In general, we will denote the set of morphisms of T_m into a $\Bbbk$-scheme Z sending 0 to a point $z \in Z$ by $\operatorname{Mor}_z(T_m, Z)$, so we have

$$\{\text{deformations of } Y \subset X \text{ over } T_m\} = \operatorname{Mor}_{[Y]}(T_m, H).$$

For simplicity we restrict ourselves for a while to the case $m = 1$, and consider deformations over T_1.

The identification of first-order deformations with morphisms from T_1 to H is the key to identifying the tangent space of H (and thus, in our case, of the Fano scheme). Indeed, for any closed $\Bbbk$-rational point z on any scheme Z we can identify the set $\operatorname{Mor}_z(T_1, Z)$ with the Zariski tangent space to Z at z. To describe the identification, recall that for any morphism $t : T_1 \to Z$ sending 0 to z we have a pullback map on functions, denoted $t^* : \mathcal{O}_{Z,z} \to R_1$. Restricting this map to $\mathfrak{m}_{Z,z}$, we get

$$t^*|_{\mathfrak{m}_{Z,z}} : \mathfrak{m}_{Z,z} \to \mathfrak{m}_{T,0} = \Bbbk\epsilon \cong \Bbbk.$$

Since t^* sends $\mathfrak{m}_{Z,z}^2$ to zero, we may identify $t^*|_{\mathfrak{m}_{Z,z}}$ with the induced map

$$t^*|_{\mathfrak{m}_{Z,z}} : \mathfrak{m}_{Z,z}/\mathfrak{m}_{Z,z}^2 \to \mathfrak{m}_{T,0} = \Bbbk\epsilon \cong \Bbbk.$$

Lemma 6.18. *Let $z \in Z$ be a $\Bbbk$-rational point on a $\Bbbk$-scheme. The map*

$$\operatorname{Mor}_z(T_1, Z) \to T_{z/Z} = \operatorname{Hom}_{\Bbbk}(\mathfrak{m}_{z/Z}/\mathfrak{m}_{z/Z}^2, \Bbbk)$$

sending a morphism t to the restriction of the pullback map on functions $t^*|_{\mathfrak{m}_{Z,z}}$ is bijective.

Proof: Giving a morphism $t : T_1 \to Z$ is equivalent to giving the local map of $\Bbbk$-algebras $t^* : \mathcal{O}_{Z,z} \to R_1$ that induces the identity map $\Bbbk \cong \mathcal{O}_{Z,z}/\mathfrak{m}_{Z,z} \to R_1/(\epsilon_1) = \Bbbk$. Thus t^* is determined by the induced map of vector spaces $\mathfrak{m}_{Z,z}/\mathfrak{m}_{Z,z}^2 \to (\epsilon_1) \subset R_1$.

Conversely, any map $\mathfrak{m}_{Z,z}/\mathfrak{m}_{Z,z}^2 \to (\epsilon)$ extends to a local algebra homomorphism $t^* : \mathcal{O}_{Z,z} \to \Bbbk[\epsilon]/(\epsilon^2) = R_1$. $\qquad\qquad\square$

As we have explained, the universal property of the Hilbert scheme of $Y \subset X$ also allows us to identify $\mathrm{Mor}_{[Y]}(T_1, H)$ with the set of first-order deformations of $Y \subset X$ over T_1. Such deformations admit another very concrete description:

Theorem 6.19. *Suppose that $Y \subset X$ are schemes. There is a one-to-one correspondence between flat families of subschemes of X over the base T_m with central fiber Y and homomorphisms of $\mathcal{O}_Y$-modules $\mathcal{I}_Y/\mathcal{I}_Y^2 \to \mathcal{O}_Y^m$. In particular, flat families of deformations of Y in X over T_1 correspond to global sections of the normal sheaf of Y in X.*

We will use the following characterization of flatness over T_m:

Lemma 6.20. *If M is a (not necessarily finitely generated) module over the ring R_m, then M is flat if and only if the map*

$$M^m \xrightarrow{\ (\epsilon_1,\ldots,\epsilon_m)\ } M$$

induces an isomorphism $(M/(\epsilon_1,\ldots,\epsilon_m)M)^m \cong (\epsilon_1,\ldots,\epsilon_m)M$.

Proof: The general criterion of Eisenbud [1995, Proposition 6.1] says that M is flat if and only if the multiplication map $\mu_I : I \otimes_R M \to IM$ is an isomorphism for all ideals I. But every nontrivial ideal of R_m is a summand of $(\epsilon_1,\ldots,\epsilon_m) = (\epsilon)$, and, since $(R/(\epsilon))^m \cong (\epsilon)$, the map $\mu_{(\epsilon)}$ may be identified with the given map $(M/(\epsilon)M)^m \to (\epsilon)M$. $\qquad\qquad\square$

Proof of Theorem 6.19: The problem is local, so we may assume that X and Y are affine. Since any homomorphism of sheaves $\mathcal{I}_Y \to \mathcal{O}_Y^m$ must annihilate $\mathcal{I}_Y^2$, we may identify a homomorphism

$$\varphi : \mathcal{I}_Y/\mathcal{I}_Y^2 \xrightarrow{\ (\varphi_1,\ldots,\varphi_m)\ } \mathcal{O}_Y^n$$

with the composition $\mathcal{I}_Y \to \mathcal{I}_Y/\mathcal{I}_Y^2 \to \mathcal{O}_Y^m$. Let $\mathcal{I}_\varphi \subset \mathcal{O}_X \otimes R_m$ be the ideal

$$\mathcal{I}_\varphi := \Big\{ g + \sum_j g_j \epsilon_j \ \Big| \ g \in \mathcal{I}_Y \text{ and } g_j \equiv \varphi_j(g) \bmod \mathcal{I}_Y \Big\},$$

and note that $\mathcal{I}_\varphi \supset \sum_j \epsilon_j \mathcal{I}_Y = (\epsilon)\mathcal{I}_Y$.

From $\mathcal{I}_\varphi$, we construct the family

$$
\begin{array}{ccccc}
Y & \lhook\joinrel\longrightarrow & Y_{T_m} & \lhook\joinrel\longrightarrow & T_m \times X \\[2pt]
\alpha \downarrow & & \beta \downarrow & \swarrow \text{\scriptsize projection} & \\[6pt]
\mathrm{Spec}\,\Bbbk & \longrightarrow & T_m & &
\end{array}
$$

where Y_{T_m} is defined by $\mathcal{I}_\varphi$. If we set all the $\epsilon_j = 0$, then $\mathcal{I}_\varphi$ becomes equal to $\mathcal{I}_Y$, so α is indeed the pullback of β.

We may identify $\mathcal{I}_\varphi / ((\epsilon)\mathcal{I}_Y)$ with the graph of $\varphi : \mathcal{I}_Y \to \mathcal{O}_Y^m$ in

$$\mathcal{I}_Y \oplus \mathcal{O}_Y^m \cong \mathcal{I}_Y \oplus \left(\bigoplus \mathcal{O}_Y \epsilon_j \right)$$
$$\subset \mathcal{O}_X \oplus \left(\bigoplus \mathcal{O}_Y \epsilon_j \right)$$
$$= \mathcal{O}_X[\epsilon]/((\epsilon)^2 + (\epsilon)\mathcal{I}_Y).$$

Thus $\mathcal{I}_\varphi \cap (\epsilon)\mathcal{O}_X = (\epsilon)\mathcal{I}_Y$, and it follows that

$$(\epsilon)(\mathcal{O}_X/\mathcal{I}_\varphi) = (\epsilon)\mathcal{O}_X/(\mathcal{I}_\varphi \cap (\epsilon)\mathcal{O}_X)$$
$$= (\epsilon)\mathcal{O}_X/(\epsilon)\mathcal{I}_Y$$
$$\cong (\mathcal{O}_X/\mathcal{I}_Y)^m \cong \mathcal{O}_Y^m.$$

By Lemma 6.20, $\mathcal{O}_X/\mathcal{I}_\varphi$ is flat over R_m.

Conversely, given an R_m-algebra of the form

$$S := \mathcal{O}_X[\epsilon]/((\epsilon)^2 + \mathcal{I}),$$

the statement that Y is the pullback of $Y_{T_m} := \operatorname{Spec} S$ over the morphism $\operatorname{Spec} \Bbbk \subset T_m$ means that $\mathcal{I}$ is congruent to $\mathcal{I}_Y$ modulo (ϵ). Multiplying by (ϵ) and using that $(\epsilon)^2 = 0$, we see that $\mathcal{I} \supset (\epsilon)\mathcal{I}_Y$. If S is flat over R_m, then we must have $\mathcal{I} \cap (\epsilon) = (\epsilon)\mathcal{I}_Y$. Putting these facts together, we see that $\mathcal{I}/(\epsilon)\mathcal{I}_Y$ is the graph of a homomorphism $\mathcal{I}_Y \to (\epsilon)\mathcal{O}_X/(\epsilon)\mathcal{I}_Y \cong \mathcal{O}_Y^m$, and this is the inverse of the construction above. $\square$

These results identify both the Zariski tangent space $T_{[Y],H}$ of the Hilbert scheme H of $Y \subset X$ at the point corresponding to Y, and the vector space of global sections of the normal sheaf, with the *set* of first-order deformations of Y in X, which we have already identified with the set $\operatorname{Mor}_{[Y]}(T_1, H)$. Since our goal is to compute the *dimension* of one of these two vector spaces in terms of the dimension of the dimension of the other, we must also ensure that the identification of sets preserves the vector space structure. This is the new content of the following result:

Theorem 6.21. *Suppose that $Y \subset X$ is a subscheme of a $\Bbbk$-scheme $X \subset \mathbb{P}^n$, and let H be the Hilbert scheme of Y. If $[Y] \in H$ denotes the point corresponding to Y, then*

$$T_{[Y]/H} \cong H^0(\mathcal{H}om_{\mathcal{O}_Y}(\mathcal{I}_{Y/X}/\mathcal{I}_{Y/X}^2, \mathcal{O}_Y))$$

as vector spaces.

Theorem 6.13 is the special case where Y is a k-plane in X.

Proof of Theorem 6.21: We will show how to give the set of morphisms $T_1 \to H$ the structure of a vector space, and prove that this third structure is compatible with the bijections we have already given.

The rules for addition and scalar multiplication in the set $\mathrm{Mor}_{[Y]}(T_1, H)$ are similar, and the one for addition is more complicated, so will define addition, and check that it is compatible with the identifications of Lemma 6.18 and Theorem 6.19. We leave the analogous treatment of scalar multiplication to the reader.

As before, we set $R_m = \Bbbk[\epsilon_1, \ldots, \epsilon_m]/(\epsilon_1, \ldots, \epsilon_m)^2$ and $T_m = \mathrm{Spec}\, R_m$ (we will only use the cases $m = 1$ and $m = 2$). A morphism of schemes $\Psi : T_2 \to H$ sending the closed point to $[Y]$ corresponds to a homomorphism $\psi : \mathfrak{m}_{H,[Y]} \to \Bbbk\epsilon_1 \oplus \Bbbk\epsilon_2$ or, equivalently, a pair of homomorphisms $\psi_1, \psi_2 : \mathfrak{m}_{H,[Y]} \to \Bbbk$, or a pair of morphisms $\Psi_1, \Psi_2 : T_1 \to H$ (in fancier language: T_2 is the coproduct of T_1 with itself in the category of pointed schemes). Moreover, there is an addition map

$$T_1 \xrightarrow{\text{(plus)}} T_2$$

that embeds T_1 as the closed subscheme with ideal $(\epsilon_1 - \epsilon_2) \subset R_2$. This map has the property that $\Psi \circ (\text{plus}) : T_1 \to H$ is the morphism corresponding to the sum $\psi_1 + \psi_2 : \mathfrak{m}_{H,[Y]} \to \Bbbk$.

Let $\mathcal{Y}_{\varphi_i}$ be the family obtained by pulling back the universal family along Ψ_i, and let $\varphi_i : \mathcal{I}_Y/\mathcal{I}_Y^2 \to \mathcal{O}_Y$ be the homomorphism corresponding to this flat family. We have a pullback diagram

$$
\begin{array}{ccccc}
\mathcal{Y}_{\varphi_1} & \longrightarrow & \mathcal{Y}_2 & \longleftarrow & \mathcal{Y}_{\varphi_2} \\
\downarrow & & \downarrow & & \downarrow \\
T_1 & \longrightarrow & T_2 & \longleftarrow & T_1
\end{array}
$$

of flat families, where $\mathcal{Y}_2 \to T_2$ is the family obtained by pulling back along Ψ. To show that the addition law on the set $\mathrm{Mor}_{[Y]}(T_1, H)$ agrees with addition in the vector space $H^0(\mathcal{H}om_{\mathcal{O}_Y}(\mathcal{I}_{Y/X}/\mathcal{I}_{Y/X}^2, \mathcal{O}_Y))$, it suffices to show that the pullback of $\mathcal{Y}_2$ along the map $(\text{plus}) : T_1 \to T_2$ is the family $\mathcal{Y}_{\varphi_1 + \varphi_2}$.

Let $\varphi : \mathcal{I}_Y/\mathcal{I}_Y^2 \to \mathcal{O}_Y^2$ be the homomorphism corresponding to $\mathcal{Y}_2$, so that the ideal of $\mathcal{Y}_2$ is the ideal $\mathcal{I}_\varphi$. If we compose φ with the map induced by the projection $R_2 \to R_1$ annihilating ϵ_2 we get the map φ_1, and similarly for ϵ_1 and φ_2. It follows that φ is in fact the map

$$\mathcal{I}_Y/\mathcal{I}_Y^2 \xrightarrow{\binom{\varphi_1}{\varphi_2}} \mathcal{O}_Y^2.$$

Thus if we pull back $\mathcal{Y}_2$ along the map (plus), that is, factor out $\epsilon_1 - \epsilon_2$ from the structure sheaf of $\mathcal{Y}_s$, the resulting algebra corresponds to the map $\varphi_1 + \varphi_2$, as required. $\quad\square$

Associated to any family

$$\mathcal{Y} \subset X \times B \xrightarrow{\pi} B$$

of subschemes of X is the *union of the schemes in the family*, defined to be the image of $\mathcal{Y}$ under the projection to X. In this spirit, if $Y \subset X$ are projective schemes, and B

is a subscheme of the Hilbert scheme of Y in X, then we define the *subscheme swept out by B* to be the union $Y' = Y'_B$ of the schemes in the restriction to B of the universal family over H.

We can now give a bound on the Zariski tangent spaces to Y' in the case where Y and X are smooth. Suppose that $p \in Y$ is a point of one of the schemes Y represented by points of B. The tangent space to Y' at p contains the tangent space to Y at p, so it is enough to bound the image of $T_p Y'$ in $T_p X / T_p Y$, which is the fiber at p of the normal bundle $(\mathcal{N}_{Y/X})_p$ of Y in X.

Intuitively, the amount the tangent space $T_p Y$ "moves" as Y moves in B is measured by the tangent space to B at $[Y]$, although some tangent vectors to B may produce trivial motions of $T_p Y$. Of course $T_{[Y]}B \subset T_{[Y]}H$, and by Theorem 6.21 the latter is $H^0(\mathcal{N}_{Y/X})$. Let $\varphi_{p,Y}$ be the evaluation map

$$\varphi_{p,Y} : H^0(\mathcal{N}_{Y/X}) \to (\mathcal{N}_{Y/X})_p = T_p X / T_p Y.$$

Proposition 6.22. *Let $Y \subset X$ be smooth projective schemes, and let $B \subset H$ be a closed subscheme of the Hilbert scheme of Y in X containing the point $[Y]$. If $p \in Y$ and Y' is the subscheme swept out by B, then*

$$T_p Y' / T_p Y \subset \varphi_{p,Y}(T_{[Y]}B).$$

This will follow directly from the following lemma:

Lemma 6.23. *Let $Z \subset Y$ be closed subschemes of a scheme X, and let $Z_\sigma, Y_\tau \subset$ Spec $\Bbbk[\epsilon]/(\epsilon^2) \times X$ be first-order deformations of Z and Y in X corresponding to the sections $\sigma \in H^0(\mathcal{N}_{Z/X})$ and $\tau \in H^0(\mathcal{N}_{Y/X})$. The scheme Z_σ is contained in Y_τ if and only if the images of σ and τ are equal under the maps*

$$\sigma \in H^0(\mathcal{N}_{Z/X}) = \mathrm{Hom}_{\mathcal{O}_X}(\mathcal{I}_{Z/X}, \mathcal{O}_Z)$$
$$\downarrow$$
$$\mathrm{Hom}_{\mathcal{O}_X}(\mathcal{I}_{Y/X}, \mathcal{O}_Z)$$
$$\uparrow$$
$$\tau \in H^0(\mathcal{N}_{Y/X}) = \mathrm{Hom}_{\mathcal{O}_X}(\mathcal{I}_{Y/X}, \mathcal{O}_Y)$$

induced by the inclusion $\mathcal{I}_{Y/X} \subset \mathcal{I}_{Z/X}$ and the projection $\mathcal{O}_Y \to \mathcal{O}_Z$. If Y and X are smooth, or more generally if $Y \subset X$ is locally a complete intersection, then $\mathcal{H}om_{\mathcal{O}_X}(\mathcal{I}_{Y/X}, \mathcal{O}_Z) \cong \mathcal{N}_{Y/X}|_Z$, and thus $\mathrm{Hom}_{\mathcal{O}_X}(\mathcal{I}_{Y/X}, \mathcal{O}_Z) \cong H^0(\mathcal{N}_{Y/X}|_Z)$.

See Figure 6.3.

Proof: The statement is local, so we can assume Z, Y and X are affine. We regard the global sections σ and τ as module homomorphisms $\mathcal{I}_{Z/X} \to \mathcal{O}_Z$ and $\mathcal{I}_{Y/X} \to \mathcal{O}_Y$.

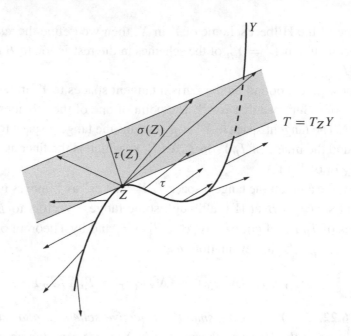

Figure 6.3 $\sigma(Z) \equiv \tau(Z)$ modulo the tangent line T to Y at Z, so the deformation of the point Z corresponding to σ keeps Z inside the deformation of Y corresponding to τ.

The schemes $Z_\sigma, Y_\tau \subset X \times T_1$ are given by the ideals

$$\mathcal{I}_\sigma = \{f + \epsilon f' \mid f \in \mathcal{I}_{Z/X} \text{ and } f' \equiv \sigma(f) \bmod \mathcal{I}_{Z/X}\}$$

and

$$\mathcal{I}_\tau = \{g + \epsilon g' \mid g \in \mathcal{I}_{Y/X} \text{ and } g' \equiv \tau(g) \bmod \mathcal{I}_{Y/X}\}$$

in $\mathcal{O}_X \otimes R_1 = \mathcal{O}_X \oplus \mathcal{O}_X \epsilon$.

Accordingly, we have $Z_\sigma \subset Y_\tau$ — that is, $\mathcal{I}_\tau \subset \mathcal{I}_\sigma$ — if and only if

$$\sigma(f) \equiv \tau(f) \bmod \mathcal{I}_{Z/X} \quad \text{for all } f \in \mathcal{I}_{Y/X},$$

which is the first statement of the lemma.

The second statement holds because, with the given hypothesis, $\mathcal{I}_{Y/X}/\mathcal{I}_{Y/X}^2$ is a vector bundle, and thus

$$\mathcal{H}om(\mathcal{I}_{Y/X}, \mathcal{O}_Z) = \mathcal{H}om(\mathcal{I}_{Y/X}/\mathcal{I}_{Y/X}^2, \mathcal{O}_Z)$$
$$= \mathcal{H}om(\mathcal{I}_{Y/X}/\mathcal{I}_{Y/X}^2, \mathcal{O}_Y) \otimes_X \mathcal{O}_Z$$
$$= \mathcal{N}_{Y/X}|_Z. \qquad \square$$

Finally, we use the notion of first-order deformation to see Corollary 6.17 geometrically, via the Gauss map $\mathcal{G}_X : X \to \mathbb{P}^{3*}$ sending $p \in X$ to the tangent plane $\mathbb{T}_p X \subset \mathbb{P}^3$ (see Section 2.1.3). The restriction of $\mathcal{G}_X$ to a line $L \subset X \subset \mathbb{P}^3$ sends L to the dual line

$$L^\perp = \{H \in \mathbb{P}^{3*} \mid L \subset H\},$$

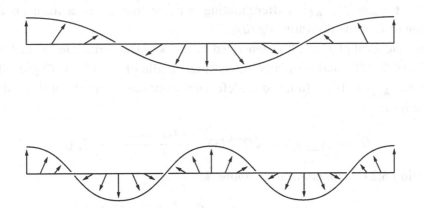

Figure 6.4 The tangent planes to a smooth quadric surface along a line wind once around the line, but in the case of a smooth cubic surface they wind around twice.

and this map, being given by the partial derivatives of the defining equation of X, has degree $d - 1$. Thus, for example, as we travel along a line on a smooth quadric surface Q, the tangent planes to Q rotate once around the line; on a smooth cubic surface X, by contrast, they wind twice around the line (see Figure 6.4). But if $\tilde{L}$ is a first-order deformation of L in $\mathbb{P}^3$, the direction of motion of a point $p \in L$ — that is, the 2-plane spanned by L and the normal vector $\sigma(p)$, where σ is the section of the normal bundle $\mathcal{N}_{L/\mathbb{P}^3}$ — is linear in p. It is thus impossible to find a first-order deformation of L on X, or on any smooth surface of higher degree.

Note that if X is singular at a point of L, the partial derivatives of the defining equation of X have a common zero along L, and so the degree of $\mathcal{G}_X : L \to L^{\perp}$ will be less than $d - 1$. Thus, for example, the tangent planes to a quadric cone are constant along a line of its ruling, and if $L \subset X$ is a line on a cubic surface with an ordinary double point on L the Gauss map will have degree 1 on L. In this case, there *will* exist first-order deformations of L on X — as we will see shortly in Section 6.7

6.4.3 Normal bundles of k-planes on hypersurfaces

In order to apply the description of the tangent space $T_L F_k(X)$ to a Fano scheme $F_k(X)$ of k-planes on a hypersurface X at a point L, we need to know something about the normal bundle of L in X.

Suppose that $L \subset X \subset \mathbb{P}^n$ is a k-plane on a (not necessarily smooth) hypersurface X of degree d in $\mathbb{P}^n$. Choose coordinates so that the ideal of L is $I_L = (x_{k+1}, \ldots, x_n)$, and let $I_X = (g) \subset I_L$. There is a unique expression

$$g = \sum_{i=k+1}^{n} x_i g_i(x_0, \ldots, x_k) + h$$

with $h \in (x_{k+1}, \ldots, x_n)^2$. Differentiating, we see that g_i, as a form on L, is the restriction to L of the derivative $\partial g/\partial x_i$.

Since the ideal of $L \subset \mathbb{P}^n$ is generated by $n - k$ linear forms, the normal bundle of L in $\mathbb{P}^n$ is $\mathcal{O}_L^{n-k}(1)$, and, similarly, the normal bundle of X in $\mathbb{P}^n$ is $\mathcal{O}_X(d)$. Thus the restriction $\mathcal{N}_{X/\mathbb{P}^n}|_L$ is $\mathcal{O}_L(d)$, and the left exact sequence of part (b) of Proposition 6.15 takes the form

$$0 \longrightarrow \mathcal{N}_{L/X} \longrightarrow \mathcal{O}_L^{n-k}(1) \xrightarrow{\ \alpha = (g_{k+1}, \ldots, g_n)\ } \mathcal{O}_L(d). \tag{6.1}$$

Proposition 6.24. *With notation as above, let*

$$\alpha = (g_{k+1}, \ldots, g_n) : \mathcal{O}_L^{n-k}(1) \to \mathcal{O}_L(d).$$

(a) *The map α is a surjection of sheaves if and only if the hypersurface X is smooth along L.*

(b) *The map α is surjective on global sections if and only if the point $[L]$ is a smooth point on $F_k(X)$ and the dimension of $F_k(X)$ at $[L]$ is equal to the "expected dimension" $(k + 1)(n - k) - \binom{k+d}{k}$.*

(c) *The map α is injective on global sections if and only if the point $[L]$ is an isolated reduced (that is, smooth) point of $F_k(X)$.*

Proof: (a) Since $L \subset X$, the derivatives of g along L are all zero, so X is smooth at a point $p \in L$ if and only if at least one of the normal derivatives $g_i = \partial g/\partial x_i$, for $i > k$, is nonzero at p. This is the condition that α is surjective as a map of sheaves.

(b) By Corollary 6.2, the dimension of $F_k(X)$ at any point is at least

$$D := (k + 1)(n - k) - \binom{k+d}{k},$$

so $F_k(X)$ is smooth of dimension D at $[L]$ if and only if the tangent space $T_{[L]}F_k(X) = H^0\mathcal{N}_{L/X}$ has dimension D. Since

$$\dim H^0(\mathcal{O}_L^{n-k}(1)) = (k + 1)(n - k) \quad \text{and} \quad \dim H^0(\mathcal{O}_L(d)) = \binom{d+k}{k},$$

we see from the exact sequence in (6.1) (before the proposition) that $\dim H^0\mathcal{N}_{L/X} = D$ if and only if α is surjective on global sections.

(c) The condition that $[L]$ is an isolated reduced point of $F_k(X)$ is the condition that $T_{[L]}F_k(X) = H^0\mathcal{N}_{L/X} = 0$, and by the argument of part (b) this happens if and only if α is injective on global sections. $\qquad\square$

We can unpack the conditions of Proposition 6.24 as follows: The condition of part (a) is equivalent to saying that the components g_i of the map α do not all vanish simultaneously at a point of L.

Using the exact sequence (6.1), and assuming that X is smooth along L so that α is a surjection of sheaves, we see that the condition of part (b) that α is surjective on global sections is equivalent to the condition $H^1(\mathcal{N}_{L/X}) = 0$. On the other hand, the global sections $x_0, \ldots, x_k$ of the i-th summand $\mathcal{O}_L(1) \subset \mathcal{O}_L^{n-1}(1)$ map by α to the sections $x_0 g_{k+i}, \ldots, x_k g_{k+i}$, so the condition of surjectivity on sections is also equivalent to the condition that the ideal $(g_{k+1}, \ldots, g_n)$ contains every form of degree d.

Similarly, it follows from the exact sequence that the condition of part (c) is equivalent to the condition $H^0(\mathcal{N}_{L/X}) = 0$. This means that there are no maps $\mathcal{O}_L$ to the kernel of α or, more concretely, that the g_i have no linear syzygies.

Although part (a) of Proposition 6.24 tells only about smoothness along L, we can do a little better: Bertini's theorem tells us that the general member of a linear series can only be singular along the base locus of the series, and it follows that the general X with a given map α is smooth except possibly along L. Thus if $\alpha : \mathcal{O}_L^{n-k}(1) \to \mathcal{O}_L(d)$ is any surjective map of sheaves, there is a smooth hypersurface X containing L such that $\mathcal{N}_{L/X} = \operatorname{Ker}\alpha$.

Example 6.25 (Cubic surfaces again). The following gives another treatment of Corollary 6.17. In the case of a cubic surfaces $X \subset \mathbb{P}^3$, we have $n = d = 3$ and the expected dimension of $F_1(X)$ is $D = 0$. If we choose

$$g_2 = x_0^2, \quad g_3 = x_1^2$$

then the conditions in all three parts of Proposition 6.24 apply: g_2 and g_3 obviously have no common zeros in $\mathbb{P}^1$; because g_2 and g_3 are relatively prime quadratic forms, they have no linear syzygies; and since

$$(x_0, x_1)(x_0^2, x_1^2) = (x_0^3, x_0^2 x_1, x_0 x_1^2, x_1^3) \subset \Bbbk[x_0, x_1, x_2, x_3],$$

the map α is surjective on global sections. Since the numbers of global sections of the source and target of α are equal, the map α is injective on global sections as well. We can see this directly, too: Because g_2, g_3 are relatively prime quadratic forms, the kernel of α is

$$\mathcal{O}_L(-1) \xrightarrow{\binom{-g_3}{g_2}} \mathcal{O}_L(1)^2,$$

so $\mathcal{N}_{L/X} = \mathcal{O}_L(-1)$, and we see again that $H^0(\mathcal{N}_{L/X}) = 0$.

From all this, we see that L will be an isolated smooth point of $F_1(X)$, where X is the hypersurface defined by the equation $x_0^2 x_2 + x_1^2 x_3 = 0$. Although this hypersurface is not smooth, Bertini's theorem, as above, shows that there are smooth cubics having the same map α. Since the rank of a linear transformation is upper-semicontinuous as the transformation varies, this will also be true for the general cubic surface containing a line. By Corollary 6.2, every cubic surface in $\mathbb{P}^3$ contains lines.

One special case of Proposition 6.24 shows that a smooth hypersurface of degree $d > 1$ cannot contain a plane of more than half its dimension:

Corollary 6.26. *Let $X \subset \mathbb{P}^n$ be a hypersurface of degree $d > 1$. If $L \subset X$ is a k-plane on X, and X is smooth along L, then*

$$k \leq \frac{n-1}{2}.$$

For example, there are no 2-planes on a smooth quadric hypersurface in $\mathbb{P}^4$ — even though the "expected dimension" $\varphi(4, 2, 2)$ is 0. This implies that all singular quadrics contain families of 2-planes of positive dimension — of course, it is easy to see this directly.

Proof: If $k > (n-1)/2$, then $k + 1 > n - k$, so $n - k$ forms on $\mathbb{P}^k$ of strictly positive degree must have a common zero, and we can apply part (a) of Proposition 6.24.   □

Remark. Corollary 6.26 is a special case of a corollary of the Lefschetz hyperplane theorem (see Appendix C), which tells us in this case that if $X \subset \mathbb{P}^n_{\mathbb{C}}$ is a smooth hypersurface and $Y \subset X$ is any subvariety of dimension $k > (n-1)/2$, then

$$\deg(X) \mid \deg(Y).$$

In the case of planes of the maximal dimension allowed by Corollary 6.26, Proposition 6.24 gives us particularly sharp information; note that this applies, in particular, to lines on surfaces in $\mathbb{P}^3$, and thus generalizes Corollary 6.17:

Corollary 6.27. *Let $X \subset \mathbb{P}^n$ be a hypersurface of degree $d \geq 3$ containing a k-plane L with $k = (n-1)/2$. If X is smooth along L then $[L]$ is an isolated smooth point of the Fano scheme $F_k(X)$. If $n = d = 3$—that is, if $X \subset \mathbb{P}^3$ is a cubic surface—then the converse is also true.*

If, in the setting of Proposition 6.24 we take an example where the g_i are general forms of degree $d-1$ in $k+1$ variables vanishing at some point of $\mathbb{P}^k$ with $d = 2, k > 1$ or $d > 3, k \geq 1$, then the g_i have no linear syzygies, so the corresponding $L \subset X$ will be a smooth point on the Fano scheme, though X is singular at a point of L. Thus the "converse" part of the corollary cannot be extended to these cases.

Proof of Corollary 6.27: If X is smooth along L, then by Proposition 6.24 the $k + 1$ forms g_i of degree $d - 1$ have no common zeros. It follows that they are a regular sequence, so all the relations among them are also of degree $d - 1 \geq 2$, so, again by Proposition 6.24, $[L]$ is a smooth point of $F_k(X)$.

In the case of a cubic surface, g_2 and g_3 are quadratic forms in two variables. If they have a zero in common then they have a linear common factor, so they have a linear syzygy.   □

Despite the nonexistence of 2-planes on smooth quadric hypersurfaces $X \subset \mathbb{P}^4$ and other examples coming from Corollary 6.26, the situation becomes uniform for hypersurfaces of degree $d \geq 3$. The proof for the general case is quite complicated, and we only sketch it. In the next section we give a complete and independent treatment for the case of lines.

Theorem 6.28. *Set* $\varphi = (k+1)(n-k) - \binom{k+d}{k}$.

(a) *If* $k = 1$ *or* $d \geq 3$ *and* $\varphi \geq 0$, *then every hypersurface of degree* d *contains* k-*planes, and the general hypersurface* X *of degree* d *in* $\mathbb{P}^n$ *has* $\dim F_k(X) = \varphi$.

(b) *If* $\varphi \leq 0$ *and* X *is a general hypersurface containing a given* k-*plane* L, *then* L *is an isolated smooth point of* $F_k(X)$.

See Exercise 6.59 for an example that can be worked out directly.

Proof: (a) The first part follows from the second using Corollary 6.2. For the second part we use Proposition 6.24. We must show that, under the given hypotheses, a general $(n-k)$-dimensional vector space of forms of degree $d-1$ generates an ideal containing all the forms of degree d.

On the other hand, for part (b) we must show that a general $(n-k)$-dimensional vector space of forms of degree $d-1$ generates an ideal without linear syzygies.

These two statements together say that if $g_{k+1}, \ldots, g_{k+n}$ is a general collection of $n-k$ forms of degree $d-1$ in $k+1$ variables, then the degree-d component of the ideal $(g_{k+1}, \ldots, g_{k+n})$ has dimension equal to $\min\{(k+1)(n-k), \binom{k+d}{k}\}$. This is a special case of the formula for the maximal Hilbert function of a homogeneous ideal with generators in given degrees conjectured in Fröberg [1985]. This particular case of Fröberg's conjecture was proved in Hochster and Laksov [1987, Theorem 1]. $\square$

6.4.4 The case of lines

The case $k = 1$ of lines is special because, very much in contrast with the general situation, we can classify vector bundles on $\mathbb{P}^1$ completely. The following result is sometimes attributed to Grothendieck, although equivalent forms go back at least to the theory of matrix pencils of Kronecker and Weierstrass:

Theorem 6.29. *Any vector bundle* $\mathcal{E}$ *on* $\mathbb{P}^1$ *is a direct sum of line bundles; that is,*

$$\mathcal{E} = \bigoplus_{i=1}^{r} \mathcal{O}_{\mathbb{P}^1}(e_i)$$

for some integers $e_1, \ldots, e_r$.

The analogous statement is false for bundles on projective space $\mathbb{P}^n$ of dimension $n \geq 2$ (see for example Exercise 5.41).

Proof: We use the Riemann–Roch theorem for vector bundles on curves. Riemann–Roch theorems in general will be discussed in Chapter 14, where we will also discuss more aspects of the behavior of vector bundles on $\mathbb{P}^1$. The reader may wish to glance ahead or, since we will not make logical use of Theorem 6.29, defer reading this proof until then.

That said, we start with a basic observation: An exact sequence of vector bundles

$$0 \longrightarrow \mathcal{E} \longrightarrow \mathcal{F} \xrightarrow{\alpha} \mathcal{G} \longrightarrow 0$$

on any variety X splits if and only if there exists a map $\beta : \mathcal{G} \to \mathcal{F}$ such that $\alpha \circ \beta = \mathrm{Id}_{\mathcal{G}}$. This will be the case whenever the map $\mathcal{H}om(\mathcal{G}, \mathcal{F}) \to \mathcal{H}om(\mathcal{G}, \mathcal{G})$ given by composition with α is surjective on global sections; from the exactness of the sequence

$$0 \longrightarrow \mathcal{H}om(\mathcal{G}, \mathcal{E}) \longrightarrow \mathcal{H}om(\mathcal{G}, \mathcal{F}) \longrightarrow \mathcal{H}om(\mathcal{G}, \mathcal{G}) \longrightarrow 0,$$

this will in turn be the case whenever $H^1(\mathcal{H}om(\mathcal{G}, \mathcal{E})) = H^1(\mathcal{G}^* \otimes \mathcal{E}) = 0$.

Now suppose $\mathcal{E}$ is a vector bundle of rank 2 on $\mathbb{P}^1$, with first Chern class of degree d. By Riemann–Roch, we have

$$h^0(\mathcal{E}) \geq d + 2;$$

from this we may deduce the existence of a nonzero global section σ of $\mathcal{E}$ vanishing at $m \geq d/2$ points of $\mathbb{P}^1$, or equivalently of an inclusion of vector bundles $\mathcal{O}_{\mathbb{P}^1}(m) \hookrightarrow \mathcal{E}$ with $m \geq d/2$. We thus have an exact sequence

$$0 \longrightarrow \mathcal{O}_{\mathbb{P}^1}(m) \longrightarrow \mathcal{E} \longrightarrow \mathcal{O}_{\mathbb{P}^1}(d - m) \longrightarrow 0,$$

and, since $2m - d \geq 0$, we have

$$H^1\big(\mathcal{H}om(\mathcal{O}_{\mathbb{P}^1}(d - m), \mathcal{O}_{\mathbb{P}^1}(m))\big) = H^1(\mathcal{O}_{\mathbb{P}^1}(2m - d)) = 0.$$

In this case, we conclude that $\mathcal{E} \cong \mathcal{O}_{\mathbb{P}^1}(m) \oplus \mathcal{O}_{\mathbb{P}^1}(d - m)$.

The case of a bundle $\mathcal{E}$ of general rank r follows by induction: If we let $\mathcal{L} \subset \mathcal{E}$ be a sub-line bundle of maximal degree m, we get a sequence

$$0 \longrightarrow \mathcal{O}_{\mathbb{P}^1}(m) \longrightarrow \mathcal{E} \xrightarrow{\alpha} \mathcal{F} \longrightarrow 0,$$

with $\mathcal{F}$ by induction a direct sum of line bundles $\mathcal{L}_i \cong \mathcal{O}_{\mathbb{P}^1}(e_i)$. Moreover, $e_i \leq m$ for all i: If $e_i > m$ for some i, then $\alpha^{-1}(\mathcal{L}_i)$ would be a bundle of rank 2 and degree $> 2m$; by the rank-2 case, this would contradict the maximality of m. Thus this sequence splits, and we are done. $\qquad\square$

We remark in passing that vector bundles on higher-dimensional projective spaces $\mathbb{P}^n$ remain mysterious, even for $n = 2$, and open problems regarding them abound. To mention just one, it is unknown whether there exist vector bundles of rank 2 on $\mathbb{P}^n$, other than direct sums of line bundles, when $n \geq 6$. Interestingly, though, Theorem 6.29

provides a tool for the study of bundles on higher-dimensional projective spaces, via the notion of *jumping lines*, which we will discuss in Section 14.4

To return to our discussion of linear spaces on hypersurfaces, suppose that $X \subset \mathbb{P}^n$ is a hypersurface of degree d and $L \subset X$ a line. We choose coordinates so that L is defined by $x_2 = \cdots = x_n = 0$. As before, we write the equation of X in the form

$$\sum_{i=2}^{n} x_i g_i(x_0, x_1) + h,$$

with $h \in (x_2, \ldots, x_n)^2$, and we let α be the map $(g_2, \ldots, g_n) : \mathcal{O}_L^{n-1} \to \mathcal{O}_L(d)$. In this situation, the expected dimension of the Fano scheme $F_1(X)$ is $\varphi := 2n - 3 - d$. We will make use of this notation throughout this subsection.

We can say exactly what normal bundles of lines in hypersurfaces are possible. Since any vector bundle on $L \cong \mathbb{P}^1$ is a direct sum of line bundles, we may write $\mathcal{N}_{L/X} \cong \bigoplus_1^{n-2} \mathcal{O}_{\mathbb{P}^1}(e_i)$.

Proposition 6.30. *Suppose that $n \geq 3$ and $d \geq 1$. There exists a smooth hypersurface X in $\mathbb{P}^n$ of degree d, and a line $\mathbb{P}^1 \cong L \subset X$ such that $\mathcal{N}_{L/X} \cong \bigoplus_{i=1}^{n-2} \mathcal{O}_{\mathbb{P}^n}(e_i)$, if and only if*

$$e_i \leq 1 \text{ for all } i \quad \text{and} \quad \sum_{i=1}^{n-2} e_i = n - 1 - d.$$

Proof: If the normal bundle is $\mathcal{N}_{L/X} \cong \bigoplus \mathcal{O}_{\mathbb{P}^1}(e_i)$, then, from the fact that there is an inclusion $\mathcal{N}_{L/X} \to \mathcal{N}_{L/\mathbb{P}^n} \cong \mathcal{O}_{\mathbb{P}^1}^{n-1}(1)$, it follows that $e_i \leq 1$ for all i. Computing Chern classes from the exact sequence of sheaves on $\mathbb{P}^1$

$$0 \longrightarrow \bigoplus_{i=1}^{n-2} \mathcal{O}_{\mathbb{P}^1}(e_i) \longrightarrow \mathcal{O}_{\mathbb{P}^1}^{n-1}(1) \longrightarrow \mathcal{O}_{\mathbb{P}^1}(d) \longrightarrow 0,$$

we get $\sum e_i = n - 1 - d$.

Conversely, suppose the e_i satisfy the given conditions. To simplify the notation, let $\mathcal{F} = \bigoplus_{i=1}^{n-2} \mathcal{O}_{\mathbb{P}^1}(e_i)$ and $\mathcal{G} = \mathcal{O}_{\mathbb{P}^1}^{n-1}(1)$. Let $\beta : \mathcal{F} \to \mathcal{G}$ be any map, and let α be the map $\mathcal{G} \to \mathcal{O}_{\mathbb{P}^1}(d)$ given by the matrix of $(n-2) \times (n-2)$ minors of the matrix of β, with appropriate signs.[1] The composition $\alpha\beta$ is zero because the i-th entry of the composite matrix is the Cauchy expansion of the determinant of a matrix obtained from β by repeating the i-th column.

---

[1] More formally and invariantly, α is the composite map

$$\mathcal{G} \cong \mathcal{O}_{\mathbb{P}^1}(n-1) \otimes \bigwedge^{n-2} \mathcal{G}^* \xrightarrow{\mathcal{O}_{\mathbb{P}^1}(n-1) \otimes \bigwedge^{n-2} \beta^*} \mathcal{O}_{\mathbb{P}^1}(n-1) \otimes \bigwedge^{n-2} \mathcal{F}^* \cong \mathcal{O}_{\mathbb{P}^1}(d),$$

where we have used an identification of $\mathcal{G}$ with $\mathcal{O}_{\mathbb{P}^1}(n-1) \otimes \bigwedge^{n-2} \mathcal{G}^*$ corresponding to a global section of $\mathcal{O}_{\mathbb{P}^1} = \mathcal{O}_{\mathbb{P}^1}(-n+1) \otimes \bigwedge^{n-1} \mathcal{G}$.

If we take β of the form

$$
\beta = \begin{pmatrix}
x_0^{1-e_1} & 0 & 0 & \cdots & 0 & 0 \\
x_1^{1-e_2} & x_0^{1-e_2} & 0 & \cdots & 0 & 0 \\
0 & x_1^{1-e_3} & x_0^{1-e_3} & \cdots & \vdots & \vdots \\
0 & 0 & x_1^{1-e_4} & \ddots & 0 & 0 \\
\vdots & \vdots & \vdots & \ddots & x_0^{1-e_{n-4}} & 0 \\
0 & 0 & 0 & \cdots & x_1^{1-e_{n-3}} & x_0^{1-e_{n-3}} \\
0 & 0 & 0 & \cdots & 0 & x_1^{1-e_{n-2}}
\end{pmatrix},
$$

then the top $(n-2) \times (n-2)$ minor will be x_0^{d-1} and the bottom $(n-2) \times (n-2)$ minor will be x_1^{d-1}. This shows that the map α will be an epimorphism of sheaves, so that the general such hypersurface X containing L will be smooth. By Eisenbud [1995, Theorem 20.9], the sequence

$$
0 \longrightarrow \mathcal{F} \xrightarrow{\beta} \mathcal{G} \xrightarrow{\alpha} \mathcal{O}_{\mathbb{P}^1}(d) \longrightarrow 0
$$

is exact, so $\mathcal{N}_{L/X} \cong \mathcal{F}$. $\square$

Corollary 6.31. *If $d \leq 2n-3$, then there exists a pair (X, L) with $X \subset \mathbb{P}^n$ a smooth hypersurface of degree d and $L \subset X$ a line such that $F_1(X)$ is smooth of dimension $2n-3-d$ in a neighborhood of $[L]$.*

Proof: Using Proposition 6.30, we observe that, if $d \leq 2n-3$, we can choose all the e_i to be ≥ -1. With this choice, $\dim H^0(\mathcal{O}_{\mathbb{P}^1}(e_i)) = e_i + 1$ for all i and hence $\dim H^0(\mathcal{N}_{L/X}) = 2n-3-d$. Since $F_1(X)$ has dimension at least $2n-3-d$ everywhere, the result follows. $\square$

Corollary 6.32. *If $d \leq 2n-3$, then every hypersurface of degree d in $\mathbb{P}^n$ contains a line.*

Proof: The universal Fano scheme $\Phi(n, d, 1)$ is irreducible of dimension $N-d+2n-3$. Moreover, Corollary 6.31 asserts that at some point $(X, L) \in \Phi$ the fiber dimension of the projection $\Phi(n, d, 1) \to \mathbb{P}^N$ is $2n-d-3$. It follows that this projection is surjective. $\square$

We have seen above that the Fano scheme of any smooth cubic surface in $\mathbb{P}^3$ is reduced and of the correct dimension. We can now say something about the higher-dimensional case as well:

Corollary 6.33. *The Fano scheme of lines on any smooth hypersurface of degree $d \leq 3$ is smooth and of dimension $2n-3-d$. But if $n \geq 4$ and $d \geq 4$, then there exist smooth hypersurfaces of degree d in $\mathbb{P}^n$ whose Fano schemes are singular or of dimension $> 2n-3-d$.*

Proof: We follow the notation of Proposition 6.30. If $d \leq 3$, then for any $e_1, \ldots, e_{n-2}$ allowed by the conditions of the proposition we have that all the $e_i \geq -1$, and thus $h^0(\mathcal{N}_{L/X}) = \chi(\mathcal{N}_{L/X}) = 2n - 3$, proving that the Fano scheme is smooth and of expected dimension at L.

On the other hand, if $n \geq 4$ and $d \geq 4$ then we can take $e_1 = \cdots = e_{n-3} = 1$ and $e_{n-2} = 2 - d \leq -2$. In this case $h^0(\mathcal{N}_{L/X}) = 2n - 6 > 2n - 3 - d$, so the Fano scheme is singular or of "too large" dimension at L. $\qquad\square$

The first statement of Corollary 6.33 is an easy case of the conjecture of Debarre and de Jong, which we will discuss further in Section 6.8.

6.5 Lines on quintic threefolds and beyond

We can now answer the first of the keynote questions of this chapter: How many lines are contained in a general quintic threefold $X \subset \mathbb{P}^4$? More generally, we can now compute the number of distinct lines on a general hypersurface X of degree $d = 2n - 3$ in $\mathbb{P}^n$, the case in which the expected dimension of the family of lines is zero.

The set-up is the same as that for the lines on a cubic surface: The defining equation g of the hypersurface X gives a section σ_g of the bundle $\mathrm{Sym}^d \, S^*$ on the Grassmannian $\mathbb{G}(1, n)$, the zero locus of σ_g is then the Fano scheme $F_1(X)$ of lines on X, and (assuming $F_1(X)$ has the expected dimension 0) the degree m of this scheme is the degree of the top Chern class $c_{d+1}(\mathrm{Sym}^d \, S^*) \in A^{d+1}(\mathbb{G}(1, n))$. If we can show in addition that $H^0(\mathcal{N}_{L/X}) = 0$ for each line $L \subset X$, then it follows as in the previous section that the Fano scheme is zero-dimensional and reduced, so the actual number of distinct lines on X is exactly m.

To calculate the Chern class we could use the splitting principle. The computation is reasonable for $n = 4$, $d = 5$, the case of the quintic threefold, but becomes successively more complicated for larger n and d. *Schubert2* (in *Macaulay2*) instead deduces it from a Gröbner basis for the Chow ring. Here is a *Schubert2* script that computes the numbers for $n = 3, \ldots, 20$, along with its output:

```
loadPackage "Schubert2"
grassmannian = (m,n) -> flagBundle({m+1, n-m})
time for n from 3 to 20 do(
    G=grassmannian(1,n);
    (S,Q) = G.Bundles;
    d = 2*n-3;
    print integral chern symmetricPower(d, dual S))
```

```
27
2875
698005
305093061
210480374951
210776836330775
```

```
289139638632755625
520764738758073845321
1192221463356102320754899
3381929766320534635615064019
11643962664020516264785825991165
47837786502063195088311032392578125
231191601420598135249236900564098773215
1298451577201796592589999161795264143531439
8386626029512440725571736265773047172289922129
61730844370508487817798328189038923397181280384657
513687287764790207960329434065844597978401438841796875
4798492409653834563672780605191070760393640761817269985515
       -- used 119.123 seconds
```

The following result gives a geometric meaning to these numbers beyond the fact that they are degrees of certain Chern classes:

Theorem 6.34. *If* $X \subset \mathbb{P}^n$ *is a general hypersurface of degree* $d \geq 1$, *then the Fano scheme* $F_1(X)$ *of lines on* X *is reduced and has the expected dimension* $2n - d - 3$.

We now have the definitive answer to Keynote Question (a):

Corollary 6.35. *A general quintic threefold* $X \subset \mathbb{P}^4$ *contains exactly 2875 lines. More generally, the numbers in the Schubert2 output above are equal to the number of distinct lines on general hypersurfaces of degrees* $3, 5, \ldots, 37$ *and dimensions* $2, 3, \ldots, 19$.

We have seen that *every* smooth cubic surface has exactly 27 distinct lines. By contrast, the hypothesis of generality in the preceding corollary is really necessary for quintic threefolds: By Corollary 6.33, the Fano scheme of a smooth quintic threefold may be singular or positive-dimensional (we will see in Exercises 6.62 and 6.67 that both possibilities actually occur).

The 2875 lines on a quintic threefold have played a significant role in algebraic geometry, and even show up in physics. For example, the Lefschetz hyperplane theorem (see for example Milnor [1963]) implies that all 2875 are homologous to each other, but one can show that they are linearly independent in the group of cycles modulo algebraic equivalence (Ceresa and Collino [1983]). On the other hand, the number of rational curves of degree d on a general quintic threefold, of which the 2875 lines are the first example, is one of the first predictions of *mirror symmetry* (see for example Cox and Katz [1999]).

Proof of Theorem 6.34: We already know that for general X of degree $d > 2n - 3$ the Fano scheme $F_1(X)$ is empty, so we henceforward assume that $d \leq 2n - 3$. We have seen in Corollary 6.31 that there exists a pair (X, L) with $X \subset \mathbb{P}^n$ a smooth hypersurface of degree d and $L \subset X$ a line such that $\dim T_L F_1(X) = 2n - 3 - d$; that is, $F_1(X)$ is smooth of the expected dimension in a neighborhood of L. We now use an incidence correspondence to deduce that, for general X, the lines $L \in F_1(X)$ with this property form an open dense subset of $F_1(X)$. In particular, if $d = 2n - 3$ then X contains just a finite number of lines, every one of which is a reduced point of $F_1(X)$.

Let $\mathbb{P}^N$ be the projective space of forms of degree d in $n+1$ variables, whose points we think of as hypersurfaces in $\mathbb{P}^n$. Consider the projection maps from the universal Fano scheme $\Phi := \Phi(n, d, 1)$:

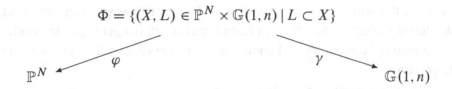

$$\Phi = \{(X, L) \in \mathbb{P}^N \times \mathbb{G}(1, n) \mid L \subset X\}$$

$$\mathbb{P}^N \xleftarrow{\quad \varphi \quad} \qquad \xrightarrow{\quad \gamma \quad} \mathbb{G}(1, n)$$

so that the fiber of φ over the point X of $\mathbb{P}^N$ is the Fano scheme $F_1(X)$ of X. As we have seen in Proposition 6.1, Φ is smooth and irreducible of dimension $N + 2n - 3 - d$. It follows that the fiber of φ through any point of Φ has dimension $\geq 2n - 3 - d$.

The set of points of $\mathbb{P}^N$ where the fiber dimension of φ is equal to $2n-3-d$ is open; within that, the set U of points where the fiber is smooth is also open. Corollary 6.31 shows that this open set is nonempty; given this, it follows that if X is a general hypersurface of degree d, then any component of $F_1(X)$ is generically reduced of dimension $N + 2n - 3 - d$. Since $F_1(X)$ is defined by the vanishing of a section of a bundle of rank $d + 1$, it is locally a complete intersection. Thus $F_1(X)$ cannot have embedded components, and the fact that it is generically reduced implies that it is reduced. $\qquad\square$

6.6 The universal Fano scheme and the geometry of families of lines

In Keynote Question (c) we asked: What is the degree of the surface S in $\mathbb{P}^3$ swept out by the lines on a cubic surface as the cubic surface moves in a general pencil? What is the genus of the curve $C \subset \mathbb{G}(1, 3)$ consisting of the points corresponding to lines on the various elements of the pencil of cubic surfaces? We can answer such questions by giving a "global" view of the universal Fano scheme as the zero locus of a section of a vector bundle, just as we have done for Fano schemes of individual hypersurfaces.

We will compute the degree of S as the number of times S intersects a general line. The task of computing this number is made easier by the fact that a general point of the surface lies on only one of the lines in question (reason: a general point that lies on two lines would have to lie on lines from different surfaces in the pencil, and thus would lie in the base locus of the pencil, contradicting the assumption that it was a general point). Thus the degree of the surface is the same as the degree of the curve $C \subset \mathbb{G}(1, 3)$ in the Plücker embedding (see Section 4.2.3 for a more general statement).

Let $\mathbb{P}^N$ be the space of hypersurfaces of degree d in $\mathbb{P}^n$. The incidence correspondence

$$\Phi = \Phi(n, d, 1) = \{(X, L) \in \mathbb{P}^N \times \mathbb{G}(1, n) \mid L \subset X\},$$

which we call the *universal* or *relative* Fano scheme of lines on such hypersurfaces, was introduced in Section 6.1. We can learn about its global geometry by realizing it as the zero locus of a section of a bundle, just as in the case of the Fano scheme of a given hypersurface.

We have seen that the maps of vector spaces

{polynomials of degree d on $\mathbb{P}^n$} $\rightarrow$ {polynomials of degree d on L}

for different $L \in \mathbb{G}(1, n)$ fit together to form a bundle map

$$V \otimes \mathcal{O}_{\mathbb{G}(1,n)} \rightarrow \mathrm{Sym}^d \, \mathcal{S}^*$$

on the Grassmannian $\mathbb{G}(1, n)$, where $V = H^0(\mathcal{O}_{\mathbb{P}^n}(d))$ is the vector space of all polynomials of degree d. Likewise, the inclusions

$$\langle f \rangle \hookrightarrow V$$

fit together to form a map of vector bundles on $\mathbb{P}V \cong \mathbb{P}^{19}$

$$\mathcal{T} = \mathcal{O}_{\mathbb{P}^{19}}(-1) \rightarrow V \otimes \mathcal{O}_{\mathbb{P}^{19}},$$

where $\mathcal{T}$ is the universal subbundle on $\mathbb{P}^{19}$.

We will put these two constructions together to understand not only $\Phi(n, d, 1)$, but also its restriction to a general linear space of forms $M \subset \mathbb{P}^N$. We denote the restriction of the universal Fano scheme to M by $\Phi(n, d, 1)|_M$.

Theorem 6.36. *The universal Fano scheme $\Phi(n, d, 1)|_M$ of lines on a general m-dimensional linear family $M = \mathbb{P}^m$ of hypersurfaces of degree d in $\mathbb{P}^n$ is reduced and of codimension $d + 1$ in the $(2n - 2 + m)$-dimensional space $\mathbb{P}^m \times \mathbb{G}(1, n)$. It is the zero locus of a section of the rank-$(d + 1)$ vector bundle $\mathcal{E} = \pi_2^* \, \mathrm{Sym}^d \, \mathcal{S}^* \otimes \pi_1^* \mathcal{O}_{\mathbb{P}^m}(1)$ on that space, so its class is $c_{d+1}(\mathcal{E})$.*

Proof: The fact that $\Phi(n, d, 1)|_M$ is reduced and of the expected dimension follows from Bertini's theorem and the corresponding statement for Φ (Proposition 6.1). To characterize $\Phi(n, d, 1)|_M$ as the zero locus of a section of a vector bundle, it likewise suffices to treat the case $M = \mathbb{P}(\mathrm{Sym}^d \, V^*)$, the space of all forms of degree d, so that $m = N := \dim V^* - 1$.

Consider the product of $\mathbb{P} \, \mathrm{Sym}^d \, V^*$ and the Grassmannian $\mathbb{G}(1, n)$, and its projections

$$\mathbb{P}(\mathrm{Sym}^d \, V^*) \xleftarrow{\pi_1} \mathbb{P}(\mathrm{Sym}^d \, V^*) \times \mathbb{G}(1, n) \xrightarrow{\pi_2} \mathbb{G}(1, n).$$

On the product, we have maps

$$\pi_1^* \mathcal{O}_{\mathbb{P} \operatorname{Sym}^d V^*}(-1) \longrightarrow \pi_1^* \operatorname{Sym}^d V^* \cong \pi_2^* \operatorname{Sym}^d V^* \longrightarrow \pi_2^* \operatorname{Sym}^d \mathcal{S}^*.$$

Restricted to the fiber over the point of $\mathbb{P} \operatorname{Sym}^d V^*$ corresponding to f, the composite map takes a generator of $\pi_1^* \mathcal{O}_{\mathbb{P} \operatorname{Sym}^d V^*}(-1)|_{(f)}$ to σ_f. Thus the zero locus of the composite map is the incidence correspondence $\Phi(n, d, 1)$.

Let σ be the corresponding global section of the bundle

$$\mathcal{E} := \mathcal{H}om(\pi_1^* \mathcal{O}_{\mathbb{P} \operatorname{Sym}^d V^*}(-1), \pi_2^* \operatorname{Sym}^d \mathcal{S}^*)$$
$$\cong \pi_2^* \operatorname{Sym}^d \mathcal{S}^* \otimes \pi_1^* \mathcal{O}_{\mathbb{P}^N}(1).$$

The zero locus of the composite map is the same as the zero locus of σ. Moreover, if we restrict to an open subset of the Grassmannian over which the universal subbundle $\mathcal{S}$ is trivial, then the vanishing of σ is given by the local equations we originally used to define the scheme structure on Φ. $\square$

Theorem 6.36 allows us to calculate the class of Φ in the Chow ring of $\mathbb{P}^N \times \mathbb{G}(1, n)$, which immediately gives the answers to Keynote Question (c). To express this, we will use the symbol ζ for the pullback to $\mathbb{P}^{19} \times \mathbb{G}(1, 3)$ of the hyperplane class on the space $\mathbb{P}^{19}$ of cubic surfaces (and for the pullback to $\mathbb{P}^{34} \times \mathbb{G}(1, 3)$ of the hyperplane class on the space $\mathbb{P}^{34}$ of quartic surfaces), and the symbols $\sigma_{i,j}$ for the pullbacks to $\mathbb{P}^{19} \times \mathbb{G}(1, 3)$ and $\mathbb{P}^{34} \times \mathbb{G}(1, 3)$ of the corresponding classes in $A(\mathbb{G}(1, 3))$.

Corollary 6.37. *The class of the universal Fano scheme $\Phi(3, 3, 1)$ of lines on cubic surfaces in $\mathbb{P}^3$ is*

$$[\Phi(3, 3, 1)] = c_4(\pi_2^* \operatorname{Sym}^3 \mathcal{S}^* \otimes \pi_1^* \mathcal{O}_{\mathbb{P}^{19}}(1))$$
$$= 27\sigma_{2,2} + 42\sigma_{2,1}\zeta + (11\sigma_2 + 21\sigma_{1,1})\zeta^2 + 6\sigma_1\zeta^3 + \zeta^4,$$

while the class of the universal Fano scheme $\Phi(3, 4, 1)$ of lines on quartic surfaces in $\mathbb{P}^3$ is

$$[\Phi(3, 4, 1)] = c_5(\pi_2^* \operatorname{Sym}^4 \mathcal{S}^* \otimes \pi_1^* \mathcal{O}_{\mathbb{P}^{34}}(1))$$
$$= 320\sigma_{2,2}\zeta + 220\sigma_{2,1}\zeta^2 + (30\sigma_2 + 55\sigma_{1,1})\zeta^3 + 10\sigma_1\zeta^4 + \zeta^5.$$

If C is the curve of lines on a general pencil of cubic surfaces, then the degree of C is 42 and the genus of C is 70. The number of quartic surfaces in a general pencil that contain a line is 320.

Restricting to a point in $\mathbb{P}^{19}$, we see again that a general cubic surface X will contain

$$[\Phi(3, 3, 1)] \cdot \zeta^{19} = 27$$

lines.

Proof of Corollary 6.37: The identifications of $[\Phi(3,3,1)]$ and $[\Phi(3,4,1)]$ with the given Chern classes is part of Theorem 6.36.

For the explicit computations of the Chern classes one can use the splitting principle or appeal to *Schubert2*. Here is the computation, via the splitting principle, for the case of $\Phi(3,3,1)$, the fourth Chern class of the bundle $\mathcal{E}$ on $\mathbb{P}^{19} \times \mathbb{G}(1,3)$:

Formally factoring the Chern class of $\pi_2^* \mathcal{S}^*$ as

$$c(\pi_2^* \mathcal{S}^*) = 1 + \sigma_1 + \sigma_{1,1} = (1+\alpha)(1+\beta),$$

we can write

$$c(\pi_1^* \mathcal{O}_{\mathbb{P}^{19}}(1) \otimes \pi_2^* \operatorname{Sym}^3 \mathcal{S}^*)$$
$$= (1 + 3\alpha + \zeta)(1 + 2\alpha + \beta + \zeta)(1 + \alpha + 2\beta + \zeta)(1 + 3\beta + \zeta),$$

and in particular the top Chern class is given by

$$c_4(\pi_1^* \mathcal{O}_{\mathbb{P}^{19}}(1) \otimes \pi_2^* \operatorname{Sym}^3 \mathcal{S}^*) = (3\alpha + \zeta)(2\alpha + \beta + \zeta)(\alpha + 2\beta + \zeta)(3\beta + \zeta)$$
$$\in A^4(\mathbb{P}^{19} \times \mathbb{G}(1,3)).$$

Evaluating, we first have

$$(3\alpha + \zeta)(3\beta + \zeta) = 9\sigma_{1,1} + 3\sigma_1 \zeta + \zeta^2,$$

and then

$$(2\alpha + \beta + \zeta)(\alpha + 2\beta + \zeta) = 2\sigma_1^2 + \sigma_{1,1} + 3\sigma_1 \zeta + \zeta^2.$$

Multiplying out, we have

$$[\Phi] = 27\sigma_{2,2} + 42\sigma_{2,1}\zeta + (11\sigma_2 + 21\sigma_{1,1})\zeta^2 + 6\sigma_1 \zeta^3 + \zeta^4.$$

Here is the corresponding *Schubert2* code:

```
n=3
d=3
m=19

P = flagBundle({1,m}, VariableNames=>{z,q1})
(Z,Q1)=P.Bundles
V = abstractSheaf(P,Rank =>n+1)
G = flagBundle({2,n-1},V,VariableNames=>{s,q})
(S,Q) = G.Bundles
p = G.StructureMap
ZG = p^*(dual Z)
chern_4 (ZG**symmetricPower_d dual S)
```

Replacing the line "d = 3" with "d = 4," we get the corresponding result for $\Phi(3,4,1)$.

From the computation of $[\Phi(3,3,1)]$, we see that the number of lines on members of a general pencil of cubics meeting a given line is

$$[\Phi] \cdot \sigma_1 \cdot \zeta^{18} = 42,$$

from which we deduce that the degree of C, which is equal to the degree of the surface swept out by the lines on our pencil of cubics, is 42. For the genus $g(C)$ of C, we use part (c) of Proposition 6.15 to conclude that the normal bundle of C is the bundle $\mathcal{E}|_C$, where $\mathcal{E}$ is the restriction to $\mathbb{P}^1 \times \mathbb{G}(1, 3)$ of the bundle $\pi_2^* \operatorname{Sym}^3 \mathcal{S}^* \otimes \pi_1^* \mathcal{O}_{\mathbb{P}^1}(1)$, whose section defines $\Phi(3, 3, 1)_{\mathbb{P}^1}$, as in Corollary 6.37. From the exact sequence

$$0 \longrightarrow \mathcal{T}_C \longrightarrow \mathcal{T}_{\mathbb{P}^1 \times \mathbb{G}(1,3)}|_C \longrightarrow \mathcal{N}_{C/\mathbb{P}^1 \times \mathbb{G}(1,3)} \longrightarrow 0,$$

we deduce that the degree of $\mathcal{T}_C$, which is $2 - 2g(C)$, is

$$\deg \mathcal{T}_C = \deg c_1(\mathcal{T}_C) = \deg([C] c_1(\mathcal{T}_{\mathbb{P}^1 \times \mathbb{G}(1,3)})) - \deg c_1(\mathcal{N}_{C/\mathbb{P}^1 \times \mathbb{G}(1,3)})$$
$$= c_4(\mathcal{E})(4\sigma_1 + 2\zeta) - c_4(\mathcal{E}) c_1(\mathcal{E}),$$

where we have used the computation $c_1(\mathcal{T}_{\mathbb{G}(1,3)}) = 4\sigma_1$ from Proposition 5.18. We can compute $c_1(\mathcal{E})$ by the splitting principle or by calling

```
chern_1 (ZG**symmetricPower_d dual S)
```

and we get $c_1(\mathcal{E}) = 6\sigma_1 + 4\zeta$. Using the fact that ζ^2 restricts to zero on the preimage of a line in $\mathbb{P}^{19}$, this gives

$$2 - 2g(C) = \deg(27\sigma_{2,2} + 42\sigma_{2,1}\zeta)(4\sigma_1 + 2\zeta - 6\sigma_1 - 4\zeta)$$
$$= \deg(-138\sigma_{2,2}\zeta) = -138,$$

whence $g = 70$. Another view of this computation is suggested in Exercise 6.54.

Finally, consider a general pencil of quartic surfaces. By Exercise 6.64, no element of the pencil will contain more than one line. It is likewise true that no line will lie on more than one element of the pencil. (If a line lay on more than one element of the pencil, it would be a component of the base locus — but, since the pencil is general, the base locus is smooth and connected.) Thus, the number of quartic surfaces that contain a line in a general pencil of quartic surfaces is the number of lines that lie on some quartic surface in the pencil, that is, the degree of $\Phi(3, 4, 1) \cap \mathbb{P}^1$. Writing σ again for the section of $\pi_2^* \operatorname{Sym}^4 \mathcal{S}^* \otimes \pi_1^* \mathcal{O}_{\mathbb{P}^1}(1)$ defined above, this is

$$\deg(\zeta^{33} c_5(\pi_2^* \operatorname{Sym}^4 \mathcal{S}^* \otimes \pi_1^* \mathcal{O}_{\mathbb{P}^1}(1))).$$

By the computation of $[\Phi(3, 4, 1)]$, this is 320.   □

The coefficients of higher powers of ζ in the class of $\Phi(3, 3, 1)$ computed above have to do with the geometry of larger linear systems of cubics: For example, we will see how to answer questions about lines on a net of cubics in Exercise 6.50.

6.6.1 Lines on the quartic surfaces in a pencil

Here is a slightly different approach to Keynote Question (b). Given that the set of quartic surfaces that contain some line is a hypersurface Γ in the projective space $\mathbb{P}^{34}$ of quartic surfaces, we are asking for the degree of Γ.

To find that number, we look again at the bundle $\mathcal{E}$ on the Grassmannian $\mathbb{G}(1,3)$, whose fiber over a point $L \in \mathbb{G}(1,3)$ is the vector space

$$\mathcal{E}_L = H^0(\mathcal{O}_L(4)),$$

that is, the fourth symmetric power $\mathrm{Sym}^4 \mathcal{S}^*$ of the dual of the universal subbundle on $\mathbb{G}(1,3)$. As before, the polynomials f and g generating the pencil define sections σ_f and σ_g of the bundle $\mathcal{E}$. The locus of lines $L \subset \mathbb{P}^3$ that lie on some element of the pencil is the locus where the values of the sections σ_f and σ_g are dependent, so the degree of this locus is the degree of the fourth Chern class $c_4(\mathcal{E}) \in A^4(\mathbb{G}(1,3)) \cong \mathbb{Z}$. As before, this can be computed either with the splitting principle or with *Schubert2*, and one finds again the number 320.

We will see another way of calculating the genus of the curve Φ in the following chapter (after we have determined the number of singular cubic surfaces in a general pencil), by expressing Φ as a 27-sheeted cover of $\mathbb{P}^1$ and using Hurwitz's theorem.

6.7 Lines on a cubic with a double point

Identifying the Fano scheme $F_1(X)$ as the Hilbert scheme of lines on X has allowed us to give a necessary and sufficient condition for its smoothness, and to show that it is indeed smooth in certain cases. But there are aspects of its geometry that we cannot get at in this way, such as the multiplicity of $F_1(X)$ at a point L where it is not smooth. We might want to know, for example, if we can find a smooth hypersurface $X \subset \mathbb{P}^n$ of degree $2n - 3$ whose Fano scheme of lines includes a point of multiplicity exactly 2, as in Harris [1979]; or, we might ask, if X has an ordinary double point, how does this affect the number of lines it will contain? To answer such questions we must go back to the local equations of $F_1(X)$ introduced (in more generality) at the beginning of this chapter.

We will describe the lines on a cubic surface with one ordinary double point. Other examples can be found in Exercise 6.55, where we will consider the case of cubic surfaces with more than one double point, and in Exercise 6.62, where we will show that it is possible to find a smooth quintic hypersurface $X \subset \mathbb{P}^4$ whose Fano scheme contains an isolated double point.

To this end, we will adapt the notation of Section 6.1.1 to the case of cubic surfaces. We work in an open neighborhood $U \subset \mathbb{G}(1,3)$ of the line

$$L : x_2 = x_3 = 0,$$

where U consists of the lines not meeting the line $x_0 = x_1 = 0$. Any line in U can be written uniquely as the row space of a matrix of the form

$$A = \begin{pmatrix} 1 & 0 & a_2 & a_3 \\ 0 & 1 & b_2 & b_3 \end{pmatrix}$$

(so $U \cong \mathbb{A}^4$, with coordinates a_2, a_3, b_2, b_3). Such a line has the parametrization

$$\mathbb{P}^1 \ni (s_0, s_1) \rightarrow (s_0, s_1)A = (s_0, s_1, a_2 s_0 + b_2 s_1, a_3 s_0 + b_3 s_1) \in \mathbb{P}^3.$$

Now let $X \subset \mathbb{P}^3$ be a cubic surface containing L, and suppose that the point $p = (1, 0, 0, 0) \in L$ is an ordinary double point of X; that is, the tangent cone to X at p is the cone over a smooth conic curve. We assume that X has no other singularities along L.

We may also suppose that the tangent cone to X at p is given by the equation $x_1 x_3 + x_2^2 = 0$. With these choices, the defining equation $g(x)$ of X can be written in the form

$$g(x) = x_0 x_1 x_3 + x_0 x_2^2 + \alpha x_1^2 x_2 + \beta x_1^2 x_3 + \gamma x_1 x_2^2 + \delta x_1 x_2 x_3 + \epsilon x_1 x_3^2 + k,$$

where $k \in (x_2, x_3)^3$. The condition that X be smooth along L except at p says that $\alpha \neq 0$; otherwise the coefficients $\alpha, \dots, \epsilon$ are arbitrary.

As we saw in Section 6.4.3, the normal bundle $\mathcal{N}_{L/X}$ can be computed from the short exact sequence

$$0 \longrightarrow \mathcal{N}_{L/X} \longrightarrow \mathcal{O}_L^2(1) \xrightarrow{(g_2 \; g_3)} \mathcal{O}_L(3),$$

where the g_3, g_3 are the coefficients of x_2, x_3 in the part of g that is not contained in (x_2, x_3); that is, $g_2 = \alpha x_1^2$ and $g_3 = x_0 x_1 + \beta x_1^2$.

Since the polynomial ring in s, t has unique factorization, the syzygies between these two forms are generated by the linear syzygy $\alpha x_1 g_3 - (x_0 + \beta x_1) g_2 = 0$, so $\mathcal{N}_{L/X} \cong \mathcal{O}_L$, and the tangent space to the Fano scheme is given by $T_{[L]} F_1(X) = H^0(\mathcal{N}_{L/X})$, which is one-dimensional. In particular, the Fano scheme is *not* "smooth of the expected dimension" at $[L]$.

We can now write down the local equations of $F_1(X)$ near L: If we substitute the four coordinates from the parametrization of a line in U into g, we get

$$g(s, t, a_2 s + b_2 t, a_3 s + b_3 t) = c_0 s^3 + c_1 s^2 t + c_2 s t^2 + c_3 t^3,$$

where the c_i are the polynomials in the $a_{i,j}$ that define the intersection of the Fano scheme with U. Writing this out, we find that, modulo terms of higher degree, the c_i are

$$c_0 = a_2^2,$$
$$c_1 = a_3 + 2a_2 b_2 + \gamma a_2^2 + \delta a_2 a_3 + \epsilon a_3^2,$$
$$c_2 = b_3 + \alpha a_2 + \beta a_3 + b_2^2 + 2\gamma a_2 b_2 + \delta(a_2 b_3 + a_3 b_2) + 2\epsilon a_3 b_3,$$
$$c_3 = \alpha b_2 + \beta b_3 + \gamma b_2^2 + \delta b_2 b_3 + \epsilon b_3^2.$$

Examining these polynomials, we see that c_1, c_2 and c_3 have independent differentials at the origin $a_2 = a_3 = b_2 = b_3 = 0$; thus, in a neighborhood of the origin the zero locus of these three is a smooth curve. Moreover, the tangent line to this curve is not contained in the plane $a_2 = 0$, so $c_0 = a_2^2$ vanishes to order exactly 2 on this curve.

Thus the component of $F_1(X)$ supported at L is zero-dimensional, and is isomorphic to Spec $\Bbbk[\epsilon]/(\epsilon^2)$. In particular, it has multiplicity 2.

Having come this far, we can answer the question: If $X \subset \mathbb{P}^3$ is a cubic surface with one ordinary double point p, and X is otherwise smooth, how many lines will X contain? We have seen that the lines $L \subset X$ passing through p count with multiplicity 2, and those not passing through p with multiplicity 1. Since we know that the total count, with multiplicity, is 27, the only question is: How many distinct lines on X pass through p?

To answer this, take $p = (1, 0, 0, 0)$ as above and expand the defining equation $g(x)$ of X around p. Since p is a double point of X, we can write

$$g(x_0, x_1, x_2, x_3) = x_0 A(x_1, x_2, x_3) + B(x_1, x_2, x_3),$$

where A is homogeneous of degree 2 and B homogeneous of degree 3. The lines on X through p then correspond to the common zeros of A and B. Moreover, if we write a line L through p as the span $L = \overline{p, q}$ with $q = (0, x_1, x_2, x_3)$, then, by Exercise 6.61, the condition that X be smooth along $L \setminus \{p\}$ is exactly the condition that the zero loci of A and B intersect transversely at (x_1, x_2, x_3). Thus there will be exactly six lines on X through p. Summarizing:

Proposition 6.38. *Let $X \subset \mathbb{P}^3$ be a cubic surface with an ordinary double point p. If X is smooth away from p, it contains exactly 21 lines: 6 through p and 15 not passing through p.*

(Compare this with the discussion starting on page 640 of Griffiths and Harris [1994].)

In Exercises 6.55–6.58, we will take up the case of cubics with more than one singularity, arriving ultimately at the statement that *a cubic surface $X \subset \mathbb{P}^3$ can have at most four isolated singular points.*

We have used the local equations of the Fano scheme only to describe the locus of lines on a single hypersurface. A similar approach gives some information about the lines on a linear system of hypersurfaces. As a sample application, we will see in Exercises 6.65 and 6.66 how to describe the singular locus of and tangent spaces to the locus $\Sigma \subset \mathbb{P}^{34}$ of quartic surfaces in $\mathbb{P}^3$ containing a line.

6.8 The Debarre–de Jong Conjecture

By Theorem 6.34, general hypersurfaces $X \subset \mathbb{P}^n$ of degree d all have Fano schemes $F_1(X)$ of the "expected" dimension $\varphi = 2n - d - 3$. On the other hand, it is easy to find smooth hypersurfaces of degree > 3 whose Fano schemes have dimension $> \varphi$; any smooth surface of degree > 3 in $\mathbb{P}^3$ that contains a line is such an example.

However, Corollary 6.33 shows that *every* smooth hypersurface of degree ≤ 3 has Fano scheme of dimension φ (that is, the open set of hypersurfaces for which $F_1(X)$ has the expected dimension contains the open set of smooth hypersurfaces when the degree is ≤ 3). Further, it was shown by Harris et al. [1998] that when $d \ll n$ every smooth hypersurface of degree d in $\mathbb{P}^n$ has a Fano scheme of lines of the correct dimension — in fact, all the $F_k(X)$ have the expected dimension when both d and k are much smaller than n. But the lower bound on n given there is very large, and examples are few. In general, we have no idea what to conjecture for the true bound required!

There is a conjecture, however, for the Fano schemes of lines. To motivate it, note that a general hypersurface $X \subset \mathbb{P}^{2m+1}$ containing an m-plane will be smooth (Exercise 6.68) and will contain a copy of the Grassmannian of lines in the m-plane, a variety of dimension $2m-2$. When $d > 2m+1$, this is larger than the expected dimension $\varphi = 2d - 3$ of $F_1(X)$. Another family of such examples is given in Exercise 6.67, but, just as in the examples above, that construction requires $d > n$.

Conjecture 6.39 (Debarre–de Jong). *If $X \subset \mathbb{P}^n$ is a smooth hypersurface of degree d with $d \le n$, then the Fano scheme $F_1(X)$ of lines on X has dimension $2n - 3 - d$.*

One striking aspect of the Debarre–de Jong conjecture is that the inequality $d \le n$ for a smooth hypersurface $X \subset \mathbb{P}^n$ is exactly equivalent to the condition that the anti-canonical bundle ω_X^* is ample, though it is not clear what role this might play in a proof.

Conjecture 6.39 has been proven for $d \le 5$ by de Jong and Debarre, and for $d \le 8$ by Beheshti (see for example Beheshti [2006]). One might worry that proving the conjecture, even for small d, would involve high-dimensional geometry, but as we will now show, it would be enough to prove the conjecture for $n = d$.

Proposition 6.40. *If $\dim F_1(X) = d - 3$ for every smooth hypersurface of degree d in $\mathbb{P}^d$, and $d \le n$, then $\dim F_1(X) = 2n - d - 3$ for every smooth hypersurface of degree d in $\mathbb{P}^n$.*

Proof: We have already treated the case of quadrics (Proposition 4.15), so we may assume that $3 \le d \le n$. Suppose that $X \subset \mathbb{P}^n$ is a smooth hypersurface and $L \subset X$ is a line. Let Λ be a general d-plane in $\mathbb{P}^n$ containing L, and let $Y = \Lambda \cap X$. By Lemma 6.41 below, Y is a smooth hypersurface of degree d in $\Lambda = \mathbb{P}^d$.

The Fano scheme $F_1(Y)$ is the intersection of $F_1(X)$ with the Schubert cycle $\Sigma_{m-n,m-n}(\Lambda) \subset \mathbb{G}(1,m)$; by the generalized principal ideal theorem,

$$d - 3 = \dim_L F_1(Y) \ge \dim_L F_1(X) - 2(n - d),$$

whence $\dim_L F_1(X) \le 2n - d - 3$, as required. $\qquad\square$

We have used a special case of the following extension of Bertini's theorem:

Lemma 6.41. *Let $k < n < m$; let $X \subset \mathbb{P}^m$ be a smooth hypersurface and $L \cong \mathbb{P}^k \subset X$ a k-plane contained in X. If $\Lambda \cong \mathbb{P}^n \subset \mathbb{P}^m$ is a general n-plane containing L, then the intersection $Y = X \cap \Lambda$ is smooth if and only if $n - 1 \geq 2k$.*

Proof: If $n - 1 < 2k$ then Y must be singular, by Corollary 6.26.

For the converse, we may assume by an obvious induction that $n = m - 1$. Bertini's theorem implies that Y is smooth away from L. On the other hand, the locus of tangent hyperplanes $\mathbb{T}_p X$ to X at points $p \in L$ is a subvariety of dimension at most k in the dual projective space $\mathbb{P}^{m*}$, while the locus of hyperplanes containing L will be the $(m - k - 1)$-plane $L^\perp \subset \mathbb{P}^{m*}$. Thus, if $n - 1 = m - 2 \geq 2k$, so that $k < m - k - 1$, then not every hyperplane containing L is tangent to X at a point of L. It follows that, for general Λ, the intersection $Y = \Lambda \cap X$ is smooth. $\square$

We can now prove Conjecture 6.39 for $d = 4$, and thereby give a negative answer for Keynote Question (d):

Theorem 6.42. *If $X \subset \mathbb{P}^n$ is a smooth hypersurface of degree 4, then the Fano scheme $F_1(X)$ has dimension $2n - 7$.*

Proof: Proposition 6.40 shows that it is enough to consider the case $n = 4$. Suppose $F \subset F_1(X)$ is an irreducible component with $\dim F \geq 2$, and let $L \in F$ be a general point. By Proposition 6.30, the normal bundle $\mathcal{N} = \mathcal{N}_{L/X}$ must be either $\mathcal{O}_L \oplus \mathcal{O}_L(-1)$ or $\mathcal{O}_L(1) \oplus \mathcal{O}_L(-2)$. Either way, all global sections of N take values in a line bundle contained in N. It follows that, for any point $p \in L$, the map $H^0(\mathcal{N}_{L/X}) \to (\mathcal{N}_{L/X})_p = T_p X / T_p L$ has rank at most 1. (Since $\dim H^0(N) \geq \dim T_L F \geq 2$, the normal bundle must in fact be $\mathcal{O}_L(1) \oplus \mathcal{O}_L(-2)$, but we do not need this.)

Let $Y \subset X$ be the subvariety swept out by the lines of $F \subset F_1(X)$. By Proposition 6.22, Y can have dimension at most 2. But by hypothesis, Y contains a two-dimensional family of lines. From Proposition 6.3 we conclude that Y is a 2-plane. Corollary 6.26 tells us this is impossible, and we are done. $\square$

6.8.1 Further open problems

The Debarre–de Jong conjecture deals with the dimension of the family of lines on a hypersurface $X \subset \mathbb{P}^n$, but we can also ask further questions about the geometry of $F_1(X)$: for example, whether it is irreducible and/or reduced. Exercises 6.70–6.73, in which we show that the Fano scheme $F_1(X)$ of lines on the Fermat quartic hypersurface $X \subset \mathbb{P}^4$ is neither, shows that the Debarre–de Jong statement cannot be strengthened for all $d \leq n$. But — based on our knowledge of examples — it does seem to be the case that the smaller d is relative to n, the better behaved $F_1(X)$ is for an arbitrary smooth hypersurface $X \subset \mathbb{P}^n$ of degree d. For example, the following questions are open:

(a) Is $F_1(X)$ is reduced and irreducible if $d \leq n - 1$ and $X \subset \mathbb{P}^n$ is a smooth hypersurface of degree d?

(b) Can we bound the dimension of the singular locus of $F_1(X)$ in terms of d? (The arguments above show that for $d = 3$ the Fano scheme $F_1(X)$ is smooth, while for $d \geq n$ it may not be reduced. What about the range $4 \leq d \leq n - 1$?)

The analogous questions for $F_k(X)$ with $k > 1$ are completely open. We can ask, for example: Given d and k, what is the largest n such that there exists a smooth hypersurface $X \subset \mathbb{P}^n$ of degree d with $\dim F_k(X) > (k + 1)(n - k) - \binom{k+d}{d}$? Again, Harris et al. [1998] says that such n are bounded, but the bound given there is probably far too large.

Finally, we can ask: Why the Fano schemes instead of other Hilbert schemes? Why not look, for example, at rational curves of any degree e on a hypersurface? Here the field is wide open. Specifically, we have an "expected" dimension: Since a rational curve $C \subset \mathbb{P}^n$ is given parametrically as the image of a map $f : \mathbb{P}^1 \to \mathbb{P}^n$, which is specified by $n + 1$ homogeneous polynomials of degree e on $\mathbb{P}^1$, and two such $(n + 1)$-tuples have the same image if and only if they differ by a scalar or by an automorphism of $\mathbb{P}^1$, the space $\mathcal{H}$ of such curves has dimension $(n+1)(e+1)-4$. On the other hand, the condition for $X = V(F)$ to contain such a curve $C \cong \mathbb{P}^1$ is that $f^*F = 0 \in H^0(\mathcal{O}_{\mathbb{P}^1}(de))$, which may be regarded as $ed + 1$ conditions on X. If we expect these conditions to be independent then we would expect the fibers of the incidence correspondence

$$\Psi = \{(X, C) \in \mathbb{P}^N \times \mathcal{H} \mid C \subset X\}$$

over $\mathcal{H}$ to have dimension $N - (de + 1)$, and Ψ correspondingly to have dimension

$$(n + 1)(e + 1) - 4 + N - (de + 1) = N + (n - d)e + n + e - 4.$$

This leads us to:

Conjecture 6.43. *If $X \subset \mathbb{P}^n$ is a general hypersurface of degree d, then X contains a rational curve of degree e if and only if*

$$\lambda(n, d, e) := (n - d)e + n + e - 4 \geq 0;$$

when this inequality is satisfied the family of such curves on X has dimension $\lambda(n, d, e)$.

We proved the conjecture in this chapter for $e = 1$, but the general case is difficult (the case $n = 4$, $d = 5$ alone is the *Clemens conjecture*, which has been the object of much study in its own right). Recently, however, there has been substantial progress: see Beheshti and Mohan Kumar [2013] and Riedl and Yang [2014].

Note that the analog of the Debarre–de Jong conjecture in this setting — that the dimension estimate of Conjecture 6.43 holds for an arbitrary smooth $X \subset \mathbb{P}^n$ of degree $d \leq n$ — is false; one counterexample is given in Exercise 6.74. But it might hold when d satisfies a stronger inequality with respect to n, perhaps for $d \leq n/e$.

6.9 Exercises

Exercise 6.44. Show that the expected number of lines on a hypersurface of degree $2n - 3$ in $\mathbb{P}^n$ (that is, the degree of $c_{2n-2}(\text{Sym}^{2n-3} \mathcal{S}^*) \in A(\mathbb{G}(1, n)))$ is always positive, and deduce that *every hypersurface of degree $2n - 3$ in $\mathbb{P}^n$ must contain a line*. (This is just a special case of Corollary 6.32; the idea here is to do it without a tangent space calculation.)

Exercise 6.45. Let $X \subset \mathbb{P}^4$ be a general quartic threefold. By Theorem 6.42, X will contain a one-parameter family of lines. Find the class in $A(\mathbb{G}(1, 4))$ of the Fano scheme $F_1(X)$, and the degree of the surface $Y \subset \mathbb{P}^4$ swept out by these lines.

Exercise 6.46. Find the class of the scheme $F_2(Q) \subset \mathbb{G}(2, 5)$ of 2-planes on a quadric $Q \subset \mathbb{P}^5$. (Do the problem first, then compare your answer to the result in Proposition 4.15.)

Exercise 6.47. Find the expected number of 2-planes on a general quartic hypersurface $X \subset \mathbb{P}^7$, that is, the degree of $c_{15}(\text{Sym}^4 \mathcal{S}^*) \in A(\mathbb{G}(2, 7))$.

Exercise 6.48. We can also use the calculation carried out in this chapter to count lines on complete intersections $X = Z_1 \cap \cdots \cap Z_k \subset \mathbb{P}^n$, simply by finding the classes of the schemes $F_1(Z_i)$ of lines on the hypersurfaces Z_i and multiplying them in $A(\mathbb{G}(1, n))$. Do this to find the number of lines on the intersection $X = Y_1 \cap Y_2 \subset \mathbb{P}^5$ of two general cubic hypersurfaces in $\mathbb{P}^5$.

Exercise 6.49. Find the Chern class $c_3(\text{Sym}^3 \mathcal{S}^*) \in A^3(\mathbb{G}(1, 3))$ as a multiple of the class $\sigma_{2,1}$. Why is this coefficient equal to the degree of the curve of lines on the cubic surfaces in a pencil? Note that this computation does not use the universal Fano scheme Φ.

Exercise 6.50. Let $\{X_t \subset \mathbb{P}^3\}_{t \in \mathbb{P}^2}$ be a general net of cubic surfaces in $\mathbb{P}^3$.

(a) Let $p \in \mathbb{P}^3$ be a general point. How many lines containing p lie on some member X_t of the net?
(b) Let $H \subset \mathbb{P}^3$ be a general plane. How many lines contained in H lie on some member X_t of the net?

Compare your answer to the second half of this question to the calculation in Chapter 2 of the degree of the locus of reducible plane cubics!

Exercise 6.51. Let $X \subset \mathbb{P}^3$ be a surface of degree $d \geq 3$. Show that if $F_1(X)$ is positive-dimensional, then either X is a cone or X has a positive-dimensional singular locus.

Exercise 6.52. Let $X \subset \mathbb{P}^4$ be a smooth cubic threefold and

$$\{S_t = X \cap H_t\}_{t \in \mathbb{P}^1}$$

a general pencil of hyperplane sections of X. What is the degree of the surface swept out by the lines on the surfaces S_t, and what is the genus of the curve parametrizing them?

Exercise 6.53. Prove Theorem 6.13 using the methods of Section 6.7, that is, by writing the local equations of $F_k(X) \subset \mathbb{G}(k, n)$

Exercise 6.54. Let $\{S_t\}_{t \in \mathbb{P}^1}$ be a general pencil of cubic surfaces, and let Φ be the incidence correspondence

$$\Phi = \{(t, L) \in \mathbb{P}^1 \times \mathbb{G}(1, 3) \mid L \subset S_t\}.$$

Using Propositions 6.38 and 7.4, show that the projection $\Phi \to \mathbb{P}^1$ has degree 27 and has six branch points over each of the 32 values of t for which S_t is singular, and deduce again the conclusion of Corollary 6.37 that the genus of Φ is 70.

Exercise 6.55. Extending the results of Section 6.7, suppose that X is a general cubic surface having two ordinary double points $p, q \in X$. Describe the scheme structure of $F_1(X)$ at the point corresponding to the line $L = \overline{p, q}$, and in particular determine the multiplicity of $F_1(X)$ at L.

Exercise 6.56. Let $X \subset \mathbb{P}^3$ be a cubic surface and $p, q \in X$ isolated singular points of X; let $L = \overline{p, q}$. Show that L is an isolated point of $F_1(X)$ and that the multiplicity $\text{mult}_L F_1(X)$ is ≥ 4.

Exercise 6.57. Let $X \subset \mathbb{P}^3$ be a cubic surface and $p_1, \ldots, p_8$ isolated singular points of X. Show that no three of the points p_i are collinear.

Exercise 6.58. Use the result of the preceding two exercises to deduce the statement that *a cubic surface $X \subset \mathbb{P}^3$ can have at most four isolated singular points.*

Exercise 6.59. Using the methods of Section 6.7, show that there exists a pair (X, Λ) with $X \subset \mathbb{P}^7$ a quartic hypersurface and $\Lambda \subset X$ a 2-plane such that Λ is an isolated, reduced point of $F_2(X)$.

Exercise 6.60. Using the result of Exercise 6.59, show that the number of 2-planes on a general quartic hypersurface $X \subset \mathbb{P}^7$ is the number calculated in Exercise 6.47 (that is, the Fano scheme $F_2(X)$ is reduced for general X).

Exercise 6.61. To complete the proof of Proposition 6.38, let $X \subset \mathbb{P}^3$ be a cubic surface with one ordinary double point $p = (1, 0, 0, 0)$, given as the zero locus of the cubic

$$F(Z_0, Z_1, Z_2, Z_3) = Z_0 A(Z_1, Z_2, Z_3) + B(Z_1, Z_2, Z_3),$$

where A is homogeneous of degree 2 and B homogeneous of degree 3. If we write a line $L \subset X$ through p as the span $L = \overline{p, q}$, with $q = (0, Z_1, Z_2, Z_3)$, show that X is smooth along $L \setminus \{p\}$ if and only if the zero loci of A and B intersect transversely at (Z_1, Z_2, Z_3).

Exercise 6.62. Show that there exists a smooth quintic threefold $X \subset \mathbb{P}^4$ whose scheme $F_1(X)$ of lines contains an isolated point of multiplicity 2.

Exercise 6.63. Let Φ be the incidence correspondence of triples consisting of a hypersurface $X \subset \mathbb{P}^n$ of degree $d = 2n - 3$, a line $L \subset X$ and a singular point p of X lying on L; that is,

$$\Phi = \{(X, L, p) \in \mathbb{P}^N \times \mathbb{G}(1, n) \times \mathbb{P}^n \mid p \in L \subset X \text{ and } p \in X_{\text{sing}}\}.$$

Show that Φ is irreducible.

Exercise 6.64. Let $\mathbb{P}^{34}$ be the space of quartic surfaces in $\mathbb{P}^3$.

(a) Show that the closure of the locus of quartics containing a pair of skew lines has dimension 32.
(b) Show that the closure of the locus of quartics containing a pair of incident lines also has dimension 32.
(c) Deduce that if $\{X_t = V(t_0 F + t_1 G)\}$ is a general pencil of quartics, then no member X_t of the pencil will contain more than one line.

Exercise 6.65. Suppose that F and G are two quartic polynomials on $\mathbb{P}^3$, and that $\{X_t = V(t_0 F + t_1 G)\}$ is the pencil of quartics they generate; let σ_F and σ_G be the sections of the bundle $\text{Sym}^4 \mathcal{S}^*$ on $\mathbb{G}(1, 3)$ corresponding to F and G. Let X_t be a member of the pencil containing a line $L \subset \mathbb{P}^3$.

(a) Find the condition on F and G for L to be a reduced point of $V(\sigma_F \wedge \sigma_G) \subset \mathbb{G}(1, 3)$.
(b) Show that this is equivalent to the condition that the point $(t, L) \in \mathbb{P}^1 \times \mathbb{G}(1, 3)$ is a simple zero of the map $\pi_1^* \mathcal{O}_{\mathbb{P}^1}(-1) \to \pi_2^* \text{Sym}^4 \mathcal{S}^*$ introduced in the proof of Theorem 6.36.

Exercise 6.66. Let $\Sigma \subset \mathbb{P}^{34}$ be the space of quartic surfaces in $\mathbb{P}^3$ containing a line. Interpret the condition of the preceding problem in terms of the geometry of the pencil $\mathcal{D}$ around the line L, and use this to answer two questions:

(a) What is the singular locus of Σ?
(b) What is the tangent hyperplane $\mathbb{T}_X \Sigma$ at a smooth point corresponding to a smooth quartic surface X containing a single line?

The following two exercises give constructions of smooth hypersurfaces containing families of lines of more than the expected dimension.

Exercise 6.67. Let $Z \subset \mathbb{P}^{n-2}$ be any smooth hypersurface. Show that the cone $\overline{p, Z} \subset \mathbb{P}^{n-1}$ over Z in $\mathbb{P}^{n-1}$ is the hyperplane section of a smooth hypersurface $X \subset \mathbb{P}^n$, and hence that for $d > n$ there exist smooth hypersurfaces $X \subset \mathbb{P}^n$ whose Fano scheme $F_1(X)$ of lines has dimension strictly greater than $2n - 3 - d$.

Exercise 6.68. Take $n = 2m + 1$ odd, and let $\Lambda \subset \mathbb{P}^n$ be an m-plane. Show that there exist smooth hypersurfaces $X \subset \mathbb{P}^n$ of any given degree d containing Λ, and deduce once more that for $d > n$ there exist smooth hypersurfaces $X \subset \mathbb{P}^n$ whose Fano scheme $F_1(X)$ of lines has dimension strictly greater than $2n - 3 - d$.

Note that the construction of Exercise 6.68 cannot be modified to provide counter-examples to the Debarre–de Jong conjecture, since by Corollary 6.26 there do not exist smooth hypersurfaces $X \subset \mathbb{P}^n$ containing linear spaces of dimension strictly greater than $(n - 1)/2$. The following exercise shows that the construction of Exercise 6.67 is similarly extremal. It requires the use of the second fundamental form (see Section 7.4.3).

Exercise 6.69. Let $X \subset \mathbb{P}^n$ be a smooth hypersurface of degree $d > 2$. Show that X can have at most finitely many hyperplane sections that are cones.

To see some of the kinds of odd behavior the variety of lines on a smooth hypersurface can exhibit, short of having the wrong dimension, the following series of exercises will look at the Fermat quartic $X \subset \mathbb{P}^4$, that is, the zero locus

$$X = V(Z_0^4 + Z_1^4 + Z_2^4 + Z_3^4 + Z_4^4).$$

The conclusion is that $F_1(X)$ has 40 irreducible components, each of which is everywhere nonreduced! We start with a useful more general fact:

Exercise 6.70. Let $S = \overline{p, C} \subset \mathbb{P}^3$ be the cone with vertex p over a plane curve C of degree $d \geq 2$, and $L \subset S$ any line. Show that the tangent space $T_L F_1(S)$ has dimension at least 2, and hence that $F_1(S)$ is everywhere nonreduced.

Exercise 6.71. Show that X has 40 conical hyperplane sections Y_i, each a cone over a quartic Fermat curve in $\mathbb{P}^2$.

Exercise 6.72. Show that the reduced locus $F_1(Y_i)_{\text{red}}$ has class $4\sigma_{3,2}$.

Exercise 6.73. Using your answer to Exercise 6.45, conclude that

$$F_1(X) = \bigcup_{i=1}^{40} F_1(Y_i);$$

in other words, $F_1(X)$ is the union of 40 double curves.

Exercise 6.74. Show that:

(a) There exist smooth quintic hypersurfaces $X \subset \mathbb{P}^5$ containing a 2-plane $\mathbb{P}^2 \subset \mathbb{P}^5$.
(b) For such a hypersurface X, the family of conic curves on X has dimension strictly greater than the number $\lambda(5, 5, 2)$ of Conjecture 6.43.

Chapter 7

Singular elements of linear series

Keynote Questions

(a) If $\{C_t = V(t_0 F + t_1 G) \subset \mathbb{P}^2\}_{t \in \mathbb{P}^1}$ is a general pencil of plane curves of degree d, how many of the curves C_t are singular? (Answer on page 253.)

(b) Let $\{C_t \subset \mathbb{P}^2\}_{t \in \mathbb{P}^2}$ be a general net of plane curves. What is the degree and geometric genus of the curve $\Gamma \subset \mathbb{P}^2$ traced out by the singular points of members of the net? What is the degree and geometric genus of the discriminant curve $\mathcal{D} = \{t \in \mathbb{P}^2 \mid C_t \text{ is singular}\}$? (Answer in Section 7.6.2.)

(c) Let $C \subset \mathbb{P}^r$ be a smooth nondegenerate curve of degree d and genus g. How many hyperplanes $H \subset \mathbb{P}^r$ have contact of order at least $r + 1$ with C at some point? (Answer on page 268.)

(d) If $\{C_t \subset \mathbb{P}^2\}_{t \in \mathbb{P}^1}$ is a general pencil of plane curves of degree d, how many of the curves C_t have hyperflexes (that is, lines having contact of order 4 with C_t)? (Answer on page 405.)

(e) If $\{C_t \subset \mathbb{P}^2\}_{t \in \mathbb{P}^4}$ is a general four-dimensional linear system of plane curves of degree d, how many of the curves C_t have a triple point? (Answer on page 257.)

In this chapter we introduce the *bundle of principal parts* associated with a line bundle $\mathcal{L}$ on a smooth variety X. This is a vector bundle on X whose fiber at a point $p \in X$ is the space of Taylor series expansions around p of sections of the line bundle, up to a given order. We will use the techniques we have developed to compute the Chern classes of this bundle, and this computation will enable us to answer many questions about singular points and other special points of varieties in families. We will start out by discussing hypersurfaces in projective space, but the techniques we develop are much more broadly applicable to families of hypersurfaces in any smooth projective variety X, and in Section 7.4.2 we will see how to generalize our formulas to that case.

In the last section (Section 7.7) we introduce a different approach to such questions, the "topological Hurwitz formula."

It is important to emphasize the standing hypothesis that the ground field $\Bbbk$ is of characteristic 0. In contrast to the preceding chapters, many of the theorems in this chapter are false as stated in characteristic $p > 0$. When it makes the geometric argument simpler, we will allow ourselves to work over the complex numbers, appealing to the "Lefschetz principle" to say that the results we obtain apply over any algebraically closed field of characteristic 0.

7.1 Singular hypersurfaces and the universal singularity

Before starting on this path, we will take a moment to talk about loci of singular plane curves, and more generally singular hypersurfaces in $\mathbb{P}^n$. Let $\mathbb{P}^N = \mathbb{P}^{\binom{d+n}{n}-1}$ be the projective space parametrizing all hypersurfaces of degree d in $\mathbb{P}^n$. Our primary object of interest is the *discriminant locus* $\mathcal{D} \subset \mathbb{P}^N$, defined as the set of singular hypersurfaces.

A central role in this chapter will be played by the *universal singular point* $\Sigma = \Sigma_{n,d}$ of hypersurfaces of degree d in $\mathbb{P}^n$, defined as follows:

$$\Sigma = \{(Y, p) \in \mathbb{P}^N \times \mathbb{P}^n \mid p \in Y_{\text{sing}}\} \xrightarrow{\;\;\pi_2\;\;} \mathbb{P}^n$$

$$\pi_1 \downarrow \qquad\qquad\qquad\qquad\qquad$$

$$\{\text{hypersurfaces } Y \text{ of degree } d \text{ in } \mathbb{P}^n\} = \mathbb{P}^N$$

If we write the general form of degree d on $\mathbb{P}^n$ as $F = \sum a_I x^I$ and think of it as a bihomogeneous form of bidegree $(1, d)$ in the coordinates a_I of $\mathbb{P}^N$ and the coordinates $x_0, \ldots, x_n$ of $\mathbb{P}^n$, then Σ is defined by the bihomogeneous equations

$$F = 0 \quad \text{and} \quad \frac{\partial F}{\partial x_i} = 0 \text{ for } i = 0, \ldots, n,$$

so Σ is an algebraic set. Note that the first of these equations is implied by the others (in characteristic 0!); given the dimension statement of Proposition 7.1 below, this means that Σ is a complete intersection of $n+1$ hypersurfaces of bidegree $(1, d-1)$ in $\mathbb{P}^N \times \mathbb{P}^n$.

The image $\mathcal{D}$ of Σ in $\mathbb{P}^N$ is the set of singular hypersurfaces, called the *discriminant*. The next proposition shows that $\mathcal{D}$ is a hypersurface, and that $\Sigma \to \mathcal{D}$ is a resolution of singularities:

Proposition 7.1. *With notation as above, suppose that $d \geq 2$.*

(a) The variety Σ is smooth and irreducible of dimension $N - 1$ (that is, codimension $n + 1$); in fact, the fibers of Σ over $\mathbb{P}^n$ are projective spaces $\mathbb{P}^{N-n-1}$.

(b) The general singular hypersurface of degree d has a unique singularity, which is an ordinary double point. In particular, Σ is birational to its image $\mathcal{D} \subset \mathbb{P}^N$.

(c) $\mathcal{D}$ is an irreducible hypersurface in $\mathbb{P}^N$.

Proof: Let $p \in \mathbb{P}^n$ be a point, and let $x_0, \ldots, x_n$ be homogeneous coordinates on $\mathbb{P}^n$ such that $p = (1, 0, \ldots, 0)$. Let $f(x_1/x_0, \ldots, x_n/x_0) = x_0^{-d} F(x_0, \ldots, x_n) = 0$ be the affine equation of the hypersurface $F = 0$. For $d \geq 1$, the $n + 1$ coefficients of the constant and linear terms f_0 and f_1 in the Taylor expansion of f at p are equal to certain coefficients of F, so the fiber of Σ over p is a projective subspace of $\mathbb{P}^N$ of codimension $n + 1$. The first part of the proposition follows from this, and implies that the discriminant $\mathcal{D} = \pi_1(\Sigma)$ is irreducible.

To prove the statements in the second part of the proposition, note that the fiber of Σ over a point $p \in \mathbb{P}^n$ contains the hypersurface that is the union of $d - 2$ hyperplanes not containing p with a cone over a nonsingular quadric in $\mathbb{P}^{n-1}$ with vertex p. This hypersurface has an ordinary double point at p, and is generically reduced. By the previous argument, the hypersurfaces corresponding to points of the fiber of Σ over p form a linear system of hypersurfaces, with no base points other than p. Bertini's theorem shows that a general member of this system is smooth away from p. Thus the fiber of the map $\pi_1 : \Sigma \to \mathcal{D} \subset \mathbb{P}^N$ over a general point of $\mathcal{D}$ consists of just one point, showing that the map is birational onto its image. Since smoothness is an open condition on a quadratic form, the general member has only an ordinary double point at p.

The fact that Σ, which has dimension $N - 1$, is birational to $\mathcal{D}$ shows that $\mathcal{D}$ also has dimension $N - 1$, completing the proof. $\qquad\square$

The defining equation of $\mathcal{D} \subset \mathbb{P}^N$ is difficult to write down explicitly, though of course it can be computed in principle by elimination theory. There are determinantal formulas in a few cases: see for example Gelfand et al. [2008] and Eisenbud et al. [2003]. Even in relatively simple cases such as $n = 1$ the discriminant locus has a lot of interesting features, as a picture of the real points of the discriminant of a quartic $f(a) = x^4 + ax^2 + bx + c$ in one variable suggests (see Figure 7.1). For a nice animation of the discriminant of a quartic polynomial, see http://youtu.be/MV2uVYqGiNc, created by Hans-Christian Graf v. Bothmer and Oliver Labs as part of their Geometrical Animations Advent Calendar.

In view of Proposition 7.1, we can rephrase the first keynote question of this chapter as asking for the number of points of intersection of a general line $L \subset \mathbb{P}^N$ with the hypersurface $\mathcal{D}$; that is, the degree of $\mathcal{D}$. How can we determine this if we cannot write down the form? As we will see below, Chern classes provide a mechanism for doing exactly this.

There is an interpretation of the discriminant hypersurface in $\mathbb{P}^N$ that relates $\mathcal{D}$ to an object previously encountered in Chapter 1. The d-th Veronese map v_d embeds $\mathbb{P}^n$ in the dual $\mathbb{P}^{N*}$ of the projective space $\mathbb{P} H^0(\mathcal{O}_{\mathbb{P}^n}(d))$, in such a way that the intersection of $v_d(\mathbb{P}^n)$ with the hyperplane corresponding to a point $F \in \mathbb{P}^N$ is isomorphic, via v_d,

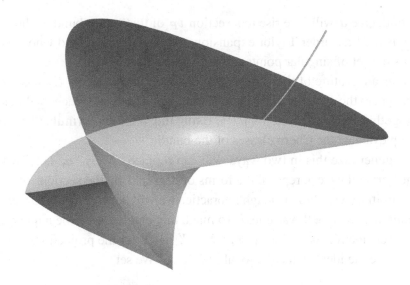

Figure 7.1 Real points of the discriminant of a quartic polynomial.

to the corresponding hypersurface $F = 0$ in $\mathbb{P}^n$. Thus the discriminant is the set of hyperplanes in $\mathbb{P}^{N*}$ that have singular intersection with $v_d(\mathbb{P}^n)$, or, equivalently, those that contain a tangent plane to $v_d(\mathbb{P}^n)$. This is the definition of the *dual variety* to $v_d(\mathbb{P}^n)$, which we first encountered in Section 2.1.3. Proposition 7.1 shows that the dual of $v_d(\mathbb{P}^n)$ is a hypersurface, and that the general tangent hyperplane is tangent at just one point, at which the intersection has an ordinary double point.

7.2 Bundles of principal parts

We can simplify the problem of describing the discriminant by linearizing it. We do not ask "is the hypersurface $X = V(F)$ singular?"; rather, we ask for each point $p \in \mathbb{P}^n$ in turn the simpler question "is X singular at p?". This is very much analogous to our approach to lines on hypersurfaces, where instead of asking "does X contains lines?" we asked for each line L "does X contain L?". As in that context, this approach converts a higher-degree equation in the coefficients of F into a family of systems of linear equations, whose solution set we can then express as the vanishing of a section of a vector bundle.

For each point $p \in \mathbb{P}^n$, we have an $(n + 1)$-dimensional vector space

$$E_p = \frac{\{\text{germs of sections of } \mathcal{O}_{\mathbb{P}^n}(d) \text{ at } p\}}{\{\text{germs vanishing to order} \geq 2 \text{ at } p\}}.$$

This space should be thought of as the vector space of first-order Taylor expansions of forms of degree d. We will see that the spaces E_p fit together to form a vector bundle, called the bundle of *first-order principal parts*, which we will write as $\mathcal{P}^1(\mathcal{O}_{\mathbb{P}^n}(d))$.

A form F of degree d will give rise to a section τ_F of this vector bundle whose value at the point p is the first-order Taylor expansion of F locally at p, and whose vanishing locus is thus the set of singular points of the hypersurface $F = 0$.

An important feature of the situation is that each vector space E_p has a naturally defined subspace: the space of germs vanishing at p. These subspaces will, as we will see, glue together into a subbundle of $\mathcal{P}^1$. Using the Whitney formula (Theorem 5.3), this will help evaluate the Chern classes of the bundle.

We can generalize this in two ways: We can replace "2" by "$m + 1$," with m any positive integer; and we can replace the forms of degree d by the sections of a coherent sheaf on an arbitrary variety X (though in practice we will be working almost exclusively with line bundles on smooth varieties). To make this precise, let $\mathcal{L}$ be a quasi-coherent sheaf on a $\Bbbk$-scheme X, and write $\pi_1, \pi_2 : X \times X \to X$ for the projections onto the two factors. Let $\mathcal{I}$ be the ideal of the diagonal in $X \times X$. We set

$$\mathcal{P}^m(\mathcal{L}) = \pi_{2*}(\pi_1^*\mathcal{L} \otimes \mathcal{O}_{X \times X}/\mathcal{I}^{m+1}),$$

which is a quasi-coherent sheaf on X. When X is smooth we call this the *bundle of principal parts*. We will parse and explain this expression below, but first we list its very useful properties:

Theorem 7.2. *The sheaves $\mathcal{P}^m(\mathcal{L})$ have the following properties:*

(a) *If $p \in X$ is a closed point, then there is a canonical identification of the fiber $\mathcal{P}^m(\mathcal{L}) \otimes \kappa(p)$ of $\mathcal{P}^m(\mathcal{L})$ at p with the sections of the restriction of $\mathcal{L}$ to the m-th-order neighborhood of p; that is,*

$$\mathcal{P}^m(\mathcal{L}) \otimes \kappa(p) = H^0(\mathcal{L} \otimes \mathcal{O}_{X,p}/\mathfrak{m}_{X,p}^{m+1})$$

as vector spaces over $\kappa(p) = \mathcal{O}_{X,p}/\mathfrak{m}_{X,p} = \Bbbk$. In other words,

$$\mathcal{P}^m(\mathcal{L}) \otimes \kappa(p) = \frac{\{\text{germs of sections of } \mathcal{L} \text{ at } p\}}{\{\text{germs vanishing to order} \geq m + 1 \text{ at } p\}}.$$

(b) *If $F \in H^0(\mathcal{L})$ is a global section, then there is a global section $\tau_F \in H^0(\mathcal{P}^m(\mathcal{L}))$ whose value at p is the class of F in $H^0(\mathcal{L} \otimes \mathcal{O}_{X,p}/\mathfrak{m}_{X,p}^{m+1})$.*

(c) *$\mathcal{P}^0(\mathcal{L}) = \mathcal{L}$, and for each $m > 0$ there is a natural right exact sequence*

$$\mathcal{L} \otimes \operatorname{Sym}^m(\Omega_X) \longrightarrow \mathcal{P}^m(\mathcal{L}) \longrightarrow \mathcal{P}^{m-1}(\mathcal{L}) \longrightarrow 0,$$

where Ω_X denotes the sheaf of $\Bbbk$-linear differential forms on X.

(d) *If X is smooth and of finite type over $\Bbbk$ and $\mathcal{L}$ is a vector bundle on X, then $\mathcal{P}^m(\mathcal{L})$ is a vector bundle on X, and the right exact sequences of part (c) are left exact as well.*

Proof: Since the constructions all commute with restriction to open sets, we may harmlessly suppose that $X = \operatorname{Spec} R$ is affine. Thus also $X \times X = \operatorname{Spec} S$, where $S := R \otimes_{\Bbbk} R$. We may think of $\mathcal{L}$ as coming from an R-module L, and then $\pi_1^* L := L \otimes_K R$. Pushing a (quasi-coherent) sheaf $\mathcal{M}$ on $X \times X$ forward by π_{2*} simply means considering the corresponding S-module as an R-module via the ring map $R \to S$ sending r to $1 \otimes_{\Bbbk} r$.

In this setting, the sheaf of ideals $\mathcal{I}$ defining the diagonal embedding of X in $X \times X$ corresponds to the ideal $I \subset S$ that is the kernel of the multiplication map $S = R \otimes_{\Bbbk} R \to R$. If R is generated as a $\Bbbk$-algebra by elements x_i, then I is generated as an ideal of S by the elements $x_i \otimes 1 - 1 \otimes x_i$.

With this notation, we see that the R-module corresponding to the sheaf $\mathcal{P}^m(\mathcal{L})$ can be written as

$$P^m(L) = (L \otimes_{\Bbbk} R)/I^{m+1}(L \otimes_{\Bbbk} R),$$

regarded as an R-module by the action $f \mapsto 1 \otimes r$ as above.

Part (a) now follows: If the $\Bbbk$-rational point p corresponds to the maximal ideal

$$\mathfrak{m} = \operatorname{Ker}\big(R \xrightarrow{\varphi} \Bbbk\big), \quad \varphi : x_i \mapsto a_i,$$

then in $R/\mathfrak{m} \otimes_R S \cong R$ the class of $x_i \otimes_{\Bbbk} 1 - 1 \otimes_{\Bbbk} x_i$ is $x_i \otimes_{\Bbbk} 1 - 1 \otimes_{\Bbbk} a_i = x_i - a_i$. Thus

$$P^m(L) \otimes_R R/\mathfrak{m} = L/(\{x_i \otimes_{\Bbbk} 1 - 1 \otimes_{\Bbbk} x_i\})^{m+1} L = L/(\{x_i - a_i\})^{m+1} L,$$

as required.

Part (b) is similarly obvious from this point of view: The section τ_F can be taken to be the image of the element $F \otimes_{\Bbbk} 1$ in $(L \otimes_{\Bbbk} R)/I^{m+1}(L \otimes_{\Bbbk} R)$. As the construction is natural, these elements will glue to a global section when we are no longer in the affine case.

Part (c) requires another important idea: The module of $\Bbbk$-linear differentials $\Omega_{R/\Bbbk}$ is isomorphic, as an R-module, to I/I^2, which has a universal derivation $\delta : R \to I/I^2$ given by $\delta(f) = f \otimes_{\Bbbk} 1 - 1 \otimes_{\Bbbk} f$. This is plausible, since when X is smooth one can see geometrically that the normal bundle of the diagonal, which is $\operatorname{Hom}(I/I^2, R)$, is isomorphic to the tangent bundle of X, which is $\operatorname{Hom}(\Omega_{R/\Bbbk}, R)$. See Eisenbud [1995, Section 16.8] for further discussion and a general proof. Given this fact, the obvious surjection $\operatorname{Sym}^m(\Omega_{R/\Bbbk}) \cong \operatorname{Sym}^m(I/I^2) \to I^m/I^{m+1}$ yields the desired right exact sequence.

Finally, it is enough to prove part (d) locally at a point $q \in X \times X$. If q is not on the diagonal then, after localizing, I is the unit ideal, and the result is trivial, so we may assume that $q = (p, p)$. Locally at p, the module L is free, so it suffices to prove the result when $L = R$.

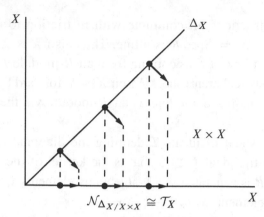

Figure 7.2 The normal bundle of the diagonal $\Delta_X \subset X \times X$ is isomorphic to the tangent bundle of X.

Write $d : R \to \Omega_{R/\Bbbk}$ for the universal $\Bbbk$-linear derivation of R. Since X is smooth, $\Omega_{R/\Bbbk}$ is locally free at p, and is generated there by elements $d(x_1), \ldots, d(x_n)$, where $x_1, \ldots, x_n$ is a system of parameters at p, and thus $\mathrm{Sym}^m(\Omega_{R/\Bbbk})$ is the free module generated by the monomials of degree m in the $d(x_i)$. Since R is a domain, I is a prime ideal.

Because $\mathrm{Sym}^m(\Omega_{R/\Bbbk})$ is free, it suffices to show that the map

$$\mathrm{Sym}^m(\Omega_{R/\Bbbk}) \to S/I^{m+1}$$

is a monomorphism (in fact, an isomorphism onto I^m/I^{m+1}) after localizing at the prime ideal I. Since $I/I^2 \cong \Omega_{R/\Bbbk}$ is free on the classes mod I^2 of the elements $x_i \otimes_{\Bbbk} 1 - 1 \otimes_{\Bbbk} x_i$ that correspond to the $d(x_i)$, Nakayama's lemma shows that, in the local ring S_I, I_I is generated by the images of the $x_i \otimes_{\Bbbk} 1 - 1 \otimes_{\Bbbk} x_i$ themselves, and it follows that these are a regular sequence. Thus the associated graded ring $S_I/I_I \oplus I_I/I_I^2 \oplus \cdots$ is a polynomial ring on the classes of the elements $x_i \otimes_{\Bbbk} 1 - 1 \otimes_{\Bbbk} x_i$, and in particular the monomials of degree m in these elements freely generate I_I^m/I_I^{m+1}. Consequently, the map $S_I \otimes_S \mathrm{Sym}^m(\Omega_{R/\Bbbk}) \to I_I^m/I_I^{m+1}$ is an isomorphism, as desired. $\qquad\square$

Remark. The name "bundle of principal parts," first used by Grothendieck and Dieudonné, was presumably suggested by the (conflicting) usage that the "principal part" of a meromorphic function of one variable at a point is the sum of the terms of negative degree in the Laurent expansion of the function around the point — a finite power series, albeit in the inverse variables. It is not the only terminology in use: $\mathcal{P}^m(\mathcal{L})$ would be called the bundle of *m-jets of sections of* $\mathcal{L}$ by those studying singularities of mappings (see for example Golubitsky and Guillemin [1973, II.2]) and some algebraic geometers (for example Perkinson [1996].) On the other hand, the m-jet terminology is in use in another conflicting sense in algebraic geometry: the "scheme of m-jets" of a scheme X is used to denote the scheme parametrizing mappings from $\mathrm{Spec}\,\Bbbk[x]/(x^{m+1})$ into X. So we have thought it best to stick to the Grothendieck–Dieudonné usage.

Example 7.3. We will not use this in any of the calculations below, but in the simplest and most interesting case, where $m = 1$, $X = \mathbb{P}^n$ and $\mathcal{L} = \mathcal{O}_{\mathbb{P}^n}(d)$, it is possible to describe the bundle $\mathcal{P}^1(\mathcal{O}_{\mathbb{P}^n}(d))$ very explicitly:

$$\mathcal{P}^1(\mathcal{O}_{\mathbb{P}^n}(d)) \cong \begin{cases} \Omega_{\mathbb{P}^n} \oplus \mathcal{O}_{\mathbb{P}^n} & \text{if } d = 0, \\ \mathcal{O}_{\mathbb{P}^n}(d-1)^{n+1} & \text{if } d \neq 0. \end{cases}$$

This curious dichotomy is explained by the answer to a more refined question: By part (d), we have a short exact sequence

$$0 \longrightarrow \Omega_{\mathbb{P}^n}(d) \longrightarrow \mathcal{P}^1(\mathcal{O}_{\mathbb{P}^n}(d)) \longrightarrow \mathcal{O}_{\mathbb{P}^n}(d) \longrightarrow 0,$$

and we can ask for its class in

$$\mathrm{Ext}^1_{\mathbb{P}^n}(\mathcal{O}_{\mathbb{P}^n}(d), \Omega_{\mathbb{P}^n}(d)) \cong \mathrm{Ext}^1_{\mathbb{P}^n}(\mathcal{O}_{\mathbb{P}^n}, \Omega_{\mathbb{P}^n}) = H^1(\Omega_{\mathbb{P}^n}) = \Bbbk.$$

More generally, for any line bundle $\mathcal{L}$ on a smooth variety X, the short exact sequence in part (d) gives us a class in

$$\mathrm{Ext}^1_X(\mathcal{L}, \Omega_X \otimes \mathcal{L}) = H^1(\Omega_X),$$

called the *Atiyah class* of $\mathcal{L}$ and denoted $\mathrm{at}(\mathcal{L})$ (Atiyah [1957] and Illusie [1972]). The formula for $\mathcal{P}^1(\mathcal{O}_{\mathbb{P}^n}(d))$ follows at once from the more refined and more uniform result that

$$\mathrm{at}(\mathcal{O}_{\mathbb{P}^n}(d)) = d \cdot \eta,$$

where $\eta \in \mathrm{Ext}^1_{\mathbb{P}^n}(\mathcal{O}_{\mathbb{P}^n}, \Omega_{\mathbb{P}^n})$ is the class of the tautological sequence

$$0 \longrightarrow \Omega_{\mathbb{P}^n} \longrightarrow \mathcal{O}_{\mathbb{P}^n}(-1)^{n+1} \xrightarrow{(x_0,\ldots,x_n)} \mathcal{O}_{\mathbb{P}^n} \longrightarrow 0.$$

See Perkinson [1996, 2.II] and Re [2012] for an analysis of all the $\mathcal{P}^m$.

7.3 Singular elements of a pencil

7.3.1 From pencils to degeneracy loci

Using the bundle of principal parts, we can tackle a slightly more general version of Keynote Question (a): How many linear combinations of general polynomials F and G of degree d on $\mathbb{P}^n$ have singular zero loci? By Proposition 7.1, none of the hypersurfaces $X_t = V(t_0 F + t_1 G)$ of the pencil will be singular at more than one point. Furthermore, no two elements of the pencil will be singular at the same point, since otherwise every member of the pencil would be singular there. Thus, the general form of the keynote question is equivalent to the question: For how many points $p \in \mathbb{P}^n$ is some element X_t of the pencil singular at p? This, in turn, amounts to asking at how many points $p \in \mathbb{P}^n$

are the values $\tau_F(p)$ and $\tau_G(p)$ in the fiber of $\mathcal{P}^1(\mathcal{O}_{\mathbb{P}^n}(d))$ at p linearly dependent, given that they are dependent at finitely many points? We can do this with Chern classes, provided that the degeneracy locus is reduced; we will establish this first.

To start, consider the behavior of the sections τ_F and τ_G around a point $p \in \mathbb{P}^n$ where they are dependent. At such a point, some linear combination $t_0 F + t_1 G$ — which we might as well take to be F — vanishes to order 2. If G were also zero at p, then the scheme $V(F, G)$ would have (at least) a double point at p. But Bertini's theorem shows that a general complete intersection such as $V(F, G)$ is smooth, so this cannot happen; thus we can assume that $G(p) \neq 0$.

To show that $V(\tau_F \wedge \tau_G)$ is reduced at p, we restrict our attention to an affine neighborhood of p where all our bundles are trivial. By Proposition 7.1, the hypersurface $C = V(F)$ has a node at p, so if we work on an affine neighborhood where the bundle $\mathcal{O}_{\mathbb{P}^n}(d)$ is trivial, and take p to be the origin with respect to coordinates $x_1, \ldots, x_n$, we may assume that the functions F and G have Taylor expansions at p of the form

$$f = f_2 + \text{(terms of order} > 2),$$
$$g = 1 + \text{(terms of order} \geq 1).$$

The sections τ_F and τ_G are then represented locally by the rows of the matrix

$$\begin{pmatrix} f & \partial f/\partial x_1 & \cdots & \partial f/\partial x_n \\ g & \partial g/\partial x_1 & \cdots & \partial g/\partial x_n \end{pmatrix}.$$

The vanishing locus of $\tau_F \wedge \tau_G$ near p is, by definition, defined by the 2×2 minors of this matrix, and to prove that it is a reduced point we need to see that it contains n functions (vanishing at p) with independent linear terms. Suppressing all the terms of the functions in the matrix that could not contribute to the linear terms of the minors, we get the matrix

$$\begin{pmatrix} 0 & \partial f_2/\partial x_1 & \cdots & \partial f_2/\partial x_n \\ 1 & 0 & \cdots & 0. \end{pmatrix}.$$

Thus there are 2×2 minors whose linear terms are $\partial f_2/\partial x_1, \ldots, \partial f_2/\partial x_n$, and these are linearly independent because $f_2 = 0$ is a smooth quadric and the characteristic is not 2.

As usual, if we assign multiplicities appropriately, we can extend the calculations to pencils whose degeneracy locus $V(\tau_F \wedge \tau_G)$ is nonreduced. In Section 7.7.2 we will see one way to calculate these multiplicities.

7.3.2 The Chern class of a bundle of principal parts

Once again, let F, G be general forms of degree d on $\mathbb{P}^n$. As we saw in the previous section, the linear combinations $t_0 F + t_1 G$ that are singular correspond exactly to points where the two sections τ_F and τ_G are dependent. The degeneracy locus of τ_F and τ_G is the n-th Chern class of the rank-$(n + 1)$ bundle $\mathcal{P}^1(\mathcal{O}_{\mathbb{P}^n}(d))$, so we turn to the computation of this class. For brevity, we will shorten $\mathcal{P}^1(\mathcal{O}_{\mathbb{P}^n}(d))$ to $\mathcal{P}^1(d)$, but the reader should keep in mind that this is *not* "a bundle $\mathcal{P}^1$ tensored with $\mathcal{O}(d)$!"

Stated explicitly, if $\zeta \in A^1(\mathbb{P}^n)$ denotes the class of a hyperplane in $\mathbb{P}^n$, we want to compute the coefficient of ζ^n in

$$c(\mathcal{P}^1(d)) \in A(\mathbb{P}^n) \cong \mathbb{Z}[\zeta]/(\zeta^{n+1}).$$

Parts (c) and (d) of Theorem 7.2 give us a short exact sequence

$$0 \longrightarrow \Omega_{\mathbb{P}^n}(d) \longrightarrow \mathcal{P}^1(d) \longrightarrow \mathcal{O}_{\mathbb{P}^n}(d) \longrightarrow 0,$$

so $c(\mathcal{P}^1(d)) = c(\mathcal{O}_{\mathbb{P}^n}(d)) \cdot c(\Omega_{\mathbb{P}^n}(d))$. (See Proposition 7.5 for the other $\mathcal{P}^m(d)$.) On the other hand, $\Omega_{\mathbb{P}^n}$ fits into a short exact sequence

$$0 \longrightarrow \Omega_{\mathbb{P}^n} \longrightarrow \mathcal{O}_{\mathbb{P}^n}(-1)^{n+1} \longrightarrow \mathcal{O}_{\mathbb{P}^n} \longrightarrow 0.$$

Tensoring with $\mathcal{O}(d)$, we get an exact sequence

$$0 \longrightarrow \Omega_{\mathbb{P}^n}(d) \longrightarrow \mathcal{O}_{\mathbb{P}^n}(d-1)^{n+1} \longrightarrow \mathcal{O}_{\mathbb{P}^n}(d) \longrightarrow 0$$

similar to the one involving $\mathcal{P}^1(d)$. This does not mean that $\mathcal{P}^1(d)$ and $\mathcal{O}_{\mathbb{P}^n}(d-1)^{n+1}$ are isomorphic (they are not), but by the Whitney formula their Chern classes agree:

$$c(\mathcal{P}^1(d)) = c(\mathcal{O}_{\mathbb{P}^n}(d-1)^{n+1}) = (1 + (d-1)\zeta)^{n+1}.$$

Putting this formula together with the idea of the previous section, we deduce:

Proposition 7.4. *The degree of the discriminant hypersurface in the space of forms of degree d on $\mathbb{P}^n$ is*

$$\deg c_n(\mathcal{P}^1(d)) = (n+1)(d-1)^n,$$

and this is the number of singular hypersurfaces in a general pencil of hypersurfaces of degree d in $\mathbb{P}^n$.

In particular, this answers Keynote Question (a): A general pencil of plane curves of degree d will have $3(d-1)^2$ singular elements.

It is pleasant to observe that the conclusion agrees with what we get from elementary geometry in the cases where it is easy to check, such as those of plane curves ($n = 2$) with $d = 1$ or $d = 2$. For $d = 1$, the statement $c_2 = 0$ simply means that there are no singular elements in a pencil of lines. The case $d = 2$ corresponds to the number of singular conics in a general pencil $\{C_t\}$ of conics. To see that this is really $3(d-1)^2 = 3$, note that the pencil $\{C_t\}$ consists of all conics passing through the four (distinct) base points, and a singular element of the pencil will thus be the union of a line joining two of the points with the line joining the other two. There are indeed three such pairs of lines (see Figure 7.3).

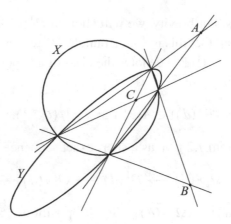

Figure 7.3 A, B, C are the singular elements of the pencil of plane conics containing X and Y.

We could also get the number 3 by viewing the pencil of conics as given by a 3×3 symmetric matrix M of linear forms on $\mathbb{P}^2$ whose entries vary linearly with a parameter t; the determinant of M will then be a cubic polynomial in t.

As the reader may have noticed, there is a simpler way to arrive at the formula of Proposition 7.4. We observed in Section 7.1 that the universal singularity

$$\Sigma = \{(Y, p) \in \mathbb{P}^N \times \mathbb{P}^n \mid p \in Y_{\text{sing}}\}$$

is a complete intersection of $n + 1$ hypersurfaces of bidegree $(1, d - 1)$ in $\mathbb{P}^N \times \mathbb{P}^n$. Denoting by α and ζ the pullbacks to $\mathbb{P}^N \times \mathbb{P}^n$ of the hyperplane classes in $\mathbb{P}^N$ and $\mathbb{P}^n$, this means that Σ has class

$$[\Sigma] = (\alpha + (d - 1)\zeta)^{n+1}$$
$$= \alpha^{n+1} + \binom{n+1}{1}\alpha^n (d - 1)\zeta + \cdots + \binom{n+1}{n}\alpha^n (d - 1)^n \zeta^n.$$

When we push this class forward to $\mathbb{P}^N$, all the terms go to 0 except the last, from which we can conclude that the class of the discriminant hypersurface $\mathcal{D}$ is $[\mathcal{D}] = (n + 1)(d - 1)^n \alpha$; that is, $\mathcal{D} = \pi_1(\Sigma)$ has degree $(n + 1)(d - 1)^n$.

Why did we adopt the approach via principal parts, given this alternative? The answer is that, as we will see in Section 7.4.2, the principal parts approach can be applied to linear series on arbitrary smooth varieties; the alternative we have just given applies *only* to projective space.

It is easy to extend the Chern class computation in Proposition 7.4 to all the $\mathcal{P}^m(\mathcal{O}_{\mathbb{P}^n}(d))$, and this will be useful in the rest of this chapter:

Proposition 7.5. $c\big(\mathcal{P}^m(\mathcal{O}_{\mathbb{P}^n}(d))\big) = (1 + (d - m)\zeta)^{\binom{n+m}{n}}.$

Proof: We will again use the exact sequences of Theorem 7.2. With the Whitney formula, they immediately give

$$c\big(\mathcal{P}^m(\mathcal{O}_{\mathbb{P}^n}(d))\big) = \prod_{j=0}^{m} c(\mathrm{Sym}^j(\Omega_{\mathbb{P}^n})(d)).$$

To derive the formula we need, we apply Lemma 7.6 below to the exact sequence

$$0 \longrightarrow \Omega_{\mathbb{P}^n} \longrightarrow \mathcal{O}_{\mathbb{P}^n}(-1)^{n+1} \longrightarrow \mathcal{O}_{\mathbb{P}^n} \longrightarrow 0$$

and the line bundle $\mathcal{L} := \mathcal{O}_{\mathbb{P}^n}(d)$. To simplify the notation, we set $\mathcal{U} = \mathcal{O}_{\mathbb{P}^n}(-1)^{n+1}$. The lemma yields

$$c(\mathrm{Sym}^j(\Omega_{\mathbb{P}^n})(d)) = c(\mathrm{Sym}^j(\mathcal{U})(d)) \cdot c(\mathrm{Sym}^{j-1}(\mathcal{U})(d))^{-1}$$

for all $j \geq 1$. Combining this with the obvious equality $\mathrm{Sym}^0(\Omega_{\mathbb{P}^n})(d) = \mathcal{O}_{\mathbb{P}^n}(d)$, we see that the product in the formula for $c\big(\mathcal{P}^m(\mathcal{O}_{\mathbb{P}^n}(d))\big)$ is

$$c(\mathcal{O}_{\mathbb{P}^n}(d)) \cdot \frac{c(\mathrm{Sym}^1(\mathcal{U})(d))}{c(\mathcal{O}_{\mathbb{P}^n}(d))} \cdot \frac{c(\mathrm{Sym}^2(\mathcal{U})(d))}{c(\mathrm{Sym}^1(\mathcal{U})(d))} \cdots,$$

which collapses to

$$c\big(\mathcal{P}^m(\mathcal{O}_{\mathbb{P}^n}(d))\big) = c(\mathrm{Sym}^m(\mathcal{U})(d)).$$

But

$$c(\mathrm{Sym}^m(\mathcal{U})(d)) = c(\mathrm{Sym}^m(\mathcal{O}_{\mathbb{P}^n}(-1)^{n+1})(d)) = c\big(\mathcal{O}_{\mathbb{P}^n}(-m)^{\binom{n+m}{n}}(d)\big)$$
$$= c\big(\mathcal{O}_{\mathbb{P}^n}(d-m)^{\binom{n+m}{n}}\big)$$
$$= (1 + (d-m)\zeta)^{\binom{n+m}{n}},$$

yielding the formula of the proposition. $\qquad\square$

Lemma 7.6. *If*

$$0 \longrightarrow \mathcal{A} \longrightarrow \mathcal{B} \longrightarrow \mathcal{C} \longrightarrow 0$$

is a short exact sequence of vector bundles on a projective variety X with rank $\mathcal{C} = 1$, *then, for any $j \geq 1$,*

$$c(\mathrm{Sym}^j(\mathcal{A}) \otimes \mathcal{L}) = c(\mathrm{Sym}^j(\mathcal{B}) \otimes \mathcal{L}) \cdot c(\mathrm{Sym}^{j-1}(\mathcal{B}) \otimes \mathcal{C} \otimes \mathcal{L})^{-1}.$$

Proof: For any right exact sequence of coherent sheaves

$$\mathcal{E} \longrightarrow \mathcal{F} \longrightarrow \mathcal{G} \longrightarrow 0,$$

the universal property of the symmetric powers (see, for example, Eisenbud [1995, Proposition A2.2.d]) shows that for each $j \geq 1$ there is a right exact sequence

$$\mathcal{E} \otimes \mathrm{Sym}^{j-1}(\mathcal{F}) \longrightarrow \mathrm{Sym}^j(\mathcal{F}) \longrightarrow \mathrm{Sym}^j(\mathcal{G}) \longrightarrow 0.$$

Since $\mathcal{A}, \mathcal{B}$ and $\mathcal{C}$ are vector bundles, the dual of the exact sequence in the hypothesis is exact, and we may apply the result on symmetric powers with $\mathcal{G} = \mathcal{A}^*, \mathcal{F} = \mathcal{B}^*$ and $\mathcal{E} = \mathcal{C}^*$.

In this case, since rank $\mathcal{E} = \text{rank} \, \mathcal{C} = 1$, the sequence

$$0 \longrightarrow \mathcal{E} \otimes \text{Sym}^{j-1}(\mathcal{F}) \longrightarrow \text{Sym}^j(\mathcal{F}) \longrightarrow \text{Sym}^j(\mathcal{G}) \longrightarrow 0$$

is left exact as well, as one sees by comparing the ranks of the three terms (this is a special case of a longer exact sequence, independent of the rank of $\mathcal{E}$, derived from the Koszul complex).

Since these are all bundles, dualizing preserves exactness, and we get an exact sequence

$$0 \longrightarrow \text{Sym}^j(\mathcal{A}^*)^* \longrightarrow \text{Sym}^j(\mathcal{B}^*)^* \longrightarrow \text{Sym}^{j-1}(\mathcal{B}^*)^* \otimes \mathcal{C} \longrightarrow 0.$$

Of course the double dual of a bundle is the bundle itself, and the dual of the j-th symmetric power is naturally isomorphic to the j-th symmetric power of the dual, so all the *'s cancel, and we can deduce the lemma from the Whitney formula. $\qquad \square$

7.3.3 Triple points of plane curves

We can adapt the preceding ideas to compute the number of points of higher order in linear families of hypersurfaces. By way of example we consider the case of triple points of plane curves.

Let $\mathbb{P}^N$ be the projective space of all plane curves of degree $d \geq 3$, and let

$$\Sigma' = \{(C, p) \in \mathbb{P}^N \times \mathbb{P}^2 \mid \text{mult}_p(C) \geq 3\}.$$

The condition that a curve C have a triple point at a given point $p \in \mathbb{P}^2$ is six independent linear conditions on the coefficients of the defining equation of C, from which we see that the fibers of the projection map $\Sigma' \to \mathbb{P}^2$ on the second factor are linear spaces $\mathbb{P}^{N-6} \subset \mathbb{P}^N$, and hence that Σ' is irreducible of dimension $N - 4$. It follows that the set of curves with a triple point is irreducible as well. An argument similar to that for double points also shows that a general curve $f = 0$ with a triple point has only one. In particular, the projection map $\Sigma' \to \mathbb{P}^N$ on the first factor is birational onto its image. It follows in turn that *the locus* $\Phi \subset \mathbb{P}^N$ *of curves possessing a point of multiplicity 3 or more is an irreducible variety of dimension* $N - 4$. We also see that if C is a general curve with a triple point at p, then p is an *ordinary* triple point of C; that is, the projectivized tangent cone $\mathbb{T}C_p X$ is smooth or, equivalently, the cubic term f_3 of the Taylor expansion of f around p has three distinct linear factors.

We ask now for the degree of the variety of curves with a triple point, or, equivalently, the answer to Keynote Question (e): If $F_0, \ldots, F_4$ are general polynomials of degree d on $\mathbb{P}^2$, for how many linear combinations $F_t = t_0 F_0 + \cdots + t_4 F_4$ (up to scalars) will the corresponding plane curve $C_t = V(F_t) \subset \mathbb{P}^2$ have a triple point?

If we write τ_F for the section defined by F in $\mathcal{P}^2(\mathcal{O}_{\mathbb{P}^2}(d))$, then C has a triple point at p if and only if τ_F vanishes at p. An argument analogous to the one given in Section 7.3.1, together with the smoothness of the tangent cone at a general triple point, shows that the 5×5 minors of the map $\mathcal{O}_{\mathbb{P}^2}^5 \to \mathcal{P}^2(\mathcal{O}_{\mathbb{P}^2}(d))$ generate the maximal ideal locally at a general point where a linear combination of the F_i defines a curve with an ordinary triple point.

Thus the number of triple points in the family is the degree of the second Chern class $c_2(\mathcal{P}^2(\mathcal{O}_{\mathbb{P}^2}(d)))$. By Proposition 7.5,

$$c(\mathcal{P}^2(\mathcal{O}_{\mathbb{P}^2}(d))) = (1 + (d-2)\zeta)^6 = 1 + 6(d-2)\zeta + 15(d^2 - 4d + 4)\zeta^2.$$

Proposition 7.7. *If $\Phi = \Psi_{d,n} \subset \mathbb{P}^N$ is the locus in the space of all curves of degree d in $\mathbb{P}^2$ of curves having a triple point, then for $d \geq 2$*

$$\deg(\Phi) = 15(d^2 - 4d + 4).$$

In case $d = 1$, the number 15 computed is of course meaningless, because the expected dimension $N - 4$ of Φ is negative — any five global sections τ_{F_i} of the bundle $\mathcal{P}^2(\mathcal{O}(1))$ are everywhere-dependent. On the other hand, the number 0 computed in the case $d = 2$, which is 0, really does reflect the fact that no conics have a triple point. For $d = 3$, the computation above gives 15, a number we already computed as the degree of the locus of "asterisks" in Section 2.2.3.

7.3.4 Cones

As we remarked, the calculation in the preceding section is a generalization of the calculation in Section 2.2.3 of the degree of the locus Φ parametrizing triples of concurrent lines ("asterisks") in the space $\mathbb{P}^9$ parametrizing plane cubic curves. There is another generalization of this problem: We can ask for the degree, in the space $\mathbb{P}^N$ parametrizing hypersurfaces of degree d in $\mathbb{P}^n$, of the locus Ψ of *cones*. We are now in a position to answer that more general problem, which we will do here.

We will not go through the steps in detail, since they are exactly analogous to the last calculation; the upshot is that the degree of Ψ is the degree of the n-th Chern class of the bundle $\mathcal{P}^{d-1}(\mathcal{O}_{\mathbb{P}^n}(d))$. By Proposition 7.5,

$$c(\mathcal{P}^{d-1}(\mathcal{O}_{\mathbb{P}^n}(d))) = (1 + \zeta)^{\binom{n+d-1}{n}},$$

and so we have:

Proposition 7.8. *If $\Psi = \Psi_{d,n} \subset \mathbb{P}^N$ is the locus of cones in the space of all hypersurfaces of degree d in $\mathbb{P}^n$, then*

$$\deg(\Psi) = \binom{\binom{n+d-1}{n}}{n}.$$

Thus, for example, in case $d = 2$ we see again that the locus of singular quadrics in $\mathbb{P}^n$ is $n + 1$, and in case $d = 3$ and $n = 2$ the locus of asterisks has degree 15. Likewise, in the space $\mathbb{P}^{14}$ of quartic plane curves, the locus of concurrent 4-tuples of lines has degree

$$\binom{\binom{5}{2}}{2} = \binom{10}{2} = 45.$$

Compare this to the calculation in Exercise 2.57!

7.4 Singular elements of linear series in general

Let X be a smooth projective variety of dimension n, and let $\mathcal{W} = (\mathcal{L}, W)$ be a linear system on X. We think of the elements of $\mathbb{P}W$ as divisors in X, and, as in Section 7.1, we introduce the incidence correspondence

$$\Sigma_{\mathcal{W}} = \{(Y, p) \in \mathbb{P}W \times X \mid p \in Y_{\text{sing}}\}$$

with projection maps $\pi_1 : \Sigma \to \mathbb{P}W$ and $\pi_2 : \Sigma \to X$. Also as in Section 7.1 we denote by $\mathcal{D} = \pi_1(\Sigma) \subset \mathbb{P}W$ the locus of singular elements of the linear series $\mathcal{W}$, which we again call the *discriminant*.

As mentioned in the introduction to this chapter, the techniques developed so far apply as well in this generality. What is missing is the analog of Proposition 7.1: We do not know in general that Σ is irreducible of codimension $n + 1$, we do not know that it maps birationally onto $\mathcal{D}$ (as we will see more fully in Section 10.6, the discriminant $\mathcal{D}$ may have dimension strictly smaller than that of Σ) and we do not know that the general singular element of $\mathcal{W}$ has one ordinary double point as its singularity. Thus the formulas we derive in this generality are only *enumerative formulas*, in the sense of Section 3.1: They apply subject to the hypothesis that the loci in question do indeed have the expected dimension, and even then only if multiplicities are taken into account.

That said, we can still calculate the Chern classes of the bundle of principal parts $\mathcal{P}^1(\mathcal{L})$, and derive an enumerative formula for the number of singular elements of a pencil of divisors (that is, the degree of $\mathcal{D} \subset \mathbb{P}W$, in case $\mathcal{D}$ is indeed a hypersurface); we will do this in Section 7.4.1 below.

We note one interpretation of $\mathcal{D}$ in case the linear series $\mathbb{P}W$ is very ample. If $X \subset \mathbb{P}^N$ is a smooth variety and

$$\mathcal{W} = (\mathcal{O}_X(1), W) \quad \text{with } W = H^0(\mathcal{O}_{\mathbb{P}^N}(1))|_X$$

is the linear series of hyperplane sections of X, then a section in W is singular if and only if the corresponding hyperplane is tangent to X. Thus the set of points in $\mathbb{P}W$

corresponding to such sections is the *dual variety* to X, and the number of singular elements in a general pencil of these sections is the degree of the dual variety. We will treat dual varieties more thoroughly in Section 10.6.

7.4.1 Number of singular elements of a pencil

Let X be a smooth projective variety of dimension n and $W = (\mathcal{L}, W)$ a pencil of divisors on X (typically, a general pencil in a larger linear series). We can use the Chern class machinery to compute the expected number of singular elements of W. To simplify the notation, we will denote the first Chern class of the line bundle $\mathcal{L}$ by $\lambda \in A^1(X)$, and the Chern classes of the cotangent bundle Ω_X of X simply by $c_1, c_2, \ldots, c_n$.

From the exact sequence

$$0 \longrightarrow \Omega_X \otimes \mathcal{L} \longrightarrow \mathcal{P}^1(\mathcal{L}) \longrightarrow \mathcal{L} \longrightarrow 0$$

and Whitney's formula, we see that the Chern class of $\mathcal{P}^1(\mathcal{L})$ is the Chern class of $\mathcal{L} \otimes (\mathcal{O}_X \oplus \Omega_X)$. Since $c_i(\mathcal{O}_X \oplus \Omega_X) = c_i(\Omega_X) = c_i$, we may apply the formula for the Chern class of a tensor product of a line bundle (Proposition 5.17) to arrive at

$$c_k(\mathcal{P}^1(\mathcal{L})) = \sum_{i=0}^{k} \binom{n+1-i}{k-i} \lambda^{k-i} c_i.$$

In particular,

$$c_n(\mathcal{P}^1(\mathcal{L})) = \sum_{i=0}^{n} (n+1-i) \lambda^{n-i} c_i$$

$$= (n+1)\lambda^n + n\lambda^{n-1} c_1 + \cdots + 2\lambda c_{n-1} + c_n. \qquad (7.1)$$

As remarked above, this represents only an enumerative formula for the number of singular elements of a pencil. But the calculations of Section 7.3.1 hold here as well: A singular element Y of a pencil corresponds to a reduced point of the relevant degeneracy locus if Y has just one ordinary double point as its singularity. Thus we have the following:

Proposition 7.9. *Let X be a smooth projective variety of dimension n. If $W = (\mathcal{L}, W)$ is a pencil of divisors on X having finitely many singular elements $D_1, \ldots, D_\delta$ such that*

(a) each D_i has just one singular point,
(b) that singular point is an ordinary double point, and
(c) that singular point is not contained in the base locus of the pencil,

then the number δ of singular elements is the degree of the class

$$\gamma(\mathcal{L}) := c_n(\mathcal{P}^1(\mathcal{L})) = (n+1)\lambda^n + n\lambda^{n-1} c_1 + \cdots + 2\lambda c_{n-1} + c_n \in A^n(X),$$

where $\lambda = c_1(\mathcal{L})$ and $c_i = c_i(\Omega_X)$.

Naturally, there will be occasions when we want to apply this formula but may not be able to verify hypotheses (a)–(c) of Proposition 7.9—for example, as we will see in Section 10.6, these hypotheses are not necessarily satisfied by a general pencil of hyperplane sections of a smooth projective variety $X \subset \mathbb{P}^r$. It is worth asking, accordingly, what can we conclude from the enumerative formula in the absence of these hypotheses.

First off, if the class $\gamma(\mathcal{L})$ is nonzero, then we can conclude that *the pencil $\mathcal{W}$ must have singular elements*; this applies to any pencil on any variety. Secondly, if $\mathcal{W}$ is a general pencil in a very ample linear series $V = (\mathcal{L}, V)$, we can form the *universal singular point*, as in Section 7.1:

$$\Sigma = \{(v, p) \in \mathbb{P}V \times X \mid p \in (D_v)_{\text{sing}}\},$$

where $D_v \subset X$ is the divisor corresponding to the element $v \in \mathbb{P}V$. As in the proof of Proposition 7.1, we see that the fibers of Σ over X are projective spaces of codimension $n+1$ in $\mathbb{P}V$, and hence that Σ has codimension $n+1$ in $\mathbb{P}V \times X$; it follows that the preimage of a general pencil $\mathbb{P}W \subset \mathbb{P}V$ is Σ will be finite. We can conclude, therefore, that in this situation *the degree of the class $\gamma(\mathcal{L})$ must be nonnegative, and if it is 0 then $\mathcal{W}$ will have no singular elements* — in other words, the locus of singular elements of the linear system V has codimension > 1 in $\mathbb{P}V$, and every singular element will have positive-dimensional singular locus. We will see an example of a situation where this is the case in Exercise 7.28 below, and investigate the question in more detail in Section 10.6.

Finally, we will see in Section 7.7.2 below a way of calculating multiplicities of the relevant degeneracy locus topologically, so that even in case the singular elements of $\mathcal{W}$ do not satisfy the hypothesis of having only one ordinary double point we can say something about the number of singular elements. (The conclusions of Section 7.7.2 are stated only for pencils of curves on a surface, but analogous statements hold in higher dimension as well.)

7.4.2 Pencils of curves on a surface

By way of an example, we will apply the results of Proposition 7.9 to pencils of curves on surfaces. For the case $d = 2$, see Exercises 7.22 and 7.23.

Suppose that $X \subset \mathbb{P}^3$ is a smooth surface of degree d and that V is the linear series of intersections of X with surfaces of degree e, so that $\mathcal{L} = \mathcal{O}_X(e)$.

We claim that the three hypotheses of Proposition 7.9 are satisfied for a general pencil $\mathcal{W} \subset V$:

(a) The fact that a general singular element of $\mathcal{W}$ (equivalently, of V) has only one singularity in the case $e = 1$ is somewhat subtle; it is equivalent to the statement that the Gauss map from the surface to its dual variety is birational. (This is sometimes false in characteristic p!) This statement is proven for all smooth hypersurfaces in Corollary 10.21.

The case $e > 1$ can be deduced from the case $e = 1$ by using Bertini's theorem and Proposition 7.10.

(b) The fact that the singularity of a general singular element of $\mathcal{W}$ is an ordinary double point is also tricky. Again, it follows for $e > 1$ from the case $e = 1$, and when $e = 1$ it can be done for a general surface $X \subset \mathbb{P}^3$ by an incidence correspondence/dimension count argument (Exercise 7.42). For an arbitrary X, however, it requires the introduction of the *second fundamental form*; we will describe this in the following section and use it to prove the statement we want in Theorem 7.11.

(c) Finally, the third hypothesis of Proposition 7.9 follows much as in the case of plane curves: By Bertini's theorem, the base locus of a general pencil in a very ample linear series is smooth — in this case a set of reduced points — and a reduced point on a smooth surface cannot be the intersection of two divisors if one of them is singular.

We have used:

Proposition 7.10. *Let $\Gamma \subset \mathbb{P}^n$ be a finite subscheme with homogeneous coordinate ring S_Γ.*

(a) *Γ imposes independent conditions on forms of degree $\deg \Gamma - 1$.*

(b) *Γ fails to impose independent conditions on forms of degree $\Gamma - 2$ if and only if Γ is contained in a line.*

(c) *Let l be a general linear form, and set $R_\Gamma := S_\Gamma/(l)S_\Gamma$. In general, Γ imposes independent conditions on forms of degree e in $\mathbb{P}^n$ if and only if R_Γ is 0 in degree e.*

Since R_Γ is generated as an S_Γ-module in degree 0, the range of integers e such that $(R_\Gamma)_e \neq 0$ is an interval in $\mathbb{Z}$ of the form $[0, \ldots, r]$, and the number r is called the *Castelnuovo–Mumford regularity* of R_Γ. See Eisenbud [2005, Chapter 4] for more on this important notion.

Proof: The condition that Γ imposes independent conditions on forms of degree s is equivalent to the statement that $(\dim S_\Gamma)_s = \deg \Gamma$.

Using the exact sequences

$$0 \longrightarrow (S_\Gamma)_{t-1} \xrightarrow{\cdot l} (S_\Gamma)_t \longrightarrow (R_\Gamma)_t \longrightarrow 0,$$

we see that $\dim_{\Bbbk}(S_\Gamma)_s = \sum_{t=0}^{s} \dim_{\Bbbk}(R_\Gamma)_t$.

By Eisenbud [1995, Section 1.9], the dimension of R_Γ is the degree of the scheme Γ, proving Part (c). Part (a) is an immediate consequence. If Γ is not contained in a line, then $(R_\Gamma)_1 \geq 2$, so $(R_\Gamma)_{\deg \Gamma - 1} = 0$, proving Part (b). See Eisenbud and Harris [1992]. $\quad\square$

Letting $\zeta \in A^1(X)$ denote the restriction of the hyperplane class, we have $c_1(\mathcal{L}) = e\zeta$, and, as we have seen,

$$
\begin{aligned}
c(\mathcal{T}_X) &= \frac{c(\mathcal{T}_{\mathbb{P}^3})}{c(\mathcal{N}_{X/\mathbb{P}^3})} \\
&= \frac{(1+\zeta)^4}{1+d\zeta} \\
&= (1 + 4\zeta + 6\zeta^2)(1 - d\zeta + d^2\zeta^2) \\
&= 1 + (4-d)\zeta + (d^2 - 4d + 6)\zeta^2.
\end{aligned}
$$

Thus $c_1 = (d-4)\zeta$ and $c_2 = (d^2 - 4d + 6)\zeta^2$. From (7.1), above we see that

$$
c_2(\mathcal{P}(\mathcal{L})) = (3e^2 + 2(d-4)e + d^2 - 4d + 6)\zeta^2.
$$

Finally, since $\deg(\zeta^2) = d$, the number of singular elements in the pencil of curves on a smooth surface $X \subset \mathbb{P}^3$ of degree d cut by a general pencil of surfaces of degree e is

$$
\deg c_2(\mathcal{P}(\mathcal{L})) = d(3e^2 + 2(d-4)e + d^2 - 4d + 6). \tag{7.2}
$$

As explained above, this will be the degree of the dual surface of the e-th Veronese image $\nu_e(X)$ of X. For example, when $e = 1$ this reduces to

$$
\deg X^* = d(d-1)^2,
$$

as calculated in Section 2.1.3.

When $e = 2$, we are computing the expected number of singular points in the intersection of X with a general pencil $\{Q_t \subset \mathbb{P}^3\}_{t \in \mathbb{P}^1}$ of quadric surfaces in $\mathbb{P}^3$, and we find that it is equal to

$$
d^3 + 2d.
$$

The reader should check the case $d = 1$ directly! We invite the reader to work out some more examples, and to derive analogous formulas in higher (and lower!) dimensions, in Exercises 7.22–7.27.

7.4.3 The second fundamental form

A useful tool in studying singularities of elements of linear series is the *second fundamental form* S_X of a smooth variety $X \subset \mathbb{P}^n$. The notion was first considered in differential geometry, and is usually described using a metric, but we give a purely algebro-geometric treatment. We will explain the definition and an application; more information can be found, for example, in Griffiths and Harris [1979].

As we shall see at the end of this section, the second fundamental form is closely related to the *Gauss map* $\mathcal{G}_X : X \to \mathbb{G}(k, n)$, which sends each point $p \in X$ to its tangent plane $\mathbb{T}_p X \subset \mathbb{P}^n$: The information S_X carries is equivalent to that of the differential

$$d\mathcal{G}_X : \mathcal{T}_X \to \mathcal{G}_X^* \mathcal{T}_{\mathbb{G}(k,n)}.$$

(See Section 2.1.3 for the definition of the Gauss map for hypersurfaces, and below for the general case.)

Since we will be dealing with both duals and pullbacks of vector bundles in this section, we will write the dual of a bundle $\mathcal{E}$ as $\mathcal{E}^\vee$ instead of our more usual $\mathcal{E}^*$.

Throughout this section X will denote a smooth subvariety of dimension k in $\mathbb{P}^n$.

We will define $S = S_X$ to be a map of sheaves on X

$$S : \mathcal{I}_X / \mathcal{I}_X^2 \to \mathrm{Sym}^2(\mathcal{T}_X^\vee).$$

We regard $\mathrm{Sym}^2(\mathcal{T}_X^\vee)$ as the bundle of quadratic forms on the tangent spaces to X. Let f be a function on an open subset of $\mathbb{P}^n$ defined in a neighborhood of $p \in X$ and vanishing on X, so that f is a local section of $\mathcal{I}_X$. When restricted to the tangent space $\mathbb{T}_p X \subset \mathbb{P}^n$ of X at p, the function f is singular at p, so the restriction $\overline{f} = f|_{\mathbb{T}_p X}$ vanishes together with all its first derivatives at p. Because of this, the quadratic part of the Taylor expansion of f at p is independent of the choice of coordinates. Via the identification $T_p(\mathbb{T}_p X) = T_p X$, we define $S(f)_p$, the value of $S(f)$ at p, to be the quadratic term in the expansion of $\overline{f}$ at p.

We claim that $S(f)$ vanishes if $f \in \mathcal{I}_X^2$, and thus that S defines a map $\mathcal{I}_X / \mathcal{I}_X^2 \to \mathrm{Sym}^2(\mathcal{T}_X^\vee)$. We work in local coordinates z_i at p. In these terms, $S(f)$ is the quadratic form defined by the Hessian matrix

$$\left(\frac{\partial^2 f}{\partial z_i \partial z_j}(p) \right).$$

If $f = gh$ is the product of two functions vanishing on X, then the locally defined function $\partial f / \partial z_i = g(\partial h / \partial z_i) + h(\partial g / \partial z_i)$ vanishes on X, so the Hessian matrix is identically 0 on X. Since S is linear, this suffices to prove the claim.

Recall from Eisenbud [1995, Chapter 16] that there is an exact sequence

$$0 \longrightarrow \mathcal{I}_{X/\mathbb{P}^n} / \mathcal{I}_{X/\mathbb{P}^n}^2 \longrightarrow \Omega_{\mathbb{P}^n}|_X \longrightarrow \Omega_X \longrightarrow 0.$$

Since X is smooth this is a short exact sequence of vector bundles. Composing the inclusion $\mathcal{I}_{X/\mathbb{P}^n} / \mathcal{I}_{X/\mathbb{P}^n}^2 \to \Omega_{\mathbb{P}^n}|_X$ with the first map of the restriction of the Euler sequence

$$0 \longrightarrow \Omega_{\mathbb{P}^n}|_X \longrightarrow \mathcal{O}_X^{n+1}(-1) \longrightarrow \mathcal{O}_X \longrightarrow 0,$$

we get an inclusion of bundles $\iota : \mathcal{I}_{X/\mathbb{P}^n} / \mathcal{I}_{X/\mathbb{P}^n}^2 \to \mathcal{O}_X^{n+1}$. The Gauss map $\mathcal{G} : X \to G(n - k, n + 1)$ may be defined as the unique map such that the pullback $\mathcal{G}^*(\mathcal{S})$ of the inclusion of the universal subbundle on $G(n - k, n + 1)$ is ι.

(The more usual definition of the Gauss map is dual to this one: Starting from the derivative $T_X \to T_{\mathbb{P}^n}|_X$ of the inclusion map, one takes the pullback of the image under the surjection

$$\mathcal{O}_X^{n+1} \to T_{\mathbb{P}^n}|_X;$$

the two descriptions are related by the duality isomorphism $G(k+1, n+1) \cong G(n-k, n+1)$.)

As explained in Section 3.2.5, the derivative of this map takes a derivation ∂ to the result of applying ∂ to the entries of a matrix representing ι and then projecting to $\mathcal{O}^{n+1}/\mathcal{E}$. We can put all these actions into the composite map

$$\mathcal{T}_X \otimes (\mathcal{I}_{X/\mathbb{P}^n}/\mathcal{I}^2_{X/\mathbb{P}^n}) \to \Omega_{\mathbb{P}^n}|_X \to \Omega_X,$$

or equivalently the map

$$\mathcal{I}_{X/\mathbb{P}^n}/\mathcal{I}^2_{X/\mathbb{P}^n} \to \Omega_{\mathbb{P}^n}|_X \otimes \Omega_X \to \Omega_X \otimes \Omega_X.$$

A local computation in coordinates x_i on X shows that the image of the class of a function $f \in \mathcal{I}_{X/\mathbb{P}^n}$ has the form

$$\sum_{i,j} \frac{\partial f}{\partial x_i \partial x_j} dx_i \otimes dx_j,$$

and is thus a symmetric tensor, an element of $\mathrm{Sym}^2(\Omega_X) \subset \Omega_X \otimes \Omega_X$; in fact, it is the quadratic term of the Taylor expansion of f.

We can now complete the proof of the result of Section 7.4.2, based on Corollary 10.21, which will be proven independently:

Theorem 7.11. *Let $X \subset \mathbb{P}^n$ be any smooth hypersurface of degree $d > 1$. The set of points $X_i \subset X$ where the rank of the quadratic form $S(f)_p$ is at most i is an algebraic subset of dimension at most i. In particular:*

(a) *There are at most finitely many points where the tangent hyperplane section $Y = X \cap \mathbb{T}_p X$ has multiplicity 3 or more at p; that is, $S(f)_p = 0$.*

(b) *If $p \in X$ is a general point, then the tangent hyperplane section $Y = X \cap \mathbb{T}_p X$ has an ordinary double point at p; that is, for general p the rank of $S(f)_p$ is equal to the dimension of X.*

Proof: Since X is a hypersurface, the ideal of X near p is generated by a single function f, so we may regard $S(f)$ as a map from X to the total space of a twist of the vector bundle $\mathrm{Sym}^2(\mathcal{T}_X^\vee)$. The locus X_i is thus the (reduced) preimage of the closed algebraic set of forms of rank $\leq i$.

It suffices to prove the general statement. Suppose that X_i had dimension $> i$ for some $i \geq 0$, and let p be a general (and in particular smooth) point of X_i. Since p is general in X_i, the null-space $T'_{X,q} \subset T_{X,q}$ of $S(f)_q$ has constant dimension $\geq \dim X - i$ for q in a neighborhood $U \subset X_i$ of p, and these null-spaces form a subbundle $\mathcal{T}'_U \subset \mathcal{T}_X|_U$. The tangent spaces to U also form a subbundle, and by our hypothesis on the dimension of X_i these two subbundles intersect in a subbundle $\mathcal{T}_U \cap \mathcal{T}'_X$ of rank ≥ 1.

We may assume that the ground field is the complex numbers. Integrating a local analytic vector field inside $\mathcal{T}_U \cap \mathcal{T}'_X$, we obtain the germ of a curve in X along which the Gauss map has derivative 0. This contradicts the assertion of Corollary 10.21 that $\mathcal{G}_X$ is a finite mapping from X to its dual. $\square$

7.5 Inflection points of curves in $\mathbb{P}^r$

Bundles of principal parts are very useful for studying maps of curves to projective space. The connection with "singular elements of linear series" comes from the fact that a hyperplane in projective space is tangent to a nondegenerate curve if and only if its intersection with the curve — an element of the linear system corresponding to the embedding — is singular. If the plane meets the curve with a higher degree of tangency — think of the tangent line at a flex point of a plane curve — then that will be reflected in a higher-order singularity. Thus the technique we developed in Section 7.2 will allow us to solve the third of the keynote questions of this chapter: How to extend the notion of flexes to curves in $\mathbb{P}^n$, and how to count them.

Recall that if $C \subset X$ is a reduced curve on a scheme X and $D \subset X$ an effective Cartier divisor on C, then for any closed point $p \in D \cap C$ we defined the multiplicity of intersection of C with D at p to be the length (or the dimension over the ground field $\Bbbk$, which will be the same since we are supposing that $\Bbbk$ is algebraically closed) of $\mathcal{O}_{C,p}/\mathcal{I}(D) \cdot \mathcal{O}_{C,p}$. Thus, for example, when $p \notin C \cap D$ the multiplicity is 0, and the multiplicity is 1 if and only if C and D are both smooth at p and meet transversely there.

For the purpose of this chapter it is convenient to expand this notion. Suppose that C is a smooth curve, $f : C \to X$ is a morphism and D is any subscheme of X such that $f^{-1}(D)$ is a finite scheme. We define the *order of contact* of D with C at $p \in C$ to be

$$\operatorname{ord}_p f^{-1}(D) := \dim_{\kappa(p)} \mathcal{O}_{C,p}/f^*(\mathcal{I}(D)).$$

Since we have assumed that C is smooth, the local ring $\mathcal{O}_{C,p}$ is a discrete valuation ring, so $\operatorname{ord}_p f^{-1}(D)$ is the minimum of the lengths of the algebras $\mathcal{O}_{C,p}/f^*(g)$, where g ranges over the local sections of $\mathcal{I}(D)$ at p, or over the generators of this ideal.

If f is the inclusion map of a smooth curve $C \subset X = \mathbb{P}^r$, and $D = \Lambda \subset \mathbb{P}^r$ is a linear subspace, then the order of contact $\operatorname{ord}_p f^{-1}(\Lambda)$ is the minimum, over the set of hyperplanes H containing Λ, of the intersection multiplicity $m_p(C, H)$.

For example, if $p \in C \subset \mathbb{P}^2$ is a smooth point of a plane curve and $L \supset p$ is any line through p, then the order of contact of L with C at p is at least 1; L is tangent to C at p if and only if it is at least 2. The line L is called a *flex tangent* if the order is at least 3, and in this case p is called a *flex* of C. Carrying this further, we say that p is a *hyperflex* if the tangent line L at p meets C with order ≥ 4. We adopt similar definitions in the situation where $f : C \to \mathbb{P}^2$ is a nonconstant morphism from a smooth curve. For a curve in 3-space we can consider both the orders of contact with lines and the orders of contact with hyperplanes.

7.5.1 Vanishing sequences and osculating planes

We will systematize these ideas by considering a linear system $\mathcal{W} = (\mathcal{L}, W)$ on a smooth curve C with $\dim W = r + 1$. Given a point $p \in C$ and a section $\sigma \in W$, the order of vanishing $\operatorname{ord}_p \sigma$ of σ at p is defined to be the length of the $\mathcal{O}_{C,p}$-module $\mathcal{L}_p/(\mathcal{O}_{C,p}\sigma)$. Again, because the ground field $\Bbbk$ is algebraically closed we have $\kappa(p) = \Bbbk$, so

$$\operatorname{ord}_p \sigma = \dim_\Bbbk \mathcal{L}_p/(\mathcal{O}_{C,p}\sigma).$$

Given $p \in C$, consider the collection of all orders of vanishing of sections $\sigma \in W$ at p.

We define the *vanishing sequence* $a(\mathcal{W}, p)$ of the linear system $\mathcal{W}$ at p to be the sequence of integers that occur as orders of vanishing at p of sections in W, arranged in strictly increasing order:

$$a(\mathcal{W}, p) := (a_0(\mathcal{W}, p) < a_1(\mathcal{W}, p) < \cdots).$$

Since sections vanishing to distinct orders are linearly independent, $a(\mathcal{W}, p)$ has at most $\dim W$ elements. On the other hand, we can find a basis for W consisting of sections vanishing to distinct orders at p (start with any basis; if two sections vanish to the same order replace one with a linear combination of the two vanishing to higher order, and repeat). It follows that the number of elements in $a(\mathcal{W}, p)$ is exactly $\dim_\Bbbk W = r + 1$:

$$a(\mathcal{W}, p) = (a_0(\mathcal{W}, p) < \cdots < a_r(\mathcal{W}, p)).$$

We set $\alpha_i = a_i - i$, and call the associated weakly increasing sequence

$$\alpha(\mathcal{W}, p) = (\alpha_0(\mathcal{W}, p) \leq \cdots \leq \alpha_r(\mathcal{W}, p))$$

the *ramification sequence* of $\mathcal{W}$ at p. When the linear system $\mathcal{W}$ or the point p we are referring to is clear from context, we will drop it from the notation and write $a_i(p)$ or just a_i in place of $a_i(\mathcal{W}, p)$, and similarly for α_i.

For example, p is a base point of $\mathcal{W}$ if and only if $a_0(p) = \alpha_0(p) > 0$, and more generally $a_0(p)$ is the multiplicity with which p appears in the base locus of $\mathcal{W}$. If p is a base point of $\mathcal{W}$ then, since C is a smooth curve, we may remove it; that is, W is in the image of the monomorphism $H^0(\mathcal{L}(-a_0 p)) \to H^0(\mathcal{L})$, and we may thus consider W as defining a linear series $\mathcal{W}' := (\mathcal{L}(-a_0 p), W)$. In this way most questions about linear systems on smooth curves can be reduced to the base point free case.

When p is not a base point of $\mathcal{W}$, so that $\mathcal{W}$ defines a morphism $f : C \to \mathbb{P}^r$ in a neighborhood of p, we have $a_1(p) = 1$ if and only if f is an embedding near p. If $r = 2$ and $\mathcal{W}$ is very ample, so that f is an embedding, we thus have $\alpha_0(p) = \alpha_1(p) = 0$ for all p and $\alpha_2(p) > 0$ for some particular p if and only if there is a line meeting the embedded curve with multiplicity > 2 at p; that is, p is an inflection point of the embedded curve. The geometric meaning of the vanishing sequence is given in general by the next result:

Proposition 7.12. *Let $\mathcal{W} = (\mathcal{L}, W)$ be a linear series on a smooth curve C, and let $p \in C$. If p is not a base point of $\mathcal{W}$, we let $\mathcal{W}' = \mathcal{W}$; in general, let $\mathcal{W}' = (\mathcal{L}(-a_0(\mathcal{W}, p)p), W)$.*

(a) $a_i(\mathcal{W}', p) = a_i(\mathcal{W}, p) - a_0(\mathcal{W}, p)$.

(b) Choose $\sigma_0, \ldots, \sigma_r \in W$ such that σ_j vanishes at p to order $a_j(\mathcal{W}', p)$, and let H_j be the hyperplane in $\mathbb{P}(W^*)$ corresponding to σ_j. The plane

$$L_i = H_{i+1} \cap \cdots \cap H_r$$

is the unique linear subspace of dimension i with highest order of contact with C at p, and that order is $a_{i+1}(\mathcal{W}, p)$.

The planes L_i are called the *osculating planes* to $f(C)$ at p. We always have $L_0 = p$. If $f(C)$ is smooth at $f(p)$ then L_1 is the tangent line, and in general it is the reduced tangent cone to the branch of $f(C)$ that is the image of an analytic neighborhood of $p \in C$.

Proof: (a) A section of $\mathcal{L}(-dp)$ that vanishes to order m as a section of $\mathcal{L}(-dp)$ will vanish to order $m + d$ at p as a section of $\mathcal{L}$.

(b) Writing f for the germ at p of the morphism defined by $\mathcal{W}'$, it follows from the definitions that $\mathrm{ord}_p L_i = a_{i+1}$. If there were an i-plane L' with higher order of contact, and we wrote

$$L' = H'_{i+1} \cap \cdots \cap H'_r$$

for some hyperplanes H'_r, then each H'_r would have order of contact with C at p strictly greater than a_{i+1}. But these would correspond to independent sections in W, and taking linear combinations of these sections we would get $r - i$ sections with vanishing orders at p strictly greater than a_{i+1}. This contradicts the assumption that the highest $r - i$ elements of the vanishing sequence are $a_{i+1}, \ldots, a_r$. $\qquad\square$

7.5.2 Total inflection: the Plücker formula

We say that p is an *inflection point* for a linear system $(\mathcal{L}, W)$ of dimension r if the ramification sequence $(\alpha_0, \ldots, \alpha_r)$ is not $(0, \ldots, 0)$, or, equivalently, if $\alpha_r > 0$, which is the same as $a_r > r$. When W arises from a morphism $f : C \to \mathbb{P}^r$ that is an embedding near p, p is an inflection point of W if and only if some hyperplane has contact $\geq r + 1$ at p.

We define the *weight* of $p \in C$ with respect to W to be

$$w(W, p) := \sum_{i=0}^{r} \alpha_i.$$

This number is a measure of what might be called the "total inflection" of W at p. We can compute the sum $\sum_{p \in C} w(W, p)$ as a Chern class of the bundle of principal parts of $\mathcal{L}$.

Theorem 7.13 (Plücker formula). *If W is a linear system of degree d and dimension r on a smooth projective curve C of genus g, then*

$$\sum_{p \in C} w(W, p) = (r + 1)d + (r + 1)r(g - 1).$$

This is our answer to Keynote Question (c). Note that it is only an enumerative formula, in the sense that each hyperplane having contact of order $r + 1$ or more with C at a point p has to be counted with multiplicity $w(W, p)$. We might expect that if C is a suitably general curve — say, one corresponding to a general point on a component of the open subset of the Hilbert scheme parametrizing smooth, irreducible, nondegenerate curves — then all inflection points of C would have weight 1, but this is actually false (see Exercises 7.40–7.41). It can be verified in some cases, such as plane curves (see Exercise 7.32), and it is true also for complete intersections with sufficiently high multidegree (see Exercise 7.39 for a step forward in that direction); it remains an open problem to say when it holds in general.

Proof: The key observation is that both sides of the desired formula are equal to the degree of the first Chern class of the bundle $\mathcal{P}^r(\mathcal{L})$. We can compute the class of this bundle from Theorem 7.2 as

$$c(\mathcal{P}^r(\mathcal{L})) = \prod_{j=0}^{r} c(\mathrm{Sym}^j(\Omega_C) \otimes \mathcal{L}).$$

Since Ω_C is a line bundle we have $\mathrm{Sym}^j(\Omega_C) = \Omega_C^j$, and thus $c((\mathrm{Sym}^j(\Omega_C) \otimes \mathcal{L}) = 1 + jc_1(\Omega_C) + c_1(\mathcal{L})$. It follows that

$$c_1(\mathcal{P}^r(\mathcal{L})) = (r + 1)c_1(\mathcal{L}) + \binom{r+1}{2}c_1(\Omega_C).$$

Since the degree of Ω_C is $2g - 2$, the degree of this class is

$$\deg c_1(\mathcal{P}^r(\mathcal{L})) = (r + 1)d + (r + 1)r(g - 1),$$

the right-hand side of the Plücker formula.

We may define a map $\varphi : \mathcal{O}_C^{r+1} \to \mathcal{P}^r(\mathcal{L})$ by choosing any basis $\sigma_0, \ldots, \sigma_r$ of W and sending the i-th basis element of $\mathcal{O}_C^{r+1}$ to the section τ_{σ_i} of $\mathcal{P}^r(\mathcal{L})$ corresponding to σ_i. We will complete the proof by showing that for any point $p \in C$ the determinant of the map φ vanishes at p to order exactly $w(W, p)$, and that there are only finitely many points $w(W, p)$ where the determinant is 0.

To this end, fix a point $p \in C$. Since the determinant of φ depends on the choice of basis $\sigma_0, \ldots, \sigma_r$ only up to scalars, we may choose the basis σ_i so that the order of vanishing $\text{ord}_p(\sigma_i) = a_i$ at p is $a_i(W, p)$. Trivializing $\mathcal{L}$ in a neighborhood of p, we may think of the section σ_i locally as a function, and φ is represented by the matrix

$$\begin{pmatrix} \sigma_0 & \sigma_1 & \cdots & \sigma_r \\ \sigma_0' & \sigma_1' & \cdots & \sigma_r' \\ \vdots & \vdots & & \vdots \\ \sigma_0^{(r)} & \sigma_1^{(r)} & \cdots & \sigma_r^{(r)} \end{pmatrix},$$

where σ_i' denotes the derivative and $\sigma_i^{(r)}$ the r-th derivative. Because σ_i vanishes to order $\geq i$ at p, the matrix evaluated at p is lower-triangular, and the entries on the diagonal are all nonzero if and only if $a_i = i$ for each i; that is, if and only if p is not an inflection point for W.

We can compute the exact order of vanishing of $\det \varphi$ at an inflection point as follows: Denote by $v(z)$ the $(r + 1)$-vector $(\sigma_0(z), \ldots, \sigma_r(z))$, so that the determinant of φ is the wedge product

$$\det(\varphi) = v \wedge v' \wedge \cdots \wedge v^{(r)}.$$

Applying the product rule, the n-th derivative of $\det(\varphi)$ is then a linear combination of terms of the form

$$v^{(\beta_0)} \wedge v^{(\beta_1+1)} \wedge \cdots \wedge v^{(\beta_r+r)},$$

with $\sum \beta_i = n$. Now, $v^{(\beta_0)}(p) = 0$ unless $\beta_0 \geq \alpha_0$; similarly, $v^{(\beta_0)}(p) \wedge v^{(\beta_1)}(p) = 0$ unless $\beta_0 + \beta_1 \geq \alpha_0 + \alpha_1$, and so on. We conclude that *any derivative of $\det(\varphi)$ of order less than $w = \sum \alpha_i$ vanishes at p*, and the expression for the w-th derivative of $\det(\varphi)$ has exactly one term nonzero at p, namely

$$v^{(\alpha_0)} \wedge v^{(\alpha_1+1)} \wedge \cdots \wedge v^{(\alpha_r+r)}.$$

Since this term appears with nonzero coefficient, we conclude that $\det(\varphi)$ vanishes to order exactly w at p.

It remains to show that not every point of C can be an inflection point for W — that is, that $\det \varphi$ is not identically zero. To prove this, suppose that $\det(\varphi)$ does vanish identically, that is, that

$$v \wedge v' \wedge \cdots \wedge v^{(k)} \equiv 0 \qquad (7.3)$$

for some $k \leq r$. Suppose in addition that k is the smallest such integer, so that at a general point $p \in C$ we have

$$v(p) \wedge v'(p) \wedge \cdots \wedge v^{(k-1)}(p) \neq 0;$$

in other words, $v(p), \ldots, v^{(k-1)}(p)$ are linearly independent, but $v^{(k)}(p)$ lies in their span Λ. Again using the product rule to differentiate the expression (7.3), we see that

$$\frac{d}{dz}(v \wedge v' \wedge \cdots \wedge v^{(k-1)} \wedge v^{(k)}) = v \wedge v' \wedge \cdots \wedge v^{(k-1)} \wedge v^{(k+1)} \equiv 0,$$

so that $v^{(k+1)}(p)$ also lies in the span of $v(p), \ldots, v^{(k-1)}(p)$. Similarly, taking the second derivative of (7.3), we see that

$$\frac{d^2}{dz^2}(v \wedge v' \wedge \cdots \wedge v^{(k-1)} \wedge v^{(k)}) = v \wedge v' \wedge \cdots \wedge v^{(k-1)} \wedge v^{(k+2)} \equiv 0,$$

where are all the other terms in the derivative are zero because they are $(k+1)$-fold wedge products of vectors lying in a k-dimensional space. Continuing in this way, we see that $v^{(m)}(p) \in \Lambda$ for all m; it follows by integration that $v(z) \in \Lambda$ for all z. This implies that the linear system W has dimension $k < r + 1$, contradicting our assumptions. $\square$

Flexes of plane curves

Theorem 7.13 gives the answer to Keynote Question (c). We do not even need to assume C is smooth; if C is singular, as long as it is reduced and irreducible we view it as the image of the map $v : \tilde{C} \to \mathbb{P}^r$ from its normalization. For example, when $r = 2$, if we apply the Plücker formula to the linear system corresponding to this map, we see that C has

$$(r + 1)d + r(r + 1)(g - 1) = 3d + 6g - 6$$

flexes, where g is the genus of $\tilde{C}$, that is, geometric genus of C. If the curve C is indeed smooth, then $2g - 2 = d(d - 3)$, and so this yields

$$3d + 6g - 6 = 3d + 3d(d - 3) = 3d(d - 2).$$

To be explicit, this formula counts points $p \in \tilde{C}$ such that, for some line $L \subset \mathbb{P}^2$, the multiplicity of the pullback divisor v^*L at p is at least 3. In particular:

(a) It does not necessarily count nodes of C, even though at a node p of C there will be lines having intersection multiplicity 3 or more with C at p.

(b) It does count singularities where the differential dv vanishes, for example cusps.

Some applications of the general Plücker formula appear in Exercises 7.35–7.37.

We mention that there is an alternative notion of a flex point of a (possibly singular) curve $C \subset \mathbb{P}^2$: a point $p \in C$ such that, for some line $L \subset \mathbb{P}^2$ through p, we have

$$m_p(C \cdot L) \geq 3.$$

In this sense, a node p of a plane curve C is a flex point, since the tangent lines to the branches of the curve at the node will have intersection multiplicity at least 3 with C at p. When we want to talk about flexes in this sense, we will refer to them as *Cartesian flexes*, since they are defined in terms of the defining equation of $C \subset \mathbb{P}^2$ rather than its parametrization by a smooth curve.

There is a classical way to calculate the number of flexes of a plane curve that does count Cartesian flexes. Briefly, if C is the zero locus of a homogeneous polynomial $F(X, Y, Z)$, we define the *Hessian* of C be the zero locus of the polynomial

$$H = \begin{vmatrix} \dfrac{\partial^2 F}{\partial X^2} & \dfrac{\partial^2 F}{\partial X \partial Y} & \dfrac{\partial^2 F}{\partial X \partial Z} \\[2mm] \dfrac{\partial^2 F}{\partial X \partial Y} & \dfrac{\partial^2 F}{\partial Y^2} & \dfrac{\partial^2 F}{\partial Y \partial Z} \\[2mm] \dfrac{\partial^2 F}{\partial X \partial Z} & \dfrac{\partial^2 F}{\partial Y \partial Z} & \dfrac{\partial^2 F}{\partial Z^2} \end{vmatrix}.$$

For a smooth plane curve C, the Cartesian flexes are exactly the points of intersection of C with its Hessian (it is even true that on a smooth curve C the weight of a flex p is equal to the intersection multiplicity of C with its Hessian at p). In Exercise 7.33, we will explore what happens to the flexes on a smooth plane curve when it acquires a node.

Hyperflexes

First, the bad news: We are not going to answer Keynote Question (d) here. The question itself is well-posed: We know that a general plane curve $C \subset \mathbb{P}^2$ of degree $d \geq 4$ has only ordinary flexes, and it is not hard to see that the locus of those curves that do have a hyperflex is a hypersurface in the space $\mathbb{P}^N$ of all such curves (see Exercise 7.38). Surely the techniques we have employed in this chapter will enable us to calculate the degree of that hypersurface? Unfortunately, they do not, and indeed the reason we included Keynote Question (d) is so that we could point out the problem.

Very much by analogy with the analysis of lines on surfaces and singular points on curves, we would like to determine the class of the "universal hyperflex:" that is, in the universal curve

$$\Phi = \{(C, p) \in \mathbb{P}^N \times \mathbb{P}^2 \mid p \in C\},$$

the locus

$$\Gamma = \{(C, p) \in \Phi \mid p \text{ is a hyperflex of } C\}.$$

Moreover, it seems as if this would be amenable to a Chern class approach: We would define a vector bundle $\mathcal{E}$ on Φ whose fiber at a point $(C, p) \in \Phi$ would be the vector space

$$\mathcal{E}_{(C,p)} = \frac{\{\text{germs of sections of } \mathcal{O}_C(1) \text{ at } p\}}{\{\text{germs vanishing to order} \geq 4 \text{ at } p\}}.$$

We would then have a map of vector bundles on Φ from the trivial bundle with fiber $H^0(\mathcal{O}_{\mathbb{P}^2}(1))$ to $\mathcal{E}$, and the degeneracy locus of this map would be the universal hyperflex Γ. Since this is the locus where three sections of a bundle of rank 4 are linearly dependent, we could conclude that

$$[\Gamma] = c_2(\mathcal{E}).$$

As we indicated, though, there is a problem with this approach. The description above of the fibers of $\mathcal{E}$ makes sense *as long as p is a smooth point of C, but not otherwise*. Reflecting this fact, if we were to try to define $\mathcal{E}$ by taking $\Delta \subset \Phi \times_{\mathbb{P}^N} \Phi$ the diagonal and setting

$$\mathcal{E} = \pi_{1*}(\pi_2^* \mathcal{O}_{\mathbb{P}^2}(1) \otimes \mathcal{O}_{\Phi \times_{\mathbb{P}^N} \Phi} / \mathcal{I}_\Delta^4),$$

the sheaf $\mathcal{E}$ would have fiber as desired over the open set $U \subset \Phi$ of (C, p) with C smooth at p, but would not even be locally free on the complement. The fact that bundles of principal parts do not behave well in families (except, of course, smooth families) is a real obstruction to carrying out this sort of calculation.

There is a way around this problem: Ziv Ran [2005a; 2005b] showed that — at least over the preimage $\mathcal{C} \subset \Phi$ of a general line $\mathbb{P}^1 \subset \mathbb{P}^N$ — the vector bundle $\mathcal{E}|_{U \cap \mathcal{C}}$ extends to a locally free sheaf on a blow-up of $\mathcal{C}$, realized as a subscheme of the relative Hilbert scheme of $\mathcal{C}$ over $\mathbb{P}^1$. This approach does yield an answer to Keynote Question (d), and indeed applies far more broadly, albeit at the expense of a level of difficulty that places it outside the range of this text.

And now, the good news: there is another way to approach Keynote Question (d), and we will explain it in Section 11.3.1.

7.5.3 The situation in higher dimension

Is there an analog of the Plücker formula for linear series on varieties of dimension greater than 1? Assuming that the linear series yields an embedding $X \subset \mathbb{P}^r$, we might ask, for a start, what sort of singularities we should expect the intersection $X \cap \Lambda$ of X with linear spaces $\Lambda \subset \mathbb{P}^r$ of a given dimension to have at a point p, and ask for the locus of points that are "exceptional" in this sense.

We do not know satisfying answers to these questions in general. One issue is that, while the singularities of subschemes of a smooth curve are simply classified by their multiplicity, there is already a tremendous variety of singularities of subschemes of surfaces. (If we have a particular class of singularities in mind, such as the A_n-singularities described in Section 11.4.1, then these questions do become well-posed; see for example the beautiful analysis of elements of a linear system having an A_n-singularity in Russell [2003].) Another problem is that the analog of the final step in the proof of Theorem 7.13 — showing that not every point on a smooth curve $C \subset \mathbb{P}^r$ can be an inflection point — may not hold. For example, a dimension count might lead us to expect that for a general point p on a smooth, nondegenerate surface $S \subset \mathbb{P}^5$ no hyperplane $H \subset \mathbb{P}^5$ intersects S in a curve $C = H \cap S$ with a triple point at p, but there are such surfaces for which this is false, and we do not know a classification of such surfaces.

We will revisit this question in Chapter 11, where we will describe the behavior of plane sections of a general surface $S \subset \mathbb{P}^3$.

7.6 Nets of plane curves

We now want to consider larger-dimensional families of plane curves, and in particular to answer the second keynote question of this chapter. A key step will be to compute the class of the universal singular point $\Sigma = \{(C, p) \mid p \in C_{\mathrm{sing}}\}$ as a subvariety of $\mathbb{P}^N \times \mathbb{P}^2$, where $\mathbb{P}^N = \mathbb{P} H^0(\mathcal{O}_{\mathbb{P}^2}(d))$.

7.6.1 Class of the universal singular point

Let $W = H^0(\mathcal{O}_{\mathbb{P}^n}(d))$, so that $\mathbb{P} W$ is the projective space of hypersurfaces of degree d in $\mathbb{P}^n$, and consider the universal m-fold point

$$\Sigma = \Sigma_{n,d,m} = \{(X, p) \in \mathbb{P} W \times \mathbb{P}^n \mid \mathrm{mult}_p(X) \geq m\},$$

and let

$$\mathbb{P} W \times \mathbb{P}^n \xrightarrow{\ \pi_2\ } \mathbb{P}^n$$
$$\pi_1 \downarrow $$
$$\mathbb{P} W$$

be the projection maps. We can express the class $[\Sigma] \in A(\mathbb{P} W \times \mathbb{P}^n)$ in terms of Chern classes:

Proposition 7.14. *$\Sigma_{n,d,m}$ is the zero locus of a section of the vector bundle*

$$\mathcal{P}^m := \pi_1^* \mathcal{O}_{\mathbb{P} W}(1) \otimes \pi_2^* \mathcal{P}^m(\mathcal{O}_{\mathbb{P}^n}(d)),$$

which has Chern class

$$c(\mathcal{P}^m) = (1 + (d - m)\zeta_n + \zeta_W)^{\binom{n+m}{n}},$$

where ζ_n and ζ_W are the pullbacks of the hyperplane classes on $\mathbb{P}^n$ and $\mathbb{P} W$ respectively.

Thus the class of $\Sigma_{n,d,m}$ in $A(\mathbb{P}W \times \mathbb{P}^n)$ is the sum of the terms of total degree $\binom{n+m}{n}$ in this expression. For example, in the case $n = 2, m = 1$ this is

$$[\Sigma] = \zeta_W^3 + 3(d-1)\zeta_2\zeta_W^2 + 3(d-1)^2\zeta_2^2\zeta_W \in A^3(\mathbb{P}W \times \mathbb{P}^2).$$

Proof: The computation is similar to the one used in the calculation of the class of the universal line in Section 6.6. Since every polynomial $F \in W$ defines a section τ_F of $\mathcal{P}^m(\mathcal{O}_{\mathbb{P}^n}(d))$, we have a map

$$W \otimes \mathcal{O}_{\mathbb{P}^n} \to \mathcal{P}^m(\mathcal{O}_{\mathbb{P}^n}(d))$$

of vector bundles on $\mathbb{P}^n$. Likewise, we have the tautological inclusion

$$\mathcal{O}_{\mathbb{P}W}(-1) \to W \otimes \mathcal{O}_{\mathbb{P}W}$$

on $\mathbb{P}W$. We pull these maps back to the product $\mathbb{P}W \times \mathbb{P}^2$ and compose them to obtain a map

$$\pi_1^*\mathcal{O}_{\mathbb{P}W}(-1) \to \pi_2^*\mathcal{P}^m(\mathcal{O}_{\mathbb{P}^n}(d)),$$

or, equivalently, a section of the bundle $\mathcal{P}^m$. The zero locus of this map is $\Sigma \subset \mathbb{P}W \times \mathbb{P}^n$, so the class of $\Sigma_{n,d,m}$ in $A(\mathbb{P}W \times \mathbb{P}^n)$ is the class of a section of $\mathcal{P}^m$, as claimed.

To compute the Chern class of $\mathcal{P}^m$, we follow the argument of Proposition 7.5, pulling back the sequences

$$0 \longrightarrow \operatorname{Sym}^i(\Omega_{\mathbb{P}^n})(d) \longrightarrow \mathcal{P}^i(\mathcal{O}_{\mathbb{P}^n}(d)) \longrightarrow \mathcal{P}^{i-1}(\mathcal{O}_{\mathbb{P}^n}(d)) \longrightarrow 0$$

and tensoring with the line bundle $\pi_1^*\mathcal{O}_{\mathbb{P}W}(1)$ to get

$$c(\mathcal{P}^m) = \prod_{j=0}^{m} c(\operatorname{Sym}^j(\pi_2^*\Omega_{\mathbb{P}^n}) \otimes \mathcal{O}(d\zeta_n + \zeta_W)),$$

where we write $\mathcal{O}(d\zeta_n + \zeta_W)$ as shorthand for the line bundle $\pi_1^*\mathcal{O}_{\mathbb{P}W}(1) \otimes \pi_2^*\mathcal{O}_{\mathbb{P}^n}(d)$. Using the exact sequences

$$0 \longrightarrow \operatorname{Sym}^i(\Omega_{\mathbb{P}^n}) \longrightarrow \operatorname{Sym}^i(\mathcal{O}_{\mathbb{P}^n}(-1)) \longrightarrow \operatorname{Sym}^{i-1}(\mathcal{O}_{\mathbb{P}^n}(-1)) \longrightarrow 0$$

we get a collapsing product as before, yielding the desired formula for the Chern class of $\mathcal{P}^m$. To deduce the special case at the end of the proposition, it suffices to remember that since ζ_2 is the pullback from a two-dimensional variety we have $\zeta_2^3 = 0$. $\square$

7.6.2 The discriminant of a net of plane curves

We return to the case of a net of plane curves of degree d. Throughout this section we fix a general net of plane curves of degree d, that is, the family of curves associated to a general linear subspace $W \subset H^0(\mathcal{O}_{\mathbb{P}^2}(d))$ of dimension 3, parametrized by $\mathcal{B} = \mathbb{P}W \cong \mathbb{P}^2$.

Let $\mathcal{D} \subset \mathcal{B}$ be the set of singular curves, called the *discriminant curve* of the net $\mathcal{B}$. Since $\mathcal{D}$ is the intersection of $\mathcal{B}$ with the discriminant hypersurface in $\mathbb{P}W$, its degree is $\deg \mathcal{D} = 3(d-1)^2$ by Proposition 7.4. Next, let $\Gamma \subset \mathbb{P}^2$ be the plane curve traced out by the singular points of members of the net, so that if we set

$$\Sigma_{\mathcal{B}} := \Sigma \cap (\mathcal{B} \times \mathbb{P}^2),$$

then the projection maps π_i on Σ restrict to surjections

$$\Sigma_{\mathcal{B}} \xrightarrow{\ \pi_2|_{\mathcal{B}}\ } \Gamma$$
$$\pi_1|_{\mathcal{B}} \Big\downarrow$$
$$\mathcal{D}$$

Since Σ is smooth of codimension 3, Bertini's theorem shows that $\Sigma_{\mathcal{B}}$ is a smooth curve in $\mathcal{B} \times \mathbb{P}^2$. Since the generic singular plane curve is singular at only one point, the map $\Sigma_{\mathcal{B}} \to \Gamma$ is birational. Since the fiber of Σ over a given point $p \in \mathbb{P}^2$ is a linear space of dimension $N - 3$, the general 2-plane $\mathcal{B}$ containing a curve singular at p will contain a unique such curve. Thus the map $\Sigma_{\mathcal{B}} \to \mathcal{D}$ is also birational, and $\Sigma_{\mathcal{B}}$ is the normalization of each of Γ and $\mathcal{D}$. In particular the geometric genus of $\mathcal{D}$ and that of Γ are the same as the genus of $\Sigma_{\mathcal{B}}$.

From the previous section, we know that $\Sigma_{\mathcal{B}}$ is the zero locus of a section of the rank-3 bundle $\mathcal{P}^1|_{\mathcal{B} \times \mathbb{P}^2}$ on $\mathcal{B} \times \mathbb{P}^2$. This makes it easy to compute the degree and genus of $\Sigma_{\mathcal{B}}$, and we will derive the degree and genus of Γ, answering Keynote Question (b):

Proposition 7.15. *With notation as above, the map $\Sigma_{\mathcal{B}} \to \Gamma$ is an isomorphism, so both curves are smooth. The curve Γ has degree $3d - 3$, and thus has genus $\binom{3d-4}{2}$. When $d \geq 2$, the curve $\mathcal{D}$ is singular.*

We will see how the singularities of $\mathcal{D}$ arise, what they look like and how many there are in Chapter 11.

Proof: We begin with the degree of Γ, the number of points of intersection of Γ with a line $L \subset \mathbb{P}^2$. Since $\Sigma_{\mathcal{B}} \to \Gamma$ is birational, this is the same as the degree of the product $[\Sigma_{\mathcal{B}}]\zeta_2 \in A^4(\mathcal{B} \times \mathbb{P}^2)$. (More formally, $\pi_{2*}[\Sigma_{\mathcal{B}}] = \Gamma$ and $\pi_{2*}[\Sigma_{\mathcal{B}}][L] = [\Sigma_{\mathcal{B}}]\zeta_2$.) Write $\zeta_{\mathcal{B}}$ for the restriction of ζ_W, the pullback of the hyperplane section from $\mathbb{P}^N$, to $\mathcal{B} \times \mathbb{P}^2$. The degree of a class in $\mathcal{B} \times \mathbb{P}^2$ is the coefficient of $\zeta_2^2 \zeta_{\mathcal{B}}^2$ in its expression in

$$A(\mathcal{B} \times \mathbb{P}^2) = \mathbb{Z}[\zeta_2, \zeta_{\mathcal{B}}]/(\zeta_2^3, \zeta_{\mathcal{B}}^3).$$

Since $\zeta_{\mathcal{B}}^3 = 0$, the last formula in Proposition 7.14 gives

$$\deg(\Gamma) = \deg \zeta_2 \big(3(d-1)\zeta_2\zeta_{\mathcal{B}}^2 + 3(d-1)^2\zeta_2^2\zeta_{\mathcal{B}}\big)$$
$$= \deg 3(d-1)\zeta_2^2\zeta_{\mathcal{B}}^2$$
$$= 3d - 3.$$

Since Γ is a plane curve, the arithmetic genus of the curve Γ is $\binom{3d-4}{2}$.

Next we compute the genus g_{Σ_B} of the smooth curve Σ_B. The normal bundle of Σ_B in $\mathbb{P}^2 \times B$ is the restriction of the rank-3 bundle $\mathcal{P}^1$, and the canonical divisor on $\mathbb{P}^2 \times B$ has class $-3\zeta_2 - 3\zeta_B$, so by the general adjunction formula (Part (c) of Proposition 6.15) the degree of the canonical class of Σ_B is the degree of the line bundle obtained by tensoring the canonical bundle of $B \times \mathbb{P}^2$ with $\wedge^3 \mathcal{P}^1$ and restricting the result to Σ_B. This is the degree of the class

$$(-3\zeta_2 - 3\zeta_B + c_1(\mathcal{P}^1))[\Sigma_B] = (-3\zeta_2 - 3\zeta_B + c_1(\mathcal{P}^1)) \cdot \big(3(d-1)\zeta_2\zeta_B^2 + 3(d-1)^2\zeta_2^2\zeta_B\big).$$

Substituting the value $c_1(\mathcal{P}^1) = 3((d-1)\zeta_2 + \zeta_W)$ from Proposition 7.14 and taking account of the fact that $\zeta_W\zeta_B = \zeta_B^2$, this becomes

$$(3d-6)\zeta_2 \cdot \big(3(d-1)\zeta_2\zeta_B^2 + 3(d-1)^2\zeta_2^2\zeta_B\big) = (3d-6)(3d-3)\zeta_2^2\zeta_B^2$$

with degree $2g_{\Sigma_B} - 2 = (3d-3)(3d-6)$, and we see that

$$g(\Sigma_B) = \frac{(3d-4)(3d-5)}{2} = \binom{3d-4}{2}.$$

Since this coincides with the arithmetic genus of Γ computed above, we see that Γ is smooth and the map $\Sigma_B \to \Gamma$ is an isomorphism. On the other hand the degree $3(d-1)^2$ of $\mathcal{D}$ is different from that of Γ for all $d \geq 2$, so in these cases the arithmetic and geometric genera of $\mathcal{D}$ differ, and $\mathcal{D}$ must be singular, completing the proof. $\quad\square$

Here is a different method for computing the degree of Γ: The net $\mathcal{B}$ of curves, having no base points, defines a regular map

$$\varphi_\mathcal{B} : \mathbb{P}^2 \to \Lambda,$$

where $\Lambda \cong \mathbb{P}^2$ is the projective plane dual to the plane parametrizing the curves in the net $\mathcal{B}$. This map expresses $\mathbb{P}^2$ as a d^2-sheeted branched cover of Λ, and the curve $\Gamma \subset \mathbb{P}^2$ is the ramification divisor of this map.

By definition,

$$\varphi^* \mathcal{O}_\Lambda(1) = \mathcal{O}_{\mathbb{P}^2}(d);$$

so that, if we denote by ζ_Λ the hyperplane class on Λ, we have $\varphi^*\zeta_\Lambda = d\zeta$.

Pulling back a 2-form via the map $\varphi : \mathbb{P}^2 \to \Lambda$ we see that

$$K_{\mathbb{P}^2} = \varphi^* K_\Lambda + \Gamma,$$

and since $K_\Lambda = -3\zeta_\Lambda$, this yields

$$-3\zeta = -3d\zeta + [\Gamma]$$

or $[\Gamma] = (3d-3)\zeta$.

These ideas work for a net $\mathcal{W} = (\mathcal{L}, W)$ on an arbitrary smooth projective surface S, as long as we know the classes $c_1(\Omega_S), c_2(\Omega_S)$ and $\lambda = c_1(\mathcal{L})$ and can evaluate the degrees of the relevant products in $A(S)$. See Exercise 7.31 for an example.

7.7 The topological Hurwitz formula

In this section we will work explicitly over the complex numbers, so that we can use the topological Euler characteristic. Using this tool, we will give a different approach to questions of singular elements of linear series. It sheds additional light on the formula of Proposition 7.4, and is applicable in many circumstances in which Proposition 7.4 cannot be used. In addition, it will allow us to describe the local structure of the discriminant hypersurface, such as its tangent planes and tangent cones. By the Lefschetz principle (see for example Harris [1995, Chapter 15]), moreover, the purely algebro-geometric consequences of this analysis, such as Propositions 7.19 and 7.20, hold more generally over an arbitrary algebraically closed field of characteristic 0. (There are also alternative ways of defining an Euler characteristic with the desired properties algebraically.)

This approach is based on the following simple observation:

Proposition 7.16. *Let X be a smooth projective variety over $\mathbb{C}$, and $Y \subset X$ a divisor. If we denote by χ_{top} the topological Euler characteristic (in the classical, or analytic, topology), then*

$$\chi_{\mathrm{top}}(X) = \chi_{\mathrm{top}}(Y) + \chi_{\mathrm{top}}(X \setminus Y).$$

Proof: This will follow from the Mayer–Vietoris sequence applied to the covering of X by $U = X \setminus Y$ and a small open neighborhood V of Y.

Let $\mathcal{L} = \mathcal{O}_X(Y)$, and let σ be the section of $\mathcal{L}$ vanishing on Y. Introducing Hermitian metrics on X and the line bundle $\mathcal{L}$, we can use the gradient of the absolute value of σ to define a C^∞ map $V \to Y$ expressing V as a fiber bundle over Y with fiber a disc D^2, and simultaneously expressing $V \cap U$ as a bundle over Y with fiber a punctured disc. It follows that

$$\chi_{\mathrm{top}}(V) = \chi_{\mathrm{top}}(Y) \quad \text{and} \quad \chi_{\mathrm{top}}(V \cap U) = 0,$$

and we deduce the desired relation.　　　　　　　　　　　　　　　　　　□

It is a surprising fact that the formula $\chi_{\mathrm{top}}(X) = \chi_{\mathrm{top}}(Y) + \chi_{\mathrm{top}}(X \setminus Y)$ applies much more generally to an arbitrary subvariety Y of an arbitrary X; see for example Fulton [1993, pp. 93–95, 142].

Now let X be a smooth projective variety, and let $f : X \to B$ be a map to a smooth curve B of genus g. This being characteristic 0, there are only a finite number of points $p_1, \ldots, p_\delta \in B$ over which the fiber X_{p_i} is singular. We can apply the relation on Euler characteristics to the divisor

$$Y = \bigcup_{i=1}^{\delta} X_{p_i} \subset X.$$

Naturally, $\chi_{\text{top}}(Y) = \sum \chi_{\text{top}}(X_{p_i})$, and on the other hand the open set $X \setminus Y$ is a fiber bundle over the complement $B \setminus \{p_1, \ldots, p_\delta\}$, so that

$$\chi_{\text{top}}(X \setminus Y) = \chi_{\text{top}}(X_\eta)\chi_{\text{top}}(B \setminus \{p_1, \ldots, p_\delta\}) = (2 - 2g - \delta)\chi_{\text{top}}(X_\eta),$$

where again η is a general point of B. Combining these, we have

$$\chi_{\text{top}}(X) = (2 - 2g - \delta)\chi_{\text{top}}(X_\eta) + \sum_{i=1}^{\delta} \chi_{\text{top}}(X_{p_i})$$

$$= \chi_{\text{top}}(B)\chi_{\text{top}}(X_\eta) + \sum_{i=1}^{\delta}(\chi_{\text{top}}(X_{p_i}) - \chi_{\text{top}}(X_\eta)).$$

In this form, we can extend the last summation over all points $q \in B$. We have proven:

Theorem 7.17 (Topological Hurwitz formula). *Let $f : X \to B$ be a morphism from a smooth projective variety to a smooth projective curve; let $\eta \in B$ be a general point. Then*

$$\chi_{\text{top}}(X) = \chi_{\text{top}}(B)\chi_{\text{top}}(X_\eta) + \sum_{q \in B}(\chi_{\text{top}}(X_q) - \chi_{\text{top}}(X_\eta)).$$

In English: The Euler characteristic of X is what it would be if X were a fiber bundle over B — that is, the product of the Euler characteristics of B and the general fiber X_η — with a "correction term" coming from each singular fiber, equal to the difference between its Euler characteristic and the Euler characteristic of the general fiber.

To see why Theorem 7.17 is a generalization of the classical Riemann–Hurwitz formula (see for example Hartshorne [1977, Section IV.2]), consider the case where X is a smooth curve of genus h and $f : X \to C$ a branched cover of degree d. For each point $p \in C$, we write the fiber X_p as a divisor:

$$f^*(p) = \sum_{q \in f^{-1}(p)} m_q \cdot q.$$

We call the integer $m_q - 1$ the *ramification index* of f at q; we define the *ramification divisor R* of f to be the sum

$$R = \sum_{q \in X}(m_q - 1) \cdot q,$$

and we define the *branch divisor B* of f to be the image of R (as a divisor, not as a scheme!) — that is,

$$B = \sum_{p \in C} b_p \cdot p, \quad \text{where } b_p = \sum_{q \in f^{-1}(p)} m_q - 1.$$

Now, since the degree of any fiber $X_p = f^{-1}(p)$ of f is equal to d, for each $p \in C$ the cardinality of $f^{-1}(p)$ will be $d - b_p$, so its contribution to the topological Hurwitz formula is $-b_p$. The formula then yields

$$2 - 2h = d(2 - 2g) - \deg(B),$$

the classical Riemann–Hurwitz formula.

7.7.1 Pencils of curves on a surface, revisited

To apply the topological Hurwitz formula to Keynote Question (a), suppose that $\{C_t = V(t_0 F + t_1 G) \subset \mathbb{P}^2\}$ is a general pencil of plane curves of degree d. Since the polynomials F and G are general, the base locus $\Gamma = V(F, G)$ of the pencil will consist of d^2 reduced points, and the total space of the pencil — that is, the graph

$$X = \{(t, p) \in \mathbb{P}^1 \times \mathbb{P}^2 \mid p \in C_t\}$$

of the rational map $[F, G] : \mathbb{P}^2 \ {-}\,{-}{\blacktriangleright} \ \mathbb{P}^1$ — is the blow-up of $\mathbb{P}^2$ along Γ. In particular, X is smooth, so Theorem 7.17 can be applied to the map $f : X \to \mathbb{P}^1$ that is the projection on the first factor.

Since X is the blow-up of $\mathbb{P}^2$ at d^2 points, we have

$$\chi_{\mathrm{top}}(X) = \chi_{\mathrm{top}}(\mathbb{P}^2) + d^2 = d^2 + 3.$$

Next, we know that a general fiber C_η of the map f is a smooth plane curve of degree d; as we saw in Example 2.17, its genus is $\binom{d-1}{2}$ and hence

$$\chi_{\mathrm{top}}(C_\eta) = -d^2 + 3d.$$

We know from Proposition 7.1 that each singular fiber C appearing in a general pencil of plane curves has a single node as singularity. By the calculation in Section 2.4.6, then, its normalization $\widetilde{C}$ will be a curve of genus $\binom{d-1}{2} - 1$ and hence Euler characteristic $-d^2 + 3d + 2$. Since C is obtained from $\widetilde{C}$ by identifying two points, we have

$$\chi_{\mathrm{top}}(C) = -d^2 + 3d + 1,$$

so the contribution of each singular fiber of f to the topological Hurwitz formula is exactly 1. It follows that the number of singular fibers is

$$\begin{aligned}
\delta &= \chi_{\mathrm{top}}(X) - \chi_{\mathrm{top}}(\mathbb{P}^1)\chi_{\mathrm{top}}(C_\eta) \\
&= d^2 + 3 - 2(-d^2 + 3d) \\
&= 3d^2 - 6d + 3,
\end{aligned}$$

as we saw before.

This same analysis can be applied to a pencil of curves on any smooth surface S. Let $\mathcal{L}$ be a line bundle on S with first Chern class $c_1(\mathcal{L}) = \lambda \in A^1(S)$, and let $W = \langle \sigma_0, \sigma_1 \rangle \subset H^0(\mathcal{L})$ be a two-dimensional vector space of sections with

$$\{C_t = V(t_0\sigma_0 + t_1\sigma_1) \subset S\}_{t \in \mathbb{P}^1}$$

the corresponding pencil of curves. We make — for the time being — two assumptions:

(a) The base locus $\Gamma = V(\{\sigma\}_{\sigma \in W})$ of the pencil is reduced; that is, it consists of $\deg(\lambda^2)$ points.

(b) Each of the finitely many singular elements of the pencil has just one node as singularity.

We also denote by $c_i = c_i(\Omega_S)$ the Chern classes of the cotangent bundle to S.

Given this, the calculation proceeds as before: we let X be the blow-up of S along Γ, and apply the topological Hurwitz formula to the natural map $f : X \to \mathbb{P}W^* \cong \mathbb{P}^1$. To start, we have

$$\chi_{\text{top}}(X) = \chi_{\text{top}}(S) + \#(\Gamma) = c_2 + \lambda^2$$

(we omit the "deg" here for simplicity). Next, by the adjunction formula, the Euler characteristic of a smooth member C_η of the pencil is given by

$$\chi_{\text{top}}(C_\eta) = -\deg(\omega_{C_\eta}) = -(c_1 + \lambda) \cdot \lambda = -\lambda^2 - c_1\lambda,$$

and, by Section 2.4.6, as in the plane curve case the Euler characteristic of each singular element of the pencil is 1 greater than the Euler characteristic of the general element. In sum, then, the number of singular fibers is

$$\begin{aligned} \delta &= \chi_{\text{top}}(X) - \chi_{\text{top}}(\mathbb{P}^1)\chi_{\text{top}}(C_\eta) \\ &= \lambda^2 + c_2 - 2(-\lambda^2 - c_1\lambda) \\ &= 3\lambda^2 + 2\lambda c_1 + c_2, \end{aligned}$$

agreeing with our previous calculation.

We will see how this may be applied in higher dimensions in Exercises 7.43–7.44.

7.7.2 Multiplicities of the discriminant hypersurface

One striking thing about this derivation of the formula for the number of singular elements in a pencil is that it gives a description of the multiplicities with which a given singular element counts that allows us to determine these multiplicities at a glance.

In the derivation of the formula, we assumed that the singular elements of the pencil had only nodes as singularities. But what if an element C of the pencil has a cusp? In that case the calculation of Section 2.4.6 says that the geometric genus of the curve — the genus of its normalization $\tilde{C}$ — is again 1 less than the genus of the smooth fiber, but this time instead of identifying two points of $\tilde{C}$ we are just "crimping" the curve at one point. (In the analytic topology, C and $\tilde{C}$ are homeomorphic.) Thus,

$$\chi_{top}(C) = \chi_{top}(\tilde{C}) = \chi_{top}(C_\eta) + 2,$$

and the fiber C "counts with multiplicity 2," in the sense that its contribution to the sum in the right-hand side of Theorem 7.17 is 2. Similarly, if C has a tacnode, we have $g(\tilde{C}) = g(C_\eta) - 2$, so that $\chi_{top}(\tilde{C}) = \chi_{top}(C_\eta) + 4$, but we identify two points of $\tilde{C}$ to form C, so in all

$$\chi_{top}(C) = \chi_{top}(C_\eta) + 3,$$

and the contribution of the fiber C to the rightmost term in Theorem 7.17 is thus 3. If C has an ordinary triple point — consisting of three smooth branches meeting at a point — then $g(\tilde{C}) = g(C_\eta) - 3$, but we identify three points of $\tilde{C}$ to form C, so

$$\chi_{top}(C) = \chi_{top}(C_\eta) + 4,$$

and the contribution of the fiber C is 4. Moreover, if a fiber has more than one isolated singularity, the same analysis shows that the multiplicity with which it appears in the formula above is just the sum of the contributions coming from the individual singularities.

In addition to giving us a way of determining the contribution of a given singular fiber to the expected number, this approach tells us something about the geometry of the discriminant locus $\mathcal{D} \subset \mathbb{P}^N$ in the space $\mathbb{P}^N$ of plane curves of degree d. To see this, suppose that $C \subset \mathbb{P}^2$ is any plane curve of degree d with isolated singularities. Let D be a general plane curve of the same degree, and consider the pencil $\mathcal{B}$ of plane curves they span — in other words, take $\mathcal{B} \subset \mathbb{P}^N$ a general line through the point $C \in \mathbb{P}^N$. By what we have said, the number of singular elements of the pencil $\mathcal{B}$ other than C will be $3(d-1)^2 - (\chi_{top}(C) - \chi_{top}(C_\eta))$, where C_η is a smooth plane curve of degree d; it follows that the intersection multiplicity $m_p(\mathcal{B}, \mathcal{D})$ of $\mathcal{B}$ and $\mathcal{D}$ at C is $\chi_{top}(C) - \chi_{top}(C_\eta)$.

Proposition 7.18. *Let $C \subset \mathbb{P}^2$ be any plane curve of degree d with isolated singularities. Then*

$$\mathrm{mult}_C(\mathcal{D}) = \chi_{top}(C) - \chi_{top}(C_\eta),$$

where C_η is a smooth plane curve of degree d

Thus a plane curve with a cusp (and no other singularities) corresponds to a double point of $\mathcal{D}$, a plane curve with a tacnode is a triple point, and so on. A curve C with one node and no other singularities is necessarily a smooth point of $\mathcal{D}$.

7.7.3 Tangent cones of the discriminant hypersurface

We can use the ideas above to describe the tangent spaces and tangent cones to the discriminant hypersurface $\mathcal{D} \subset \mathbb{P}^N$. To do this, we have to remove the first assumption in our application of the topological Hurwitz formula to pencils of curves, and deal with pencils whose base loci are not reduced.

To consider the simplest such situation, suppose that $p \in L \subset \mathbb{P}^2$ are a point and a line in the plane, and that F, G are general forms of degree d such that $V(F)$ and $V(G)$ pass through p and are tangent to L at p. Let $\Gamma = V(F, G)$ be the base locus of the pencil $C_t = V(t_0 F + t_1 G)$, so that Γ will be a scheme of degree d^2 consisting of $d^2 - 2$ reduced points and one scheme of degree 2 supported at p. Since being singular at p is one linear condition on the elements of the pencil, exactly one member of the pencil (which we may take to be C_0 after re-parametrizing the pencil) will be singular at p.

We could arrive at such a pencil by taking F to be the equation of a general curve with a node p and G a general polynomial vanishing at p; thus for the general pencil above, the singular element C_0 of the pencil will have a node at p, with neither branch tangent to L, while all the others elements are smooth at p and have a common tangent line $\mathbb{T}_p(C_t) = L$ at p.

Let X be the minimal smooth blow-up resolving the indeterminacy of the rational map φ from $\mathbb{P}^2$ to $\mathbb{P}^1$ associated to the pencil — that is, X is obtained by blowing up $\mathbb{P}^2$ at Γ_{red} and then blowing up the resulting surface at the point p' on the exceptional divisor corresponding to the common tangent line L to the smooth members of the pencil at p. (This is *not* the blow-up of S along the scheme Γ, which is singular! See for example Eisenbud and Harris [2000, IV.2.3].) Note that we are blowing up $\mathbb{P}^2$ a total of d^2 times, so that the Euler characteristic $\chi_{\text{top}}(X)$ is equal to $3 + d^2$, just as in the general case.

What is different is the fiber of the map $X \to \mathbb{P}^1$ over $t = 0$: Rather than being a copy of the curve $C_0 = V(F)$, it is the union of the proper transform of C_0 and the proper transform E of the first exceptional divisor, that is, the union of the normalization $\tilde{C}_0$ of C_0 and a copy E of $\mathbb{P}^1$, meeting at the two points of $\tilde{C}_0$ lying over the node p (See Figure 7.4).

In sum, the Euler characteristic of the fiber over $t = 0$ is

$$\chi_{\text{top}}(\tilde{C}_0) + \chi_{\text{top}}(E) - 2 = \chi_{\text{top}}(\tilde{C}_0) = \chi_{\text{top}}(C_\eta) + 2,$$

and the fiber counts with multiplicity 2. We can use this to analyze the tangent planes to $\mathcal{D}$ at its simplest points:

Proposition 7.19. *Let C be a plane curve with a node at p and no other singularities. The tangent plane $\mathbb{T}_C \mathcal{D} \subset \mathbb{P}^N$ is the hyperplane $H_p \subset \mathcal{D}$ of curves containing the point p.*

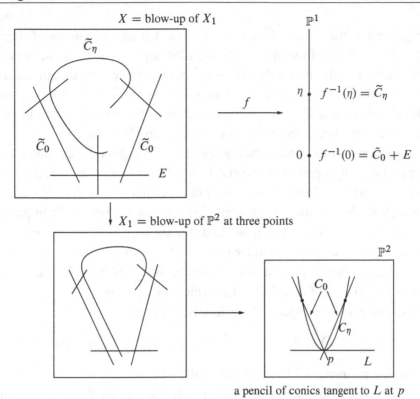

Figure 7.4 The morphism $f : X \to \mathbb{P}^1$ coming from the pencil of conics tangent to L at p.

Proof: If $C \subset \mathbb{P}^2$ is a plane curve with one node p and no other singularities, then, by Proposition 7.18, C is a smooth point of $\mathcal{D}$. It thus suffices to show that H_p is contained in the tangent space to $\mathcal{D}$ at C_0. But, as we have seen, if $\mathcal{B} \subset \mathbb{P}^N$ is a general pencil including C and having p as a base point, $\mathcal{B}$ will meet $\mathcal{D}$ in exactly $3(d-1)^2 - 1$ points — in other words, a general line $\mathcal{B} \subset \mathbb{P}^N$ through C and lying in H_p will be tangent to $\mathcal{D}$ somewhere.

As above, we may suppose that $\mathcal{B}$ is spanned by a curve $F = 0$ with a node at p and no other singularities and a smooth curve $G = 0$ that passes through p.

To complete the argument — to show that such a line is indeed tangent to $\mathcal{D}$ specifically at C, and not somewhere else — we have to do two things: We have to relate the pencil $\mathcal{B}$ to nearby general pencils, and we have to localize the Euler characteristic. For the first, choose a general polynomial G', and consider the family of pencils $\{\mathcal{B}_s\}$ with $\mathcal{B}_0 = \mathcal{B}$ and $\mathcal{B}_s$ the pencil spanned by F and a linear combination $G_s = G + sG'$; that is,

$$\mathcal{B}_s = \{V(F + t(G + sG')) \mid t \in \mathbb{P}^1\}.$$

For each s, we let X_s be the total space of the pencil $\mathcal{B}_s$ and $f_s : X_s \to \mathbb{P}^1$ the map $[F, G + sG']$.

For general s, the pencil $\mathcal{B}_s$ will be a general pencil of curves of degree d; in particular, if $\mu > 0$ is sufficiently small, then for any $0 < |s| < \mu$ the pencil $\mathcal{B}_s$ will intersect $\mathcal{D}$ transversely in exactly $3(d-1)^2$ points. Moreover, by our description of $\mathcal{B}_0 \cap \mathcal{D}$, we know that as s approaches 0, two of these $3(d-1)^2$ points will approach a particular element $C_{t_0} \in \mathcal{B}$ — the point of tangency of $\mathcal{B}$ with $\mathcal{D}$ — and the remaining $3(d-1)^2 - 2$ will remain distinct from each other and from C_{t_0}.

For the second component of the argument (localizing the Euler characteristic), we cover the t-line $\mathbb{P}^1_t$ by a pair of open sets: $U = (|t| < \epsilon)$ a disc around the point $t = 0$, and $V = \mathbb{P}^1 \setminus (|t| \geq \epsilon/2)$ the complement of a smaller closed disc. We can choose ϵ small enough so that no singular fiber of $\mathcal{B}$ other than C_0 lies in $\overline{U}$; in particular, no singular fiber lies in $U \cap V$. It follows that, for some $\mu > 0$, the same is true for all $\mathcal{B}_s$ with $|s| < \mu$: none of the singular fibers of $\mathcal{B}_s$ lie in the overlap $U \cap V$. For any $0 < |s| < \mu$, accordingly, the number of singular fibers of $\mathcal{B}_s$ in U is the intersection multiplicity $m_{C_0}(\mathcal{B}, \mathcal{D})$ of $\mathcal{B}$ and $\mathcal{D}$ at C_0, which we claim is 2.

Now consider the total space of our family of pencils:

$$\Phi = \{(s, t, p) \in \Delta \times \mathbb{P}^1 \times \mathbb{P}^2 \mid F(p) + t(G(p) + sG'(p)) = 0\}.$$

Let Φ^V be the preimage of V in Φ. Since the fiber Φ^V_s of Φ^V over each $s \in \Delta$ is smooth, Φ^V is a fiber bundle over Δ and, in particular, all the Φ^V_s have the same Euler characteristic. We know that Φ^V_0 has exactly $3(d-1)^2 - 2$ singular fibers, each a curve with a single node, so that by Theorem 7.17

$$\chi_{\text{top}}(\Phi^V_0) = -d(d-3) + 3(d-1)^2 - 2;$$

since for $s \neq 0$ the Φ^V_s have the same Euler characteristic, the same logic tells us that they also have exactly $3(d-1)^2 - 2$ singular fibers over V. It follows that Φ^V_s has two singular fibers for $0 < |s| < \mu$, completing the argument. □

This argument shows that, more generally, if C is a plane curve with a unique singular point p, the tangent cone to $\mathcal{D}$ at C will be a multiple of the hyperplane H_p, and, more generally still, if C has isolated singularities $p_1, \ldots, p_\delta$, the tangent cone $\mathbb{T}_C \mathcal{D}$ is supported on the union of the planes H_{p_i}.

There is also a sort of converse to Proposition 7.18:

Proposition 7.20. *The smooth locus of $\mathcal{D}$ consists exactly of those curves with a single node and no other singularity.*

Proof: Proposition 7.18 gives one inclusion: if C has a node and no other singularity, it is a smooth point of $\mathcal{D}$. Moreover, if C has more than one (isolated) singular point, then the projection map $\Sigma \to \mathcal{D}$ is finite but not one-to-one over C; it is intuitively clear (and follows from Zariski's main theorem) that $\mathcal{D}$ is analytically reducible and hence singular at C. Moreover, we observe that if $d \geq 3$ any curve with multiple components is a limit of curves with isolated singularities and at least three nodes — just deform each multiple component $m C_0 \subset C$ to a union of m general translates of C_0 — so these must also lie in the singular locus of $\mathcal{D}$.

It remains to see that if C is a singular curve having a singularity p other than a node, then $\mathcal{D}$ is singular at C. This follows from an analysis of plane curve singularities: If C has isolated singularities including a point p of multiplicity $k \geq 3$, then, as we saw in Section 2.4.6, the genus of the normalization $\widetilde{C}$ is at most

$$g(\tilde{D}) \leq \binom{d-1}{2} - \frac{k(k-1)}{2}$$

and, since at most k points of the normalization lying over p are identified in C,

$$\chi_{\text{top}}(C) \geq 2 - 2g(\widetilde{C}) + k - 1 \geq -d(d-3) + (k-1)^2.$$

As for double points p other than a node, we have already done the case of a cusp; other double points will drop the genus of the normalization by 2 or more, and since we have at most two points of the normalization lying over p, we must have $\chi_{\text{top}}(C) \geq -d(d-3) + 3$. $\qquad\square$

Finally, note that the techniques of this section can be applied in exactly the same way in one dimension lower!

Proposition 7.21. *Let* $\mathbb{P}^d = \mathbb{P} H^0(\mathcal{O}_{\mathbb{P}^1}(d))$ *be the space of polynomials of degree d on* $\mathbb{P}^1$, *and* $\mathcal{D} \subset \mathbb{P}^d$ *the discriminant hypersurface, that is, the locus of polynomials with a repeated root. If $F \in \mathcal{D}$ is a point corresponding to a polynomial with exactly one double root p and $d - 2$ simple roots, then $\mathcal{D}$ is smooth at F with tangent space the space of polynomials vanishing at p.*

We leave the proof via the topological Hurwitz formula as an exercise; for an algebraic proof, see Proposition 8.6.

We add that there are many, many problems having to do with the local geometry of $\mathcal{D}$ and its stratification by singularity type, only a small fraction of which we know how to answer. The statements above barely scratch the surface; for more, see for example Brieskorn and Knörrer [1986] or Teissier [1977].

7.8 Exercises

Exercise 7.22. Let $S = \mathbb{P}^1 \times \mathbb{P}^1$, and let $\{C_t \subset S\}_{t \in \mathbb{P}^1}$ be a general pencil of curves of type (a, b) on S, where $a, b > 0$. What is the expected number of curves C_t that are singular? (Make sure your answer agrees with (7.2) in the case $(a, b) = (1, 1)$!)

Exercise 7.23. Prove that the number found in the previous exercise is the actual number of singular elements; that is, prove the three hypotheses of Proposition 7.9 in the case of $S = \mathbb{P}^1 \times \mathbb{P}^1$ and the line bundle $\mathcal{O}(a, b)$.

Exercise 7.24. Let $S \subset \mathbb{P}^3$ be a smooth cubic surface and $L \subset S$ a line. Let $\{C_t\}_{t \in \mathbb{P}^1}$ be the pencil of conics on S cut out by the pencil of planes $\{H_t \subset \mathbb{P}^3\}$ containing L. How many of the conics C_t are singular? Use this to answer the question of how many other lines on S meet L.

Exercise 7.25. Let $p \in \mathbb{P}^2$ be a point, and let $\{C_t \subset \mathbb{P}^2\}_{t \in \mathbb{P}^1}$ be a general pencil of plane curves singular at p — in other words, let F and G be two general polynomials vanishing to order 2 at p, and take $C_t = V(t_0 F + t_1 G)$. How many of the curves C_t will be singular somewhere else as well?

Exercise 7.26. Let $S = X_1 \cap X_2 \subset \mathbb{P}^4$ be a smooth complete intersection of hypersurfaces of degrees e and f. If $\{H_t \subset \mathbb{P}^4\}_{t \in \mathbb{P}^1}$ is a general pencil of hyperplanes in $\mathbb{P}^4$, find the expected number of singular hyperplane sections $S \cap H_t$. (Equivalently: if $\Lambda \cong \mathbb{P}^2 \subset \mathbb{P}^4$ is a general 2-plane, how many tangent planes to S intersect Λ in a line?)

Exercise 7.27. Let $X \subset \mathbb{P}^4$ be a smooth hypersurface of degree d. Using formula (7.1), find the expected number of singular hyperplane sections of X in a pencil. Again, compare your answer to the result of Section 2.1.3.

Exercise 7.28. Let $X \cong \mathbb{P}^1 \times \mathbb{P}^2 \subset \mathbb{P}^5$ be the Segre threefold. Using formula (7.1), find the number of singular hyperplane sections of X in a pencil.

Exercise 7.29. Let $S = X_1 \cap X_2 \subset \mathbb{P}^4$ be a smooth complete intersection of hypersurfaces of degrees e and f. What is the expected number of hyperplane sections of S having a triple point? (Check this in the case $e = f = 2$!)

Exercise 7.30. Let $S \subset \mathbb{P}^n$ be a smooth surface of degree d whose general hyperplane section is a curve of genus g; let e and f be the degrees of the classes $c_1(\mathcal{T}_S)^2, c_2(\mathcal{T}_S) \in A^2(S)$. Find the class of the cycle $T_1(S) \subset \mathbb{G}(1, n)$ of lines tangent to S in terms of d, e, f and g; from Exercise 4.21, we need only the intersection number $[T_1(S)] \cdot \sigma_3$. *Hint:* Consider instead the variety of tangent planes $T_2(S) \subset \mathbb{G}(2, n)$, and find the intersection with σ_2 as the intersection with $(\sigma_1)^2$ minus the intersection with $\sigma_{1,1}$.

Exercise 7.31. Let $S \subset \mathbb{P}^3$ be a general surface of degree d and $\mathcal{B}$ a general net of plane sections of S (that is, intersections of X with planes containing a general point $p \in \mathbb{P}^3$). What are the degree and genus of the curve $\Gamma \subset S$ traced out by singular points of this net? What are the degree and genus of the discriminant curve? Use this to describe the geometry of the finite map $\pi_p : S \to \mathbb{P}^2$ given by projection from p.

Exercise 7.32. Verify that for a general curve $C \subset \mathbb{P}^2$ of degree d the number $3d(d-2)$ is the actual number of flexes of C, that is, that all inflection points of C have weight 1.

Exercise 7.33. Let $\{C_t \subset \mathbb{P}^2\}_{t \in \mathbb{P}^1}$ be a general pencil of plane curves of degree $d \geq 3$; suppose C_0 is a singular element of C (so that in particular by Proposition 7.1 C_0 will have just one node as singularity). By our formula, C_0 will have six fewer flexes than the general member C_t of the pencil. Where do the other six flexes go? If we consider the incidence correspondence

$$\Phi = \{(t, p) \in \mathbb{P}^1 \times \mathbb{P}^2 \mid C_t \text{ is smooth and } p \text{ is a flex of } C_t\},$$

what is the geometry of the closure of Φ near $t = 0$? Bonus question: Describe the geometry of

$$\widetilde{\Phi} = \{(t, p, L) \in \mathbb{P}^1 \times \mathbb{P}^2 \times \mathbb{P}^{2*} \mid C_t \text{ smooth}, p \text{ a flex of } C_t \text{ and } L = \mathbb{T}_p C_t\}$$

near $t = 0$.

Exercise 7.34. Find the points on $\mathbb{P}^1$, if any, that are ramification points for the maps $\mathbb{P}^1 \to \mathbb{P}^3$ given by

$$(s, t) \mapsto (s^3, s^2 t, s t^2, t^3) \in \mathbb{P}^3 \quad \text{and} \quad (s, t) \mapsto (s^4, s^3 t, s t^3, t^4) \in \mathbb{P}^3.$$

Exercise 7.35. Show that the only smooth, irreducible and nondegenerate curve $C \subset \mathbb{P}^r$ with no inflection points is the rational normal curve.

Exercise 7.36. We define an *elliptic normal curve* to be a smooth irreducible nondegenerate curve of genus 1 and degree $r + 1$ in $\mathbb{P}^r$. Observe that for an elliptic normal curve E the Plücker formula yields the number $(r + 1)^2$ of inflection points. Show that these are exactly the images of any one under the group of translations of order $r + 1$ on E, each having weight 1.

Exercise 7.37. Let C be a smooth curve of genus $g \geq 2$. A point $p \in C$ is called a *Weierstrass point* if there exists a nonconstant rational function on C with a pole of order g or less at p and regular on $C \setminus \{p\}$.

(a) Show that the Weierstrass points of C are exactly the inflection points of the canonical map $\varphi : C \to \mathbb{P}^{g-1}$.

(b) Use this to count the number of Weierstrass points on C.

Exercise 7.38. Let $\mathbb{P}^N$ be the space of all plane curves of degree $d \geq 4$, and let $H \subset \mathbb{P}^N$ be the closure of the locus of smooth curves with a hyperflex. Show that H is a hypersurface. (We will be able to calculate the degree of this hypersurface once we have developed the techniques of Chapter 11.)

Exercise 7.39. To prove that a general complete intersection $C \subset \mathbb{P}^3$ does not have weight-2 inflection points, we need to prove that it does not have flex lines (lines with multiplicity-3 intersection with the curve) or planes with a point of contact of order 5. Prove the first statement: that a general complete intersection of two surfaces S_1 and S_2 of degrees $d_1 \geq d_2 > 1$ does not have a flex line.

The following two exercises show how to construct an example of a component of the Hilbert scheme whose general member is a smooth, irreducible, nondegenerate curve having inflection points of weight > 1.

Exercise 7.40. Let $S = \overline{p, E} \subset \mathbb{P}^n$ be a cone over an elliptic normal curve $E \subset \mathbb{P}^{n-1}$ (that is, a smooth curve of genus 1 embedded by a complete linear system of degree n), and let $L_1, \ldots, L_{n-1} \subset S$ be lines of the ruling. Show that, for $n > 9$ and $m \gg 0$:

(a) The residual intersection C of S with a general hypersurface $X \subset \mathbb{P}^n$ of degree m containing $L_1, \ldots, L_{n-1}$ is a smooth, irreducible and nondegenerate curve.

(b) Every deformation of C also lies on a cone over an elliptic normal curve. (The condition $n > 9$ is necessary to ensure that the surface S has itself no deformations other than cones. This follows from the classification of *del Pezzo surfaces*; see for example Beauville [1996].)

Thus the smooth, irreducible and nondegenerate curves C constructed in this fashion form an open subset of the Hilbert scheme of curves in $\mathbb{P}^n$.

Exercise 7.41. Let $C \subset S \subset \mathbb{P}^n$ be a curve as constructed in the preceding problem. Show that C has inflection points of weight > 1 (look at points where C is tangent to a line of the ruling of S).

Exercise 7.42. Let $S \subset \mathbb{P}^3$ be a general surface of degree $d \geq 2$, $p \in S$ a general point and $H = \mathbb{T}_p S \subset \mathbb{P}^3$ the tangent plane to S at p. Show by an elementary dimension count (not using the second fundamental form or quoting Theorem 7.11) that the intersection $H \cap S$ has an ordinary double point at p.

Exercise 7.43. Let $S = \mathbb{P}^1 \times \mathbb{P}^1$, and let $\{C_t \subset S\}_{t \in \mathbb{P}^1}$ be a general pencil of curves of type (a, b) on S. Use the topological Hurwitz formula to say how many of the curves C_t are singular. (Compare this with your answer to Exercise 7.22.)

Exercise 7.44. Let $p \in \mathbb{P}^2$ be a point, and let $\{C_t \subset \mathbb{P}^2\}_{t \in \mathbb{P}^1}$ be a general pencil of plane curves of degree d singular at p, as in Exercise 7.25. Use the topological Hurwitz formula to count the number of curves in the pencil singular somewhere else.

Exercise 7.45. Let $\mathbb{P}^5$ be the space of conic plane curves and $\mathcal{D} \subset \mathbb{P}^5$ the discriminant hypersurface. Let $C \in \mathcal{D}$ be a point corresponding to a double line. What is the multiplicity of $\mathcal{D}$ at C, and what is the tangent cone?

Exercise 7.46. Now, let $\mathbb{P}^{14}$ be the space of quartic plane curves and $\mathcal{D} \subset \mathbb{P}^{14}$ the discriminant hypersurface. Let $C \in \mathcal{D}$ be a point corresponding to a double conic. What is the multiplicity of $\mathcal{D}$ at C, and what is the tangent cone?

Chapter 8

Compactifying parameter spaces

Keynote Questions

(a) (The five conic problem) Given five general plane conics $C_1, \ldots, C_5 \subset \mathbb{P}^2$, how many smooth conics $C \subset \mathbb{P}^2$ are tangent to all five? (Answer on page 308.)

(b) Given 11 general points $p_1, \ldots, p_{11} \in \mathbb{P}^2$ in the plane, how many rational quartic curves $C \subset \mathbb{P}^2$ contain them all? (Answer on page 321.)

All the applications of intersection theory to enumerative geometry exploit the fact that interesting classes of algebraic varieties — lines, hypersurfaces and so on — are themselves parametrized by the points of an algebraic variety, the *parameter space*, and our efforts have all been toward counting intersections on these spaces. But to use intersection theory to count something, the parameter space must be projective (or at least proper) so that we have a degree map, as defined in Chapter 1. In the first case we treated in this book, that of the family of planes of a certain dimension in projective space, the natural parameter space was the Grassmannian, and the fact that it is projective is what makes the Schubert calculus so useful for enumeration. When we studied the questions about linear spaces on hypersurfaces, we were similarly concerned with parameter spaces that were projective — the Grassmannian $\mathbb{G}(k, n)$ and, in connection with questions involving families of hypersurfaces, the projective space $\mathbb{P}^N$ of hypersurfaces itself. These spaces have an additional feature of importance: a universal family of the geometric objects we are studying, or (amounting to the same thing) the property of representing a functor we understand. This property is useful in many ways, first of all for understanding tangent spaces, and thus transversality questions.

In many interesting cases, however, the "natural" parameter space for a problem is *not* projective. To use the tools of intersection theory to count something, we must add points to the parameter space to complete it to a projective (or at least proper) variety. It is customary to call these new points the *boundary*, although this is not a topological boundary in any ordinary sense — the boundary points may look like any other point of the space — and (more reasonably) to call the enlarged space a *compactification* of the original space. If we are lucky, the boundary points of the compactification still parametrize some sort of geometric object we understand. In such cases we can use this structure to solve geometric problems. But as we shall see, the boundary can also get in the way, even when it seems quite natural. In such cases, we might look for a "better" compactification... but just how to do so is a matter of art rather than of science.

Perhaps the first problem in enumerative geometry where this tension became clear is the five conic problem, which was solved in a naive way, not taking the difficulty into account (and therefore getting the wrong answer) by Steiner [1848], and again, with the necessary subtlety (and correct answer!) by Chasles [1864]. In this case there is a very beautiful and classical construction of a good parameter space, the space of *complete conics*. In this chapter we will explore the construction, and briefly discuss two more general constructions: Hilbert schemes and Kontsevich spaces.

8.1 Approaches to the five conic problem

To reiterate the problem: Given five general plane conics $C_1, \ldots, C_5$, how many smooth conics are tangent to all five? Here is a naive approach:

(a) The set of plane conics is parametrized by $\mathbb{P}^5$. The locus of conics tangent to each given C_i is an irreducible hypersurface $Z_i \subset \mathbb{P}^5$, as one sees by considering the incidence correspondence

$$\{(C, p) \in \mathbb{P}^5 \times C_i \mid C \text{ a conic tangent to } C_i \text{ at } p\}$$

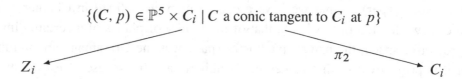

$$Z_i \qquad\qquad\qquad\qquad \pi_2 \qquad\qquad C_i$$

and noting that the fibers of π_2 are linear subspaces of $\mathbb{P}^5$ of dimension 3. (Here, "tangent to C_i at p" means $m_p(C \cdot C_i) \geq 2$, that is, the restriction to C_i of the defining equation of C vanishes to order at least 2 at p.)

(b) The degree of Z_i is 6. To see this, we intersect Z_i with a general line in $\mathbb{P}^5$ — that is, we take a general pencil of conics and count how many are tangent to C_i. The conic C_i may be thought of as the embedding of $\mathbb{P}^1$ in $\mathbb{P}^2$ by the complete linear system of degree 2. Thus a general pencil of conics cuts out a general linear series on C_i of degree 4, and the degree of Z_i is the number of divisors in this family with fewer than four distinct points. The linear series defines a general map $C_i \to \mathbb{P}^1$ of degree 4 with distinct branch points, and by the Riemann–Hurwitz theorem (Section 7.7) the number of branch points of this map is six.

(c) Thus the number of points of intersection of $Z_1, \ldots, Z_5$, *assuming they intersect transversely,* will be $6^5 = 7776$.

Alas, 7776 is *not* the answer to the question we posed. The problem is not hard to spot: far from being transverse, the hypersurfaces Z_i do not even meet in a finite set! To be sure, the part of the intersection within the open set $U \subset \mathbb{P}^5$ of smooth conics (which is what we wanted to count) *is* finite, and even transverse, as we will verify below. The trouble is with the compactification: we used the space of all (possibly singular) conics, and "excess" intersection of the Z_i takes place along the boundary.

In more detail: the hypersurface Z_i is the *closure* in $\mathbb{P}^5$ of the locus of smooth conics C tangent to C_i. A smooth conic C is tangent to C_i exactly when the defining equation F of C, restricted to $C_i \cong \mathbb{P}^1$ and viewed as a quartic polynomial on $\mathbb{P}^1$, has a multiple root. When we extend this characterization to arbitrary conics C we see that *a double line is tangent to every conic.* Thus the five hypersurfaces $Z_1, \ldots, Z_5 \subset \mathbb{P}^5$ will all contain the locus $S \subset \mathbb{P}^5$ of double lines, which is a Veronese surface in the $\mathbb{P}^5$ of conics. As we shall see, the intersection $\bigcap Z_i$ is the union of S and the finite set of smooth conics tangent to the five C_i. The presence of this extra component S means that the number we seek has little to do with the intersection product $\prod [Z_i] \in A^5(\mathbb{P}^5)$.

There are at least three successful approaches to dealing with this issue:

Blowing up the excess locus

Suppose we are interested in intersections inside some quasi-projective variety U and we have a compactification V of U; in the example above, U is the space of smooth conics and V the space of all conics. We could blow up some locus in the boundary $V \setminus U$ to obtain a new compactification. This is the classical way of separating subvarieties of a given variety that we do not want to meet. In the five conic problem, we would blow up the surface S in $\mathbb{P}^5$ and consider the proper transforms $\tilde{Z}_i$ of the hypersurfaces Z_i in the blow-up $X = \mathrm{Bl}_S \mathbb{P}^5$. If we are lucky (and in this case we are), we will have eliminated the excess intersection — that is, the $\tilde{Z}_i$ will not intersect anywhere in the exceptional divisor $E \subset X$ of the blow-up. (If this were not the case we would have to blow up again, along the common intersection $\bigcap \tilde{Z}_i \cap E$.) In our case, the $\tilde{Z}_i$ intersect transversely, and only inside U. To finish the argument, we could determine the Chow ring $A(X)$ of the blow-up, find the class $\zeta \in A^1(X)$ of the $\tilde{Z}_i$ (as members of a family parametrized by an open subset of $\mathbb{P}^5$, they all have the same class) and evaluate the product $\zeta^5 \in A^5(X)$.

Readers who want to carry this out themselves can find a description of the Chow ring of a blow-up in Section 13.6; there is also a complete account of this approach in Griffiths and Harris [1994, Section 6.1].

This approach has the virtue of being universally applicable, at least in theory: Any component of any intersection of cycles can be eliminated by blowing up repeatedly. But often we cannot recognize the blow-up as the parameter space of any nice geometric objects, and this makes the computations less intuitive and sometimes unwieldy. For example, this approach to the problem of counting cubics satisfying nine tangency conditions (solved heuristically by Maillard and Zeuthen in the 19th century and rigorously in Aluffi [1990] and Kleiman and Speiser [1991]) requires multiple blow-ups of the space $\mathbb{P}^9$ of cubics and complex calculations.

Excess intersection formulas

Excess intersection problems were already considered by Salmon in 1847, and were generalized greatly by Cayley around 1868. The excess intersection formula of Fulton and MacPherson (see Fulton [1984, Chapter 9]) subsumes them all: It is a general formula that assigns to every connected component of an intersection $\bigcap Z_i \subset X$ a class in the appropriate dimension, in such a way that the sum of these classes (viewed as classes on the ambient variety X via the inclusion) equals the product of the classes of the intersecting cycles. This applies whenever all but at most one of the subvarieties Z_i are locally complete intersections in X; in our case all are hypersurfaces. We will give an exposition of the formula in Chapter 13, and show in Section 13.3.5 how it may be applied to the five conic problem, as was originally carried out in Fulton and MacPherson [1978].

As a general method, excess intersection formulas are often an improvement on blowing up. But, as with the blow-up approach, they require some knowledge of the normal bundles (or, more generally, normal cones) of the various loci involved.

Changing the parameter space

To understand what sort of compactification is "right" for a given problem is, as we have said, an art. In the case of the five conic problem, we can take a hint from the fact that the problem is about tangencies. The set of lines tangent to a nonsingular conic is again a conic in the dual space (we will identify it explicitly below). But when a conic degenerates to the union of two lines or a double line, the dual conic seems to disappear — the dual of a line is only a point! This leads us to ask for a compactification of the space of smooth conics that keeps track of information about limiting positions of tangents.

There are at least two ways to make a compactification that encodes the necessary information. One is to use the *Kontsevich space*. It parametrizes not subschemes of $\mathbb{P}^2$, but rather maps $f : C \to \mathbb{P}^2$ with C a nodal curve of arithmetic genus 0. This is an important construction, which generalizes to a parametrization of curves of any degree and genus in any variety. We will discuss it informally in the second half of this chapter. But proving even the existence of Kontsevich spaces requires a considerable development, and we will not take this route; the reader will find an exposition in Fulton and Pandharipande [1997].

The other way to describe a compactification of the space of smooth conics that preserves the tangency information is through the idea of *complete conics*. The space of complete conics is very well-behaved, and we will spend the first half of this chapter on this beautiful construction. It turns out that the space we will construct is isomorphic to the Kontsevich space for conics (and, for that matter, to the blow-up $\mathrm{Bl}_S \, \mathbb{P}^5$ of $\mathbb{P}^5$ along the surface of double lines), but generalizes in a different direction: There are analogs for quadric hypersurfaces of any dimension, for linear transformations ("complete colineations") and, more generally, for symmetric spaces (see De Concini and Procesi [1983; 1985], De Concini et al. [1988] and Bifet et al. [1990]), but not for curves of higher degree or genus. (There is an analogous construction but, as we will remark at the end of Section 8.2.2, in general the space constructed is highly singular and not well-understood.)

8.2 Complete conics

We begin with an informal discussion. Later in this section we will provide a rigorous foundation for what we describe. Recall that the *dual* of a smooth conic $C \subset \mathbb{P}^2$ is the set of lines tangent to C, regarded as a curve $C^* \subset \mathbb{P}^{2*}$. As we shall see, C^* is also a smooth conic (this would not be true in characteristic 2!).

8.2.1 Informal description

Degenerating the dual

Consider what happens to the dual conic as a smooth conic degenerates to a singular conic — either two distinct lines or a double line. That is, let $C \to B$ be a one-parameter family of conics with parameter t, with C_t smooth for $t \neq 0$. Associating to each curve C_t the dual conic $C_t^* \subset \mathbb{P}^{2*}$ we get a regular map from the punctured disc $B \setminus \{0\}$ to the space $\mathbb{P}^{5*}$ of conics in $\mathbb{P}^{2*}$. (If $\mathbb{P}^2 = \mathbb{P}V$ and $\mathbb{P}^{2*} = \mathbb{P}V^*$, the space of conics on each are respectively $\mathbb{P}\,\mathrm{Sym}^2 V^*$ and $\mathbb{P}\,\mathrm{Sym}^2 V$ — in particular, they are naturally dual to one another, so if we write the former as $\mathbb{P}^5$ it makes sense to write the latter as $\mathbb{P}^{5*}$.) Since the space $\mathbb{P}^{5*}$ of all conics in $\mathbb{P}^{2*}$ is proper, this extends to a regular map on all of B — in other words, there is a well-defined conic $C_0^* = \lim_{t \to 0} C_t^*$. However, as we will see, this limit depends in general on the family C and not just on the curve C_0: in other words, the limit of the duals C_t^* is *not* determined by the limit of the curves C_t.

To provide a compactification of the space U of smooth conics that captures this phenomenon, we realize U as a locally closed subset of $\mathbb{P}^5 \times \mathbb{P}^{5*}$: As we will see in the following section, the map $C \mapsto C^*$ is regular on smooth conics, so U is isomorphic to the graph of the map $U \to \mathbb{P}^{5*}$ sending a smooth conic C to its dual. That is, we set

$$U = \{(C, C^*) \in \mathbb{P}^5 \times \mathbb{P}^{5*} \mid C \text{ a smooth conic in } \mathbb{P}^2 \text{ and } C^* \subset \mathbb{P}^{2*} \text{ its dual}\}.$$

The desired compactification, the *variety of complete conics*, is the closure

$$X = \overline{U} \subset \mathbb{P}^5 \times \mathbb{P}^{5*}.$$

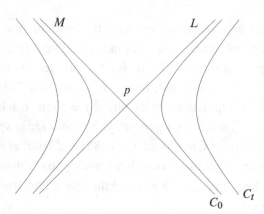

Figure 8.1 Conics specializing to a conic $C_0 = L \cup M$ of rank 2.

The dual of the dual of a smooth conic is the original conic, as we shall soon see (in fact, the same statement holds for varieties much more generally, and will be proven in Section 10.6), so the set U is symmetric under exchanging $\mathbb{P}^5$ and $\mathbb{P}^{5*}$. It follows that X is symmetric too. (As one consequence of this symmetry, note that if $(C, C^*) \in X$ and either C or C^* is smooth, then the other is too.) The set $U \subset \mathbb{P}^5$ of smooth conics is by definition dense in X, and it follows that X is irreducible and of dimension 5 as well.

What happens to C^* when C becomes singular? Let us first consider the case of a family $\{C_t\}$ of smooth conics approaching a conic C_0 of rank 2, that is, $C_0 = L \cup M$ is the union of a pair of distinct lines; for example, the family given (in affine coordinates on $\mathbb{P}^2$) as

$$\mathcal{C} = \{(t, x, y) \in B \times \mathbb{P}^2 \mid y^2 = x^2 - t\},$$

as shown in Figure 8.1. The picture makes it easy to guess what happens: Any collection $\{L_t\}$ of lines with L_t tangent to C_t for $t \neq 0$ approaches a line L_0 through the point $p = L \cap M$, and conversely any line L_0 through p is a limit of lines L_t tangent to C_t. (Actually, the second statement follows from the first, given that the limit $C_0' = \lim_{t \to 0} C_t^*$ is one-dimensional.) Since C_0' is by definition a conic, it must be the double of the line in $\mathbb{P}^{2*}$ dual to the point p, irrespective of the family $\{C_t\}$ used to construct it or of the positions of the lines L and M.

Things are much more interesting when we consider a family of smooth conics $\{C_t \subset \mathbb{P}^2\}$ specializing to a double line $C_0 = 2L$, and ask what the limit $\lim_{t \to 0} C_t^*$ of the dual conics $C_t^* \subset \mathbb{P}^{2*}$ may be. One way to realize such a family of conics is as the images of a family of maps $\varphi_t : \mathbb{P}^1 \to \mathbb{P}^2$. Such a family of maps is given by a triple of polynomials $(f_t(x), g_t(x), h_t(x))$, homogeneous of degree 2 in $x = (x_0, x_1)$, whose coefficients are regular functions in t. In our present circumstances, our hypotheses say that for $t \neq 0$ the polynomials f_t, g_t and h_t are linearly independent (and so span $H^0(\mathcal{O}_{\mathbb{P}^1}(2))$), but for $t = 0$ they span only a two-dimensional vector space $W \subset H^0(\mathcal{O}_{\mathbb{P}^1}(2))$. For now, we will make the additional assumption that the linear system $\mathcal{W} = (\mathcal{O}_{\mathbb{P}^1}(2), W)$ is base point free; the case where it is not will be dealt with below.

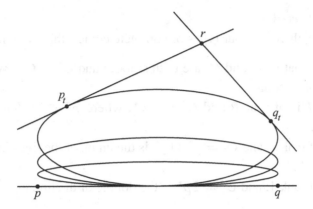

Figure 8.2 The family of conics $y^2 = t(x^2 - 2y)$.

To see what the limit of the dual conics C_t^* will be in this situation, let $u, v \in \mathbb{P}^1$ be the ramification points of the map $\varphi_W : \mathbb{P}^1 \to \mathbb{P}^1$ associated to W (note that the map φ_0 is just the composition of this map with the inclusion of the target $\mathbb{P}^1$ as the line $L \subset \mathbb{P}^2$), and let $p = \varphi_0(u)$ and $q = \varphi_0(v) \in L$ be their images. We claim that in this case the limit $\lim_{t \to 0} C_t^*$ of the dual conics is the conic $C_0^* = p^* + q^* \subset \mathbb{P}^{2*}$ consisting of lines through p and lines through q.

To prove this, let $r \in \mathbb{P}^2$ be any point not in L and not in any curve C_t, and let $\pi : \mathbb{P}^2 \to L$ be the projection from r to L. The composition $\pi \circ \varphi_t : \mathbb{P}^1 \to L \cong \mathbb{P}^1$ has degree 2; let $u_t, v_t \in \mathbb{P}^1$ be the ramification points of this map and $p_t, q_t \in L$ the corresponding branch points. Suppose that $\pi \circ \varphi_t$ is the map associated to the pencil $W_t = (\mathcal{O}_{\mathbb{P}^1}(2), W_t)$ for a two-dimensional vector space $W_t \subset H^0(\mathcal{O}_{\mathbb{P}^1}(2))$. As $t \to 0$, the linear system W_t approaches the linear system W; correspondingly, the divisor $u_t + v_t$ approaches $u + v$ and $p_t + q_t$ approaches $p + q$. In other words, the tangent lines to C_t passing through r — which are exactly the lines $\overline{r, \varphi_t(u_t)} = \overline{r, p_t}$ and $\overline{r, \varphi_t(v_t)} = \overline{r, q_t}$ — approach the lines $\overline{r, p}$ and $\overline{r, q}$, *independently of* r. Thus every line through p or q is a limit of tangent lines to C_t, and conversely.

It is important to note that in this situation, unlike in the case where C_0 is the union of two distinct lines, the limit of the dual conics C_t^* is not determined by the conic C_0. As we will see in Section 8.2.2, the points p and q may be any pair of points of L, depending on the path along which C_t approaches C_0.

The remaining case to consider is when the branch points $p_t, q_t \in L$ of the maps $\pi \circ \varphi_t$ approach the same point $p \in L$. (Typically, this corresponds to the case where W has a base point: When W has a base point u, the ramification of W is concentrated at this point, which must then be the limit as $t \to 0$ of both the ramification points u_t and v_t of W_t.) In this case, the same logic shows that the limit of the dual conics C_t^* will be the double $2p^*$ of the line $p^* \subset \mathbb{P}^{2*}$ dual to the image point $p = \varphi_0(u)$.

Types of complete conics

In conclusion, there are four types of complete conics, that is, points $(C, C') \in X$:

(a) $(C, C') \in U$; that is, C and C' are both smooth and $C' = C^*$. We will call these *smooth* complete conics.

(b) $C = L \cup M$ is of rank 2 and $C' = 2p^*$, where $p^* \subset \mathbb{P}^{2*}$ is the line dual to $p = L \cap M$.

(c) $C = 2L$ is of rank 1, and $C' = p^* \cup q^*$ is the union of the lines in $\mathbb{P}^{2*}$ dual to two points $p, q \in L$.

(d) $C = 2L$ is of rank 1, and $C' = 2p^*$ is the double of the line in $\mathbb{P}^{2*}$ dual to a point $p \in L$.

Note that the description is exactly the same if we reverse the roles of C and C', except that the second and third types are exchanged. Note also that the points of each type form a locally closed subset of X, with the first open and the last closed, and all four are orbits of the action of PGL_3 on $\mathbb{P}^5 \times \mathbb{P}^{5*}$.

As we have already explained, the locus of complete conics of type (a) is isomorphic to U; in particular, it has dimension 5. It is easy to see that those of type (b) are determined by the pair of lines L, M, and thus form a set of dimension 4. By symmetry (or inspection) the same is true for type (c). Finally, those of type (d) are determined by the line L and the point $p \in L$; thus these form a set of dimension 3, which is in fact the intersection of the closures of the sets of points described in (b) and (c).

8.2.2 Rigorous description

Let us now verify all these statements, using the equations defining the locus $X \subset \mathbb{P}^5 \times \mathbb{P}^{5*}$. We could do this explicitly in coordinates, but it will save a great deal of ink if we use a little multilinear algebra. The reader to whom this is new will find more than enough background in Appendix 2 of Eisenbud [1995]. The multilinear algebra allows us to treat some basic properties in all dimensions with no extra effort, so we begin with some general results about duality for quadrics.

Duals of quadrics

Let V be a vector space. Recall that since we are assuming the characteristic of the ground field $\Bbbk$ is not 2 the following three notions are equivalent:

- A symmetric linear map $\varphi : V \to V^*$.
- A quadratic map $q : V \to \Bbbk$.
- An element $q' \in \mathrm{Sym}^2(V^*)$.

Explicitly, if we start with a symmetric map $\varphi : V \to V^*$ then we take $q(x) = \langle \varphi(x), x \rangle$, and the element $q' \in \mathrm{Sym}^2(V^*)$ comes about from the identification of $\mathrm{Sym}(V^*)$ with the ring of polynomial functions on V.

Any one of these objects, if nonzero, defines a quadric hypersurface $Q \subset \mathbb{P}V$, defined as the zero locus $Q = V(q)$ of q, or equivalently the locus

$$\{v \in \mathbb{P}V \mid \langle \varphi(v), v \rangle = 0\}.$$

(Here, and in the remainder of this discussion, we will abuse notation and use the same symbol v to denote both a nonzero vector $v \in V$ and the corresponding point in $\mathbb{P}V$.) The quadric $Q \subset \mathbb{P}V$ will be smooth if and only if φ is an isomorphism; more generally, the singular locus of Q will be the (projectivization of the) kernel of φ. The *rank* of Q is defined to be, equivalently, the rank of the linear map φ, or $n - \dim(Q_{\text{sing}})$ (where we adopt, for the present purposes only, the convention that $\dim(\varnothing) = -1$); another way to characterize it is to say that a quadric of rank k is the cone with vertex a linear space $Q_{\text{sing}} \cong \mathbb{P}^{n-k} \subset \mathbb{P}^n$ over a smooth quadric hypersurface $\bar{Q} \subset \mathbb{P}^{k-1}$.

Now, the dual of any variety $X \subset \mathbb{P}^n$ is defined to be the closure in $\mathbb{P}^{n*}$ of the locus of hyperplanes tangent to X (that is, containing the tangent space $\mathbb{T}_p X$ at a smooth point $p \in X$). (We will describe this construction in far more detail in Section 10.6.) Given the description in the last paragraph of a quadric Q of rank k as a cone, we see that the dual of a quadric of rank k has dimension $k - 2$. That said, we ask: what, in these terms, is the dual to Q?

To state the result, recall that if $\varphi : V \to W$ is any map of vector spaces of dimension $n + 1$, then there is a *cofactor map* $\varphi^c : W \to V$, represented by a matrix whose entries are signed $n \times n$ minors of φ, satisfying $\varphi \circ \varphi^c = \det(\varphi) \operatorname{Id}_W$ and $\varphi^c \circ \varphi = \det(\varphi) \operatorname{Id}_V$. In invariant terms, φ^c is the composite

$$W \cong \wedge^n W^* \xrightarrow{\wedge^n \varphi^*} \wedge^n V^* \cong V,$$

where the identifications $W \cong \wedge^n W^*$ and $\wedge^n V^* \cong V$ are defined by choices of nonzero vectors in the one-dimensional spaces $\wedge^{n+1} W^*$ and $\wedge^{n+1} V^*$ respectively. Note that when the rank of φ is $< n$ the map φ^c is zero.

Proposition 8.1. *Let $Q \subset \mathbb{P}(V) = \mathbb{P}^n$ be the quadric corresponding to the symmetric map $\varphi : V \to V^*$, and let $v \in V$ be a nonzero vector such that $\langle \varphi(v), v \rangle = 0$, so that $v \in Q$. The tangent hyperplane to Q at v is*

$$\mathbb{T}_v Q = \{w \in \mathbb{P}(V) \mid \langle \varphi(v), w \rangle = 0\}.$$

The dual of Q is thus

$$Q^* = \{\varphi(v) \in \mathbb{P}(V^*) \mid v \in Q \text{ and } \varphi(v) \neq 0\}.$$

In particular, if Q is nonsingular (that is, if the rank of φ is $n + 1$), then Q^* is the image $\varphi(Q)$ of Q under the induced map $\varphi : \mathbb{P}V \to \mathbb{P}V^*$, and Q^* is the quadric corresponding to the cofactor map φ^c.

On the other hand, if the rank of Q is n, and Q^c is the quadric corresponding to the cofactor map φ^c, then Q^c is the unique double hyperplane containing Q^*; that is, the support of Q^c is the hyperplane corresponding to the annihilator of the singular point of Q.

Proof: For any $w \in V$, the line $\overline{v, w} \subset \mathbb{P}V$ spanned by v and w is tangent to Q at v if and only if

$$\langle \varphi(v + \epsilon w), v + \epsilon w \rangle = 0 \bmod (\epsilon^2).$$

Expanding this out, we get

$$\langle \varphi(w), v \rangle + \langle \varphi(v), w \rangle = 0,$$

and, by the symmetry of φ and the assumption that we are not in characteristic 2, this is the case if and only if

$$\langle \varphi(v), w \rangle = 0,$$

proving the first statement and identifying the dual variety as $Q^* = \varphi(Q)$.

Suppose the rank of Q is n or $n + 1$. Let φ^c be the matrix of cofactors of φ, so that $\varphi^c \varphi = \det \varphi \circ I$, where I is the identity map. Since rank Q = rank $\varphi \geq n$, the map φ^c is nonzero. The quadric Q^c is by definition the set of all $w \in V^*$ such that $\langle w, \varphi^c(w) \rangle = 0$. If $v \in Q$ then

$$\langle \varphi(v), \varphi^c \varphi(v) \rangle = (\det \varphi)\langle \varphi(v), v \rangle = 0,$$

so $\varphi(Q)$ is contained in Q^c.

If rank $\varphi = n + 1$, so that φ is an isomorphism, then $Q^* = \varphi(Q)$ is again a quadric hypersurface, and we must have $Q^* = Q^c$. If rank $\varphi = n$, then since $\varphi^c \varphi = 0$ the rank of φ^c is 1, and the associated quadric is a double plane. On the other hand, Q is the cone over a nonsingular quadric in $\mathbb{P}^{n-1}$, and Q^* is the dual of that quadric inside a hyperplane (corresponding to the vertex of Q) in $\mathbb{P}^{n*}$. Thus Q^* spans the plane contained in Q^c. $\square$

The following easy consequence will be useful for the five conic problem:

Corollary 8.2. *If Q and Q' are smooth quadrics, then Q and Q' have the same tangent hyperplane $l = 0$ at some point of intersection $v \in Q \cap Q'$ if and only if Q^* and Q'^* have the common tangent hyperplane $v = 0$ at the point of intersection $l \in Q^* \cap Q'^*$.*

In particular, it follows that if D is a smooth plane conic then the divisor $Z_D \subset X$, which is the closure of the set of complete conics (C, C') such that C is smooth and tangent to D, is equal to the divisor defined similarly starting from the dual conic D^*, that is, the closure of the set of (C, C') such that $C' \subset \mathbb{P}^{2*}$ is smooth and tangent to the dual conic D^*.

Proof: Suppose that Q and Q' correspond to symmetric maps φ and ψ. Since the tangent planes at v are the same, Proposition 8.1 shows that $\varphi(v) = \psi(v) \in Q^* \cap Q'^*$. Since $v = \varphi^{-1}(\varphi(v)) = \psi^{-1}(\psi(v)) \sim \psi^{-1}(\varphi(v))$, we see that Q^* and Q'^* are in fact tangent at $\varphi(v)$. (In addition to the fact that the duality interchanges points and planes, we are really proving that the dual of Q^* is Q, and similarly for Q'. Such a thing is actually true for any nondegenerate variety, as we will see in Section 10.6.) $\quad\square$

Equations for the variety of complete conics

We now return to the case of conics in $\mathbb{P}^2$, and suppose that V is three-dimensional.

Proposition 8.3. *The variety*

$$X \subset \mathbb{P}(\mathrm{Sym}^2 V^*) \times \mathbb{P}(\mathrm{Sym}^2 V) = \mathbb{P}^5 \times \mathbb{P}^{5*}$$

of complete conics is smooth and irreducible. Thinking of $(\varphi, \psi) \in \mathbb{P}^5 \times \mathbb{P}^{5*}$ as coming from a pair of symmetric matrices $\varphi : V \to V^*$ and $\psi : V^* \to V$, the scheme X is defined by the ideal I generated by the eight bilinear equations specifying that the product $\psi \circ \varphi$ has its diagonal entries equal to one another (two equations) and its off-diagonal entries equal to zero (six equations).

(For the experts: it follows from the proposition that the ideal I has codimension 5, and that its saturation, in the bihomogeneous sense, is prime. Computation shows that the polynomial ring modulo I is Cohen–Macaulay. With the proposition, this implies that I is prime. In particular, I is preserved under the interchange of factors φ and ψ, which does not seem evident from the form given.)

Proof: Let Y be the subscheme defined by the given equations. We first show that Y agrees set-theoretically with X on at least the locus of those points (φ, ψ) where rank $\varphi \geq 2$ or rank $\psi \geq 2$, that is, where φ or ψ corresponds to a smooth conic or the union of two distinct lines. On the locus of smooth conics, φ has rank 3 and $(\varphi, \psi) \in Y$ if and only if $\psi = \varphi^{-1}$ up to scalars, so Proposition 8.1 shows that the dual conic is defined by ψ. Moreover, if the rank of φ is 2 and $(\varphi, \psi) \in Y$, then we see from the equations that $\psi \circ \varphi = 0$. Up to scalars, $\psi = \varphi^c$ is the unique possibility, and again Proposition 8.1 shows that the corresponding conic C' is the dual of C. To see the uniqueness (up to scalars) in terms of matrices, note that in suitable bases

$$\varphi = \begin{pmatrix} 1 & 0 & 0 \\ 0 & 1 & 0 \\ 0 & 0 & 0 \end{pmatrix}$$

and the symmetric matrices annihilating the image have the form

$$\psi = \begin{pmatrix} 0 & 0 & 0 \\ 0 & 0 & 0 \\ 0 & 0 & a \end{pmatrix} = a\varphi^c.$$

The same arguments show that when rank $\psi \geq 2$, $\varphi = \psi^c$ and again they correspond to dual conics. (Note that since rank $\psi^c = 1$ on this locus we do *not* have $\psi = \varphi^c$ there.)

Since X is defined as the closure in $\mathbb{P}^5 \times \mathbb{P}^{5*}$ of the locus U of pairs (C, C^*) with C smooth, we see now in particular that $X \subset Y$. We will show next that Y is smooth of dimension 5 locally at any point $(\varphi, \psi) \in Y$ where both φ and ψ have rank 1. We will use this to show that Y is everywhere smooth of dimension 5.

To this end, suppose that $(\varphi, \psi) \in Y$ and that both φ and ψ have rank 1. The tangent space to Y at the point (φ, ψ) may be described as the locus of pairs of symmetric matrices $\alpha : V \to V^*$, $\beta : V^* \to V$ such that

$$(\psi + \epsilon \beta) \circ (\varphi + \epsilon \alpha) \mod (\epsilon^2)$$

has equal diagonal entries and zero off-diagonal entries. Since both φ and ψ have rank 1, the rank of $\psi \circ \alpha + \beta \circ \varphi$ is at most 2, so this is equivalent to saying that

$$\psi \circ \alpha + \beta \circ \varphi = 0.$$

We must show that this linear condition on the entries of the pair (α, β) is equivalent to five independent linear conditions. In suitable coordinates the maps φ, ψ will be represented by the matrices

$$\varphi = \begin{pmatrix} 1 & 0 & 0 \\ 0 & 0 & 0 \\ 0 & 0 & 0 \end{pmatrix} \quad \text{and} \quad \psi = \begin{pmatrix} 0 & 0 & 0 \\ 0 & 1 & 0 \\ 0 & 0 & 0 \end{pmatrix}.$$

Multiplying out, we see that

$$\psi \circ \alpha = \begin{pmatrix} 0 & 0 & 0 \\ \alpha_{2,1} & \alpha_{2,2} & \alpha_{2,3} \\ 0 & 0 & 0 \end{pmatrix} \quad \text{and} \quad \beta \circ \varphi = \begin{pmatrix} \beta_{1,1} & 0 & 0 \\ \beta_{2,1} & 0 & 0 \\ \beta_{3,1} & 0 & 0 \end{pmatrix}.$$

Thus the equation $\psi \alpha + \beta \varphi = 0$ is equivalent to the equations $\alpha_{2,1} + \beta_{2,1} = 0$ and $\alpha_{2,2} = \alpha_{2,3} = \beta_{1,1} = \beta_{3,1} = 0$: five independent linear conditions, as required.

To complete the proof of smoothness, note that Y is preserved scheme-theoretically by the action of the orthogonal group G. (Proof: If $(\varphi, \psi) \in Y$ and α is orthogonal, then $(\alpha \varphi \alpha^*, \alpha \psi \alpha^*) \in Y$ since $\alpha^* \alpha = 1$.) Any closed point on Y where rank $\varphi \geq 2$ degenerates under the action of G to a point where rank $\varphi = 1$. (Proof: If α is orthogonal, that is, $\alpha \alpha^* = 1$, then the matrix $\varphi \psi$ is diagonal if and only if $\alpha \varphi \alpha^* \alpha \psi \alpha^* = \alpha \varphi \psi \alpha^*$ is diagonal. Thus, in a basis for which φ is diagonal, stretching one of the coordinates will make the corresponding entry of φ approach zero, and $\psi = \varphi^c$ moves at the same time; a similar argument works when rank $\psi \geq 2$.) Consequently, if the singular locus of Y were not empty it would have to intersect the locus of pairs of matrices of rank 1, and we have seen that this is not the case.

Finally, to see Y is equal to X scheme-theoretically it is enough to show that the open subset U of Y is dense in Y. We use the fact that each point (φ, ψ) of Y corresponds to a unique pair of quadrics (Q, Q'). When φ has rank 2, Q corresponds to a pair of distinct lines, and Q' is uniquely determined. Thus this set is four-dimensional. The same goes for the case where ψ has rank 2. On the other hand, when both φ and ψ have rank 1, Q is the double of a line L and Q' is the double of a line corresponding to one of the points of L; thus this set is only three-dimensional. Since Y is everywhere smooth of dimension 5, any component of Y must intersect the set where φ and ψ have rank 3, as required. $\square$

The classification of the points of X into the four types on page 296 follows from Proposition 8.3:

Corollary 8.4. *If* $(\varphi, \psi) \in X$, *then one of the following holds:*

(a) *(Smooth complete conics) If* φ *is of rank 3, then* ψ *must be its inverse.*

(b) *If* φ *is of rank 2, then (since X is symmetric) the products* $\psi \circ \varphi$ *and* $\varphi \circ \psi$ *must both be zero; it follows that* ψ *is the unique (up to scalars) symmetric map* $V^* \to V$ *whose kernel is the image of* φ *and whose image is the kernel of* φ.

(c) *If* φ *is of rank 1,* ψ *may have rank 1 or 2; in the latter case, it may be any symmetric map* $V^* \to V$ *whose kernel is the image of* φ *and whose image is the kernel of* φ.

(d) *If* φ *and* ψ *both have rank 1, they simply have to satisfy the condition that the kernel of* ψ *contains the image of* φ *and vice versa.*

Relations with the blow-up of $\mathbb{P}^5$

We mentioned at the beginning of this chapter that another approach to the problem of excess intersection in the five conic problem would be to blow up the excess component — that is, to pass to the blow-up $Z = \mathrm{Bl}_S \mathbb{P}^5$ of $\mathbb{P}^5$ along the surface $S \subset \mathbb{P}^5$ of double lines. It is natural to ask: what is the relation of the space X of complete conics to the blow-up Z?

The answer is that they are in fact the same space. To see this, it is helpful to recall the characterization of blow-ups given in Eisenbud and Harris [2000, Proposition IV-22]: For an affine scheme Y and subscheme $A \subset Y$ with ideal $(f_1, \ldots, f_k)$, the blow-up $\mathrm{Bl}_A Y \to Y$ of Y along A is the closure in $Y \times \mathbb{P}^{k-1}$ of the graph of the map $Y \setminus A \to \mathbb{P}^{k-1}$ given by $[f_1, \ldots, f_k]$. We can globalize this: Let Y be any scheme and $A \subset Y$ a subscheme. If $\mathcal{L}$ is a line bundle on Y and $\sigma_1, \ldots, s_k \in H^0(\mathcal{L})$ sections generating the subsheaf $\mathcal{I}_{A/Y} \otimes \mathcal{L}$, then the closure of the graph of the map $Y \setminus A \to \mathbb{P}^{k-1}$ given by $[f_1, \ldots, f_k]$ is the blow-up $\mathrm{Bl}_A Y \to Y$ of Y along A.

This is exactly what we have in the construction of the space X of complete conics. Again, we think of the space $\mathbb{P}^5$ of conics as the space of symmetric 3×3 matrices M, and the Veronese surface $S \subset \mathbb{P}^5$ of double lines as the locus of matrices of rank 1. The six minors σ_i of M are then sections of $\mathcal{O}_{\mathbb{P}^5}(2)$ generating $\mathcal{I}_{S/\mathbb{P}^5}(2)$, so that the blow-up

$\mathrm{Bl}_S \, \mathbb{P}^5$ is the closure of the graph of the map $[\sigma_1, \ldots, \sigma_6] : \mathbb{P}^5 \setminus S \to \mathbb{P}^5$. But as we have just seen this is the map sending a conic to its dual, so the closure of the graph is the variety X of complete conics.

One note: We could construct an analogous compactification of the space $U \subset \mathbb{P}^N$ of smooth plane curves of any degree d by associating to each smooth $C \subset \mathbb{P}^2$ its dual curve. This defines a regular map $U \to \mathbb{P}^M$, where $\mathbb{P}^M$ is the space of plane curves of degree $d(d-1)$, and we can compactify U by taking the closure in $\mathbb{P}^N \times \mathbb{P}^M$ of the graph of this map. The resulting spaces are highly singular—already in the case $d = 3$, Aluffi [1990] showed it takes five blow-ups of $\mathbb{P}^9$ to resolve the singularities—so in general this is not a useful approach.

8.2.3 Solution to the five conic problem

Now that we have established that the space X of complete conics is smooth and projective, we will show how to solve the five conic problem. To any smooth conic $D \subset \mathbb{P}^2$ we associate a divisor $Z = Z_D \subset X$, which we define to be the closure in X of the locus of pairs $(C, C^*) \in X$ with C smooth and tangent to D, and let $\zeta = [Z_D] \in A^1(X)$ be its class. We will address each of the following issues:

- We have to show that in passing from the "naive" compactification $\mathbb{P}^5$ of the space U of smooth conics to the more sensitive compactification X, we have in fact eliminated the problem of extraneous intersection; in other words, we have to show that for five general conics C_i the corresponding divisors $Z_{C_i} \subset X$ intersect only in points $(C, C') \in X$ with C and $C' = C^*$ smooth.
- We have to show that the five divisors Z_{C_i} are transverse at each point where they intersect.
- We have to determine the Chow ring of the space X, or at least the structure of a subring of $A(X)$ containing the class ζ of the hypersurfaces Z_{C_i} we wish to intersect.
- We have to identify the class ζ in this ring and find the degree of the fifth power $\zeta^5 \in A^5(X)$.

Complete conics tangent to five general conics are smooth

We begin by recalling that X is symmetric under the operation of interchanging the factors $\mathbb{P}^5$ and $\mathbb{P}^{5*}$.

Let us start by showing that no complete conic (C, C') of type (b) lies in the intersection of the divisors associated to five general conics. The first thing we need to do is to describe the points (C, C') of type (b) lying in Z_D for a smooth conic D. This is straightforward: If $C = L \cup M$ is a conic of rank 2 which is a limit of smooth conics tangent to D, then C also must have a point of intersection multiplicity 2 or more with D; thus either L or M is a tangent line to D, or the point $p = L \cap M$ lies on D. (Note that by symmetry a similar description holds for the points of type (c): the complete conic $(2L, p^* + q^*)$ will lie on Z_D only if L is tangent to D, or p or q lie on D.)

Now, suppose that (C, C') is a complete conic of type (b) lying in the intersection of the divisors $Z_i = Z_{C_i}$ associated to five general conics C_i. Write $C = L \cup M$, and set $p = L \cap M$. We note that since the C_i are general, no three are concurrent; thus p can lie on at most two of the conics C_i. We will proceed by considering three cases in turn:

- p *lies on none of the conics* C_i. This is the most immediate case: Since the conics C_i^* are also general, it is likewise the case that no three of them are concurrent. In other words, no line in the plane is tangent to more than two of the C_i, and correspondingly $(L \cup M, p) \in Z_{C_i}$ for at most four of the C_i.

- p *lies on two of the conics* C_i, say C_1 and C_2. Since C_3, C_4 and C_5 are general with respect to C_1 and C_2, none of the finitely many lines tangent to two of them passes through a point of $C_1 \cap C_2$; thus L and M can each be tangent to at most one of the conics C_3, C_4 and C_5, and again we see that $(L \cup M, p) \in Z_{C_i}$ for at most four of the C_i.

- p *lies on exactly one of the conics* C_i, say C_1. Now, since C_1 is general with respect to C_2, C_3, C_4 and C_5, it will not contain any of the finitely many points of pairwise intersection of lines tangent to two of them. Thus L and M cannot each be tangent to two of the conics $C_2, \ldots, C_5$, and once more we see that $(L \cup M, p) \in Z_{C_i}$ for at most four of the C_i.

Thus no complete conic of type (b) can lie in the intersection of the Z_{C_i}; by symmetry, no complete conic of type (c) can either.

It remains to verify that no complete conic (C, C') of type (d) can lie in the intersection $\bigcap Z_{C_i}$, and again we have to start by characterizing the intersection of a cycle $Z = Z_D$ with the locus of complete conics of type (d).

To do this, write an arbitrary complete conic of type (d) as $(2M, 2q^*)$, with $q \in M$. If $(2M, 2q^*) \in Z_D$, then there is a one-parameter family $(C_t, C_t') \in Z_D$ with C_t smooth, $C_t' = C_t^*$ for $t \neq 0$ and $(C_0, C_0') = (2M, 2q^*)$; let $p_t \in C_t \cap D$ be the point of tangency of C_t with D, and set $p = \lim_{t \to 0} p_t \in M$. The tangent line $\mathbb{T}_{p_t} C_t = \mathbb{T}_{p_t} D$ to C_t at p_t will have as limit the tangent line L to D at p, so $L \in q^*$. Thus both p and q are in both L and M. If $p = q$ then in particular $q \in D$. On the other hand, if $p \neq q$, then we must have $M = \overline{p, q} = L$, so $M \in D^*$. We conclude, therefore, that *a complete conic $(2M, 2q^*)$ of type (d) can lie in Z_D only if either $q \in D$ or $M \in D^*$.*

Given this, we see that the first condition ($q \in C_i$) can be satisfied for at most two of the C_i, and the latter ($M \in C_i^*$) likewise for at most two; thus no complete conic $(2M, 2q^*)$ of type (d) can lie in Z_{C_i} for all $i = 1, \ldots, 5$.

Transversality

In order to prove that the cycles $Z_{C_i} \subset X$ intersect transversely when the conics $C_1, \ldots, C_5$ are general, we need a description of the tangent spaces to the Z_{C_i} at points of $\bigcap Z_{C_i}$. We have just shown that such points are represented by smooth conics, and the open subscheme parametrizing smooth conics is the same whether we are working

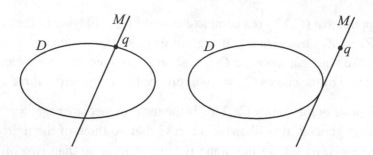

Figure 8.3 The two types of complete conics $(2M, 2q^*)$ of type (d) tangent to D.

in $\mathbb{P}^5$ or in the space of complete conics, so we may express the answer in terms of the geometry of $\mathbb{P}^5$.

Lemma 8.5. *Let* $D \subset \mathbb{P}^2$ *be a smooth conic curve and* $Z_D^\circ \subset \mathbb{P}^5$ *the variety of smooth plane conics* C *tangent to* D.

(a) *If* C *has a point* p *of simple tangency with* D *and is otherwise transverse to* D, *then* Z_D° *is smooth at* $[C]$.

(b) *In this case, the projective tangent plane* $\mathbb{T}_{[C]} Z_D^\circ$ *to* Z_D° *at* $[C]$ *is the hyperplane* $H_p \subset \mathbb{P}^5$ *of conics containing the point* p.

Proof: First, identify D with $\mathbb{P}^1$, and consider the restriction map

$$H^0(\mathcal{O}_{\mathbb{P}^2}(2)) \to H^0(\mathcal{O}_D(2)) = H^0(\mathcal{O}_{\mathbb{P}^1}(4)).$$

This map is surjective, with kernel the one-dimensional subspace spanned by the section representing D itself. In terms of projective spaces, the restriction induces a rational map

$$\pi_D : \mathbb{P}^5 = \mathbb{P}H^0(\mathcal{O}_{\mathbb{P}^2}(2)) \to \mathbb{P}H^0(\mathcal{O}_{\mathbb{P}^1}(4)) = \mathbb{P}^4$$

(this is just the linear projection of $\mathbb{P}^5$ from the point $D \in \mathbb{P}^5$ to $\mathbb{P}^4$). The closure Z_D° in $\mathbb{P}^5$ is thus the cone with vertex $D \in \mathbb{P}^5$ over the hypersurface $\mathcal{D} \subset \mathbb{P}^4$ of singular divisors in the linear system $|\mathcal{O}_{\mathbb{P}^1}(4)|$; Lemma 8.5 will follow directly from the next result:

Proposition 8.6. *Let* $\mathbb{P}^d = \mathbb{P}H^0(\mathcal{O}_{\mathbb{P}^1}(d))$ *be the space of polynomials of degree* d *on* $\mathbb{P}^1$ *and* $\mathcal{D} \subset \mathbb{P}^d$ *the discriminant hypersurface, that is, the locus of polynomials with a repeated root. If* $F \in \mathcal{D}$ *is a point corresponding to a polynomial with exactly one double root* p *and* $d - 2$ *simple roots, then* $\mathcal{D}$ *is smooth at* F *with tangent space the space of polynomials vanishing at* p.

Proof: Note that we have already seen this statement: it is the content of Proposition 7.21 (stated in Section 7.7.3 as a consequence of the topological Hurwitz formula). For another proof, this time in local coordinates, we can introduce the incidence correspondence

$$\Psi = \{(F, p) \in \mathbb{P}^d \times \mathbb{P}^1 \mid \operatorname{ord}_p(F) \geq 2\},$$

and write down its equations in local coordinates (a, x) in $\mathbb{P}^d \times \mathbb{P}^1$: Ψ is the zero locus of the polynomials

$$R(a, t) = a_d x^d + a_{d-1} x^{d-1} + \cdots + a_1 x + a_0$$

and

$$S(a, t) = d a_d x^{d-1} + (d-1) a_{d-1} x^{d-2} + \cdots + 2 a_2 x + a_1.$$

Evaluated at a general point (a, x) where $a_1 = a_0 = x = 0$, all the partial derivatives of R and S vanish except

$$\begin{pmatrix} \dfrac{\partial R}{\partial a_1} & \dfrac{\partial R}{\partial a_0} & \dfrac{\partial R}{\partial x} \\[2mm] \dfrac{\partial S}{\partial a_1} & \dfrac{\partial S}{\partial a_0} & \dfrac{\partial S}{\partial x} \end{pmatrix} = \begin{pmatrix} 0 & 1 & 0 \\ 1 & 0 & 2a_2 \end{pmatrix}.$$

The fact that the first 2×2 minor is nonzero assures us that Ψ is smooth at the point, and the fact that $a_2 \neq 0$ and the characteristic is not 2 assures us that the differential $d\pi : T_{(a,0)}\Psi \to T_a \mathbb{P}^d$ of the projection $\pi : \mathcal{D} \to \mathbb{P}^d$ is injective, with image the plane $a_0 = 0$. Finally, the fact that π is one-to-one at such a point tells us the image $\mathcal{D} = \pi(\Psi)$ is smooth at the image point. $\qquad\square$

Getting back to the statement of Lemma 8.5, if $C \subset \mathbb{P}^2$ is a conic with a point p of simple tangency with D and is otherwise transverse to D, then, by Proposition 8.6, $\mathcal{D}$ is smooth at the image point in $\mathbb{P}^4$, with tangent space the space of polynomials vanishing at p. Since Z_D is the cone over $\mathcal{D}$ it follows that Z_D is smooth at C; the tangent space statement follows as well. $\qquad\square$

In order to apply Lemma 8.5, we need to establish some more facts about a conic tangent to five general conics:

Lemma 8.7. *Let $C_1, \ldots, C_5 \subset \mathbb{P}^2$ be general conics, and $C \subset \mathbb{P}^2$ any smooth conic tangent to all five. Each conic C_i is simply tangent to C at a point p_i and is otherwise transverse to C, and the points $p_i \in C$ are distinct.*

Proof: Let U be the set of smooth conics, and consider incidence correspondences

$$\Phi = \{(C_1, \ldots, C_5; C) \in (U^5 \times U) \mid \text{each } C_i \text{ is tangent to } C\}$$
$$\subset \Phi' = \{(C_1, \ldots, C_5; C) \in ((\mathbb{P}^5)^5 \times U) \mid \text{each } C_i \text{ is tangent to } C\}.$$

The set Φ is an open subset of the set Φ'. Since U is irreducible of dimension 5 and the projection map $\Phi' \to U$ on the last factor has irreducible fibers $(Z_C)^5$ of dimension 20, we see that Φ', and thus also Φ, is irreducible of dimension 25.

There are certainly points in Φ where the conditions of the lemma are satisfied: simply choose a conic C and five general conics C_i tangent to it. Thus the set of $(C_1, \ldots, C_5; C) \in \Phi$ where the conditions of the lemma are not satisfied is a proper closed subset, and as such it can have dimension at most 24, and cannot dominate U^5 under the projection to the first factor. This proves the lemma. $\qquad\square$

To complete the argument for transversality, let $[C] \in \bigcap Z_{C_i}$ be a point corresponding to the conic $C \subset \mathbb{P}^2$. By Lemma 8.7 the points p_i of tangency of C with the C_i are distinct points on C. Since C is the unique conic through these five points, the intersection of the tangent spaces to Z_{C_i} at $[C]$

$$\bigcap \mathbb{T}_{[C]} Z_{C_i} = \bigcap H_{p_i} = \{[C]\}$$

is zero-dimensional, proving transversality.

8.2.4 Chow ring of the space of complete conics

Having confirmed that the intersection $\bigcap Z_{C_i}$ indeed behaves well, let us turn now to computing the intersection number. We start by describing the relevant subgroup of the Chow group $A(X)$.

First, let $\alpha, \beta \in A^1(X)$ be the pullbacks to $X \subset \mathbb{P}^5 \times \mathbb{P}^{5*}$ of the hyperplane classes on $\mathbb{P}^5$ and $\mathbb{P}^{5*}$. These are respectively represented by the divisors

$$A_p = \{(C, C^*) \mid p \in C\}$$

(for any point $p \in \mathbb{P}^2$) and

$$B_L = \{(C, C^*) \mid L \in C^*\}$$

(for any point $L \in \mathbb{P}^{2*}$).

Also, let $\gamma, \varphi \in A^4(X)$ be the classes of the curves Γ and Φ that are the pullbacks to X of general lines in $\mathbb{P}^5$ and $\mathbb{P}^{5*}$. These are, respectively, the classes of the loci of complete conics (C, C^*) such that C contains four general points in the plane, and such that C^* contains four points $L_i \in \mathbb{P}^{2*}$ (that is, C is tangent to four lines in $\mathbb{P}^2$).

Lemma 8.8. *The group $A^1(X)$ of divisor classes on X has rank 2, and is generated over the rationals by α and β. The intersection number of these classes with γ and φ are given by the table*

$$\begin{array}{c} \quad \alpha \quad \beta \\ \begin{array}{c} \gamma \\ \varphi \end{array} \begin{pmatrix} 1 & 2 \\ 2 & 1 \end{pmatrix} \end{array}$$

Proof: We first show that the rank of $A^1(X)$ is at most 2. The open subset $U \subset X$ of pairs (C, C^*) with C and C^* smooth is isomorphic to the complement of a hypersurface in $\mathbb{P}^5$, and hence has torsion Picard group: Any line bundle on U extends to a line bundle on $\mathbb{P}^5$, a power of which is represented by a divisor supported on the complement $\mathbb{P}^5 \setminus U$.

Thus, if L is any line bundle on X, a power of L is trivial on U and so is represented by a divisor supported on the complement $X \setminus U$. But the complement of U in X has just two irreducible components: the closures D_2 and D_3 of the loci of complete conics of type (b) and (c). Any divisor class on X is thus a rational linear combination of the classes of D_2 and D_3, from which we see that the rank of the Picard group of X is at most 2.

Since passing through a point is one linear condition on a quadric, we have $\deg(\alpha\gamma) = 1$ and dually $\deg(\beta\varphi) = 1$. Similarly, since a general pencil of conics will cut out on a line $L \subset \mathbb{P}^2$ a pencil of degree 2, which will have two branch points, we see that $\deg(\alpha\varphi) = 2$ and again by duality $\deg(\beta\gamma) = 2$. Since the matrix of intersections between α, β and γ, φ is nonsingular, we conclude that α and β generate $\mathrm{Pic}(X) \otimes \mathbb{Q}$. $\qquad\square$

In fact, α and β generate $A^1(X)$ over $\mathbb{Z}$ as well, as we could see from the description of X as a blow-up of $\mathbb{P}^5$.

The class of the divisor of complete conics tangent to C

It follows from Lemma 8.8 that we can write $\zeta = p\alpha + q\beta \in A^1(X) \otimes \mathbb{Q}$ for some $p, q \in \mathbb{Q}$. To compute p and q, we use the fact that, restricted to the open set $U \subset X$, the divisor Z is a sextic hypersurface; it follows that $\deg \zeta\gamma = p + 2q = 6$, and since ζ is symmetric in α and β we get $\deg \zeta\varphi = q + 2p = 6$ as well. Thus

$$\zeta = 2\alpha + 2\beta \in A^1(X) \otimes \mathbb{Q}.$$

From this we see that $\deg(\zeta^5) = 32 \deg(\alpha + \beta)^5$, and it suffices to evaluate the degree of the class $\alpha^{5-i}\beta^i \in A^5(X)$ for $i = 0, \ldots, 5$. By symmetry, it is enough to do this for $i = 0$, 1 and 2.

To do this, observe first that the projection of $X \subset \mathbb{P}^5 \times \mathbb{P}^{5*}$ onto the first factor is an isomorphism on the set U_1 of pairs (C, C') such that rank $C \geq 2$ (the map $U \to \mathbb{P}^{5*}$ sending a smooth conic C to its dual in fact extends to a regular map on U_1 sending a conic $C = L \cup M$ of rank 2 to the double line $2p^* \in \mathbb{P}^{5*}$, where $p = L \cap M$). Since all conics passing through three given general points have rank ≥ 2, the intersections needed will occur only in U_1. Since the degree of a zero-dimensional intersection is equal to the degree of the intersection scheme, this implies that we can make the computations on $\mathbb{P}^5$ instead of on X. For this we will use Bézout's theorem:

- $i = 0$: Passing through a point is a linear condition on quadrics. There is a unique quadric through five general points, and the intersection of five hyperplanes in $\mathbb{P}^5$ has degree 1, so $\deg(\alpha^5) = 1$.
- $i = 1$: The quadrics tangent to a given line form a quadric hypersurface in $\mathbb{P}^5$. Since not all conics in the one-dimensional linear space of conics through four general points will be tangent to a general line, $\deg(\alpha^4\beta) = 2$.

- $i = 2$: Similarly, we see that the conics passing through three given general points and tangent to a general line form a conic curve in $U_1 \subset \mathbb{P}^5$. Not all these conics are tangent to another given general line. (For example, after fixing coordinates we may think of circles as the conics passing through the points $\pm\sqrt{-1}$ on the line at ∞. Certainly there are circles through a given point and tangent to a given line that are not tangent to another given line.) It follows that $\deg(\alpha^3\beta^2)$ is the degree of the zero-dimensional intersection of a plane with two quadrics, that is, 4.

Thus

$$\deg((\alpha + \beta)^5) = \binom{5}{0} + 2\binom{5}{1} + 4\binom{5}{2} + 4\binom{5}{3} + 2\binom{5}{4} + \binom{5}{5}$$
$$= 1 + 10 + 40 + 40 + 10 + 1$$
$$= 102$$

and, correspondingly,

$$\zeta^5 = 2^5 \cdot 102 = 3264.$$

This proves:

Theorem 8.9. *There are 3264 plane conics tangent to five general plane conics.*

Of course, the fact that we are imposing the condition of being tangent to a conic is arbitrary; we can use the space of complete conics to count conics tangent to five general plane curves of any degree, as Exercises 8.11 and 8.12 show, and indeed we can extend this to the condition of tangency with singular curves, as Exercises 8.14–8.16 indicate. See Fulton et al. [1983] for a general formula enumerating members of a k-dimensional families of varieties tangent to k given varieties.

Other divisor classes on the space of complete conics

We will take a moment here to describe as well the classes of two other important divisors on the space X of complete conics: the closures E and G of the strata of complete conics of types (b) and (c). As we mentioned in the initial section of this chapter, the space X can also be realized, via the projection map $X \subset \mathbb{P}^5 \times \mathbb{P}^{5*} \to \mathbb{P}^5$, as the blow-up of $\mathbb{P}^5$ along the Veronese surface $S \subset \mathbb{P}^5$ of double lines, or dually as a blow-up of $\mathbb{P}^{5*}$; in these descriptions of X, the divisors G and E are the exceptional divisors of the blow-up maps.

We can describe the classes ϵ and ξ of E and G by the same method we used to determine the class of Z_d, that is, by calculating their intersection numbers with the curves Γ and Φ. For E, we see that a general pencil of plane conics will have three singular elements, so that $\deg(\epsilon\gamma) = 3$ (that is, the image of E in $\mathbb{P}^5$ is a cubic hypersurface), while the image of E in $\mathbb{P}^{5*}$ has codimension 3, and so will not meet a general line in $\mathbb{P}^{5*}$ at all, so that $\deg(\epsilon\varphi) = 0$; solving, we obtain

$$\epsilon = 2\alpha - \beta, \quad \text{and dually} \quad \xi = 2\beta - \alpha.$$

Alternatively, we can argue that in the space $\mathbb{P}^5$ of conics the closure B_L of the locus of smooth conics tangent to a given line $L \subset \mathbb{P}^2$ is a quadric hypersurface containing the Veronese surface of double lines. It necessarily has multiplicity 1 along S (the singular locus of a quadric hypersurface in $\mathbb{P}^n$ is contained in a proper linear subspace of $\mathbb{P}^n$), so that its proper transform has class $\beta = 2\alpha - \xi$, and the relations above follow.

8.3 Complete quadrics

There are beautiful generalizations of the construction of the space of complete conics to the case of quadrics in $\mathbb{P}^n$ and more general bilinear forms or homomorphisms. The paper Laksov [1987] gives an excellent account and many references. Here is a sketch of the beginning of the story. As usual we restrict ourselves to characteristic 0, though the description holds more generally as long as the ground field has characteristic $\neq 2$.

As in the case of conics, we represent a quadric in $\mathbb{P}^n = \mathbb{P}V$ by a symmetric transformation $\varphi : V \to V^*$, or equivalently a symmetric bilinear form in $\operatorname{Sym}^2 V^*$. To this transformation we associate the sequence of symmetric transformations

$$\varphi_i : \wedge^i V \to \wedge^i (V^*) = (\wedge^i V)^* \quad \text{for } i = 1, \ldots, n.$$

Here the identification $\wedge^i (V^*) = (\wedge^i V)^*$ is canonical — see for example Eisenbud [1995, Section A2.3].

We think of φ_i as an element of $\operatorname{Sym}^2(\wedge^i V^*)$, and we define *the variety of complete quadrics in* $\mathbb{P}^n$, which we will denote by Φ, to be the closure in

$$\prod_{i=1}^{n} \mathbb{P}(\operatorname{Sym}^2(\wedge^i V^*))$$

of the image of the set of smooth quadrics under the map $\varphi \mapsto (\varphi_1, \ldots, \varphi_n)$.

The space $\mathbb{P}(\wedge^i V^*)$ in which the quadric corresponding to φ_i lies is the ambient space of the Grassmannian $G_i = \mathbb{G}(i - 1, n)$ of $(i - 1)$-planes in $\mathbb{P}^n$, and in fact an $(i - 1)$-plane $\Lambda \subset \mathbb{P}^n$ is tangent to Q if the point $[\Lambda] \in G_i$ lies in this quadric.

From the definition we see that Φ has an open set U isomorphic to the open set corresponding to quadrics in the projective space of quadratic forms on $\mathbb{P}^n$. As with the case of complete conics, there is a beautiful description of the points that are not in U.

To start, let $\mathbb{P}^n = \mathbb{P}V$ and consider a flag $\mathcal{V}$ of subspaces of V of arbitrary length r and dimensions $k = \{k_1 < \cdots < k_r\}$:

$$0 \subset V_{k_1} \subset V_{k_2} \subset \cdots \subset V_{k_r} \subset V.$$

Now consider the variety F_k of pairs $(\mathcal{V}, Q)$, where $\mathcal{V}$ is a flag as above and $Q = (Q_1, \ldots, Q_{r+1})$, where the Q_i are smooth quadric hypersurfaces in the projective space $\mathbb{P}(V_{k_i}/V_{k_{i-1}})$; this is an open subset of a product of projective bundles over the variety of flags $\mathcal{V}$. We then have:

Proposition 8.10. *There is a stratification of* Φ *whose strata are the varieties* F_k*, where* k *ranges over all strictly increasing sequences* $0 < k_1 < \cdots < k_r < r$.

One can also describe the limit of a family of smooth quadrics $q_t \in U = F_\varnothing$ when the family approaches a quadric q_0 of rank $n + 1 - k$, as in

$$\varphi_t := \begin{pmatrix} t \cdot I_k & 0 \\ 0 & I_{n+1-k} \end{pmatrix}.$$

The limit lies in the stratum $F_{\{k\}}$, where the flag consists of one intermediate space $0 \subset V_k \subset V$; the k-plane V_k will be the kernel of φ_0, the quadric Q_2 on $\mathbb{P}(V/V_k)$ will be the quadric induced by Q_0 on the quotient, and Q_1 will be the quadric on $\mathbb{P}V_k$ associated to the limit

$$\lim_{t \to 0} \frac{\varphi_t|_{V_k}}{t}.$$

In general, the stratum F_k lies in the closure of F_l exactly when $l \subset k$; the specialization relations can be defined inductively, using the above example.

8.4 Parameter spaces of curves

So far in this chapter we have been studying compactifications of parameter spaces of smooth conics. The most obvious is perhaps $\mathbb{P}^5$, which we can identify as the space of all subschemes of $\mathbb{P}^2$ having pure dimension 1 and degree 2 (and thus arithmetic genus 0), and we have shown how the compactification by complete conics was more useful for dealing with problems involving tangencies. Here we have used the fact that the dual of a smooth conic is again a smooth conic. It would have been a different story if the problem had involved twisted cubics in $\mathbb{P}^3$ rather than conics in $\mathbb{P}^2$ — if we had asked, for example, for the number of twisted cubic curves meeting each of 12 lines, or tangent to each of 12 planes, or, as in one classical example, the number of twisted cubic curves tangent to each of 12 quadrics. In that case it is not so clear how to make any parameter space and compactification at all!

In this section, we will discuss two general approaches to constructing parameter spaces for curves in general: the Hilbert scheme of curves and the Kontsevich space of stable maps. (In specific cases, other approaches may be possible as well; for example, see Cavazzani [2016] in the case of twisted cubics.) The Hilbert scheme and the Kontsevich space each have advantages and disadvantages, as we will see.

8.4.1 Hilbert schemes

Recall from Section 6.3 that the Hilbert scheme $\mathcal{H}_P(\mathbb{P}^n)$ is a parameter space for subschemes of $\mathbb{P}^n$ with Hilbert polynomial P; in the case of curves (one-dimensional subschemes), this means all subschemes with fixed degree and arithmetic genus. We start by describing the Hilbert schemes parametrizing conic and cubic curves in $\mathbb{P}^2$ and $\mathbb{P}^3$; when we come to Kontsevich spaces, we will describe these cases in that setting for contrast.

The Hilbert schemes of conics and cubics in $\mathbb{P}^2$

As we have observed, these are just the projective spaces $\mathbb{P}^5$ and $\mathbb{P}^9$ associated to the vector spaces of homogeneous quadratic and cubic polynomials on $\mathbb{P}^2$; they parametrize subschemes of $\mathbb{P}^2$ with Hilbert polynomial $2m + 1$ and $3m$ respectively.

The Hilbert scheme of plane conics in $\mathbb{P}^3$

We will discuss this space at much greater length in the following chapter (where, in particular, we will prove all the assertions made here!). Briefly, the story is this: Any subscheme of $\mathbb{P}^3$ with Hilbert polynomial $2m + 1$ is necessarily the complete intersection of a plane and a quadric surface; the plane, naturally, is unique. This means that the Hilbert scheme admits a map to the dual projective space $\mathbb{P}^{3*}$; the fiber over a point $H \in \mathbb{P}^{3*}$ is the $\mathbb{P}^5$ of conics in $H \cong \mathbb{P}^2$. (This $\mathbb{P}^5$-bundle structure is what allows us to calculate its Chow ring; we will use this information to solve the enumerative problem of counting the conics meeting each of eight general lines in $\mathbb{P}^3$.) In any event, the Hilbert scheme of plane conics in $\mathbb{P}^3$ is irreducible and smooth of dimension 8.

The Hilbert scheme of twisted cubics

In the case of the Hilbert scheme parametrizing twisted cubic curves in $\mathbb{P}^3$ (that is, parametrizing subschemes of $\mathbb{P}^3$ with Hilbert polynomial $3m + 1$) we begin to see some of the pathologies that affect Hilbert schemes in general. It has one component of dimension 12 whose general point corresponds to a twisted cubic curve. But it also has a second component, whose general point corresponds to the union of a plane cubic $C \subset \mathbb{P}^2 \subset \mathbb{P}^3$ and a point $p \in \mathbb{P}^3$. Moreover, this second component has dimension 15 (the choice of plane has three degrees of freedom, the cubic inside the plane nine more, and the point gives an additional three). These two components meet along the 11-dimensional subscheme of singular plane cubics C with an embedded point at the singularity, not contained in the plane spanned by C (see Piene and Schlessinger [1985]).

8.4.2 Report card for the Hilbert scheme

The Hilbert scheme is from some points of view the most natural parameter space that is generally available for projective schemes. Among its advantages: As shown in Section 6.3, it represents a functor that is easy to understand. There is a useful cohomological description of the tangent spaces to the Hilbert scheme, and, beyond that, a deformation theory that in some cases can describe its local structure. It was shown to be connected in characteristic 0 by Hartshorne [1966] and in finite characteristic by Pardue [1996] (see Peeva and Stillman [2005] for another proof). Reeves [1995] showed that the radius of the graph of its irreducible components is at most one more than the dimension of the varieties being parametrized. And, of course, associated to a point on the Hilbert scheme is all the rich structure of a homogeneous ideal in the ring $\Bbbk[x_0, \ldots, x_n]$ and its free resolution.

However, as a compactification of the space of smooth curves, the Hilbert scheme has drawbacks that sometimes make it difficult to use:

(a) *It has extraneous components, often of differing dimensions.* We see this phenomenon already in the case of twisted cubics, above. Of course we could take just the closure $\mathcal{H}^\circ$ in the Hilbert scheme of the locus of smooth curves, but we would lose some of the nice properties, like the description of the tangent space. (Thus while it is relatively easy to describe the singular locus of $\mathcal{H}$, we do not know in general how to describe the singular locus of $\mathcal{H}^\circ$ along the locus where it intersects other components; in the case of twisted cubics it was not known until Piene and Schlessinger [1985] that $\mathcal{H}^\circ$ is smooth.)

In fact, we do not know for curves of higher degree how many such extraneous components there are, nor their dimensions: For $r \geq 3$ and large d, the Hilbert scheme of zero-dimensional subschemes of degree d in $\mathbb{P}^r$ will have an unknown number of extraneous components of unknown dimensions, and this creates even more extraneous components in the Hilbert schemes of curves.

(b) *No one knows what is in the closure of the locus of smooth curves.* If we do choose to deal with the closure $\mathcal{H}^\circ$ of the locus of smooth curves rather than the whole Hilbert scheme — as it seems we must — we face another problem: except in a few special cases, we cannot tell if a given point in the Hilbert scheme is in this closure. That is, we may not know how to tell whether a given singular one-dimensional scheme $C \subset \mathbb{P}^r$ is smoothable.

(c) *It has many singularities.* The singularities of the Hilbert scheme are, in a precise sense, arbitrarily bad: Vakil [2006b] has shown that the completion of every affine local $\Bbbk$-algebra appears (up to adding variables) as the completion of a local ring on a Hilbert scheme of curves.

8.4.3 The Kontsevich space

In the case of curves in a variety, the *Kontsevich space* is an alternative compactification. A precise treatment of this object is given in Fulton and Pandharipande [1997]; here we will treat it informally, sketch some of its properties, and indicate how it is used, with the hope that this will give the interested reader a taste of what to expect.

The Kontsevich space $\overline{M}_{g,0}(\mathbb{P}^r, d)$ parametrizes what are called *stable maps* of degree d and genus g to $\mathbb{P}^r$. These are morphisms

$$f : C \to \mathbb{P}^r,$$

with C a connected curve of arithmetic genus g having only nodes as singularities, such that the image $f_*[C]$ of the fundamental class of C is equal to d times the class of a line in $A_1(\mathbb{P}^r)$, and satisfying the one additional condition that the automorphism group of the map f — that is, automorphisms φ of C such that $f \circ \varphi = f$ — is finite. (This last condition is automatically satisfied if the map f is finite; it is relevant only for maps that are constant on an irreducible component of C, and amounts to saying that any smooth, rational component C_0 of C on which f is constant must intersect the rest of the curve C in at least three points.) Two such maps $f : C \to \mathbb{P}^r$ and $f' : C' \to \mathbb{P}^r$ are said to be the same if there exists an isomorphism $\alpha : C \to C'$ with $f' \circ \alpha = f$. There is an analogous notion of a family of stable maps, and the Kontsevich space $\overline{M}_{g,0}(\mathbb{P}^r, d)$ is a coarse moduli space for the functor of families of stable maps. Note that we are taking the quotient by automorphisms of the source, but not of the target, so that $\overline{M}_{g,0}(\mathbb{P}^r, d)$ shares with the Hilbert scheme $\mathcal{H}_{dm-g+1}(\mathbb{P}^r)$ a common open subset parametrizing smooth curves $C \subset \mathbb{P}^r$ of degree d and genus g.

There are natural variants of this: the space $\overline{M}_{g,n}(\mathbb{P}^r, d)$ parametrizes maps $f : C \to \mathbb{P}^r$ with C a nodal curve having n marked distinct smooth points $p_1, \ldots, p_n \in C$. (Here an automorphism of f is an automorphism of C fixing the points p_i and commuting with f; the condition of stability is thus that any smooth, rational component C_0 of C on which f is constant must have at least three distinguished points, counting both marked points and points of intersection with the rest of the curve C.) More generally, for any projective variety X and numerical equivalence class $\beta \in \mathrm{Num}_1(X)$, we have a space $\overline{M}_{g,n}(X, \beta)$ parametrizing maps $f : C \to X$ with fundamental class $f_*[C] = \beta$, again with C nodal and f having finite automorphism group.

It is a remarkable fact that the Kontsevich space is proper: In other words, if $\mathcal{C} \subset \mathcal{D} \times \mathbb{P}^r$ is a flat family of subschemes of $\mathbb{P}^r$ parametrized by a smooth, one-dimensional base $\mathcal{D}$, and the fiber C_t is a smooth curve for $t \neq 0$, then no matter what the singularities of C_0 are there is a unique stable map $f : \tilde{C}_0 \to \mathbb{P}^r$ which is the limit of the inclusions $\iota_t : C_t \hookrightarrow \mathbb{P}^r$. Note that this limiting stable map $f : \tilde{C}_0 \to \mathbb{P}^r$ depends on the family, not just on the scheme C_0; the import of this in practice is that the Kontsevich space is often locally a blow-up of the Hilbert scheme along loci of curves with singularities worse than nodes. (This is not to say we have in general a regular map from the Kontsevich space to the Hilbert scheme; as we will see in the examples below, the limiting stable map $f : \tilde{C}_0 \to \mathbb{P}^r$ does not determine the flat limit C_0 either.) We will see how this plays out in the four relatively simple cases discussed above in connection with the Hilbert scheme:

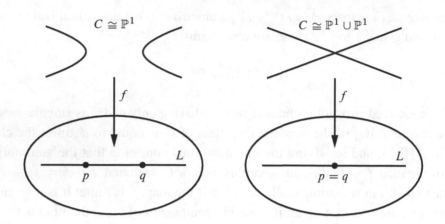

Figure 8.4 Stable maps of degree 2 with image a line.

Plane conics

One indication of how useful the Kontsevich space can be is that, in the case of $\overline{M}_0(\mathbb{P}^2, 2)$ (that is, plane conics), the Kontsevich space is actually equal to the space of complete conics.

To begin with, if $C \subset \mathbb{P}^2$ is a conic of rank 2 or 3 — that is, anything but a double line — then the inclusion map $\iota : C \hookrightarrow \mathbb{P}^2$ is a stable map; thus the open set $W \subset \mathbb{P}^5$ of such conics is likewise an open subset of the Kontsevich space $\overline{M}_0(\mathbb{P}^2, 2)$.

But when a one-parameter family $C \subset \mathcal{D} \times \mathbb{P}^2$ of conics specializes to a double line $C_0 = 2L$, the limiting stable map is a finite, degree-2 map $f : C \to L$, with C isomorphic to either $\mathbb{P}^1$ or two copies of $\mathbb{P}^1$ meeting at a point. Such a map is characterized, up to automorphisms of the source curve, by its branch divisor $B \subset L$, a divisor of degree 2. (If B consists of two distinct points, then $C \cong \mathbb{P}^1$, while if $B = 2p$ for some $p \in L$, the curve C is reducible.) Thus the data of the limiting stable map is equivalent to specifying the limiting dual curve.

This suggests what is in fact the case: The identification of the common open subset W of the Kontsevich space $\overline{M}_0(\mathbb{P}^2, 2)$ and the Hilbert scheme $\mathcal{H}_{2m+1}(\mathbb{P}^2) = \mathbb{P}^5$ extends to a regular morphism, and to a biregular isomorphism of $\overline{M}_0(\mathbb{P}^2, 2)$ with the space X of complete conics, commuting with the projection $X \to \mathbb{P}^5$:

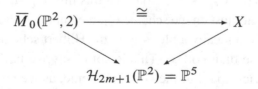

We will not verify these assertions, but they are not hard to prove given the analysis of limits of conics and their duals in Section 8.2.1.

Plane conics in $\mathbb{P}^3$

By contrast, there is not a natural regular map in either direction between the Hilbert scheme of conics in space and the Kontsevich space $\overline{M}_0(\mathbb{P}^3, 2)$. Of course there is a common open set U: its points correspond to reduced connected curves of degree 2 embedded in $\mathbb{P}^3$ (such a curve is either a smooth conic in a plane or the union of two coplanar lines). To see that the identification of this open set does not extend to a regular map in either direction, note first that, as before, if $C \subset \mathcal{D} \times \mathbb{P}^3$ is a family of conics specializing to a double line C_0, then the limiting stable map is a finite, degree-2 cover $f : C \to L$, and this cover is not determined by the flat limit C_0 of the schemes $C_t \subset \mathbb{P}^3$. Thus the identity map on U does not extend to a regular map from the Hilbert scheme to the Kontsevich space. On the other hand, the scheme C_0 is contained in a plane — the limit of the unique planes containing the C_t. Since it has degree 2, the plane containing it is unique. But this plane is not determined by the data of the map f. Thus the identity map on U does not extend to a regular map from the Kontsevich space to the Hilbert scheme either.

The birational equivalence between the Hilbert scheme and the Kontsevich space is of a type that appears often in higher-dimensional birational geometry: the Kontsevich space is obtained from the Hilbert scheme $\mathcal{H}$ by blowing up the locus of double lines, and then blowing down the exceptional divisor along another ruling. (The blow-up of $\mathcal{H}$ along the double line locus is isomorphic to the blow-up of $\overline{M}_0(\mathbb{P}^3, 2)$ along the locus of stable maps of degree 2 onto a line; both can be described as the space of pairs $(H; (C, C^*))$, where $H \subset \mathbb{P}^3$ is a plane and (C, C^*) a complete conic in $H \cong \mathbb{P}^2$.) The birational isomorphism between the Hilbert scheme and Kontsevich space in this case is an example of what is known as a *flip* in higher-dimensional birational geometry.

Plane cubics

Here, we do have a regular map from the Kontsevich space $\overline{M}_1(\mathbb{P}^2, 3)$ to the Hilbert scheme $\mathcal{H}_{3m}(\mathbb{P}^2) \cong \mathbb{P}^9$, and it does some interesting things: It blows up the locus of triple lines, much as in the example of plane conics, and the locus of cubics consisting of a double line and a line as well. But it also blows up the locus of cubics with a cusp, and cubics consisting of a conic and a tangent line, and these are trickier: The blow-up along the locus of cuspidal cubics, for example, can be obtained either by three blow-ups with smooth centers or by one blow-up along a nonreduced scheme supported on this locus.

But what we really want to illustrate here is that the Kontsevich space $\overline{M}_{1,0}(\mathbb{P}^2, 3)$ is not irreducible — in fact, it is not even nine-dimensional! For example, maps of the form $f : C \to \mathbb{P}^2$ with C consisting of the union of an elliptic curve E and a copy of $\mathbb{P}^1$, where f maps $\mathbb{P}^1$ to a nodal plane cubic C_0 and maps E to a smooth point of C_0, form a 10-dimensional family of stable maps; in fact, these form an open subset of a second irreducible component of $\overline{M}_1(\mathbb{P}^2, 3)$, as illustrated in Figure 8.5.

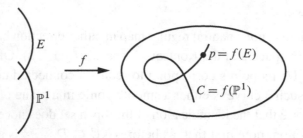

Figure 8.5 A typical point in the 10-dimensional component of $\overline{M}_{1,0}(\mathbb{P}^2, 3)$.

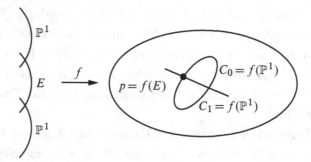

Figure 8.6 General member of a third component of $\overline{M}_{1,0}(\mathbb{P}^2, 3)$.

And there is also a third component, whose general member is depicted in Figure 8.6.

Twisted cubics

Here the shoe is on the other foot. The Hilbert scheme $\mathcal{H} = \mathcal{H}_{3m+1}$ has, as we saw, a second irreducible component besides the closure $\mathcal{H}_0$ of the locus of actual twisted cubics, and the presence of this component makes it difficult to work with. For example, it takes quite a bit of analysis to see that $\mathcal{H}_0$ is smooth, since we have no simple way of describing its tangent space; see Piene and Schlessinger [1985] for details. By contrast, the Kontsevich space is irreducible, and has only relatively mild (finite quotient) singularities.

8.4.4 Report card for the Kontsevich space

As with the Hilbert scheme, there are difficulties in using the Kontsevich space:

(a) *It has extraneous components.* These arise in a completely different way from the extraneous components of the Hilbert scheme, but they are there. A typical example of an extraneous component of the Kontsevich space $\overline{M}_g(\mathbb{P}^r, d)$ consists of maps $f : C \to \mathbb{P}^r$ in which C was the union of a rational curve $C_0 \cong \mathbb{P}^1$, mapping to a rational curve of degree d in $\mathbb{P}^r$, and C_1 an arbitrary curve of genus g meeting C_0 in one point and on which f was constant; if the curve C_1 does not itself admit a nondegenerate map of degree d to $\mathbb{P}^r$, this map cannot be smoothed.

So, using the Kontsevich space rather than the Hilbert scheme does not solve this problem, but it does provide a frequently useful alternative: There are situations where the Kontsevich space has extraneous components and the Hilbert scheme does not—like the case of plane cubics described above—and also situations where the reverse is true, such as the case of twisted cubics.

(b) *No one knows what is in the closure of the locus of smooth curves.* This, unfortunately, remains an issue with the Kontsevich space. Even in the case of the space $\overline{M}_g(\mathbb{P}^2, d)$ parametrizing plane curves, where it might be hoped that the Kontsevich space would provide a better compactification of the Severi variety parametrizing reduced and irreducible plane curves of degree d and geometric genus g than simply its closure in the space $\mathbb{P}^N$ of all plane curves of degree d, the fact that we do not know which stable maps are smoothable represents a real obstacle to its use.

(c) *It has points corresponding to highly singular schemes, and these tend to be in turn highly singular points of $\overline{M}_g(\mathbb{P}^r, d)$.* Still true, but in this respect, at least, it might be said that the Kontsevich space represents an improvement over the Hilbert scheme: Even when the image $f(C)$ of a stable map $f : C \to \mathbb{P}^r$ is highly singular, the fact that the source of the map is at worst nodal makes the deformation theory of the map relatively tractable.

Finally, we mention one other virtue of the Kontsevich space: It allows us to work with tangency conditions, without modifying the space and without excess intersection. The reason is simple: If $X \subset \mathbb{P}^r$ is a smooth hypersurface, the closure Z_X in $\overline{M}_g(\mathbb{P}^r, d)$ of the locus of embedded curves tangent to X is contained in the locus of maps $f : C \to \mathbb{P}^r$ such that the preimage $f^{-1}(X)$ is nonreduced or positive-dimensional. Thus, for example, a point in $\overline{M}_g(\mathbb{P}^2, d)$ corresponding to a multiple curve—that is, a map $f : C \to \mathbb{P}^2$ that is multiple-to-one onto its image—is not necessarily in Z_X.

8.5 How the Kontsevich space is used: rational plane curves

One case in which the Kontsevich space is truly well-behaved is the case $g = 0$. Here the space $\overline{M}_0(\mathbb{P}^r, d)$ is irreducible—it has no extraneous components—and, moreover, its singularities are at worst finite quotient singularities (in fact, it is the coarse moduli space of a smooth Deligne–Mumford stack). Indeed, the use of the Kontsevich space has been phenomenally successful in answering enumerative questions about rational curves in projective space. We will close out this chapter with an example of this; specifically, we will answer the second keynote question, and, more generally, the question of how many rational curves $C \subset \mathbb{P}^2$ of degree d are there passing through $3d - 1$ general points in the plane.

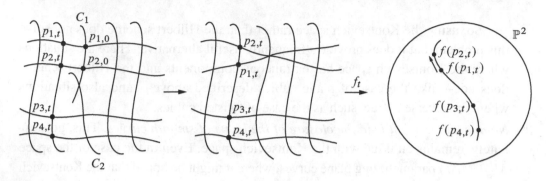

Figure 8.7 A family of maps that blows down C_1.

Since we have not even defined the Kontsevich space, this analysis will be far from complete. The paper Fulton and Pandharipande [1997] provides enough background to complete it.

Before starting the calculation, let us check that we do in fact expect a finite number. Maps of degree d from $\mathbb{P}^1$ to $\mathbb{P}^2$ are given by triples $[F, G, H]$ of homogeneous polynomials of degree d on $\mathbb{P}^1$ with no common zeros; since the vector space of polynomials of degree d on $\mathbb{P}^1$ has dimension $d + 1$, the space U of all such triples has dimension $3d + 3$. Now look at the map $U \to \mathbb{P}^N$ from U to the space $\mathbb{P}^N$ of plane curves of degree d, sending such a triple to the image (as divisor) of the corresponding map $\mathbb{P}^1 \to \mathbb{P}^2$. This has four-dimensional fibers (we can multiply F, G and H by a common scalar, or compose the map with an automorphism of $\mathbb{P}^1$), so we conclude that the image has dimension $3d - 1$. In particular, we see that there are no rational curves of degree d passing through $3d$ general points of $\mathbb{P}^2$, and we expect a finite number (possibly 0) through $3d - 1$. We will denote the number by $N(d)$.

We will work on the space $M_d := \overline{M}_{0,4}(\mathbb{P}^2, d)$ of stable maps from curves with four marked points. This is convenient, since on M_d we have a rational function φ, given by the *cross-ratio*: at a point of M_d corresponding to a map $f : (C; p_1, p_2, p_3, p_4) \to \mathbb{P}^2$ with $C \cong \mathbb{P}^1$ irreducible, it is the cross-ratio of the points $p_1, p_2, p_3, p_4 \in \mathbb{P}^1$; that is, in terms of an affine coordinate z on $\mathbb{P}^1$,

$$\varphi = \frac{(z_1 - z_2)(z_3 - z_4)}{(z_1 - z_3)(z_2 - z_4)},$$

where $z_i = z(p_i)$. The cross-ratio takes on the values 0, 1 and ∞ only when two of the points coincide, which in our setting corresponds to when the curve C is reducible: For example, if C has two components C_1 and C_2, with $p_1, p_2 \in C_1$ and $p_3, p_4 \in C_2$, then by blowing down the curve C_1 in the total space of the source family, we can realize $(C, p_1, \ldots, p_4)$ as a limit of pointed curves $(C_t, p_1(t), \ldots, p_4(t))$ with C_t irreducible and $\lim_{t \to 0} p_1(t) = \lim_{t \to 0} p_2(t)$ (see Figure 8.7). Thus φ has a zero at such a point. Similarly, if three of the p_i, or all four, lie on one component of C, then φ will be equal to the cross-ratio of four distinct points on $\mathbb{P}^1$, and so will not be 0, 1 or ∞.

We now introduce a curve $B \subset M_d$ on which we will make the calculation. Fix a point $p \in \mathbb{P}^2$ and two lines $L, M \subset \mathbb{P}^2$ passing through p; fix two more general points $q, r \in \mathbb{P}^2$ and a collection $\Gamma \subset \mathbb{P}^2$ of $3d - 4$ general points. We consider the locus

$$B = \left\{ f : (C; p_1, p_2, p_3, p_4) \to \mathbb{P}^2 \;\middle|\; \begin{array}{c} f(p_1) = q, \; f(p_2) = r, \\ f(p_3) \in L, \; f(p_4) \in M, \\ \Gamma \subset f(C) \end{array} \right\} \subset M_d.$$

Since, as we said, the space of rational curves of degree d in $\mathbb{P}^2$ has dimension $3d - 1$, and we are requiring the curves in our family to pass through $3d - 2$ points (the points q and r, and the $3d - 4$ points of Γ), our locus B will be a curve.

There may be points in B for which the source C of the corresponding map $f : (C; p_1, p_2, p_3, p_4) \to \mathbb{P}^2$ is reducible. But in these cases C *will have no more than two components*. To see this, note that if the image of C has components $D_1, \ldots, D_k$ of degrees $d_1, \ldots, d_k$, by the above the curve D_i can contain at most $3d_i - 1$ of the $3d - 2$ points $\Gamma \cup \{q, r\}$. Thus

$$3d - 2 \geq \sum_{i=1}^{k} (3d_i - 1) = 3d - k,$$

whence $k \leq 2$. As a consequence, we see that the map f cannot be constant on any component: By the stability condition, if f were constant on a component C_0, then C_0 would have to meet at least three other components — but f can be nonconstant on only two, and it follows that the stability condition is violated on some component.

This argument also shows that there are only finitely many points in B for which the source C is reducible: If $D = D_1 \cup D_2 \subset \mathbb{P}^2$, with D_i a rational curve of degree d_i, and $\Gamma \cup \{q, r\} \subset D$, then by the above D_i must contain exactly $3d_i - 1$ of the $3d - 2$ points $\Gamma \cup \{q, r\}$. The number of such plane curves D is thus

$$\binom{3d - 2}{3d_1 - 1} N(d_1) N(d_2).$$

Moreover, for each such plane curve D there are $d_1 d_2$ stable maps $f : C \to \mathbb{P}^2$ with image D: We can take C the normalization of D at all but any one of the points of intersection $D_1 \cap D_2$. (By Exercise 8.18, D_1 and D_2 will intersect transversely.)

On with the calculation! We equate the number of zeros and the number of poles of φ on B. To begin with, we consider points $f : (C; p_1, p_2, p_3, p_4) \to \mathbb{P}^2$ of B with C irreducible. Since $f(p_1) = q$ and $f(p_2) = r$ are fixed and lie off the lines L and M, the only way any two of the points p_i can coincide on such a curve is if

$$f(p_3) = f(p_4) = p, \quad \text{where } L \cap M = \{p\}.$$

Such points are zeros of φ; the number of these zeros is the number of rational plane curves of degree d through the $3d - 1$ points p, q, r and Γ, that is to say, $N(d)$. (Of course, to make a rigorous argument we would have to determine the multiplicities of these zeros; since we are just sketching the calculation, we will omit the verification that all multiplicities are 1, here and in the following.)

What about zeros and poles of φ coming from points

$$f : (C; p_1, p_2, p_3, p_4) \to \mathbb{P}^2$$

in B with $C = C_1 \cup C_2$ reducible? As we have observed, we get a zero of φ at such a point if and only if the points p_1 and p_2 lie on one component of C and p_3 and p_4 lie on the other. How many such points are there? Well, letting d_1 be the degree of the component C_1 of C containing p_1 and p_2, and $d_2 = d - d_1$ the degree of the other component C_2, we see that $f(C_1)$ must contain q, r and $3d_1 - 3$ of the points of Γ, while C_2 contains the remaining $3d - 4 - (3d_1 - 3) = 3d_2 - 1$ points of Γ. For any subset of $3d_1 - 3$ points of Γ, the number of such plane curves is $N(d_1)N(d_2)$, and for each such plane curve there are d_2 choices of the point $p_3 \in C_2 \cap f^{-1}(L)$ and d_2 choices of the point $p_4 \in C_2 \cap f^{-1}(M)$, as well as $d_1 d_2$ choices of the point $f(C_1 \cap C_2) \in f(C_1) \cap f(C_2)$. We thus have a total of

$$\sum_{d_1=1}^{d-1} d_1 d_2^3 \binom{3d-4}{3d_1-3} N(d_1)N(d_2)$$

zeros of φ arising in this way.

The poles of φ are counted similarly. These can occur only at points

$$f : (C; p_1, p_2, p_3, p_4) \to \mathbb{P}^2$$

in B with $C = C_1 \cup C_2$ reducible, specifically with the points p_1 and p_3 lying on one component of C and p_2 and p_4 on the other. Again letting d_1 be the degree of the component C_1 of C containing p_1 and p_3, and $d_2 = d - d_1$ the degree of the other component C_2, we see that $f(C_1)$ must contain q and $3d_1 - 2$ points of Γ, and $f(C_2)$ the remaining $3d - 4 - (3d_1 - 2) = 3d_2 - 2$ points of Γ, plus r. For any subset of $3d_1 - 2$ points of Γ, the number of such plane curves is $N(d_1)N(d_2)$, and for each such plane curve there are d_1 choices of the point $p_3 \in C_2 \cap f^{-1}(L)$ and d_2 choices of the point $p_4 \in C_2 \cap f^{-1}(M)$, as well as $d_1 d_2$ choices of the point $f(C_1 \cap C_2) \in f(C_1) \cap f(C_2)$. We thus have a total of

$$\sum_{d_1=1}^{d-1} d_1^2 d_2^2 \binom{3d-4}{3d_1-2} N(d_1)N(d_2)$$

poles of φ arising in this way. Now, adding up the poles and zeros, we conclude that

$$N(d) = \sum_{d_1=1}^{d-1} d_1 d_2 \left[d_1 d_2 \binom{3d-4}{3d_1-2} - d_2^2 \binom{3d-4}{3d_1-3} \right] N(d_1)N(d_2),$$

a recursive formula that allows us to determine $N(d)$ if we know $N(d')$ for $d' < d$.

To see how this works, we start with the fact that there is a unique line through two points, and a unique conic through five general points, so $N(d_1) = N(d_2) = 1$. Next, if we take $d = 3$ we see that

$$N(3) = 2\left[2\binom{5}{1} - 4\binom{5}{0}\right] + 2\left[2\binom{5}{4} - \binom{5}{3}\right] = 12.$$

In fact, we have already seen this in Proposition 7.4: The set of all cubics containing eight general points $p_1, \ldots, p_8 \in \mathbb{P}^2$ is a general pencil, and we are counting the number of singular elements of that pencil.

Continuing to $d = 4$, we have

$$N(4) = 3 \cdot 12\left[3\binom{8}{1} - 9\binom{8}{0}\right] + 4\left[4\binom{8}{4} - 4\binom{8}{3}\right] + 3 \cdot 12\left[3\binom{8}{7} - \binom{8}{6}\right] = 620.$$

Always ignoring the question of multiplicity, this answers Keynote Question (b): There are 620 rational quartic curves through 11 general points of $\mathbb{P}^2$.

Exercises 8.19 and 8.20 suggest some additional problems that can be solved using spaces of stable maps.

8.6 Exercises

Exercise 8.11. Let $D \subset \mathbb{P}^2$ be a smooth curve of degree d, and let $Z_D \subset X$ be the closure, in the space X of complete conics, of the locus of smooth conics tangent to D. Find the class $[Z_D] \in A^1(X)$ of the cycle Z_D.

Exercise 8.12. Now let $D_1, \ldots, D_5 \subset \mathbb{P}^2$ be general curves of degrees $d_1, \ldots, d_5$. Show that the corresponding cycles $Z_{D_i} \subset X$ intersect transversely, and that the intersection is contained in the open set U of smooth conics.

Exercise 8.13. Combining the preceding two exercises, find the number of smooth conics tangent to each of five general curves $D_i \subset \mathbb{P}^2$.

Exercise 8.14. Let $D \subset \mathbb{P}^2$ be a curve of degree d with δ nodes and κ ordinary cusps (for a definition of cusps, see Section 11.4.1), and smooth otherwise. Let $Z_D \subset X$ be the closure, in the space X of complete conics, of the locus of smooth conics tangent to D at a smooth point of D. Find the class $[Z_D] \in A^1(X)$ of the cycle Z_D.

Exercise 8.15. Let $\{D_t\}$ be a family of plane curves of degree d, with D_t smooth for $t \neq 0$ and D_0 having a node at a point p. What is the limit of the cycles Z_{D_t} as $t \to 0$?

Exercise 8.16. Here is a very 19th century way of deriving the result of Exercise 8.11 above. Let $\{D_t\}$ be a pencil of plane curves of degree d, with D_t smooth for general t and D_0 consisting of the union of d general lines in the plane. Using the description of the limit of the cycles Z_{D_t} as $t \to 0$ in the preceding exercise, find the class of the cycle Z_{D_t} for t general.

Exercise 8.17. True or false: There are only finitely many PGL_4-orbits in the Kontsevich space $\overline{M}_0(\mathbb{P}^3, 3)$.

Exercise 8.18. Let Γ_1 and Γ_2 be collections of $3d_1 - 1$ and $3d_2 - 1$ general points in $\mathbb{P}^2$, and $D_i \subset \mathbb{P}^2$ any of the finitely many rational curves of degree d_i passing through Γ_i. Show that D_1 and D_2 intersect transversely.

Exercise 8.19. Let $p_1, \ldots, p_7 \in \mathbb{P}^2$ be general points and $L \subset \mathbb{P}^2$ a general line. How many rational cubics pass through $p_1, \ldots, p_7$ and are tangent to L?

Exercise 8.20. (a) Let $M = \overline{M}_0(\mathbb{P}^2, d)$ be the Kontsevich space of rational plane curves of degree d, and let $U \subset M$ be the open set of immersions $f : \mathbb{P}^1 \to \mathbb{P}^2$ that are birational onto their images. For $D \subset \mathbb{P}^2$ a smooth curve, let $Z_D^\circ \subset U$ be the locus of maps $f : \mathbb{P}^1 \to \mathbb{P}^2$ such that $f(\mathbb{P}^1)$ is tangent to D at a smooth point of $f(\mathbb{P}^1)$, and $Z_D \subset M$ its closure. Show that Z_D is contained in the locus of maps $f : C \to \mathbb{P}^r$ such that the preimage $f^{-1}(D)$ is nonreduced or positive-dimensional.

(b) Given this, show that for $D_1, \ldots, D_{3d-1}$ general curves the intersection $\bigcap Z_{D_i}$ is contained in U.

Chapter 9

Projective bundles and their Chow rings

Keynote Questions

(a) Given eight general lines $L_1, \ldots, L_8 \subset \mathbb{P}^3$, how many plane conic curves in $\mathbb{P}^3$ meet all eight? (Answer on page 354.)

(b) Can a ruled surface (that is, a $\mathbb{P}^1$-bundle over a curve) contain more than one curve of negative self-intersection? (Answer on page 341.)

Many interesting varieties, such as scrolls, blow-ups of linear subspaces of projective spaces, and some natural parameter spaces for enumerative problems can be described as projective bundles over simpler varieties. In this chapter we will investigate such varieties and compute their Chow rings. This is a tremendously useful tool, and in particular will allow us to answer the first of the keynote questions above. It will also help us to describe the Chow ring of a blow-up, which we will do in Chapter 13.

9.1 Projective bundles and the tautological divisor class

Definition 9.1. A *projective bundle* over a scheme X is a map $\pi : Y \to X$ such that for any point $p \in X$ there is a Zariski open neighborhood $U \subset X$ of p in X with $Y_U := \pi^{-1}(U) \cong U \times \mathbb{P}^r$ as U-schemes; that is, there are commuting maps

$$
\begin{array}{ccc}
\pi^{-1}(U) & \xrightarrow{\ \cong\ } & U \times \mathbb{P}^r \\
& \searrow{\scriptstyle \pi} \quad \swarrow{\scriptstyle \pi_1} & \\
& U &
\end{array}
$$

where $\pi_1 : U \times \mathbb{P}^r \to U$ is projection on the first factor.

One can make projective bundles from vector bundles as follows: First, if $\mathcal{E} \cong \mathcal{O}_X^{r+1}$ is a trivial vector bundle, then

$$X \times \mathbb{P}^r = \operatorname{Proj}(\mathcal{O}_X[x_0, \ldots, x_r]) = \operatorname{Proj}(\operatorname{Sym} \mathcal{E}^*),$$

and the structure map $\mathcal{O}_X \to \operatorname{Sym} \mathcal{E}^*$ corresponds to the projection $\pi : X \times \mathbb{P}^r \to X$. By definition, any vector bundle $\mathcal{E}$ becomes trivial on an open cover of X, so $\mathbb{P}\mathcal{E} := \operatorname{Proj}(\operatorname{Sym} \mathcal{E}^*) \to X$ is a projective bundle, called the *projectivization of* $\mathcal{E}$. In fact, every projective bundle can be written as $\mathbb{P}\mathcal{E}$ for some vector bundle $\mathcal{E}$. Before we can prove this, we need to know a little more about projectivizations of vector bundles.[1]

From the local description of $\mathbb{P}\mathcal{E}$ as a product, it follows that the points of $\mathbb{P}\mathcal{E}$ correspond to pairs (x, ξ) with $x \in X$ and ξ a one-dimensional subspace $\xi \subset \mathcal{E}_x$ of the fiber $\mathcal{E}_x$ of $\mathcal{E}$. The bundle $\pi^*\mathcal{E}$ on $\mathbb{P}\mathcal{E}$ thus comes equipped with a *tautological subbundle* of rank 1, whose fiber at a point $(x, \xi) \in \mathbb{P}\mathcal{E}$ is the subspace $\xi \subset \mathcal{E}_x$ of the fiber $\mathcal{E}_x$ corresponding to the point $\xi \in \mathbb{P}\mathcal{E}_x$. This subbundle is denoted by $\mathcal{O}_{\mathbb{P}\mathcal{E}}(-1) \subset \pi^*\mathcal{E}$. On an open set $U \subset X$ where $\mathcal{E}$ becomes trivial, so that $\pi^{-1}U = U \times \mathbb{P}^r$, the bundle $\mathcal{O}_{\mathbb{P}\mathcal{E}}(-1)$ is the pullback of $\mathcal{O}_{\mathbb{P}^r}(-1)$ from the second factor. We write $\mathcal{O}_{\mathbb{P}\mathcal{E}}(1) = \mathcal{H}om(\mathcal{O}_{\mathbb{P}\mathcal{E}}(-1), \mathcal{O}_{\mathbb{P}\mathcal{E}})$ for the dual bundle. Dualizing the inclusion of the tautological bundle, we get a surjection $\pi^*\mathcal{E}^* \to \mathcal{O}_{\mathbb{P}\mathcal{E}}(1)$.

We can get an idea of the relation between $\mathbb{P}\mathcal{E}$ and $\mathcal{E}$ from the case where $\mathcal{E}$ is a line bundle. In this case $\mathbb{P}\mathcal{E}$ is locally $X \times \mathbb{P}^0$, so the projection $\pi : \mathbb{P}\mathcal{E} \to X$ is an isomorphism. Identifying $\mathbb{P}\mathcal{E}$ with X via π, we see that $\pi^*(\mathcal{E}) = \mathcal{E}$, and moreover $\mathcal{O}_{\mathbb{P}\mathcal{E}}(-1) = \mathcal{E}$.

From this example we see that the bundles $\mathcal{E}$ and $\mathcal{O}_{\mathbb{P}\mathcal{E}}(-1)$ are not determined by the scheme $\mathbb{P}\mathcal{E}$ or even by the map $\pi : \mathbb{P}\mathcal{E} \to X$ — rather, the bundle $\mathcal{E}$ is an additional piece of data that determines the bundle $\mathcal{O}_{\mathbb{P}\mathcal{E}}(-1)$. We shall soon see that, in general, the projective bundle $\mathbb{P}\mathcal{E} \to X$ alone determines $\mathcal{E}$ up to tensor product with a line bundle (Proposition 9.4), and that the line bundle $\mathcal{O}_{\mathbb{P}\mathcal{E}}(-1)$ on $\mathbb{P}\mathcal{E}$ determines $\mathcal{E}$ completely (Proposition 9.3).

9.1.1 Example: rational normal scrolls

Before continuing with the general theory we pause to work out the case of projective bundles over $\mathbb{P}^1$. As we saw in Theorem 6.29, vector bundles on $\mathbb{P}^1$ are particularly simple: Each one is a direct sum of line bundles $\bigoplus_{i=0}^r \mathcal{O}_{\mathbb{P}^1}(a_i)$.

Write $\mathbb{P}^1 = \mathbb{P}V$, where V is a vector space of dimension 2, with homogeneous coordinates $s, t \in V^*$. Recall that for $a \geq 1$ the *rational normal curve* of degree a is the

---

[1] There is a conflicting definition that is also in use. Some sources, following Grothendieck, define the projectivization of $\mathcal{E}$ to be what we would call the projectivization of $\mathcal{E}^*$, that is, $\pi : \operatorname{Proj}(\operatorname{Sym} \mathcal{E}) \to X$. Its points correspond to 1-quotients of fibers of $\mathcal{E}$. We are following the classical tradition, which is also the convention adopted in Fulton [1984]. Grothendieck's convention is better adapted to the generalization from vector bundles to arbitrary coherent sheaves, which we will not use.

image of the morphism

$$\varphi : \mathbb{P}^1 \to \mathbb{P}^a, \quad (s,t) \mapsto (s^a, s^{a-1}t, \ldots, t^a).$$

When $a = 0$ we take $\varphi : \mathbb{P}^1 \to \mathbb{P}^0$ to be the constant map. More invariantly, for $a \geq 0$, we can think of φ as the map $\mathbb{P}V \to \mathbb{P}^a = \mathbb{P}W^*$ given by the complete linear series

$$|\mathcal{O}_{\mathbb{P}^1}(a)| := (\mathcal{O}_{\mathbb{P}^1}(a), H^0(\mathcal{O}_{\mathbb{P}^1}(a))).$$

Fix a sequence of $r + 1$ nonnegative integers $a_0, \ldots, a_r$, and let $\mathcal{E} = \bigoplus_i \mathcal{O}_{\mathbb{P}^1}(-a_i)$. We will analyze the projective bundle $\mathbb{P}\mathcal{E}$ by mapping it to a projective space $\mathbb{P}^N$ using the line bundle $\mathcal{O}_{\mathbb{P}\mathcal{E}}(1)$.

Set $W_i = H^0(\mathcal{O}_{\mathbb{P}^1}(a_i)) = \mathrm{Sym}^i V^*$ and $W = H^0(\mathcal{E}^*) = \bigoplus W_i$, and write

$$N = \dim W - 1 = \sum (a_i + 1) - 1 = r + \sum a_i.$$

Inside $\mathbb{P}\mathcal{E}$, we consider the $r + 1$ rational curves

$$C_i = \mathbb{P}(\mathcal{O}_{\mathbb{P}^1}(-a_i)) \cong \mathbb{P}^1.$$

There are natural maps

$$W = H^0(\mathcal{E}^*) \to H^0(\pi^*\mathcal{E}^*) \to H^0(\mathcal{O}_{\mathbb{P}\mathcal{E}}(1)),$$

and from the commutative diagrams

$$
\begin{array}{ccc}
\bigoplus W_i = W & \longrightarrow & H^0(\mathcal{O}_{\mathbb{P}\mathcal{E}}(1)) \\
\text{projection} \downarrow & & \downarrow \text{restriction to } C_i \\
W_i = H^0(\mathcal{O}_{\mathbb{P}^1}(a_i)) & \overset{\cong}{\longrightarrow} & H^0(\mathcal{O}_{\mathbb{P}\mathcal{O}_{\mathbb{P}^1}(-a_i)}(1))
\end{array}
$$

we see that $W \to H^0(\mathcal{O}_{\mathbb{P}\mathcal{E}}(1))$ is a monomorphism and that its restriction to C_i is the complete linear series $|\mathcal{O}_{\mathbb{P}^1}(a_i)|$. Let $\varphi_i : \mathbb{P}^1 \to \mathbb{P}W_i^* \subset \mathbb{P}W^*$ be the corresponding morphism, which embeds C_i as the rational normal curve of degree a_i as above.

For each $p \in \mathbb{P}^1$, the restriction of the linear series $\mathcal{W} := (\mathcal{O}_{\mathbb{P}\mathcal{E}}(1), W)$ to the fiber $\mathbb{P}^r = \pi^{-1}(p)$ is a subseries of $|\mathcal{O}_{\mathbb{P}^r}(1)|$. Since the image contains the $r + 1$ linearly independent points $\varphi_i(p)$, it is the complete linear series, and this fiber is mapped isomorphically to the $\mathbb{P}^r$ that is the linear span of the points $\varphi_i(p)$. Thus the linear series $\mathcal{W}$ is base point free, and defines a morphism $\varphi : \mathbb{P}\mathcal{E} \to \mathbb{P}^N$.

We define the *rational normal scroll*

$$S(a_0, \ldots, a_r) \subset \mathbb{P}\left(\bigoplus W_i\right) = \mathbb{P}^N$$

to be the image $\varphi(\mathbb{P}\mathcal{E})$ of this morphism. It is the union of the r-dimensional planes spanned by $\varphi_0(p), \ldots, \varphi_r(p)$ as p runs over $\mathbb{P}^1$:

$$S := S(a_0, \ldots, a_r) = \bigcup_{p \in \mathbb{P}^1} \overline{\varphi_0(p), \ldots, \varphi_r(p)}.$$

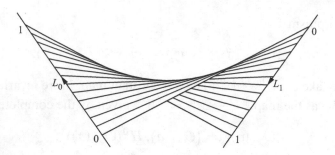

Figure 9.1 $S(1,1)$, the union of lines joining corresponding points on the parametrized skew lines L_0 and L_1, is a nonsingular quadric in $\mathbb{P}^3$.

Since each $\mathbb{P}\mathcal{O}_{\mathbb{P}^1}(-a_i)$ is embedded by the restriction of $\mathcal{W}$, and the distinct $\varphi_i(p)$ are linearly independent for every $p \in \mathbb{P}^1$, it is already clear that φ is set-theoretically an injection.

In the next section, we will show that $\mathcal{W}$ is the complete linear series $|\mathcal{O}_{\mathbb{P}\mathcal{E}}(1)|$, and that when all $a_i > 0$ the map φ induces an isomorphism $\mathbb{P}\mathcal{E} \cong S$. The ideal of forms vanishing on a rational normal scroll is also easy to describe (Exercises 9.27–9.29).

We will also show that

$$\mathbb{P}\left(\bigoplus_{i=0}^{r} \mathcal{O}_{\mathbb{P}^1}(a_i)\right) \cong \mathbb{P}\left(\bigoplus_{i=0}^{r} \mathcal{O}_{\mathbb{P}^1}(b_i)\right)$$

if and only if there is an integer b such that (after possibly reordering the indices) $b_i = a_i + b$ for all i; thus the description above can also be applied to describe the bundles $\mathbb{P}(\bigoplus_{i=0}^{r} \mathcal{O}_{\mathbb{P}^1}(a_i))$ even when some of the a_i are negative.

Some examples of this construction are already familiar. In the case $r = 0$, we have already noted that $S(a_0)$ is the rational normal curve of degree a_0 (or a point, if $a_0 = 0$). From the construction of $S(1, 1) \subset \mathbb{P}^3$ above as the union of lines joining corresponding points on two given disjoint lines, the images of φ_0 and φ_1, we see that $S(1, 1)$ is the nonsingular quadric in $\mathbb{P}^3$: the lines in the union are the lines in one of the two rulings, while the images of φ_0 and φ_1 are two of the lines in the other ruling (see Figure 9.1). Another instance is the scroll $S(1, 1, 1) \subset \mathbb{P}^5$, which is the *Segre threefold*, that is, the image of the Segre embedding $\mathbb{P}^1 \times \mathbb{P}^2 \to \mathbb{P}^5$.

If $a_r = 0$, then from the construction we see that $S(a_0, \dots, a_r)$ is a cone over $S(a_0, \dots, a_{r-1})$, and similarly for the other a_i. This remark allows us to reduce most questions about scrolls to the case where all $a_i > 0$ for all i. For example, the quadric in $\mathbb{P}^3$ with an isolated singularity, that is, the cone over a nonsingular conic in $\mathbb{P}^2$, can be described as $S(2, 0)$ or $S(0, 2)$.

To describe the first example beyond these, the scroll $S(1, 2) \subset \mathbb{P}^4$, we choose an isomorphism between a line L and a nonsingular conic C lying in a plane disjoint from L. The scroll is then the union of the lines joining the points of L to the corresponding points of C.

There is much more to say about the geometry of rational normal scrolls, some of which will be deduced from the more general situation of projective bundles in the next sections, some in Exercises 9.27–9.29. For more information see Eisenbud and Harris [1987] or Harris [1995].

9.2 Maps to a projective bundle

One of our goals is to show that every projective bundle $\pi : Y \to X$ is the projectivization of a vector bundle $\mathcal{E}$ on X, as stated above. In fact, we will construct the bundle $\mathcal{E}$ from the geometry of π, as the dual of the direct image of a suitably chosen line bundle $\mathcal{L}$ on Y. To construct the isomorphism $Y \to \mathbb{P}\mathcal{E}$, we will use the following universal property, which generalizes the one for projective spaces:

Proposition 9.2 (Universal property of Proj). *Given a vector bundle $\mathcal{E}$ on a scheme X, commutative diagrams of maps of schemes*

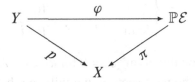

are in natural one-to-one correspondence with line subbundles $\mathcal{L} \subset p^*\mathcal{E}$.

Proof: Given φ, we pull back the inclusion $\mathcal{O}_{\mathbb{P}\mathcal{E}}(-1) \subset \pi^*\mathcal{E}$ via φ and get

$$\varphi^*\mathcal{O}_{\mathbb{P}\mathcal{E}}(-1) \subset \varphi^*\pi^*\mathcal{E} = p^*\mathcal{E}.$$

Conversely, given $\mathcal{L} \subset p^*\mathcal{E}$, we may cover X by open sets on which $\mathcal{E}$ and $\mathcal{L}$ are trivial, and get a unique map over each of these using the universal property of ordinary projective space. By uniqueness, these maps glue together to give a map over all of X. □

To prepare for the next step we need at the least to know how to reconstruct $\mathcal{E}$ from a line bundle on $\mathbb{P}\mathcal{E}$. For future use, we will treat an easy generalization. Write $\mathcal{O}_{\mathbb{P}\mathcal{E}}(m)$ for the m-th tensor power of $\mathcal{O}_{\mathbb{P}\mathcal{E}}(1)$. Thus (for any integer m) the sheaf $\mathcal{O}_{\mathbb{P}\mathcal{E}}(m)$ is the sheaf on $\mathrm{Proj}(\mathrm{Sym}\,\mathcal{E}^*)$ associated to the sheaf of $\mathrm{Sym}\,\mathcal{E}^*$-modules $(\mathrm{Sym}\,\mathcal{E}^*)(m)$ on X, obtained by shifting the grading of $\mathrm{Sym}\,\mathcal{E}^*$. For any quasi-coherent sheaf $\mathcal{F}$ on $\mathbb{P}\mathcal{E}$ we write $\mathcal{F}(m)$ to denote $\mathcal{F} \otimes \mathcal{O}_{\mathbb{P}\mathcal{E}}(m)$.

The surjection $\pi^*\mathcal{E}^* \to \mathcal{O}_{\mathbb{P}\mathcal{E}}(1)$, restricted to the fiber over a point $(x, \xi) \in \mathbb{P}\mathcal{E}$, sends a linear form on $\mathcal{E}_x$ to its restriction to the subspace $\xi \subset \mathcal{E}_x$. Thus any global section σ of $\mathcal{E}^*$ gives rise to a global section $\tilde{\sigma}$ of $\mathcal{O}_{\mathbb{P}\mathcal{E}}(1)$. The following result strengthens and extends this observation:

Proposition 9.3. *If* $\pi : \mathbb{P}\mathcal{E} \to X$ *is a projectivized vector bundle on* X *then for* $m \geq 0$

$$\pi_* \mathcal{O}_{\mathbb{P}\mathcal{E}}(m) = \operatorname{Sym}^m \mathcal{E}^*,$$

and $R^i \pi_* \mathcal{O}_{\mathbb{P}\mathcal{E}}(m) = 0$ *for* $i > 0$.

Taking $m = 1$, we see that the map $\pi : \mathbb{P}\mathcal{E} \to X$, together with the tautological line bundle $\mathcal{O}_{\mathbb{P}\mathcal{E}}(1)$, determines $\mathcal{E}$.

Proof: Suppose that $\mathcal{E}$ has rank $r + 1$. Over an affine open set $U \subset X$ where $\mathcal{E}|_U \cong \mathcal{O}_U^{r+1}$, the natural maps $H^0(\pi^* \operatorname{Sym}^m \mathcal{E}|_U) \to H^0(\mathcal{O}_{\mathbb{P}\mathcal{E}}(m)|_U)$ are isomorphisms, while $H^i(\mathcal{O}_{\mathbb{P}\mathcal{E}}(m)|_U) = 0$ for $i > 0$, so the proposition follows immediately from the definition of the direct image functors. $\qquad\square$

Remark. Proposition 9.3 is a direct generalization of the standard computation of $H^0(\mathcal{O}_{\mathbb{P}^r}(m))$ — the case when X is a point. Though we will not make use of these facts, the rest of the computation of the cohomology of line bundles on a projective space, and Serre duality, also generalize, and one can show that

$$R^i \pi_* \mathcal{O}_{\mathbb{P}\mathcal{E}}(m) = \begin{cases} \operatorname{Sym}^m \mathcal{E}^* & \text{for } i = 0, \\ 0 & \text{for } 0 < i < r - 1, \\ \operatorname{Sym}^{-m-r-1} \mathcal{E} & \text{for } i = r. \end{cases}$$

(Here we adopt the convention that $\operatorname{Sym}^k \mathcal{E} = 0$ for $k < 0$.) As a part of our computation of the Chow ring of $\mathbb{P}\mathcal{E}$ in the next section, we will see that every line bundle on $\mathbb{P}\mathcal{E}$ has the form $\pi^* \mathcal{L} \otimes \mathcal{O}_{\mathbb{P}\mathcal{E}}(m)$ for a unique line bundle $\mathcal{L}$ on X and integer m; that is, $\operatorname{Pic}(\mathbb{P}\mathcal{E}) \cong \operatorname{Pic} X \oplus \mathbb{Z}$. From the push-pull formula of Proposition B.7, we get a computation of the direct images of any line bundle:

$$R^i \pi_* (\pi^* \mathcal{L} \otimes \mathcal{O}_{\mathbb{P}\mathcal{E}}(m)) = \mathcal{L} \otimes R^i \pi_* (\mathcal{O}_{\mathbb{P}\mathcal{E}}(m)).$$

See Dieudonné [1969, p. 308] for equivalent material, with references to EGA.

Serre duality also generalizes to a *relative duality*. For example, setting

$$\omega_{\mathbb{P}\mathcal{E}/B} = \wedge^r \mathcal{E}(-r - 1)$$

we have $R^r \pi_* (\omega_{\mathbb{P}\mathcal{E}/B}) = \mathcal{O}_B$, and more generally

$$R^r \pi_* (\mathcal{M}) = \mathcal{H}om(\pi_* (\omega \otimes \mathcal{M}^{-1}), \mathcal{O}_B)$$

for any line bundle $\mathcal{M}$ on $\mathbb{P}\mathcal{E}$.

See Altman and Kleiman [1970], in particular Theorem 3.8, for most of this.

Supposing that $\mathcal{E}^*$ has a global section $\sigma \neq 0$, the proof of Proposition 9.3 shows that the corresponding section $\tilde{\sigma}$ of $\mathcal{O}_{\mathbb{P}\mathcal{E}}(1)$ vanishes on the locus of pairs (x, ξ) such that σ_x vanishes on ξ; thus the divisor $(\tilde{\sigma})$ meets a general fiber of $\pi : \mathbb{P}\mathcal{E} \to X$ in a hyperplane. It will not in general meet *every* fiber of $\mathbb{P}\mathcal{E} \to X$ in a hyperplane, however;

the section σ of $\mathcal{E}^*$ may have zeros $x \in X$, and the divisor $(\tilde{\sigma}) \subset \mathbb{P}\mathcal{E}$ will contain the corresponding fibers $(\mathbb{P}\mathcal{E})_x = \pi^{-1}(x)$.

Using these ideas, we can characterize the schemes over X that are projective bundles:

Proposition 9.4. *Let* $\pi : Y \to X$ *be a smooth morphism of projective schemes whose (scheme-theoretic) fibers are all isomorphic to* $\mathbb{P}^r$. *The following are equivalent:*

(a) $Y = \mathbb{P}\mathcal{E}$ *is the projectivization of a vector bundle* $\mathcal{E}$ *on* X.

(b) $\pi : Y \to X$ *is a projective bundle; that is, it is locally isomorphic to a product in the Zariski topology on* X.

(c) *There exists a line bundle* $\mathcal{L}$ *on* Y *whose restriction to each fiber* $Y_x \cong \mathbb{P}^r$ *of* π *is isomorphic to* $\mathcal{O}_{\mathbb{P}^r}(1)$.

(d) *There exists a Cartier divisor* $D \subset Y$ *intersecting a general fiber* $Y_x \cong \mathbb{P}^r$ *of* π *in a hyperplane.*

Proof: Condition (a) clearly implies (b) and (c): The projectivization of a vector bundle is locally trivial in the Zariski topology, since a vector bundle is, and comes with the line bundle $\mathcal{O}_{\mathbb{P}\mathcal{E}}(1)$.

It is clear that (b) implies (d): Just take an isomorphism $\pi^{-1}(U) \cong U \times \mathbb{P}^r$ for any Zariski open $U \subset X$, choose a hyperplane $H \cong \mathbb{P}^{r-1} \subset \mathbb{P}^r$ and take D the closure in Y of $U \times H$.

Also, it is easy to see that (c) and (d) are equivalent: If D is a divisor as in (d), the line bundle $\mathcal{L} = \mathcal{O}_Y(D)$, restricted to a general fiber, is $\mathcal{O}_{\mathbb{P}^r}(1)$. By the constancy of the Euler characteristic of a sheaf in a flat family (Corollary B.12), the restriction of $\mathcal{L}$ to any fiber is $\mathcal{O}_{\mathbb{P}^r}(1)$.

Conversely, if $\mathcal{L}$ is a line bundle as in (c), tensoring with the pullback of an ample line bundle from X we can assume the existence of a nonzero global section of $\mathcal{L}$, whose zero locus will be the divisor of part (d).

To complete the argument we take $\mathcal{L}$ as in part (c), and we must prove that Y is as in part (a). For any $p \in X$ we have $H^1(\mathcal{L}_p) = H^1(\mathcal{O}_{\mathbb{P}^r}(1)) = 0$, so Theorem B.5 shows that $\mathcal{E} := \pi_*\mathcal{L}$ is a vector bundle whose fiber at p is $H^0(\mathcal{O}_{\mathbb{P}^r}(1))$.

We claim that there is an isomorphism $\alpha : Y \to \mathbb{P}\mathcal{E}$ commuting with the projections to X. By Proposition 9.2 we can define the morphism α by giving a line bundle that is a subbundle of $\pi^*\mathcal{E}$, or equivalently a line bundle that is a homomorphic image of $\pi^*\mathcal{E}^* = \pi^*\pi_*\mathcal{L}$.

There is a natural map $\pi^*\pi_*\mathcal{L} \to \mathcal{L}$ coming from the definitions of π^* and π_*. Restricted to the fiber over a point p this map becomes the surjection $\mathcal{E}_p \otimes \mathcal{O}_{\mathbb{P}(\mathcal{E}_p)} \to \mathcal{O}_{\mathbb{P}(\mathcal{E}_p)}(1)$, so $\pi^*\pi_*\mathcal{L} \to \mathcal{L}$ is surjective. Let $\alpha : Y \to \mathbb{P}\mathcal{E}$ be the corresponding morphism.

The map α is an isomorphism on each fiber of π because it restricts to the map $\pi^{-1}(p) \cong \mathbb{P}^r \to \mathbb{P}^r$ given by the complete linear series $|\mathcal{O}_{\mathbb{P}^r}(1)|$. This shows that α is a set-theoretic isomorphism.

To prove that α is a scheme-theoretic isomorphism, we need to show that if α carries $y \in Y$ to a point $q \in \mathbb{P}\mathcal{E}$ then the map of local rings $\alpha^* : \mathcal{O}_{\mathbb{P}\mathcal{E},q} \to \mathcal{O}_{Y,y}$ is an isomorphism. Of course it is enough to prove this after completing both rings. Set $p = \pi(y)$. By smoothness, the completions of both local rings are isomorphic to $\hat{\mathcal{O}}_{X,p}[\![z_0, \ldots, z_r]\!]$. Since α commutes with the projections, it induces the identity modulo the maximal ideal of $\mathcal{O}_{X,p}$, and thus induces an isomorphism. $\square$

We can also use Proposition 9.2 to see when two vector bundles give the same projective bundle:

Corollary 9.5. *Let X be a scheme. Two projective bundles $\pi : \mathbb{P}\mathcal{E} \to X$ and $\pi' : \mathbb{P}\mathcal{E}' \to X$ are isomorphic as X-schemes if and only if there is a line bundle $\mathcal{L}$ on X such that $\mathcal{L} \otimes \mathcal{E}' = \mathcal{E}$. In this case the line bundle $\mathcal{O}_{\mathbb{P}\mathcal{E}}(-1)$ corresponds under the isomorphism to $\pi'^*(\mathcal{L}) \otimes \mathcal{O}_{\mathbb{P}\mathcal{E}'}(-1)$.*

Proof: Let $\mathcal{L}$ be a line bundle, and set $\mathcal{E}' = \mathcal{L} \otimes \mathcal{E}$. Tensoring the tautological subbundle $\mathcal{O}_{\mathbb{P}\mathcal{E}'}(-1) \to \pi'^*\mathcal{E}' = \pi'^*\mathcal{L} \otimes \pi'^*\mathcal{E}$ with $\pi'^*(\mathcal{L}^{-1})$, we get a subbundle $\pi'^*(\mathcal{L}^{-1}) \otimes \mathcal{O}_{\mathbb{P}\mathcal{E}'}(-1) \to \pi'^*\mathcal{E}$. By Proposition 9.2 this determines a unique morphism of X-schemes $\varphi : \mathbb{P}\mathcal{E}' \to \mathbb{P}\mathcal{E}$ such that

$$\varphi^*\mathcal{O}_{\mathbb{P}\mathcal{E}}(-1) = \pi'^*(\mathcal{L}^{-1}) \otimes \mathcal{O}_{\mathbb{P}\mathcal{E}'}(-1).$$

The inverse map is defined similarly. The proof that they are inverse to each other is that the composite $\mathbb{P}\mathcal{E} \to \mathbb{P}(\mathcal{L} \otimes \mathcal{E}) \to \mathbb{P}\mathcal{E}$ corresponds to the original subbundle $\mathcal{O}_{\mathbb{P}\mathcal{E}}(-1) \subset \pi^*\mathcal{E}$.

Conversely, suppose that $\mathcal{E}'$ is a vector bundle on X, and let $\pi' : \mathbb{P}\mathcal{E}' \to X$ be the projection. If $\varphi : \mathbb{P}\mathcal{E}' \to \mathbb{P}\mathcal{E}$ is an isomorphism commuting with the projections to X, then since any isomorphism from $\mathbb{P}^n$ to itself preserves the bundle $\mathcal{O}_{\mathbb{P}^n}(1)$ it follows that $\varphi^*\mathcal{O}_{\mathbb{P}\mathcal{E}}(1)$ restricts on each fiber $\mathbb{P}(\mathcal{E}'_x) \cong \mathbb{P}^n$ to the bundle $\mathcal{O}_{\mathbb{P}^n}(1)$. By Corollary B.6, $\mathcal{O}_{\mathbb{P}\mathcal{E}'}(1) = \pi'^*(\mathcal{L}) \otimes \varphi^*\mathcal{O}_{\mathbb{P}\mathcal{E}}(1)$ for some line bundle $\mathcal{L}$ on X. Thus

$$\begin{aligned}
\mathcal{E}'^* = \pi'_*(\mathcal{O}_{\mathbb{P}\mathcal{E}'}(1)) &= \pi'_*(\pi'^*\mathcal{L} \otimes \varphi^*\mathcal{O}_{\mathbb{P}\mathcal{E}}(1)) \\
&= \mathcal{L} \otimes \pi'_*\varphi^*\mathcal{O}_{\mathbb{P}\mathcal{E}}(1) \\
&= \mathcal{L} \otimes \pi'_*\varphi_*^{-1}\mathcal{O}_{\mathbb{P}\mathcal{E}}(1) \\
&= \mathcal{L} \otimes \pi_*\mathcal{O}_{\mathbb{P}\mathcal{E}}(1) \\
&= \mathcal{L} \otimes \mathcal{E}^*,
\end{aligned}$$

and also $\varphi^*\mathcal{O}_{\mathbb{P}\mathcal{E}}(-1) = \pi^*(\mathcal{L}) \otimes \mathcal{O}_{\mathbb{P}\mathcal{E}'}(-1)$, as claimed. $\square$

9.3 Chow ring of a projective bundle

We now turn to the central problem of this chapter: to describe the Chow ring of a projective bundle $Y = \mathbb{P}\mathcal{E} \to X$. We will see that the Chow groups of Y depend only on the rank of $\mathcal{E}$, but the ring structure reflects the Chern classes of $\mathcal{E}$.

As we mentioned in Section 2.1.4, the Künneth theorem holds for the Chow ring of the product of any smooth variety with a projective space. Thus, if

$$Y = \mathbb{P}(\mathcal{O}_{\mathbb{P}^r}^{r+1}) = X \times \mathbb{P}^r$$

then

$$A(Y) \cong A(X) \otimes_{\mathbb{Z}} A(\mathbb{P}^r)$$
$$\cong A(X) \otimes_{\mathbb{Z}} \mathbb{Z}[\zeta]/(\zeta^{r+1})$$
$$\cong A(X)[\zeta]/(\zeta^{r+1}),$$

where ζ is the pullback of the hyperplane class on $\mathbb{P}^r$. In particular,

$$A(Y) = \bigoplus_{i=0}^{r} \zeta^i A(X)$$

as groups. (Given that the pullback map $A(X) \to A(Y)$ is injective, here and in what follows we think of $A(X)$ as a subalgebra of $A(Y)$, suppressing the "π^*;" for example, when we write products of the form $\alpha\beta$ with $\alpha \in A(X)$ and $\beta \in A(Y)$, we mean $(\pi^*\alpha)\beta \in A(Y)$.)

The general case is not much more complicated:

Theorem 9.6. *Let $\mathcal{E}$ be a vector bundle of rank $r + 1$ on a smooth projective scheme X, and let $\zeta = c_1(\mathcal{O}_{\mathbb{P}\mathcal{E}}(1)) \in A^1(\mathbb{P}\mathcal{E})$. Let $\pi : \mathbb{P}\mathcal{E} \to X$ be the projection. The map $\pi^* : A(X) \to A(\mathbb{P}\mathcal{E})$ is an injection of rings, and via this map*

$$A(\mathbb{P}\mathcal{E}) \cong A(X)[\zeta]/(\zeta^{r+1} + c_1(\mathcal{E})\zeta^r + \cdots + c_{r+1}(\mathcal{E})).$$

In particular, the group homomorphism $A(X)^{\oplus r+1} \to A(\mathbb{P}\mathcal{E})$ given by $(\alpha_0, \ldots, \alpha_r) \mapsto \sum \zeta^i \pi^*(\alpha_i)$ is an isomorphism, so that

$$A(\mathbb{P}\mathcal{E}) \cong \bigoplus_{i=0}^{r} \zeta^i A(X)$$

as groups.

It is worth remarking that much of the statement of Theorem 9.6 remains true without the assumption that X is smooth: If $\mathcal{E}$ is a vector bundle of rank $r + 1$ over an arbitrary scheme X and $\mathbb{P}\mathcal{E} = \mathrm{Proj}(\mathrm{Sym}\,\mathcal{E}^*)$ its associated projective bundle, then we have a well-defined line bundle $\mathcal{O}_{\mathbb{P}\mathcal{E}}(1)$ on $\mathbb{P}\mathcal{E}$ such that $\zeta = c_1(\mathcal{O}_{\mathbb{P}\mathcal{E}}(1))$ restricts to the hyperplane class on each fiber, and we can show that

$$A(\mathbb{P}\mathcal{E}) \cong \bigoplus_{i=0}^{r} \zeta^i A(X)$$

as groups, just as in the smooth case (see Fulton [1984, Chapter 3]). (Note that in this setting we do not have a ring structure on $A(X)$ or $A(\mathbb{P}\mathcal{E})$, but multiplication by the class ζ is still well-defined since it is the Chern class of a line bundle.)

It was one of the insights of Grothendieck [1958] that Theorem 9.6 could be inverted and used to *define* the Chern classes of $\mathcal{E}$ as the coefficients in the unique expression of ζ^{r+1} as a linear combination of the classes $1, \zeta, \ldots, \zeta^r$ (or rather to prove the existence of classes satisfying the axioms of Theorem 5.3). The original definitions of Chern and Stiefel–Whitney classes in the 1930s came from topology. They did not mention degeneracy loci, but could be directly related to that characterization of the classes; as we have seen in Chapters 6 and 7, this is closer to the way Chern classes are thought of and used in practice. As a definition, however, it has the drawback of depending on the existence of global sections. (This is a problem only in the algebro-geometric context; in the continuous or C^∞ settings, thanks to partitions of unity there is never a shortage of sections.) While it is possible to define Chern classes for bundles with enough sections via degeneracy loci, and even (as we illustrate in Section 5.9.1) to prove basic properties such as the Whitney formula in that setting, in order to have a full toolkit of techniques for calculating Chern classes it is necessary to extend the definition to arbitrary bundles, and for this the Grothendieck–Serre definition is better.

We isolate part of the proof of Theorem 9.6 that will be useful elsewhere:

Lemma 9.7. *Let the hypotheses be as in Theorem 9.6. If $\alpha \in A(X)$, then*

$$\pi_*(\zeta^i \alpha) = \begin{cases} \alpha & \text{if } i = r, \\ 0 & \text{if } i < r. \end{cases}$$

Proof: By the push-pull formula (Proposition B.7), $\pi_*(\zeta^i \alpha) = \pi_*(\zeta^i)\alpha$. If $i < r$, then $\pi_*(\zeta^i)$ is zero for dimension reasons. If $i = r$, we see similarly that $\pi_*(\zeta^r)$ must be a multiple $m[X] \in A^0(X)$ of the fundamental class of X. Let η be the class of a point $x \in X$ and $f = \pi^*(\eta) = [\mathbb{P}\mathcal{E}_x]$ the class of the fiber $\mathbb{P}\mathcal{E}_x \cong \mathbb{P}^r$. Intersecting both sides of the equality $\pi_*(\zeta^r) = m[X]$ with η and taking degrees, we have

$$m = \deg(\pi_*(\zeta^r) \cdot \eta) = \deg(\zeta^r \cdot [\mathbb{P}^r]) = 1,$$

since the restriction of ζ to a fiber is the hyperplane class. $\qquad\qquad\square$

In fact, we have encountered this construction before, in the proof of Lemma 5.12.

Proof of Theorem 9.6: Let $\psi : A(\mathbb{P}\mathcal{E}) \to \bigoplus_{i=0}^{r} A(X)\zeta^i$ be the map

$$\beta \mapsto \sum_i \pi_*(\zeta^{r-i}\beta)\zeta^i,$$

and let $\varphi : \bigoplus_{i=0}^{r} A(X) \to A(\mathbb{P}\mathcal{E})$ be the sum of the multiplications by powers of ζ:

$$\varphi : (\alpha_0, \ldots, \alpha_r) \mapsto \sum_i \zeta^i \alpha_i.$$

By Lemma 9.7, the composite $\psi\varphi$ is upper-triangular with ones on the diagonal; in particular, φ is a monomorphism.

To prove the additive part of Theorem 9.6, it now suffices to show that the subgroups $\zeta^i A(X)$ generate $A(\mathbb{P}\mathcal{E})$ additively. This is a relative version of the fact that the linear subspaces of a projective space generate its Chow ring, and the proof runs along the same lines. In the case of a single projective space, we used the technique of dynamic projection to degenerate a given subvariety $Z \subset \mathbb{P}^n$ to a multiple of a linear space; we do the same thing here, but in a family of projective spaces.

We start with a definition. If $Z \subset \mathbb{P}\mathcal{E}$ is a k-dimensional subvariety, we say that Z has *footprint* l if the image $W = \pi(Z)$ has dimension l, or equivalently if the general fiber of the map $\pi : Z \to W$ has dimension $k - l$.

Lemma 9.8. *If $Z \subset \mathbb{P}\mathcal{E}$ is a subvariety of dimension k and footprint l, then*

$$Z \sim Z' + \sum n_i B_i$$

for some subvarieties $B_i \subset \mathbb{P}\mathcal{E}$ such that:

(a) $[Z'] = \zeta^{r-k+l} \alpha$ for a class $\alpha \in A(X)$.

(b) Each B_i has footprint strictly less than l.

Applying the lemma repeatedly, we can express the class of an arbitrary subvariety as a sum of classes of the form $\zeta^i \alpha$, establishing the group isomorphism $A(\mathbb{P}\mathcal{E}) \cong \bigoplus \zeta^i A(X)$.

Proof of Lemma 9.8: By Corollary 9.5, replacing $\mathcal{E}$ with its tensor product with a line bundle $\mathcal{L}$ does not change $\mathbb{P}\mathcal{E}$, but has the effect of replacing the class ζ by $\zeta - \pi^* c_1(\mathcal{L})$. In particular, it does not affect the truth of our assertion, so we can assume from the outset that $\mathcal{E}^*$ is generated by global sections.

This done, we choose a point $x \in \pi(Z) \subset X$ and a general collection $\tau_0, \ldots, \tau_r$ of global sections of $\mathcal{E}^*$, making sure that the τ_i satisfy two conditions:

(a) $\tau_0(x), \ldots, \tau_r(x)$ are independent, that is, they span the fiber $\mathcal{E}_x$.
(b) The zero locus $(\tau_0(x) = \cdots = \tau_{k-l}(x) = 0) \subset \mathbb{P}\mathcal{E}_x$ is disjoint from the fiber $Z_x = Z \cap \mathbb{P}\mathcal{E}_x$ of Z over x.

These are both open conditions; let $U \subset X$ be the locus of $x \in X$ where they hold. Note in particular that, by the first condition, the bundle $\mathbb{P}\mathcal{E}$ is trivial over U, the sections $\tau_0, \ldots, \tau_r$ giving an isomorphism $\mathbb{P}\mathcal{E}_U \cong U \times \mathbb{P}^r$.

Now consider the one-parameter group of automorphisms A_t of $\mathbb{P}\mathcal{E}_U \cong U \times \mathbb{P}^r$ given, in terms of this trivialization, by the matrix

$$\begin{pmatrix} I_{k-l+1} & 0 \\ 0 & t \cdot I_{r-k+l} \end{pmatrix}.$$

Let $\tilde{Z} = Z \cap \mathbb{P}\mathcal{E}_U$ be the preimage of U in Z (note that $\pi(Z) \cap U \neq \varnothing$, since $x \in \pi(Z)$); let Z_t be the closure of the image $A_t(\tilde{Z})$ and let Z_0 be the limiting cycle, as $t \to 0$, of the subvarieties Z_t. In other words, let $\Phi^\circ \subset \mathbb{A}^1 \times \mathbb{P}\mathcal{E}$ be the incidence correspondence

$$\Phi^\circ = \{(t, p) \in \mathbb{A}^1 \times \mathbb{P}\mathcal{E} \mid t \neq 0 \text{ and } p \in A_t(\tilde{Z})\};$$

let Φ be the closure of Φ° and let Z_0 be the fiber of Φ over $t = 0$.

What does Z_0 look like? Over the open subset $U \subset X$ the original cycle Z has been flattened to a multiple of the zero locus $\tau_{k-l+1} = \cdots = \tau_r = 0$. There is thus a unique component Z' of Z_0 dominating $W = \pi(Z)$, and it is the closure of the intersection of the common zero locus $\tau_{k-l+1} = \cdots = \tau_r = 0$ with the preimage $\pi^{-1}(W \cap U)$.

Now, we have arranged for $\mathcal{E}^*$ to be generated by global sections, so that the linear series $|\mathcal{O}_{\mathbb{P}\mathcal{E}}(1)|$ has no base locus. Since the τ_i are general sections of $\mathcal{O}_{\mathbb{P}\mathcal{E}}(1)$, by Bertini the common zero locus $\tau_{k-l+1} = \cdots = \tau_r = 0$ of $r - k + l$ of them intersects the subvariety $\pi^{-1}(W)$ generically transversely, in a k-dimensional subvariety of $\mathbb{P}\mathcal{E}$ with class $[W] \cdot \zeta^{r-k+l}$; moreover, since this intersection is fibered over W with fibers $\mathbb{P}^{k-l}$, it is irreducible. In sum,

$$[Z'] = m[W] \cdot \zeta^{r-k+l}$$

for some multiplicity m.

To complete the proof we note that we do not need to know what happens over the complement of $U \cap W$ in W, because any component of Z_0 not dominating W necessarily has footprint smaller than l.   □

From this description of the Chow groups we see that we can write ζ^{r+1} as a linear combination of products of (pullbacks of) classes in $A(X)$ with lower powers of ζ — that is, ζ satisfies a monic polynomial f of degree $r + 1$ over $A(X)$. Thus the ring homomorphism $A(X)[\zeta] \to A(\mathbb{P}\mathcal{E})$ factors through the quotient $A(X)[\zeta]/(f)$. Since $A(X)[\zeta]/(f) \cong \bigoplus \zeta^i A(X)$ as groups, it follows that the map $A(X)[\zeta]/(f) \to A(\mathbb{P}\mathcal{E})$ is an isomorphism of rings.

It remains to identify the polynomial f. Let $\mathcal{S} = \mathcal{O}_{\mathbb{P}\mathcal{E}}(-1)$, and let $\mathcal{Q}$ be the cokernel of the natural inclusion $\mathcal{S} \to \pi^*\mathcal{E}$, a bundle of rank r. We have an exact sequence

$$0 \longrightarrow \mathcal{S} \longrightarrow \pi^*\mathcal{E} \longrightarrow \mathcal{Q} \longrightarrow 0.$$

Identifying $A(X)$ with $\pi^*A(X)$ as before, we have

$$c(\mathcal{S}) \cdot c(\mathcal{Q}) = c(\mathcal{E})$$

by the Whitney formula (Theorem 5.3).

We defined the class ζ to be the first Chern class of the line bundle $\mathcal{O}_{\mathbb{P}\mathcal{E}}(1)$, which is the dual of $\mathcal{S}$; thus $c(\mathcal{S}) = 1 - \zeta$, and we can write this as

$$c(\mathcal{Q}) = c(\mathcal{E}) \cdot c(\mathcal{S})^{-1} = c(\mathcal{E})(1 + \zeta + \zeta^2 + \cdots).$$

Since $\mathcal{Q}$ is a vector bundle of rank r, we conclude that

$$0 = c_{r+1}(\mathcal{Q}) = \zeta^{r+1} + c_1(\mathcal{E})\zeta^r + c_2(\mathcal{E})\zeta^{r-1} + \cdots + c_r(\mathcal{E})\zeta + c_{r+1}(\mathcal{E}),$$

so the polynomial f is given by the formula in the theorem. $\square$

If $\mathcal{L}$ is a line bundle on X then Corollary 9.5 shows that $\mathbb{P}\mathcal{E} \cong \mathbb{P}(\mathcal{E} \otimes \mathcal{L})$, but the class ζ is different in the two representations; the two classes differ by multiplication with the pullback of $\mathcal{L}$. The relation between the two resulting descriptions of the Chow ring is addressed in Exercises 9.30 and 9.31.

Using Theorem 9.6, we can immediately compute the degrees of rational normal scrolls, or, more generally, of any projectivized vector bundle $\mathbb{P}\mathcal{E}$ over a curve X, embedded by $|\mathcal{O}_{\mathbb{P}\mathcal{E}}(1)|$:

Corollary 9.9. *If $a_0, \ldots, a_r$ are positive integers, then the degree of the rational normal scroll $S(a_0, \ldots, a_r)$ is $\sum a_i$. More generally, if $\mathcal{E}$ is a vector bundle on a smooth curve X and the line bundle $\mathcal{O}_{\mathbb{P}\mathcal{E}}(1)$ on $\mathbb{P}\mathcal{E}$ is very ample, then the degree of the image of $\mathbb{P}\mathcal{E}$ under the embedding given by $|\mathcal{O}_{\mathbb{P}\mathcal{E}}(1)|$ is $-\deg c_1(\mathcal{E})$.*

Note that degree and codimension of a scroll S satisfy the equation

$$\deg S = 1 + \operatorname{codim} S.$$

This is the minimal degree for any subvariety of projective space not contained in a hyperplane. The Veronese surface in $\mathbb{P}^5$, and any cone over it, also satisfy this equation, but these are the only "varieties of minimal degree." See Harris [1995, Theorem 19.19].

Proof: If the rank of $\mathcal{E}$ is $r + 1$ then the dimension of $\mathbb{P}\mathcal{E}$ is $r + 1$, so the degree of the image of $\mathbb{P}\mathcal{E}$ under the embedding given by $|\mathcal{O}_{\mathbb{P}\mathcal{E}}(1)|$ is $\deg \zeta^{r+1}$. Since X is one-dimensional, we have $c_i(\mathcal{E}) = 0$ for $i > 1$, so $\zeta^{r+1} = -c_1(\mathcal{E})$. If $X = \mathbb{P}^1$ and

$$\mathcal{E} = \mathcal{O}_{\mathbb{P}^1}(-a_0) \oplus \cdots \oplus \mathcal{O}_{\mathbb{P}^1}(-a_r),$$

then $\deg c_1(\mathcal{E}) = -\sum a_i$ and $S(a_0, \ldots, a_r)$ is the embedding of $\mathbb{P}\mathcal{E}$ by $\mathcal{O}_{\mathbb{P}\mathcal{E}}(1)$. $\square$

9.3.1 The universal k-plane over $\mathbb{G}(k, n)$

In this section and the next, we will use Theorem 9.6 to give a description of the Chow ring of some varieties that arise often in algebraic geometry: the universal k-plane over the Grassmannian $\mathbb{G}(k, n)$ and the blow-up of $\mathbb{P}^n$ along a linear space.

For the first of these, let $G = \mathbb{G}(k, \mathbb{P}V)$ be the Grassmannian parametrizing k-planes $\Lambda \subset \mathbb{P}V$ in the projectivization of an $(n + 1)$-dimensional vector space V, and let Φ be the universal plane

$$\Phi = \{(\Lambda, p) \in G \times \mathbb{P}V \mid p \in \Lambda\},$$

initially introduced in Section 3.2.3. We can recognize Φ, via the projection $\pi : \Phi \to G$ on the first factor, as the projectivization $\mathbb{P}\mathcal{S}$ of the universal subbundle on G, and use Theorem 9.6 to describe $A(\Phi)$. We will use the notation introduced above: We will identify $A(G)$ with its image in $A(\Phi)$ via the pullback map π^*, and denote the first Chern class of the tautological bundle $\mathcal{O}_{\mathbb{P}\mathcal{S}}(1)$ by $\zeta \in A^1(\Phi)$.

Note that a linear form $l \in V^*$ on V gives rise to a section of $\mathcal{S}^*$ by restriction in turn to each subspace of V, hence to a section of $\pi^*\mathcal{S}^*$, and ultimately to a section of $\mathcal{O}_{\mathbb{P}\mathcal{S}}(1)$ via the surjection $\pi^*\mathcal{S}^* \to \mathcal{O}_{\mathbb{P}\mathcal{S}}(1)$ dual to the tautological inclusion $\mathcal{O}_{\mathbb{P}\mathcal{S}}(-1) \hookrightarrow \pi^*\mathcal{S}$. Simply put, if we think of $\Phi = \mathbb{P}\mathcal{S}$ as the variety of pairs $(\tilde{\Lambda}, \xi)$ with $\tilde{\Lambda} \subset V$ a $(k + 1)$-dimensional subspace and $\xi \subset \tilde{\Lambda}$ a one-dimensional subspace, then we can define a section σ_l of $\mathcal{O}_{\mathbb{P}\mathcal{S}}(1)$ by setting

$$\sigma_l(\tilde{\Lambda}, \xi) = l|_\xi.$$

In particular, we see that the zero locus of the section σ_l is just the locus of $(\tilde{\Lambda}, \xi)$ such that ξ is contained in the hyperplane $\mathrm{Ker}(l) \subset V$, and hence *the tautological class $\zeta = c_1(\mathcal{O}_{\mathbb{P}\mathcal{S}}(1)) \in A^1(\Phi)$ is just the pullback of the hyperplane class on $\mathbb{P}V$ via the projection map $\eta : \Phi \to \mathbb{P}V$ on the second factor.*

Recalling the calculation of the Chern classes of the universal bundles on $\mathbb{G}(k, n)$ from Section 5.6.2 and applying Theorem 9.6, we conclude:

Proposition 9.10. *Let $G = \mathbb{G}(k, n)$ be the Grassmannian of k-planes in $\mathbb{P}^n$ and $\Phi \subset G \times \mathbb{P}^n$ the universal k-plane as above, with $\pi : \Phi \to G$ and $\eta : \Phi \to \mathbb{P}^n$ the projection maps. We have then*

$$A(\Phi) = A(G)[\zeta]/(\zeta^{k+1} - \sigma_1\zeta^k + \sigma_{1,1}\zeta^{k-1} + \cdots + (-1)^{k+1}\sigma_{1,1,\ldots,1}),$$

where $\zeta \in A^1(\Phi)$ is the tautological class, or equivalently the pullback via η of the hyperplane class in $\mathbb{P}^n$.

The two special cases occurring most commonly are the cases $k = n - 1$ of the universal hyperplane and the case $k = 1$ of the universal line. In the first case,

$$\Phi = \{(H, p) \in \mathbb{P}^{n*} \times \mathbb{P}^n \mid p \in H\},$$

and if we let ω be pullback to Φ of the hyperplane class in $\mathbb{P}^{n*}$, we have

$$A(\Phi) = \mathbb{Z}[\omega, \zeta]/(\omega^{n+1}, \zeta^{n+1}, \zeta^n - \omega\zeta^{n-1} + \cdots + (-1)^n\omega^n).$$

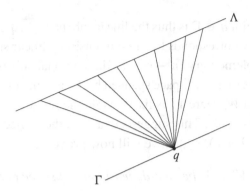

Figure 9.2 The fiber over a point under the projection of $\mathbb{P}^3$ from the line Λ.

We have written the ideal of relations in this way to emphasize the symmetry, but it is redundant: we could drop either ω^{n+1} or ζ^{n+1}. Note that when $a + b = \dim(\Phi) = 2n - 1$, we have

$$\deg(\omega^a \zeta^b) = \begin{cases} 1 & \text{if } (a, b) = (n, n - 1) \text{ or } (n - 1, n), \\ 0 & \text{otherwise,} \end{cases}$$

which we could also see from the fact that $\Phi \subset \mathbb{P}^{n*} \times \mathbb{P}^n$ is a hypersurface of bidegree $(1, 1)$.

The universal line will also come up a lot in the following chapters; in this case we have

$$A(\Phi) = A(\mathbb{G}(1, n))[\zeta]/(\zeta^2 - \sigma_1 \zeta + \sigma_{1,1}).$$

We will leave it to the reader to calculate the degrees of monomials $\sigma_1^a \sigma_{1,1}^b \zeta^c$ of top degree $a + 2b + c = \dim(\Phi) = 2n - 1$ in Exercise 9.33.

9.3.2 The blow-up of $\mathbb{P}^n$ along a linear space

In Section 2.1.9 we saw how to describe the Chow ring of the blow-up of projective space at a point. We can now analyze much more generally and systematically the Chow ring of the blow-up $Z = \mathrm{Bl}_\Lambda \mathbb{P}^n$ of projective space $\mathbb{P}^n = \mathbb{P}V$ along any linear subspace $\Lambda \cong \mathbb{P}^{r-1}$. The key is to realize Z as the total space of a projective bundle.

To understand the picture, first recall that the blow-up is the graph of the rational map $\pi_\Lambda : \mathbb{P}^n \dashrightarrow \mathbb{P}^{n-r}$ given by projection from Λ. Thus $Z \subset \mathbb{P}^n \times \mathbb{P}^{n-r}$. We will show that the projection $Z \to \mathbb{P}^{n-r}$ to the second factor makes Z into a projective bundle. Certainly, each fiber of the projection is an r-dimensional projective space (see Figure 9.2). Concretely, if we choose an $(n - r)$-plane $\Gamma \subset \mathbb{P}^n$ disjoint from Λ, we can write

$$Z = \{(p, q) \in \mathbb{P}^n \times \Gamma \mid p \in \overline{\Lambda, q}\}.$$

The fiber over a point $q \in \Gamma$ is thus the linear subspace $\overline{\Lambda, q} \cong \mathbb{P}^{n-r+1} \subset \mathbb{P}^n$. If we write $\mathbb{P}^n$ as $\mathbb{P}V$, then Λ corresponds to an r-dimensional linear subspace $V' \subset V$ and Γ corresponds to a complementary $(n - r + 1)$-dimensional subspace W. The fiber of Z over $q \in \Gamma$ corresponds to the subspace spanned by V' and the one-dimensional subspace $\tilde{q}$ corresponding to q in W. Here V' is fixed, while the one-dimensional subspace varies over all such subspaces of W. This suggests that Z is the projectivization of the bundle $\mathcal{O}_{\mathbb{P}^{n-r}}(-1) \oplus (V' \otimes \mathcal{O}_{\mathbb{P}^{n-r}})$, which we will now prove:

Proposition 9.11. *Let $V' \subset V$ be an r-dimensional subspace of an $(n + 1)$-dimensional vector space V, and let*

$$\mathcal{E} = \mathcal{O}_{\mathbb{P}^{n-r}}(-1) \oplus (V' \otimes \mathcal{O}_{\mathbb{P}^{n-r}}),$$

so that $\mathcal{E}$ is a vector bundle of rank $r + 1$ on $\mathbb{P}^{n-r} = \mathbb{P}(V/V')$. The blow-up Z of $\mathbb{P}(V)$ along the $(r - 1)$-dimensional subspace $\mathbb{P}(V')$, together with its projection to $\mathbb{P}^{n-r}$, is isomorphic to the projective bundle $\pi : \mathbb{P}\mathcal{E} \to \mathbb{P}^{n-r}$. Under this isomorphism, the blow-up map $Z \to \mathbb{P}^n$ corresponds to the complete linear series $|\mathcal{O}_{\mathbb{P}\mathcal{E}}(1)|$.

Proof: Choose a complement $V/V' \cong W \subset V$ to V', so that $V = W \oplus V'$. With $\mathcal{E}$ as in the proposition, the natural inclusion $\mathcal{O}_{\mathbb{P}W}(-1) \subset (W \otimes \mathcal{O}_{\mathbb{P}W})$ induces an inclusion

$$\mathcal{E} \subset (W \otimes \mathcal{O}_{\mathbb{P}W}) \oplus (V' \otimes \mathcal{O}_{\mathbb{P}W}) = V \otimes \mathcal{O}_{\mathbb{P}W}.$$

The dual map, which is a surjection, induces an isomorphism $V^* \to H^0(\mathcal{E}^*) = V'^* \oplus W^*$. Thus $\mathcal{E}^*$ is generated by its global sections and the complete linear series $|\mathcal{O}_{\mathbb{P}\mathcal{E}}(1)|$ corresponds to a map $\mathbb{P}\mathcal{E} \to \mathbb{P}V$.

The fiber of $\mathcal{E}$ over a point $q \in \mathbb{P}W$ is, as a subspace of V, equal to $V' \oplus \tilde{q}$, whose projectivization is the fiber over q of the blow-up Z of $\mathbb{P}V'$ in $\mathbb{P}V$. Thus, together with the projection map $\pi : \mathbb{P}\mathcal{E} \to \mathbb{P}W$, we get a closed immersion $\varphi : \mathbb{P}\mathcal{E} \to \mathbb{P}V \times \mathbb{P}W$ that maps the fiber of $\mathbb{P}\mathcal{E}$ isomorphically to Z. $\qquad \square$

Corollary 9.12. *Let $Z \subset \mathbb{P}^n \times \mathbb{P}^{n-r}$ be the blow-up of an $(r - 1)$-plane Λ in $\mathbb{P}^n$. Writing $\alpha, \zeta \in A^1(Z)$ for the pullbacks of the hyperplane classes on $\mathbb{P}^{n-r}$ and $\mathbb{P}^n$ respectively, we have*

$$A(Z) = \mathbb{Z}[\alpha, \zeta]/(\alpha^{n-r+1}, \zeta^{r+1} - \alpha \zeta^r).$$

With this notation the class of the exceptional divisor $E \subset Z$, the preimage of Λ in Z, is

$$[E] = \zeta - \alpha.$$

Proof: The Chern class of $\mathcal{E} = \mathcal{O}_{\mathbb{P}^{n-r}}(-1) \oplus (V' \otimes \mathcal{O}^r_{\mathbb{P}^{n-r}})$ is $1 - \alpha$, so the formula for $A(Z)$ follows at once from Theorem 9.6. Since ζ is the class of the preimage of a hyperplane $H \subset \mathbb{P}^n$ (which could contain Λ), and α is represented by the proper transform of a hyperplane containing Λ, we have $[E] = \zeta - \alpha$ as claimed. $\qquad \square$

For example, in the case of the blow-up of the plane at a point we have

$$[E]^2 = (\zeta - \alpha)^2 = \zeta^2 - 2\alpha\zeta + \alpha^2 = -\zeta^2,$$

that is, minus the class of a point, as we already knew. But we can now compute $\deg[E]^n$ in general (Exercise 9.38).

9.3.3 Nested pairs of divisors on $\mathbb{P}^1$ revisited

We start by introducing two vector bundles that arise often in studying the geometry of rational curves; in particular, they will be a central object of study in Section 10.4.2.

To begin with, let $\mathbb{P}^d = \mathbb{P}H^0(\mathcal{O}_{\mathbb{P}^1}(d))$ be the projective space of polynomials of degree d in two variables modulo scalars — that is, divisors of degree d on $\mathbb{P}^1$. For any $e \geq d$, then, we can define a vector bundle $\mathcal{F}$ on $\mathbb{P}^d$ informally by associating to each divisor $D \in \mathbb{P}^d$ the vector space

$$\mathcal{F}_D = H^0(\mathcal{I}_D(e))$$

of polynomials of degree e on $\mathbb{P}^1$ vanishing on D. Similarly, we can define a bundle $\mathcal{E}$ on $\mathbb{P}^d$ informally by associating to each divisor $D \in \mathbb{P}^d$ the quotient vector space

$$\mathcal{E}_D = H^0(\mathcal{O}_{\mathbb{P}^1}(e))/H^0(\mathcal{I}_D(e)) = H^0(\mathcal{O}_D(e))$$

of polynomials of degree e modulo those vanishing on D. To define these bundles precisely, let $\mathcal{D} \subset \mathbb{P}^d \times \mathbb{P}^1$ be the universal divisor of degree d, that is

$$\mathcal{D} = \{(D, p) \in \mathbb{P}^d \times \mathbb{P}^1 \mid p \in D\},$$

and let $\mu : \mathbb{P}^d \times \mathbb{P}^1 \to \mathbb{P}^d$ and $\nu : \mathbb{P}^d \times \mathbb{P}^1 \to \mathbb{P}^1$ be the projection maps. We can then take

$$\mathcal{F} = \mu_*(\nu^*\mathcal{O}_{\mathbb{P}^1}(e) \otimes \mathcal{I}_{\mathcal{D}})$$

and

$$\mathcal{E} = \alpha_*(\nu^*\mathcal{O}_{\mathbb{P}^1}(e) \otimes \mathcal{O}_{\mathcal{D}});$$

an application of the theorem on cohomology and base change shows that these have the fibers indicated, and that the exact sequence of sheaves on $\mathbb{P}^d \times \mathbb{P}^1$

$$0 \longrightarrow \mathcal{I}_{\mathcal{D}} \longrightarrow \mathcal{O}_{\mathbb{P}^d \times \mathbb{P}^1} \longrightarrow \mathcal{O}_{\mathcal{D}} \longrightarrow 0,$$

tensored with the line bundle $\nu^*\mathcal{O}_{\mathbb{P}^1}(e)$ and pushed forward to $\mathbb{P}^d$, gives the expected exact sequence

$$0 \longrightarrow \mathcal{F} \longrightarrow H^0(\mathcal{O}_{\mathbb{P}^1}(e)) \otimes \mathcal{O}_{\mathbb{P}^d} \longrightarrow \mathcal{E} \longrightarrow 0 \qquad (9.1)$$

of bundles on $\mathbb{P}^d$.

Consider now the projectivization $\Phi = \mathbb{P}\mathcal{E}$ of the bundle $\mathcal{E}$. This is a variety we have encountered before, in Section 2.1.8: We can realize it as the subvariety

$$\Phi = \{(D, E) \in \mathbb{P}^d \times \mathbb{P}^e \mid E \geq D\}$$

of *nested pairs* of divisors of degrees d and e on $\mathbb{P}^1$. Moreover, under the inclusion of $\Phi = \mathbb{P}\mathcal{E}$ in $\mathbb{P}^d \times \mathbb{P}^e$, the pullback τ of the hyperplane class from $\mathbb{P}^e$ restricts to the tautological class $\zeta = c_1(\mathcal{O}_{\mathbb{P}\mathcal{E}}(1))$ on $\mathbb{P}\mathcal{E}$.

We can use this to describe the Chow ring of Φ, and correspondingly the Chern classes of $\mathcal{E}$. The key, as it was in Section 2.1.8, is to observe that $\Phi \cong \mathbb{P}^d \times \mathbb{P}^{e-d}$ abstractly, via the map

$$\alpha : \mathbb{P}^d \times \mathbb{P}^{e-d} \to \mathbb{P}^d \times \mathbb{P}^e, \quad (D, D') \mapsto (D, D + D').$$

Let σ, τ and μ be the pullbacks of the hyperplane classes on $\mathbb{P}^d$, $\mathbb{P}^e$ and $\mathbb{P}^{e-d}$, respectively. As we saw in Section 2.1.8, the pullback of the class τ to Φ is the sum $\sigma + \mu$. We can then rewrite the relation $\mu^{e-d+1} = 0$ in $A(\mathbb{P}\mathcal{E})$ as

$$0 = (\zeta - \sigma)^{e-d+1} = \sum (-1)^i \binom{e-d+1}{i} \sigma^i \zeta^{e-d+1-i},$$

and we conclude that

$$c_i(\mathcal{E}) = (-1)^i \binom{e-d+1}{i} \sigma^i.$$

To express this more compactly, we can write the total Chern class as

$$c(\mathcal{E}) = (1 - \sigma)^{e-d+1}.$$

In this form, it follows from the exact sequence (9.1) that

$$c(\mathcal{F}) = \frac{1}{(1-\sigma)^{e-d+1}} = \sum \binom{e-d+i}{i} \sigma^i,$$

so we have the Chern classes of $\mathcal{F}$ as well.

9.4 Projectivization of a subbundle

If $\mathcal{E}$ is a vector bundle on a smooth variety X and $\mathcal{F} \subset \mathcal{E}$ a subbundle then $\mathbb{P}\mathcal{F}$ is naturally a subvariety of $\mathbb{P}\mathcal{E}$, and we can ask for its class in the Chow ring $A(\mathbb{P}\mathcal{E})$. This will be a crucial element in understanding the Chow ring of a blow-up in general (Section 13.6); for now, it will allow us to answer Keynote Question (b).

Let $\pi : \mathbb{P}\mathcal{E} \to X$ be the projection and let $\mathcal{O}_{\mathbb{P}\mathcal{E}}(-1) \subset \pi^*\mathcal{E}$ be the universal subbundle. A point $p \in \mathbb{P}\mathcal{E}$ lying over a point $x \in X$ corresponds to the one-dimensional space that is the fiber of $\mathcal{O}_{\mathbb{P}\mathcal{E}}(-1)$ at p. Thus $p \in \mathbb{P}\mathcal{F}$ if and only if this space is contained in the fiber of $\mathcal{F}$. In other words, $p \in \mathbb{P}\mathcal{F}$ if and only if the composite map

$$\varphi : \mathcal{O}_{\mathbb{P}\mathcal{E}}(-1) \to \pi^*\mathcal{E} \to \pi^*(\mathcal{E}/\mathcal{F})$$

vanishes at p. We can view φ as a global section of the bundle

$$\mathcal{H}om(\mathcal{O}_{\mathbb{P}\mathcal{E}}(-1), \pi^*(\mathcal{E}/\mathcal{F})) \cong \mathcal{O}_{\mathbb{P}\mathcal{E}}(1) \otimes \pi^*(\mathcal{E}/\mathcal{F}).$$

If we write everything in local coordinates then we see that $\mathbb{P}\mathcal{F}$ is scheme-theoretically defined by the vanishing of φ. Since the codimension of $\mathbb{P}\mathcal{F}$ is the same as the rank of $\mathcal{E}/\mathcal{F}$, it follows that $[\mathbb{P}\mathcal{F}] \in A(\mathbb{P}\mathcal{E})$ is given by a Chern class, which we can compute using the formula for the Chern class of the tensor product of a bundle with a line bundle (Proposition 5.17):

Proposition 9.13. *If X is a smooth projective variety and $\mathcal{F} \subset \mathcal{E}$ are vector bundles on X of ranks s and r respectively, then*

$$[\mathbb{P}\mathcal{F}] = c_{r-s}(\mathcal{O}_{\mathbb{P}\mathcal{E}}(1) \otimes \pi^*(\mathcal{E}/\mathcal{F}))$$
$$= \zeta^{r-s} + \gamma_1 \zeta^{r-s-1} + \cdots + \gamma_{r-s} \in A^{r-s}(\mathbb{P}\mathcal{E}),$$

where $\zeta = c_1(\mathcal{O}_{\mathbb{P}\mathcal{E}}(1))$ and $\gamma_k = c_k(\mathcal{E}/\mathcal{F})$. Moreover, the normal bundle of $\mathbb{P}\mathcal{F}$ in $\mathbb{P}\mathcal{E}$ is $\mathcal{O}_{\mathbb{P}\mathcal{E}}(1) \otimes \pi^*(\mathcal{E}/\mathcal{F})$.

This formula will be useful to us in many settings; for an immediate application, see Exercises 9.43 and 9.44.

An important reason to consider projectivized subbundles is suggested by the following characterization of sections. Giving a section — that is, a map $\alpha : X \to \mathbb{P}\mathcal{E}$ such that $\pi \circ \alpha$ is the identity — is the same as giving the image of the section; and we will therefore refer to the image as a section as well.

Proposition 9.14. *If $\mathcal{L} \subset \mathcal{E}$ is a line subbundle of a vector bundle $\mathcal{E}$ on a variety X, then $\mathbb{P}\mathcal{L} \subset \mathbb{P}\mathcal{E}$ is the image of a section $X \to \mathbb{P}\mathcal{E}$ of the projection $\mathbb{P}\mathcal{E} \to X$, and every section has this form.*

Informally: giving a section is the same as specifying point of $\mathbb{P}\mathcal{E}$ over each point of X, that is, giving a one-dimensional subspace of each fiber of $\mathcal{E}$.

Proof: By the universal property of $\pi : \mathbb{P}\mathcal{E} \to X$, giving a map $\alpha : X \to \mathbb{P}\mathcal{E}$ that "commutes with" the identity map $X \to X$ is the same as giving a line subbundle of $\mathcal{E}$. $\square$

9.4.1 Ruled surfaces

Recall that a *ruled surface* is by definition the projectivization of a vector bundle of rank 2 over a smooth curve. We can now answer Keynote Question (b):

Proposition 9.15. *A ruled surface can contain at most one irreducible and reduced curve of negative self-intersection.*

Proof: Let X be a smooth curve, let $\pi : \mathbb{P}\mathcal{E} \to X$ be a ruled surface, and suppose that $C_1, C_2 \subset \mathbb{P}\mathcal{E}$ are two irreducible curves of strictly negative self-intersection. A fiber $\pi^{-1}(x)$ satisfies $[\pi^{-1}(x)]^2 = \pi^*([x]^2) = 0$, so the induced maps $\pi : C_i \to X$ are finite. Let $C_1' \to C_1$ be the normalization of C_1, and let $\alpha : C_1' \to C_1 \subset X$ be the corresponding map. Consider the pullback diagram

$$\begin{array}{ccc} \mathbb{P}\alpha^*\mathcal{E} = C_1' \times_X \mathbb{P}\mathcal{E} & \xrightarrow{\ \beta\ } & \mathbb{P}\mathcal{E} \\ \downarrow & & \downarrow{\scriptstyle \pi} \\ C_1' & \xrightarrow[\ \alpha\ ]{} & X \end{array}$$

The preimage $\beta^{-1}(C_1) = C_1' \times_X C_1$ represents a cycle $m\Sigma_1 + D_1$, where Σ_1 is a section, D_1 has no component in common with Σ_1 and $m > 0$. Hence

$$\begin{aligned} m^2 \deg[\Sigma_1]^2 &= \deg[\Sigma_1][\beta^*C_1] - \deg[\Sigma_1][D_1] \\ &\leq \deg[\Sigma_1][\beta^*C_1] \\ &= \deg[\beta_*\Sigma_1][C_1] \\ &= \deg[C_1]^2, \end{aligned}$$

so $\deg[\Sigma_1]^2 < 0$.

Since a section pulls back to a section with the same self-intersection, we can repeat the process with a component of $\beta^{-1}C_2$ to obtain two sections Σ_1 and Σ_2 of negative self-intersection. We can analyze this case using Proposition 9.14. Suppose that $\Sigma_i = \mathbb{P}\mathcal{L}_i \subset \mathbb{P}\mathcal{E}$.

By Theorem 9.6, we have

$$A(\mathbb{P}\mathcal{E}) = A(X)[\zeta]/(\zeta^2 + c_1(\mathcal{E})\zeta),$$

where $\zeta = c_1(\mathcal{O}_{\mathbb{P}\mathcal{E}}(1))$. Now $\deg(c_1(\mathcal{E})\zeta) = \deg \pi_*(c_1(\mathcal{E})\zeta) = \deg c_1(\mathcal{E})$ because ζ meets each fiber of π in degree 1. It then follows that $\deg \zeta^2 = -\deg c_1(\mathcal{E})$. By Proposition 9.13,

$$[\Sigma_i] = \zeta + c_1(\mathcal{E}) - c_1(\mathcal{L}_i),$$

so

$$0 > \deg[\Sigma_i]^2 = \deg \zeta^2 + 2 \deg c_1(\mathcal{E}) - 2 \deg \mathcal{L}_i.$$

Thus $2 \deg \mathcal{L}_i > \deg c_1(\mathcal{E})$. (Exercise 9.50 strengthens this conclusion slightly.)

Supposing now that $\Sigma_1 \neq \Sigma_2$, we get an exact sequence

$$0 \longrightarrow \mathcal{L}_1 \oplus \mathcal{L}_2 \longrightarrow \mathcal{E} \longrightarrow \mathcal{G} \longrightarrow 0,$$

where $\mathcal{G}$ is a sheaf with finite support; it follows that $\deg \mathcal{E} \geq \deg \mathcal{L}_1 + \deg \mathcal{L}_2 > \deg \mathcal{E}$, a contradiction. $\qquad\square$

By contrast, it is possible for a (nonruled) smooth projective surface to contain infinitely many irreducible curves of negative self-intersection; Exercises 9.45–9.47 show how to construct an example. It is an open problem (in characteristic 0) whether the self-intersections of irreducible curves on a surface S are bounded below, that is, whether a surface can contain a sequence $C_1, C_2, \dots$ of irreducible curves with $\deg(C_n \cdot C_n) \to -\infty$. (In characteristic $p > 0$, János Kollár has shown us an example, described in Exercise 9.49.)

9.4.2 Self-intersection of the zero section

The total space of a vector bundle $\mathcal{E}$ on a scheme X may itself be considered as a scheme $\mathbb{A}\mathcal{E} := \mathrm{Spec}(\mathrm{Sym}\,\mathcal{E}^*)$ over X. For various purposes it is useful to have a compactification of $\mathbb{A}\mathcal{E}$, that is, a variety proper over X that includes $\mathbb{A}\mathcal{E}$ as an open subset, and we will describe the simplest such construction here.

It is natural to try to compactify each fiber by putting it inside a projective space of the same dimension, and we can do this globally by taking the projectivization of the direct sum $\mathcal{E} \oplus \mathcal{O}_X$; that is, we set

$$\overline{\mathcal{E}} := \mathbb{P}(\mathcal{E} \oplus \mathcal{O}_X).$$

Let r be the rank of $\mathcal{E}$. Since $c(\mathcal{E} \oplus \mathcal{O}_X) = c(\mathcal{E})$, we have

$$A(\overline{\mathcal{E}}) = A(X)[\zeta]/(\zeta^{r+1} + c_1(\mathcal{E})\zeta^r + \cdots + c_r(\mathcal{E})\zeta).$$

In terms of coordinates, $\mathbb{A}\mathcal{E} \subset \overline{\mathcal{E}}$ is "the locus where the last coordinate is nonzero." Its complement is the divisor $\mathbb{P}\mathcal{E} \subset \mathbb{P}(\mathcal{E} \oplus \mathcal{O}_X)$, which we therefore call the "hyperplane at infinity." Since this is the locus where the section of $\mathcal{O}_{\overline{\mathcal{E}}}(1)$ corresponding to $1 \in \mathcal{O}_X \subset (\mathcal{E} \oplus \mathcal{O}_X)^*$ vanishes, we get

$$\zeta := c_1(\mathcal{O}_{\overline{\mathcal{E}}}(1)) = [\mathbb{P}\mathcal{E}].$$

(One can also see this from Proposition 9.13.)

The section $\mathbb{P}\mathcal{O}_X \subset \overline{\mathcal{E}}$ is the locus where all the coordinates in $\mathcal{E}^*$ vanish; it is thus the zero section of $\mathbb{A}\mathcal{E}$, which we will call Σ_0. By Proposition 9.13, we have $[\Sigma_0] = \zeta^r + c_1(\mathcal{E})\zeta^{r-1} + \cdots + c_{r-1}(\mathcal{E})\zeta + c_r(\mathcal{E})$. More generally, if τ is a global section of $\mathcal{E}$, then $(\tau, 1)$ is a nowhere-vanishing section of $\mathcal{E} \oplus \mathcal{O}_X$, and the line subbundle it generates corresponds to a section of $\overline{\mathcal{E}}$, which we will call Σ_τ. Using Proposition 9.13 or the family $\Sigma_{t\tau}$, which gives a rational equivalence between Σ_τ and Σ_0, we see that $[\Sigma_\tau] = [\Sigma_0]$. If τ vanishes in codimension r, then

$$\pi_*([\Sigma_0]^2) = \pi_*([[\Sigma_0][\Sigma_\tau]]) = [(\tau)_0] = c_r(\mathcal{E}).$$

We claim that this formula holds in general:

Proposition 9.16. *Let $\mathcal{E}$ be a vector bundle of rank r on a smooth variety X, and let $\pi : \overline{\mathcal{E}} = \mathbb{P}(\mathcal{E} \oplus \mathcal{O}_X) \to X$ be the projection. Let $\iota : X \to \mathbb{A}(\mathcal{E}) \subset \overline{\mathcal{E}}$ be the zero section, with image $\Sigma_0 = \mathbb{P}(\mathcal{O}_X)$. We have*

$$\pi_*([\Sigma_0]^2) = c_r(\mathcal{E}),$$

and, for any class $\alpha \in A(X)$,

$$\iota^* \iota_* \alpha = \alpha c_r(\mathcal{E}).$$

Proof: By Proposition 9.13,

$$[\Sigma_0] = \zeta^r + c_1(\mathcal{E})\zeta^{r-1} + \cdots + c_{r-1}(\mathcal{E})\zeta + c_r(\mathcal{E}).$$

Since Σ_0 is disjoint from the hyperplane at infinity $\mathbb{P}\mathcal{E} \subset \mathbb{P}(\mathcal{E} \oplus \mathcal{O}_X)$, which has class ζ, we get $[\Sigma_0]\zeta = 0$. (This also follows from the computation of $A(\overline{\mathcal{E}})$.) Thus

$$\begin{aligned}[\Sigma_0]^2 &= [\Sigma_0](\zeta^r + c_1(\mathcal{E})\zeta^{r-1} + \cdots + c_{r-1}(\mathcal{E})\zeta + c_r(\mathcal{E})) \\ &= [\Sigma_0]c_r(\mathcal{E}) \in A(\overline{\mathcal{E}}).\end{aligned}$$

From the push-pull formula we get $\pi_*([\Sigma_0]^2) = (\pi_*[\Sigma_0])c_r(\mathcal{E}) = c_r(\mathcal{E})$, proving the first assertion.

For the second assertion, we use the fact that π induces an isomorphism from Σ_0 to X, and thus $\iota^*\beta = \pi_*(\beta \cap [\Sigma_0])$ for any cycle β on $\overline{\mathcal{E}}$. Thus

$$\iota^* \iota_* \alpha = \iota^*(\iota_*\alpha[\Sigma_0]) = \pi_*(\iota_*\alpha[\Sigma_0]^2) = \alpha c_r(\mathcal{E}),$$

as required. $\qquad\square$

See Theorem 13.7 for a generalization.

9.5 Brauer–Severi varieties

We defined a projective bundle to be a morphism $\pi : Y \to X$ that is isomorphic to a product with projective space over Zariski open subsets covering the target X. Interestingly, if we had weakened the condition to saying that π was a product locally in the étale, or analytic, topology on X, we would get in general a larger class of morphisms! In this section, we will illustrate the difference with an example of a morphism that satisfies the weaker condition but not the stronger.

We start with a definition: A *Brauer–Severi variety* over a variety X is a variety Y together with a proper, smooth map $\pi : Y \to X$ such that all the (scheme-theoretic) fibers of π are isomorphic to $\mathbb{P}^r$, for some fixed r. Thus any projective bundle $\pi : Y \to X$ is a Brauer–Severi variety. But, as we will see, the converse is false.

It is in fact the case that such a morphism π will be trivial locally in the étale (or, in case the ground field is $\mathbb{C}$, the analytic) topology, in the sense that every point $x \in X$ will have an étale or analytic neighborhood U such that $\pi^{-1}(U) \cong U \times \mathbb{P}^r$. This is a

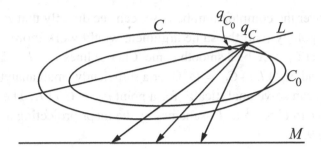

Figure 9.3 Local analytic triviality of the universal family of conics in the plane: $Y|_U \cong C \in U \times \mathbb{P}^1$ via projection from $q_C \in C \in U$.

consequence of the fact that $\mathbb{P}^r$ has no nontrivial deformations. But it may not be trivial locally in the Zariski topology. Here is an example:

Example 9.17. Let $\mathbb{P}^5$ be the space of conics in $\mathbb{P}^2 = \mathbb{P}V$, and consider the universal conic

$$\Phi = \{(C, p) \in \mathbb{P}^5 \times \mathbb{P}^2 \mid p \in C\} \xrightarrow{\;\pi_2\;} \mathbb{P}^2$$
$$\downarrow{\scriptstyle \pi_1}$$
$$\mathbb{P}^5$$

with its projections π_i to the two factors. We can realize Φ as the total space of a $\mathbb{P}^4$-bundle over $\mathbb{P}^2$ via π_2: Indeed, Φ is the projectivization of the rank-5 subbundle $\mathcal{E} \subset \mathrm{Sym}^2 V^*$ whose fiber $\mathcal{E}_p$ at a point p is the subspace of quadratic polynomials vanishing at p. (In particular, Φ is smooth.) In these terms, the tautological class $\zeta = c_1(\mathcal{O}_{\mathbb{P}\mathcal{E}}(1)) \in A^1(\Phi)$ is the pullback of the hyperplane class $\pi_1^*(\mathcal{O}_{\mathbb{P}^5}(1))$. By Theorem 9.6, the divisor class group $A^1(\Phi) \cong \mathbb{Z}^2$ is generated by the pullbacks of the hyperplane classes from $\mathbb{P}^2$ and $\mathbb{P}^5$. Note that these classes restrict to classes of degrees 2 and 0 on any fiber of π_1. Thus *the intersection of the fiber of π_1 with any divisor on Φ has even degree.*

We now consider the projection π_1. To obtain a map whose fibers are all isomorphic to $\mathbb{P}^1$, we let $X \subset \mathbb{P}^5$ be the open subset corresponding to smooth conics and let $\pi : Y = \Phi_X \to X$ be the restriction of π_1 to the preimage of X in Φ. By definition, the fibers of π are smooth conics, and in particular isomorphic to $\mathbb{P}^1$, so Φ_X is a Brauer–Severi variety over X.

We claim that $\pi : Y \to X$ is not a projective bundle. Indeed, if there were a nonempty Zariski open $U \subset X \subset \mathbb{P}^5$ such that $\pi : Y_U \to U$ were isomorphic to the projection to U of the product $U \times \mathbb{P}^1$, then we could take a section of Y_U and take its closure in Φ, obtaining a divisor in Φ meeting the general fiber of $\Phi \to \mathbb{P}^5$ in a reduced point. This contradicts the computation above. Thus $\pi : Y \to X$ is not a projective bundle.

If we work over the complex numbers, we can see directly that π is locally trivial in the analytic topology (and the same argument would work more generally for the étale topology). Let $C_0 \in X$ be a smooth conic. Choose lines $L, M \subset \mathbb{P}^2$ such that L is transverse to C_0 and $M \cap L \cap C_0 = \varnothing$. Over a sufficiently small analytic neighborhood U of $C_0 \in X$ we can solve analytically for a point $q_C \in C \cap L$. The restriction of Y to U is isomorphic to $U \times \mathbb{P}^1$ as U-schemes by the maps projecting a fiber C from q_C to M (see Figure 9.3).

The conclusion of this example may be interpreted as a theorem in polynomial algebra: It says that *there does not a exist a rational solution to the general quadratic polynomial*. In other words, there do not exist rational functions $X(a, b, c, d, e, f)$, $Y(a, b, c, d, e, f)$ and $Z(a, b, c, d, e, f)$ such that

$$aX^2 + bY^2 + cZ^2 + dXY + eXZ + fYZ \equiv 0.$$

This is a generalization of the statement that the roots of a quadratic polynomial in one variable are not expressible as rational functions of its coefficients, though much stronger: Polynomials in several variables have many more solutions than polynomials in one variable! The same is true of polynomials of any degree $d > 1$ in any number of variables (Exercise 9.51).

The set of Brauer–Severi varieties over a given variety X, modulo an equivalence relation that makes the projective bundles trivial, can be given the structure of a group, called the *Brauer group* of X. There is another avatar of this group, as the group of *Azumaya algebras* over $\mathcal{O}_X$ modulo those that are the endomorphism algebras of vector bundles. Understanding the Brauer groups of varieties is an important goal of arithmetic geometry. See for example Artin [1982] for more about Brauer–Severi varieties, and Grothendieck [1966a] or Serre [1979] for more on the Brauer group.

9.6 Chow ring of a Grassmannian bundle

Suppose that X is any smooth variety and $\mathcal{E}$ is a vector bundle of rank n on X. Generalizing the projective bundle associated to $\mathcal{E}$, we can form the *Grassmannian bundle* $G(k, \mathcal{E})$ of k-planes in the fibers of $\mathcal{E}$; that is,

$$G(k, \mathcal{E}) = \{(x, V) \mid x \in X, V \subset \mathcal{E}_x\} \xrightarrow{\pi} X.$$

(As with a single Grassmannian, we can realize $G(k, \mathcal{E})$ as a subvariety of the projectivization $\mathbb{P}(\wedge^k \mathcal{E})$.) There is a description of the Chow ring of $G(k, \mathcal{E})$ that extends both the description of the Chow ring of a projective bundle above and the description of the Chow ring of $G(k, n)$ given in Theorem 5.26; we will explain it here without proof.

As in the projective bundle case, there is a *tautological subbundle* $\mathcal{S} \subset \pi^* \mathcal{E}$ defined on $G(k, \mathcal{E})$; this is a rank-k bundle whose fiber over a point (x, V) is the vector space

$V \subset \mathcal{E}_x$. Let $\mathcal{Q} = \pi^*(\mathcal{E})/\mathcal{S}$ be the *tautological quotient bundle*. As in the case of projective bundles, the Chow ring $A(G(k, \mathcal{E}))$ is generated as an $A(X)$-algebra by the Chern classes $c_i(\mathcal{S})$, and also by the classes $c_i(\mathcal{Q})$. To understand the relations they satisfy, consider the exact sequence

$$0 \longrightarrow \mathcal{S} \longrightarrow \pi^*\mathcal{E} \longrightarrow \mathcal{Q} \longrightarrow 0.$$

By the Whitney formula

$$c(\mathcal{Q}) = \frac{c(\mathcal{E})}{c(\mathcal{S})}.$$

Since $\mathcal{Q}$ has rank $n - k$, the Chern classes $c_l(\mathcal{Q})$ vanish for $l > n - k$, and as in the projective bundle case (above) or the case of $G(k, n)$ (Theorem 5.26) this gives all the relations:

Theorem 9.18. *Let X be a smooth variety, and let $\mathcal{E}$ be a vector bundle of rank n on X. If $G = G(k, \mathcal{E}) \to X$ is the bundle of k-planes in the fibers of $\mathcal{E}$ then*

$$A(G) = A(X)[\zeta_1, \ldots, \zeta_k] \Big/ \left(\left\{ \frac{c(\mathcal{E})}{1 - \zeta_1 + \zeta_2 + \cdots \pm \zeta_k} \right\}^l, l > \dim G - n + k \right),$$

where $\{\eta\}^l$ denotes the component of η of codimension l and ζ_k has degree k.

In fact, the same formula holds without the assumption that X is smooth, as long as one has developed the theory of Chern classes on singular varieties, as in Fulton [1984, Chapter 3]

One can go further and, fixing a sequence of ranks $0 < r_1 < \cdots < r_m < \operatorname{rank} \mathcal{E}$, consider the *flag bundle* $\mathbb{F}(r_1, \ldots, r_m, \mathcal{E})$ whose fiber over a point of X is the set of all flags of subspaces of the given ranks in $\mathcal{E}$. There is again an analogous description of the Chow ring of this space. See Grayson et al. [2012] for this result and an interesting proof that is in some ways more explicit than the one we have given, even in the case of $A(G(k, n))$.

9.7 Conics in $\mathbb{P}^3$ meeting eight lines

The family of plane conics in $\mathbb{P}^3$ is naturally a projective bundle, and we will now use this fact, together with Theorem 9.6, to compute the number of such conics intersecting each of eight general lines $L_1, \ldots, L_8 \subset \mathbb{P}^3$.

We start by checking that we should expect a finite number. There is a three-parameter family of planes in $\mathbb{P}^3$, and a five-parameter family of conics in each. Since two distinct planes intersect only in a line, the space of conics, whatever it is, should have dimension $3 + 5 = 8$.

Next, the locus D_L of conics meeting a given line $L \subset \mathbb{P}^3$ has codimension 1 in the space of conics: If $C \subset \mathbb{P}^3$ is the image of the map given by (F_0, F_1, F_2, F_3), the condition that C meet the line $Z_0 = Z_1 = 0$ is that F_0 and F_1 have a common zero. More geometrically: A one-parameter family of conics sweeps out a surface that meets L in a finite set, so a curve in the space of conics will intersect the locus of conics meeting L a finite number of times. It is reasonable, then, to ask whether there is only a finite number of conics that meet each of eight general lines and, if so, how many there are.

We will proceed as follows. First, as a parameter space for conics in $\mathbb{P}^3$, we will use a projective bundle $\mathcal{Q} \to \mathbb{P}^{3*}$, whose points correspond to pairs (H, C) with H a plane in $\mathbb{P}^3$ and C a conic in H; we will use the theory developed earlier in this chapter to calculate in its Chow ring. In particular, we will identify the class $\delta \in A(\mathcal{Q})$ of the cycle $D_L \subset \mathcal{Q}$ of conics meeting a given line L, and compute the number $\deg \delta^8$, our candidate for the number of conics meeting eight given general lines L_i.

To prove that this number is correct, we must show that the cycles D_{L_i} meet transversely, and this requires a tangent space calculation. To do this, we will show that our bundle $\mathcal{Q}$ is in fact isomorphic to the Hilbert scheme $\mathcal{H} = \mathcal{H}_{2m+1}(\mathbb{P}^3)$ of subschemes of $\mathbb{P}^3$ having Hilbert polynomial $p(m) = 2m + 1$. This will allow us to prove the necessary transversality by describing the tangent spaces to D_L in terms of the general description of the tangent spaces to Hilbert schemes from Theorem 6.21; this is a special case of an important general principle explained in Exercise 9.60.

9.7.1 The parameter space as projective bundle

Since the conics in a given plane naturally form a $\mathbb{P}^5$, and each conic is contained in a unique plane, it is plausible that the set of all conics in $\mathbb{P}^3$ is a $\mathbb{P}^5$-bundle over $\mathbb{P}^{3*}$, the projective space of planes in $\mathbb{P}^3$.

To make this structure explicit, consider the tautological exact sequence on $\mathbb{P}^{3*}$, which we may write as

$$0 \longrightarrow \mathcal{S} \longrightarrow \mathcal{O}_{\mathbb{P}^{3*}}^4 \xrightarrow{\;(x_0, x_1, x_2, x_3)\;} \mathcal{O}_{\mathbb{P}^{3*}}(1) \longrightarrow 0.$$

The projective bundle $\mathbb{P}\mathcal{S} \subset \mathbb{P}(\mathcal{O}_{\mathbb{P}^{3*}}^4) = \mathbb{P}^{3*} \times \mathbb{P}^3$ is the family of 2-planes in $\mathbb{P}^3$: the fiber of $\mathbb{P}\mathcal{S}$ over a point $a = (a_0, \dots, a_3) \in \mathbb{P}^{3*}$ is the plane $H_a \subset \mathbb{P}^3$ defined by $\sum a_i x_i = 0$. The dual $\mathcal{S}_a^*$ is thus the space of linear forms on this plane, and, setting $\mathcal{E} := \mathrm{Sym}^2(\mathcal{S}^*)$, the fiber of $\mathbb{P}\mathcal{E}$ over the point a may be identified with the set of conics in H_a. We will therefore take $\mathcal{Q} = \mathbb{P}\mathcal{E}$ as our parameter space for conics in $\mathbb{P}^3$. Note that there is a *tautological family of conics in $\mathbb{P}^3$*

$$\mathcal{X} \subset \mathbb{P}\mathcal{E} \times_{\mathbb{P}^{3*}} \mathbb{P}\mathcal{S} \subset \mathbb{P}\mathcal{E} \times \mathbb{P}^3$$

whose points are pairs consisting of a conic in a 2-plane and a point on that conic, with projections both to $\mathbb{P}^{3*}$ and to $\mathbb{P}^3$.

From the dual of the exact sequence above, we derive an exact sequence

$$0 \longrightarrow \mathcal{O}^4_{\mathbb{P}^{3*}} \otimes \mathcal{O}_{\mathbb{P}^{3*}}(-1) \longrightarrow \mathrm{Sym}^2(\mathcal{O}^4_{\mathbb{P}^{3*}}) \longrightarrow \mathcal{E} \longrightarrow 0.$$

If we denote the tautological class on $\mathbb{P}^{3*}$ by ω, then, taking into account that $\omega^4 = 0$, the Whitney formula (Theorem 5.3) yields

$$c(\mathcal{E}) = 1/(1 - \omega)^4 = 1 + 4\omega + 10\omega^2 + 20\omega^3.$$

We can now apply Theorem 9.6 to describe the Chow ring of $\mathcal{Q}$. Letting $\zeta \in A^1(\mathcal{Q})$ be the first Chern class of the tautological quotient $\mathcal{O}_{\mathbb{P}\mathcal{E}}(1)$ of the pullback of $\mathcal{E}^*$ to $\mathcal{Q}$, we get

$$A(\mathcal{Q}) = A(\mathbb{P}^{3*})[\zeta]/(\zeta^6 + 4\omega\zeta^5 + 10\omega^2\zeta^4 + 20\omega^3\zeta^3)$$
$$= \mathbb{Z}[\omega, \zeta]/(\omega^4, \zeta^6 + 4\omega\zeta^5 + 10\omega^2\zeta^4 + 20\omega^3\zeta^3).$$

9.7.2　The class δ of the cycle of conics meeting a line

We next compute the class $\delta \in A^1(\mathcal{Q})$ of the divisor $D = D_L$ using the technique of undetermined coefficients. We know that $\delta = a\omega + b\zeta$ for some pair of integers a and b, and restricting to curves in $\mathcal{Q}$ gives us linear relations on a and b. Let $\Gamma \subset \mathcal{Q}$ be the curve corresponding to a general pencil $\{C_\lambda \subset H\}$ of conics in a general plane $H \subset \mathbb{P}^3$ and let $\Phi \subset \mathcal{Q}$ be the curve consisting of a general pencil of plane sections $\{H_\lambda \cap Q\}$ of a fixed quadric Q. We denote their classes in $A_1(\mathcal{Q})$ by γ and φ respectively.

We claim that the following table gives the intersection numbers between our divisor classes ω, ζ, δ, and the curves Γ, Φ:

	ω	ζ	δ
γ	0	1	1
φ	1	0	2

The calculation of the five intersection numbers other than $\zeta\varphi$ is easy, and we leave to the reader the pleasure of working them out (Exercise 9.54).

We can compute $\zeta\varphi$ as the degree of the restriction of the bundle $\mathcal{O}_{\mathbb{P}\mathcal{E}}(1)$ to the curve Φ; equivalently, to show that $\zeta\varphi = 0$ we must show that $\mathcal{T} = \mathcal{O}_{\mathbb{P}\mathcal{E}}(-1)$ is trivial on Φ. To see this, recall that a point of $\mathcal{Q}$ is a pair (H, ξ), with H a plane in $\mathbb{P}^3$ and ξ a one-dimensional subspace of $H^0(\mathcal{O}_H(2))$; the fiber of $\mathcal{T}$ over the point (H, ξ) is the vector space ξ. Now, if $F \in H^0(\mathcal{O}_{\mathbb{P}^3}(2))$ is the homogeneous quadratic polynomial defining Q, we see that the restrictions of F to the planes H_λ give an everywhere-nonzero section of $\mathcal{T}$ over Φ, proving that $\mathcal{T}|_\Phi$ is the trivial bundle, as required.

Given the intersection numbers in the table above, we conclude that

$$\delta = 2\omega + \zeta.$$

There is also a direct way to arrive at this class, which we will describe in Exercise 9.55.

9.7.3 The degree of δ^8

To compute δ^8, we need to know the degrees of the monomials $\omega^i \zeta^j$ of degree 8. To start with, we have $\omega^4 = 0$, and, since ω^3 is the class of a fiber of $\mathcal{Q} \to \mathbb{P}^{3*}$ and ζ restricts to the hyperplane class on this fiber, we have

$$\deg(\omega^3 \zeta^5) = 1.$$

To evaluate the next monomial $\omega^2 \zeta^6$, we use the relation

$$\zeta^6 = -4\omega\zeta^5 - 10\omega^2\zeta^4 - 20\omega^3\zeta^3$$

of Theorem 9.6, which gives

$$\deg(\omega^2 \zeta^6) = \deg \omega^2(-4\omega\zeta^5 - 10\omega^2\zeta^4 - 20\omega^3\zeta^3) = -4.$$

The same idea yields

$$\deg(\omega\zeta^7) = 6 \quad \text{and} \quad \deg \zeta^8 = -4.$$

Putting these together we obtain

$$\deg((2\omega + \zeta)^8) = \deg\left(\zeta^8 + 2\binom{8}{1}\omega\zeta^7 + 4\binom{8}{2}\omega^2\zeta^6 + 8\binom{8}{3}\omega^3\zeta^5\right) = 92.$$

Writing $\pi : \mathcal{Q} \to \mathbb{P}^3$ for the projection, the numbers $\deg(\omega^i \zeta^j)$ computed above may be interpreted (via the push-pull formula) as the degrees of the classes $\pi_* \zeta^k$, which are called *Segre classes* of the bundle $\mathcal{E}$. See Definition 10.1 and, for an alternative computation, Proposition 10.3.

9.7.4 The parameter space as Hilbert scheme

If $C \subset \Lambda$ is a smooth plane conic then the Hilbert polynomial of C is $p(m) = 2m + 1$. Let $\mathcal{H} := \mathcal{H}_{2m+1}$ be the Hilbert scheme of subschemes of $\mathbb{P}^3$ with this Hilbert polynomial, and let $\mathcal{C} \to \mathcal{H} \times \mathbb{P}^3$ be the universal family. We have already described the tautological family of plane conics $\mathcal{X} \to \mathcal{Q} \times \mathbb{P}^3$, and by the universal property of the Hilbert scheme there is a unique map $\psi : \mathcal{Q} \to \mathcal{H}$ such that $\mathcal{X} = (\psi \times 1)^*\mathcal{C}$.

Theorem 9.19. $\mathcal{Q}$ *with its universal family* $\mathcal{X} \to \mathcal{Q} \times \mathbb{P}^3$ *is isomorphic to* $\mathcal{H}$ *with its universal family* $\mathcal{C} \to \mathcal{H} \times \mathbb{P}^3$ *via the map* ψ.

We postpone the proof to develop a few necessary facts about subschemes C with Hilbert polynomial $p(m) = 2m + 1$. To show that C is really a conic, we first want to show that C is contained in a plane Λ — that is, there is a linear form vanishing on C. Since the number of independent linear forms on $\mathbb{P}^3$ is $4 = p(1) + 1$, it suffices to show that the value $h_C(1)$ of the Hilbert function of C — that is, the dimension $(S_C)_1$ of the degree-1 part of the homogeneous coordinate ring of C — is equal to $p(1)$.

Once this is established we must show that a nonzero quadratic form on Λ vanishes on C, and it suffices, for similar reasons as above, to show that $h_C(2) = \dim(S_C)_2 = 5 = p(2)$. In fact, we will prove that if $C \subset \mathbb{P}^3$ is any subscheme with Hilbert polynomial $p(m) = 2m + 1$, then the Hilbert function $h_C(m)$ of C is equal to $p(m)$ for all m. This is contained in the following result:

Proposition 9.20. *Let $C \subset \mathbb{P}^n$ be a subscheme, and let $\mathcal{I}_C$ be its ideal sheaf and $S_C = \Bbbk[x_0, \dots, x_n]/I$ its homogeneous coordinate ring.*

(a) *If the Hilbert polynomial of S_C is $p_C(m) = 2m + 1$, then the Hilbert function of S_C is also equal to $2m + 1$.*

(b) *C is the complete intersection of a unique 2-plane and a (non-unique) quadric hypersurface.*

(c) *$H^1(\mathcal{I}_C(m)) = 0$ for all $m \geq 0$.*

Proof: The form of the Hilbert polynomial implies that C has dimension 1 and degree 2. Thus a general plane section $\Gamma = \{x = 0\} \cap C$ is a subscheme of degree 2 in the plane, either two distinct points or one double point. In either case, the Hilbert function of Γ is $h_\Gamma(m) = 2$ for all $m \geq 1$. Writing S_C for the homogeneous coordinate ring of C, we have a surjective map $S_C \to S_\Gamma$ whose kernel contains $x S_C$, whence

$$h_C(m) - h_C(m-1) \geq h_\Gamma(m) = 2$$

for $m \geq 1$. Since $h_C(0) = 1$, it follows that $h_C(m) \geq 2m + 1$ for all $m \geq 0$, and that a strict inequality for any value of m implies the same for all larger values. Since $h_C(m) = p_C(m) = 2m + 1$ for large m, the inequality above must be an equality for all $m \geq 1$, proving the first statement.

The second statement follows. From $h_C(1) = 3$, we see that C is be contained in a unique plane Λ. From $h_C(2) = 5$, we see that C lies on five linearly independent quadrics; since at most four of these can contain Λ, we see that C lies on a quadric $Q \subset \mathbb{P}^3$ not containing Λ. The subscheme $C' := \Lambda \cap Q$ also has Hilbert function $2m + 1$, and since $C \subset C'$ they are equal.

To prove the last statement, we use the long exact sequence in cohomology

$$H^0(\mathcal{I}_C(m)) \longrightarrow H^0(\mathcal{O}_{\mathbb{P}^n}(m)) \longrightarrow H^0(\mathcal{O}_C(m)) \longrightarrow H^1(\mathcal{I}_C(m)) \longrightarrow H^1(\mathcal{O}_{\mathbb{P}^n}(m)).$$

Since the last term is zero and the cokernel of the map $H^0(\mathcal{I}_C(m)) \to H^0(\mathcal{O}_{\mathbb{P}^3}(m))$ is the component of degree m in S_C, it suffices to show that $h^0(\mathcal{O}_C(m)) = 2m + 1$. But as C is defined in the plane by a quadratic hypersurface, we have also a sequence

$$0 \longrightarrow H^0(\mathcal{O}_{\mathbb{P}^2}(m-2)) \longrightarrow H^0(\mathcal{O}_{\mathbb{P}^2}(m)) \longrightarrow H^0(\mathcal{O}_C(m)) \longrightarrow H^1(\mathcal{O}_{\mathbb{P}^2}(m-2)),$$

and, since the twists of $\mathcal{O}_{\mathbb{P}^2}$ have no intermediate cohomology, we get

$$h^0(\mathcal{O}_C(m)) = h^0(\mathcal{O}_{\mathbb{P}^2}(m)) - h^0(\mathcal{O}_{\mathbb{P}^2}(m-2)) = \binom{m+2}{2} - \binom{m}{2} = 2m + 1,$$

as required. $\qquad\square$

Proof of Theorem 9.19: By Proposition 9.20, the fibers of $\mathcal{C} \subset \mathcal{H} \times \mathbb{P}^3$ over closed points of $\mathcal{H}$ are precisely the distinct conics in $\mathbb{P}^3$. Since this is also true for $\mathcal{X} \subset \mathcal{Q} \times \mathbb{P}^3$, the map $\psi : \mathcal{Q} \to \mathcal{H}$ is bijective on closed points.

Since $\mathcal{Q}$ is smooth, it now suffices to prove that $\mathcal{H}$ is smooth. From the bijectivity of ψ, we see that $\dim \mathcal{H} = \dim \mathcal{Q} = 8$, so it suffices, in fact, to prove that the tangent space to $\mathcal{H}$ at each point $[C]$ has dimension 8. By Theorem 6.21, there is an isomorphism $T_{[C]/\mathcal{H}} \cong H^0(\mathcal{N}_{C/\mathbb{P}^3})$. Using Proposition 9.20 again, we know that C is a complete intersection of a linear form and a quadric. Thus $\mathcal{N}_{C/\mathbb{P}^3} = (\mathcal{O}_{\mathbb{P}^3}(1) \oplus \mathcal{O}_{\mathbb{P}^3}(2))|_C$, and the dimension of the tangent space is $h^0(\mathcal{O}_C(1)) + h^0(\mathcal{O}_C(2))$.

By Proposition 9.20, $H^1(\mathcal{I}_{C/\mathbb{P}^3}(m)) = 0$ for all m, so the desired value is the sum of the values of the Hilbert function of C at 1 and at 2. Putting this together, we get

$$\dim T_{[C]/\mathcal{H}} = (2 \cdot 1 + 1) + (2 \cdot 2 + 1) = 8$$

as required. $\qquad\square$

9.7.5 Tangent spaces to incidence cycles

To prove that the D_{L_i} intersect transversely we need to compute their tangent spaces at the points of intersection. This task is made easier by the fact that, for general L_i, the intersection of the D_{L_i} takes place in the locus U of smooth conics, as we shall now prove:

Lemma 9.21. *For a general choice of lines $L_1, \ldots, L_8 \subset \mathbb{P}^3$, no singular conic meets all eight.*

Proof: The family of singular conics has dimension 7, and the family of lines meeting a line, or a singular conic, has dimension 3. Thus the family consisting of 8-tuples of lines meeting a singular conic has dimension $7 + 3 \cdot 8 = 31$, while the family of 8-tuples of lines has dimension $8 \cdot 4 = 32$. $\qquad\square$

Next we describe the tangent spaces to the cycles D_L at points in U. Again, we use the computation of the tangent space to $\mathcal{Q} \cong \mathcal{H}$ at a point $[C]$ corresponding to a conic C as $T_{[C]/\mathcal{H}} = H^0(\mathcal{N}_{C/\mathbb{P}^3})$.

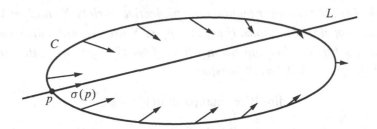

Figure 9.4 If C is a conic meeting a line L at a point p, then a deformation of C corresponding to a normal section σ remains in D_L if and only if $\sigma(p)$ is tangent to L.

Proposition 9.22. *Let $L \subset \mathbb{P}^3$ be a line and $D_L \subset \mathcal{H}$ the locus of conics meeting L. If $C \subset \mathbb{P}^3$ is a smooth plane conic such that $C \cap L = \{p\}$ is a single reduced point, then D_L is smooth at $[C]$, and its tangent space at $[C]$ is the space of sections of the normal bundle whose value at p lies in the normal direction spanned by L; that is,*

$$T_{[C]}D_L = \left\{ \sigma \in H^0(\mathcal{N}_{C/\mathbb{P}^3}) \mid \sigma(p) \in \frac{T_p L + T_p C}{T_p C} \right\}.$$

See Figure 9.4 for an illustration.

Proof: We prove Proposition 9.22 by introducing an incidence correspondence: For $L \subset \mathbb{P}^3$ a line, we let

$$\Phi_L = \{(p, C) \in L \times \mathcal{H} \mid p \in C\}.$$

The image of Φ_L under the projection π_2 to the second factor is the cycle $D_L \subset \mathcal{H}$ of conics meeting L. By Lemma 6.23, the tangent space to Φ_L at the point (p, C) is

$$T_{(p,C)}\Phi_L = \{(v, \sigma) \in T_p L \times H^0(\mathcal{N}_{C/\mathbb{P}^3}) \mid \sigma(p) \equiv v \bmod T_p C\}.$$

In particular, Φ_L will be smooth at (p, C), and the projection π_2 will carry its tangent space injectively to the space of sections $\sigma \in H^0(\mathcal{N}_{C/\mathbb{P}^3})$ such that $\sigma(p) \in (T_p L + T_p C)/T_p C$. Since the map π_2 is one-to-one over p, it follows that D_L is smooth at $[C]$ with this tangent space. $\qquad\square$

This argument also applies to Hilbert schemes in a more general context; see Exercise 9.60.

Corollary 9.23. *Let C be a smooth conic in $\mathbb{P}^3$. If $L_1, \dots, L_8$ are general lines meeting C at general points, then the cycles $D_{L_1}, \dots, D_{L_8} \subset \mathcal{Q} \cong \mathcal{H}$ meet transversely at $[C]$.*

Proof: By Proposition 9.22, it suffices to show that the eight linear conditions specifying that a global section of the normal bundle of C lie in specified one-dimensional subspaces at eight points of C are independent, for a general choice of the points and the subspaces. Since the rank of the normal bundle is 2, this is a special case of Lemma 9.24, proved below. $\qquad\square$

Lemma 9.24. *Let $\mathcal{E}$ be a vector bundle on a projective variety X, and let $V \subset H^0(\mathcal{E})$ be a vector space of global sections. If $p_1, \ldots, p_k \in X$ are general points and $V_i \subset \mathcal{E}_{p_i}$ a general linear subspace of codimension 1 in the fiber $\mathcal{E}_{p_i}$ of $\mathcal{E}$ at p_i, then the subspace $W = \{\sigma \in V \mid \sigma(p_i) \in V_i\}$ has dimension*

$$\dim W = \max\{0, \dim(V) - k\}.$$

The obvious analog of this result fails if we allow codim $V_i > 1$; see Exercise 9.53.

Proof: Proceeding inductively, it suffices to show the case $k = 1$, and note that if the general section in V had value in every hyperplane $V_i \subset \mathcal{E}_p$ at a dense set of points $p \in X$ then $V = 0$. $\qquad\square$

9.7.6 Proof of transversality

Proposition 9.25. *If $L_1, \ldots, L_8 \subset \mathbb{P}^3$ are eight general lines, then the cycles $D_{L_i} \subset \mathcal{Q}$ intersect transversely.*

Proof: To start, we introduce the incidence correspondence

$$\Sigma = \{(L_1, \ldots, L_8; C) \in \mathbb{G}(1,3)^8 \times \mathcal{Q} \mid C \cap L_i \neq \varnothing \text{ for all } i\}.$$

Since the locus of lines $L \subset \mathbb{P}^3$ meeting a given smooth conic C is an irreducible hypersurface in the Grassmannian $\mathbb{G}(1,3)$, we see via projection to $\mathcal{Q}$ that Σ is irreducible of dimension 32.

Now, let $\Sigma_0 \subset \Sigma$ be the locus of $(L_1, \ldots, L_8; C)$ such that the cycles D_{L_i} fail to intersect transversely at $[C]$; this is a closed subset of Σ. By Corollary 9.23, $\Sigma_0 \neq \Sigma$, so $\dim \Sigma_0 < 32$. It follows that Σ_0 does not dominate $\mathbb{G}(1,3)^8$, so for a general point $(L_1, \ldots, L_8) \in \mathbb{G}(1,3)^8$ the cycles D_{L_i} are transverse at every point of their intersection. $\qquad\square$

In sum, we have proved:

Theorem 9.26. *There are exactly 92 distinct plane conics in $\mathbb{P}^3$ meeting eight general lines, and each of them is smooth.*

As with any enumerative formula that applies to the general form of a problem, the computation still tells us something in the case of eight arbitrary lines. For one thing, it says that if $L_1, \ldots, L_8 \subset \mathbb{P}^3$ are *any* eight lines, there will be at least one conic meeting all eight (here we have to include degenerate conics as well as smooth), and, if we assume that the number of conics meeting all eight (again including degenerate ones) is finite, then, assigning to each such conic C a multiplicity (equal to the scheme-theoretic degree of the component of the intersection $\bigcap D_{L_i} \subset \mathcal{H}$ supported at $[C]$, since the cycles D_L are Cohen–Macaulay), the total number of conics will be 92. In particular, as long as the number is finite, there cannot be *more* than 92 distinct conics meeting all eight lines.

In Exercises 9.56–9.68 we will look at some other problems involving conics in $\mathbb{P}^3$, including some problems involving calculations in $A(\mathcal{H})$, some other applications of the techniques we have developed here and some problems that require other parameter spaces for conics.

9.8 Exercises

Exercise 9.27. Choosing coordinates $x_0, x_1, \ldots, x_a$ on $\mathbb{P}^a$ corresponding to the monomials $s^a, s^{a-1}, \ldots, t^a$, show that the 2×2 minors of the matrix

$$\begin{pmatrix} x_0 & x_1 & \cdots & x_{a-1} \\ x_1 & x_2 & \cdots & x_a \end{pmatrix}$$

vanish identically on the rational normal curve $S(a)$. By working in local coordinates, show that the ideal I generated by the minors defines the curve scheme-theoretically. Find a set of monomials forming a basis for the ring $\Bbbk[x_0, x_1, \ldots, x_a]/I$, and show that in degree d it has dimension $ad + 1$. By comparing this with the Hilbert function of $\mathbb{P}^1$, prove that I is the saturated ideal of the rational normal curve.

Exercise 9.28. In order to do the same as we did in the previous exercise for surface scrolls, prove that the Hilbert polynomial $f_S(d)$ of the surface scroll $S(a, b) \subset \mathbb{P}^{a+b+1}$ satisfies

$$f_S(d) \geq (a + b)\binom{d+1}{2} + d + 1.$$

Exercise 9.29. Let $x_0, \ldots, x_{a+b+1}$ be coordinates in $\mathbb{P}^{a+b+1}$. Prove that the 2×2 minors of the matrix

$$\begin{pmatrix} x_0 & x_1 & \cdots & x_{a-1} & x_{a+1} & x_{a+2} & \cdots & x_{a+b} \\ x_1 & x_2 & \cdots & x_a & x_{a+2} & x_{a+3} & \cdots & x_{a+b+1} \end{pmatrix}$$

vanish on a surface scroll $S(a, b)$. As in Exercise 9.27, show that the ideal I generated by the minors defines the surface scheme-theoretically. Then, using Exercise 9.28, prove that I is the saturated ideal of the surface scroll.

Exercise 9.30. Let X be a smooth projective variety, $\mathcal{E}$ a vector bundle on X and $\mathbb{P}\mathcal{E} \to X$ its projectivization. Let $\mathcal{L}$ be any line bundle on X; as we have seen, there is a natural isomorphism $\mathbb{P}\mathcal{E} \cong \mathbb{P}(\mathcal{E} \otimes \mathcal{L})$, such that

$$\mathcal{O}_{\mathbb{P}(\mathcal{E}\otimes\mathcal{L})}(1) \cong \mathcal{O}_{\mathbb{P}\mathcal{E}}(1) \otimes \pi^*\mathcal{L}^*.$$

Using the results of Section 5.5.1, show that the two descriptions of the Chow ring of $\mathbb{P}\mathcal{E} = \mathbb{P}(\mathcal{E} \otimes \mathcal{L})$ agree.

Exercise 9.31. Let $\pi : Y \to X$ be a projective bundle.

(a) Show that the direct sum decomposition of the group $A(X)$ given in Theorem 9.6 depends on the choice of vector bundle $\mathcal{E}$ with $Y \cong \mathbb{P}\mathcal{E}$.

(b) Show that if we define group homomorphisms $\psi_i : A(Y) \to A(X)^{\oplus i+1}$ by

$$\psi_i : \alpha \mapsto (\pi_*(\alpha), \pi_*(\zeta\alpha), \dots, \pi_*(\zeta^i\alpha)),$$

then the filtration of $A(Y)$ given by

$$A(Y) \supset \mathrm{Ker}(\psi_0) \supset \mathrm{Ker}(\psi_1) \supset \cdots \supset \mathrm{Ker}(\psi_{r-1}) \supset \mathrm{Ker}(\psi_r) = 0$$

is independent of the choice of $\mathcal{E}$.

Hint: Give a geometric characterization of the cycles in each subspace of $A(Y)$.

Exercise 9.32. In Example 9.17, we used intersection theory to show that there does not exist a rational solution to the general quadratic polynomial; that is, there do not exist rational functions $X(a, \dots, f)$, $Y(a, \dots, f)$ and $Z(a, \dots, f)$ such that

$$aX^2 + bY^2 + cZ^2 + dXY + eXZ + fYZ \equiv 0.$$

To gain some appreciation of the usefulness of intersection theory, give an elementary proof of this assertion.

Exercise 9.33. Let

$$\Phi = \{(L, p) \in \mathbb{G}(1, n) \times \mathbb{P}^n \mid p \in L\}$$

be the universal line in $\mathbb{P}^n$, and let σ_1, $\sigma_{1,1}$ and ζ be the pullbacks of the Schubert classes $\sigma_1 \in A^1(\mathbb{G}(1, n))$, $\sigma_{1,1} \in A^2(\mathbb{G}(1, n))$ and the hyperplane class $\zeta \in A^1(\mathbb{P}^n)$ respectively. Find the degree of all monomials $\sigma_1^a \sigma_{1,1}^b \zeta^c$ of top degree $a + 2b + c = \dim(\Phi) = 2n - 1$.

Exercise 9.34. Consider the flag variety $\mathbb{F}$ of pairs consisting of a point $p \in \mathbb{P}^3$ and a line $L \subset \mathbb{P}^3$ containing p; that is,

$$\mathbb{F} = \{(p, L) \in \mathbb{P}^3 \times \mathbb{G}(1, 3) \mid p \in L \subset \mathbb{P}^3\}.$$

$\mathbb{F}$ may be viewed as a $\mathbb{P}^1$-bundle over $\mathbb{G}(1, 3)$, or as a $\mathbb{P}^2$-bundle over $\mathbb{P}^3$. Calculate the Chow ring $A(\mathbb{F})$ via each map, and show that the two descriptions agree.

Exercise 9.35. By Theorem 9.6, the Chow ring of the product $\mathbb{P}^3 \times \mathbb{G}(1, 3)$ is just the tensor product of their Chow rings; that is

$$A(\mathbb{P}^3 \times \mathbb{G}(1, 3)) = A(\mathbb{G}(1, 3))[\zeta]/(\zeta^4).$$

In these terms, find the class of the flag variety $\mathbb{F} \subset \mathbb{P}^3 \times \mathbb{G}(1, 3)$ of Exercise 9.34.

Exercise 9.36. Generalizing the preceding problem, let

$$\mathbb{F}(0, k, r) = \{(p, \Lambda) \in \mathbb{P}^r \times \mathbb{G}(1, r) \mid p \in \Lambda\}.$$

Find the class of $\mathbb{F}(0, 1, r) \subset \mathbb{P}^r \times \mathbb{G}(1, r)$.

Exercise 9.37. Generalizing Exercise 9.35 in a different direction, let

$$\Phi_r = \{(L, M) \in \mathbb{G}(1, r) \times \mathbb{G}(1, r) \mid L \cap M \neq \varnothing\}.$$

Given that by Theorem 9.18 we have

$$A(\mathbb{G}(1, r) \times \mathbb{G}(1, r)) \cong A(\mathbb{G}(1, r)) \otimes A(\mathbb{G}(1, r)),$$

find the class of Φ_r in $A(\mathbb{G}(1, r) \times \mathbb{G}(1, r))$ for:

(a) $r = 3$.
(b) $r = 4$.
(c) General r.

Exercise 9.38. Let Z be the blow-up of $\mathbb{P}^n$ along an $(r - 1)$-plane, and let $E \subset Z$ be the exceptional divisor. Find the degree of the top power $[E]^n \in A(Z)$.

Exercise 9.39. Again let $Z = \mathrm{Bl}_\Lambda \mathbb{P}^n$ be the blow-up of $\mathbb{P}^n$ along an $(r - 1)$-plane Λ. In terms of the description of the Chow ring of Z given in Corollary 9.12, find the classes of the following:

(a) The proper transform of a linear space $\mathbb{P}^s$ containing Λ, for each $s > r$.
(b) The proper transform of a linear space $\mathbb{P}^s$ in general position with respect to Λ (that is, disjoint from Λ if $s \leq n - r$, and transverse to Λ if $s > n - r$).
(c) In general, the proper transform of a linear space $\mathbb{P}^s$ intersecting Λ in an l-plane.

Exercise 9.40. Let $Z = \mathrm{Bl}_L \mathbb{P}^3$ be the blow-up of $\mathbb{P}^3$ along a line. In terms of the description of the Chow ring of Z given in Corollary 9.12, find the classes of the proper transform of a smooth surface $S \subset \mathbb{P}^3$ of degree d containing L.

Exercise 9.41. Now let $Z = \mathrm{Bl}_L \mathbb{P}^4$ be the blow-up of $\mathbb{P}^4$ along a line, and let $S \subset \mathbb{P}^4$ be a smooth surface of degree d containing L. Show by example that the class of the proper transform of S in Z is not determined by this data. For example, try taking $S = S(1, 2) \subset \mathbb{P}^4$ a cubic scroll, with L either

(a) a line of the ruling of S, or
(b) the *directrix* of S, that is, the unique curve of negative self-intersection,

and observe that you get different answers.

Exercise 9.42. Let $Z = \mathrm{Bl}_\Lambda \, \mathbb{P}^n$ be the blow-up of $\mathbb{P}^n$ along an $(r-1)$-plane Λ; that is, if we consider the subspace $\mathbb{P}^{n-r} \subset \mathbb{G}(r, n)$ of r-planes containing Λ, we have

$$Z = \{(p, \Gamma) \in \mathbb{P}^n \times \mathbb{P}^{n-r} \mid p \in \Gamma\}.$$

Using the description of the Chow ring of Z given in Corollary 9.12, find the class of $Z \subset \mathbb{P}^n \times \mathbb{P}^{n-r}$.

Exercise 9.43. Let C be a smooth curve, $\mathcal{E}$ a vector bundle of rank r on C and $\mathcal{F}, \mathcal{G} \subset \mathcal{E}$ two subbundles of complementary ranks s and $r - s$ such that for general $p \in C$ the fibers $\mathcal{F}_p$ and $\mathcal{G}_p$ are complementary in $\mathcal{E}_p$. In terms of the Chern classes of the three bundles, describe the locus of $p \in C$ where $\mathcal{F}_p \cap \mathcal{G}_p \neq 0$:

(a) By using the result of Proposition 9.13 to calculate the class of the intersection $\mathbb{P}\mathcal{F} \cap \mathbb{P}\mathcal{G}$ in $\mathbb{P}\mathcal{E}$.

(b) By considering the bundle map $\mathcal{F} \oplus \mathcal{G} \to \mathcal{E}$.

Exercise 9.44. To generalize the preceding problem: Let X be a smooth projective variety of any dimension, $\mathcal{E}$ a vector bundle of rank r on X and $\mathcal{F}, \mathcal{G} \subset \mathcal{E}$ subbundles of ranks a and b with $a + b \leq r$. Describe the locus of $p \in C$ where $\mathcal{F}_p \cap \mathcal{G}_p \neq 0$, assuming this locus has the expected codimension $r + 1 - a - b$.

We will see how to generalize this calculation substantially using the *Porteous formula* of Chapter 12; see Exercise 12.11.

The following three exercises show one way to construct a surface with infinitely many reduced and irreducible curves of negative self-intersection.

Exercise 9.45. Let F and G be two general polynomials of degree 3 in $\mathbb{P}^2$, and let $\{C_t = V(t_0 F + t_1 G)\}_{t \in \mathbb{P}^1}$ be the associated pencil of curves; let $p_1, p_2, \ldots, p_9$ be the base points of this pencil. Show that for very general $t \in \mathbb{P}^1$ (that is, for all but countably many t) the line bundle $\mathcal{O}_{C_t}(p_1 - p_2)$ is not torsion in $\mathrm{Pic}(C_t) = A^1(C_t)$.

Exercise 9.46. Now let S be the blow-up of the plane at the points $p_1, \ldots, p_9$ — that is, the graph of the rational map $\mathbb{P}^2 \dashrightarrow \mathbb{P}^1$ given by (F, G) — and let $E_1, \ldots, E_9$ be the exceptional divisors. Show that there is a biregular automorphism $\varphi : S \to S$ that commutes with the projection $S \to \mathbb{P}^1$ and carries E_1 to E_2.

Exercise 9.47. Using the result of Exercise 9.45, show that the automorphism φ of Exercise 9.46 has infinite order, and deduce that the surface S contains infinitely many irreducible curves of negative self-intersection.

Exercise 9.48. An amusing enumerative problem: In the circumstances of the preceding exercises, for how many $t \in \mathbb{P}^1$ will $\mathcal{O}_{C_t}(p_1 - p_2)$ be torsion of order 2 — that is, for how many t will $\mathcal{O}_{C_t}(2p_1) \cong \mathcal{O}_{C_t}(2p_2)$?

Exercise 9.49. Let C be a smooth curve of genus $g \geq 2$ over a field of characteristic $p > 0$; let $\varphi : C \to C$ be the Frobenius morphism. If $\Gamma_n \subset C \times C$ is the graph of φ^n and $\gamma_n = [\Gamma_n] \in A^1(C \times C)$ its class, show that the self-intersection $\deg(\gamma_n^2)$ goes to $-\infty$ as $n \to \infty$.

Exercise 9.50. Show that if $\mathcal{E}$ is a vector bundle of rank 2 and degree e on a smooth projective curve X, and $\mathcal{L}$ and $\mathcal{M}$ sub-line bundles of degrees a and b corresponding to sections of $\mathbb{P}\mathcal{E}$ with classes σ and τ, then

$$\sigma\tau = e - a - b \quad \text{and} \quad \sigma^2 + \tau^2 = 2e - 2a - 2b.$$

In particular, if $\mathcal{L}$ and $\mathcal{M}$ are distinct then $\deg \sigma^2 + \deg \tau^2 \geq 0$, with equality holding if and only if $\mathcal{E} = \mathcal{L} \oplus \mathcal{M}$.

Exercise 9.51. Let $\mathbb{P}^N$ be the space of hypersurfaces of degree d in $\mathbb{P}^n$. Using the analysis of Example 9.17 as a template, show that for $d > 1$ the universal hypersurface

$$\Phi_{d,n} = \{(X, p) \in \mathbb{P}^N \times \mathbb{P}^n \mid p \in X\} \to \mathbb{P}^N$$

admits no rational section.

Exercise 9.52. Consider the flag variety $\mathbb{F}$ of pairs consisting of a point $p \in \mathbb{P}^4$ and a 2-plane $\Lambda \subset \mathbb{P}^4$ containing p; that is,

$$\mathbb{F} = \{(p, L) \mid p \in \Lambda \subset \mathbb{P}^4\} \subset \mathbb{P}^4 \times \mathbb{G}(2, 4).$$

$\mathbb{F}$ may be viewed as a $\mathbb{P}^2$-bundle over $\mathbb{G}(2, 4)$, or as a $\mathbb{G}(1, 3)$-bundle over $\mathbb{P}^4$. Calculate the Chow ring $A(\mathbb{F})$ via each map, and show that the two descriptions agree.

Exercise 9.53. Show that the analog of Lemma 9.24 is false if we allow the V_i to have codimension >1: in other words, if $V_i \subset \mathcal{E}_{p_i}$ is a general linear subspace of codimension m_i, then the corresponding subspace $W \subset H^0(\mathcal{E})$ need not have dimension $\max\{0, h^0(\mathcal{E}) - \sum m_i\}$.
Hint: Consider a bundle whose sections all lie in a proper subbundle.

Exercise 9.54. Calculate the remaining five intersection numbers in the table of intersection numbers on page 349 of Section 9.7.2.

Exercise 9.55. To find the class $\delta = [D_L] \in A^1(\mathcal{H})$ of the cycle of conics meeting a line directly, restrict to the open subset $U \subset \mathcal{H}$ of pairs $(H, \xi) \in \mathcal{H}$ such that H does not contain L (since the complement of this open subset of $\mathcal{H}$ has codimension 2, any relation among divisor classes that holds in U will hold in $\mathcal{H}$). Show that we have a map $\alpha : U \to L$ sending a pair (H, ξ) to the point $p = H \cap L$, and that in U the divisor D_L is the zero locus of the map of line bundles

$$T \to \alpha^* \mathcal{O}_L(2)$$

sending a quadric $Q \in \xi$ to $Q(p)$.

Exercise 9.56. Let $\Delta \subset \mathcal{H}$ be the locus of singular conics.

(a) Show that Δ is an irreducible divisor in $\mathcal{H}$.

(b) Express the class $\delta \in A^1(\mathcal{H})$ as a linear combination of ω and ζ.

(c) Use this to calculate the number of singular conics meeting each of seven general lines in $\mathbb{P}^3$.

(d) Verify your answer to the last part by calculating this number directly.

Exercise 9.57. Let $p \in \mathbb{P}^3$ be a point and $F_p \subset \mathcal{H}$ the locus of conics containing the point p. Show that F_p is six-dimensional, and find its class in $A^2(\mathcal{H})$.

Exercise 9.58. Use the result of the preceding exercise to find the number of conics passing through a point p and meeting each of six general lines in $\mathbb{P}^3$, the number of conics passing through two points p, q and meeting each of four general lines in $\mathbb{P}^3$, and the number of conics passing through three points p, q, r and meeting each of two general lines in $\mathbb{P}^3$. Verify your answers to the last two parts by direct examination.

Exercise 9.59. Find the class in $A^3(\mathcal{H})$ of the locus of double lines (note that this is five-dimensional, not four!).

Exercise 9.60. Suppose that $X \subset \mathbb{P}^n$ is a subscheme of pure dimension l and $\mathcal{H}$ a component of the Hilbert scheme parametrizing subschemes of $\mathbb{P}^n$ of pure dimension $k < n - l$ in $\mathbb{P}^n$; let $[Y] \in \mathcal{H}$ be a smooth point corresponding to a subscheme $Y \subset \mathbb{P}^n$ such that $Y \cap X = \{p\}$ is a single reduced point, and suppose moreover that p is a smooth point of both X and Y. Finally, let $\Sigma_X \subset \mathcal{H}$ be the locus of subschemes meeting X.

Use the technique of Proposition 9.22 to show that $\Sigma_X \subset \mathcal{H}$ is smooth at $[Y]$, of the expected codimension $n - k - l$, with tangent space

$$T_{[Y]}\Sigma_X = \left\{ \sigma \in H^0(\mathcal{N}_{Y/\mathbb{P}^n}) \mid \sigma(p) \in \frac{T_p X + T_p Y}{T_p Y} \right\}.$$

The next few problems deal with an example of a phenomenon encountered in the preceding chapter: the possibility that the cycles in our parameter space corresponding to the conditions imposed in fact do not meet transversely, or even properly.

Exercise 9.61. Let $H \subset \mathbb{P}^3$ be a plane, and let $\mathcal{E}_H \subset \mathcal{H}$ be the closure of the locus of smooth conics $C \subset \mathbb{P}^3$ tangent to H. Show that this is a divisor, and find its class $[\mathcal{E}_H] \in A^1(\mathcal{H})$.

Exercise 9.62. Find the number of smooth conics in $\mathbb{P}^3$ meeting each of seven general lines $L_1, \ldots, L_7 \subset \mathbb{P}^3$ and tangent to a general plane $H \subset \mathbb{P}^3$. More generally, for $k = 1, 2$ and 3 find the number of smooth conics in $\mathbb{P}^3$ meeting each of $8 - k$ general lines $L_1, \ldots, L_{8-k} \subset \mathbb{P}^3$ and tangent to k general planes $H_1, \ldots, H_k \subset \mathbb{P}^4$.

Exercise 9.63. For $k \geq 4$, why do the methods developed here not work to calculate the number of smooth conics in $\mathbb{P}^3$ meeting each of $8 - k$ general lines $L_1, \ldots, L_{8-k} \subset \mathbb{P}^3$ and tangent to k general planes $H_1, \ldots, H_k \subset \mathbb{P}^3$? What can you do to find these numbers? (In fact, we have seen one way to deal with this in Chapter 8.)

Next, some problems involving conics in $\mathbb{P}^4$:

Exercise 9.64. Now let $\mathcal{K}$ be the space of conics in $\mathbb{P}^4$ (again, defined to be complete intersections of two hyperplanes and a quadric). Use the description of $\mathcal{K}$ as a $\mathbb{P}^5$-bundle over the Grassmannian $\mathbb{G}(2, 4)$ to determine its Chow ring.

Exercise 9.65. In terms of your answer to the preceding problem, find the class of the locus D_Λ of conics meeting a 2-plane Λ, and of the locus $\mathcal{E}_L$ of conics meeting a line $L \subset \mathbb{P}^4$.

Exercise 9.66. Find the expected number of conics in $\mathbb{P}^4$ meeting each of 11 general 2-planes $\Lambda_1, \ldots, \Lambda_{11} \subset \mathbb{P}^4$.

Exercise 9.67. Prove that your answer to the preceding problem is in fact the actual number of conics by showing that for general 2-planes $\Lambda_1, \ldots, \Lambda_{11} \subset \mathbb{P}^4$ the corresponding cycles D_{Λ_i} intersect transversely.

Finally, here is a challenge problem:

Exercise 9.68. Let $\{S_t \subset \mathbb{P}^3\}_{t \in \mathbb{P}^1}$ be a general pencil of quartic surfaces (that is, take A and B general homogeneous quartic polynomials, and set $S_t = V(t_0 A + t_1 B) \subset \mathbb{P}^3$). How many of the surfaces S_t contain a conic?

Chapter 10

Segre classes and varieties of linear spaces

Keynote Questions

(a) Let $v_1, \dots, v_{2n}$ be general tangent vector fields on $\mathbb{P}^n$. At how many points of $\mathbb{P}^n$ is there a nonzero cotangent vector annihilated by all the v_i? (Answer on page 366.)

(b) If f is a general polynomial of degree $d = 2m - 1$ in one variable over a field of characteristic 0, then there is a unique way to write f as a sum of m d-th powers of linear forms (Proposition 10.15). If f and g are general polynomials of degree $d = 2m$ in one variable, how many linear combinations of f and g are expressible as a sum of m d-th powers of linear forms? (Answer on page 377.)

(c) If $C \subset \mathbb{P}^4$ is a general rational curve of degree d, how many 3-secant lines does C have? (Answer on page 379.)

(d) If $C \subset \mathbb{P}^3$ is a general rational curve of degree d, what is the degree of the surface swept out by the 3-secant lines to C? (Answer on page 380.)

10.1 Segre classes

Our understanding of the Chow rings of projective bundles makes accessible the computation of the classes of another natural series of loci associated to a vector bundle.

We start with a naive question. Suppose that $\mathcal{E}$ is a vector bundle on a scheme X and that $\mathcal{E}$ is generated by global sections. How many global sections does it actually take to generate $\mathcal{E}$? More generally, what sort of locus is it where a given number of general global sections fail to generate $\mathcal{E}$ locally?

We can get a feeling for these questions as follows. First, consider the case where $\mathcal{E}$ is a line bundle. In this case, each regular section corresponds to a divisor of class $c_1(\mathcal{E})$. If $\mathcal{E}$ is generated by its global sections, the linear series of these divisors is base point free, so a general collection of i of them will intersect in a codimension-i locus of class

$c_1(\mathcal{E})^i$. That is, the locus where i general sections of $\mathcal{E}$ fail to generate $\mathcal{E}$ has "expected" codimension i and class $c_1(\mathcal{E})^i$.

Now suppose that $\mathcal{E}$ has rank $r > 1$; again, suppose that it is generated by global sections. Choose r general sections, and let X' be the codimension-1 subset of $\mathcal{E}$ consisting of points p where the sections do not generate $\mathcal{E}$. One can hope that at a general point of X' the sections have only one degeneracy relation, so that on some open set $U \subset X'$ they generate a corank-1 subbundle of $\mathcal{E}' \subset \mathcal{E}$, and the quotient $\mathcal{E}/\mathcal{E}'$ is a line bundle on U. The sections of $\mathcal{E}$ yield sections of $\mathcal{E}/\mathcal{E}'$, so if it is a line bundle they will vanish in codimension 1 in U; that is, we should expect $r + 1$ general sections of $\mathcal{E}$ to generate $\mathcal{E}$ away from a codimension-2 subset of X. Continuing in this way (and assuming that $r \geq i$), it seems that $r + i - 1$ sections of $\mathcal{E}$ might generate $\mathcal{E}$ away from a codimension-i locus. In particular, $r + \dim X$ sections might generate $\mathcal{E}$ locally everywhere.

A case beloved by algebraists is that of $\mathcal{E} = \mathcal{O}_{\mathbb{P}V}(1)^r$. Here a collection of $r + i - 1$ general sections is a general map

$$\mathcal{O}_{\mathbb{P}V}^{r+i-1} \xrightarrow{\varphi} \mathcal{O}_{\mathbb{P}V}(1)^r,$$

that is, a general $r \times (r + i - 1)$ matrix of linear forms. The locus where the sections fail to generate is the support of the cokernel, which is defined by the $r \times r$ minors of the matrix. By the generalized principal ideal theorem (Theorem 0.2), the codimension of this locus is at most i, and in fact equality holds (as we shall soon see) whenever $r + 1 \geq i$. In fact, the support of the cokernel is exactly the scheme defined by the ideal of minors in this case (see Buchsbaum and Eisenbud [1977]).

It turns out that the construction of projective bundles gives us an effective way of reducing this question (and many others) about vector bundles to the case of line bundles, passing from $\mathcal{E}$ to the line bundle $\mathcal{O}_{\mathbb{P}\mathcal{E}}(1)$ on $\mathbb{P}\mathcal{E}$. To relate this line bundle to classes on X, we push forward its self-intersections:

Definition 10.1. Let X be a smooth projective variety, let $\mathcal{E}$ be a vector bundle of rank r on X and $\pi : \mathbb{P}\mathcal{E} \to X$ its projectivization, and let $\zeta = c_1(\mathcal{O}_{\mathbb{P}\mathcal{E}}(1))$. The *i*-*th Segre class* of $\mathcal{E}$ is the class

$$s_i(\mathcal{E}) = \pi_*(\zeta^{r-1+i}) \in A^i(X),$$

and the (total) *Segre class* of $\mathcal{E}$ is the sum

$$s(\mathcal{E}) = 1 + s_1(\mathcal{E}) + s_2(\mathcal{E}) + \cdots.$$

(For a more general definition of the Segre classes, see Fulton [1984, Chapter 4].)

The Segre classes give the answer to our question about generating vector bundles:

Proposition 10.2. *Let $\mathcal{E}$ be a vector bundle of rank r on a smooth variety X that is generated by global sections, and let $\tau_1, \ldots, \tau_{r+i-1}$ be general sections. If X_i is the scheme where $\tau_1, \ldots, \tau_{r+i-1}$ fail to generate $\mathcal{E}$, then X_i has pure codimension i and the class $[X_i]$ is equal to $(-1)^i s_i(\mathcal{E})$.*

We will prove here only the weaker statement that $(-1)^i s_i(\mathcal{E})$ is represented by a positive linear combination of the components of X_i; the stronger version is a special case of *Porteous' formula* (Theorem 12.4), which will be proved in full in Chapter 12.

The proposition shows an interesting parallel between the Chern classes and the Segre classes of a bundle:

- The i-th Chern class $c_i(\mathcal{E})$ is the locus of fibers where a suitably general bundle map

$$\mathcal{O}_X^{\oplus r-i+1} \to \mathcal{E}$$

  fails to be injective.

- The i-th Segre class $s_i(\mathcal{E})$ is $(-1)^i$ times the locus of fibers where a suitably general bundle map

$$\mathcal{O}_X^{\oplus r+i-1} \to \mathcal{E}$$

  fails to be surjective.

The Segre classes may seem to give a way of defining new cycle class invariants of a vector bundle, but in fact they are essentially a different way of packaging the information contained in the Chern classes. Postponing the proof of Proposition 10.2 for a moment, we explain the remarkable relationship:

Proposition 10.3. *The Segre and Chern classes of a bundle $\mathcal{E}$ on X are reciprocals of one another in the Chow ring of X:*

$$s(\mathcal{E})c(\mathcal{E}) = 1 \in A(X).$$

Using the formula $c_i(\mathcal{E}^*) = (-1)^i c_i(\mathcal{E})$, we deduce that

$$s_i(\mathcal{E}^*) = (-1)^i s_i(\mathcal{E}).$$

Also, for any exact sequence $0 \longrightarrow \mathcal{E} \longrightarrow \mathcal{F} \longrightarrow \mathcal{G} \longrightarrow 0$ of vector bundles, the Whitney formula gives $c(\mathcal{F}) = c(\mathcal{E})c(\mathcal{G})$, whence

$$s(\mathcal{F}) = s(\mathcal{E})s(\mathcal{G}).$$

Proof of Proposition 10.3: If $\mathcal{S}$ and $\mathcal{Q}$ are the tautological sub- and quotient bundles on $\mathbb{P}\mathcal{E}$ and $\zeta = c_1(\mathcal{S}^*)$ is the tautological class, then $c(\mathcal{S}) = 1 - \zeta$, so by the Whitney formula

$$c(\mathcal{Q}) = \frac{c(\pi^*\mathcal{E})}{c(\mathcal{S})} = c(\pi^*\mathcal{E})(1 + \zeta + \zeta^2 + \cdots) \in A(\mathbb{P}\mathcal{E}).$$

We now push this equation forward to X. Considering first the left-hand side, we see that for $i < r-1$ the Chern class $c_i(Q)$ is represented by a cycle of dimension $> \dim X$, so it maps to 0, while the top Chern class $c_{r-1}(Q)$ maps to a multiple of the fundamental class of X — in fact, we saw in Lemma 9.7 that the multiple is 1. Thus $\pi_*(c(Q)) = 1 \in A(X)$. On the other hand, the push-pull formula tells us that

$$\pi_*(c(\pi^*\mathcal{E})(1 + \zeta + \zeta^2 + \cdots)) = c(\mathcal{E}) \cdot \pi_*(1 + \zeta + \zeta^2 + \cdots)$$
$$= c(\mathcal{E})s(\mathcal{E}),$$

completing the argument. $\square$

For example, if $X = \mathbb{P}^n$ and $\mathcal{E} = (\mathcal{O}_{\mathbb{P}^n}(1))^r$, then

$$s(\mathcal{E}) = \frac{1}{c(\mathcal{E})} = \frac{1}{(1 + \zeta)^r} = 1 - r\zeta + \binom{r+1}{2}\zeta^2 - \binom{r+2}{3}\zeta^3 + \cdots .$$

Proof of Proposition 10.2: Let $V = H^0(\mathcal{E})$; suppose that $\dim V = n$. Since $\mathcal{E}$ is generated by global sections, we have a natural map $\varphi : X \to G(n - r, n)$ sending each point $p \in X$ to the kernel of the evaluation map $V \to \mathcal{E}_p$, that is, the subspace of sections of $\mathcal{E}$ vanishing at p. Via this map, $\mathcal{E}$ is the pullback φ^*Q of the universal quotient bundle on $G(n - r, V)$, and by Section 5.6.2 we have correspondingly

$$c_i(\mathcal{E}) = \varphi^*(c(Q)) = \varphi^*(\sigma_i).$$

In fact, we can see this directly: If $\tau_1, \ldots, \tau_{r-i+1} \in V$ are general sections of $\mathcal{E}$, the locus where they fail to be independent will be the preimage of the Schubert cycle $\Sigma_i(W)$, where $W \subset V$ is the span of $\tau_1, \ldots, \tau_{r-i+1}$, and, since the plane $W \subset V$ is general, by Kleiman transversality the class $[\varphi^{-1}(\Sigma_i(W))]$ of the preimage is the pullback of the class $[\Sigma_i(W)] = \sigma_i$.

In the same way, if $\tau_1, \ldots, \tau_{r+i-1} \in V$ are general sections of $\mathcal{E}$, the scheme X_i where they fail to span will be the preimage of the Schubert cycle $\Sigma_{1^i}(W) = \Sigma_{1,1,\ldots,1}(W)$, where $W \subset V$ is the span of $\tau_1, \ldots, \tau_{r+i-1}$; again, we can invoke Kleiman to deduce that the X_i have pure codimension i and that

$$[X_i] = \varphi^*(\sigma_{1^i}).$$

Finally, we saw in Corollary 4.10 that in the Chow ring of the Grassmannian we have

$$(1 + \sigma_1 + \sigma_2 + \cdots)(1 - \sigma_1 + \sigma_{1,1} - \cdots) = 1,$$

and combining these we have

$$\sum(-1)^i[X_i] = \varphi^*\left(\sum(-1)^i\sigma_{1^i}\right) = \varphi^*\frac{1}{\sum\sigma_i} = \frac{1}{c(\mathcal{E})} = s(\mathcal{E}),$$

as desired. $\square$

We can now answer Keynote Question (a). A tangent vector field on $\mathbb{P}^n$ is a section of $\mathcal{T}_{\mathbb{P}^n} = \Omega^*_{\mathbb{P}^n}$, so the question can be rephrased as: At how many points of $\mathbb{P}^n$ do $2n$ general sections of $\mathcal{T}_{\mathbb{P}^n}$ fail to generate $\mathcal{T}_{\mathbb{P}^n}$? By Proposition 10.2, this is $(-1)^n$ times the degree of the Segre class $s_n(\mathcal{T}_{\mathbb{P}^n})$. By Proposition 10.3, $s(\mathcal{T}_{\mathbb{P}^n}) = 1/c(\mathcal{T}_{\mathbb{P}^n})$. And, as we have seen (in Section 5.7.1), $c(\mathcal{T}_{\mathbb{P}^n}) = (1 + \zeta)^{n+1}$, where ζ is the hyperplane class on $\mathbb{P}^n$. Putting this together,

$$s(\mathcal{T}_{\mathbb{P}^n}) = \frac{1}{(1+\zeta)^{n+1}} = 1 - (n+1)\zeta + \binom{n+2}{2}\zeta^2 + \cdots,$$

so the answer is $\binom{2n}{n}$.

10.2 Varieties swept out by linear spaces

We can use Segre classes to calculate the degrees of some interesting varieties "swept out" by linear spaces in the following sense. Let B be a smooth variety of dimension m and $\alpha : B \to G = \mathbb{G}(k, n)$ a map to the Grassmannian of k-planes in $\mathbb{P}^n$, and let

$$X = \bigcup_{b \in B} \Lambda_{\alpha(b)} \subset \mathbb{P}^n$$

be the union of the planes in $\mathbb{P}^n$ corresponding to the points of the image of B. Let $\mathcal{S}$ be the universal subbundle on G and

$$\Phi = \mathbb{P}\mathcal{S} = \{(\Lambda, p) \in G \times \mathbb{P}^n \mid p \in \Lambda\}$$

the universal k-plane. Form the fiber product

$$\Phi_B = B \times_G \Phi = \{(b, p) \in B \times \mathbb{P}^n \mid p \in \Lambda_{\alpha(b)}\},$$

with projection maps

$$B \xleftarrow{\pi} \Phi_B \xrightarrow{\eta} \mathbb{P}^n,$$

so that we can write

$$X = \eta(\Phi_B).$$

Since Φ_B is necessarily a variety of dimension $m + k$, we see from this that X will be a subvariety of $\mathbb{P}^n$ of dimension at most $m + k$. In case it has dimension equal to $m + k$ — that is, the map η is generically finite of some degree d, or in other words a general point of X lies on d of the planes $\Lambda_{\alpha(b)}$ — we will say that X is *swept out d times* by the planes $\Lambda_{\alpha(b)}$.

Assuming now that X has the "expected" dimension $m + k$, we can ask for its degree in $\mathbb{P}^n$. This can be conveniently expressed as the degree of a Segre class:

Proposition 10.4. *Let $B \subset \mathbb{G}(k, n)$ be a smooth projective variety of dimension m, $\alpha : B \to G = \mathbb{G}(k, n)$ any morphism and $\mathcal{E} = \alpha^* S$ the pullback of the universal subbundle on G. If*

$$X = \bigcup_{b \in B} \Lambda_{\alpha(b)} \subset \mathbb{P}^n$$

is swept out d times by the planes corresponding to points of B, then

$$\deg(X) = \deg(s_m(\mathcal{E}))/d.$$

Proof: If $L \in H^0(\mathcal{O}_{\mathbb{P}^n}(1))$ is a homogeneous linear form on $\mathbb{P}^n$, then L defines a section of $\mathcal{E}^*$ by restriction to each fiber of $\mathcal{E} = S_B$, and hence a section σ_L of $\mathcal{O}_{\mathbb{P}\mathcal{E}}(1)$. The preimage $\eta_B^{-1}(H)$ of the hyperplane $H = V(L) \subset \mathbb{P}^n$ given by L is the zero locus of σ_L. Thus the pullback of the hyperplane class on $\mathbb{P}^n$ under the map η_B is the tautological class $\zeta = c_1(\mathcal{O}_{\mathbb{P}\mathcal{E}}(1))$ on $\mathbb{P}\mathcal{E}$, and it follows that $d \cdot \deg(X) = \deg \zeta^{m+k} = \deg s_m(\mathcal{E})$, as required. $\square$

Alternatively, we could argue that the degree of X is the number of points of its intersection with a general $(n - m - k)$-plane $\Gamma \subset \mathbb{P}^n$; since the class of the cycle $\Sigma_m(\Gamma) \subset G$ of k-planes meeting Γ is the Schubert class $\sigma_m \in A^m(G)$, this is $1/d$ times the degree of the pullback $\alpha^* \sigma_m$. Thus we have

$$d \cdot \deg(X) = \deg \alpha^* \sigma_m$$
$$= \deg c_m(\alpha^* \mathcal{Q})$$
$$= \deg s_m(\mathcal{E})$$

since $s(S) = 1/c(S) = c(\mathcal{Q})$.

10.3 Secant varieties

The study of secant varieties to projective varieties $X \subset \mathbb{P}^n$ is a rich one, with a substantial history and many fundamental open problems. In this section, we will discuss some of the basic questions. In the following sections we will use Segre classes to compute the degrees of secant varieties to rational curves.

10.3.1 Symmetric powers

A k-secant m-plane to a variety X in $\mathbb{P}^n$ is a linear space $\Lambda \cong \mathbb{P}^m \subset \mathbb{P}^n$ of dimension m that meets X in k points, so it will be useful to introduce a classical construction of a variety whose points are (unordered) k-tuples of points of X: the k-th *symmetric power* $X^{(k)}$ of X.

Formally, we define $X^{(k)}$ to be the quotient of the ordinary k-fold product X^k by the action of the symmetric group on k letters $\mathfrak{S}_k$, acting on X^k by permuting the factors. If $X = \operatorname{Spec} A$ is any affine scheme, this means that

$$X^{(k)} := \operatorname{Spec}((A \otimes A \otimes \cdots \otimes A)^{\mathfrak{S}_k}).$$

When X is quasi-projective, $X^{(k)}$ is defined by patching together symmetric powers of affine open subsets of X. The main theorem of Galois theory shows that when X is a variety the extension of rational function fields $\Bbbk(X^k)/\Bbbk(X^{(k)})$ is Galois, and of degree $k!$.

One can show that such quotients are *categorical*: Any morphism $X^{(k)} \to Y$ determines an $\mathfrak{S}_k$-invariant morphism $X^k \to Y$, and this is a one-to-one correspondence. Further, the closed points of X correspond naturally to the effective 0-cycles on X: they are usually denoted additively by $p_1 + \cdots + p_k$, where the $p_i \in X$ need not be distinct. For these results, see Mumford [2008, Chapter 12].

Since the natural map $X^k \to X^{(k)}$ is finite, $X^{(k)}$ is affine or projective if and only if X is.

A familiar example is the case $X = \mathbb{A}^1$: Here, $X = \operatorname{Spec} \Bbbk[t]$, so

$$(\mathbb{A}^1)^{(k)} = \operatorname{Spec}(\Bbbk[t_1, \ldots, t_k]^{\mathfrak{S}_k}).$$

This ring of invariants is a polynomial ring on the k elementary symmetric functions (see for example Eisenbud [1995, Section 1.3] for an algebraic proof), so $(\mathbb{A}^1)^{(k)} = \mathbb{A}^k$. Set-theoretically, this is the statement that a monic polynomial is determined by the set of its roots, counting multiplicity.

A similar result holds for $\mathbb{P}^1$. We could deduce it from the case of $\mathbb{A}^1$, but instead we give a geometric proof:

Proposition 10.5. $(\mathbb{P}^1)^{(k)} \cong \mathbb{P}H^0(\mathcal{O}_{\mathbb{P}^1}(k)) = \mathbb{P}^k$.

Proof: We think of $\mathbb{P}^1$ as $\mathbb{P}H^0(\mathcal{O}_{\mathbb{P}^1}(1))$, the space of linear forms in 2 variables modulo scalars. The product of k linear forms is a form of degree k, which is independent of the order in which the product is taken. Thus multiplication defines a morphism $\varphi : (\mathbb{P}^1)^k \to \mathbb{P}H^0(\mathcal{O}_{\mathbb{P}^1}(k))$ that is invariant under the group $\mathfrak{S}_k$. The morphism φ is finite and generically $k!$-to-one, so it has degree $k!$.

Since φ is invariant, it factors through a morphism $\psi : (\mathbb{P}^1)^{(k)} \to \mathbb{P}^k$, and, since the degree of the quotient map $(\mathbb{P}^1)^k \to (\mathbb{P}^1)^{(k)}$ is $k!$, we see that ψ is birational. Since $\mathbb{P}^k$ is normal and ψ is finite and birational, ψ is an isomorphism. $\qquad\square$

The construction of $X^{(k)}$ is most useful when X is a smooth curve. One reason is given by the following result:

Proposition 10.6. *If X is a variety and $k > 1$, then $X^{(k)}$ is smooth if and only if X is smooth and $\dim X \leq 1$.*

Proof: If $\dim X = 0$, then X consists of a single reduced point, and $X^{(k)}$ is also a single reduced point. Thus we may assume that $\dim X > 0$.

Away from the subsets where at least two factors are equal, the quotient map $X^k \to X^{(k)}$ is an unramified covering. Thus if X is singular at a point p, and $p, q_1, \ldots, q_{k-1} \in X$ are distinct points, then near $p + q_1 + \cdots + q_{k-1}$ the variety $X^{(k)}$ looks like the product X^k near $(p, q_1, \ldots, q_{k-1})$; in particular, it is singular. Thus if X is singular then $X^{(k)}$ is singular.

Now suppose that X is smooth and of dimension ≥ 2. If $X^{(k)}$ were smooth as well, the quotient map $\pi : X^k \to X^{(k)}$ would be étale away from the diagonal in X^k, a locus of codimension at least 2. But the differential $d\pi : \mathcal{T}_{X^k} \to \pi^* \mathcal{T}_{X^{(k)}}$, being a map between vector bundles of equal rank, would necessarily be singular in codimension 1, a contradiction.

It remains to see that if X is a smooth curve then $X^{(k)}$ will be smooth. This in fact follows from the special case $X = \mathbb{P}^1$ described in Proposition 10.5: in the analytic topology, any collection of points p_i on any smooth curve X have neighborhoods isomorphic to open subsets of $\mathbb{P}^1$, and it follows that any point of $X^{(k)}$ has a neighborhood isomorphic to an open subset of $\mathbb{P}^k$. $\qquad\square$

The symmetric powers of a smooth curve C are central to the analysis of the geometry of C, as we will see illustrated in Appendix D. We can think of a point of $C^{(k)}$ as a subscheme $D \subset C$, and use notation such as $D \cup D'$ and $D \cap D'$ accordingly. In fact, $C^{(k)}$ is isomorphic to the Hilbert scheme of subschemes of C with constant Hilbert polynomial k — that is, zero-dimensional subschemes of degree k (see Arbarello et al. [1985] for a proof). When $\dim X > 1$ or X is singular, a point on $X^{(k)}$ does *not* in general determine a subscheme of X, and the Hilbert schemes $\mathcal{H}_k(X)$ are often more useful.

10.3.2 Secant varieties in general

In this subsection we will prove a basic result related to the dimension of secant varieties. Then we will state without proof some general results that may help to orient the reader. In the following two sections we will prove a number of results about the secant varieties of rational curves.

Let $X \subset \mathbb{P}^r$ be a projective variety of dimension n not contained in a hyperplane. Since $m \leq r + 1$ general points of X are linearly independent, for any $m \leq r$ we have a rational map

$$\tau : X^{(m)} \dashrightarrow \mathbb{G}(m - 1, r),$$

called the *secant plane map*, sending a general m-tuple $p_1 + \cdots + p_m$ to the span $\overline{p_1, \ldots, p_m} \cong \mathbb{P}^{m-1} \subset \mathbb{P}^r$. (In coordinates: If $p_i = (x_{i,0}, \ldots, x_{i,r})$, then τ is given by the maximal minors of the matrix $(x_{i,j})$.) We define the *locus of secant $(m-1)$-planes* to X to be the image $\Psi_m(X) \subset \mathbb{G}(m - 1, r)$ of the rational map τ — that is, the closure in $\mathbb{G}(m-1, r)$ of the locus of $(m-1)$-planes spanned by m linearly independent points of X. Finally, the variety

$$\mathrm{Sec}_m(X) = \bigcup_{\Lambda \in \Psi_m(X)} \Lambda \subset \mathbb{P}^r$$

is called the *m-th secant variety of* X.

Caution: If $\Lambda \in \Psi_m$ and $\Lambda \cap X$ is finite, then $\deg(\Lambda \cap X) \geq m$, but the converse is false; Exercise 10.24 suggests an example of this.

If $n > 1$ and $m > 1$ then the secant plane map $\tau : X^{(m)} \dashrightarrow \mathbb{G}(m-1, r)$ is never regular: When a point $p \in X$ on a variety of dimension 2 or more approaches another point $q \in X$, the limiting position of the secant line $\overline{p, q}$ necessarily depends on the direction of approach. (When X is a curve and q a smooth point of X, the limit is always the tangent line $\mathbb{T}_q X$.) This illustrates the point that — in this context, at least — the Hilbert scheme $\mathcal{H}_m(X)$ may be a better compactification of the space of unordered m-tuples of points on X than the symmetric power: When $m = 2$, for example, the map $\tilde{\tau} : \mathcal{H}_2(X) \to \mathbb{G}(1, r)$ sending a subscheme of length 2 to its span is always regular. Further, if we fix m and replace the embedding $X \subset \mathbb{P}^r$ by a sufficiently high Veronese re-embedding, then every length-m subscheme of X will span an $m-1$ plane, so the map $\mathcal{H}_m(X) \to \mathbb{G}(m-1, r)$ will be regular. In this chapter, we will care only about the image of τ, so it does not matter which we use.

We begin with the dimension of $\mathrm{Sec}_m(X)$:

Proposition 10.7. *If* $m \leq r - n$, *then the map* τ *is birational onto its image; in particular,* $\Psi_m(X)$ *has dimension* $\dim X^{(m)} = mn$.

This is slightly more subtle than it might at first appear. The first case would be the statement that if $C \subset \mathbb{P}^3$ is a nondegenerate curve then the line joining two general points of C does not meet C a third time. Though intuitively plausible, this is tricky to prove, and requires the hypothesis of characteristic 0. For the proof we will use the following general position result:

Lemma 10.8 (General position lemma). *If* $X \subset \mathbb{P}^r$ *is a nondegenerate variety of dimension* n *and* $\Gamma \cong \mathbb{P}^{r-n} \subset \mathbb{P}^r$ *a general linear subspace of complementary dimension, then the points of* $\Gamma \cap X$ *are in linear general position; that is, any* $r-n+1$ *of them span* Γ.

We will not prove this here; a good reference is the discussion of the *uniform position lemma* in Arbarello et al. [1985, Section III.1]).

Proof of Proposition 10.7: The proposition amounts to the claim that if $p_1, \ldots, p_m \in X$ are general points, then the plane $\overline{p_1, \ldots, p_m}$ they span contains no other points of X.

To prove this, let $U \subset X^{(m)}$ be the open subset of m-tuples of distinct, linearly independent points, and consider the incidence correspondence

$$\Psi = \{(p_1 + \cdots + p_m, \Gamma) \in U \times \mathbb{G}(r - n, r) \mid p_1, \ldots, p_m \in \Gamma\}.$$

Via projection on the first factor, we see that Ψ is irreducible, and by Lemma 10.8 it dominates $\mathbb{G}(r - n, r)$; it follows that *a general $(r - n)$-plane Γ containing m general points $p_1, \ldots, p_m \in X$ is a general $(r - n)$-plane in $\mathbb{P}^r$*, and applying Lemma 10.8 again we deduce that the $(m - 1)$-plane $\overline{p_1, \ldots, p_m}$ contains no other points of X. $\quad\square$

Let

$$\Phi = \{(\Lambda, p) \in \mathbb{G}(m - 1, r) \times \mathbb{P}^r \mid p \in \Lambda\}$$

be the universal $(m - 1)$-plane in $\mathbb{P}^r$, with projection maps

$$
\begin{array}{ccc}
\Phi & \xrightarrow{\quad \eta \quad} & \mathbb{P}^r \\
{\scriptstyle \pi} \downarrow & & \\
\mathbb{G}(m - 1, r). & &
\end{array}
$$

Set

$$\Phi_m(X) = \pi^{-1}(\Psi_m(X)) \quad \text{and} \quad \eta_X = \eta|_{\Phi_m(X)},$$

so that the m-th secant variety $\mathrm{Sec}_m(X)$ is the image of η_X. We will call $\Phi_m(X)$ the *abstract secant variety*.

Projection on the first factor shows $\Phi_m(X)$ is irreducible of dimension $mn + m - 1$, so that $\dim \mathrm{Sec}_m(X) \leq mn + m - 1$, with equality holding when a general point on $\mathrm{Sec}_m(X)$ lies on only finitely many m-secant $(m - 1)$-planes to X. By way of language, if $X \subset \mathbb{P}^r$ has dimension n we will call $\min(mn + m - 1, r)$ the *expected dimension* of the secant variety $\mathrm{Sec}_m(X)$; we will say that X is *m-defective* if $\dim \mathrm{Sec}_m(X) < \min(mn + m - 1, r)$, and *defective* if it is m-defective for some m.

Everyone's favorite example of a defective variety is the Veronese surface in $\mathbb{P}^5$:

Proposition 10.9. *The Veronese surface $X = \nu_2(\mathbb{P}^2) \subset \mathbb{P}^5$ is 2-defective.*

In fact, the Veronese surface is the *only* 2-defective smooth projective surface! This much more difficult theorem was asserted, and partially proven, by Severi. The proof was completed by Moishezon in characteristic 0; see Dale [1985] for a modern treatment that works in all characteristics.

Proof: The Veronese surface may be realized as the locus where a symmetric 3×3 matrix

$$M = \begin{pmatrix} z_0 & z_1 & z_2 \\ z_1 & z_3 & z_4 \\ z_2 & z_4 & z_5 \end{pmatrix}$$

has rank 1. But if M has rank 1 at two points $p, q \in \mathbb{P}^5$, then M has rank at most 2 at any point of the form $\lambda p + \mu q$. Thus the determinant of M vanishes on the whole line spanned by p and q, so the cubic form $\det M$ vanishes on the secant locus $\mathrm{Sec}_2(X)$. Thus $\dim \mathrm{Sec}_2(X) \leq 5 - 1 = 4$, not $2 \times 2 + 1 = 5$. $\square$

We can give a more geometric proof using a basic result introduced by Terracini [1911]:

Proposition 10.10 (Terracini's lemma). *Let $X \subset \mathbb{P}^r$ be a variety and $p_1, \ldots, p_m \in X$ linearly independent smooth points of X. If $p \in \Gamma = \overline{p_1, \ldots, p_m}$ is any point in their span not in the span of any proper subset, then the image of the differential $d\eta_X$ at the point $(\Gamma, p) \in \Phi_m(X)$ is the span*

$$\mathrm{Im}\, d\eta_X = \overline{\mathbb{T}_{p_1} X, \ldots, \mathbb{T}_{p_m} X}$$

of the tangent planes to X at the points p_i. In particular, if X has dimension n and $r \geq mn + m - 1$, then X is m-defective if and only if its tangent spaces at m general points are dependent.

For a proof, see Landsberg [2012].

We can use Terracini's lemma to see that the Veronese surface $X \subset \mathbb{P}^5$ is 2-defective as follows: A hyperplane $H \subset \mathbb{P}^5$ contains the tangent plane to X at a point p if and only if the curve $H \cap X$ is singular at p. Of course we can consider $H \cap X$ as a conic in $\mathbb{P}^2 \cong X$, and, from the definition of the Veronese surface, we see that every conic appears in this way. Now, two planes in $\mathbb{P}^5$ are dependent if and only if they are both contained in a hyperplane. Putting this together with Terracini's lemma, we see that to show that X is 2-defective we must show that given any two points in $\mathbb{P}^2$ there is a conic in $\mathbb{P}^2$ that is singular at both these points; of course, the double line passing through the points is such a conic.

We can also use Terracini's lemma to show that there are no defective curves:

Proposition 10.11. *If $C \subset \mathbb{P}^r$ is a nondegenerate reduced irreducible curve, then $\dim \mathrm{Sec}_m(X) = \min(2m - 1, r)$ for every m.*

Proof: By Terracini's lemma, it suffices to show that if $p_1, \ldots, p_m \in C$ are general points then the tangent lines $\mathbb{T}_{p_i} C$ are linearly independent when $2m - 1 \leq r$ and span $\mathbb{P}^r$ when $2m - 1 \geq r$.

We have already seen in the proof of Theorem 7.13 that for a general point $p \in C$ the divisor $(r + 1)p$ spans $\mathbb{P}^r$ (that is, a general point $p \in C$ is not inflectionary); it follows that the divisor $2m \cdot p$ spans a $\mathbb{P}^{2m-1}$ when $2m \leq r + 1$ and spans $\mathbb{P}^r$ when $2m \geq r + 1$. By lower-semicontinuity of rank, it follows that for general $p_1, \ldots, p_m$ the divisor $2p_1 + \cdots + 2p_m$ has span of the same dimension. $\square$

The general question of which nondegenerate varieties are defective is a fascinating one, with a long history. Perhaps because of Terracini's lemma, which relates the issue to the question of when multiples of general points impose independent conditions on polynomials (the *interpolation problem*), the case of Veronese embeddings of projective spaces has attracted a great deal of attention. The following is a result of Alexander and Hirschowitz [1995]. The proof was later simplified by Karen Chandler, and an exposition of this version, with a further simplification, can be found in Brambilla and Ottaviani [2008].

Theorem 10.12. *The defective Veronese varieties are the following:*

- $v_2(\mathbb{P}^n)$ *is 2-defective for any n.*
- $v_4(\mathbb{P}^2)$ *is 5-defective.*
- $v_4(\mathbb{P}^3)$ *is 9-defective.*
- $v_3(\mathbb{P}^4)$ *is 7-defective.*
- $v_4(\mathbb{P}^4)$ *is 14-defective.*

We will see in Exercises 10.26–10.29 that the Veronese varieties listed in the theorem are indeed defective (the hard part is the converse!). Note that by Terracini's lemma Theorem 10.12 implies (and indeed is equivalent to) the following corollary:

Corollary 10.13. *Let $p_1, \ldots, p_m$ be general points in $\mathbb{A}^n$, and let d be any positive integer such that $\binom{d+n}{n} \geq m(n+1)$. There exists a polynomial f of degree d on $\mathbb{A}^n$ with specified values and derivatives at the points p_i, except in the cases $d = 2$ and $(n, d, m) = (2, 4, 5), (3, 4, 9), (4, 3, 7)$ and $(4, 4, 14)$.*

10.4 Secant varieties of rational normal curves

We turn now from secant varieties in general to the special case of rational curves. Every rational curve is the projection of a rational normal curve, and its secant varieties are correspondingly projections of the secant varieties of the rational normal curve, so we will focus initially on that case.

10.4.1 Secants to rational normal curves

We begin with the observation that finite sets of points on a rational normal curve are always "as independent as possible." (This property actually characterizes rational normal curves, as we invite the reader to show in Exercise 10.31.)

Lemma 10.14. *Let $C \subset \mathbb{P}^d$ be a rational normal curve. If $D \subset C$ is a divisor of degree $m \leq d + 1$, then D is not contained in any linear subspace of $\mathbb{P}^d$ of dimension $< m - 1$.*

Informally: Any finite subscheme $D \subset C$ of length $\leq d + 1$ is linearly independent, in the sense that the map $H^0(\mathcal{O}_{\mathbb{P}^d}(1)) \to H^0(\mathcal{O}_D(1))$ is surjective. On an affine subset of $\mathbb{P}^1$ the parametrization of the rational normal curve looks like $t \mapsto (1, t, t^2, \ldots, t^d)$, so the independence of the images of any $d + 1$ points $a_0, \ldots, a_d$ is given by the nonvanishing of the Vandermonde determinant

$$\det \begin{pmatrix} 1 & a_0 & \cdots & a_0^d \\ \vdots & \vdots & \ddots & \vdots \\ 1 & a_d & \cdots & a_d^d \end{pmatrix} = \prod_{i < j} (a_i - a_j).$$

Proof: If D were contained in a linear subspace L of dimension $n < m - 1$, then adding $d - n - 1$ general points to D we would arrive at a divisor $D' \subset C$ of degree $m + d - n - 1$ contained in a hyperplane H. Since C is not contained in a hyperplane, the intersection $H \cap C$ is finite, and we deduce the contradiction $\deg C > d$. $\qquad\qquad \square$

In Section 10.3.2 we described the secant map as a regular map on an open set, that is, as a rational map

$$\tau : C^{(m)} \dashrightarrow \mathbb{G}(m - 1, d).$$

One consequence of Lemma 10.14 is that the secant plane map has a natural extension to an injective map of sets. It is not hard to show that τ is actually a morphism, and in fact an embedding. We can thus regard the restriction Φ_C to the image of τ of the universal $\mathbb{P}^{m-1}$-bundle over $\mathbb{G}(m - 1, d)$ as a $\mathbb{P}^{m-1}$-bundle over $(\mathbb{P}^1)^{(m)} = \mathbb{P}^m$, and $\mathrm{Sec}_m(C)$ is the image of this bundle.

Proposition 10.15. *Let $C \subset \mathbb{P}^d$ be a rational normal curve. When $2m - 1 \leq d$, the map $\eta_C : \Phi_m(C) \to \mathbb{P}^d$ is birational onto its image $\mathrm{Sec}_m(C)$; more precisely, it is one-to-one over the complement of $\mathrm{Sec}_{m-1}(C)$ in $\mathrm{Sec}_m(C)$.*

Proof: Suppose a point $p \in \mathbb{P}^d$ is the image of two different points of $\Phi_m(C)$, say $(\overline{D}, p)$ and $(\overline{D}', p)$. Let $k = \dim(\overline{D} \cap \overline{D}')$; note that $0 < k < m - 1$. Since the span of $\overline{D}$ and $\overline{D}'$ has dimension $2m - 2 - k$, by Lemma 10.14 the union (as subschemes of C) of D and D' can have degree at most $2m - k - 1$. It follows that the intersection $D \cap D'$ (again, as subschemes of C) has degree at least $k + 1$. Thus $\overline{D} \cap \overline{D}'$ is a secant k-plane, and $p \in S_k(C) \subset S_{m-1}(C)$. $\qquad\qquad \square$

Proposition 10.15 is not particularly remarkable in case $2m - 1 < d$: all irreducible, nondegenerate curves $C \subset \mathbb{P}^d$ have the property that when $2m - 1 < d$ a general point on the m-secant variety $\mathrm{Sec}_m(C)$ lies on a unique m-secant $(m - 1)$-plane to C (see Exercises 10.33–10.35). In case $2m - 1 = d$, however, it is striking. For example, the twisted cubic curve $C \subset \mathbb{P}^3$ is the *unique* nondegenerate space curve whose secant lines sweep out $\mathbb{P}^3$ only once (see Exercise 10.30).

10.4.2 Degrees of the secant varieties

Let $C \subset \mathbb{P}^d$ be a rational normal curve, and m any integer with $2m - 1 \leq d$. Since the secant plane map $\tau : C^{(m)} \to \mathbb{G}(m - 1, d)$ is regular, and $\Phi_C \to \mathrm{Sec}_m(C)$ is birational, it is reasonable to hope that we can answer enumerative questions about the geometry of the varieties $\mathrm{Sec}_m(C)$. We will do this in the remainder of this section and the next, starting with the calculation of the degree of $\mathrm{Sec}_m(C)$.

There are a few cases that we can do without any machinery; for example:

(a) $\mathrm{Sec}_1(C) = C$, so $\deg \mathrm{Sec}_1(C) = d$.
(b) The case $m = 2$ is not quite as trivial, but is readily done: The variety $\mathrm{Sec}_2(C)$ has dimension 3, so its degree is the number of points in which it intersects a general $(d - 3)$-plane Λ. As we saw in Exercise 3.34, the projection of C from Λ to $\mathbb{P}^2$ will map C birationally onto a plane curve C_0 with nodes, and the points of $\Lambda \cap S_2(C)$ correspond to these nodes. Since C_0 has arithmetic genus $\binom{d-1}{2}$ and geometric genus 0, we conclude that $\deg \mathrm{Sec}_2(C) = \binom{d-1}{2}$.
(c) Finally, if d is odd and $m = (d + 1)/2$, then the secant locus is all of $\mathbb{P}^d$, so $\deg \mathrm{Sec}_m(C) = 1$.

In order to go further, we use the Segre class technique of Proposition 10.4. To begin with, the map

$$\Phi_m(C) \xrightarrow{\pi} \Psi_m(C) = \tau((\mathbb{P}^1)^{(m)}) \cong (\mathbb{P}^1)^{(m)} \cong \mathbb{P}^m$$

has the form $\mathbb{P}\mathcal{H} \to \mathbb{P}^m$, where $\mathcal{H}$ is the pullback $\tau^* S$ of the tautological subbundle on $\mathbb{G}(m - 1, d)$ to $\mathbb{P}^m$.

In fact, we have already seen this bundle before, in Section 9.3.3! To see this, let $V = H^0(\mathcal{O}_{\mathbb{P}^1}(d))$. The rational normal curve lives naturally in $\mathbb{P}V^*$, as the locus C of linear functionals on V given by evaluation at a point $p \in \mathbb{P}^1$. The span of a divisor D of degree m on C is the space of linear functionals vanishing on those points, that is, the annihilator in V^* of the subspace $V_D = H^0(\mathcal{I}_{D,\mathbb{P}^1}(d))$. Thus the map $\tau : \mathbb{P}^m \to \mathbb{G}(m - 1, d)$ sends $D \in \mathbb{P}^m$ to the subspace $\mathrm{Ann}(V_D) \subset V^*$, and it follows that the pullback $\tau^* S$ is the dual $\mathcal{F}^*$ of the bundle $\mathcal{F}$ introduced in Section 9.3.3. We have from the results of that section that

$$c(\mathcal{F}) = \frac{1}{(1 - \sigma)^{d-m+1}},$$

where $\sigma \in A^1(\mathbb{P}^m)$ is the hyperplane class. The Segre class is the inverse, and taking the dual we have

$$s(\mathcal{F}^*) = (1 + \sigma)^{d-m+1}.$$

Finally, we can deduce:

Theorem 10.16. *If $C \subset \mathbb{P}^d$ is a rational normal curve, then, for $2m - 1 \leq d$,*

$$\deg \operatorname{Sec}_m(C) = \binom{d-m+1}{m}.$$

Note that in the case $d = 2m - 1$, the calculation reaffirms the conclusion of Proposition 10.15 that the m-secant planes to C sweep out $\mathbb{P}^d$ exactly once.

10.4.3 Expression of a form as a sum of powers

We can now answer Keynote Question (b): If f and g are general polynomials of degree $d = 2m$ in one variable, how many linear combinations of f and g are expressible as a sum of m d-th powers of linear forms?

This question is related to secants of rational normal curves, because if we realize $\mathbb{P}^d$ as the projective space of forms of degree d on $\mathbb{P}^1$ then the curve of pure d-th powers is a rational normal curve — it is the image of the morphism

$$\mu : \mathbb{P}^1 \ni (s, t) \mapsto \left(s^d, ds^{d-1}t, \binom{d}{2}s^{d-2}t, \ldots, t^d \right) \in \mathbb{P}^d.$$

(Note that we are relying here on the hypothesis of characteristic 0: If, for example, d is equal to the characteristic, then μ is a purely inseparable map whose image is a line!)

A point $p \in \mathbb{P}^d$ lies on the plane spanned by distinct points $q_1, \ldots, q_m \in C$ if and only if the homogeneous coordinates of p can be expressed as a linear combination of the homogeneous coordinates of $q_1, \ldots, q_m$. Thus a form of degree d is a linear combination of m d-th powers of linear forms if and only if the corresponding point in $\mathbb{P}^d$ lies in the union of the m-secant $(m - 1)$-planes to $\mu(\mathbb{P}^1)$, and questions about the expression of a polynomial as a sum of powers become questions about the secants.

There is an important subtlety: It is *not* the case that every point of $\operatorname{Sec}_m(C)$ corresponds to a polynomial that is expressible as a sum of m d-th powers! For example, the tangent lines to C are contained in $\operatorname{Sec}_2(C)$. If $d \geq 3$, then no 2-plane in $\mathbb{P}^d$ meets the rational normal curve in four points, so a tangent line to C cannot meet any other secant line at a point off C. Thus the points on the tangent lines away from C are points of $\operatorname{Sec}_2(C)$ that cannot be expressed as the sum of two pure d-th powers.

(The points on the tangent lines do have an interesting characterization, however: At the point corresponding to the polynomial $f(t) = (t - \lambda)^d$, the tangent line is the set of linear combinations of f and $\partial f / \partial t$, or equivalently the set of polynomials that have $d - 1$ roots equal to λ.)

By definition, $\operatorname{Sec}_m(C)$ contains an open set consisting of points on secant $(m - 1)$-planes spanned by m distinct points of C. Further, by Proposition 10.15 a point in the open subset $\operatorname{Sec}_m(C) \setminus \operatorname{Sec}_{m-1}(C)$ of $\operatorname{Sec}_m(C)$ lies on the span of a unique divisor of degree m. Thus Theorem 10.16 yields the answer to Keynote Question (b), and even a generalization:

Corollary 10.17. *If $d \geq 2m - 1$, then the number of linear combinations of $d - 2m + 2$ general forms of degree d that can be expressed as the sum of m pure d-th powers is* $\deg \mathrm{Sec}_m(C) = \binom{d-m+1}{m}$.

10.5 Special secant planes

For a curve $C \subset \mathbb{P}^r$ other than a rational normal curve, it is interesting to consider the subspaces that meet C in a dependent set of points; these are called *special secant planes*. Examples of this that we will investigate below include trisecant and quadrisecant lines to a curve $C \subset \mathbb{P}^3$, and trisecant lines to a curve $C \subset \mathbb{P}^4$.

We start, as usual, with the question of dimension: When would we expect a curve $C \subset \mathbb{P}^r$ to contain m points lying in a $\mathbb{P}^{m-1-k}$-plane? What would be the expected dimension of the locus $C_k^{(m)} \subset C^{(m)}$ of such m-tuples?

There are many ways to set this up. One would be to express the locus of such m-tuples as a determinantal variety: If the coordinates of points $p_1, \ldots, p_m \in C$ are the rows of the matrix

$$
M = \begin{pmatrix} x_{1,0} & x_{1,1} & \cdots & x_{1,r} \\ \vdots & \vdots & \ddots & \vdots \\ x_{m,0} & x_{m,1} & \cdots & x_{m,r} \end{pmatrix},
$$

then $C_k^{(m)}$ is just the locus where this matrix has rank $m - k$ or less. Now, in the space of $m \times (r + 1)$ matrices, those of rank $m - k$ or less have codimension $k(r + 1 - m + k)$, so we would expect the locus $C_k^{(m)}$ of m-tuples spanning only a $\mathbb{P}^{m-1-k}$ to have dimension

$$
m - k(r + 1 - m + k) = (k + 1)(m - r - k) + r.
$$

An alternative in the case C is a rational curve would be to express C as the projection $\pi_\Lambda : \tilde{C} \to C$ of a rational normal curve $\tilde{C} \subset \mathbb{P}^d$ from a plane $\Lambda \cong \mathbb{P}^{d-r-1}$. The m-secant $(m - 1 - k)$-planes to C then correspond to the m-secant $(m - 1)$-planes to $\tilde{C}$ that intersect Λ in a $(k - 1)$-plane, that is, the preimage under the secant plane map $\tau : C^{(m)} \to \mathbb{G}(m - 1, d)$ of the Schubert cycle

$$
\Sigma_{(r+1-m+k)^k}(\Lambda) = \{\Gamma \in \mathbb{G}(m - 1, d) \mid \dim(\Gamma \cap \Lambda) \geq k - 1\}.
$$

This Schubert cycle has codimension $k(r + 1 - m + k)$, so again we would expect the preimage to have dimension $m - k(r + 1 - m + k)$.

The first three cases (with $k > 0$) are:

(a) trisecants to a curve $C \subset \mathbb{P}^3$ (that is, $r = 3$, $m = 3$ and $k = 1$), where we expect a one-parameter family;

(b) quadrisecants to a curve $C \subset \mathbb{P}^3$ (that is, $r = 3$, $m = 4$ and $k = 2$), where we expect finitely many; and

(c) trisecants to a curve $C \subset \mathbb{P}^4$ (that is, $r = 4$, $m = 3$ and $k = 1$), where, again, we expect finitely many.

That our expectations for the dimensions of these loci are indeed the case for general rational curves is shown in Exercise 10.36, though it is *not* necessarily true of a general point on any component of the Hilbert scheme of curves of higher genus, as shown in Exercise 10.37.

In this section, we will show how to count the trisecants to a general rational curve in $\mathbb{P}^4$, answering Keynote Question (c), and we will determine the degree of the trisecant surface of a general rational curve in $\mathbb{P}^3$, answering Keynote Question (d). We leave the case of quadrisecants to a rational curve in $\mathbb{P}^3$ to Exercise 10.38 for now; it will also be a direct application of Porteous' formula in Section 12.4.4.

10.5.1 The class of the locus of secant planes

In case the curve C is rational, the answers to all of the above questions come directly from the answer to a question we have not yet addressed directly: If $C \subset \mathbb{P}^d$ is a rational normal curve, $\tau : C^{(m)} \cong \mathbb{P}^m \to \mathbb{G}(m-1, d)$ the secant plane map and $\Psi_m(C) \subset \mathbb{G}(m-1, d)$ the image of τ, *what is the class* $[\Psi_m(C)] \in A_m(\mathbb{G}(m-1, d))$?

We have all the tools to answer this question at hand: We know that the pullback $\tau^* \mathcal{S}^*$ of the dual of the universal subbundle $\mathcal{S}$ on $\mathbb{G}(m-1, d)$ is the bundle $\mathcal{E}^*$ whose Chern classes we gave in Section 10.4.2. We know that $c_i(\mathcal{S}^*) = \sigma_{1^i}$, so this says that

$$\tau^* \sigma_{1^i} = \binom{d - m + i}{i} \zeta^i \in A^i(\mathbb{P}^m),$$

where ζ as usual is the hyperplane class in $C^{(m)} \cong \mathbb{P}^m$. Equivalently, since we also know that the Segre class of $\mathcal{S}$ is $s(\mathcal{S}) = 1 + \sigma_1 + \sigma_2 + \cdots + \sigma_{d-m}$, we have

$$\tau^* \sigma_i = \binom{d - m + 1}{i} \zeta^i \in A^i(\mathbb{P}^m). \tag{10.1}$$

Since the classes σ_i generate the Chow ring of $\mathbb{G}(m-1, d)$ (and τ is an embedding), this determines the class of the image. We will use this idea to compute $[\Psi_m(C)]$ explicitly in the cases below. It is an interesting fact that the map $\tau : C^{(m)} \cong \mathbb{P}^m \to \mathbb{G}(m-1, d)$ composed with the Plücker embedding of the Grassmannian is the d-th Veronese map on $\mathbb{P}^m$; see Exercise 10.32

Trisecants to a rational curve in $\mathbb{P}^4$

How many trisecant lines does a general rational curve of degree d in $\mathbb{P}^4$ possess? We already know the answer in at least two cases. First, there are no trisecant lines to a rational normal curve in the case $d = 4$. If $d = 5$, Proposition 10.15 says that a general point $p \in \mathbb{P}^5$ lies on a unique 3-secant 2-plane to a rational normal curve $\tilde{C} \subset \mathbb{P}^5$, and thus the projection of that curve from a general point has just one trisecant line.

The general case may be handled similarly to the case $d = 5$ above: We use the fact that a rational curve $C \subset \mathbb{P}^4$ of degree d is the projection of a rational normal curve $\tilde{C} \subset \mathbb{P}^d$ from a $(d - 5)$-plane $\Lambda \subset \mathbb{P}^d$. The trisecant lines to C then correspond to 3-secant 2-planes to $\tilde{C}$ of degree d that meet Λ. The trisecant lines to C correspond to the intersection of the Schubert cycle $\Sigma_3(\Lambda) \subset \mathbb{G}(2, d)$ of 2-planes meeting Λ with the cycle $\Psi_3(\tilde{C}) \subset \mathbb{G}(2, d)$ of 3-secant 2-planes to $\tilde{C}$.

This gives the answer to Keynote Question (c):

Proposition 10.18. *If $C \subset \mathbb{P}^4$ is a general rational curve of degree d, then C has $\binom{d-2}{3}$ trisecant lines.*

Proof: Since C is general, it is the projection of the rational normal curve $\tilde{C}$ from a general $(d - 5)$-plane Λ. The number of trisecant lines is the number of points in which $\Sigma_3(\Lambda)$ meets $\Psi_3(\tilde{C})$. By Kleiman transversality, this is the degree of the intersection class $[\Psi_3(C)]\sigma_3$, or equivalently the degree of the pullback $\tau^*\sigma_3$; by the above, this is $\binom{d-2}{3}$. $\qquad\square$

Trisecants to a rational curve in $\mathbb{P}^3$

We next turn to Keynote Question (d): If $C \subset \mathbb{P}^3$ is a general rational curve of degree d, what is the degree of the surface $S \subset \mathbb{P}^3$ swept out by the 3-secant lines to C?

Again we already know the answer in the simplest cases: 0 in the case $d = 3$ (a rational normal curve has no trisecants); and 2 in the case $d = 4$, since a smooth rational quartic is a curve of type $(1, 3)$ on a quadric surface $Q \subset \mathbb{P}^3$, and the trisecants of C comprise one ruling of Q.

To set up the general case, let C be the projection $\pi_\Lambda(\tilde{C})$ of a rational normal curve $\tilde{C} \subset \mathbb{P}^d$ from a general plane $\Lambda \cong \mathbb{P}^{d-4}$; let $L \subset \mathbb{P}^3$ be a general line, and let $\Gamma = \pi_\Lambda^{-1}(L) \subset \mathbb{P}^d$ be the corresponding $(d - 2)$-plane containing Λ. The points of intersection of L with S correspond to 3-secant 2-planes to $\tilde{C} \subset \mathbb{P}^d$ that

(a) meet Λ, and

(b) intersect Γ in a line.

These are the points of intersection of $\Psi_3(\tilde{C})$ with the Schubert cycle $\Sigma_{2,1}(\Lambda, \Gamma)$. Kleiman transversality shows that the cardinality of this intersection is the degree of the pullback $\tau^*(\sigma_{2,1})$.

To evaluate this we express $\sigma_{2,1}$ as a polynomial in $\sigma_1, \sigma_2, \dots$ and evaluate each term using (10.1). Giambelli's formula (Proposition 4.16) tells us that

$$\sigma_{2,1} = \begin{vmatrix} \sigma_2 & \sigma_3 \\ \sigma_0 & \sigma_1 \end{vmatrix} = \sigma_1\sigma_2 - \sigma_3$$

(an equality we could readily derive by hand). By (10.1),

$$\deg \tau^*(\sigma_1\sigma_2) = \binom{d-2}{1}\binom{d-2}{2} \quad \text{and} \quad \deg \tau^*(\sigma_3) = \binom{d-2}{3}.$$

Putting these things together, we have the answer to the question:

Proposition 10.19. *If $C \subset \mathbb{P}^3$ is a general rational curve of degree d, then the degree of the surface swept out by trisecant lines to C is*

$$\binom{d-2}{1}\binom{d-2}{2} - \binom{d-2}{3} = 2\binom{d-1}{3}.$$

10.5.2 Secants to curves of positive genus

It is instructive to ask whether we could extend the computations of Sections 10.4 and 10.5 to curves other than rational ones. There is one key problem: in treating rational curves, we made essential use of the fact that the space of effective divisors of degree m on $\mathbb{P}^1$ is the variety $\mathbb{P}^m$, whose Chow ring we know. But when C has positive genus, the Chow rings $A(C^{(k)})$ of symmetric powers of C are unknown.

In the case of $g = 1$ this is not an insurmountable problem; it is the content of Exercises 10.48 and 10.49. For genera $g \geq 2$, however, it typically necessitates the use of a coarser equivalence relation on cycles, such as homology rather than rational equivalence. Given this framework, however, it is indeed possible to extend the results of this chapter to curves of arbitrary genus; see (Arbarello et al. [1985, Chapter 8]), where there are explicit formulas generalizing all the above formulas to arbitrary genus.

10.6 Dual varieties and conormal varieties

We next turn to a remarkable property of projective varieties called *reflexivity*. A corollary of reflexivity is the deep fact that the dual of the dual of a variety is the variety itself. See Kleiman [1986] for a comprehensive account of the history of these matters (our account is based on that in Kleiman [1984]). We emphasize that the statements below are very much dependent on the hypothesis of characteristic 0; see the references above for the characteristic p case.

Let $X \subset \mathbb{P}^n$ be a subvariety of dimension k. If X is smooth, we define the *conormal variety* $CX \subset \mathbb{P}^n \times \mathbb{P}^{n*}$ to be the incidence correspondence

$$CX = \{(p, H) \in \mathbb{P}^n \times \mathbb{P}^{n*} \mid p \in X \text{ and } \mathbb{T}_p X \subset H\}.$$

If X is singular, we define CX to be the closure in $\mathbb{P}^n \times \mathbb{P}^{n*}$ of the locus CX° of such pairs (p, H), where p is a smooth point of X. Whatever the dimension of X, the conormal variety CX will have dimension $n - 1$: it is the closure of the locus CX°, which maps onto the smooth locus of X with fibers of dimension $n - k - 1$. The *dual variety* $X^* \subset \mathbb{P}^{n*}$ of X is the image of CX under projection on the second factor.

In these terms, we can state the main theorem of this section:

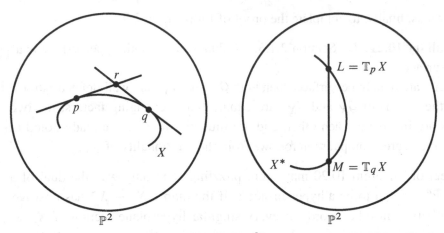

Figure 10.1 The tangent line to $X^* \subset \mathbb{P}^{2*}$ at $L = \mathbb{T}_p X$ is the line dual to p.

Theorem 10.20 (Reflexivity). *If $X \subset \mathbb{P}^n$ is any variety and $X^* \subset \mathbb{P}^{n*}$ its dual, then the conormal variety $CX \subset \mathbb{P}^n \times \mathbb{P}^{n*}$ is equal to $C(X^*) \subset \mathbb{P}^{n*} \times \mathbb{P}^n$ with the factors reversed. It follows that $(X^*)^* = X$ — that is, the dual of the dual of X is X.*

For example, if X is a plane curve, then the statement $X^{**} = X$ says that if $p \in X$ is a smooth point then the tangent line to X^* at the point $L = \mathbb{T}_p X$ is the line $p^* \subset \mathbb{P}^{2*}$ of lines through p. More picturesquely put, the tangent lines to points near $x \in X$ "roll" on the point x. It is true more generally that the osculating k-planes to a smooth curve $X \subset \mathbb{P}^n$ at points near $p \in X$ move, to first order, by rotating around the osculating $(k - 1)$-plane to X at p while staying in the osculating $(k + 1)$-plane to X at p (see Exercise 10.47).

This picture, for plane curves, can be made precise as follows. Observe that if $p \in X$ is a smooth point, then the tangent line $\mathbb{T}_p X \subset \mathbb{P}^2$ is the limit of the secant lines $\overline{p, q}$ as $q \in X$ approaches p. Applied to the dual curve $X^* \subset \mathbb{P}^{2*}$, this says that the tangent line $\mathbb{T}_L X^* \subset \mathbb{P}^{2*}$ to the curve X^* at a point L is the limit of the secant lines $\overline{L, M}$ as $M \in X^*$ approaches L. But the line $\overline{L, M} \subset \mathbb{P}^{2*}$ joining two points $L, M \in \mathbb{P}^{2*}$ corresponding to lines $L, M \subset \mathbb{P}^2$ is the line in $\mathbb{P}^{2*}$ dual to the point $L \cap M$ in $\mathbb{P}^2$.

Now, the equality $X^{**} = X$ means that the tangent line to X^* at the point $L = \mathbb{T}_p X$ corresponding to $p \in X$ is p itself; this amounts to saying that the limit as $q \in X$ approaches p of the point of intersection $r = \mathbb{T}_p X \cap \mathbb{T}_q X$ is just the point p itself, which is clear from Figure 10.1.

Combining Theorem 10.20 with the argument at the beginning of Section 2.1.3, we see that the Gauss map of a smooth hypersurface is birational as well as finite:

Corollary 10.21. *If X is a hypersurface whose dual is also a hypersurface, then the Gauss map $\mathcal{G}_X : X \to X^*$ is birational, with inverse $\mathcal{G}_{X^*} : X^* \to X$. Thus if $X \subset \mathbb{P}^n$ is a smooth hypersurface of degree d then the Gauss map is finite and birational, and X^* is a hypersurface of degree $d(d - 1)^{n-1}$.*

This allows us, finally, to complete the proof of Proposition 2.9.

Proof of Corollary 10.21: By Section 2.1.3, the dual of a smooth hypersurface is always a hypersurface.

If X and X^* are both hypersurfaces then both $\mathcal{G}_X$ and $\mathcal{G}_{X^*}$ are well-defined rational maps. Since the graphs of $\mathcal{G}_X$ and $\mathcal{G}_{X^*}$ are equal after exchanging factors, the two rational maps are inverse to each other, and are thus birational. As already noted in Section 2.1.3, the degree computation follows from the birationality of $\mathcal{G}_X$. $\qquad\square$

One aspect of Theorem 10.20 may seem puzzling. The only way the dual of a variety $X \subset \mathbb{P}^n$ can fail to be a hypersurface is if the map $CX \to X^*$ has positive-dimensional fibers — in other words, if every singular hyperplane section of X has positive-dimensional singular locus. This is a rare circumstance; as we will see in Exercise 10.42, it can never be the case for a smooth complete intersection, and, as we will see in Exercise 10.44, it can only happen for X swept out by positive-dimensional linear spaces. But if we have a one-to-one correspondence between varieties $X \subset \mathbb{P}^n$ and their dual varieties, we seem to be suggesting that there are as many hypersurfaces as varieties of arbitrary dimension in $\mathbb{P}^n$! The discrepancy is due to the fact that the duals of smooth varieties tend to be highly singular — see, for example, Exercise 10.45.

There are many fascinating results about the geometry of dual varieties and conormal varieties. We recommend in particular Kleiman [1986], and the surprising and beautiful theorems of Ein and Landman (see Ein [1986]) and Zak [1991]. Ein and Landman proved, for example, that for any smooth variety $X \subset \mathbb{P}^n$ of dimension d in characteristic 0 the difference $(n-1) - \dim X^*$ is congruent to $\dim X$ modulo 2!

As we mentioned earlier, it is relatively rare for the dual X^* of a smooth variety $X \subset \mathbb{P}^n$ to not be a hypersurface. Exercises 10.43 and 10.46 give two circumstances where it does happen.

10.6.1 The universal hyperplane as projectivized cotangent bundle

The proof of the reflexivity theorem will make use of the *universal hyperplane*

$$\Psi = \{(p, H) \in \mathbb{P}^n \times \mathbb{P}^{n*} \mid p \in H\},$$

a special case of the universal k-plane introduced in Section 3.2.3 and analyzed further in Section 9.3.1. To express it another way, let V be an $(n+1)$-dimensional vector space and $W = V^*$ its dual; we can then write

$$\Psi = \{(v, w) \in \mathbb{P}V \times \mathbb{P}W \mid w(v) = 0\}.$$

Write the tautological sequence on $\mathbb{P}V$ as

$$0 \longrightarrow \mathcal{O}_{\mathbb{P}V}(-1) \longrightarrow V \otimes \mathcal{O}_{\mathbb{P}V} \longrightarrow \mathcal{Q} \longrightarrow 0.$$

Thus $Q^* \subset W \otimes \mathcal{O}_{\mathbb{P}V}$. From the inclusion it follows that the line bundle $\mathcal{O}_{\mathbb{P}Q^*}(-1)$ on $\mathbb{P}Q^*$ is the restriction from $\mathbb{P}^n \times \mathbb{P}^{n*}$ of the bundle $\pi_2^* \mathcal{O}_{\mathbb{P}^{n*}}(-1)$. In Section 6.1.1 we observed that $\Psi \subset \mathbb{P}V \times \mathbb{P}W$ may be regarded as the projectivization $\mathbb{P}Q^*$ inside $\mathbb{P}W \otimes \mathcal{O}_{\mathbb{P}V} = \mathbb{P}V \times \mathbb{P}W$.

To simplify notation, we write π_V and π_W for the projections from $Z := \mathbb{P}V \times \mathbb{P}W$ to $\mathbb{P}V$ and $\mathbb{P}W$ respectively, and we set $\mathcal{O}_Z(a, b) := \pi_V^* \mathcal{O}_{\mathbb{P}V}(a) \otimes \pi_W^* \mathcal{O}_{\mathbb{P}W}(b)$. In this language, $\Psi = \pi_V^*(Q^*)$ and $\mathcal{O}_{\mathbb{P}Q^*}(-1)$ is the restriction of $\mathcal{O}_Z(0, -1)$. For our present purpose, we want to give a more symmetric description.

Proposition 10.22. *The map* $\pi_V : \Psi \to \mathbb{P}V$ *may be described as the projectivization of the cotangent bundle* $\mathbb{P}\Omega_{\mathbb{P}V}$ *of* $\mathbb{P}V$*. The tautological subbundle*

$$\mathcal{O}_{\mathbb{P}\Omega_{\mathbb{P}V}}(-1) \subset \pi_V^*(\Omega_{\mathbb{P}V})$$

is the restriction to $\Psi \subset \mathbb{P}V \times \mathbb{P}W = Z$ *of* $\mathcal{O}_Z(-1, -1)$*.*

Proof: The Euler sequence

$$0 \longrightarrow \Omega_{\mathbb{P}V} \longrightarrow W \otimes \mathcal{O}_{\mathbb{P}V}(-1) \longrightarrow \mathcal{O}_{\mathbb{P}V} \longrightarrow 0$$

that may be taken as the definition of $\Omega_{\mathbb{P}V}$ is the tautological sequence twisted by $\mathcal{O}_{\mathbb{P}V}(-1)$, and in particular $\Omega_{\mathbb{P}V} = Q^* \otimes \mathcal{O}_{\mathbb{P}V}(-1)$. By Corollary 9.5, we have $\mathbb{P}Q^* \cong \mathbb{P}\Omega$, with

$$\mathcal{O}_{\mathbb{P}\Omega}(-1) = \mathcal{O}_{\mathbb{P}Q^*} \otimes \pi_V^* \mathcal{O}_{\mathbb{P}V}(-1),$$

and this is the restriction to Ψ of $\mathcal{O}_Z(0, -1) \otimes \pi_V^* \mathcal{O}_{\mathbb{P}V}(-1) = \mathcal{O}_Z(-1, -1)$, as required. $\square$

Proof of Theorem 10.20: If $X \subset \mathbb{P}V$ is any subvariety then, over the open set where X is smooth, the conormal variety $CX \subset \Psi = \mathbb{P}\Omega_{\mathbb{P}V} = \text{Proj Sym } \mathcal{T}_{\mathbb{P}V}$ is the projectivized conormal bundle $\mathbb{P}K = \text{Proj Sym } K^*$, where

$$K := \text{Ker}(\Omega_{\mathbb{P}V}|_X \to \Omega_X).$$

The conormal variety of X itself is defined as the closure of this set. Over the open set where X is smooth, K^* is the cokernel of the map of bundles

$$\mathcal{T}_X \to \mathcal{T}_{\mathbb{P}V}|_X,$$

so $\text{Sym } K^*$ is equal to $\text{Sym } \mathcal{T}_{\mathbb{P}V}|_X$ modulo the ideal generated by $\mathcal{T}_X$, thought of as being contained in the degree-1 part $\mathcal{T}_{\mathbb{P}V}|_X$ of the graded algebra $\text{Sym } \mathcal{T}_{\mathbb{P}V}|_X$. Sheafifying, this means that the ideal sheaf of CX in $\mathcal{O}_{\mathbb{P}\Omega_{\mathbb{P}V}}$ is the image of the composite map u in the diagram

$$\pi_V^* \mathcal{T}_X \otimes \mathcal{O}_{\Omega_{\mathbb{P}^n}}(-1) \longrightarrow \pi_V^* \mathcal{T}_{\mathbb{P}^n} \otimes \mathcal{O}_{\Omega_{\mathbb{P}^n}}(-1)$$

$$\downarrow$$

$$\overset{u}{\searrow} \quad \mathcal{O}_{\mathbb{P}\Omega_{\mathbb{P}^n}}$$

where the vertical map is the dual of

$$\mathcal{O}_{\mathbb{P}\Omega_{\mathbb{P}^n}}(-1) \xrightarrow{\ \sum A_i\, dZ_i\ } \pi_V^* \Omega_{\mathbb{P}^n},$$

the tautological inclusion, tensored with $\mathcal{O}_{\Omega_{\mathbb{P}^n}}(-1)$.

With these equations, we can tell whether a given subvariety C of $\Psi \cap \pi_V^{-1}(X)$ is a subset of CX. Let $\iota : C \to \Psi$ be the inclusion. Set

$$v = u^* \otimes \mathcal{O}_{\mathbb{P}\Omega_{\mathbb{P}^n}}(-1) : \mathcal{O}_{\mathbb{P}\Omega_{\mathbb{P}^n}}(-1) \to \pi_V^* \Omega_X,$$

and consider the diagram

$$
\begin{array}{ccccc}
\mathcal{O}_{\mathbb{P}\Omega_{\mathbb{P}^n}}(-1)|_C & \xrightarrow{\ \sum A_i\, dZ_i\ } & \pi_V^*\Omega_{\mathbb{P}^n}|_C & \xrightarrow{\ d\pi_V\ } & \Omega_\Psi|_C \\
& {\scriptstyle v|_C}\searrow & \downarrow & & \downarrow {\scriptstyle d\iota} \\
& & \pi_V^*\Omega_X|_C & \xrightarrow{\ d(\pi_1|_C)\ } & \Omega_C
\end{array}
$$

From what we have said about the equations of the conormal variety, we see that $C \subset CX$ if and only if $v|_C = 0$; since $d\pi_V|_C$ is generically injective, we see that $C \subset CX$ if and only if the composition $d\pi_V|_C \circ v|_C$ is zero.

We will show that this condition is symmetric, so that $C \subset CX$ if and only if $C \subset C(X^*)$ (with the factors reversed). Since Ψ is defined by a hypersurface of bidegree $(1, 1)$, we have an exact sequence

$$0 \longrightarrow \mathcal{O}_{\mathbb{P}^n \times \mathbb{P}W}(-1, -1) \xrightarrow{\ \varphi\ } \Omega_{\mathbb{P}^n \times \mathbb{P}W}|_\Psi \longrightarrow \Omega_\Psi \longrightarrow 0.$$

In coordinates, using the decomposition

$$\Omega_{\mathbb{P}^n \times \mathbb{P}W} = \pi_V^*(\Omega_{\mathbb{P}^n}) \oplus \pi_W^*(\Omega_{\mathbb{P}W})$$

this becomes

$$0 \longrightarrow \mathcal{O}_{\mathbb{P}^n \times \mathbb{P}W}(-1, -1)|_\Psi \xrightarrow{\ \sum A_i\, dZ_i,\, \sum Z_i\, dA_i\ } \pi_V^*(\Omega_{\mathbb{P}^n})|_\Psi \oplus \pi_W^*(\Omega_{\mathbb{P}W})|_\Psi$$
$$\xrightarrow{\ (d\pi_V,\, d\pi_W)\ } \Omega_\Psi \longrightarrow 0.$$

Noting that $\mathcal{O}_{\mathbb{P}\Omega_{\mathbb{P}^n}}(-1) = \mathcal{O}_{\mathbb{P}^n \times \mathbb{P}W}(-1, -1)|_\Psi$, we see that if $\iota : C \to \Psi$ is the inclusion of any subvariety, then the composition

$$\mathcal{O}_{\mathbb{P}\Omega_{\mathbb{P}^n}}(-1) \xrightarrow{\ \sum A_i\, dZ_i\ } \pi_V^*(\Omega_{\mathbb{P}^n}) \xrightarrow{\ d\pi_V\ } \Omega_\Psi$$

is the negative of the composition

$$\mathcal{O}_{\mathbb{P}\Omega_{\mathbb{P}^n}}(-1) \xrightarrow{\ \sum Z_i\, dA_i\ } \pi_W^*(\Omega_{\mathbb{P}W}) \xrightarrow{\ d\pi_W\ } \Omega_\Psi.$$

It follows that the composition

$$\mathcal{O}_{\mathbb{P}\Omega_{\mathbb{P}^n}}(-1)|_C \xrightarrow{\sum A_i \, dZ_i} \pi_V^* \Omega_{\mathbb{P}^n}|_C \xrightarrow{d\pi_V} \Omega_\Psi|_C$$

$$\downarrow d\iota$$

$$\Omega_C$$

is zero if and only if the composition

$$\mathcal{O}_{\mathbb{P}\Omega_{\mathbb{P}W}}(-1)|_C \xrightarrow{\sum Z_i \, dA_i} \pi_W^* \Omega_{\mathbb{P}W}|_C \xrightarrow{d\pi_W} \Omega_\Psi|_C$$

$$\downarrow d\iota$$

$$\Omega_C$$

is zero.

If $C \subset C(X^*)$ then the composite above is zero, and it follows that $C \subset CX$. If $C \subset \Psi$, then $C \subset CX$ if and only if $C \subset CX'$. Applying this argument to $C = CX$ and $C = C(X^*)$, we obtain the desired equality. $\square$

10.7 Exercises

Exercise 10.23. Use the result of Exercise 9.36 (describing the class of the universal k-plane in $\mathbb{G}(k, r) \times \mathbb{P}^r$) to give an alternative proof of Proposition 10.4.

Exercise 10.24. (Improper secants) Let $X \subset \mathbb{P}^r$ be a variety, and $\Psi_m(X) \subset \mathbb{G}(m-1, r)$ the image of the secant plane map $\tau : X^{(m)} \dashrightarrow \mathbb{G}(m-1, r)$. Show by example that not every $(m-1)$-plane Λ such that $\deg(\Lambda \cap X) \geq m$ lies in $\Psi_m(X)$. (For example, try X a curve in $\mathbb{P}^5$ with a trisecant line, with $m = 3$.)

Exercise 10.25. Prove Proposition 10.7 in the case of a nondegenerate space curve $C \subset \mathbb{P}^3$ — that is, that the line joining two general points of C does not meet the curve a third time — without using the general position lemma (Lemma 10.8).

Exercises 10.26–10.29 verify that the Veronese varieties listed in Theorem 10.12 are indeed defective.

Exercise 10.26. Show that for $p, q \in \mathbb{P}^n$ the subspace $H^0(\mathcal{I}_p^2 \mathcal{I}_q^2(2)) \subset H^0(\mathcal{O}_{\mathbb{P}^n}(2))$ of quadrics singular at p and q has codimension $2n + 1$ (rather than the expected $2n + 2$). Deduce that any two tangent planes to the quadratic Veronese variety $v_2(\mathbb{P}^n)$ meet, and thus that $v_2(\mathbb{P}^n)$ is 2-defective for any n.

Exercise 10.27. Show that for any five points $p_1, \ldots, p_5 \in \mathbb{P}^2$ there exists a quartic curve double at all five; deduce that the tangent planes $\mathbb{T}_{p_i} S$ to the quartic Veronese surface $S = v_4(\mathbb{P}^2) \subset \mathbb{P}^{14}$ are dependent (equivalently, fail to span $\mathbb{P}^{14}$), and hence that S is 5-defective.

Exercise 10.28. Show that for any nine points $p_1, \ldots, p_9 \in \mathbb{P}^3$ there exists a quartic surface double at all nine; deduce that the tangent planes $\mathbb{T}_{p_i} X$ to the quartic Veronese threefold $X = \nu_4(\mathbb{P}^3) \subset \mathbb{P}^{34}$ are dependent (equivalently, fail to span $\mathbb{P}^{34}$), and hence that X is 9-defective.

Exercise 10.29. Finally, show that for any seven points $p_1, \ldots, p_7 \in \mathbb{P}^4$ there exists a cubic threefold double at all seven; deduce that the tangent planes $\mathbb{T}_{p_i} X$ to the cubic Veronese fourfold $X = \nu_3(\mathbb{P}^4) \subset \mathbb{P}^{34}$ are dependent (equivalently, fail to span $\mathbb{P}^{34}$), and hence that X is 7-defective.
Hint: This problem is harder than the preceding three; you have to use the fact that through seven general points in $\mathbb{P}^4$ there passes a rational normal quartic curve.

The following exercises can be solved using the following fact, the *completeness of the adjoint series* for plane curves: if C is a nodal curve of degree d in $\mathbb{P}^2$, and $\tilde{C}$ its normalization, then we obtain the entire canonical series $H^0(K_{\tilde{C}})$ by pulling back polynomials of degree $d - 3$ on $\mathbb{P}^2$ vanishing on the nodes of C (see Arbarello et al. [1985, Appendix A]).

Exercise 10.30. Show that the twisted cubic curve is the unique nondegenerate curve $C \subset \mathbb{P}^3$ such that a general point $p \in \mathbb{P}^3$ lies on a unique secant line to C. (Note: This can be done without it, but it is easy if you apply the *Castelnuovo bound* on the genus of a curve in $\mathbb{P}^3$; see Chapter 3 of Arbarello et al. [1985] for a statement and proof.)

Exercise 10.31. Show that the rational normal curve and the elliptic normal curve of degree $d + 1$ are the only nondegenerate curves $C \subset \mathbb{P}^d$ with the property that every divisor of degree d on C spans a hyperplane.

Exercise 10.32. Let $C \subset \mathbb{P}^d$ be a rational normal curve. Show that the map $\tau : C^{(m)} \cong \mathbb{P}^m \to \mathbb{G}(m - 1, d)$ sending a divisor of degree m on C to its span composed with the Plücker embedding of the Grassmannian is the d-th Veronese map on $\mathbb{P}^m$.
Hint: Show that the hypersurface in $\mathbb{P}^m$ associated to any monomial of degree d is the preimage of a hyperplane section of $\mathbb{G}(m - 1, d)$.

For the following three exercises, $C \subset \mathbb{P}^d$ will be an irreducible, nondegenerate curve and $2m - 1 < d$. The exercises will prove the assertion made in the text that a general point on the m-secant variety $\mathrm{Sec}_m(C)$ lies on a *unique* m-secant $(m - 1)$-plane to C.

Exercise 10.33. Show by a dimension count that a general point of $\mathrm{Sec}_m(C)$ lies on only *proper* secants, that is, $m - 1$ planes spanned by m distinct points of C.

Exercise 10.34. Using Lemma 10.8, show that the variety of $2m$-secant $(2m - 2)$-planes to C (equivalently, the locus $C_1^{(2m)}$ of divisors of degree $2m$ on C contained in a $(2m - 2)$-plane) has dimension at most $2m - 2$.

Exercise 10.35. Now suppose that a general point of $\mathrm{Sec}_m(C)$ lay on two or more m-secant planes. Show that the dimension of the variety of $2m$-secant $(2m - 2)$-planes to C would be at least $2m - 1$.

Exercise 10.36. Show that if $C \subset \mathbb{P}^r$ is a general rational curve of degree d, and k is a number such that $d \geq r+k$ and $m-1 \geq k$, then the locus of m-secant $(m-k-1)$-planes has the expected dimension $m - k(r + 1 - m + k)$.

Exercise 10.37. By contrast with the last exercise, show that there exist components $\mathcal{H}$ of the Hilbert scheme of curves in $\mathbb{P}^3$ whose general point corresponds to a smooth, nondegenerate curve $C \subset \mathbb{P}^3$ with a positive-dimensional family of quadrisecant lines, or with a quintisecant line.

Exercise 10.38. Compute the number of quadrisecant lines to a general rational curve $C \subset \mathbb{P}^3$ of degree d.

Hint: In the notation of Section 10.5, the answer is the degree of the class $\deg \tau^*(\sigma_{2,2}) \in A^4(\mathbb{P}^4)$. Express the class $\sigma_{2,2}$ in terms of the special Schubert classes σ_i and use (10.1) to evaluate it.

Exercise 10.39. Let $S \subset \mathbb{P}^n$ be a smooth surface of degree d, and let g be the genus of a general hyperplane section of S; let e and f be the degrees of the classes $c_1(\mathcal{T}_S)^2$ and $c_2(\mathcal{T}_S) \in A^2(S)$. Find the class of the cycle $T_1(S) \subset \mathbb{G}(1, n)$ of lines tangent to S in terms of d, e, f and g. (Note: From Exercise 4.21, we need only the intersection number $\deg([T_1(S)] \cdot \sigma_3)$; do this using Segre classes.)

Exercise 10.40. Let $C \subset \mathbb{P}^3$ be a smooth curve of degree n and genus g, and S and $T \subset \mathbb{P}^3$ two smooth surfaces containing C, of degrees d and e. At how many points of C are S and T tangent?

Exercise 10.41. Show the conclusion of Corollary 10.21 fails in characteristic $p > 0$:

(a) Let $\Bbbk$ be a field of characteristic 2, and consider the plane curve

$$C = V(X^2 - YZ) \subset \mathbb{P}^2.$$

Show that C is smooth, but the dual curve $C^* \subset \mathbb{P}^{2*}$ is a line, so that $C^{**} \neq C$.

(b) Now suppose that the ground field $\Bbbk$ has characteristic $p > 0$, set $q = p^e$ and consider the plane curve

$$C = V(YZ^q + Y^q Z - X^{q+1}) \subset \mathbb{P}^2.$$

Show that C is smooth, and that the double dual curve C^{**} is equal to C, but that $\mathcal{G}_C : C \to C^*$ is not birational!

Exercise 10.42. We saw in Section 2.1.3 that if $X \subset \mathbb{P}^n$ is a smooth hypersurface of degree $d > 1$ then the dual variety $X^* \subset \mathbb{P}^{n*}$ must again be a hypersurface. Show more generally that if $X \subset \mathbb{P}^n$ is any smooth complete intersection of hypersurfaces of degrees $d_i > 1$ then X^* will be a hypersurface.

Exercise 10.43. Let $X \subset \mathbb{P}^n$ be a k-dimensional *scroll*, that is, a variety given as the union

$$X = \bigcup \Lambda_b$$

of a one-parameter family of $(k-1)$-planes $\{\Lambda_b \cong \mathbb{P}^{k-1} \subset \mathbb{P}^n\}$; suppose that $k \geq 2$ (see Section 9.1.1).

(a) Show that if $H \subset \mathbb{P}^n$ is a general hyperplane containing the tangent plane $\mathbb{T}_p X$ to X at a smooth point p then the hyperplane section $H \cap X$ is reducible.
(b) Deduce that $\dim X^* \leq n - k + 2$ when $k \geq 3$.

Exercise 10.44. This is a sort of partial converse to Exercise 10.43 above. Let $X \subset \mathbb{P}^n$ be any variety. Use Theorem 10.20 to deduce that if the dual X^* is not a hypersurface, then X must be swept out by positive-dimensional linear spaces.

Exercise 10.45. Let $X \subset \mathbb{P}^n$ be a smooth hypersurface of degree $d > 2$. Show that the dual variety X^* is necessarily singular.

Exercise 10.46. Let $X = \mathbb{G}(1, 4) \subset \mathbb{P}^9$ be the Grassmannian of lines in $\mathbb{P}^4$, embedded in $\mathbb{P}^9$ by the Plücker embedding. Show that the dual of X is projectively equivalent to X itself!

Exercise 10.47. Let $X \subset \mathbb{P}^n$ be a smooth curve, and for any $k = 1, \ldots, n - 1$ let

$$v_k : X \rightarrow \mathbb{G}(k, n)$$

be the map sending a point $p \in X$ to its osculating k-plane. Show that the tangent line to the curve $v_k(X) \subset \mathbb{G}(k, n)$ at $v_k(p)$ is the (tangent line to the) Schubert cycle of k-planes containing the osculating $(k-1)$-plane to X at p and contained in the osculating $(k+1)$-plane to X at p; in other words, to first order the osculating k-planes move by rotating around the osculating $(k-1)$-plane to X at p while staying in the osculating $(k+1)$-plane to X at p.

Exercise 10.48. If E is a smooth elliptic curve (over an algebraically closed field this means a curve of genus 1 with a chosen point), the addition law on E expresses the k-th symmetric power E_k as a $\mathbb{P}^{k-1}$-bundle over E. Verify this, and use it to give a description of $A(E_k)$.

Exercise 10.49. Using the preceding exercise, find the degrees of the secant varieties of an elliptic normal curve $E \subset \mathbb{P}^d$.

Chapter 11

Contact problems

Keynote Questions

(a) Given a general quintic surface $S \subset \mathbb{P}^3$, how many lines $L \subset \mathbb{P}^3$ meet S in only one point? (Answer on page 391.)

(b) If $\{C_t = V(t_0 F + t_1 G + t_2 H) \subset \mathbb{P}^2\}$ is a general net of cubic plane curves, how many of the curves C_t will have cusps? (Answer on page 416.)

(c) If $\{C_t = V(t_0 F + t_1 G) \subset \mathbb{P}^2\}$ is a general pencil of quartic plane curves, how many of the curves C_t will have hyperflexes? (Answer on page 405.)

(d) If $\{C_t\}$ is again a general pencil of quartic plane curves, what are the degree and genus of the curve traced out by flexes of members of the pencil? (Answer in Section 11.3.2.)

Problems such as these, dealing with orders of contact of varieties with linear spaces, are known as *contact problems*. Their solution can often be reduced to the computation of the Chern classes of associated bundles. The most important of the bundles involved is a relative version of the bundle of principal parts introduced in Chapter 7 and described by Theorem 7.2. We will begin with an illustration showing how these arise.

One point of terminology: We define the *order of contact* of a curve C on a smooth variety X with a Cartier divisor $D \subset X$ at $p \in C$ to be the length of the component of the scheme of intersection $C \cap D$ supported at p, or (equivalently) if $\nu : \tilde{C} \to C$ is the normalization, the sum of the orders of vanishing of the defining equation of D at points of $\tilde{C}$ lying over p. If p is an isolated point of $C \cap D$, this is the same as the intersection multiplicity $m_p(C \cdot D)$, and we will use this to denote the order of contact; however, we will also adopt the convention that if $C \subset D$ then the order of contact is ∞, so that the condition $m_p(C \cdot D) \geq m$ is a closed condition on C, D and p.

Finally, we reiterate our standing hypothesis that our ground field has characteristic 0. As with most questions involving derivatives, the content of this chapter is much more complicated in characteristic p, and many of the results derived here are false in that setting.

11.1 Lines meeting a surface to high order

Consider a general quintic surface $S \subset \mathbb{P}^3$. A general line meets S in five points; to require them all to coincide is four conditions, and there is a four-dimensional family of lines in $\mathbb{P}^3$. Thus we would "expect" there to be only finitely many lines meeting S in just one point. On this basis we would expect, more generally, that for a general surface $S \subset \mathbb{P}^3$ of any degree $d \geq 5$ there will be a finite number of lines having a point of contact of order 5 with S.

As we shall show, this expectation is fulfilled, and we can compute the number. To verify the dimension statement, we introduce the flag variety

$$\Phi = \{(L, p) \in \mathbb{G}(1, 3) \times \mathbb{P}^3 \mid p \in L\},$$

which we think of as the universal line over $\mathbb{G}(1, 3)$; we can also realize Φ as the projectivization $\mathbb{P}\mathcal{S}$ of the universal subbundle $\mathcal{S}$ on $\mathbb{G}(1, 3)$. Next, we fix $d \geq 4$ and look at pairs consisting of a point $(L, p) \in \Phi$ and a surface $S \subset \mathbb{P}^3$ of degree d such that the line L has contact of order at least 5 with S at p (or is contained in S):

$$\Gamma = \{(L, p, S) \in \Phi \times \mathbb{P}^N \mid m_p(S \cdot L) \geq 5\},$$

where $\mathbb{P}^N$ is the space of surfaces of degree d in $\mathbb{P}^3$.

Assuming $d \geq 4$, the fiber of Γ over any point $(L, p) \in \Phi$ is a linear subspace of codimension 5 in the space $\mathbb{P}^N$ of surfaces of degree d. Since Φ is irreducible of dimension 5, it follows that Γ is irreducible of dimension N, and hence that the fiber of Γ over a general point $[S] \in \mathbb{P}^N$ is finite. Note that this includes the possibility that the fiber over a general point is empty, as in fact will be the case when $d = 4$: any line with a point of contact of order 5 with a quartic surface S must lie in S, but, as we saw in Chapter 6, a general quartic surface contains no lines. In the case $d = 4$, correspondingly, Γ projects with one-dimensional fibers to the hypersurface $\Sigma \subset \mathbb{P}^{34}$ of quartics that do contain a line. By contrast, we will see (as a consequence of Theorem 11.1) that if $d \geq 5$ then the projection $\Gamma \to \mathbb{P}^N$ is generically finite and surjective.

To linearize the problem, we consider for each pair $(L, p) \in \Phi$ the five-dimensional vector space

$$E_{(L,p)} = \frac{\{\text{germs of sections of } \mathcal{O}_L(d) \text{ at } p\}}{\{\text{germs vanishing to order} \geq 5 \text{ at } p\}}.$$

To say that $m_p(S \cdot L) \geq 5$ means exactly that the defining equation F of S is in the kernel of the map

$$H^0(\mathcal{O}_{\mathbb{P}^3}(d)) \to E_{(L,p)}$$

given by restriction of F to a neighborhood of p in L.

To compute the number of lines with five-fold contact, we will define a vector bundle $\mathcal{E}$ on Φ whose fiber at a point $(L, p) \in \Phi$ is the vector space $E_{(L,p)} = H^0(\mathcal{O}_L(d)/\mathfrak{m}_p^5(d))$ defined above, so that a polynomial F of degree d on $\mathbb{P}^3$ will give a global section σ_F of $\mathcal{E}$ by restriction in turn to each pointed line (L, p). The zeros of the section σ_F will then be the points $(L, p) \in \Phi$ such that $m_p(S \cdot L) \geq 5$, and — assuming that there are no unforeseen multiplicities — the answer to our enumerative problem will be the degree of the top Chern class $c_5(\mathcal{E}) \in A^5(\Phi)$. The necessary theory and computation will occupy the next two sections, and will prove:

Theorem 11.1. *If S is a general quintic surface, then there are exactly 575 lines meeting S in only one point. More generally, if $S \subset \mathbb{P}^3$ is a general surface of degree $d \geq 4$, then there are exactly $35d^3 - 200d^2 + 240d$ lines having a point of contact of order 5 with S.*

Note that this does return the correct answer 0 in the case $d = 4$! (In case $d \leq 3$, the number is meaningless, since the locus of such pairs (L, p) is positive-dimensional.)

11.1.1 Bundles of relative principal parts

The desired bundle $\mathcal{E}$ on Φ is a *bundle of relative principal parts* associated to the map

$$\pi : \Phi \to \mathbb{G}(1, 3), \quad (L, p) \mapsto [L].$$

The construction is a relative version of that of Section 7.2; the reader may wish to review that section before proceeding. The facts we need are the analogs of some of the properties spelled out in Theorem 7.2.

Suppose more generally that $\pi : Y \to X$ is a proper smooth map of schemes, and let $\mathcal{L}$ be a vector bundle on Y. Set $Z = Y \times_X Y$, the fiber product of Y with itself over X, with projection maps $\pi_1, \pi_2 : Z \to Y$, and let $\Delta \subset Z$ be the diagonal, so that we have a diagram

$$
\begin{array}{ccc}
\Delta \longrightarrow Z = Y \times_X Y & \xrightarrow{\ \pi_1\ } & Y \\
\Big\downarrow{\scriptstyle \pi_2} & & \Big\downarrow{\scriptstyle \pi} \\
Y & \xrightarrow{\ \pi\ } & X
\end{array}
$$

The *bundle of relative m-th order principal parts* $\mathcal{P}_{Y/X}^m(\mathcal{L})$ is by definition

$$\mathcal{P}_{Y/X}^m(\mathcal{L}) = \pi_{2*}(\pi_1^* \mathcal{L} \otimes \mathcal{O}_Z/\mathcal{I}_\Delta^{m+1}).$$

Theorem 11.2. *With $\pi : Y \to X$ and projections $\pi_i : Y \times_X Y \to Y$ as above:*

(a) *The sheaf $\mathcal{P}^m_{Y/X}(\mathcal{L})$ is a vector bundle on Y, and its fiber at a point $y \in Y$ is the vector space*

$$\mathcal{P}^m_{Y/X}(\mathcal{L})_y = \frac{\{\text{germs of sections of } \mathcal{L}|_{F_y} \text{ at } y\}}{\{\text{germs vanishing to order} \geq m + 1 \text{ at } y\}},$$

where $F_y = \pi^{-1}(\pi(y)) \subset Y$ is the fiber of π through y.

(b) *We have an isomorphism $\pi^* \pi_* \mathcal{L} \cong \pi_{2*} \pi_1^* \mathcal{L}$.*

(c) *The quotient map $\pi_1^* \mathcal{L} \to \pi_1^* \mathcal{L} \otimes \mathcal{O}_Z / \mathcal{I}_\Delta^{m+1}$ pushes forward to give a map*

$$\pi^* \pi_* \mathcal{L} \cong \pi_{2*} \pi_1^* \mathcal{L} \to \mathcal{P}^m_{Y/X}(\mathcal{L}),$$

and the image of a global section $G \in H^0(\mathcal{L})$ is the section σ_G of $\mathcal{P}^m_{Y/X}(\mathcal{L})$ whose value at a point $y \in Y$ is the restriction of G to a neighborhood of y in F_y.

(d) *$\mathcal{P}^0_{Y/X}\mathcal{L} = \mathcal{L}$. For $m > 1$, the filtration of the fiber $\mathcal{P}^m_{Y/X}(\mathcal{L})_y$ by order of vanishing at y corresponds to a filtration of $\mathcal{P}^m_{Y/X}(\mathcal{L})$ by subbundles that are the kernels of surjections $\mathcal{P}^m_{Y/X}(\mathcal{L}) \to \mathcal{P}^k_{Y/X}(\mathcal{L})$ for $k < m$. The graded pieces of this filtration are identified by the exact sequences*

$$0 \longrightarrow \mathcal{L} \otimes \mathrm{Sym}^m(\Omega_{Y/X}) \longrightarrow \mathcal{P}^m_{Y/X}(\mathcal{L}) \longrightarrow \mathcal{P}^{m-1}_{Y/X}(\mathcal{L}) \longrightarrow 0. \qquad (11.1)$$

The exact sequences in (11.1) allow us to express the Chern classes of the bundles $\mathcal{P}^m_{Y/X}(\mathcal{L})$ in terms of the Chern classes of $\mathcal{L}$ and those of $\Omega_{Y/X}$. We will compute the latter in the case where Y is a projectivized vector bundle over X in the next section, and this will allow us to complete the answer to Keynote Question (a).

Proof: Just as in the absolute case (Theorem 7.2), part (a) is an application of the theorem on cohomology and base change (Theorem B.5). Similarly, part (b) follows from statement (2) on page 525 in the appendix on cohomology and base change (Section B.2).

Part (c) is also a direct analog of the absolute case. For part (d), consider the diagonal $\Delta := \Delta_{Y/X} \subset Y \times_X Y$ and its ideal sheaf $\mathcal{I}_\Delta$. As in the absolute case, we have $\pi_{1*}(\mathcal{I}_\Delta / \mathcal{I}_\Delta^2) = \Omega_{Y/X}$ (see Eisenbud [1995, Theorem 16.24] for the affine case, to which the problem reduces). The sheaf $\Omega_{Y/X}$ is a vector bundle on Y because π is smooth. Since Δ is locally a complete intersection in $Y \times_X Y$, it follows (see, for example, Eisenbud [1995, Exercise 17.14]) that

$$\mathcal{I}_\Delta^m / \mathcal{I}_\Delta^{m+1} = \mathrm{Sym}^m(\mathcal{I}_\Delta / \mathcal{I}_\Delta^2).$$

The desired exact sequences are derived from this exactly as in the absolute case. $\qquad \square$

11.1.2 Relative tangent bundles of projective bundles

To use the sequences (11.1) to calculate the Chern class of $\mathcal{P}^m_{Y/X}(\mathcal{L})$, we need to understand the relative tangent bundle $\mathcal{T}_{Y/X}$. Recall first the definition: If $\pi : Y \to X$ is a smooth map, then the differential $d\pi : \mathcal{T}_Y \to \pi^*\mathcal{T}_X$ is surjective. Its kernel is called the *relative tangent bundle* of π, and denoted by $\mathcal{T}_\pi$ or, when there is no ambiguity, by $\mathcal{T}_{Y/X}$; its local sections are the vector fields on Y that are everywhere tangent to a fiber. Thus, for example, if $x \in X$ then the restriction $\mathcal{T}_{Y/X}|_{\pi^{-1}(x)}$ is the tangent bundle to the smooth variety $\pi^{-1}(x)$.

One special case in which we can describe the relative tangent bundle explicitly is when $\pi : Y = \mathbb{P}\mathcal{E} \to X$ is a projective bundle (as was the case in the example of Keynote Question (a), discussed in Section 11.1 above); in this section we will show how.

Recall from Section 3.2.4 that if $\xi \in \mathbb{P}V$ is a point in the projectivization $\mathbb{P}V$ of a vector space V, then we can identify the tangent space $T_\xi\mathbb{P}V$ with the vector space $\text{Hom}(\xi, V/\xi)$. As we showed, these identifications fit together to give an identification of bundles

$$\mathcal{T}_{\mathbb{P}V} = \mathcal{H}om(\mathcal{S}, \mathcal{Q}),$$

where $\mathcal{S} = \mathcal{O}_{\mathbb{P}V}(-1)$ and $\mathcal{Q}$ are the universal sub- and quotient bundles.

This identification extends to families of projective spaces. Explicitly, suppose $\mathcal{E}$ is a vector bundle on X and $\mathbb{P}\mathcal{E}$ its projectivization, with universal sub- and quotient bundles $\mathcal{S} = \mathcal{O}_{\mathbb{P}\mathcal{E}}(-1)$ and $\mathcal{Q}$. At every point $(x, \xi) \in \mathbb{P}\mathcal{E}$, with $x \in X$ and $\xi \subset \mathcal{E}_x$, we have an identification $T_\xi\mathbb{P}\mathcal{E}_x = \text{Hom}(\xi, \mathcal{E}_x/\xi) = \text{Hom}(\mathcal{S}_{(x,\xi)}, \mathcal{Q}_{(x,\xi)})$, and these agree on overlaps of such open sets to give a global isomorphism:

Proposition 11.3. $\mathcal{T}_{\mathbb{P}\mathcal{E}/X} \cong \mathcal{H}om(\mathcal{S}, \mathcal{Q}).$

Proof: This is a special case of the statement that with notation as in the proposition the relative tangent bundle of the Grassmannian bundle $G(k, \mathcal{E}) \to X$ is

$$\mathcal{T}_{G(k,\mathcal{E})/X} = \mathcal{H}om(\mathcal{S}, \mathcal{Q}).$$

Over an open subset where $\mathcal{E}$ is trivial this is an immediate consequence of the isomorphism described in Section 3.2.4 between the tangent bundle of a Grassmannian and the bundle $\mathcal{H}om(\mathcal{S}, \mathcal{Q})$, and as in that setting the fact that these isomorphisms are independent of choices says they fit together to give the desired isomorphism $\mathcal{T}_{G(k,\mathcal{E})/X} = \mathcal{H}om(\mathcal{S}, \mathcal{Q})$. $\square$

Using the exact sequence

$$0 \longrightarrow \mathcal{S} \longrightarrow \pi^*\mathcal{E} \longrightarrow \mathcal{Q} \longrightarrow 0,$$

Proposition 11.3 yields an exact sequence

$$0 \longrightarrow \mathcal{O}_{\mathbb{P}\mathcal{E}} \longrightarrow \pi^*\mathcal{E} \otimes \mathcal{O}_{\mathbb{P}\mathcal{E}}(1) \longrightarrow \mathcal{T}_{\mathbb{P}\mathcal{E}/X} \longrightarrow 0,$$

the *relative Euler sequence*. Applying the formula for the Chern classes of the tensor product of a vector bundle with a line bundle (Proposition 5.17), we arrive at the following theorem:

Theorem 11.4. *If $\mathcal{E}$ is a vector bundle of rank $r + 1$ on the smooth variety X, then the Chern classes of the relative tangent bundle $\mathcal{T}_{\mathbb{P}\mathcal{E}/X}$ are*

$$c_k(\mathcal{T}_{\mathbb{P}\mathcal{E}/X}) = \sum_{i=0}^{k} \binom{r+1-i}{k-i} c_i(\mathcal{E})\zeta^{k-i},$$

where $\zeta = c_1(\mathcal{O}_{\mathbb{P}\mathcal{E}}(1)) \in A^1(\mathbb{P}\mathcal{E})$ and we identify $A(X)$ with its image in $A(\mathbb{P}\mathcal{E})$ via the pullback map.

11.1.3 Chern classes of contact bundles

Returning to Keynote Question (a), we again let

$$\Phi = \{(L, p) \in \mathbb{G}(1, 3) \times \mathbb{P}^3 \mid p \in L\}$$

be the universal line over $\mathbb{G}(1, 3)$. Via the projection $\pi : \Phi \to \mathbb{G}(1, 3)$, this is the projectivization $\mathbb{P}\mathcal{S}$ of the universal subbundle $\mathcal{S}$ on $\mathbb{G}(1, 3)$. Let $\mathcal{E}$ be the bundle on Φ given by

$$\mathcal{E} = \mathcal{P}^4_{\Phi/\mathbb{G}(1,3)}(\beta^*\mathcal{O}_{\mathbb{P}^3}(d)),$$

where $\beta : \Phi \to \mathbb{P}^3$ is the projection $(L, p) \mapsto p$ on the second factor. By Theorem 11.2, this has fiber $\mathcal{E}_{(L,p)} = H^0(\mathcal{O}_L(d)/\mathfrak{m}_p^5(d))$ at a point $(L, p) \in \Phi$. Thus, counting multiplicities, the number of lines having a point of contact of order at least 5 with a general surface of degree d is the degree of the Chern class $c_5(\mathcal{E})$.

To find the degree of $c_5(\mathcal{E})$, we recall first the description of the Chow ring of Φ given in Section 9.3.1: Since

$$\Phi = \mathbb{P}\mathcal{S} \to \mathbb{G}(1, 3)$$

is the projectivization of the universal subbundle on $\mathbb{G}(1, 3)$, and

$$c_1(\mathcal{S}) = -\sigma_1 \quad \text{and} \quad c_2(\mathcal{S}) = \sigma_{11},$$

Theorem 9.6 yields

$$A(\Phi) = A(\mathbb{G}(1, 3))[\zeta]/(\zeta^2 - \sigma_1\zeta + \sigma_{11}),$$

where $\zeta \in A^1(\Phi)$ is the first Chern class of the line bundle $\mathcal{O}_{\mathbb{P}\mathcal{S}}(1)$. Recall, moreover, that the class ζ can also be realized as the pullback $\zeta = \beta^*\omega$, where $\beta : \Phi \to \mathbb{P}^3$ is the projection $(L, p) \mapsto p$ on the second factor and $\omega \in A^1(\mathbb{P}^3)$ is the hyperplane class.

The relative tangent bundle $\mathcal{T}_{\Phi/\mathbb{G}(1,3)}$ is a line bundle on Φ. By Theorem 11.4, its first Chern class is

$$c_1(\mathcal{T}_{\Phi/\mathbb{G}(1,3)}) = 2\zeta - \sigma_1.$$

By Theorem 11.2, the bundle $\mathcal{E} = \mathcal{P}^4_{\Phi/\mathbb{G}(1,3)}(\beta^*\mathcal{O}_{\mathbb{P}^3}(d))$ has a filtration with successive quotients

$$\beta^*\mathcal{O}_{\mathbb{P}^3}(d), \ \beta^*\mathcal{O}_{\mathbb{P}^3}(d) \otimes \Omega_{\Phi/\mathbb{G}(1,3)}, \ldots, \ \beta^*\mathcal{O}_{\mathbb{P}^3}(d) \otimes \mathrm{Sym}^4 \Omega_{\Phi/\mathbb{G}(1,3)}.$$

The bundle $\Omega_{\Phi/\mathbb{G}(1,3)}$ is dual to the relative tangent bundle $\mathcal{T}_{\Phi/\mathbb{G}(1,3)}$, so its m-th symmetric power has Chern class

$$c(\mathrm{Sym}^m \Omega_{\Phi/\mathbb{G}(1,3)}) = 1 + m(\sigma_1 - 2\zeta).$$

With the formula $c_1(\beta^*\mathcal{O}_{\mathbb{P}^3}(d)) = d\zeta$, this gives

$$c(\beta^*\mathcal{O}_{\mathbb{P}^3}(d) \otimes \mathrm{Sym}^m \Omega_{\Phi/\mathbb{G}(1,3)}) = 1 + (d - 2m)\zeta + m\sigma_1,$$

and altogether

$$c(\mathcal{E}) = \prod_{m=0}^{4} (1 + (d - 2m)\zeta + m\sigma_1).$$

In particular,

$$c_5(\mathcal{E}) = d\zeta \cdot ((d-2)\zeta + \sigma_1) \cdot ((d-4)\zeta + 2\sigma_1) \cdot ((d-6)\zeta + 3\sigma_1) \cdot ((d-8)\zeta + 4\sigma_1).$$

We can evaluate the degrees of monomials of degree 5 in ζ and σ_1 by using the Segre classes introduced in Section 10.1, and in particular Proposition 10.3: We have

$$\deg(\zeta^a \sigma_1^b) = \deg \pi_*(\zeta^a \sigma_1^b) = \deg(s_{a-1}(\mathcal{S})\sigma_1^b)),$$

where $s_k(\mathcal{S})$ is the k-th Segre class of $\mathcal{S}$. Combining Proposition 10.3 and the Whitney formula, we have

$$s(\mathcal{S}) = \frac{1}{c(\mathcal{S})} = c(\mathcal{Q}) = 1 + \sigma_1 + \sigma_2,$$

and so we have

$$\deg(\zeta\sigma_1^4)_\Phi = \deg(\sigma_1^4)_{\mathbb{G}(1,3)} = 2,$$
$$\deg(\zeta^2\sigma_1^3)_\Phi = \deg(\sigma_1^4)_{\mathbb{G}(1,3)} = 2,$$
$$\deg(\zeta^3\sigma_1^2)_\Phi = \deg(\sigma_2\sigma_1^2)_{\mathbb{G}(1,3)} = 1.$$

The remaining monomials of degree 5 in ζ and σ_1 are all zero: $\sigma_1^5 = 0$ since the Grassmannian $\mathbb{G}(1,3)$ is four-dimensional, while $\zeta^4\sigma_1 = \zeta^5 = 0$ because the Segre classes of $\mathcal{S}$ vanish above degree 2 (alternatively, since $\zeta = \beta^*\omega$ is the pullback of a class on $\mathbb{P}^3$ we see immediately that $\zeta^4 = 0$).

Putting this together with the formula above for $c_5(\mathcal{E})$, we get

$$\deg c_5(\mathcal{E})$$
$$= \deg\big(d\zeta((d-2)\zeta + \sigma_1)((d-4)\zeta + 2\sigma_1)((d-6)\zeta + 3\sigma_1)((d-8)\zeta + 4\sigma_1)\big)$$
$$= \deg\big(24d\zeta\sigma_1^4 + d(50d - 192)\zeta^2\sigma_1^3 + d(35d^2 - 200d + 240)\zeta^3\sigma_1^2\big)$$
$$= 35d^3 - 200d^2 + 240d.$$

Assuming there are only finitely many and counting multiplicities, this is the number of lines having a point of contact of order at least 5 with S.

To answer the keynote question, we need to know in addition that for a general surface $S \subset \mathbb{P}^3$ of degree $d \geq 5$ all the lines having a point of contact of order 5 with S "count with multiplicity 1" — that is, all the zeros of the corresponding section of the bundle $\mathcal{E}$ on Φ are simple zeros. To do this, we invoke the irreducibility of the incidence correspondence

$$\Gamma = \{(L, p, S) \mid m_p(S \cdot L) \geq 5\} \subset \Phi \times \mathbb{P}^N,$$

introduced in Section 11.1. By virtue of the irreducibility of Γ, it is enough to show that at just one point $(L, p, S) \in \Gamma$ the section of $\mathcal{E}$ corresponding to S has a simple zero at $(L, p) \in \Phi$: Given this, the locus of (L, p, S) for which this is not the case, being a proper closed subvariety of Γ, will have strictly smaller dimension, and so cannot dominate $\mathbb{P}^N$. As for locating such a triple (L, p, S), Exercise 11.17 suggests one. We should also check that for S general, no line has a point of contact of order at least 6 with S, or more than one point of contact of order at least 5; this is implied by Exercise 11.18. This completes the proof of Theorem 11.1.

11.2 The case of negative expected dimension

In this section, we will describe a rather different application of the contact calculus developed so far: We will use it to bound the maximum number of occurrences of some phenomena that occur in negative "expected dimension."

We begin by explaining an example. We do not expect a surface $S \subset \mathbb{P}^3$ of degree $d \geq 4$ to contain any lines. But some smooth quartics do contain a line and some contain several. Thus we can ask: How many lines can a smooth surface of degree d contain?

We observe to begin with that the number of lines a smooth surface of degree d can contain is certainly bounded: If we let $\mathbb{P}^N$ be the space of surfaces of degree $d \geq 4$, and write

$$\Sigma = \{(S, L) \in \mathbb{P}^N \times \mathbb{G}(1, 3) \mid L \subset S\}$$

for the incidence correspondence, then the set of points of $\mathbb{P}^N$ over which the fiber of the map $\pi : \Sigma \to \mathbb{P}^N$ is finite of degree $\geq m$ is a locally closed subset of $\mathbb{P}^N$ for any m. Since, as we saw in Section 2.4.2, a smooth surface in $\mathbb{P}^3$ of degree > 2 cannot contain a positive-dimensional family lines, by the Noetherian property the degrees of the fibers over the open set $U \subset \mathbb{P}^N$ of smooth surfaces are bounded. We can thus ask in particular:

Question 11.5. *What is the largest number $M(d)$ of lines that a smooth surface $S \subset \mathbb{P}^3$ of degree d can have?*

Remarkably, we do not know the answer to this in general!

The situation here is typical: there is a large range of quasi-enumerative problems where the actual number is indeterminate because the expected dimension of the solution set is negative. In general, almost every time we have an enumerative problem there are analogous "negative expected dimension" variants. For example, we can ask:

Question 11.6. *(a) How many isolated singular points can a hypersurface $X \subset \mathbb{P}^n$ of degree d have?*
(b) How many tritangents can a plane curve $C \subset \mathbb{P}^2$ of degree d have? How many hyperflexes?
(c) How many cuspidal curves can a pencil of plane curves of degree d have? How many reducible ones? How many totally reducible ones (that is, unions of lines)?

We can even go all the way back to Bézout, and ask:

Question 11.7. *How many isolated points of intersection can $n + k$ linearly independent hypersurfaces of degree d in $\mathbb{P}^n$ have?*

Here there is at least a conjecture, described in Eisenbud et al. [1996] and proved in the case $k = 1$ for reduced sets of points by Lazarsfeld [1994, Exercise 4.12]. For a general discussion of these questions (and a more general conjecture), see Eisenbud et al. [1996].

All of these problems are attractive (especially Question 11.7). But we will not pursue them here; rather, we will focus on the original problem of bounding the number of lines on a smooth surface in $\mathbb{P}^3$, in order to illustrate how we can use enumerative methods to find such a bound.

11.2.1 Lines on smooth surfaces in $\mathbb{P}^3$

Since the number of lines on a smooth surface S of degree $d \geq 4$ is variable, it cannot be the solution to an enumerative problem of the sort we have been considering. But we can still use enumerative geometry to bound the number. What we will do is to find a curve F on S whose degree is determined enumeratively and such that F contains all the lines on S. In this we follow a line of argument proposed in Clebsch [1861, p. 106].

A natural approach is to relax the condition that a line L be contained in S to the condition that L meets S with multiplicity $\geq m$ at some point $p \in S$. We can adjust m so that the set of points p for which some line satisfies this condition has expected dimension 1, defining a curve on the surface (as we will see, the right multiplicity is 4). Since this curve must contain all the lines lying on the surface, its degree — which we can compute by enumerative means — is a bound for the number of such lines.

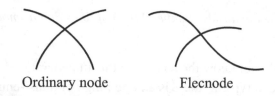

Ordinary node   Flecnode

Figure 11.1 A flecnode is a node in which one branch has a flex at the node.

First of all, we say that a point $p \in S$ is *flecnodal* if there exists a line $L \subset \mathbb{P}^3$ having contact of order 4 or more with S at p; let $F \subset S$ be the locus of such points. (The reason for the name comes from another characterization of such points: for a general surface S, a general flecnodal point $p \in S$ will be one such that the intersection $S \cap \mathbb{T}_p S$ has a *flecnode* at p, that is, a node such that one branch has contact of order at least 3 with its tangent line; see Figure 11.1.) As we will show in Proposition 11.8, the flecnodal locus $F \subset S$ of a smooth surface of degree $d \geq 3$ will always have dimension 1.

As we have observed, any line lying in S is contained in the flecnodal locus F. Of course when $d = 3$ any line meeting S with multiplicity ≥ 4 must lie in S, so the flecnodal locus is exactly the union of the 27 lines in S. To describe the locus of flecnodes on S more generally, we again write Φ for the incidence correspondence

$$\Phi = \{(L, p) \in \mathbb{G}(1, 3) \times \mathbb{P}^3 \mid p \in L\},$$

and we let $\zeta, \sigma_1 \in A^1(\Phi)$ be the pullbacks of the corresponding classes on $\mathbb{P}^3$ and $\mathbb{G}(1, 3)$. Given a surface S, we wish to find the class of the locus

$$\Gamma = \{(L, p) \in \Phi \mid m_p(L \cdot S) \geq 4\}.$$

Since the flecnodal locus $F \subset S$ is the image of Γ under the projection of Φ to $\mathbb{P}^3$, knowledge of this class will determine in particular the degree of F.

To compute the class of Γ, consider the bundle $\mathcal{F} = \mathcal{P}^3_{\Phi/\mathbb{G}(1,3)}(\pi_2^* \mathcal{O}_{\mathbb{P}^3}(d))$ of third-order relative principal parts of $\pi_2^* \mathcal{O}_{\mathbb{P}^3}(d)$. It is a bundle of rank 4 on Φ whose fiber at a point (L, p) is the vector space of germs of sections of $\mathcal{O}_L(d)$ at p, modulo those vanishing to order at least 4 at p:

$$\mathcal{F}_{(L,p)} = H^0(\mathcal{O}_L(d)/\mathcal{I}_p^4(d)).$$

If $A \in H^0(\mathcal{O}_{\mathbb{P}^3}(d))$ is a homogeneous polynomial of degree d defining a surface S, then the restrictions of A to each line $L \subset \mathbb{P}^3$ yield a global section σ_A of the bundle $\mathcal{F}$, whose zeros are the pairs (L, p) such that L meets S with multiplicity ≥ 4 at p.

Since $\dim \Phi = 5$ and $\mathcal{F}$ has rank 4, the locus Γ (if not empty) is at least one-dimensional; if it has dimension exactly 1 then its class is the top Chern class

$$[\Gamma] = c_4(\mathcal{F}) \in A^4(\Phi).$$

We can calculate this class as before: We can filter the bundle $\mathcal{F}$ by order of vanishing — that is, invoke the exact sequences (11.1) — and apply the Whitney formula to arrive at

$$c_4(\mathcal{F}) = d\zeta \cdot ((d-2)\zeta + \sigma_1) \cdot ((d-4)\zeta + 2\sigma_1) \cdot ((d-6)\zeta + 3\sigma_1).$$

Of course, none of this will help us bound the number of lines on S if every point of S is a flecnode! The following result, which was assumed by Clebsch, is thus crucial for this approach. A proof can be found in McCrory and Shifrin [1984, Lemma 2.10]. For partial results in finite characteristic see Voloch [2003].

Proposition 11.8. *If $S \subset \mathbb{P}^3$ is a smooth surface of degree $d \geq 3$ over a field of characteristic 0, the locus*

$$\Gamma = \{(L, p) \in \Phi \mid m_p(L \cdot S) \geq 4\}$$

has dimension 1. In particular, the general point of S is not flecnodal.

We will defer the proof of this proposition to the next section, and continue to derive our bound on the number of lines. By the proposition, the flecnodal locus $F \subset S$ of S is a curve, whose degree is the degree of the intersection of Γ with the class ζ. We can evaluate this as before:

$$\begin{aligned} \deg(F) &= d\zeta^2 \cdot ((d-2)\zeta + \sigma_1) \cdot ((d-4)\zeta + 2\sigma_1) \cdot ((d-6)\zeta + 3\sigma_1) \\ &= d(11d - 24). \end{aligned}$$

Putting this together, we have proven a bound on the number of lines in S:

Proposition 11.9. *The maximum number $M(d)$ of lines lying on a smooth surface $S \subset \mathbb{P}^3$ of degree $d \geq 3$ is at most $d(11d - 24)$.*

In the case $d = 3$ this gives the exact answer since $d(11d - 24) = 27$. But for $d \geq 4$ the bound is not sharp: Segre [1943] proved the slightly better bound $M(d) \leq d(11d - 28) + 12$. Even this is not sharp; for example with $d = 4$ we have $d(11d - 28) + 12 = 76$, but Segre also showed that the maximum number of lines on a smooth quartic surface is exactly $M(4) = 64$.

Of course, we can give a lower bound for $M(d)$ simply by exhibiting a surface with a large number of lines. The Fermat surface $V(x^d + y^d + z^d + w^d)$, for example, has exactly $3d^2$ lines (Exercise 11.25), whence $M(d) \geq 3d^2$. This is still the record-holder for general d. More is known for some particular values of d; Exercises 11.26 and 11.27 exhibit surfaces with more lines in the cases $d = 4, 6, 8, 12$ and 20 (respectively, 64, 180, 256, 864, and 1600 lines), and Boissière and Sarti [2007] find an octic with 352 (the current champion!).

11.2.2 The flecnodal locus

It remains to prove that for any smooth surface $S \subset \mathbb{P}^3$ of degree $d \geq 3$ the locus $\Gamma \subset \Phi$ of pairs (L, p) with $m_p(L \cdot S) \geq 4$ has dimension 1. The following proof was shown us by Francesco Cavazzani:

Proof of Proposition 11.8: Suppose on the contrary that the locus $\Gamma \subset \Phi$ has a component Γ_0 of dimension 2 or more, and let (L_0, p_0) be a general point of Γ_0. Since the fiber of Γ over a point $p \in S$ consists of lines through p in $\mathbb{T}_p S$, it has dimension at most 1. By Theorem 7.11(a), there are only finitely many points p over which the fiber has positive dimension. Thus Γ_0 must dominate S, so p_0 is a general point of S. By Theorem 7.11(b), the intersection $S \cap \mathbb{T}_{p_0} S$ has a node at p_0.

We will proceed by introducing local coordinates on Φ and writing down the defining equations of the subset Γ. To start with, we can find an affine open $\mathbb{A}^3 \subset \mathbb{P}^3$ and choose coordinates (x, y, z) on $\mathbb{A}^3$ so that the point p_0 is the origin $(0, 0, 0) \in \mathbb{A}^3$ and the line $L_0 = \{(x, 0, 0)\}$ is the x-axis; we can also take the tangent plane $\mathbb{T}_{p_0} S$ to be the plane $z = 0$, and, given that the tangent plane section $S \cap \mathbb{T}_{p_0} S$ has a node at p_0, we can take the tangent cone at p to the intersection $S \cap \mathbb{T}_{p_0} S$ to be the union $V(z, xy)$ of the x- and y-axes.

We can take coordinates (a, b, c, d, e) in a neighborhood U of $(L_0, p_0) \in \Phi$ so that if (L, p) is the pair corresponding to (a, b, c, d, e) then

$$p = (a, b, c) \quad \text{and} \quad L = \{(a + t, b + dt, c + et)\}.$$

Let $f(x, y, z)$ be the defining equation of S in $\mathbb{A}^3$. If we write the restriction of f to L as

$$f|_L = f(a + t, b + dt, c + et) = \sum_{i \geq 0} \alpha_i(a, b, c, d, e) t^i,$$

the four functions $\alpha_0, \alpha_1, \alpha_2$ and α_3 will be the defining equations of Γ in U. We want to show their common zero locus has codimension 4 in Φ; we will actually prove the stronger fact that their differentials at (L_0, p_0) are independent.

By the specifications above of $p_0, L_0, \mathbb{T}_{p_0} S$ and $\mathbb{T}C_{p_0}(S \cap \mathbb{T}_{p_0} S)$, we can write

$$f(x, y, z) = z \cdot u(x, y, z) + xy \cdot v(x, y) + y^3 \cdot l(y) + x^4 \cdot m(x).$$

Note that since S is smooth at p_0 we have $u(0, 0, 0) \neq 0$, and since the tangent plane section $S \cap \mathbb{T}_{p_0} S$ has multiplicity 2 at p_0 we have $v(0, 0) \neq 0$; rescaling the coordinates, we can assume $u(0, 0, 0) = v(0, 0) = 1$. Note by contrast that we may have $m(0) = 0$; this will be the case exactly when $m_{p_0}(L_0 \cdot S) \geq 5$.

Now, we can just plug $(a + t, b + dt, c + et)$ in for (x, y, z) in this expression to write out $f|_L$, and hence the coefficients $\alpha_i(a, b, c, d, e)$. This is potentially messy, but in fact it will be enough to evaluate the differentials of the α_i at $(L_0.p_0)$ — that is, at

$(a, b, c, d, e) = (0, 0, 0, 0, 0)$ — and so we can work modulo the ideal $(a, b, c, d, e)^2$. That said, we have

$$f|_L = f(a + t, b + dt, c + et) =$$
$$(c + et)u + (a + t)(b + dt)v + (b + dt)^3 l(b + dt) + (a + t)^4 m(a + t),$$

and thus

$$\alpha_0 \equiv c \mod (a, b, c, d, e)^2,$$
$$\alpha_1 \equiv e + b \mod (c) + (a, b, c, d, e)^2,$$
$$\alpha_2 \equiv d \mod (b, c, e) + (a, b, c, d, e)^2,$$

and finally

$$\alpha_3 \equiv 4a \cdot m(0) \mod (b, c, d, e) + (a, b, c, d, e)^2.$$

What we see from this is that *the differentials of $\alpha_0, \ldots, \alpha_3$ at (L_0, p_0) are linearly independent, unless $m(0) = 0$;* or in other words, *if there is a two-dimensional locus $\Sigma \subset \Phi$ of pairs (L, p) such that $m_p(L \cdot S) \geq 4$, then in fact we must have $m_p(L \cdot S) \geq 5$ for all $(L, p) \in \Sigma$.* But we can carry out exactly the same argument again to show that if there is a two-dimensional family of lines having contact of order 5 or more with S, then all these lines in fact have contact of order 6 or more with S, and so on. We conclude that if Γ has dimension 2 or more, then S must be ruled by lines; in other words, S must be a plane or a quadric. $\square$

11.3 Flexes via defining equations

In our initial discussion of flexes in Section 7.5, we gave the curve $C \subset \mathbb{P}^2$ in question *parametrically* — that is, as the image of a map $\nu : \widetilde{C} \to \mathbb{P}^2$ from a smooth curve $\widetilde{C}$, the normalization of C, to $\mathbb{P}^2$. We defined flexes as the points $p \in \widetilde{C}$ such that for some line $L \subset \mathbb{P}^2$ the multiplicity $m_p(\nu^* L)$ is at least 3.

This definition does not work well in families of curves. As we shall see, when a smooth plane curve degenerates to one with a node, a certain number of the flexes approach the node; but, according to the definition in Section 7.5, the nodal point will generally not be a flex, since in general neither branch of the node will have contact of order 3 or more with its tangent line. Thus to track the behavior of flexes in families we need a different way of describing them, related to the notion of Cartesian flexes described in Section 7.5.2. We will call the objects described below "flex lines" rather than flexes.

We define a *flex line* of $C \subset \mathbb{P}^2$ to be a pair (L, p) with $L \subset \mathbb{P}^2$ a line and $p \in L$ a point such that C and L intersect at p with multiplicity ≥ 3; that is, the set Γ of flex lines is the locus

$$\Gamma = \{(L, p) \in \mathbb{P}^{2*} \times \mathbb{P}^2 \mid m_p(C \cdot L) \geq 3\}.$$

Thus if C is the vanishing locus of a polynomial F, then (L, p) is a flex line if and only if the restriction of F to L vanishes to order at least 3 at p; in other words, instead of taking the defining equation of L and restricting to C (or, more precisely, pulling back to the normalization of C), we are restricting the defining equation of C to L. The two are the same when C is smooth, but different in general: For example, if C is a general curve with a node at p, the tangent lines to the two branches are flex lines at the node.

To compute the number of flexes on a curve defined by a homogeneous form F, we define Ψ to be the incidence correspondence

$$\Psi = \{(L, p) \in \mathbb{P}^{2*} \times \mathbb{P}^2 \mid p \in L\},$$

thought of as the universal line over $\mathbb{P}^{2*}$, and consider

$$\mathcal{E} = \mathcal{P}^2_{\Psi/\mathbb{P}^{2*}}(\pi_2^* \mathcal{O}_{\mathbb{P}^2}(d)),$$

a rank-3 vector bundle on the three-dimensional variety Ψ, whose fiber at a point (L, p) is

$$\mathcal{E}_{(L,p)} = H^0(\mathcal{O}_L(d)/\mathcal{I}_p^3(d)).$$

The homogeneous polynomial F gives rise to a section σ_F of $\mathcal{E}$, and the zeros of this section correspond to the flex lines of the corresponding plane curve $C = V(F)$. Thus the number of flex lines — when this number is finite, of course, and counting multiplicity — is the degree of $c_3(\mathcal{E}) \in A^3(\Psi)$.

Since the projection on the first factor expresses Ψ as the projectivization

$$\Psi = \mathbb{P}\mathcal{S} \to \mathbb{P}^{2*}$$

of the universal subbundle $\mathcal{S}$ on $\mathbb{P}^{2*}$, we can give a presentation of the Chow ring exactly as in the case of the universal line Φ over $\mathbb{G}(1, 3)$ in Section 11.1.3. Letting $\sigma \in A^1(\mathbb{P}^{2*})$ be the hyperplane class, we have

$$A(\Psi) = A(\mathbb{P}^{2*})[\zeta]/(\zeta^2 - \sigma\zeta + \sigma^2),$$

where $\zeta \in A^1(\Phi)$ is the first Chern class of the line bundle $\mathcal{O}_{\mathbb{P}\mathcal{S}}(1)$. Recall from Section 9.3.1, moreover, that the class ζ can also be realized as the pullback $\zeta = \beta^*\omega$, where $\beta : \Phi \to \mathbb{P}^2$ is the projection $(L, p) \mapsto p$ on the second factor and $\omega \in A^1(\mathbb{P}^2)$ is the hyperplane class.

We can also evaluate the degrees of monomials of degree 3 in ζ and σ as before by using the Segre classes introduced in Section 10.1, and in particular Proposition 10.3: We have

$$\deg(\zeta\sigma^2) = \deg(\zeta^2\sigma) = 1 \quad \text{and} \quad \deg(\zeta^3) = \deg(\sigma^3) = 0.$$

(We could also see these directly by observing that $\Psi \subset \mathbb{P}^{2*} \times \mathbb{P}^2$ is a hypersurface of bidegree $(1, 1)$, and the classes σ and ζ are the pullbacks of the hyperplane classes in the two factors.)

We can now calculate the Chern classes of $\mathcal{E}$ by applying the exact sequences (11.1) and using Whitney's formula, and we get

$$c_3(\mathcal{E}) = d\zeta \cdot ((d-2)\zeta + \sigma) \cdot ((d-4)\zeta + 2\sigma).$$

Hence

$$\deg(c_3(\mathcal{E})) = 2d(d-2) + d(d-4) + 2d$$
$$= 3d(d-2).$$

This shows that the number of flex lines, counted with multiplicity, is the same in the singular case as in the smooth case, whenever the number is finite. (Note that if $F = 0$ defines a nonreduced curve, or a curve containing a straight line as a component, the section defined by F vanishes in the wrong codimension.) The present derivation allows us to go further in two ways, both having to do with the behavior of flexes in families. In particular, it will permit us to solve Keynote Question (c).

11.3.1 Hyperflexes

We define a *hyperflex line* to a plane curve C similarly: It is a pair (L, p) such that L and C meet with multiplicity at least 4 at p. As with ordinary flex lines (and for the same reason), this definition is equivalent to the definition of a hyperflex given in Section 7.5 when the point p is a smooth point of C, but not in general: If a curve $C \subset \mathbb{P}^2$ has an ordinary flecnode at p (that is, two branches, one not a flex and the other a flex that is not a hyperflex), then the tangent line to the flexed branch of C at p will be a hyperflex line, though p is not a hyperflex in the sense of Section 7.5. Since a general pencil of plane curves will not include any elements possessing a flecnode (Exercise 11.29), this will not affect our answer to Keynote Question (c).

To describe the locus of hyperflex lines in a family of curves, we denote by $\mathbb{P}^N$ the space of plane curves of degree d, and consider the incidence correspondence

$$\Sigma = \{(L, p, C) \in \Psi \times \mathbb{P}^N \mid m_p(L \cdot C) \geq 4\}.$$

When $d \geq 3$, the fibers of the projection $\Sigma \to \Psi$ are linear spaces of dimension $N - 4$, from which we see that Σ is irreducible of dimension $N - 1$; in particular, it follows that a general curve $C \subset \mathbb{P}^2$ of degree $d > 1$ has no hyperflexes. Furthermore, since for $d \geq 4$ the general fiber of the projection $\Sigma \to \mathbb{P}^N$ is finite (see the proof of Theorem 7.13), the locus $\Xi \subset \mathbb{P}^N$ of curves that do admit a hyperflex is a hypersurface in $\mathbb{P}^N$ in this case. Keynote Question (c) is equivalent to asking for the degree of this hypersurface in the case $d = 4$. We will actually compute it for all d.

To do this, we consider the three-dimensional variety $\Psi \subset \mathbb{P}^{2*} \times \mathbb{P}^2$ as above, and introduce the rank-4 bundle

$$\mathcal{E} = \mathcal{P}^3_{\Psi/\mathbb{P}^{2*}}(\pi_2^* \mathcal{O}_{\mathbb{P}}^2(d))$$

whose fiber at a point (L, p) is

$$\mathcal{E}_{(L,p)} = H^0(\mathcal{O}_L(d)/\mathcal{I}_p^4(d)).$$

With this definition in hand, we consider a general pencil $\{t_0 F + t_1 G\}_{t \in \mathbb{P}^1}$ of homogeneous polynomials of degree d on $\mathbb{P}^2$. The polynomials F, G give rise to sections σ_F, σ_G of $\mathcal{E}$, and the set of pairs (L, p) that are hyperflexes of some element of our pencil is the locus where these sections fail to be linearly independent. Thus the number of hyperflex lines, counted with multiplicities, is the degree of $c_3(\mathcal{E}) \in A^3(\Psi)$.

We do this as before: Filtering the bundle $\mathcal{E}$ by order of vanishing, we arrive at the expression

$$c(\mathcal{E}) = (1 + d\zeta)(1 + (d-2)\zeta + \sigma)(1 + (d-4)\zeta + 2\sigma)(1 + (d-6)\zeta + 3\sigma).$$

Thus

$$\begin{aligned}
c_3(\mathcal{E}) &= (18d^2 - 88d + 72)\zeta^2\sigma + (22d - 36)\zeta\sigma^2 \\
&= 18d^2 - 66d + 36 \\
&= 6(d-3)(3d-2).
\end{aligned}$$

This gives zero when $d = 3$, as it should: a cubic with a hyperflex is necessarily reducible, and a general pencil of plane cubics will not include any reducible ones. We also remark that the number is meaningless in the cases $d = 1$ and $d = 2$, since every point on a line is a hyperflex and a pencil of conics will contain reducible conics equal to the union of two lines.

To show that the actual number of elements of a general pencil possessing hyperflexes is equal to the number predicted, we have to verify that for general polynomials F and G the degeneracy locus $V(\sigma_F \wedge \sigma_G) \subset \Psi$ is reduced. We do this, as in the argument carried out in Section 7.3.1, in two steps: We first use an irreducibility argument to reduce the problem to exhibiting a single pair F, G of polynomials and a point $(L, p) \in \Psi$ such that $V(\sigma_F \wedge \sigma_G)$ is reduced at (L, p), then use a local calculation to show that there do indeed exist such F, G and (L, p).

For the first, a standard incidence correspondence suffices: We let $\mathbb{P}^N$ be the space of plane curves of degree d and $\mathbb{G} = \mathbb{G}(1, N)$ the Grassmannian of pencils of such curves, and consider the locus

$$\Upsilon = \{(\mathcal{D}, L, p) \in \mathbb{G} \times \Psi \mid \text{some } C \in \mathcal{D} \text{ has a hyperflex line at } (L, p)\}.$$

The fiber of Υ over (L, p) is irreducible of dimension $2N - 5$: It is the Schubert cycle $\Sigma_3(\Lambda) \subset \mathbb{G}$, where $\Lambda = \{C \in \mathbb{P}^N \mid m_p(L \cdot C) \geq 4\}$ is the codimension-4 subspace of $\mathbb{P}^N$ consisting of curves with a hyperflex line at (L, p). It follows that Υ is irreducible of dimension $2N - 2 = \dim \mathbb{G}$. Now, if $\Upsilon' \subset \Upsilon$ is the locus of $(\mathcal{D}, L, p)$ such that $V(\sigma_F \wedge \sigma_G)$ is *not* reduced of dimension 0 at (L, p) (where $\mathcal{D}$ is the pencil spanned by F and G), then, since $\Upsilon' \subset \Upsilon$ is closed,

$$\Upsilon' \neq \Upsilon \implies \dim \Upsilon' < 2N - 2,$$

and it follows that if $\Upsilon' \neq \Upsilon$ then Υ' cannot dominate $\mathbb{G}$.

Thus we need only exhibit a single F, G and (L, p) such that $V(\sigma_F \wedge \sigma_G)$ is reduced at (L, p). We do this using local coordinates. Choose $\mathbb{A}^2 \subset \mathbb{P}^2$ with coordinates x, y so that $p = (0, 0)$ is the origin and $L \subset \mathbb{A}^2$ is the line $y = 0$. Set $f(x, y) = F(x, y, 1)$ and $g(x, y) = G(x, y, 1)$.

For local coordinates on Ψ in a neighborhood of the point (L, p), we can take the functions x, y and b, where

$$p = (x, y) \quad \text{and} \quad L = \{(x + t, y + bt)\}_{t \in \Bbbk}.$$

We can trivialize the bundle $\mathcal{E}$ in this neighborhood of (L, p), so that the section σ_F of $\mathcal{E}$ is given by the first four terms in the Taylor expansion of the polynomial $f(x + t, y + bt)$ around $t = 0$. Thus, for example, the section associated to the polynomial $f(x, y) = y + x^4$ (that is, $F(x, y, z) = yz^{d-1} + x^4 z^{d-4}$) is represented by the first four terms in the expansion of $y + bt + (x + t)^4$:

$$\sigma_F = (y + x^4, b + 4x^3, 6x^2, 4x),$$

and the general polynomial $g(x, y) = \sum a_{i,j} x^i y^j$ gives rise to the section σ_G represented by the vector

$$(a_{0,0} + a_{1,0}x + a_{0,1}y + \cdots, a_{1,0} + a_{0,1}b + a_{1,1}y + 2a_{2,0}x + \cdots, a_{2,0} + \cdots, a_{3,0} + \cdots).$$

(Here we are omitting terms in the ideal $(x, y, b)^2$.) The section $\sigma_F \wedge \sigma_G$ is given by the 2×2 minors of the matrix

$$\begin{pmatrix} y + x^4 & b + 4x^3 & 6x^2 & 4x \\ a_{0,0} + \cdots & a_{1,0} + \cdots & a_{2,0} + \cdots & a_{3,0} + \cdots \end{pmatrix}.$$

We have minors with linear terms $a_{1,0}y - a_{0,0}b$, $a_{3,0}y - 4a_{0,0}x$ and $a_{3,0}b - 4a_{2,0}x$, and for general values of the $a_{i,j}$ these are independent. This shows that the section $\sigma_F \wedge \sigma_G$ vanishes simply at p, as required. Thus:

Proposition 11.10. *In a general pencil of degree-d plane curves, exactly $6(d-3)(3d-2)$ will have hyperflexes; in particular, in a general pencil of quartic plane curves, exactly 60 members will have hyperflexes.*

11.3.2 Flexes on families of curves

We can also use the approach via defining equations to answer another question about flexes in pencils, one that sheds some more light on how flexes behave in families. Again, suppose that $\{C_t = V(t_0 F + t_1 G)\}_{t \in \mathbb{P}^1}$ is a general pencil of plane curves of degree d. The general member C_t of the pencil will have, as we have seen, $3d(d-2)$ flex points, and as t varies these points will sweep out another curve B in the plane. We can ask: What are the degree and genus of this curve? What is the geometry of this curve

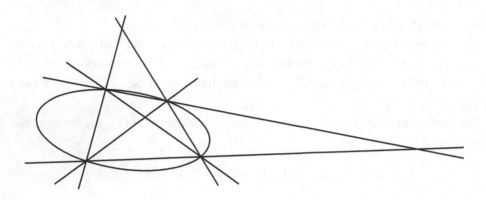

Figure 11.2 The singular elements of a pencil of conics are the pairs of lines joining the four base points.

around singular points of curves in the pencil? We will answer these questions in this section and the next.

To this end, we again write

$$\Psi = \{(L, p) \in \mathbb{P}^{2*} \times \mathbb{P}^2 \mid p \in L\},$$

and set

$$\Gamma = \{(t, L, p) \in \mathbb{P}^1 \times \Psi \mid m_p(L \cdot C_t) \geq 3\}.$$

We will describe Γ as the zero locus of a section of a rank-3 vector bundle on the four-dimensional variety $\mathbb{P}^1 \times \Psi$. For $d \geq 2$, we will show that Γ has the expected dimension 1, and we will ask the reader to show that in fact Γ is smooth by completing the sketch given in Exercise 11.33. This will allow us to determine not only the class of Γ (which will give us the degree of its image B under the projection $\mathbb{P}^1 \times \Psi \to \Psi \to \mathbb{P}^2$) but its genus as well. We will also describe the projection of Γ to $\mathbb{P}^1$, which tells how the flexes may come together as the curve moves in the pencil.

The case of a pencil of conics, $d = 2$, is easy to analyze directly, and already exhibits some of the phenomena involved. As we saw in Proposition 7.4 and the discussion immediately following, a general pencil of conics will have three singular elements, each consisting of two of the straight lines through two of the four base points of the pencil.

A smooth conic has no flexes, while the flex lines of a singular conic C are the pairs (L, p) with $p \in L \subset C$. Thus the curve B, consisting of points $p \in \mathbb{P}^2$ such that some (L, p) is a flex line of some member of the pencil, is the union of the singular members of the pencil — that is, the union of the six lines joining two of the four base points. As such it has degree 6, four triple points, and three additional double points. However, the points of the curve Γ "remember" the flex line to which they belong, so Γ is the *disjoint* union of the six lines — a smooth curve, which is the normalization of B. The singularities of B are typical of the situation of pencils of curves of higher degree, as we shall see: In general, B will have triple points at the base points of the pencil and nodes at the nodes of the singular elements of the pencil. In the case of conics, the projection map $\Gamma \to \mathbb{P}^1$ has three nonempty fibers, each consisting of one of the singular members of the pencil. For general pencils of degree > 2 we shall see that the projection is a finite cover.

Returning to the general case, we again write $\mathcal{E}$ for the rank-3 vector bundle

$$\mathcal{E} = \mathcal{P}^2_{\Psi/\mathbb{P}^2*}(\pi_2^*\mathcal{O}_{\mathbb{P}^2}(d)).$$

Writing V for the two-dimensional vector space spanned by F and G, the sections σ_F and σ_G define a map of bundles

$$V \otimes \mathcal{O}_\Psi \to \mathcal{E}.$$

We now pull this map back to $\mathbb{P}^1 \times \Psi$ via the projection $\nu : \mathbb{P}^1 \times \Psi \to \Psi$. If $\mathbb{P}^1 = \mathbb{P}V$ is the projective line parametrizing our pencil, we also have a natural inclusion

$$\mathcal{O}_{\mathbb{P}V}(-1) \hookrightarrow V \otimes \mathcal{O}_{\mathbb{P}V},$$

which we can pull back to the product $\mathbb{P}^1 \times \Psi$ via the projection $\mu : \mathbb{P}^1 \times \Psi \to \mathbb{P}^1$. Composing these, we arrive at a map

$$\rho : \mu^*\mathcal{O}_{\mathbb{P}^1}(-1) \to V \otimes \mathcal{O}_{\mathbb{P}^1 \times \Psi} \to \nu^*\mathcal{E};$$

over the point $(t, L, p) \in \mathbb{P}^1 \times \Psi$, this is the map that takes a scalar multiple of $t_0 F + t_1 G$ to its restriction to L (modulo sections of $\mathcal{O}_L(d)$ vanishing to order 3 at p). In particular, *the zero locus of this map is the incidence correspondence* Γ.

Tensoring with the line bundle $\mu^*\mathcal{O}_{\mathbb{P}^1}(1)$, we can think of ρ as a section of the bundle $\mu^*\mathcal{O}_{\mathbb{P}^1}(1) \otimes \nu^*\mathcal{E}$; the class of Γ is thus given by the Chern class

$$[\Gamma] = c_3(\mu^*\mathcal{O}_{\mathbb{P}^1}(1) \otimes \nu^*\mathcal{E}) \in A^3(\mathbb{P}^1 \times \Psi).$$

We denote by η the class of a point in $A^1(\mathbb{P}^1)$, or its pullback to $\mathbb{P}^1 \times \Psi$. Similarly, we use the notation ζ and σ, introduced as classes in $A(\Psi)$ above, to denote the pullbacks of these classes to $\mathbb{P}^1 \times \Psi$. With this notation we have

$$c(\mu^*\mathcal{O}_{\mathbb{P}^1}(1) \otimes \nu^*\mathcal{E}) = (1 + \eta + d\zeta)(1 + \eta + (d-2)\zeta + \sigma)(1 + \eta + (d-4)\zeta + 2\sigma).$$

Collecting the terms of degree 3, we get

$$c_3(\mu^*\mathcal{O}_{\mathbb{P}^1}(1) \otimes \nu^*\mathcal{E})$$
$$= (3d^2 - 8d)\zeta^2\sigma + 2d\zeta\sigma^2 + \eta((3d^2 - 12d + 8)\zeta^2 + (6d - 8)\zeta\sigma + 2\sigma^2).$$

To find the degree of the curve $B \subset \mathbb{P}^2$ swept out by the flex points of members of the family, we intersect with the (pullback of the) class ζ of a line $L \subset \mathbb{P}^2$; we get

$$\deg(B) = \zeta \cdot c_3(\mu^*\mathcal{O}_{\mathbb{P}^1}(1) \otimes \nu^*\mathcal{E}) = 6d - 6.$$

Note that this yields the answer 6 in the case $d = 2$, consistent with our previous analysis.

We can use the same constructions to find the geometric genus of the curve Γ. As we observed in Proposition 6.15, the normal bundle to Γ in the product $\mathbb{P}^1 \times \Psi$ is the restriction to Γ of the bundle $\mu^* \mathcal{O}_{\mathbb{P}^1}(1) \otimes \nu^* \mathcal{E}$. Since $\Psi \subset \mathbb{P}^{2*} \times \mathbb{P}^2$ is a hypersurface of bidegree $(1, 1)$, its canonical class is

$$-c_1(\mathcal{T}_\Psi) = K_\Psi = K_{\mathbb{P}^{2*} \times \mathbb{P}^2} + \zeta + \sigma = -2\zeta - 2\sigma;$$

it follows that

$$K_{\mathbb{P}^1 \times \Psi} = -2\eta - 2\zeta - 2\sigma.$$

By the calculation above,

$$c_1(\mathcal{E}) = 3\eta + (3d - 6)\zeta + 3\sigma,$$

and so we have

$$K_\Gamma = (\eta + (3d - 8)\zeta + \sigma)|_\Gamma.$$

We have seen that the degree of $\eta|_\Gamma$ is $3d(d - 2)$, and $\deg(\zeta|_\Gamma) = 6d - 6$; similarly, we can calculate

$$\deg(\sigma|_\Gamma) = (3d^2 - 12d + 8) + (6d - 8) = 3d^2 - 6d.$$

Altogether, we have

$$2g(\Gamma) - 2 = \deg(K_\Gamma) = 24d^2 - 78d + 48,$$

and so

$$g(\Gamma) = 12d^2 - 39d + 25.$$

Note that when $d = 2$ this yields $g(\Gamma) = -5$, as it should: As we saw, in this case Γ consists of the disjoint union of six copies of $\mathbb{P}^1$.

11.3.3 Geometry of the curve of flex lines

We will leave the proofs of most of the assertions in this section to Exercises 11.34–11.38; here, we simply outline the main points of the analysis.

We begin with the geometry of the plane curve B traced out by the flex points of the curves C_t — that is, the image of the curve Γ under projection to $\mathbb{P}^2$. We have already seen that the degree of B is $6d - 6$.

The singularities of B can be located as follows: At each base point p of the pencil, all members of the pencil are smooth. We will see in Exercise 11.34 that three members of the pencil have flexes at p, so that B has a triple point at each base point of the pencil. The only other singularities of B occur at points $p \in \mathbb{P}^2$ that are nodes of the curve C_t containing them. As we have seen, at such a point the tangent lines to the two branches are each flex lines to C_t, so that map $\Gamma \to B$ is two-to-one there; as we will verify in Exercise 11.35, the curve B will have a node there.

Since the projection $\Gamma \to B$ is the normalization, these observations give another way to derive the formula for the genus of Γ: There are in general d^2 base points of the pencil, and as we saw in Chapter 7 there will be $3(d-1)^2$ nodes of elements C_t of our pencil, so that the genus of Γ is

$$\begin{aligned} g(\Gamma) &= p_a(B) - 3d^2 - 3(d-1)^2 \\ &= \tfrac{1}{2}(6d-7)(6d-8) - 3d^2 - 3(d-1)^2 \\ &= 12d^2 - 39d + 25. \end{aligned}$$

We can study the geometry of the curve Γ in another way as well: via the projection $\Gamma \to \mathbb{P}^1$ on the first factor. Since a general member of our pencil has $3d(d-2)$ flexes, Γ is a degree $3d(d-2)$ cover of the line $\mathbb{P}^1$ parametrizing our pencil. Where is this cover branched? The Plücker formula of Section 7.5.2 shows that if C_t is smooth it can fail to have exactly $3d(d-2)$ flexes only if it has a hyperflex, in which case the hyperflex counts as two ordinary flexes. Such hyperflexes are thus ordinary ramification points of the cover $\Gamma \to \mathbb{P}^1$.

That leaves only the singular elements of the pencil to consider, and this is where it gets interesting. By the formula of Section 7.5.2, a curve of degree d with a node has genus one lower, and hence six fewer flexes (in the sense of that section), than a smooth curve of the same degree. If C_{t_0} is a singular element of a general pencil of plane curves, then as $t \to t_0$ three of the flex lines of the curves C_t approach each of the tangent lines to the branches of C_{t_0} at the node (Exercise 11.38). Thus each of the tangent lines to the branches of C_{t_0} at the node is a ramification point of index 2 of the cover $\Gamma \to \mathbb{P}^1$.

We can put this all together with the Riemann–Hurwitz formula to compute the genus of Γ yet again: Since there are $6(d-3)(3d-2)$ hyperflexes in the pencil, and $3(d-1)^2$ singular elements,

$$2g(\Gamma) - 2 = -2 \cdot 3d(d-2) + 6(d-3)(3d-2) + 4 \cdot 3(d-1)^2,$$

and so

$$\begin{aligned} g(\Gamma) &= -3d(d-2) + 3(d-3)(3d-2) + 2 \cdot 3(d-1)^2 + 1 \\ &= 12d^2 - 39d + 25. \end{aligned}$$

11.4 Cusps of plane curves

As a final application we will answer the second keynote question of this chapter: How many curves in a general net of cubics in $\mathbb{P}^2$ have cusps? This will finally complete our calculation, begun in Section 2.2, of the degrees of loci in the space $\mathbb{P}^9$ of plane cubics corresponding to isomorphism classes of cubic curves. Solving this problem requires the introduction of a new class of vector bundles that further generalize the idea of the bundles of principal parts.

We start by saying what we mean by a cusp. An *ordinary cusp* of a plane curve C over the complex numbers is a point p such that, in an analytic neighborhood of p in the plane, the equation of C can be written as $y^2 - x^3 = 0$ in suitable (analytic) coordinates. If we were working over an algebraically closed field $\Bbbk$ other than the complex numbers, we could say instead that the completion of the local ring of C at p is isomorphic to $\Bbbk[[x, y]]/(y^2 - x^3)$, and this is equivalent when $\Bbbk = \mathbb{C}$. Similar generalizations can be made for many of the remarks below.

It is inconvenient to do enumerative geometry with ordinary cusps directly, because the locus of ordinary cusps in a family of curves is not in general closed: ordinary cusps can degenerate to various other sorts of singularities (as in the family $y^2 - tx^3 + x^n$ when $t \to 0$). For this reason we will define a *cusp* of a plane curve to be point where the Taylor expansion of the equation of the curve has no constant or linear terms, and where the quadratic term is a square (possibly zero). As will become clearer in the next section, this means that a cusp is a point at which the completion of the local ring of the curve, in some local analytic coordinate system, has the form

$$\widehat{\mathscr{O}}_{C,p} = \Bbbk[[x, y]]/(ay^2 + \text{terms of degree at least 3}),$$

where a is a constant that may be equal to 0. From the point of view of a general net of curves of degree at least 3, the difference between an ordinary cusp and a general cusp, in our sense, is immaterial: Proposition 11.13 will show that no cusps other than ordinary ones appear.

It is interesting to ask questions about curves on other smooth surfaces besides $\mathbb{P}^2$. Most of the results of this section can be carried over to general nets of curves in any sufficiently ample linear series on any smooth surface, but we will not pursue this generalization.

11.4.1 Plane curve singularities

Before plunging into the enumerative geometry of cusps, we pause to explain a little of the general picture of curve singularities.

Let $p \in C$ be a point on a reduced curve. In an analytic neighborhood of the point, C looks like the union of finitely many branches, each parametrized by a one-to-one map from a disc. Over the complex numbers these maps can be taken to be parametrizations by holomorphic functions of one variable; in general, this statement should be interpreted to mean simply that the completion $\widehat{\mathscr{O}}_{C,p}$ of the local ring $\mathscr{O}_{C,p}$ of C at p is reduced and the normalization of each of its irreducible components (the branches) has the form $\Bbbk[[t]]$, where $\Bbbk$ is the ground field (if our ground field were not algebraically closed then the coefficient field might be a finite extension of the ground field). These statements are part of the theory of completions; see Eisenbud [1995, Chapter 7].

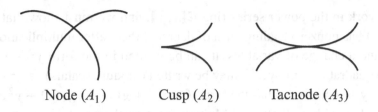

Node (A_1)   Cusp (A_2)   Tacnode (A_3)

Figure 11.3 The simplest double points.

It is a consequence of the Weierstrass preparation theorem that over the complex numbers two reduced germs of analytic curves are isomorphic if and only if the completions of their local rings are isomorphic, so we will use the analytic language although we will work with the completions. See Greuel et al. [2007] for more details.

The theory of singularities in general is vast. But what will concern us here are double points of curves, and in that very limited setting we can actually give a classification, which we will do now.

To begin with, we have already defined the notion of the *multiplicity* of a variety X at a point $p \in X$ in Section 1.3.8. One consequence is that if X has multiplicity 2 at p ("p is a *double point* of X"), then the Zariski tangent space $T_p X$ has dimension $\dim X + 1$. In particular, if p is a double point of a curve C then $\dim T_p C = 2$, so that an analytic neighborhood of p in C is embeddable in the plane, and hence the completion $\hat{\mathcal{O}}_{C,p}$ of the local ring $\mathcal{O}_{C,p}$ of C at p has the form $\Bbbk[\![x, y]\!]/(g(x, y))$, where g has leading term of degree exactly 2.

Definition 11.11. Let $p \in C$ be a double point of a reduced curve C. We say that p is *an A_n-singularity* of C (or that C *has an A_n-singularity at p*) if $\hat{\mathcal{O}}_{C,p} \cong \Bbbk[\![x, y]\!]/(y^2 - x^{n+1})$, that is, if in suitable (analytic) coordinates, C has equation $y^2 - x^{n+1} = 0$ around p.

For example, a double point $p \in C$ is a *node* (C has two smooth branches meeting transversely at p) if and only if C has, in suitable analytic coordinates, equation $y^2 - x^2 = 0$, and is thus an A_1-singularity. Similarly, an *ordinary cusp* is a point $p \in C$ with local analytic equation $y^2 - x^3 = 0$ (A_2-singularity), and an *ordinary tacnode* is a point with local analytic equation $y^2 - x^4 = 0$ (A_3-singularity); this looks like two smooth curves simply tangent to one another at p. In general, if $p \in C$ is an A_n-singularity and $n = 2m + 1$ is odd, then a neighborhood of p in C consists of two smooth branches meeting with multiplicity $m + 1$ at p (we can write $y^2 - x^{2m+2} = (y - x^{m+1})(y + x^{m+1})$), while if n is even then C is analytically irreducible at p.

Proposition 11.12. *Over an algebraically closed field any double point of a plane curve C is an A_n-singularity for some $n \geq 1$.*

Proof: We work in the power series ring $\mathbb{C}[\![x, y]\!]$, and we must show that if a power series $f(x, y)$ has nonzero leading term of degree 2 then, after multiplication by a unit of $\mathbb{C}[\![x, y]\!]$ and a change of variables, it can be written in the form $y^2 - x^{n+1}$. Since any nonzero quadratic form over $\mathbb{C}$ may be written (modulo scalars) as $y^2 + ax^2$ with $a \in \mathbb{C}$, we may assume that f has the form $f = y^2 + g(x) + yg_1(x) + y^2g_2(x, y)$ for some g, g_1 and g_2 with $g_2(0, 0) = 0$. Multiplying f by the unit $1 - g_2(x, y)$, we reduce to the case $g_2 = 0$. Making a change of variable of the form $y' = y - g_1(x)$ (called a *Tschirnhausen transformation*), we can raise the order of vanishing of g_1; repeating this operation and taking the limit we may assume that $g_1 = 0$ as well. But if g has order $n + 1$, then, by Hensel's lemma (Eisenbud [1995, Theorem 7.3]), g has an $(n + 1)$-st root of the form $x + ax^2 + \cdots$. We may take this power series to be a new variable, and after this change of variables we get $f(x, y) = y^2 - x^{n+1}$, as required. $\qquad\square$

In the space $\mathbb{P}^N$ parametrizing all plane curves of given degree d, we can estimate the dimension of the locus of curves having certain types of singularities, at least when the degree of the curves is large compared with the complexity of the singularity (this is an open problem when the degree is small; see Greuel et al. [2007] for more information):

Proposition 11.13. *Let $\mathbb{P}^N$ be the space of plane curves of degree $d \geq k$, and let $\Delta_k \subset \mathbb{P}^N \times \mathbb{P}^2$ be the set of pairs (C, p) such that C has an A_k-singularity at p. Δ_k is locally closed and has codimension $k + 2$ in $\mathbb{P}^N \times \mathbb{P}^2$. Its closure is irreducible, and contains in addition the locus $\Phi \subset \mathbb{P}^N \times \mathbb{P}^2$ of pairs (C, p) such that C has multiplicity 3 or more at p and the locus $\Xi \subset \mathbb{P}^N \times \mathbb{P}^2$ of pairs (C, p) such that p lies on a multiple component of C; in fact,*

$$\overline{\Delta_k} = \Phi \cup \Xi \cup \bigcup_{l \geq k} \Delta_l.$$

Note that since the projection $\Delta_k \to \mathbb{P}^N$ on the first factor is generically finite, the image of Δ_k will have codimension k in $\mathbb{P}^N$. Thus, among all plane curves we will see curves with a node in codimension 1, curves with a cusp in codimension 2 and curves with a tacnode in codimension 3; all other singularities should occur in codimension 4 and higher.

Finally, note that the situation is much less clear when k is large relative to d; for example, as we mentioned in Section 2.2, it is not known for all d and k whether there exists an irreducible plane curve of degree d with an A_k-singularity. In particular, it is not known for $d > 6$ whether there exists an irreducible plane curve $C \subset \mathbb{P}^2$ of degree d with an $A_{(d-1)(d-2)}$-singularity (this is the largest value allowed by the genus formula).

11.4.2 Characterizing cusps

As in the case of the simpler problem of counting singular elements of a pencil of curves, the first thing we need to do to study the cusps in a net of plane curves is to linearize the problem. The difficulty arises from the fact that even after we specify a point $p \in \mathbb{P}^2$ it is not a linear condition on the curves in our linear system to have a cusp at p. It becomes linear, though, if we specify both the point p and a line $L \subset \mathbb{P}^2$ through p with which we require our curve to have intersection multiplicity at least 3. Thus we will work on the universal line over $\mathbb{P}^{2*}$

$$\Psi = \{(L, p) \in \mathbb{P}^{2*} \times \mathbb{P}^2 \mid p \in L\},$$

which we used in Section 11.3 above. In the present circumstances, we also want to think of Ψ as a subscheme of the Hilbert scheme $\mathcal{H}_2(\mathbb{P}^2)$ parametrizing subschemes of $\mathbb{P}^2$ of degree 2. Specifically, it is the locus in $\mathcal{H}_2(\mathbb{P}^2)$ of subschemes of $\mathbb{P}^2$ supported at a single point: We associate to $(L, p) \in \Psi$ the subscheme $\Gamma = \Gamma_{L,p} \subset \mathbb{P}^2$ of degree 2 supported at p with tangent line $\mathbb{T}_p \Gamma = L \subset \mathbb{P}^2$.

For a given point $(L, p) \in \Psi$, we want to express the condition that the curve $C = V(\sigma)$ associated to a section σ of the line bundle $\mathcal{L} = \mathcal{O}_{\mathbb{P}^2}(d)$ on $\mathbb{P}^2$ have a cusp at p with $m_p(C \cdot L) \geq 3$. This suggests that we introduce for each (L, p) the ideal $J_{L,p}$ of functions whose zero locus has such a cusp; that is, we set

$$J_{L,p} = \mathfrak{m}_p^3 + I_\Gamma^2,$$

where $\Gamma = \Gamma_{L,p} \subset \mathbb{P}^2$ is the subscheme of degree 2 supported at p with tangent line L.

We want to construct a vector bundle $\mathcal{E}$ on Ψ whose fiber at a point (L, p) is

$$\mathcal{E}_{(L,p)} = H^0(\mathbb{P}^2, \mathcal{L}/\mathcal{L} \otimes J_{L,p}).$$

To do this, consider the product $\Psi \times \mathbb{P}^2$, with projection maps π_1 and π_2 to Ψ and $\mathbb{P}^2$. Let $\Delta \subset \Psi \times \mathbb{P}^2$ be the graph of the projection map $\Psi \subset \mathbb{P}^{2*} \times \mathbb{P}^2 \to \mathbb{P}^2$ — in other words,

$$\Delta = \{((L, p), q) \in \Psi \times \mathbb{P}^2 \mid p = q\}.$$

Likewise, let $\Gamma \subset B \times \mathbb{P}^2$ be the universal scheme of degree 2 over $\Psi \subset \mathcal{H}_2(\mathbb{P}^2)$. We then take

$$\mathcal{E} = \pi_{1*}\big(\pi_2^*\mathcal{L}/\pi_2^*\mathcal{L} \otimes \big(\mathcal{I}_{\Delta/\Psi \times \mathbb{P}^2}^3 + \mathcal{I}_{\Gamma/\Psi \times \mathbb{P}^2}^2\big)\big);$$

by the theorem on cohomology and base change (Theorem B.5), this is the bundle we want.

A global section of the line bundle $\mathcal{L}$ gives rise to a section of $\mathcal{E}$ by restriction. Given a net $\mathcal{D}$ corresponding to a three-dimensional vector space $V \subset H^0(\mathcal{L})$, we get three sections of $\mathcal{E}$, and the locus in B where they fail to be independent — that is, where some linear combination is zero — is the locus of (p, ξ) such that some element of the net has a cusp at p in the direction ξ. In sum, observing that two elements of a general net cannot have cusps at the same point, and that a general cuspidal curve of degree $d > 2$ has a unique cusp (we leave the verification of this fact to the reader), the (enumerative) answer to our question is the degree of the Chern class $c_3(\mathcal{E})$. In the remainder of this section we will calculate this.

One remark before we launch into the calculation. We are using here the fact that we can characterize the condition that a curve C have a cusp at p by saying that C contains a scheme isomorphic to $\operatorname{Spec} \Bbbk[x, y]/J_{L,p}$; the parameter space Ψ can be viewed as parametrizing such subschemes of $\mathbb{P}^2$. We could use the same technique to count curves with tacnodes; this is sketched in Exercises 11.46–11.48.

If we try to apply the same techniques to count curves with other singularities, however, we run into trouble. For example, the condition that C have an A_5-singularity (an *oscnode*, in the classical terminology) is that C contain a scheme isomorphic to $\operatorname{Spec} \Bbbk[x, y]/(y, x^3)^2$. But the parameter space for such subschemes of the plane is not complete (schemes isomorphic to $\operatorname{Spec} \Bbbk[x, y]/(y, x^3)$ can specialize to the "fat point" scheme $\operatorname{Spec} \Bbbk[x, y]/(x, y)^2$), and if we try to complete it in the most natural way, by taking the closure in the Hilbert scheme, the relevant bundle $\mathcal{E}$ does not extend as a bundle to the closure. This problem is addressed and largely solved in Russell [2003].

11.4.3 Solution to the enumerative problem

We start by recalling the description of the Chow ring $A(\Psi)$ of $\Psi \subset \mathbb{P}^{2*} \times \mathbb{P}^2$ from Section 11.3: We have

$$A(\Psi) = A(\mathbb{P}^{2*})[\zeta]/(\zeta^2 - \sigma\zeta + \sigma^2) = \mathbb{Z}[\sigma, \zeta]/(\sigma^3, \zeta^2 - \sigma\zeta + \sigma^2),$$

where $\sigma \in \Psi$ is the pullback of the hyperplane class in $\mathbb{P}^{2*}$ and ζ is the pullback of the hyperplane class in $\mathbb{P}^2$ (equivalently, if we view Ψ as the projectivization of the universal subbundle $\mathcal{S}$ on $\mathbb{P}^{2*}$, the first Chern class of the line bundle $\mathcal{O}_{\mathbb{P}\mathcal{S}}(1)$). The degrees of monomials of top degree 3 in σ and ζ are

$$\deg(\zeta\sigma^2) = \deg(\zeta^2\sigma) = 1 \quad \text{and} \quad \deg(\zeta^3) = \deg(\sigma^3) = 0.$$

Now, in order to find the Chern class of $\mathcal{E}$ we want to relate it to more familiar bundles. To this end, we observe that the inclusions

$$\mathfrak{m}_p^3 \hookrightarrow J_{L,p} \hookrightarrow \mathfrak{m}_p^2$$

and the corresponding quotients

$$\frac{\mathcal{L}_p}{\mathfrak{m}_p^3 \mathcal{L}_p} \to \frac{\mathcal{L}_p}{J_{L,p}\mathcal{L}_p} \to \frac{\mathcal{L}_p}{\mathfrak{m}_p^2 \mathcal{L}_p}$$

globalize to give us surjections of sheaves

$$\pi_2^* \mathcal{P}_{\mathbb{P}^2}^2(\mathcal{L}) \xrightarrow{\alpha} \mathcal{E} \quad \text{and} \quad \mathcal{E} \longrightarrow \pi_2^* \mathcal{P}_{\mathbb{P}^2}^1(\mathcal{L});$$

the composition

$$\pi_2^* \mathcal{P}_{\mathbb{P}^2}^2(\mathcal{L}) \xrightarrow{\beta} \pi_2^* \mathcal{P}_{\mathbb{P}^2}^1(\mathcal{L})$$

is the standard quotient map of Theorem 7.2.

Consider the corresponding inclusion

$$\mathrm{Ker}(\alpha) \hookrightarrow \mathrm{Ker}(\beta) = \pi_2^*(\mathrm{Sym}^2 \, \mathcal{T}_{\mathbb{P}^2}^* \otimes \mathcal{L}).$$

What is the image? It is the tensor product of $\mathcal{L}$ with the sub-line bundle of $\pi_2^* \, \mathrm{Sym}^2 \, \mathcal{T}_{\mathbb{P}^2}^*$ whose fiber at each point (L, p) is the subspace spanned by the square of the linear form on $T_p \mathbb{P}^2$ vanishing on $T_p L \subset T_p \mathbb{P}^2$. In other words, the inclusion $T_p L \hookrightarrow T_p \mathbb{P}^2$ at each point $(L, p) \in \Psi$ gives rise to a sequence

$$0 \longrightarrow \mathcal{N} \longrightarrow \beta^* \mathcal{T}_{\mathbb{P}^2}^* \longrightarrow \mathcal{T}_{\Psi/\mathbb{P}^{2*}}^* \longrightarrow 0, \tag{11.2}$$

where $\mathcal{N}$ is the sub-line bundle of $\beta^* \mathcal{T}_{\mathbb{P}^2}^*$ whose fiber at (L, p) is the space of linear forms on $T_p \mathbb{P}^2$ vanishing on $T_p L \subset T_p \mathbb{P}^2$ (we can think of $\mathcal{N}$ as the "relative conormal bundle" of the family $\Psi \subset \mathbb{P}^{2*} \times \mathbb{P}^2 \to \mathbb{P}^{2*}$). Taking symmetric squares, we have an inclusion

$$\mathrm{Sym}^2 \, \mathcal{N} \hookrightarrow \beta^* \, \mathrm{Sym}^2 \, \mathcal{T}_{\mathbb{P}^2}^*,$$

and tensoring with the pullback of $\mathcal{L}$ we arrive at an inclusion

$$\mathrm{Sym}^2 \, \mathcal{N} \otimes \beta^* \mathcal{L} \hookrightarrow \beta^*(\mathrm{Sym}^2 \, \mathcal{T}_{\mathbb{P}^2}^* \otimes \mathcal{L}),$$

whose image is exactly $\beta^* \mathcal{P}_{\mathbb{P}^2}^2(\mathcal{L})/\mathcal{E}$.

We can put this all together to calculate the Chern class $c(\mathcal{E})$. To begin with, we know the classes of the bundle $\mathcal{P}_{\mathbb{P}^2}^2(\mathcal{L})$ from Proposition 7.5: We have

$$c(\beta^* \mathcal{P}_{\mathbb{P}^2}^2(\mathcal{L})) = (1 + (d - 2)\zeta)^6 = 1 + 6(d - 2)\zeta + 15(d - 2)^2 \zeta^2.$$

Next, the Chern class of the line bundle $\mathcal{N}$ can be found from the sequence (11.2): We have

$$c_1(\mathcal{N}) = c_1(\beta^* \mathcal{T}_{\mathbb{P}^2}^*) - c_1(\mathcal{T}_{\Psi/\mathbb{P}^{2*}}^*)$$

$$= -3\zeta - (-2\zeta + \sigma)$$

$$= -\sigma - \zeta,$$

where the equality $c_1(\mathcal{T}_{\Psi/\mathbb{P}^{2*}}^*) = -2\zeta + \sigma$ comes from Theorem 11.4. Thus

$$c(\mathrm{Sym}^2 \, \mathcal{N} \otimes \beta^* \mathcal{L}) = 1 + (d - 2)\zeta - 2\sigma,$$

and since $\beta^* \mathcal{P}^2_{\mathbb{P}^2}(\mathcal{L})/\mathcal{E} \cong \mathrm{Sym}^2 \mathcal{N} \otimes \beta^* \mathcal{L}$ the Whitney formula gives

$$c(\mathcal{E}) = \frac{c(\beta^* \mathcal{P}^2_{\mathbb{P}^2}(\mathcal{L}))}{c(\mathrm{Sym}^2 \mathcal{N} \otimes \beta^* \mathcal{L})}$$

$$= \frac{1 + 6(d-2)\zeta + 15(d-2)^2 \zeta^2}{1 - (2\sigma - (d-2)\zeta)}$$

$$= (1 + 6(d-2)\zeta + 15(d-2)^2 \zeta^2) \sum_{k=0}^{3} (2\sigma - (d-2)\zeta)^k.$$

In particular, the third Chern class $c_3(\mathcal{E})$ is

$$c_3(\mathcal{E}) = (2\sigma - (d-2)\zeta)^3 + 6(d-2)\zeta(2\sigma - (d-2)\zeta)^2 + 15(d-2)^2 \zeta^2 (2\sigma - (d-2)\zeta),$$

and taking degrees we have

$$\deg c_3(\mathcal{E}) = 30(d-2)^2 + 24(d-2) - 24(d-2)^2 - 12(d-2) + 6(d-2)^2$$

$$= 12d^2 - 36d + 24.$$

We have thus proven the enumerative formula:

Proposition 11.14. *The number of cuspidal elements of a net $\mathcal{D}$ of curves of degree d on $\mathbb{P}^2$, assuming there are only finitely many and counting multiplicities, is*

$$12d^2 - 36d + 24 = 12(d-1)(d-2).$$

Of course, to answer Keynote Question (b) we have to verify that for a general net there are indeed only finitely many cusps, and that they all count with multiplicity 1. The first of these statements follows easily from the dimension count of Proposition 11.13. The second can be verified by explicit calculation in local coordinates, analogous to what we did to verify, for example, that hyperflexes in a general pencil occur with multiplicity 1; alternatively, we can use the method described in Section 11.4.4 below.

Note that the formula yields 0 in the cases $d = 1$ and 2, as it should. And, in the case $d = 3$, we see that a general net of plane cubics will have 24 cuspidal members, answering Keynote Question (b). Equivalently, we see that the locus of cuspidal plane cubics has degree 24 in the space $\mathbb{P}^9$ of all plane cubics, completing the analysis begun in Section 2.2.

Note that there was no need to restrict ourselves to nets of curves in $\mathbb{P}^2$; a similar analysis could be made for the number of cusps (possibly with multiplicities) in a sufficiently general net of divisors associated to a sufficiently ample line bundle $\mathcal{L}$ on any surface S. (Here the role of Ψ would be played by the projectivized tangent bundle $\mathbb{P}\mathcal{T}_S$.) We leave this version of the calculation to the reader; the answer is that the number of cuspidal elements in a net of curves $\mathcal{D} = (\mathcal{L}, V)$ on a surface S is

$$\deg(12\lambda^2 - 12\lambda c_1 + 2c_1^2 + 2c_2),$$

where $\lambda = c_1(\mathcal{L})$ and $c_i = c_i(\mathcal{T}_S)$. As always, this number is subject to the usual caveats: it is meaningful only if the number of cuspidal curves in the net is in fact finite; in this case, it represents the number of cuspidal curves counted with multiplicity (with multiplicity defined as the degree of the component of the zero scheme of the corresponding section of $\mathcal{E}$ supported at (p, ξ)).

11.4.4 Another approach to the cusp problem

There is another approach to the problem of counting cuspidal curves in a linear system, one that gives a beautiful picture of the geometry of nets. It is not part of the overall logical structure of this book, so we will run through the sequence of steps involved without proof; the reader who is interested can view supplying the verifications as an extended exercise.

To begin with, let S be a smooth projective surface and $\mathcal{L}$ a very ample line bundle; let $\mathcal{D} \subset |\mathcal{L}|$ be a general two-dimensional subseries, corresponding to a three-dimensional vector subspace $V \subset H^0(\mathcal{L})$. We have a natural map

$$\varphi : S \to \mathbb{P}^2 = \mathbb{P}V^*$$

to the projectivization of the dual V^*; the preimages $\varphi^{-1} \subset S$ of the lines $L \subset \mathbb{P}V^*$ are the divisors $C \subset S$ of the linear system $\mathcal{D}$. If we want, we can think of the complete linear system $|\mathcal{L}|$ as giving an embedding of S in the larger projective space $\mathbb{P}^n = \mathbb{P}H^0(\mathcal{L})^*$, and the map φ as the projection of S corresponding to a general $(n-3)$-plane.

Now, the geometry of generic projections of smooth varieties is well understood in low dimensions. Mather [1971; 1973] showed that these are the same in the algebro-geometric setting as in the differentiable; in the latter context the singularities of general projections of surfaces are described in Golubitsky and Guillemin [1973, Section 6.2]. The upshot is that if $\varphi : S \to \mathbb{P}^2$ is the projection of a smooth surface $S \subset \mathbb{P}^n$ from a general $(n-3)$-plane, then:

- The ramification divisor $R \subset S$ of the map φ is a smooth curve.
- The branch divisor $B \subset \mathbb{P}V^*$ is the birational image of R, and has only nodes and ordinary cusps as singularities.

In fact, étale locally around any point $p \in S$, one of three things is true. Either:

(i) The map is étale (if $p \notin R$).
(ii) The map is simply ramified, that is, of the form $(x, y) \mapsto (x, y^2)$ (if p is a point of R not lying over a cusp of B).
(iii) The surface S is given, in terms of local coordinates (x, y) on $\mathbb{P}V^*$ around $\varphi(p)$, by the equation

$$z^3 - xz - y = 0.$$

(This is the picture around a point where three sheets of the cover come together; in a neighborhood of $\varphi(p)$ the branch curve is the zero locus of the discriminant $4x^3 - 27y^2$, and in particular has a cusp at $\varphi(p)$.)

The interesting thing about this set-up is that we have two plane curves associated to it, lying in dual projective planes:

(a) The branch curve $B \subset \mathbb{P}V^*$ of the map φ.
(b) In the dual space $\mathbb{P}V$ parametrizing divisors in the net $\mathcal{D}$, we have the *discriminant curve* $\Delta \subset \mathbb{P}V$, that is, the locus of singular elements of the net.

What ties everything together is the observation that *the discriminant curve $\Delta \subset \mathbb{P}V$ is the dual curve of the branch curve $B \subset \mathbb{P}V^*$*. To see this, note that if $L \subset \mathbb{P}V^*$ is a line transverse to B (in particular, not passing through any of the singular points of B), then the preimage $\varphi^{-1}(L) \subset S$ will be smooth: This is certainly true away from points of B, where the map φ is étale, and at a point $p \in L \cap B$ we can take local coordinates (x, y) on $\mathbb{P}V^*$ with L given by $y = 0$ and B by $x = 0$; at a point of $\varphi^{-1}(p)$ the cover $S \to \mathbb{P}V^*$ will either be étale or given by $z^2 = x$. A similar calculation shows conversely that if L is tangent to B at a smooth point then $\pi^{-1}(L)$ will be singular.

At this point, we invoke the classical *Plücker formulas for plane curves*. These say that if $C \subset \mathbb{P}^2$ is a plane curve of degree $d > 1$ and geometric genus g having δ nodes and κ cusps as its only singularities, and the dual curve C^* has degree d^* and δ^* nodes and κ^* cusps as singularities, then

$$d^* = d(d - 1) - 2\delta - 3\kappa,$$
$$d = d^*(d^* - 1) - 2\delta^* - 3\kappa^*,$$
$$g = \tfrac{1}{2}(d - 1)(d - 2) - \delta - \kappa = \tfrac{1}{2}(d^* - 1)(d^* - 2) - \delta^* - \kappa^*.$$

See, for example, Griffiths and Harris [1994, p. 277ff.]. Given these, all we have to do is write down everything we know about the curves R, B and Δ. To begin with, we invoke the Riemann–Hurwitz formula for finite covers $f : X \to Y$: If η is a rational canonical form on Y with divisor D, the divisor of the pullback $f^*\eta$ will be the preimage of D plus the ramification divisor $R \subset X$; thus

$$K_X = f^* K_Y + R \in A^1(X).$$

In our present circumstances, this says that

$$K_S = \varphi^* K_{\mathbb{P}V^*} \otimes \mathcal{O}_S(R);$$

since the pullback $\varphi^* \mathcal{O}_{\mathbb{P}V^*}(1)$ is equal to $\mathcal{L}$, we can write this as

$$K_S = \mathcal{L}^{-3}(R),$$

or, in terms of the notation $c_1 = c_1(T_S^*)$ and $\lambda = c_1(\mathcal{L})$, the class of R is

$$[R] = c_1 + 3\lambda \in A^1(S).$$

Among other things, this tells us the genus g of the curve R: Since R is smooth, by adjunction we have

$$\begin{aligned} g &= \tfrac{1}{2} R \cdot (R + K_S) + 1 \\ &= \tfrac{1}{2}(c_1 + 3\lambda)(2c_1 + 3\lambda) + 1 \\ &= \tfrac{1}{2}(9\lambda^2 + 9\lambda c_1 + 2c_1^2) + 1. \end{aligned}$$

It also tells us the degree d of the branch curve $B = \varphi(R) \subset \mathbb{P}V^*$: This is the intersection of R with the preimage of a line, so that

$$d = \lambda(c_1 + 3\lambda) = 3\lambda^2 + \lambda c_1.$$

Finally, we also know the degree e of the discriminant curve $\Delta \subset \mathbb{P}V$: This is the number of singular elements in a pencil, which we calculated back in Chapter 7; we have

$$e = 3\lambda^2 + 2\lambda c_1 + c_2.$$

We now have enough information to determine the number of cusps of Δ. Let δ and κ denote the number of nodes and cusps of Δ respectively. First off, the geometric genus of Δ is given by

$$g = \tfrac{1}{2}(e-1)(e-2) - \delta - \kappa,$$

and the degree d of the dual curve is

$$d = e(e-1) - 2\delta - 3\kappa.$$

Subtracting twice the first equation from the second yields

$$\begin{aligned} \kappa &= 2g - d + 2(e-1) \\ &= 9\lambda^2 + 9\lambda c_1 + 2c_1^2 + 2 - (3\lambda^2 + \lambda c_1) + 2(3\lambda^2 + 2\lambda c_1 + c_2 - 1) \\ &= 12\lambda^2 + 12\lambda c_1 + 2c_1^2 + 2c_2, \end{aligned}$$

agreeing with result stated at the end of Section 11.4.3. Note that this method also gives us a geometric sense of when a cusp "counts with multiplicity one;" in particular, if all the hypotheses above about the geometry of the map φ are satisfied, the count is exact.

This also gives us a formula for the number of curves C in the net with two nodes. This is the number δ of nodes of the curve Δ, which we get by subtracting three times the equation for g above from the equation for d: this yields

$$\delta = d - 3g - e(e-1) + \tfrac{3}{2}(e-1)(e-2),$$

where d, e and g are given in terms of the classes λ, c_1 and c_2 by the equations above.

Note that the formula returns 0 in the cases $d = 1$ and $d = 2$, as it should, and in the case $d = 3$ it gives 21 — the degree of the locus of reducible cubics in the $\mathbb{P}^9$ of all cubics, as calculated in Section 2.2.

Exercises 11.49–11.51 describe an alternative (and perhaps cleaner) way of deriving the formula for the number of binodal curves in a net, via linearization.

11.5 Exercises

Exercise 11.15. Let $X \subset \mathbb{P}^4$ be a general hypersurface of degree $d \geq 6$. How many lines $L \subset \mathbb{P}^4$ will have a point of contact of order 7 with X?

Exercise 11.16. Let $S \subset \mathbb{P}^3$ be a general surface of degree $d \geq 2$. Using the dimension counts of Proposition 11.13 and incidence correspondences, show that:

(a) For p in a dense open subset $U \subset S$, the intersection $S \cap \mathbb{T}_p S$ has an ordinary double point (a node) at p.

(b) There is a one-dimensional locally closed locus $Q \subset S$ such that for $p \in Q$ the intersection $S \cap \mathbb{T}_p S$ has a cusp at p.

(c) There will be a finite set Γ of points $p \in S$, lying in the closure of Q, such that the intersection $S \cap \mathbb{T}_p S$ has a tacnode at p.

(d) S is the disjoint union of U, Q and Γ; that is, no singularities other than nodes, cusps and tacnodes appear among the plane sections of S.

Exercise 11.17. Let Φ be the universal line over $\mathbb{G}(1, 3)$ and $\mathcal{E}$ the bundle on Φ introduced in Section 11.1. Let $L \subset \mathbb{P}^3$ be the line $X_2 = X_3 = 0$, and let $p \in L$ be the point $[1, 0, 0, 0]$. By trivializing the bundle $\mathcal{E}$ in a neighborhood of $(L, p) \in \Phi$ and writing everything in local coordinates, show that the section of $\mathcal{E}$ coming from the polynomial $X_1^5 + X_0^4 X_2 + X_0^2 X_1^2 X_3$ has a simple zero at (L, p).

Exercise 11.18. Let $S \subset \mathbb{P}^3$ be a general surface of degree $d \geq 4$. Show that, for any line $L \subset \mathbb{P}^3$ and any pair of distinct points $p, q \in L$:

(a) $m_p(S \cdot L) \leq 5$.

(b) $m_p(S \cdot L) + m_q(S \cdot L) \leq 6$.

Exercise 11.19. A point p on a smooth surface $S \subset \mathbb{P}^3$ is called an *Eckhart point* of S if the intersection $S \cap \mathbb{T}_p S$ has a triple point at p. Recall that in Exercise 7.42 we saw that a general surface $S \subset \mathbb{P}^3$ of degree d has no Eckhart points.

(a) Show that the locus of smooth surfaces that do have an Eckhart point is an open subset of an irreducible hypersurface $\Psi \subset \mathbb{P}^{\binom{d+3}{3}-1}$ in the space of all surfaces.

(b) Show that a general surface $S \subset \mathbb{P}^3$ that does have an Eckhart point has only one.

(c) Find the degree of the hypersurface Ψ.

Exercise 11.20. Consider a smooth surface $S \subset \mathbb{P}^4$. Show that we would expect there to be a finite number of hyperplane sections $H \cap S$ of S with triple points, and count the number in terms of the hyperplane class $\zeta \in A^1(S)$ and the Chern classes of the tangent bundle to S.

Exercise 11.21. Applying your answer to the preceding exercise, find the number of hyperplane sections of $S \subset \mathbb{P}^4$ with triple points in each of the following cases:

(a) S is a complete intersection of two quadrics in $\mathbb{P}^4$.

(b) S is a cubic scroll (Section 9.1.1).

(c) S is a general projection of the Veronese surface $v_2(\mathbb{P}^2) \subset \mathbb{P}^5$.

In each case, can you check your answer directly?

Exercise 11.22. For $S \subset \mathbb{P}^3$ a general surface of degree d, find the degree of the surface swept out by the lines in $\mathbb{P}^3$ having a point of contact of order at least 4 with S.

The following exercise describes in some more detail the geometry of the *flecnodal locus* $\Gamma \subset S$ of a smooth surface $S \subset \mathbb{P}^3$, introduced in Section 11.2.1; we will use the notation of that section.

Exercise 11.23. Let $S \subset \mathbb{P}^3$ be a general surface of degree d.

(a) Find the first Chern class of the bundle $\mathcal{F}$.

(b) Show that the curve Γ is smooth, and that the projection $\Gamma \to C$ is generically one-to-one.

(c) Using the preceding parts, find the genus of the curve Γ.

(d) Show, on the other hand, that the flecnodal curve of S is the intersection of S with a surface of degree $11d - 24$, and use this to calculate the arithmetic genus of C.

(e) Can you describe the singularities of the curve C? Do these account for the discrepancy between the genera of Γ and of C?

Exercise 11.24. Let $\mathbb{P}^N$ be the space of surfaces of degree $d \geq 4$ in $\mathbb{P}^3$ and $\Psi \subset \mathbb{P}^N$ the locus of surfaces containing a line. Show that the maximum possible number $M(d)$ of lines on a smooth surface $S \subset \mathbb{P}^3$ of degree d is at most the degree of Ψ by considering the pencil spanned by S and a general second surface T. Is this bound better or worse than the one derived in Section 11.2.1?

Exercise 11.25. Show that for $d \geq 3$ the Fermat surface $S_d = V(x^d + y^d + z^d + w^d) \subset \mathbb{P}^3$ contains exactly $3d^2$ lines.

Exercise 11.26. For $F(x, y)$ any homogeneous polynomial of degree d, consider the surface $S \subset \mathbb{P}^3$ given by the equation

$$F(x, y) - F(z, w) = 0.$$

If α is the order of the group of automorphisms of $\mathbb{P}^1$ preserving the polynomial F (that is, carrying the set of roots of F to itself), show that S contains at least $d^2 + \alpha d$ lines. *Hint:* if L_1 and L_2 are the lines $z = w = 0$ and $x = y = 0$ respectively, and $\varphi : L_1 \to L_2$ any isomorphism carrying the zero locus $F(x, y) = 0$ to $F(z, w) = 0$, consider the intersection of S with the quadric

$$Q_\varphi = \bigcup_{p \in L_1} \overline{p, \varphi(p)}.$$

Exercise 11.27. Using the preceding exercise, exhibit smooth surfaces $S \subset \mathbb{P}^3$ of degrees 4, 6, 8, 12 and 20 having at least 64, 180, 256, 864 and 1600 lines, respectively.

Exercise 11.28. Verify that the Fermat quartic curve $C = V(x^4 + y^4 + z^4) \subset \mathbb{P}^2$ has 12 hyperflexes and no ordinary flexes.

Exercise 11.29. Recall that a node $p \in C$ of a plane curve is called a *flecnode* if one of the branches of C at p has contact of order 3 or more with its tangent line. Show that the closure, in the space $\mathbb{P}^N$ of all plane curves of degree $d \geq 4$, of the locus of curves with a flecnode is irreducible of dimension $N - 2$.

Exercise 11.30. How many elements of a general net of plane curves of degree d will have flecnodes?

Exercise 11.31. Verify that for a general pencil $\{C_t = V(t_0 F + t_1 G)\}$ of plane curves of degree d, if (L, p) is a hyperflex of some element C_t of the pencil, then:

(a) $m_p(C_t \cdot L) = 4$; that is, no line has a point of contact of order 5 or more with any element of the pencil.
(b) p is a smooth point of C_t.
(c) p is not a base point of the pencil.

Using these facts, show that the degeneracy locus of the sections σ_F and σ_G of the bundle $\mathcal{E}$ introduced in Section 11.3.1 is reduced.

Exercise 11.32. Let $\{C_t = V(t_0 F + t_1 G)\}$ be a general pencil of plane curves of degree d. If $p \in \mathbb{P}^2$ is a general point, how many flex lines to members of the pencil $\{C_t\}$ pass through p?

For Exercises 11.33–11.38, we let $\{C_t = V(t_0 F + t_1 G)\}$ be a general pencil of plane curves of degree d,

$$\Psi = \{(L, p) \in \mathbb{P}^{2*} \times \mathbb{P}^2 \mid p \in L\}$$

be the universal line and

$$\Gamma = \{(t, L, p) \in \mathbb{P}^1 \times \Phi \mid m_p(L \cdot C_t) \geq 3\}.$$

Let $B \subset \mathbb{P}^2$ be the image of Γ under the projection $\Gamma \to \Phi \to \mathbb{P}^2$; that is, the curve traced out by flex points of members of the pencil.

Exercise 11.33. First, show that Γ is indeed smooth, by showing that the "universal flex"

$$\Sigma = \{(C, L, p) \mid m_p(C \cdot L) \geq 3\} \subset \mathbb{P}^N \times \Phi$$

(where $\mathbb{P}^N$ is the space parametrizing all degree-d plane curves) is smooth and invoking Bertini. Can you give explicit conditions on the pencil equivalent to the smoothness of Γ?

Exercise 11.34. If $p \in \mathbb{P}^2$ is a base point of the pencil, show that exactly three members of the pencil have a flex point at p, and that the curve B has an ordinary triple point at p.

Exercise 11.35. If a point $p \in \mathbb{P}^2$ is a node of the curve C_t containing it, the tangent lines to the two branches are each flex lines to C_t, so that the map $\Gamma \to B$ is two-to-one there. Show that the curve B has correspondingly a node at p.

Exercise 11.36. Finally, show that the triple points and nodes of B described in the preceding two exercises are the only singularities of B.

Exercise 11.37. Let C_t be an element of our pencil with a hyperflex (L, p). Show that the map $\Gamma \to \mathbb{P}^1$ is simply ramified at (t, L, p), and simply branched at t.

Exercise 11.38. Let C_t be an element of our pencil with a node p; let L_1 and L_2 be the tangent lines to the two branches of C_t at p. Show that $(t, p, L_i) \in \Gamma$, and that these are ramification points of weight 2 of the map $\Gamma \to \mathbb{P}^1$ (that is, each of the lines L_i is a limit of three flex lines of nearby smooth curves in our pencil, and these three are cyclically permuted by the monodromy in the family). Conclude that t is a branch point of multiplicity 4 for the cover $\Gamma \to \mathbb{P}^1$.

Exercise 11.39. Let $\{C_t\}$ be a general pencil of plane curves of degree d including a cuspidal curve C_0. (That is, let $C_0 = V(F)$ be a general cuspidal curve, $C_\infty = V(G)$ a general curve and $\{C_t = V(F + tG)\}$ the pencil they span.) As $t \to 0$, how many flexes of C_t approach the cusp of C_0? How about if C_0 has a tacnode?

The following series of exercises (Exercises 11.40–11.44) sketches a proof of Proposition 11.13.

Exercise 11.40. Suppose that $p \in C$ is an A_n-singularity for $n \geq 3$. Show that the blow-up $C' = \mathrm{Bl}_p\, C$ of C at p has a unique point q lying over p, and that $q \in C'$ is an A_{n-2}-singularity. Conclude in particular that the normalization $\tilde{C} \to C$ of C at p has genus

$$p_a(\tilde{C}) = p_a(C) - \lfloor \tfrac{1}{2}(n + 1) \rfloor.$$

Exercise 11.41. Let S be a smooth surface and $C \subset S$ a curve with an A_{2n-1}-singularity at p.

(a) Show that there is a unique curvilinear subscheme $\Gamma \subset S$ of degree n supported at p such that a local defining equation of $C \subset S$ at p lies in the ideal $\mathcal{I}_\Gamma^2$.

(b) If $\tilde{S} = \mathrm{Bl}_\Gamma\, S$ is the blow-up of S along Γ, show that the proper transform $\tilde{C}$ of C in $\tilde{S}$ is smooth over p and intersects the exceptional divisor E transversely twice at smooth points of $\tilde{S}$.

(c) Conversely, show that if $D \subset \tilde{S}$ is any such curve then the image of D in S has an A_{2n-1}-singularity at p.

Exercise 11.42. Prove the analog of Exercise 11.41 for A_{2n}-singularities. This is the same statement, except that in the second and third parts the phrase "intersects the exceptional divisor E transversely twice at smooth points of $\tilde{S}$" should be replaced with "is simply tangent to the exceptional divisor E at a smooth point of $\tilde{S}$ and does not meet E otherwise."

Exercise 11.43. Let $\mathcal{L}$ be a line bundle on a smooth surface S, and assume that for any curvilinear subscheme $\Gamma \subset S$ of degree n supported at a single point we have

$$H^1(\mathcal{L} \otimes \mathcal{I}_\Gamma^2) = 0.$$

Show that the locus $\Delta_k \subset \mathbb{P}H^0(\mathcal{L})$ of curves in the linear series $|\mathcal{L}|$ with an A_k-singularity is locally closed and irreducible of codimension k in $\mathbb{P}H^0(\mathcal{L})$ for all $k \leq 2n - 2$.

Exercise 11.44. Deduce from the above exercises the statement of Proposition 11.13.

Exercise 11.45. Show that if $\mathcal{L}$ is the n-th power of a very ample line bundle, then the condition $H^1(\mathcal{L} \otimes \mathcal{I}_\Gamma^2) = 0$ is satisfied for any curvilinear subscheme $\Gamma \subset S$ of degree $n/2$ or less. Conclude in particular that if $\mathcal{D} \subset |\mathcal{L}|$ is a general net in the complete linear series $|\mathcal{L}|$ associated to the fourth or higher power of a very ample bundle then no curve $C \in \mathcal{D}$ has singularities other than nodes and ordinary cusps.

The following three exercises sketch out a calculation of the number of curves $C \subset S$ with a tacnode in a suitably general three-dimensional linear system. (Here, as in the case of cusps, when we use the term "tacnode" without the adjective "ordinary" we include as well singularities that are specializations of ordinary tacnodes, that is, A_n-singularities for any $n \geq 3$, triple points or points on multiple components.)

Exercise 11.46. Let S be a smooth surface and $\mathcal{L}$ a line bundle on S. Let $B = \mathbb{P}\mathcal{T}_S$ be the projectivization of the tangent bundle of S, which we may think of as a parameter space for subschemes $\Gamma \subset S$ of degree 2 supported at a single point. Construct a vector bundle $\mathcal{E}$ on B whose fiber at a point $\Gamma \in B$ may be naturally identified with the vector space

$$\mathcal{E}_\Gamma = H^0(\mathcal{L}/\mathcal{L} \otimes \mathcal{I}_\Gamma^2)$$

Exercise 11.47. In terms of the description of the Chow ring $A(B)$ of $B = \mathbb{P}\mathcal{T}_S$ given in Section 11.4.2, calculate the top Chern class of the bundle constructed in Exercise 11.46.

Exercise 11.48. Using the preceding two exercises, find an enumerative formula for the number of curves in a three-dimensional linear series $\mathcal{D} \subset |\mathcal{L}|$ that have a tacnode. If $S \subset \mathbb{P}^3$ is a smooth surface of degree d, apply this to find the expected number of plane sections with a tacnode. Check your answer by calculating the number directly in the cases $d = 2$ and 3.

The following three exercises describe a way of deriving the formula for the number of binodal curves in a net via linearization. We begin by introducing a smooth, projective compactification of the space of unordered pairs of points $p, q \in \mathbb{P}^2$: We set

$$\widetilde{\Phi} = \{(L, p, q) \mid p, q \in L\} \subset \mathbb{P}^{2*} \times \mathbb{P}^2 \times \mathbb{P}^2,$$

and let Φ be the quotient of $\widetilde{\Phi}$ by the involution $(L, p, q) \mapsto (L, q, p)$. To put it differently, Φ consists of pairs (L, D) with $L \subset \mathbb{P}^2$ a line and $D \subset L$ a subscheme of degree 2; or, differently still, Φ is the Hilbert scheme of subschemes of $\mathbb{P}^2$ with Hilbert polynomial 2. (Compare this with the description in Section 9.7.4 of the Hilbert scheme of conic curves in $\mathbb{P}^3$ — this is the same thing, one dimension lower.)

Exercise 11.49. Observe that the projection $\Phi \to \mathbb{P}^{2*}$ expresses Φ as a projective bundle over $\mathbb{P}^{2*}$, and use this to calculate its Chow ring.

Exercise 11.50. Viewing Φ as the Hilbert scheme of subschemes of $\mathbb{P}^2$ of dimension 0 and degree 2, construct a vector bundle $\mathcal{E}$ on Φ whose fiber at a point D is the space

$$\mathcal{E}_{(L,p,q)} = H^0(\mathcal{O}_{\mathbb{P}^2}(d)/\mathcal{I}_D^2(d)).$$

(What would go wrong if instead of using the Hilbert scheme Φ as our parameter space we used the Chow variety — that is, the symmetric square of $\mathbb{P}^2$?) Express the condition that a curve $C = V(F) \subset \mathbb{P}^2$ be singular at p and q in terms of the vanishing of an associated section σ_F of $\mathcal{E}$ on $\mathcal{H}$ at (L, p, q)

Exercise 11.51. Calculate the Chern classes of this bundle, and derive accordingly the formula for the number of binodal curves in a net.

Chapter 12

Porteous' formula

12.1 Degeneracy loci

We saw in Chapter 5 that the Chern class $c_i(\mathcal{F})$ of a vector bundle $\mathcal{F}$ of rank f on a smooth variety X can be characterized, when $\mathcal{F}$ is generated by global sections, as the class of the scheme where $e = f - i + 1$ general sections of $\mathcal{F}$ become dependent: Specifically, if the locus where a map

$$\varphi : \mathcal{O}_X^e \to \mathcal{F}$$

fails to have maximal rank has the expected codimension i, then $c_i(\mathcal{F})$ is the class of the scheme that is locally defined by the $e \times e$ minors of a matrix representing φ. A similar result holds for Segre classes.

We can substantially extend the usefulness of this characterization in two ways: by considering the rank-k locus of a map $\mathcal{O}_X^e \to \mathcal{F}$ for arbitrary $k \leq \min(e, f)$, and by replacing $\mathcal{O}_X^e$ with an arbitrary vector bundle $\mathcal{E}$. In this chapter we will do both: We henceforth consider the class of the scheme $M_k(\varphi)$ where a map of vector bundles $\varphi : \mathcal{E} \to \mathcal{F}$ has rank $\leq k$, locally defined by the ideal of $(k+1) \times (k+1)$ minors of a matrix representation of φ. Such loci are called *degeneracy loci*. We write e and f for the ranks of $\mathcal{E}$ and $\mathcal{F}$, respectively.

In the "generic" case, where X is an affine space of dimension ef and φ_{gen} is the map defined by an $f \times e$ matrix of variables, the codimension of the locus $M_k(\varphi_{\text{gen}})$ is $(e - k)(f - k)$ (Harris [1995, Proposition 12.2] or Eisenbud [1995, Exercise 10.10]).

In general we say that $(e - k)(f - k)$ is the *expected codimension* of $M_k(\varphi)$. In this chapter we will give a formula for the class of $M_k(\varphi)$ under the assumption that it has the expected dimension

Such a formula was first found by Giambelli in 1904, in the special case where $\mathcal{E}$ and $\mathcal{F}$ are both direct sums of line bundles. René Thom observed more generally in the context of differential geometry that when $M_k(\varphi)$ has the expected codimension its class (suitably construed) depends only on the Chern classes of $\mathcal{E}$ and $\mathcal{F}$. This was made explicit by Porteous (see Porteous [1971], which reproduces notes from 1962), giving the expression now called Porteous' formula. (The formula might more properly be called the Giambelli–Thom–Porteous formula; we have chosen to call it the Porteous formula for brevity and because that is how it appears in much of the literature.) The result was proven (in a more general form, in which one specifies the ranks of the restriction of φ to a flag of subbundles of $\mathcal{E}$) in the context of algebraic geometry by Kempf and Laksov [1974].

The form of the expression is interesting in itself: Porteous' formula expresses $[M_k(\varphi)]$ as a polynomial in the components of the ratio $c(\mathcal{F})/c(\mathcal{E})$.

To get an idea of what is to come, consider the case $k = 0$, and suppose that the locus $M_0(\varphi)$, where the map φ induces the zero map on the fibers, has the expected codimension ef. The map φ may be regarded as a global section of the bundle $\mathcal{E}^* \otimes \mathcal{F}$, and the locus $M_0(\varphi)$ is the locus where this global section vanishes; thus its class is $c_{ef}(\mathcal{E}^* \otimes \mathcal{F})$.

The splitting principle makes it easy to understand $c_{ef}(\mathcal{E}^* \otimes \mathcal{F})$: If $\mathcal{E} = \bigoplus \mathcal{L}_i$ and $\mathcal{F} = \bigoplus \mathcal{M}_i$ were sums of line bundles, then $\mathcal{E}^* \otimes \mathcal{F}$ would be the sum of the $\mathcal{L}_i^* \otimes \mathcal{M}_j$. If we write $c(\mathcal{L}_i) = 1 + \alpha_i$ and $c(\mathcal{M}_j) = 1 + \beta_j$, then $c(\mathcal{L}_i^* \otimes \mathcal{M}_j) = 1 + \beta_j - \alpha_i$, so, by Whitney's formula,

$$c(\mathcal{E}^* \otimes \mathcal{F}) = \prod_{i,j} (1 + \beta_j - \alpha_i),$$

and in particular

$$c_{ef}(\mathcal{E}^* \otimes \mathcal{F}) = \prod_{i,j} (\beta_j - \alpha_i).$$

This expression is symmetric in each of the two sets of variables α_i and β_j, so it can be written in terms of the elementary symmetric functions of these variables, which are the Chern classes of $\mathcal{E}$ and $\mathcal{F}$. If we think of the α_i as the roots of the *Chern polynomial* $c_t(\mathcal{E}) := 1 + c_1(\mathcal{E})t + c_2(\mathcal{E})t^2 + \cdots$, and similarly for $c_t(\mathcal{F})$, then $c_{ef}(\mathcal{E}^* \otimes \mathcal{F})$ is the classical *resultant* of $c_t(\mathcal{E})$ and $c_t(\mathcal{F})$, written $\text{Res}_t(c_t(\mathcal{E}), c_t(\mathcal{F}))$. (See for example

Eisenbud [1995, Section 14.1] for more about resultants and their role in algebraic geometry.) By the splitting principle, the result we have obtained holds for all maps of vector bundles:

Proposition 12.1. *If $\mathcal{E}$ and $\mathcal{F}$ are vector bundles of ranks e and f on a smooth variety X, then*

$$c_{ef}(\mathcal{E}^* \otimes \mathcal{F}) = \mathrm{Res}_t(c_t(\mathcal{E}), c_t(\mathcal{F})).$$

The polynomials $c_t(\mathcal{E})$ and $c_t(\mathcal{F})$ each have constant coefficient 1, and in this case we can express the resultant differently.

We first introduce some notation. For any sequence of elements $\gamma := (\gamma_0, \gamma_1, \dots)$ in a commutative ring and any natural numbers e, f, we set $\Delta^e_f(\gamma) = \det \mathbb{D}^e_f(\gamma)$, where

$$
\mathbb{D}^e_f(\gamma) := \begin{pmatrix}
\gamma_f & \gamma_{f+1} & \cdots & \cdots & \gamma_{e+f-1} \\
\gamma_{f-1} & \gamma_f & \cdots & \cdots & \gamma_{e+f-2} \\
\vdots & \vdots & \ddots & & \vdots \\
\vdots & \vdots & & \ddots & \vdots \\
\gamma_{f-e+1} & \gamma_{f-e+2} & \cdots & \cdots & \gamma_f
\end{pmatrix}.
$$

Proposition 12.2. *If $a(t) = 1 + a_1 t + \cdots + a_e t^e$ and $b(t) = 1 + b_1 t + \cdots + b_f t^f$ are polynomials with constant coefficient 1, then*

$$\mathrm{Res}_t(a(t), b(t)) = \Delta^e_f\left[\frac{b(t)}{a(t)}\right] = (-1)^{ef} \Delta^f_e\left[\frac{a(t)}{b(t)}\right],$$

where $[b(t)/a(t)]$ denotes the sequence of coefficients $(1, c_1, c_2, \dots)$ of the formal power series $b(t)/a(t) = 1 + c_1 t + c_2 t^2 + \cdots$, and similarly for $[a(t)/b(t)]$.

We will give the proof in Section 12.2.

For any element $\gamma \in A(X)$, we write $[\gamma]$ for the sequence $(\gamma_0, \gamma_1, \dots)$, where γ_i is the component of γ of degree i. The next corollary gives the expression of the top Chern class of a tensor product that we will use:

Corollary 12.3. *If $\mathcal{E}$ and $\mathcal{F}$ are vector bundles of ranks e and f on a smooth variety X, then*

$$c_{ef}(\mathcal{E}^* \otimes \mathcal{F}) = \mathrm{Res}_t(c_t(\mathcal{E}), c_t(\mathcal{F})) = \Delta^e_f\left[\frac{c(\mathcal{F})}{c(\mathcal{E})}\right].$$

In particular, if $\varphi : \mathcal{E} \to \mathcal{F}$ is a homomorphism that vanishes in expected codimension ef, then

$$[M_0(\varphi)] = \Delta^e_f\left[\frac{c(\mathcal{F})}{c(\mathcal{E})}\right].$$

Porteous' formula for the class of an arbitrary degeneracy locus follows the same pattern:

Theorem 12.4 (Porteous' formula). *Let $\varphi : \mathcal{E} \to \mathcal{F}$ be a map of vector bundles of ranks e and f on a smooth variety X. If the scheme $M_k(\varphi) \subset X$ has codimension $(e-k)(f-k)$, then its class is given by*

$$[M_k(\varphi)] = \Delta_{f-k}^{e-k}\left[\frac{c(\mathcal{F})}{c(\mathcal{E})}\right].$$

The formula is easiest to interpret in the case $k = e - 1 < f$; in this case $\Delta_{e-k}^{f-k}(\gamma)$ is the determinant of the 1×1 matrix

$$\mathbb{D}_{f-e+1}^1\left[\frac{c(\mathcal{F})}{c(\mathcal{E})}\right] = \left\{\frac{c(\mathcal{F})}{c(\mathcal{E})}\right\}^{f-e+1},$$

where we write $\{\alpha\}^k$ for the codimension-k part of a Chow class $\alpha \in A(X)$. Specializing further, if $\mathcal{E} = \mathcal{O}_X^e$ then $\{c(\mathcal{F})/c(\mathcal{E})\}^{f-e+1} = c_{f-e+1}(\mathcal{F})$, so we recover the characterization of Chern classes as degeneracy loci (Theorem 5.3). If instead $\mathcal{F} = \mathcal{O}_X^f$, then $\{c(\mathcal{F})/c(\mathcal{E})]\}^{f-e+1} = \{1/c(\mathcal{E})\}^{f-e+1}$, the Segre class $s_{f-e+1}(\mathcal{E}^*)$, so we recover the characterization of the Segre class as degeneracy locus of a map $\mathcal{O}_X^f \to \mathcal{E}^*$.

More generally, $\Delta_{f-e+1}^1[c(\mathcal{F})/c(\mathcal{E})]$ represents an obstruction to the existence of an inclusion of vector bundles $\varphi : \mathcal{E} \to \mathcal{F}$: If φ were an inclusion of vector bundles, then $\mathcal{F}/\mathcal{E}$ would be a vector bundle of rank equal to rank $\mathcal{F}$ − rank $\mathcal{E} = f - e$, so $\Delta_{f-e+1}^1[c(\mathcal{F})/c(\mathcal{E})] = \{c(\mathcal{F})/c(\mathcal{E})\}^{f-e+1} = 0$.

The proof of Theorem 12.4 will be given in Section 12.3: After a reduction to a "generic case," we will express $[M_k(\varphi)]$ as the image of $[M_0(\psi)]$, where ψ is a map from a bundle $\mathcal{S}$ of rank $e - k$ to a bundle $\mathcal{F}'$ of rank f on a Grassmannian bundle over X; under the pushforward to X, the entry $\{c(\mathcal{F}')/c(\mathcal{S})\}^i$ in the matrix $\mathbb{D}_f^{e-k}[c(\mathcal{F})/c(\mathcal{E})]$ will be replaced by $\{c(\mathcal{F})/c(\mathcal{E})\}^{i-k}$, yielding Porteous' formula.

12.2 Porteous' formula for $M_0(\varphi)$

In this section we will prove the resultant formula of Proposition 12.2, and thus complete the proof of Corollary 12.3.

Proof of Proposition 12.2: Since the polynomial $\prod(\beta_i - \alpha_j)$ has no repeated factors, it divides any polynomial in the α_i and β_j that vanishes when one of the α_i is equal to one of the β_j. We first show that $\Delta_f^e(b(t)/a(t))$ has this vanishing property. Indeed, if $a(t)$ and $b(t)$ have a common factor $1 + \gamma$, then dividing $a(t)$ by this root gives a polynomial $\bar{a}(t)$ such that

$$\bar{a}(t)\frac{b(t)}{a(t)} = g(t)$$

is a polynomial of degree $< f$. If we write the ratio b/a as a power series

$$\frac{b(t)}{a(t)} = 1 + c_1 t + c_2 t^2 + \cdots,$$

and substitute the power series $1 + c_1 t + \cdots$ into this expression, we get a power series

$$g(t) = \bar{a}(t) c(t) = 1 + \bar{c}_1 t + \bar{c}_2 t^2 + \cdots$$

whose coefficients $\bar{c}_i = c_i + \bar{a}_1 c_{i-1} + \cdots + \bar{a}_{e-1} c_{i-e+1}$ vanish for $i \geq f$. It follows that the vector $(\bar{a}_{e-1}, \ldots, \bar{a}_1, -1)$ is annihilated by $\mathbb{D}_f^e(b(t)/a(t))$, and thus $\Delta_f^e(b(t)/a(t)) = \det \mathbb{D}_f^e(b(t)/a(t)) = 0$.

It follows that

$$\Delta_f^e(b(t)/a(t)) = d \operatorname{Res}_t(a(t), b(t)) = d \prod_{i,j} (\beta_j - \alpha_i)$$

for some polynomial d in the α_i and β_j.

Writing $a(t) = \prod(1 + \alpha_i t)$ and $b(t) = \prod_j (1 + \beta_j t)$, we see that the coefficient of t^k in the power series $\sum c_k t^k = b(t)/a(t)$ is homogeneous of degree k in the variables α_i, β_j, and thus every term in the determinant $\Delta_f^e(b(t)/a(t))$ has degree ef, as does $\operatorname{Res}_t(a(t), b(t))$. It follows that d is a constant.

If we take all the $a_i = 0$, we see that $b(t)/a(t) = b(t)$, and $\mathbb{D}_f^e(b(t)/a(t))$ becomes lower-triangular; in this case its determinant is $(b_f)^e = \left(\prod_j \beta_j\right)^e = \operatorname{Res}_t(a(t), b(t))$, so $d = 1$. □

12.3 Proof of Porteous' formula in general

12.3.1 Reduction to a generic case

We first explain how to reduce the proof to a case where a slightly stronger hypothesis holds:

(a) $M_k(\varphi)$ is of the expected dimension $(e - k)(f - k)$.

(b) $M_k(\varphi)$ is reduced.

(c) The points $x \in X$ where the map φ_x has rank exactly k are dense in $M_k(\varphi)$; equivalently, $M_{k-1}(\varphi)$ has codimension $> (e - k)(f - k)$.

To do this, consider the map

$$\psi : \mathcal{E} \xrightarrow{\binom{1}{\varphi}} \mathcal{E} \oplus \mathcal{F}$$

taking $\mathcal{E}$ onto the graph $\Gamma_\varphi \subset \mathcal{E} \oplus \mathcal{F}$ of φ. The original map φ is the composition of ψ with the projection to $\mathcal{F}$.

We now form the Grassmannian bundle $\pi : X' := G(e, \mathcal{E} \oplus \mathcal{F}) \to X$, and we write $\mathcal{S} \to \pi^*(\mathcal{E} \oplus \mathcal{F})$ for the tautological subbundle of rank e. Since ψ is an inclusion of bundles, the universal property of the Grassmannian guarantees that there is a unique map $u : X \to X'$ such that the pullback under u of the tautological inclusion map $\mathcal{S} \to \pi^*(\mathcal{E} \oplus \mathcal{F})$ on the Grassmannian is $\psi : \mathcal{E} \to \mathcal{E} \oplus \mathcal{F}$, and thus the pullback of the composite map $\varphi' : \mathcal{S} \to \pi^*(\mathcal{E} \oplus \mathcal{F}) \to \pi^*\mathcal{F}$ is φ. It follows that $M_k(\varphi) = u^{-1}(M_k(\varphi'))$.

Since $\mathcal{E} \subset \mathcal{E} \oplus \mathcal{F}$ is the kernel of the projection to $\mathcal{F}$, the points of $M_k(\varphi')$ are the points $x \in X'$ such that the fiber of $\mathcal{S}$ meets the fiber of $\pi^*\mathcal{E}$ in dimension at least $e - k$. With notation parallel to that of Chapter 4, this is the Schubert cycle $\Sigma := \Sigma_{(e-k)^{f-k}}(\mathcal{E})$: It is defined over any open subset of X where the bundles in question are trivial, by the same determinantal formula that defines the corresponding Schubert cycle in the case of vector spaces. A look at this formula shows that $\Sigma = M_k(\varphi)$ as schemes.

Over an open set in X where the bundles $\mathcal{E}$ and $\mathcal{F}$ are trivial, the Grassmannian $X' = G(e, \mathcal{E} \oplus \mathcal{F})$ is the product of X with the ordinary Grassmannian $G(e, e + f)$ and the Schubert cycle Σ is the product of X with the corresponding Schubert cycle in $G(e, e + f)$. As was mentioned in the discussion of the equations of Schubert varieties after Theorem 4.3, these varieties are reduced, irreducible and Cohen–Macaulay (Hochster [1973] or De Concini et al. [1982]). In particular, $M_k(\varphi') = \Sigma$ is reduced, irreducible and Cohen–Macaulay of codimension $(e-k)(f-k)$. Moreover, $M_{k-1}(\varphi') = \Sigma_{(e-k+1)^{f-k}}(\mathcal{E})$ has codimension $(e-k+1)(f-k) > \mathrm{codim}\,\Sigma$, so the points $x' \in X'$ where $\varphi'_{x'}$ has rank exactly k are dense in $M_k(\varphi')$.

Because $M_k(\varphi') = \Sigma$ is a Cohen–Macaulay subvariety, we can apply the Cohen–Macaulay case of Theorem 1.23 and conclude that $[M_k(\varphi)] = u^*[M_k(\varphi')]$. Since $u^* : A(X') \to A(X)$ is a ring homomorphism, and since the Chern classes of $\mathcal{S}$ and $\pi^*\mathcal{F}$ pull back to the Chern classes of $\mathcal{E}$ and $\mathcal{F}$ respectively, we see that it suffices to prove Porteous' formula for the map φ'.

12.3.2 Relation to the case $k = 0$

Replacing φ by φ' as above, we may assume that φ satisfies hypotheses (a), (b) and (c) of the previous section.

We next linearize the problem by introducing more data. To say that $x \in M_k(\varphi)$ means that there is some k-dimensional subspace of $\mathcal{F}_x$ that contains $\varphi_x(\mathcal{E}_x)$, and by assumptions (a) and (c) the subspace is equal to $\varphi_x(\mathcal{E}_x)$ when x is a general point of $M_k(\varphi)$. To make use of this idea, we introduce the Grassmannian $\pi : G(e - k, \mathcal{E}) \to X$. We write $\mathcal{S} \subset \pi^*\mathcal{E}$ for the tautological rank-k subbundle. Let $\mu : \mathcal{S} \to \pi^*\mathcal{E} \to \pi^*\mathcal{F}$ be the composite map. The locus in $G(e - k, \mathcal{F})$ where $\pi^*(\varphi) : \pi^*\mathcal{E} \to \pi^*\mathcal{F}$ factors through the tautological rank-k quotient $\pi^*(\mathcal{E})/\mathcal{S}$ may also be described as $M_0(\mu)$, the locus of points x where μ vanishes. It follows that the map from $M_0(\mu)$ to $M_k(\varphi)$ is surjective and generically one-to-one. Since we have assumed that $M_k(\varphi)$ is reduced, we can compute the class $[M_k(\varphi)]$ as $\pi_*[M_0(\mu)]$.

From the fact that π is generically one-to-one on $M_0(\mu)$, we have that $\dim M_0(\mu) = \dim M_k(\varphi) = \dim X - (e-k)(f-k)$. Since $\dim G(e-k, \mathcal{E}) = \dim X + k(e-k)$, it follows that $M_0(\mu)$ has the expected codimension $(e-k)f$, and thus by Corollary 12.3

$$[M_k(\varphi)] = \pi_*[M_0(\mu)] = \pi_* \Delta_f^{e-k}\left[\frac{c(\pi^*\mathcal{F})}{c(\mathcal{S})}\right].$$

12.3.3 Pushforward from the Grassmannian bundle

Completion of the Proof of Theorem 12.4: It remains to compute

$$\pi_* \Delta_f^{e-k}\left[\frac{c(\pi^*\mathcal{F})}{c(\mathcal{S})}\right].$$

Let $\mathcal{Q} = (\pi^*\mathcal{E})/\mathcal{S}$. By Whitney's formula,

$$c(\mathcal{S}) = \frac{c(\pi^*\mathcal{E})}{c(\mathcal{Q})},$$

so

$$\frac{c(\pi^*\mathcal{F})}{c(\mathcal{S})} = \frac{c(\pi^*\mathcal{F})}{c(\pi^*\mathcal{E})}c(\mathcal{Q}) = \pi^*\left(\frac{c(\mathcal{F})}{c(\mathcal{E})}\right)c(\mathcal{Q}).$$

The point is that we have isolated the factors in the entries of the matrix

$$\mathbb{D}_f^{e-k}[\pi^*(c(\mathcal{F})/c(\mathcal{E}))c(\mathcal{Q})]$$

that are pullbacks from X. To take advantage of this, we expand the determinant $\Delta_f^{e-k}[\pi^*(c(\mathcal{F})/c(\mathcal{E}))c(\mathcal{Q})]$ into a sum of classes of the form $\pi^*(\delta)\epsilon$, where ϵ is a product of $e-k$ Chern classes of $\mathcal{Q}$. From the push-pull formula (Theorem 1.23), we see that π_* takes $\pi^*(\delta)\epsilon$ to $\delta\pi_*(\epsilon)$.

The fibers of the morphism π are all isomorphic to the Grassmannian $G(e-k, e)$, which has dimension $(e-k)k$, so any class of codimension $< (e-k)k$ pushes forward to 0. Since $\mathcal{Q}$ has rank k, the only product of $e-k$ Chern classes of $\mathcal{Q}$ that has nonzero pushforward is $c_k(\mathcal{Q})^{e-k}$, and this class will push forward to some multiple $d[X]$ of the fundamental class of X. The coefficient d is the intersection number of $c_k(\mathcal{Q})^{e-k}$ with the general fiber of π.

We can compute d by first restricting $\mathcal{Q}$ to a general fiber of π, obtaining the tautological quotient bundle $\overline{\mathcal{Q}}$ on $G(e-k, e)$. The number d is the degree of $c_k(\overline{\mathcal{Q}})^{e-k}$, which is 1 by Corollary 4.2.

It follows that if ϵ is any k-fold product of Chern classes of $\mathcal{Q}$ then

$$\pi_*(\pi^*(\delta)\epsilon) = \begin{cases} \delta & \text{if } \epsilon = c_k(\mathcal{Q})^{e-k}, \\ 0 & \text{otherwise.} \end{cases}$$

In particular, since each term in the usual expansion of an $(e - k) \times (e - k)$ determinant is the product of $e - k$ factors,

$$
\pi_* \Delta_f^{e-k} \left[\frac{c(\pi^* \mathcal{F})}{c(\mathcal{S})} \right] = \pi_* \det \begin{pmatrix} c_f & c_{f+1} & \cdots & c_{e-k+f-1} \\ c_{f-1} & c_f & \cdots & c_{e-k+f-2} \\ \vdots & \vdots & & \vdots \\ \vdots & \vdots & & \vdots \\ \vdots & \vdots & & \vdots \\ c_{f-e+k+1} & \cdots & \cdots & c_f \end{pmatrix},
$$

where

$$
c_j = \pi^* \left\{ \frac{c(\mathcal{F})}{c(\mathcal{E})} \right\}^{j-k} c_k(\mathcal{Q}).
$$

Since π^* is a ring homomorphism, the coefficient of $c_k(\mathcal{Q})^{e-k}$ in this expansion is

$$
\pi^* \Delta_{f-k}^{e-k} \left[\frac{c(\mathcal{F})}{c(\mathcal{E})} \right],
$$

and we see that

$$
\pi_* \Delta_f^{e-k} \left[\frac{c(\pi^* \mathcal{F})}{c(\mathcal{S})} \right] = \Delta_{f-k}^{e-k} \left[\frac{c(\mathcal{F})}{c(\mathcal{E})} \right],
$$

as required.   □

12.4 Geometric applications

12.4.1 Degrees of determinantal varieties

As a direct application, we use Porteous' formula to determine the degrees of the varieties of $e \times f$ matrices of rank $\leq k$. Another approach is to form an explicit basis for the component of each degree in the homogeneous coordinate ring of this variety, and thus to compute the Hilbert function. This seems first to have been done in Hodge [1943]; for a modern treatment related to his ideas, see De Concini et al. [1982]. For other approaches, generalizations and references, see Abhyankar [1984] and Herzog and Trung [1992].

Theorem 12.5. *Let A be an $e \times f$ matrix of linear forms on $\mathbb{P}^r$, and let $M_k :=$ $M_k(A) \subset \mathbb{P}^r$ be the scheme defined by its $(k + 1) \times (k + 1)$ minors. If M_k has the expected codimension $(e - k)(f - k)$ in $\mathbb{P}^r$, then its degree is*

$$
\deg(M_k) = \prod_{i=0}^{e-k-1} \frac{i! (f + i)!}{(k + i)! (f - k + i)!}.
$$

Note that as a special case we could take $\mathbb{P}^r = \mathbb{P}^{ef-1}$ the space of all nonzero $e \times f$ matrices modulo scalars and A the matrix of homogeneous coordinates on $\mathbb{P}^{ef-1}$. Indeed, the general case follows from this one: Any matrix A as in the statement of the theorem corresponds to a linear map $\mathbb{P}^r \to \mathbb{P}^{ef-1}$, and if the preimage M_k of the locus $\Phi_k \subset \mathbb{P}^{ef-1}$ of matrices of rank k or less has the expected codimension its degree must be equal to the degree of Φ_k.

The formula simplifies in the case $k = e - 1$, with $f \geq e$. Here we see that

$$\deg(M_{e-1}) = \frac{0!\,(f)!}{(e-1)!\,(f-(e-1))!} = \binom{f}{e-1};$$

if in addition $e = f$, the degree is f, the degree of the determinant.

On the other hand, when $k = 1$ the formula telescopes: We have

$$\prod_{i=0}^{e-2} \frac{i!}{(i+1)!} = \frac{1}{(e-1)!}$$

and

$$\prod_{i=0}^{e-2} \frac{(f+i)!}{(f-1+i)!} = \frac{(e+f-2)!}{(f-1)!},$$

so we get

$$\deg(M_1) = \frac{1}{(e-1)!} \cdot \frac{(e+f-2)!}{(f-1)!} = \binom{e+f-2}{e-1}.$$

Note that when $\mathbb{P}^r = \mathbb{P}^{ef-1}$ is the space of all nonzero $e \times f$ matrices modulo scalars and A the matrix of homogeneous coordinates on $\mathbb{P}^{ef-1}$, the scheme $M_1(A)$ is the Segre variety $\mathbb{P}^{e-1} \times \mathbb{P}^{f-1} \subset \mathbb{P}^{ef-1}$ and the formula gives the degree of that variety, agreeing with the computation made by other means in Section 2.1.5.

Proof of Theorem 12.5: Multiplication by the matrix A defines a vector bundle map

$$(\mathcal{O}_{\mathbb{P}^r})^{\oplus e} \to (\mathcal{O}_{\mathbb{P}^r}(1))^{\oplus f},$$

and we are asking for the class of the locus where this map has rank k or less. Letting $\zeta \in A^1(\mathbb{P}^r)$ be the hyperplane class, we have

$$c(\mathcal{F}) = (1+\zeta)^f = \sum_{r=0}^{f} \binom{f}{r} \zeta^r,$$

from which we conclude that the class of M_k is given by

$$[M_k] = \begin{vmatrix} \binom{f}{f-k}\zeta^{f-k} & \cdots & \binom{f}{f+e-2k-1}\zeta^{f+e-2k-1} \\ \vdots & & \vdots \\ \binom{f}{f-e+1}\zeta^{f-e+1} & \cdots & \binom{f}{f-k}\zeta^{f-k} \end{vmatrix}$$

$$= \begin{vmatrix} \binom{f}{f-k} & \cdots & \binom{f}{f+e-2k-1} \\ \vdots & & \vdots \\ \binom{f}{f-e+1} & \cdots & \binom{f}{f-k} \end{vmatrix} \zeta^{(e-k)(f-k)}.$$

The degree of M_k is thus the last determinant, which may be simplified as follows. To begin with, we make a series of column operations: First, we replace each column, starting with the second, with the sum of it and the column to its left, arriving at

$$\begin{vmatrix} \binom{f}{f-k} & \binom{f}{f-k+1} & \cdots & \binom{f}{f+e-2k-1} \\ \vdots & \vdots & & \vdots \\ \vdots & \vdots & & \vdots \\ \binom{f}{f-e+1} & \binom{f}{f-e+2} & \cdots & \binom{f}{f-k} \end{vmatrix}$$

$$= \begin{vmatrix} \binom{f}{f-k} & \binom{f+1}{f-k+1} & \cdots & \binom{f+1}{f+e-2k-1} \\ \vdots & \vdots & & \vdots \\ \vdots & \vdots & & \vdots \\ \binom{f}{f-e+1} & \binom{f+1}{f-e+2} & \cdots & \binom{f+1}{f-k} \end{vmatrix}.$$

Now we do the same thing again, this time starting with the third column, then again, starting with the fourth, and so on, obtaining the determinant

$$\begin{vmatrix} \binom{f}{f-k} & \binom{f+1}{f-k+1} & \cdots & \binom{f+e-k-1}{f+e-2k-1} \\ \vdots & \vdots & & \vdots \\ \vdots & \vdots & & \vdots \\ \binom{f}{f-e+1} & \binom{f+1}{f-e+2} & \cdots & \binom{f+e-k-1}{f-k} \end{vmatrix}$$

$$= \begin{vmatrix} \dfrac{f!}{k!(f-k)!} & \dfrac{(f+1)!}{k!(f-k+1)!} & \cdots & \dfrac{(f+e-k-1)!}{k!(f+e-2k-1)!} \\ \vdots & \vdots & & \vdots \\ \vdots & \vdots & & \vdots \\ \dfrac{f!}{(e-1)!(f-e+1)!} & \dfrac{(f+1)!}{(e-1)!(f-e+2)!} & \cdots & \dfrac{(f+e-k-1)!}{(e-1)!(f-k)!} \end{vmatrix}.$$

We can pull a factor of $f!$ from the first column, $(f + 1)!$ from the second, and so on; similarly, we can pull a $k!$ from the denominators in the first row, a $(k + 1)!$ from the denominators in the second row, and so on. We arrive at the product

$$\prod_{i=0}^{e-k-1} \frac{(f + i)!}{(k + i)!} \begin{vmatrix} \frac{1}{(f-k)!} & \frac{1}{(f-k+1)!} & \cdots & \frac{1}{(f+e-2k-1)!} \\ \vdots & \vdots & & \vdots \\ \vdots & \vdots & & \vdots \\ \frac{1}{(f-e+1)!} & \frac{1}{(f-e+2)!} & \cdots & \frac{1}{(f-k)!} \end{vmatrix}.$$

Next, we multiply the first column by $(f - k)!$, the second by $(f - k + 1)!$, and so on, obtaining the expression

$$\prod_{i=0}^{e-k-1} \frac{(f + i)!}{(k + i)!(f - k + i)!} \begin{vmatrix} 1 & 1 & \cdots \\ f - k & f - k + 1 & \cdots \\ (f - k)(f - k - 1) & (f - k + 1)(f - k) & \cdots \\ \vdots & \vdots & \end{vmatrix}.$$

Finally, we can recognize the columns of this matrix as the series of monic polynomials $1, x, x(x - 1), x(x - 1)(x - 2), \ldots$ of degrees $0, 1, 2, \ldots, m - k - 1$, applied to the integers $f - k, f - k + 1, \ldots, f + e - 2k - 1$. Its determinant is thus equal to the Vandermonde determinant

$$\begin{vmatrix} 1 & 1 & 1 & \cdots \\ f - k & f - k + 1 & f - k + 2 & \cdots \\ (f - k)^2 & (f - k + 1)^2 & (f - k + 2)^2 & \cdots \\ \vdots & \vdots & \vdots & \end{vmatrix},$$

which is equal to $\prod_{i=0}^{e-k-1} i!$. Putting this all together, we have established Theorem 12.5. $\square$

12.4.2 Pinch points of surfaces

Let $C \subset \mathbb{P}^n$ be a smooth curve. A classical theorem (Exercise 3.34) describes the projection $\pi = \pi_\Lambda : C \to \mathbb{P}^2$ from a general $(n - 3)$-plane: it is an immersion whose image has only ordinary nodes as singularities. We would now like to describe in similar fashion the geometry of the projection $\pi : S \to \mathbb{P}^3$ of a smooth surface $S \subset \mathbb{P}^n$ from a general plane $L \cong \mathbb{P}^{n-4}$.

In general, Mather [1971; 1973] gave normal forms for the singularities of general projections of a smooth variety of dimension ≤ 7 to a hypersurface. For the case of a smooth surface $S \subset \mathbb{P}^n$ and a general projection $\pi : S \to \mathbb{P}^3$ this is classical and easy to describe; see for example Griffiths and Harris [1994, p. 611]. There are precisely three local analytic types of singular points of the image $\pi(S) \subset \mathbb{P}^3$:

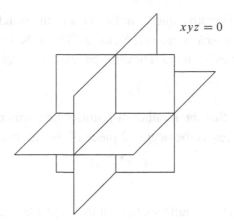

Figure 12.1 Three branches of the double curve meeting in a triple point.

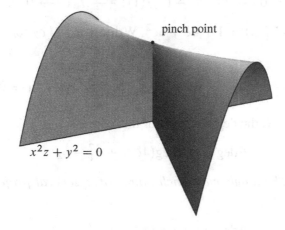

Figure 12.2 Double curve passing through a pinch point.

- There is a curve in $\pi(S)$ over which the map is two-to-one, and in an analytic neighborhood of a general point on this curve the surface S_0 is the union of two smooth sheets crossing transversely.
- There are a finite number of points in $\pi(S)$ with preimage of length 3, and at each such point $\pi(S)$ is the union of three smooth sheets intersecting transversely (Figure 12.1).
- There are finitely many points of S where the differential of π is not injective, and in suitable local coordinates near such a point the map is $(s, t) \mapsto (s, st, t^2)$, so that the image satisfies the equation $x^2 z = y^2$. Such points in the image are called *pinch points* (Figure 12.2). In the given coordinates the double curve is the z-axis, and the two sheets at the point $(0, 0, z_0)$ will have tangent planes $y = \pm\sqrt{z_0} \cdot x$.

The geometry of the map, and of the image, is beautiful: pinch points are points of the double curve of $\pi(S)$ where the local monodromy interchanges the two sheets.

We now ask the enumerative question: In terms of the standard invariants of the surface $S \subset \mathbb{P}^n$, *how many pinch points will S_0 have?* This is Keynote Question (b), and we can answer it with Porteous' formula. The differential of the map π is a vector bundle map

$$d\pi : \mathcal{T}_S \to \pi^* \mathcal{T}_{\mathbb{P}^3},$$

and the formula tells us that the number of points where this map fails to be injective, counted with multiplicities, is the degree-2 piece of the quotient

$$\frac{c(\pi^* \mathcal{T}_{\mathbb{P}^3})}{c(\mathcal{T}_S)}.$$

Denote by $\zeta = c_1(\mathcal{O}_S(1))$ the pullback to S of the hyperplane class on $\mathbb{P}^3$ (equivalently, the restriction to S of the hyperplane class on $\mathbb{P}^n$). Pulling back the Euler sequence

$$0 \longrightarrow \mathcal{O}_{\mathbb{P}^3} \longrightarrow \mathcal{O}_{\mathbb{P}^3}(1)^4 \longrightarrow \mathcal{T}_{\mathbb{P}^3} \longrightarrow 0$$

to S, we see that $c(\pi^* \mathcal{T}_{\mathbb{P}^3}) = 1 + 4\zeta + 6\zeta^2$. Writing c_1 and c_2 for $c_1(\mathcal{T}_S^*)$ and $c_2(\mathcal{T}_S^*)$, we have

$$\frac{c(\pi^* \mathcal{T}_{\mathbb{P}^3})}{c(\mathcal{T}_S)} = \frac{1 + 4\zeta + 6\zeta^2}{1 - c_1 + c_2} = (1 + 4\zeta + 6\zeta^2)(1 + c_1 + (c_1^2 - c_2)).$$

Since $\deg(\zeta^2) = \deg S$, the degree-2 part of this expression is

$$6 \deg(S) + \deg(4\zeta c_1 + c_1^2 - c_2).$$

Proposition 12.6. *The number of pinch points of a general projection of a smooth surface $S \subset \mathbb{P}^n$ to $\mathbb{P}^3$ is*

$$6 \deg(S) + \deg(4\zeta c_1 + c_1^2 - c_2),$$

where the $c_i = c_i(\mathcal{T}_S^*)$ are the Chern classes of the cotangent bundle of S.

Proof: The map π fails to be an immersion at a point $s \in S \subset \mathbb{P}^n$ if and only if the projection center $\Lambda \cong \mathbb{P}^{n-4} \subset \mathbb{P}^n$ meets the projective tangent plane $\mathbb{T}_p S$ to S at p. The union of the tangent planes to S is clearly at most four-dimensional, so for general Λ the number of such points is finite.

It remains to prove that each pinch point counts with multiplicity 1 in the degeneracy locus of the differential $d\pi : \mathcal{T}_S \to \pi^* \mathcal{T}_{\mathbb{P}^3}$. This is equivalent to saying that at each point where the differential drops rank the 2×2 minors of the Jacobian matrix representing it generate the maximal ideal. From the local form of the map given above, we see that the Jacobian is

$$
\begin{array}{c}
 \\
s \\
st \\
t^2
\end{array}
\begin{array}{cc}
\partial/\partial s & \partial/\partial t \\
\left(\begin{array}{cc}
1 & 0 \\
t & s \\
0 & 2t
\end{array} \right),
\end{array}
$$

with ideal of 2×2 minors (s, t) as required. $\qquad \square$

A beautiful example of Proposition 12.6 is the projection $\pi : S \to \mathbb{P}^3$ of the Veronese surface $S = \nu_2(\mathbb{P}^2) \subset \mathbb{P}^5$ from a general line $L \subset \mathbb{P}^5$. We will sketch the geometry of the map and its image briefly.

To begin with, the secant variety of the Veronese surface is a hypersurface $X \subset \mathbb{P}^5$. This is a reflection of the fact that a point $x \in \mathbb{P}^5$ lying on a secant line $\overline{p,q}$ of S in fact lies on a one-dimensional family of secants: the line in $\mathbb{P}^2$ through the points $p, q \in S = \mathbb{P}^2$ is carried, under the Veronese map, to a conic curve $C \subset S \subset \mathbb{P}^5$, and, since the point x lies in the plane spanned by C, every line through x in that plane will be a secant line to S.

Now, since X is in fact a cubic hypersurface, a general line $L \subset \mathbb{P}^5$ will meet X in three points, corresponding to three lines in $\mathbb{P}^2$. Each of these lines will be carried into a conic in $\mathbb{P}^5$ and then under projection from L will be mapped onto a line with degree 2 (and two branch points). Thus the double curve of $\pi(S)$ will consist of three lines, meeting in a single point in $\mathbb{P}^3$ (the unique triple point of $\pi(S)$), and each line will have two pinch points, accounting for the six pinch points predicted by Proposition 12.6.

12.4.3 Pinch points and the tangential variety of S

There is an alternative derivation of the formula of Proposition 12.6, based on the observation that the number of pinch points of a general projection of a smooth surface $S \subset \mathbb{P}^n$ to $\mathbb{P}^3$ is related to the degree of the *tangential surface* $T(S)$ of S, that is, the union

$$T(S) = \bigcup_{p \in S} \mathbb{T}_p S \subset \mathbb{P}^n$$

of the projective tangent planes to S. More precisely, the number of pinch points of a general projection of $S \subset \mathbb{P}^n$ to $\mathbb{P}^3$ is the degree of $T(S)$ times the number d of tangent planes $\mathbb{T}_p S$ containing a general point of $T(S)$.

To carry this out, recall from Section 7.4.3 that the *Gauss map* $\mathcal{G} : S \to \mathbb{G}(2, n)$ is the map sending each point $p \in S$ to its projective tangent plane $\mathbb{T}_p S$; assuming $T(S)$ does have the expected dimension 4, we can apply Proposition 10.4 on the degree of a variety swept out by linear spaces to express this degree as the second Segre class of the pullback $\mathcal{G}^*(\mathcal{S})$ via $\mathcal{G}$ of the universal subbundle $\mathcal{S}$ on $\mathbb{G}(2, n)$. To express this, we recall also from Section 7.4.3 that we have an exact sequence

$$0 \longrightarrow \mathcal{O}_X(-1) \longrightarrow \mathcal{G}^*\mathcal{S} \longrightarrow \mathcal{T}_X(-1) \longrightarrow 0$$

relating $\mathcal{G}^*(\mathcal{S})$ to the hyperplane bundle of $S \subset \mathbb{P}^n$ and a twist of the tangent bundle. Applying the formula for the Chern class of a tensor product with a line bundle (Proposition 5.17) and the Whitney formula, we arrive at the conclusion of Proposition 12.6 again.

There is one advantage of this approach: Once we show that the tangential variety $T(S)$ is indeed four-dimensional, we may conclude that *if $S \subset \mathbb{P}^n$ is any smooth, irreducible nondegenerate surface and $n \geq 4$, the number of pinch points of a general projection of S to $\mathbb{P}^3$ is positive.* We will sketch a proof that $T(S)$ is indeed four-dimensional in Exercises 12.16–12.18.

On the other hand, the derivation of Proposition 12.6 via Porteous carried out here has an advantage as well: It applies to any map of a surface S to $\mathbb{P}^3$, assuming only that the singularities of the map and its image are as described on page 436 for a general projection. Indeed, the same method may be applied to a map of a surface S to any smooth threefold X, again assuming the singularities of the map are as described.

12.4.4 Quadrisecants to rational curves

As a final application of Porteous' formula we will count the number of *quadrisecant lines* to a rational space curve $C \subset \mathbb{P}^3$ (that is, lines meeting C in four points). This will conclude our discussion, began in Section 10.5, of special secant planes to rational curves; other cases that can similarly be dealt with using the Porteous formula are suggested in Exercise 12.23.

The question we will address here is: Is there an enumerative formula for the number of quadrisecant lines to a curve $C \subset \mathbb{P}^3$, say in terms of the degree d and genus g of C? We discuss a formula in the general case at the end of the subsection, but first we treat the rational case, and we suppose that $C \cong \mathbb{P}^1$.

Instead of looking at all lines in $\mathbb{P}^3$ and imposing the condition of meeting C four times, we will look at 4-tuples of points on C and impose the condition that they span only a line. We will use the set-up of Section 10.4.2: We identify the space of subschemes Γ of degree 4 in $C \cong \mathbb{P}^1$ with the symmetric power $C^{(4)} \cong \mathbb{P}^4$ of C, and introduce the bundle $\mathcal{E}^*$ on $C^{(4)}$ with fibers

$$\mathcal{E}_\Gamma^* = H^0(\mathcal{O}_\Gamma(d)).$$

As in Section 10.4.2, a global section σ of $\mathcal{O}_C(1) = \mathcal{O}_{\mathbb{P}^1}(d)$ gives rise to a global section of $\mathcal{E}^*$ by restriction to each subscheme of C in turn. The restriction map

$$H^0(\mathcal{O}_{\mathbb{P}^3}(1)) \hookrightarrow H^0(\mathcal{O}_{\mathbb{P}^1}(d)) \to H^0(\mathcal{O}_\Gamma(d))$$

gives us a map $\varphi : \mathcal{F} \to \mathcal{E}^*$ of vector bundles on $C^{(4)}$, where $\mathcal{F}$ is the trivial bundle with fiber $H^0(\mathcal{O}_{\mathbb{P}^3}(1))$, and we see that *the locus $M_2(\varphi) \subset \mathbb{P}^4$ of subschemes $\Gamma \subset \mathbb{P}^1$ of degree 4 with $\dim \overline{\Gamma} = 1$ is the locus where the map φ has rank 2.* (Note that φ can never have rank < 2.) Porteous' formula will then give us an expression for the class of this locus, valid in the case it has the expected dimension $4 - 2 \times 2 = 0$.

To carry this out, recall from Theorem 10.16 that the Chern classes of $\mathcal{E}^*$ are

$$c_i(\mathcal{E}^*) = \binom{d-4+i}{i} \zeta^i \in A^i(\mathbb{P}^4),$$

where $\zeta \in A^1(\mathbb{P}^4)$ is as always the hyperplane class. Now we apply Porteous' formula, which tells us that the class of the locus of subschemes $\Gamma \subset \mathbb{P}^1$ contained in a line is

$$[M_2(\varphi)] = \begin{vmatrix} c_2(\mathcal{E}^*) & c_3(\mathcal{E}^*) \\ c_1(\mathcal{E}^*) & c_2(\mathcal{E}^*) \end{vmatrix}$$

$$= \begin{vmatrix} \binom{d-2}{2}\zeta^2 & \binom{d-1}{3}\zeta^3 \\ \binom{d-3}{1}\zeta & \binom{d-2}{2}\zeta^2 \end{vmatrix}$$

$$= \begin{vmatrix} \binom{d-2}{2} & \binom{d-1}{3} \\ \binom{d-3}{1} & \binom{d-2}{2} \end{vmatrix} \zeta^4$$

$$= \left(\tfrac{1}{4}(d-2)^2(d-3)^2 - \tfrac{1}{6}(d-1)(d-2)(d-3)^2\right)\zeta^4$$

$$= \tfrac{1}{12}(d-3)^2(d-2)(3(d-2)-2(d-1))\zeta^4$$

$$= \tfrac{1}{12}(d-2)(d-3)^2(d-4)\zeta^4.$$

This gives us the enumerative formula:

Proposition 12.7. *If* $C \subset \mathbb{P}^3$ *is a rational space curve of degree* d *possessing only finitely many quadrisecant lines then the number of such lines, counted with multiplicities, is*

$$\tfrac{1}{12}(d-2)(d-3)^2(d-4).$$

Note as a check that this number is 0 in the cases $d = 2, 3$ and 4, as it should be.

We will see in Exercise 12.24 a condition for a given quadrisecant line to be simple, that is, to count with multiplicity 1. We will also see in Exercise 12.25 that for $C \subset \mathbb{P}^3$ a *general* rational curve of degree d — that is, a general projection of a rational normal curve from $\mathbb{P}^d$ to $\mathbb{P}^3$ — all quadrisecants are simple, so this is the actual number of quadrisecant lines.

Quadrisecants to curves of higher genus

If we try to generalize the arguments above to the case where C has higher genus, a new issue arises. The Hilbert scheme parametrizing subschemes of degree 4 in $\mathbb{P}^1$ is $\mathbb{P}^4$, whose Chow ring we know. But the space of subschemes of degree 4 of a smooth curve C of higher genus — again, the fourth symmetric power $C^{(4)}$ of C — is more complex; in particular, its Chow ring is much harder to determine explicitly (see Collino [1975]). The most general formula, for the number of d-secant $(d - r - 1)$-planes to a curve of degree n and genus g in $\mathbb{P}^s$, is derived (using topological cohomology groups rather than Chow groups) in Chapter 8 of Arbarello et al. [1985]; the formula for the number q of quadrisecant lines to a curve $C \subset \mathbb{P}^3$ of degree d and genus g is

$$q = \tfrac{1}{12}(d-2)(d-3)^2(d-4) - \tfrac{1}{2}g(d^2 - 7d + 13 - g).$$

We could also approach the problem of counting quadrisecant lines to a space curve of arbitrary genus via the classical theory of correspondences. This is described in Chapter 2 of Griffiths and Harris [1994].

12.5 Exercises

Exercise 12.8. Let $A = (P_{i,j})$ be a 2×3 matrix whose entries $P_{i,j}$ are general polynomials of degree $a_{i,j}$ on $\mathbb{P}^3$. Assuming that $a_{1,j} + a_{2,k} = a_{1,k} + a_{2,j}$ for all j and k — so that the minors of A are homogeneous — what is the degree of the curve $M_1(A)$ where A has rank 1?

Exercise 12.9. In Exercise 2.32, we introduced the variety of triples of collinear points, that is,

$$\Psi = \{(p, q, r) \in \mathbb{P}^n \times \mathbb{P}^n \times \mathbb{P}^n \mid p, q \text{ and } r \text{ are collinear in } \mathbb{P}^n\}.$$

Calculate the class $\psi = [\Psi] \in A^{n-1}(\mathbb{P}^n \times \mathbb{P}^n \times \mathbb{P}^n)$ by applying Porteous to the evaluation map $\mathcal{E} \to \mathcal{F}$, where $\mathcal{E}$ is the trivial bundle on $\mathbb{P}^n \times \mathbb{P}^n \times \mathbb{P}^n$ with fiber $H^0(\mathcal{O}_{\mathbb{P}^n}(1))$ and

$$\mathcal{F} = \pi_1^* \mathcal{O}_{\mathbb{P}^n}(1) \oplus \pi_2^* \mathcal{O}_{\mathbb{P}^n}(1) \oplus \pi_3^* \mathcal{O}_{\mathbb{P}^n}(1),$$

with $\pi_i : \mathbb{P}^n \times \mathbb{P}^n \times \mathbb{P}^n \to \mathbb{P}^n$ projection on the i-th factor.

Exercise 12.10. In Exercise 9.37, we introduced

$$\Phi_r = \{(L, M) \in \mathbb{G}(1, r) \times \mathbb{G}(1, r) \mid L \cap M \neq \varnothing\}.$$

Find the class of Φ_r in $A(\mathbb{G}(1, r) \times \mathbb{G}(1, r))$ using Porteous' formula.

The following exercise uses Porteous to generalize the result of Exercise 9.44.

Exercise 12.11. Let X be a smooth projective variety, $\mathcal{E}$ a vector bundle of rank r on X and $\mathcal{F}, \mathcal{G} \subset \mathcal{E}$ subbundles of ranks a and b. For any k, find the class of the locus

$$\Sigma = \{p \in X \mid \dim(\mathcal{F}_p \cap \mathcal{G}_p) \geq k\},$$

assuming this locus has the expected (positive) codimension.

Exercise 12.12. Verify Proposition 12.6 directly in case S is a smooth surface in $\mathbb{P}^3$ to begin with.

Exercise 12.13. Verify Proposition 12.6 directly in case $S = S(1, 2) \subset \mathbb{P}^4$ is a cubic scroll (Section 9.1.1). What is the double curve of the image S_0?

Exercise 12.14. Verify Proposition 12.6 directly in case $S = S(2, 2) \subset \mathbb{P}^5$ is a rational normal surface scroll. What does the double curve of S_0 look like in this case, and how many triple points will S_0 have?

Exercise 12.15. Let $S \subset \mathbb{P}^n$ be a smooth surface.

(a) Show that we have a map from the projective bundle $\mathbb{P}(\mathcal{T}_S \oplus \mathcal{O}_S)$ to $\mathbb{P}^n$ with image the tangential variety X of S (specifically, carrying the fiber over p to the tangent plane $\mathbb{T}_p(S)$).

(b) Show that the pullback of $\mathcal{O}_{\mathbb{P}^n}(1)$ under this map is the line bundle

$$\mathcal{O}_{\mathbb{P}(\mathcal{T}_S \oplus \mathcal{O}_S)}(1) \otimes \mathcal{O}_S(1).$$

(c) Use this and our description of the Chow ring of the projective bundle $\mathbb{P}(\mathcal{T}_S \oplus \mathcal{O}_S)$ to re-derive the formula of Proposition 12.6 for the degree of X.

Exercises 12.16–12.18 suggest a proof of the assertion made in Section 12.4.3 that if $S \subset \mathbb{P}^n$ is a nondegenerate surface and $n \geq 4$ then the union of the tangent planes to S is a fourfold.

Exercise 12.16. Let $B \subset \mathbb{G}(2, n)$ be an irreducible surface, and let

$$X = \bigcup_{\Lambda \in B} \Lambda \subset \mathbb{P}^n$$

be the variety swept out by the corresponding 2-planes. Show that if X is three-dimensional then

(a) a general point $p \in X$ lies on a one-dimensional family of planes $\Lambda \in B$, and hence
(b) any two planes $\Lambda, \Lambda' \in B$ meet in a line.

Exercise 12.17. Let $B \subset \mathbb{G}(2, n)$ be an irreducible surface such that any pair of planes $\Lambda, \Lambda' \in B$ meet in a line. Show that either all the planes Λ lie in a fixed 3-plane or all the planes Λ contain a fixed line.

Exercise 12.18. Using the preceding two exercises, conclude that if $S \subset \mathbb{P}^n$ is a nondegenerate surface and $n \geq 4$ then the union of the tangent planes to S is a fourfold.

Exercise 12.19. Let $X \subset \mathbb{P}^n$ be a smooth sixfold and $\pi : X \to \mathbb{P}^7$ a general projection. Find the number of points where the differential $d\pi$ has rank 4 or less.

Exercise 12.20. Let $S \subset \mathbb{P}^n$ be a smooth surface of degree d whose general hyperplane section is a curve of genus g; let e and f be the degrees of the classes $c_1(\mathcal{T}_S)^2$ and $c_2(\mathcal{T}_S) \in A^2(S)$. Find the class of the cycle $T_1(S) \subset \mathbb{G}(1, n)$ of lines tangent to S in terms of d, e, f and g. From Exercise 4.21, we need only the intersection number $[T_1(S)] \cdot \sigma_3$; find it as the number of pinch points of a projection of S from a general $\mathbb{P}^{n-4}$.

Exercise 12.21. Let $C \subset \mathbb{P}^3$ be a smooth, nondegenerate curve. Show that the general secant line to C is not trisecant, and deduce that the locus of trisecant lines to C, if nonempty, has dimension 1. (For extra credit, show that it is empty only in case C is a twisted cubic or an elliptic quartic.)

Exercise 12.22. Check the conclusion of Proposition 12.7 for general rational curves of degrees $d = 5$ and 6 by independently counting the number of quadrisecant lines to such curves.

Hint: Show that such a curve C will lie on a smooth cubic surface S, and observe that the quadrisecant lines to C will be contained in S.

Exercise 12.23. Use Porteous' formula to find the expected number of:

(a) Trisecant lines to a rational curve $C \subset \mathbb{P}^4$.
(b) 6-secant 2-planes to a rational curve $C \subset \mathbb{P}^4$.
(c) 8-secant 3-planes to a rational curve $C \subset \mathbb{P}^5$.
(d) 4-secant 2-planes to a rational curve $C \subset \mathbb{P}^6$.

Exercise 12.24. Let $C \subset \mathbb{P}^3$ be a smooth curve and $L \subset \mathbb{P}^3$ a line meeting C in exactly four points $p_1, \ldots, p_4$ and not tangent to C at any of them. Suppose that the tangent lines $\mathbb{T}_{p_i}(C)$ to C at the p_i are pairwise independent mod L (that is, they span distinct planes with L), and that the cross-ratio of the four points $p_1, \ldots, p_4 \in L$ is *not* equal to the cross-ratio of the four planes $\mathbb{T}_{p_1}(C) + L, \ldots, \mathbb{T}_{p_4}(C) + L$. Show that $\Gamma = p_1 + \cdots + p_4 \in \mathbb{P}^4$ counts as a quadrisecant line with multiplicity 1 (that is, the 3×3 minors of a matrix representative of φ near Γ generate the maximal ideal $\mathfrak{m}_\Gamma \subset \mathcal{O}_{\mathbb{P}^4, \Gamma}$).

Exercise 12.25. Let C now be a general rational curve of degree d in $\mathbb{P}^3$.

(a) Show that C has no 5-secant lines.
(b) Show that if $L \subset \mathbb{P}^3$ is any quadrisecant line to C, then L meets C in four distinct points.
(c) Finally, show that every quadrisecant line to C satisfies the conditions of the preceding exercise, and deduce that the number of quadrisecant lines to C is exactly $\frac{1}{12}(d-2)(d-3)^2(d-4)$. (Note: The two preceding parts are straightforward dimension counts; this one is a little more subtle.)

Exercise 12.26. Let $\{Q_\mu \subset \mathbb{P}^3\}_{\mu \in \mathbb{P}^3}$ be a general web of quadrics in $\mathbb{P}^3$, that is, the three-dimensional linear series corresponding to a general four-dimensional vector space $V \subset H^0(\mathcal{O}_{\mathbb{P}^3}(2))$.

(a) Find the number of 2-planes $\Lambda \subset \mathbb{P}^3$ that are contained in some quadric of the net.
(b) A line $L \subset \mathbb{P}^3$ is said to be a *special line* for the web if it lies on a pencil of quadrics in the web. Find the class of the locus $\Sigma \subset \mathbb{G}(1,3)$ of special lines.

Chapter 13

Excess intersections and the Chow ring of a blow-up

Keynote Questions

(a) Suppose that S_1, S_2 and $S_3 \subset \mathbb{P}^3$ are three surfaces of degrees s_1, s_2 and s_3 whose intersection consists of the disjoint union of a reduced line L and a zero-dimensional scheme $\Gamma \subset \mathbb{P}^3$. What is the degree of Γ? More generally, what happens when we replace L with a smooth curve of genus g and degree d? (Answer on page 450.)

(b) Suppose that S_1 and $S_2 \subset \mathbb{P}^4$ are two surfaces whose intersection consists of the disjoint union of a reduced line L and a zero-dimensional scheme $\Gamma \subset \mathbb{P}^4$. In terms of the geometry of S_1 and S_2, can we say what the degree of Γ is? Can we say what the degree of Γ is in terms of the degrees of S_1 and S_2 alone? As in the preceding question, what happens when we replace L with a smooth curve of genus g and degree d? (Answers on page 459.)

(c) Let $\Lambda \cong \mathbb{P}^{n-c} \subset \mathbb{P}^n$ be a codimension-c linear subspace, and let $Q_1, \ldots, Q_n \subset \mathbb{P}^n$ be general quadric hypersurfaces containing Λ. If we write the intersection $\bigcap Q_i$ as the union

$$\bigcap_{i=1}^{n} Q_i = \Lambda \cup \Gamma,$$

what is the degree of Γ? (Answer on page 460.)

(d) Is it the case that every smooth curve $C \subset \mathbb{P}^3$ is the scheme-theoretic intersection of three surfaces? (Answer on page 452.)

(e) Let $C \subset \mathbb{P}^3$ be a smooth curve of degree d and genus g. If $S, T \subset \mathbb{P}^3$ are smooth surfaces of degrees s and t containing C, at how many points of C are S and T tangent? (Answer on page 476.)

Let X be a smooth projective variety and $S_1, \ldots, S_k \subset X$ subvarieties. Basic intersection theory, for example in the form of Bézout's theorem for proper intersections (Theorem 1.1), tells us that the sum of the classes of the irreducible components of the intersection, with appropriate multiplicities, is $\prod [S_i]$, but *only* under the hypothesis that the intersection has the expected dimension. The first three keynote questions of this chapter are examples of what are called *excess intersection* problems: situations in which we wish to describe something about improper intersections, where the intersection has components of dimension greater than expected.

A remarkable discovery of Fulton and MacPherson is that this is possible in surprising generality: Given a collection of cycles $S_i \subset X$, subject to mild hypotheses there is a canonical way of assigning a class $\gamma_{C_\alpha} \in A(C_\alpha)$ of the right dimension to each connected component C_α of the intersection $\bigcap S_i$ so that the sum of the pushforwards of these classes in $A(X)$ is equal to $\prod [S_i]$. Moreover, the γ_{C_α} are determined by local geometry. The result is sometimes called the *excess intersection formula*.

The first part of this chapter will be devoted to an exposition of this formula. We begin with some elementary examples, including Keynote Question (a), worked out by hand, which should make it at least plausible that such a formula should exist. We then present the general statement, Theorem 13.3. (An excellent account, with proofs, from which some of the material in this chapter is taken, is Fulton and MacPherson [1978]; see also Fulton [1984, Section 6.3].) We also give a heuristic argument which may help to explain the form that the excess intersection formula takes. As an application of the excess intersection formula, we answer Keynote Questions (b) and (c), and explain how the excess intersection formula applies to the problem of finding the number of conics tangent to five given conics, as suggested in Chapter 8.

The rest of the chapter is devoted to several related topics: the technique of specialization to the normal cone, the "key formula" for intersections in a subvariety, and a description of the Chow ring of the blow-up of a smooth variety along a smooth subvariety.

One note: We do not have the tools to give a proof of either the excess intersection formula or the Grothendieck Riemann–Roch formula, the subjects of this chapter and the next. Nonetheless, given their beauty and their importance in modern intersection theory, we wanted to give an exposition of both topics. Thus in the final two chapters of this book, we will not attempt to prove all the assertions made, focusing rather on what they say and heuristically why they might be true.

One aspect of this is that while up to now we have introduced Chern classes only on smooth varieties, here we want to invoke the fact that they may be defined much more generally for locally free sheaves on any scheme, and that they continue to satisfy Theorem 5.3. In particular, we can use part (d) of that theorem to define the product of an arbitrary Chow class with the Chern class of a bundle; in other words, a Chern class $c_i(\mathcal{E})$ of a vector bundle on an arbitrary scheme X defines an operation $A_k(X) \to A_{k-i}(X)$. This is the point of view taken by Fulton; the reader can find proofs of these assertions in Fulton [1984, Chapter 3].

13.1 First examples

13.1.1 The intersection of a divisor and a subvariety

Suppose that X is a smooth projective variety, $D \subset X$ is a Cartier divisor and $\iota : C \hookrightarrow X$ the inclusion of a subvariety C in X. If C intersects D generically transversely, then the intersection class $[C][D]$ of D and C is $[C \cap D]$; more generally, since D is Cohen–Macaulay, this holds as long as C is not contained in D. But what if C *is* contained in D? In this case $D \cap C = C$, so we would like to find a class γ_C on C that pushes forward to the class $[D][C]$ on X.

Since D is an effective divisor, it is the zero locus of a global section σ of the line bundle $\mathcal{L} = \mathcal{O}_X(D)$ on X, and is equivalent to $(\sigma')_0 - (\sigma')_\infty$, the divisor of zeros minus poles, of any rational section σ' of $\mathcal{L}$. We can find a rational section σ' of $\mathcal{L}$ such that $(\sigma')_0$ and $(\sigma')_\infty$ are both generically transverse to C. (Reason: If $\mathcal{L}'$ is a very ample bundle then $\mathcal{L} \otimes \mathcal{L}'^n$ is very ample for large n, and thus both $\mathcal{L} \otimes \mathcal{L}'^n$ and $\mathcal{L}'^n$ have sections without poles that vanish on divisors generically transverse to C. If we call these sections τ and τ', then τ/τ' is a rational section of $\mathcal{L}$ with the desired property.) If we take γ_C to be the class of the cycle $\langle (\sigma')_0 \cap C \rangle - \langle (\sigma')_\infty \cap C \rangle$, then since intersection products are well-defined on rational equivalence classes we have

$$\iota_*(\gamma_C) = [(\sigma')_0 \cap C] - [(\sigma')_\infty \cap C] = [D][C] \in A(X),$$

as required.

We may think of the class γ_C as the class of a rational section of the line bundle $\mathcal{L}|_C$, or equivalently as the first Chern class of $\mathcal{L}|_C$. Anticipating what is to come, we note that

$$\mathcal{N}_{D/X} := \mathrm{Hom}_X(\mathcal{I}_{D/X}, \mathcal{O}_D) = \mathrm{Hom}_X(\mathcal{L}^{-1}, \mathcal{O}_D) = \mathcal{L}|_D.$$

Thus we have proven:

Proposition 13.1. *Suppose that X is a smooth projective variety. Let $\iota : C \hookrightarrow X$ be the inclusion of a subvariety of codimension k in a smooth variety X, and let $D \subset X$ be an effective Cartier divisor containing C. We have*

$$[C][D] = \iota_* \gamma_C \in A^{k+1}(X),$$

where

$$\gamma_C = c_1(\mathcal{N}_{D/X}|_C).$$

Under suitable hypotheses we can give a geometric interpretation of Proposition 13.1 that shows why the normal bundle of D, restricted to C, is relevant. For simplicity we will take C to be a smooth curve, and suppose that there is a rational deformation of the divisor $D \subset X$, that is, a one-parameter family of divisors $\mathcal{D} \subset \Delta \times X$, parametrized by $\mathbb{A}^1$, such that the special fiber D_0 is D. (This will be the case in particular whenever

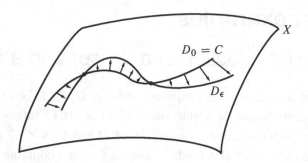

Figure 13.1 The limits of the points of intersection of C with the deformed divisor D_ϵ are zeros of the corresponding section of the normal bundle.

$h^0(\mathcal{O}_X(D)) > 1$.) Suppose moreover that a general member D_t of the deformation is transverse to C; that is,

$$D_t \cap C = \{p_1(t), \ldots, p_k(t)\} \quad \text{for } t \neq 0,$$

with $k = \deg([D_t] \cdot [C]) = \deg([D] \cdot [C])$. Finally, suppose that the deformation is nontrivial to first order along C; that is, its restriction to the scheme $\operatorname{Spec} \Bbbk[t]/(t)^2 \subset \Delta$ supported at 0 corresponds to a section σ of the normal bundle $\mathcal{N}_{D/X}$ that is not identically 0 along C.

In this situation, we claim that the limit as $t \to 0$ of the divisor $D_t \cap C$ is the zero locus of the section σ, and thus represents the class $c_1(\mathcal{N}_{D/X}|_C)$. Heuristically, away from zeros of σ the deformation moves D away from itself (and hence away from C), while the points where σ is zero are stationary to first order.

To see this, take local analytic coordinates $(z_1, \ldots, z_n)$ on X near a point of C such that, locally,

$$D = (z_n = 0) \quad \text{and} \quad C = (z_2 = \cdots = z_n = 0).$$

Locally, z_1 is a coordinate on C, and the normal bundle of D is trivial. We can write the family $\mathcal{D} \subset \Delta \times X$ as the zero locus of the function

$$z_n + t f_1(z_1, \ldots, z_{n-1}) + t^2 f_2(z_1, \ldots, z_{n-1}) + \cdots,$$

with $f_1(z_1, \ldots, z_{n-1})$ representing the corresponding section of the normal bundle. Restricting to $\Delta \times C$, this becomes

$$t f_1|_C + t^2 f_2|_C + \cdots = t(f_1|_C + t f_2|_C + \cdots),$$

from which we see that a zero of $f_1|_C$ of order m will be a limit of exactly m points of intersection of C with D_t as $t \to 0$.

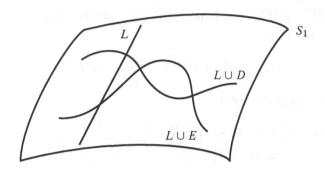

Figure 13.2 S_2 and S_3 intersect S_1 in $L \cup D$ and $L \cup E$.

13.1.2 Three surfaces in $\mathbb{P}^3$ containing a curve

Next we consider Keynote Question (a): Given three surfaces S_1, S_2 and $S_3 \subset \mathbb{P}^3$ of degrees s_i whose intersection consists of the disjoint union of a reduced line L and a zero-dimensional scheme Γ, we wish to find the degree of Γ. We will approach the question naively, solving it the way it might have been solved in the 19th century. We will then explain the modification necessary if L is replaced with an arbitrary smooth curve. This leads us to a formula exhibiting the main features of the general case.

By way of preparation, we would like to be able to assume that S_1 is smooth without changing the intersection $S_1 \cap S_2 \cap S_3$. We can do this if we reorder the surfaces so that the degree s_1 of S_1 is maximal among the s_i. With this choice we can replace S_1 by the zero locus of a general linear combination $F_1 + AF_2 + BF_3$, where F_i is the defining equation of S_i and A and B are general polynomials of degrees $s_1 - s_2$ and $s_1 - s_3$. The surface S_1 is then the general element of a linear system in $\mathbb{P}^3$ whose base locus is smooth of dimension less than 3/2, and smoothness follows from Proposition 5.6.

Because S_1 is smooth, we can simplify the problem of computing deg Γ by working in the intersection ring $A(S_1)$. By hypothesis, the intersection of the two surfaces S_1 and S_2 may be written as

$$S_1 \cap S_2 = L + D \in Z_1(S_1)$$

for some divisor D on S_1. Similarly, write

$$S_1 \cap S_3 = L + E \in Z_1(S_1)$$

for some divisor E on S_1. (Note that D, E or even both may be 0.) If $p \in D \cap L$, then p is a singular point of $S_1 \cap S_2$. If also $p \in S_3$, then, since S_3 is a Cartier divisor, p is a singular point of $S_1 \cap S_2 \cap S_3$ as well, contradicting our hypothesis that the intersection is smooth along L. Thus $S_1 \cap S_2 \cap S_3$ is the disjoint union of L and $D \cap E$, so $\Gamma = D \cap E$. In particular the degree of Γ is the intersection number of the classes of D and E in $A(S_1)$.

Denote by $H \in A^1(S_1)$ the hyperplane class on S_1; since $S_1 \cap S_2 \sim s_2 H$ and $S_1 \cap S_3 \sim s_3 H$, we have

$$D \sim s_2 H - L \quad \text{and} \quad E \sim s_3 H - L$$

in $A^1(S_1)$. Thus, in $A(S_1)$ we have

$$
\begin{aligned}
[D][E] &= [s_2 H - L][s_3 H - L] \\
&= s_2 s_3 [H]^2 - (s_2 + s_3)[H][L] + [L]^2 \\
&= s_2 s_3 [H]^2 - (s_2 + s_3)[H][L] + \iota_* c_1 \mathcal{N}_{L/S_1},
\end{aligned}
$$

where ι denotes the inclusion of L in S_1, and $[L]^2 = \iota_* c_1 \mathcal{N}_{L/S_1}$ by the reasoning of Section 13.1.1.

To obtain a numerical result, we note that $\deg([H]^2) = \deg S_1 = s_1$, while $\deg([H][L]) = \deg L = 1$. We can compute $\deg(c_1(\mathcal{N}_{L/S_1})) = \deg([L]^2)$ as in Section 2.4 by using the adjunction formula twice. First, $K_L = [L]|_L + [K_{S_1}]|_L$ and $K_{S_1} = \mathcal{O}_{S_1}(s_1 - 4)$. Thus

$$\deg([L][K_{S_1}]) = s_1 - 4,$$

and since $\deg K_L = 2g(L) - 2 = -2$ we get

$$\deg([L]^2) = 2 - s_1.$$

This gives

$$\deg(\Gamma) = \deg([D][E]) = \prod s_i - \sum s_i + 2,$$

which is the answer to our first keynote question. Note that in this formula the number $\prod s_i = \deg[S_1][S_2][S_3]$ is the degree that the intersection would have if there were no curve component, so we can think of the remaining terms $\sum s_i - 2$ as representing "the contribution γ_L of the line to the intersection."

For example, if $s_1 = s_2 = s_3 = 1$ we get $\deg \Gamma = 0$, corresponding to the fact that three planes meet in a linear space. More generally, if $s_1 = 1$ then we get $\deg \Gamma = (s_2 - 1)(s_3 - 1)$, corresponding to the fact that the residual curves D and E are plane curves of degrees $s_2 - 1$ and $s_3 - 1$, respectively.

We can make a similar computation if we replace the line L by any smooth curve C. If D has degree d and genus g then the adjunction formula, applied in the same way, gives

$$\deg(c_1(\mathcal{N}_{C/S_1})) = \deg([C]^2) = 2g - 2 - d(s_1 - 4).$$

It follows that in this more general case

$$\deg \Gamma = \deg([D][E]) = \prod s_i - d\left(\sum s_i\right) + 4d + 2g - 2.$$

We can interpret this formula as saying that the class $\prod[S_i]$ can be decomposed into the actual number of isolated points of intersection (with multiplicities) and a class, supported on the positive-dimensional component of $\bigcap S_i$, expressed in terms of various normal bundles. First, note that

$$\deg c_1(\mathcal{N}_{S_i/\mathbb{P}^3}|_C) = ds_i.$$

Next consider the natural exact sequence of ideal sheaves

$$0 \longrightarrow \mathcal{I}_{S_1/\mathbb{P}^3} \longrightarrow \mathcal{I}_{C/\mathbb{P}^3} \longrightarrow \mathcal{I}_{C/S_1} \longrightarrow 0.$$

Since S_1 is a Cartier divisor in $\mathbb{P}^3$ and C is a Cartier divisor on S_1, the left-hand term is a line bundle on $\mathbb{P}^3$ and the right-hand term is the line bundle on S_1. Thus the sequence restricts to an exact sequence of vector bundles on C and, applying the functor $\mathcal{H}om_{\mathcal{O}_S}(-, \mathcal{O}_C)$, we derive the exact sequence of restricted normal bundles

$$0 \longrightarrow \mathcal{N}_{C/S_1} \longrightarrow \mathcal{N}_{C/\mathbb{P}^3} \longrightarrow \mathcal{N}_{S_1/\mathbb{P}^3}|_C \longrightarrow 0.$$

In particular,

$$\begin{aligned}
\deg c_1(\mathcal{N}_{C/\mathbb{P}^3}) &= \deg c_1(\mathcal{N}_{C/S_1}) + \deg c_1(\mathcal{N}_{S_1/\mathbb{P}^3}|_C) \\
&= 2g - 2 - d(s_1 - 4) + ds_1 \\
&= 4d + 2g - 2.
\end{aligned}$$

Putting this together, we have proven an excess intersection formula for our case:

Proposition 13.2. *Let S_1, S_2 and $S_3 \subset \mathbb{P}^3$ be surfaces of degrees s_1, s_2 and s_3 whose intersection consists of the disjoint union of a smooth curve C of degree d and genus g and a zero-dimensional scheme Γ. We have*

$$\begin{aligned}
\deg\left(\prod[S_i]\right) &= \deg(\Gamma) + d\left(\sum s_i\right) - (4d + 2g - 2) \\
&= \deg(\Gamma) + \sum \deg c_1(\mathcal{N}_{S_i/\mathbb{P}^3}|_C) - \deg c_1(\mathcal{N}_{C/\mathbb{P}^3}).
\end{aligned}$$

Another way to view this result is to imagine that we have one-parameter families $S_1, S_2, S_3 \subset \Delta \times \mathbb{P}^3$ specializing to S_1, S_2 and S_3 such that the fibers $(S_1)_t$, $(S_2)_t$ and $(S_3)_t$ intersect transversely in $s_1s_2s_3$ points $p_1(t), \ldots, p_{s_1s_2s_3}(t)$ for $t \neq 0$. Proposition 13.2 tells us that the number of points $p_i(t)$ that tend toward C as $\lambda \to 0$ is

$$\sum \deg c_1(\mathcal{N}_{S_i/\mathbb{P}^3}|_C) - \deg c_1(\mathcal{N}_{C/\mathbb{P}^3});$$

in other words, this is "the contribution of C to the total degree $s_1s_2s_3$ of the intersection." We will see in Section 13.3.1 a heuristic way to interpret this expression.

We can apply the formula of Proposition 13.2 to answer Keynote Question (d): If $C \subset \mathbb{P}^3$ is a smooth curve of degree d and genus g, can C necessarily be expressed as the (scheme-theoretic) intersection of three surfaces? By the formula of Proposition 13.2, the degrees s_i of the three surfaces would have to satisfy the equality

$$\prod s_i - d\left(\sum s_i\right) + 4d + 2g - 2 = 0.$$

For example, if $C \subset \mathbb{P}^3$ is an elliptic quintic curve, then C lies on no planes or quadrics (this follows from the genus formula for smooth curves on a plane and quadric), but when all s_i are ≥ 3 the quantity $\prod s_i - 5(\sum s_i) + 20$ is positive. (In Exercise 13.16 the reader is asked to answer the corresponding question for quintic curves $C \subset \mathbb{P}^3$ of genera 0 and 2.) This question was first answered by Peskine and Szpiro [1974] in a different way; the method given here is from Fulton [1984, Example 9.1.2]. Exercises 13.17 and 13.18 give further examples of such applications.

We will now explain how to restate Proposition 13.2 in forms that match the general expression of the excess intersection formula given in Theorem 13.3. The expression $\sum_i \deg c_1(\mathcal{N}_{S_i/\mathbb{P}^3}|_C) - \deg c_1(\mathcal{N}_{C/\mathbb{P}^3})$ that appears in Proposition 13.2 may be thought of as the degree of the component in $A_0(C)$ of the ratio of Chern classes

$$\frac{\prod c(\mathcal{N}_{S_i/\mathbb{P}^3}|_C)}{c(\mathcal{N}_{C/\mathbb{P}^3})}.$$

This expression also works for the components of $\bigcap S_i$ that are of the correct dimension — that is, of dimension 0. For if $p \in \bigcap S_i$ is an isolated zero-dimensional (possibly nonreduced) component, then the multiplicity of the reduced point p_{red} in the intersection is equal to the degree of p, as per the discussion in Section 1.3.8, and the normal bundle $\mathcal{N}_{p/\mathbb{P}^3}$ is isomorphic to the sum of the normal bundles of the S_i restricted to p, so that

$$\frac{\prod c(\mathcal{N}_{S_i/\mathbb{P}^3}|_p)}{c(\mathcal{N}_{p/\mathbb{P}^3})} = 1,$$

representing the fundamental class $[p] \in A_0(p)$. If for each connected component C_α of the intersection we define the class

$$\gamma C_\alpha = \left\{ \frac{\prod_{i=1}^3 c(\mathcal{N}_{S_i/\mathbb{P}^3}|_{C_\alpha})}{c(\mathcal{N}_{C_\alpha/\mathbb{P}^3})} \right\}_0 \in A_0(C_\alpha),$$

then we have shown

$$\prod_{i=1}^3 [S_i] = \sum \iota_{C_\alpha *}(\gamma C_\alpha),$$

where the sum is taken over all connected components C_α of $\bigcap S_i$.

This expression for γ_{C_α} is symmetric in the S_i but, as we shall soon see, there is sometimes an advantage in using the following nonsymmetric form. We single out S_1, and note that since S_1 is smooth, C is actually a Cartier divisor on S_1. Thus it makes sense to speak of the normal bundle of C in S_1, and we have an exact sequence of normal bundles

$$0 \longrightarrow \mathcal{N}_{S_1/\mathbb{P}^3} \longrightarrow \mathcal{N}_{C_\alpha/\mathbb{P}^3} \longrightarrow \mathcal{N}_{C_\alpha/S_1} \longrightarrow 0.$$

Using Whitney's formula (Theorem 5.3), we get

$$c(\mathcal{N}_{C_\alpha/\mathbb{P}^3}) = c(\mathcal{N}_{S_1/\mathbb{P}^3})c(\mathcal{N}_{C_\alpha/S_1}).$$

Substituting the right side for the left in the formula for γ_{C_α}, we get

$$\gamma_{C_\alpha} = \left\{ \frac{\prod_{i=2}^{3} c(\mathcal{N}_{S_i/\mathbb{P}^3}|C_\alpha)}{c(\mathcal{N}_{C_\alpha/S_1})} \right\}_0.$$

Bearing in mind that the Segre class of a bundle is the inverse of the Chern class, we get the expressions

$$\gamma_{C_\alpha} = \left\{ s(\mathcal{N}_{C_\alpha/\mathbb{P}^3}) \prod_{i=1}^{3} c(\mathcal{N}_{S_i/\mathbb{P}^3}|C_a) \right\}_0$$

$$= \left\{ s(\mathcal{N}_{C_\alpha/S_1}) \prod_{i=2}^{3} c(\mathcal{N}_{S_i/\mathbb{P}^3}|C_\alpha) \right\}_0.$$

13.2 Segre classes of subvarieties

One might wonder why we have bothered to replace the inverse Chern class in the formula above with the Segre class. The reason is that Segre classes can be defined in a much more general context; indeed, this is the approach of Fulton [1984], where the Segre classes are used to define the Chern classes. What we need for the definition of Segre classes is really just a projective morphism $\pi : E \to C$ and a distinguished Cartier divisor class $\zeta \in A(E)$. Since we can compute the intersection of a Cartier class with any subvariety, we can define a codimension-i class on E as the i-th self-intersection ζ^i, and push $\sum \zeta^i$ forward to C. In the case where E is the projectivization of a bundle $\mathcal{N}$ on C and $\zeta = c_1(\mathcal{O}_{\mathbb{P}\mathcal{N}}(1))$, the result is by definition the Segre class $s(\mathcal{N})$. More generally we say that $\pi : E \to C$ is a *cone* if $E = \mathrm{Proj}\,\mathcal{S}$, where $\mathcal{S}$ is a graded, locally finitely generated sheaf of algebras over $\mathcal{O}_C$ with $\mathcal{S}_0 = \mathcal{O}_C$, and π is the morphism corresponding to the inclusion. Taking ζ to be the line bundle on E associated to the sheaf of $\mathcal{S}$-modules $\mathcal{S}(1)$, we define the Segre class of the cone E to be

$$s(E) = \pi_* \left(\sum_k \zeta^k \right) \in A(C).$$

For our purposes it suffices to take the case where X is a variety, $C \subset X$ is a proper subscheme, and $\mathcal{S} = \bigoplus_n (\mathcal{I}_{C/X}^n / \mathcal{I}_{C/X}^{n+1})$, that is, the case where $\pi : E \to C$ is the exceptional divisor of the blow-up $B = \mathrm{Bl}_C X$ of X along C and ζ is the negative of the class of the exceptional divisor $E \subset B$, restricted to E. Following the pattern above we define the *Segre class of C in X* to be

$$s(C, X) := \pi_* \left(\sum_{k \geq 0} c_1(\mathcal{O}_E(1))^k \right),$$

where the intersections are taken in $A(E)$.

For example, in the case where X is smooth and E is equidimensional (in particular, when C is locally a complete intersection), since the blow-up has the same dimension as X the dimension of E is $\dim X - 1$, so the relative dimension of E over C is $\mathrm{codim}\, C - 1$; the codimension-k component $s_k(C, X)$ of $s(C, X)$ is thus

$$s_k(C, X) := \pi_*(\zeta^{\mathrm{codim}\, C - 1 + k}).$$

Indeed, if C is locally a complete intersection in X then

$$\mathcal{I}_{C/X}^n / \mathcal{I}_{C/X}^{n+1} \cong \mathrm{Sym}^n(\mathcal{I}_{C/X}/\mathcal{I}_{C/X}^2) = \mathrm{Sym}^n(\mathcal{N}_{C/X}^*),$$

so $\mathrm{Proj}(\bigoplus_n \mathcal{I}_{C/X}^n / \mathcal{I}_{C/X}^{n+1}) = \mathrm{Proj}(\mathcal{N})$ and thus

$$s(C, X) = s(\mathcal{N}_{C/X}) = c(\mathcal{N}_{C/X})^{-1},$$

as before.

13.3　The excess intersection formula

Putting together these ideas, and recalling that on any variety C we can take the product of an arbitrary class in $A(C)$ with the Chern class of a bundle on C, we can express a very general form of Bézout's theorem. It suffices to treat the case where we intersect just two subvarieties:

Theorem 13.3 (Excess intersection formula). *If $S \subset X$ is a subvariety of a smooth variety X and T is a locally complete intersection subvariety of X, then*

$$[S][T] := \sum_C (\iota_C)_*(\gamma_C),$$

where:

- *The sum is taken over the connected components C of $S \cap T$.*
- *$\iota_C : C \to X$ denotes the inclusion morphism.*
- *$\gamma_C = \{s(C, S)c(\mathcal{N}_{T/X}|_C)\}_d \in A_d(C)$, where $d = \dim X - \mathrm{codim}\, S - \mathrm{codim}\, T$ is the "expected dimension" of the intersection.*

If the subvariety S is locally a complete intersection as well, then we have a symmetric form

$$\gamma_C = \{s(C, X)c(\mathcal{N}_{S/X}|_C)c(\mathcal{N}_{T/X}|_C)\}_d.$$

Two notes on this statement. First, it should be emphasized that each connected component C of $S \cap T$ is to be taken with the scheme structure inherited from $S \cap T$; this is important, because it may affect the classes $s(C, S)$ and $s(C, X)$. Secondly, it might appear that when $S \subset X$ is a locally complete intersection we have an exact sequence of normal bundles

$$0 \longrightarrow \mathcal{N}_{C/S} \longrightarrow \mathcal{N}_{C/X} \longrightarrow \mathcal{N}_{S/X}|_C \longrightarrow 0,$$

from which it would follow by Whitney that

$$s(\mathcal{N}_{C/S}) = \frac{1}{c(\mathcal{N}_{C/S})} = \frac{c(\mathcal{N}_{S/X}|_C)}{c(\mathcal{N}_{C/X})} = s(\mathcal{N}_{C/X})c(\mathcal{N}_{S/X}|_C),$$

accounting for the difference in the two expressions above for γ_C. Sadly, the equality $s(C, S) = s(C, X)c(\mathcal{N}_{S/X})$ is not true in general, even when both S and T are locally complete intersections, as for example in the case $X = \mathbb{P}^3$, C and T are both equal to a line $L \subset \mathbb{P}^3$ and S is the union of two planes containing L. So the equality of the two expressions in this case is even subtler than it looks.

By induction, we could extend the formula to a formula for the intersection of S with an arbitrary number of locally complete intersection subvarieties $S_i \subset X$. In the case where the subvarieties S_i to be intersected are all hypersurfaces, there is a form of the excess intersection formula due to Wolfgang Vogel that is sometimes more easily adapted to computation; see Vogel [1984] and, for a comparison with Theorem 13.3, van Gastel [1990]. We will give a brief description of Vogel's approach in Section 13.3.6 below.

Theorem 13.3 is of great theoretical importance in at least three circumstances:

Intersection products on nonsmooth varieties: The formula can serve as a definition of the intersection product of locally complete intersection subvarieties of an *arbitrary* variety X; there is no smoothness used in defining the terms on the right-hand side of the formula. This is, in fact, the way that Fulton [1984] defined intersections in this case.

Intersection products on smooth varieties: Defining intersections with locally complete intersection subvarieties actually suffices to define all intersection products on a smooth variety! Indeed, suppose $T_1, T_2 \subset Y$ are arbitrary subvarieties. Since Y is smooth the diagonal $Y \cong \Delta \subset Y \times Y$ is a locally complete intersection subvariety, so we can use the asymmetric form of the formula on $X = Y \times Y$ to define classes on the connected components of $\Delta \cap (T_1 \times T_2) \cong T_1 \cap T_2$. Pushing these classes forward to $\Delta \cong X$ and adding, we get the intersection class of T_1 and T_2 in X. Again, this is the path followed in Fulton [1984].

General pullbacks: In Theorem 1.23, we characterized the pullback φ^* of equivalence classes of cycles along a projective morphism $\varphi : X \to Y$ of smooth varieties by taking the class of a cycle A on X to the class of $\varphi^{-1}(A)$, in the case where A is generically transverse to φ, in the sense that A is a linear combination of subvarieties of Y whose preimages in X are generically reduced of the same codimension; the moving lemma shows that we can find a cycle A with this property in any rational equivalence class, so that specifying the pullback class in this case determines φ^*. Theorem 13.3 gives a different way of defining $\varphi^*([A])$, without smoothness hypotheses and without the moving lemma, in the cases where $A \subset X$ is a closed locally complete intersection subscheme of X or the graph of φ is locally a complete intersection in $X \times Y$ (a situation that holds whenever both X and Y are smooth): In these cases one can apply Theorem 13.3 to compute the product of the class $[X \times A]$ and the class of the graph Γ_φ of φ as a class on the intersection $(X \times A) \cap \Gamma_\varphi \cong \varphi^{-1}(A)$ inside $A(X \times Y)$, and push the result forward along the projection to X. (Note that, as in the case of intersections, the pullback class $\varphi^*[A]$ is expressed as the pushforward of a class on the preimage $\varphi^{-1}(A)$; this is occasionally very useful, as in the proof of Proposition 13.12.) One must, of course, prove that this agrees with $[\varphi^{-1}(A)] \in A(X)$ in the case where A is generically transverse to φ. Once more, this is the route taken in Fulton [1984].

For the proof of Theorem 13.3 and the treatment of its consequences as above, we refer to Fulton [1984, Section 6.3], where all this is worked out. We will give only the proof of a special case, Theorem 13.7; however, this proof contains one of the major new ideas, from Fulton and MacPherson [1978], that went into the general theorem.

In the following section, we will give a heuristic argument that may help to explain the form of Theorem 13.3. Following that, we work out some examples, including the second and third keynote questions.

13.3.1 Heuristic argument for the excess intersection formula

The expression given in Theorem 13.3 for the class γ_C may seem to come out of nowhere. The following calculation may help explain where it is coming from; though it is not a suitable framework for a proof — it involves far too many extra hypotheses, and we have omitted the multiplicity calculations that would be necessary, even subject to those hypotheses, to make it into a proof — it will hopefully at least make the form of Theorem 13.3 more plausible.

To begin with, we make the following assumptions:

- X will be a smooth, projective variety of dimension n.
- S and $T \subset X$ will be smooth subvarieties of codimensions k and l respectively.
- The intersection $C = S \cap T$ will be smooth, with connected components C_α of codimension $k + l - m_\alpha$.

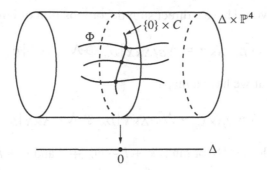

Figure 13.3 Limits of points of intersection of the deformed varieties S_λ and T_λ are singular points of $\mathcal{S} \cap \mathcal{T} = \Phi \cup (\{0\} \times C)$.

Again, we want to assign to each C_α a cycle class $\gamma_\alpha \in A^{m_\alpha}(C_\alpha)$ of dimension $k + l - n$, which represents the contribution of C_α to the total intersection $S \cap T$, that is, such that if we denote by $i_\alpha : C_\alpha \to X$ the inclusion, then

$$\sum_\alpha (i_\alpha)_*(\gamma_\alpha) = [S] \cdot [T] \in A^{k+l}(X).$$

We will do this by imagining that we can deform S and T to cycles S_λ and T_λ on X intersecting transversely, and asking for the limiting position of the intersection $S_\lambda \cap T_\lambda$ as $\lambda \to 0$. We therefore make a crucial (and very frequently counterfactual) fourth hypothesis:

- There exist families $\mathcal{S}, \mathcal{T} \subset \Delta \times X$, flat over a smooth rational curve Δ, with fibers $S_0 = S$ and $T_0 = T$, such that $S_\lambda \cap T_\lambda$ is transverse for $\lambda \neq 0$.

Given all this, we can express the intersection $\mathcal{S} \cap \mathcal{T}$ of the $(k+1)$- and $(l+1)$-folds $\mathcal{S}$ and $\mathcal{T}$ in the $(n+1)$-fold $\Delta \times X$ as a union

$$\mathcal{S} \cap \mathcal{T} = \Phi \cup D,$$

where Φ is flat over Δ, consisting of the components of the intersection $S_\lambda \cap T_\lambda$ for $\lambda \neq 0$ and their limits — that is, the closure in $\Delta \times X$ of the intersection of $\mathcal{S} \cap \mathcal{T}$ with $(\Delta \setminus \{0\}) \times X$ — and $D = \{0\} \times C$, as in Figure 13.3 above.

Now, we know that the cycle $\Xi = \Phi \cap (\{0\} \times X)$ has dimension $k + l - n$, and has class $[S] \cdot [T] \in A^{k+l}(X)$ as a cycle on X. Moreover, Ξ consists of the sum of its intersections $\Xi_\alpha = \Phi \cap D_\alpha$ with the connected components $D_\alpha = \{0\} \times C_\alpha$ of $\mathcal{S} \cap \mathcal{T}$. To find the class $[S] \cdot [T]$, accordingly, we have to figure out the class of the intersection $\Phi \cap D_\alpha$ for each connected component.

Finally, one way to characterize the points of $\Phi \cap D_\alpha$ is to observe that *they are the points $p \in D_\alpha$ where the tangent spaces $T_p \mathcal{S}$ and $T_p \mathcal{T} \subset T_p(\Delta \times X)$ fail to intersect in $T_p D_\alpha$.* Now, if we had

$$T_p D_\alpha = T_p \mathcal{S} \cap T_p \mathcal{T}$$

for all $p \in D_\alpha$, we would have a direct sum decomposition of bundles

$$\mathcal{N}_{D_\alpha/\Delta \times X} = \mathcal{N}_{S/\Delta \times X}|_{D_\alpha} \oplus \mathcal{N}_{T/\Delta \times X}|_{D_\alpha}.$$

In general, we see that we have a map

$$\mathcal{N}_{D_\alpha/\Delta \times X} \to \mathcal{N}_{S/\Delta \times X}|_{D_\alpha} \oplus \mathcal{N}_{T/\Delta \times X}|_{D_\alpha}$$

between vector bundles on D_α of ranks $k + l - m_\alpha + 1$ and $k + l$, and the locus where this map fails to be injective (as a map of vector bundles) is exactly the cycle Ξ_α.

We have

$$c(\mathcal{N}_{D_\alpha/\Delta \times X}) = c(\mathcal{N}_{C_\alpha/X})$$

and

$$\mathcal{N}_{S/\Delta \times X}|_{D_\alpha} = \mathcal{N}_{S/X}|_{C_\alpha},$$

and likewise

$$\mathcal{N}_{T/\Delta \times X}|_{D} = \mathcal{N}_{T/X}|_{C_\alpha}.$$

Now we can apply Porteous to deduce that the class of $\Xi = \Phi \cap D_\alpha$ is

$$\gamma_\alpha = [\Xi_\alpha] = \left\{ \frac{c(\mathcal{N}_{S/X}|_{C_\alpha}) \cdot c(\mathcal{N}_{T/X}|_{C_\alpha})}{c(\mathcal{N}_{C_\alpha/X})} \right\}^{m_\alpha} \in A^{m_\alpha}(D_\alpha),$$

from which we arrive at the statement of Theorem 13.3.

13.3.2 Connected components versus irreducible components

One further remark is in order before we get to the examples. One might ask if it is possible to refine the excess intersection formula to associate a class γ_C to each *irreducible* component of the intersection, as opposed to each connected component, in such a way that when we push forward each γ_C into $A(X)$ and sum the results, we get the class $[S][T]$. The following example shows that this is not in fact possible.

Example 13.4. Let $L_1, \ldots, L_4 \subset \mathbb{P}^2$ be four concurrent (but distinct) lines, and consider two reduced effective cycles S and T on $\mathbb{P}^2$, where S is the sum $L_1 + L_2 + L_3$ and $T = L_1 + L_2 + L_4$, as in Figure 13.4. Since S and T are cubic curves, the product $[S][T] \in A(\mathbb{P}^2)$ is the class of (any) cycle consisting of nine points. On the other hand, the intersection of the underlying algebraic sets is $C := L_1 \cup L_2$, and the two lines L_1 and L_2 play completely symmetric roles. Since 9 is an odd number, there is no canonical way of dividing the intersection cycle into a cycle on L_1 and a cycle on L_2.

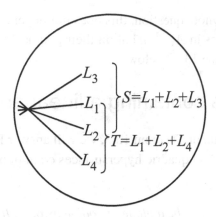

Figure 13.4 The product $[S][T]$ is a cycle of nine points on $L_1 \cup L_2$.

13.3.3 Two surfaces in $\mathbb{P}^4$ containing a curve

We can use Theorem 13.3 to answer the second of our keynote questions: Given smooth surfaces S and $T \subset \mathbb{P}^4$ of degrees s and t whose intersection consists of a smooth curve C of genus g and degree d and a collection Γ of reduced points, we ask, as before, what we can say about the degree of Γ.

Because S and T are assumed smooth, they are locally complete intersections. Moreover, since Γ is reduced each point $p \in \gamma$ satisfies $\gamma_p = [p]$. Thus the asymmetric form of the expression for γ_L in Theorem 13.3 gives

$$st = \deg \Gamma + \deg \left\{ \frac{c(N_{T/\mathbb{P}^4}|_C)}{c(\mathcal{N}_{C/S})} \right\}_0$$

$$= \deg \Gamma + \deg c_1(N_{T/\mathbb{P}^4}|_C) - \deg c_1(\mathcal{N}_{C/S}).$$

Now $\deg c_1(\mathcal{N}_{C/S})$ is the degree of the class $[C]^2$, computed in $A(S)$, which we write $[C_S]^2$. On the other hand, the adjunction formula gives $\deg K_C = 2g - 2 = \deg K_T|_C + \deg[C_T]^2$, and

$$K_T = c_1(\mathcal{N}_{T/\mathbb{P}^4}) + K_{\mathbb{P}^4}|_T = c_1(\mathcal{N}_{T/\mathbb{P}^4}) + 5H,$$

where H is the hyperplane class in $A^1(T)$ (Hartshorne [1977, Proposition II.8.20]). Thus

$$\deg c_1(\mathcal{N}_{T/\mathbb{P}^4}|_C) = 2g - 2 + 5d - \deg[C_T]^2.$$

Putting this all together, we get

$$st = \deg \Gamma + 2g - 2 + 5d - \deg[C_S]^2 - \deg[C_T]^2,$$

which becomes $\deg \Gamma = st - 3 + \deg[L_S]^2 + \deg[L_T]^2$ when $C = L$, a line. For example, if $S, T \subset \mathbb{P}^4$ are 2-planes containing the line L, then Γ is empty and indeed the formula gives $\deg \Gamma = 0$. For other examples, see Exercise 13.19. In contrast to the answer to our first keynote question, this does not depend only on the degrees of S and T (that is, their classes in $A(\mathbb{P}^4)$), but on their geometry; the simplest example of this is described in Exercise 13.20 below.

13.3.4 Quadrics containing a linear space

As a second application of Theorem 13.3, we will answer Keynote Question (c). Let $Q_1, \ldots, Q_n \subset \mathbb{P}^n$ be general quadric hypersurfaces containing a codimension-c plane $\Lambda \cong \mathbb{P}^{n-c} \subset \mathbb{P}^n$.

Proposition 13.5. *Let $\Lambda \subset \mathbb{P}^n$ be a plane of codimension c. If $Q_1, \ldots, Q_n$ are quadric hypersurfaces that are general among those containing Λ, then*

$$\bigcap_{i=1}^{n} Q_i = \Lambda \cup \Gamma$$

as schemes, where Γ is a set of

$$\binom{n}{0} + \binom{n}{1} + \cdots + \binom{n}{c-1} = 2^n - \binom{n}{c} - \cdots - \binom{n}{n}$$

reduced points, disjoint from Λ.

At one extreme, when $c = 1$ each Q_i is the union of Λ with a generic hyperplane, and the intersection of these hyperplanes is a single point outside Λ. At the other extreme, when $c = n$, so that Λ is a point, the set Γ consists of all but one of the 2^n points in the complete intersection of the Q_i.

More interesting geometrically is the case where $c = 2$. Suppose Λ is the zero locus of the linear forms X_0 and X_1. Then each of the quadrics Q_i can be written as the zero locus of a linear combination of X_0 and X_1:

$$Q_i = V(F_i), \quad \text{where } F_i = X_0 L_i + X_1 M_i$$

for some linear forms L_i and M_i. Now consider the $2 \times (k+1)$ matrix

$$\Phi = \begin{pmatrix} X_0 & M_1 & M_2 & \cdots & M_k \\ X_1 & L_1 & L_2 & \cdots & L_k \end{pmatrix}.$$

Away from the locus $\Lambda = V(X_0, X_1)$ where X_0 and X_1 both vanish, the rank-1 locus of Φ is just the intersection of the quadrics Q_i; in other words,

$$\Gamma = \{X \in \mathbb{P}^n \mid \text{rank}(\Phi(X)) \leq 1\}.$$

The degree of Γ is then given by Porteous' formula; as we worked out in Section 12.4.1, this has degree $k + 1$. This is a special case of what is sometimes called the *Steiner construction*; see Griffiths and Harris [1994, Section 4.3] for more information.

Proof of Proposition 13.5: We first claim that Λ is a reduced connected component of the intersection and Γ consists of a collection of reduced points. In particular, since Λ is a connected component of the intersection, we may apply Theorem 13.3. Since the ideal of Λ is generated by c linear forms l_i, the conormal bundle $\mathcal{N}^* := \mathcal{I}_\Lambda/\mathcal{I}_\Lambda^2$ of Λ is $\mathcal{O}_\Lambda^c(-1)$. The equation of a quadric hypersurface vanishing on Λ is a linear combination of the l_i with linear coefficients, and thus defines a general section of $\mathcal{N}(-1)$. Thus if we fix equations for the Q_i we get a general map

$$\mathcal{O}_\Lambda^n(-2) \to \mathcal{O}_\Lambda^c(-1).$$

Since a general $c \times n$ matrix has minors vanishing in codimension $n - c + 1$, this map is locally a surjection everywhere on Λ, which is to say that the Q_i cut out Λ scheme-theoretically. Since $\bigcap Q_i$ is smooth along Λ, no other component of $\bigcap Q_i$ meets Λ. In addition, since the linear series spanned by the Q_i has no base points outside of Λ, it follows that the "residual" scheme Γ consists of reduced points, as claimed.

The normal bundle of Q_i in $\mathbb{P}^n$ is $\mathcal{O}_{\mathbb{P}^n}(2)|_{Q_i} = \mathcal{O}_{Q_i}(2)$. Thus, writing $\zeta \in A^1(\Lambda)$ for the hyperplane class, we have

$$c(\mathcal{N}_{Q_i/\mathbb{P}^n}|_\Lambda) = 1 + 2\zeta.$$

From $\mathcal{N}_{\Lambda/\mathbb{P}^n} \cong \mathcal{O}_\Lambda(1)^{\oplus c}$, we get

$$c(\mathcal{N}_{\Lambda/\mathbb{P}^n}) = (1 + \zeta)^c.$$

Thus, in the notation of Theorem 13.3, the contribution γ_Λ of the component Λ to the intersection is given by

$$\gamma_\Lambda = \left\{ \frac{(1 + 2\zeta)^n}{(1 + \zeta)^c} \right\}_0 \in A_0(\Lambda) \cong \mathbb{Z}.$$

This is the coefficient of z^{n-c} in

$$\frac{(1 + 2z)^n}{(1 + z)^c},$$

which is

$$x := \sum_{j=0}^{n-c} (-1)^j 2^{n-c-j} \binom{n}{n-c-j} \binom{c+j-1}{j},$$

and we must show that it is equal to $\sum_{i=c}^{n} \binom{n}{i}$, the number of subsets with $\geq c$ elements in a set S of cardinality n.

Now

$$2^{n-c-j} \binom{n}{n-c-j} \binom{c+j-1}{j} = \#\{A \subset B \subset S \mid \#B = n - c - j\} \binom{c+j-1}{j}.$$

Thus we may regard the alternating sum x as counting subsets by inclusion-exclusion: Setting $k = n - c$, a given subset A of cardinality a is counted a total of

$$\sum (-1)^j \binom{n-a}{k-a-j} \binom{n-k+j-1}{j}$$

times, and we recognize the last expression as the coefficient of z^{k-a} in the power series expansion of

$$\frac{(1+z)^{n-a}}{(1+z)^{n-k}},$$

which is 1 since the quotient is $(1+z)^{k-a}$.   □

We can also verify the statement of Proposition 13.5 by specializing to the case where each quadric Q_i is a union of a general hyperplane H_i containing Λ and a second general hyperplane H_i'. Each point of $\bigcap Q_i$ is the point of intersection of a subset of d hyperplanes $\{H_i\}_{i \in D}$ and $n - d$ hyperplanes $\{H_i'\}_{i \notin D}$, and the point is outside Λ if and only if $d < c$. This establishes a bijection between the points of intersection outside Λ and subsets of $\{1, \ldots, n\}$ of cardinality $< c$, yielding the result. Indeed, given that the number of points depends only on the rational equivalence class of the Q_i and the fact that Λ is a component of their intersection (the *principle of specialization*), this argument gives an alternative proof of Proposition 13.5.

13.3.5 The five conic problem

Theorem 13.3 gives another way to solve the five conic problem treated in Chapter 8: How many conics $C \subset \mathbb{P}^2$ are tangent to each of five given conics $C_1, \ldots, C_5$. The problem was first solved by this method in Fulton and MacPherson [1978]; a short version appears in Fulton [1984, p. 158]. Here we simply sketch the necessary ideas.

As we saw in Chapter 8, the set of conics tangent to C_i is a sextic hypersurface $Z_i \subset \mathbb{P}^5$ in the space $\mathbb{P}^5$ parametrizing all plane conics. As we also saw there, the hypersurfaces $Z_1, \ldots, Z_5 \subset \mathbb{P}^5$ do not intersect properly; rather, they all contain the Veronese surface $S \subset \mathbb{P}^5$ corresponding to double lines. Thus, if T denotes the component of the (scheme-theoretic) intersection $\bigcap Z_i$ supported on S, we have

$$\bigcap_{i=1}^{5} Z_i = T \cup \Gamma,$$

and the problem is to determine the cardinality of Γ. In Chapter 8 we did this by replacing $\mathbb{P}^5$ by the space of *complete conics*. Now we can apply Theorem 13.3 directly in $\mathbb{P}^5$. To do this, we need:

(a) *The Chern class of the restriction to S of the normal bundle of $Z_i \subset \mathbb{P}^5$.* This is the easiest part: Let $\zeta \in A^1(S)$ be the class of a line in $S \cong \mathbb{P}^2$, and let $\eta \in A^1(\mathbb{P}^5)$ be the hyperplane class on $\mathbb{P}^5$; note that the restriction of η to S is 2ζ. Since the Z_i are sextic hypersurfaces, $\mathcal{N}_{Z_i/\mathbb{P}^5} = \mathcal{O}_{Z_i}(6)$ and so

$$c(\mathcal{N}_{Z_i/\mathbb{P}^5}|_S) = 1 + 12\zeta.$$

(b) *The multiplicity of Z_i along S.* By Riemann–Hurwitz, a general pencil of plane conics including a double line $2L$ will have four other elements tangent to C_i, so that $\text{mult}_S(Z_i) = 2$. (See Exercise 13.31.)

(c) *The Chern classes of the normal bundle of $S \subset \mathbb{P}^5$.* In terms of the hyperplane classes $\eta \in A^1(\mathbb{P}^5)$ and $\zeta \in A^1(S)$, we have

$$c(\mathcal{T}_S) = (1 + \zeta)^3 = 1 + 3\zeta + 3\zeta^2$$

and

$$c(\mathcal{T}_{\mathbb{P}^5}|_S) = (1 + \eta)^6|_S = 1 + 12\zeta + 60\zeta^2.$$

Applying the Whitney formula to the sequence

$$0 \longrightarrow \mathcal{T}_S \longrightarrow \mathcal{T}_{\mathbb{P}^5}|_S \longrightarrow \mathcal{N}_{S/\mathbb{P}^5} \longrightarrow 0,$$

we conclude that

$$c(\mathcal{N}_{S/\mathbb{P}^5}) = \frac{1 + 12\zeta + 60\zeta^2}{1 + 3\zeta + 3\zeta^2}$$
$$= 1 + 9\zeta + 30\zeta^2,$$

and inverting this we have

$$s(\mathcal{N}_{S/\mathbb{P}^5}) = 1 - 9\zeta + 51\zeta^2.$$

(d) *The scheme-theoretic intersection of the hypersurfaces Z_i.* This is easy to state: The component of $\bigcap Z_i$ supported on S is exactly the scheme $T = V(\mathcal{I}^2_{S/\mathbb{P}^5})$ defined by the square of the ideal $\mathcal{I}_{S/\mathbb{P}^5}$. We will not give a proof here; given part (b) above, the statement is equivalent to the statement that the proper transforms of the Z_i in the blow-up of $\mathbb{P}^5$ along S have no common intersection in the exceptional divisor, which is proved Griffiths and Harris [1994, Chapter 6]. Alternatively, via the isomorphism of the blow-up with the space of complete conics, it is tantamount to the statement, proved in Section 8.2.3, that every complete conic tangent to each of $C_1, \ldots, C_5$ is smooth.

Given part (d), the blow-up of $\mathbb{P}^5$ along T is the same as the blow-up along S, but with the exceptional divisor doubled. Applying the definition of Section 13.2, the k-th graded piece of the Segre class $s(T, \mathbb{P}^5)$ is 2^{k+3} times the corresponding graded piece of $s(S, \mathbb{P}^5)$, so that

$$s(T, \mathbb{P}^5) = 8 - 144\zeta + 1632\zeta^2.$$

Thus the contribution of S to the degree of the intersection $\bigcap Z_i$ is

$$
\deg \left(\prod c(\mathcal{N}_{Z_i/\mathbb{P}^5}|_S) \cdot s(T, \mathbb{P}^5) \right) = \deg((1 + 12\zeta)^5 (8 - 144\zeta + 1632\zeta^2))
$$
$$
= 1632 - 60 \cdot 144 + 1440 \cdot 8
$$
$$
= 4512,
$$

and the degree of Γ is correspondingly $7776 - 4512 = 3264$.

13.3.6 Intersections of hypersurfaces in general: Vogel's approach

We take a moment here to describe Vogel's approach to the problem of excess intersection, in the case where the varieties being intersected are all hypersurfaces. In fact, we have already seen this approach carried out in a special case: The method described here is exactly what we did in the case of the intersection of three surfaces in $\mathbb{P}^3$ considered in Section 13.1.2.

Briefly, let X be a smooth, n-dimensional projective variety and $D_1, \ldots, D_k$ a collection of hypersurfaces in X. Assume the intersection $Y = \bigcap D_i$ has components of dimension strictly greater than $n - k$, as well as components of the expected dimension $n - k$. For simplicity, say

$$
\bigcap_{i=1}^{k} D_i = \Phi \cup \Gamma,
$$

with Φ and Γ smooth and disjoint, Φ of pure dimension $n - k + m$ and Γ of pure dimension $n - k$. Can we find the class of the sum Γ of the components of the expected dimension $n - k$?

In fact, we can, if we know something about the geometry of the divisors D_i and their intersection. What we want to do here is intersect the hypersurfaces D_i one at a time, and focus on the first step where the intersection fails to have the expected dimension; if we allow ourselves to change the order of the divisors D_i, we can assume that this occurs after we have intersected $k - m + 1$ of the divisors, at which point Φ appears as a component of excess intersection. The point is, if we back up one step, we see that *the previous intersection $D_1 \cap \cdots \cap D_{k-m}$ must have been reducible*: if $\Gamma \neq \varnothing$, then we must have

$$
D_1 \cap \cdots \cap D_{k-m} = \Phi \cup B_0.
$$

Now back up one further step, and consider the intersection $S = D_1 \cap \cdots \cap D_{k-m-1}$, which has dimension $n - k + m + 1$. We have an equation of divisors on S

$$
D_{k-m} \cap S = A + B_0,
$$

and similarly we can write

$$D_{k-m+\alpha} \cap S = A + B_\alpha$$

for each $\alpha = 0, \ldots, m$. We can then express the cycle Γ as a proper intersection of $m + 1$ divisors on the variety S:

$$\Gamma = B_0 \cap \cdots \cap B_m,$$

and if we know enough about the geometry of Φ—specifically, if we can evaluate the products of powers $[\Phi]^l \in A^l(S)$ with products of the classes $[D_\alpha]|_S \in A^1(S)$—we can then evaluate the class of this intersection as

$$[\Gamma] = \prod_{\alpha=0}^{m} ([D_{k-m+\alpha}]|_S - [\Phi]) \in A^{m+1}(S) \to A^{n-k}(X).$$

As we said, this approach was the one we used in Section 13.1.2; for some other examples, try Exercises 13.17 and 13.18.

13.4 Intersections in a subvariety

One frequently occurring situation in which excess intersection arises is the case of cycles A and B on a smooth variety X that happen to both lie on a proper subvariety $Z \subsetneq X$; in this case, the generalized principal ideal theorem (Theorem 0.2) says that their intersection has dimension at least $\dim A + \dim B - \dim Z > \dim A + \dim B - \dim X$. Thus as cycles on X their intersection cannot even be dimensionally transverse. Nevertheless we can relate their intersection class $[A][B] \in A(Z)$ in Z to the intersection of their classes on X. (To avoid confusion, we will denote the classes of A and B, viewed as cycles on X, by $\iota_*[A], \iota_*[B] \in A(X)$.)

Proposition 13.6 (Key formula). *Let $\iota : Z \to X$ be an inclusion of smooth projective varieties of codimension m, and let $\mathcal{N} = \mathcal{N}_{Z/X}$ be the normal bundle of Z in X. If $\alpha \in A^a(Z)$ and $\beta \in A^b(Z)$, then*

$$\iota_*\alpha \cdot \iota_*\beta = \iota_*(\alpha \cdot \beta \cdot c_m(\mathcal{N}_{Z/X})) \in A^{a+b+2m}(X).$$

Proposition 13.6 follows easily from the important special case where $B = Z$. Using the fact that $\alpha[Z] = \alpha \in A(Z)$, this takes the following form:

Theorem 13.7. *Let $\iota : Z \to X$ be an inclusion of smooth projective varieties of codimension m, and let $\mathcal{N}_{Z/X}$ be the normal bundle of Z in X. For any class $\alpha \in A(Z)$ we have*

$$\iota^*(\iota_*\alpha) = \alpha \cdot c_m(\mathcal{N}_{Z/X}) \in A^{a+m}(Z).$$

Proof of Proposition 13.6 from Theorem 13.7: From $\iota^*(\iota_*\alpha) = \alpha \mathcal{N}_{Z/X}$, we use the push-pull formula to get

$$\iota_*(\alpha \mathcal{N}_{Z/X}\beta) = \iota_*(\iota^*(i_*\alpha)\beta) = (\iota_*\alpha)(\iota_*\beta). \qquad \square$$

Note that Theorem 13.7 and Proposition 13.6 may be deduced from the more general Theorem 13.3. The result is easier to visualize in the special case, however, and it will give us an occasion to describe a key technique, that of *specialization to the normal cone*.

One way to see the plausibility of Theorem 13.7 is to consider the special case where the normal bundle $\mathcal{N}_{Z/X}$ extends to a bundle $\mathcal{N}$ with enough sections on all of X (this happens, for example, when Z is a suitably positive divisor): In that case, $[Z] = c_m(\mathcal{N})$, so at least

$$\iota_*(\iota^*(\iota_*\alpha)) = \iota_*[Z]\alpha = c_m(\mathcal{N})\alpha.$$

The situation above can actually be realized topologically: Consider the complex case, and suppose that $\mathcal{N}_{Z/X}$ has enough sections. Topologically, the normal bundle looks like a tubular neighborhood of Z in X, and again $c_m(\mathcal{N}_{Z/X})$ is the class of the zero locus of a general section σ. Thinking of the image σ as a perturbation Z' of Z within the tubular neighborhood, the set where $\sigma = 0$ corresponds to the intersection $Z' \cap Z$ — that is, the self-intersection of Z as a subvariety of X.

Neither of these ideas suffice to prove Theorem 13.7, even in the complex analytic case: Rational equivalence is more subtle than homological equivalence, and the tubular neighborhood theorem that we used is false in the category of complex analytic or algebraic varieties. (For example, no analytic neighborhood of a conic curve $C \subset \mathbb{P}^2$ is biholomorphic to any neighborhood of the zero section in the normal bundle $\mathcal{N}_{C/\mathbb{P}^2}$; see Exercise 13.23.) However, the technique of "specialization to the normal cone" (also called "deformation to the normal cone"), introduced in Fulton and MacPherson [1978] (see also the references at the end of Fulton [1984, Chapter 5]), provides a flat degeneration from the neighborhood of Z in X to the neighborhood of Z in its normal bundle, which suffices. We will take up this technique in the next section, and then return to the proof of Theorem 13.7 in the following one.

Note that as a special case of Theorem 13.7, we see that if $\iota : Z \subset X$ is an inclusion of smooth projective varieties of codimension m then the square of the class $[Z] \in A^m(X)$ is the pushforward $\iota_*(c_m(\mathcal{N}_{Z/X}))$ of the top Chern class of the normal bundle. (We can use this, for example, to determine the self-intersection of a linear space $\Lambda \cong \mathbb{P}^m \subset X$ on a smooth hypersurface $X \subset \mathbb{P}^{2m+1}$, as suggested in Exercise 13.22.) More generally, associating to a class $\zeta \in A^m(X)$ the pushforwards $\iota_*(c_k(\mathcal{N}_{Z/X}))$ of the other Chern classes of the normal bundle of a smooth representative $\iota : Z \to X$ (that is, a smooth subvariety $Z \subset X$ with $[Z] = \zeta$) represents the analog, in the Chow setting, of the *Steenrod squares* in algebraic topology.

13.4.1 Specialization to the normal cone

Suppose again that X is a smooth projective variety of dimension n and $Z \subset X$ is a smooth subvariety of codimension m. Let

$$\mu : \mathcal{X} = \mathrm{Bl}_{\{0\} \times Z}(\mathbb{P}^1 \times X) \to \mathbb{P}^1 \times X$$

be the blow-up of $\mathbb{P}^1 \times X$ along the subvariety $\{0\} \times Z$, and write $E \subset \mathcal{X}$ for the exceptional divisor. As a variety, E is the projectivization of the normal bundle

$$\mathcal{N}_{\{0\} \times Z / \mathbb{P}^1 \times X} \cong \mathcal{N}_{Z/X} \oplus \mathcal{O}_Z.$$

Thus E is the compactification of the total space of the normal bundle $\mathcal{N} = \mathcal{N}_{Z/X}$ described in Section 9.4.2.

We think of $\mathcal{X}$ as a family of projective varieties over $\mathbb{P}^1$ via the composition $\alpha = \pi_1 \circ \mu : \mathcal{X} \to \mathbb{P}^1 \times X \to \mathbb{P}^1$. Because $\mathcal{X}$ is a variety, the family is flat. The fibers X_t of $\mathcal{X}$ over $t \neq 0$ are all isomorphic to X, while the fiber X_0 of $\mathcal{X}$ over $t = 0$ consists of two irreducible components: the proper transform $\tilde{X}$ of $\{0\} \times X$ in $\mathcal{X}$ (isomorphic to the blow-up $\mathrm{Bl}_Z(X)$) and the exceptional divisor $E \cong \mathbb{P}(\mathcal{N} \oplus \mathcal{O}_Z)$, with the two intersecting along the "hyperplane at infinity" $\mathbb{P}\mathcal{N} \subset \mathbb{P}(\mathcal{N} \oplus \mathcal{O}_X)$ in E, which is the exceptional divisor in the first component $\tilde{X} \cong \mathrm{Bl}_Z(X)$.

Now consider the subvariety $\mathbb{P}^1 \times Z \subset \mathbb{P}^1 \times X$, and let $\mathcal{Z}$ be its proper transform in $\mathcal{X}$. Since $\{0\} \times Z$ is a Cartier divisor in $\mathbb{P}^1 \times Z$, the morphism μ carries $\mathcal{Z}$ isomorphically to Z.

Write $|\mathcal{N}| \subset \mathbb{P}(\mathcal{N} \oplus \mathcal{O}_Z)$ for the open subset that is isomorphic to the total space of the normal bundle $\mathcal{N}$ of Z in X. Because $\mathbb{P}^1 \times Z$ intersects $\{0\} \times X$ transversely in $\{0\} \times Z$, the intersection $\mathcal{Z} \cap X_0$ is the zero section $\mathcal{N}_0 \subset |\mathcal{N}|$. In particular, $\mathcal{Z}$ does not meet the component $\tilde{X}$ of the zero fiber $\mathcal{X}_0$.

Let

$$\nu = \pi_2 \circ \mu : \mathcal{X} \to X$$

be the composition of the blow-up map $\mu : \mathcal{X} \to \mathbb{P}^1 \times X$ with the projection on the second factor. For $t \in \mathbb{P}^1$ other than 0, ν carries X_t isomorphically to X. As for the fiber X_0, on the component $\tilde{X} \subset X_0$ the map ν is the blow-down map $\tilde{X} = \mathrm{Bl}_Z X \to X$, while on the component E the map ν is the composition $i \circ \pi$ of the bundle map $\pi : E \cong \mathbb{P}(\mathcal{N} \oplus \mathcal{O}_Z) \to Z$ with the inclusion $i : Z \hookrightarrow X$. The situation is summarized in Figure 13.5.

We may describe the situation by saying that, in the family $\mathcal{X} \to \mathbb{P}^1$, as $t \in \mathbb{P}^1$ approaches 0 the neighborhood of Z in the fibers specializes from the neighborhood of Z in X to the neighborhood of Z in the total space of its normal bundle in X. More formally:

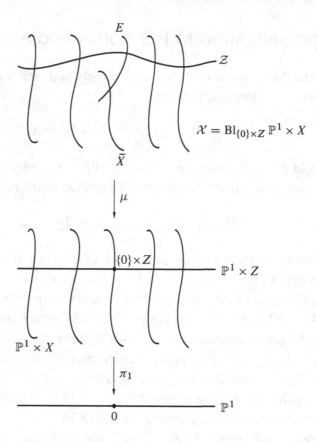

Figure 13.5 Specialization to the normal cone.

Theorem 13.8. *There is a flat family* $\alpha : \mathcal{X} \to \mathbb{P}^1$ *containing a subvariety* $\mathcal{Z} \cong Z \times \mathbb{P}^1$ *such that the restriction to* $\mathbb{P}^1 \setminus \{0\}$ *is isomorphic to*

$$(\mathbb{P}^1 \setminus \{0\}) \times X \hookleftarrow (\mathbb{P}^1 \setminus \{0\}) \times Z$$

and such that an open neighborhood of $Z \cong \alpha^{-1}(0) \cap \mathcal{Z}$ *in* $\alpha^{-1}(0)$ *is isomorphic to the total space of the normal bundle of* Z *in* X.

If we have a family $\{A_t\}_{t \in \mathbb{P}^1}$ of cycles in X that we would like to intersect with Z, then we can use this construction to transform the intersection of A_0 with Z into the intersection of the fiber of the proper transform of $\mathcal{A}$ in $\mathcal{X}$ with the zero section in the compactified normal bundle $E \cong \mathbb{P}(\mathcal{N} \oplus \mathcal{O}_Z)$. We can use our knowledge of the Chow rings of projective bundles to analyze this intersection.

13.4.2 Proof of the key formula

We now return to the proof of Theorem 13.7. Using Theorem 13.8 we will reduce the general case to the one treated in Proposition 9.16, where X is the total space of the compactification $\mathbb{P}(\mathcal{N}_{Z/X} \oplus \mathcal{O}_X)$ of the normal bundle of Z.

We will use the notation introduced in Section 13.4.1. The idea of the proof is that under the specialization to the normal cone the class $i^* i_* \alpha$ is deformed into the rationally equivalent class $j^* j_* \alpha$, where

$$j : Z \to \mathbb{P}(\mathcal{N}_{Z/X} \oplus \mathcal{O}_Z) = E$$

is the section sending Z to the zero section $\mathcal{N}_0 \subset |\mathcal{N}| \subset \overline{\mathcal{N}} := \mathbb{P}(\mathcal{N}_{Z/X} \oplus \mathcal{O}_Z)$. By Proposition 9.16, $j^* j_* \alpha = \alpha \cdot c_m(\mathcal{N})$.

Proof of Theorem 13.7: We may assume that α is the class of an irreducible subvariety $A \subset Z$. Let $\mathcal{A}$ and $\mathcal{Z}$ be the proper transforms of the subvarieties $A \times \mathbb{P}^1$ and $Z \times \mathbb{P}^1$. Since $Z \times \mathbb{P}^1$ meets $X \times \{0\}$ transversely, $\mathcal{Z} \cong Z \times \mathbb{P}^1$ via the projection, and this isomorphism also induces an isomorphism $\mathcal{A} \cong A \times \mathbb{P}^1$. To simplify notation, we will write $A_t \subset Z_t \subset X_t$ for the copies in the general fiber of $\mathcal{X}$, but we will write $A \subset Z \subset E$ instead of $A_0 \subset Z_0 \subset E$ for the fibers $A \subset Z$ contained in $E \subset X_0$.

By the moving lemma, we can find a cycle $\mathcal{C}$ on $\mathcal{X}$ linearly equivalent to $\mathcal{A}$ and generically transverse to $\mathcal{Z}$, to Z, to E, to X_t and to Z_t. The family $\mathcal{A}$ meets X_t and E generically transversely in A_t and A, respectively, so the equality $[\mathcal{C}] = [\mathcal{A}] \in A(\mathcal{X})$ restricts to equalities $i_* [A_t] = i_* [\mathcal{A} \cap X_t] = [\mathcal{C} \cap X_t] \in A(X_t)$ and $j_* [A] = [\mathcal{C} \cap E] \in A(E)$. Since $\mathcal{C}$ meets Z_t and Z generically transversely as well, we have $i^* i_* [A_t] = [\mathcal{C} \cap Z_t]$ and $j^* j_* [A] = [\mathcal{C} \cap E]$.

By generic transversality, neither Z_t nor Z can be contained in $\mathcal{C}$. It follows that after removing any components that do not dominate $\mathbb{P}^1$ the cycle $\mathcal{C}$ in $\mathcal{Z} \cong Z \times \mathbb{P}^1$ is a rational equivalence between $i^* i_* [A]$ and $j^* j_* [A]$. By Proposition 9.16, $j^* j_* [A] = [A] \cdot c_m(\mathcal{N})$, as required. $\qquad\qquad \square$

13.5 Pullbacks to a subvariety

Theorem 13.7 has an important extension which we will explain here and use in Section 13.6 to compute the relations in the Chow ring of a blow-up. Suppose that $\pi : X' \to X$ is a morphism of smooth varieties and $Z \subset X$ is a subvariety. Set $Z' = \pi^{-1}(Z)$ and let $\pi' : Z' \to Z$ be the restriction of π. If $A \subset Z \subset X$, then the expected dimension of $\pi^{-1}(A)$ is $\dim A + \dim X' - \dim X$; however — at least when Z and Z' are smooth — the actual dimension of $\pi^{-1}(A) = \pi'^{-1}(A)$ is at least $\dim A + \dim Z' - \dim Z$. Since fiber dimension is only semicontinuous, it may well happen that $\dim Z' - \dim Z > \dim X' - \dim X$. When this occurs, A cannot be transverse to π, in the sense that $\pi^{-1}(A)$ is generically reduced of the right codimension, even when it is transverse to π'. As we shall see, this is exactly the situation when π is the blow-up of a smooth (or locally complete intersection) subvariety Z. Thus it is interesting to try to compute $\pi^*[A] \in A(X')$; more precisely, writing $i : Z \to X$ and $i' : Z' \to X'$ for the inclusions, we wish to compute $\pi^*(i_*[A])$ in terms of $\pi'^*([A])$ or perhaps $i'_* \pi'^*([A])$. The following picture may help keep track of the notation:

$$
\begin{array}{ccc}
\pi^{-1}(A) \lhook\joinrel\longrightarrow & Z' & \xrightarrow{\ i'\ } X' \\
& \downarrow{\scriptstyle \pi'} & \downarrow{\scriptstyle \pi} \\
A \lhook\joinrel\longrightarrow & Z & \xrightarrow{\ i\ } X
\end{array}
$$

Theorem 13.7 does exactly what we want in the special case where $\pi = i : Z \hookrightarrow X$ is an inclusion of smooth varieties: then $\pi' : Z' \to Z$ is the identity, $i' = i$, and we saw that

$$
\begin{aligned}
\pi^*(i_*[A]) &= i^*(i_*[A]) \\
&= [A] c_m \mathcal{N}_{Z/X} \\
&= i'_*(\pi'^*([A]) \pi'^*(c_m \mathcal{N}_{Z/X})).
\end{aligned}
$$

The next result is a direct generalization. In order to state it, we have to assume that Z' is smooth; this is unnecessarily restrictive, but to state the theorem in its correct generality requires the formalism of Fulton [1984].

Theorem 13.9. *Suppose that $Z \subset X$ is a smooth subvariety of a smooth variety X, and that $\pi : X' \to X$ is a morphism from another smooth variety. Let $Z' = \pi^{-1}(Z)$ and assume that Z' is smooth, with connected components C_α of dimension c_α. Write $i : Z \to X$ and $i'_\alpha : C_\alpha \to X'$ for the inclusion maps, and likewise π_α for the restriction of π to C_α. For any class $\beta \in A_b(Z)$,*

$$
\pi^*(i_* \beta) = \sum_\alpha (i'_\alpha)_* \{ \pi_\alpha^*(\beta c(\mathcal{N}_{Z/X})) s(C_\alpha, X') \}_{b + \dim X' - \dim X}.
$$

For the proof, see Fulton [1984, Chapter 6].

13.5.1 The degree of a generically finite morphism

Let $\varphi : X \to Y$ be a generically finite projective morphism to a smooth variety Y. If $q \in Y$ is a point such that $\varphi^{-1}(q)$ is finite, then the degree of φ is the number of points of $\varphi^{-1}(q)$, counted with appropriate multiplicity. By the moving lemma, any point $q \in Y$ is rationally equivalent to a cycle of points that is transverse to the map φ in the sense of Definition 1.22, so $\deg \varphi = \deg \varphi^*[q]$ for *any* point q. Using Theorem 13.9 we can express this in terms of the local geometry of X near $\varphi^{-1}(q)$:

Corollary 13.10. *Let $\varphi : X \to Y$ be a generically finite surjective map of smooth projective varieties. If $q \in Y$ is any point, then*

$$
\deg(\varphi) = \deg \varphi^*[q] = \deg \{ s(\varphi^{-1}(q), X) \}_0.
$$

Proof: The normal bundle of q in Y is trivial, so the given formula follows from Theorem 13.9. $\qquad\square$

As a trivial example, consider the degree-1 map $\varphi : X = \mathrm{Bl}_q Y \to Y$ that is the blow-up of an n-dimensional variety Y at a point q. Since the normal bundle to the exceptional divisor $E = \varphi^{-1}(q) \cong \mathbb{P}^{n-1}$ is $\mathcal{O}_E(-1)$, we have

$$s(\mathcal{N}_{E/X}) = \frac{1}{1-\zeta},$$

and the coefficient of ζ^{n-1} is indeed 1.

A nontrivial example where the principle of Corollary 13.10 is decisive is the beautiful calculation by Donagi and Smith [1980] of the degree of the "Prym map" in genus 6. Here φ is the map from the space R_6 of unramified covers of curves of genus 6 to the space A_5 of abelian varieties of dimension 5 (both of dimension 15) defined by the "Prym construction;" while it does not seem possible to enumerate the points of a general fiber, Donagi and Smith were able to calculate its degree by looking at a very special point (the Prym of a double cover of a smooth plane quintic) over which the fiber has three components: a point, a curve and a surface!

13.6 The Chow ring of a blow-up

We can now describe the Chow ring of a blow-up of a smooth projective variety along a smooth subvariety. After reviewing some basic facts about blow-ups, we give a set of generators and calculate their products. We illustrate the results in Section 13.6.3. In the last section we complete the story by describing the relations among the generators.

Throughout this section we will use the following notation. Let X be a smooth projective variety and $Z \subset X$ a smooth subvariety of codimension m. Write $i : Z \to X$ for the inclusion map. Let $\pi : W = \mathrm{Bl}_Z X \to X$ be the blow-up of X along Z. Let $E \subset W$ be the exceptional divisor and $j : E \hookrightarrow W$ the inclusion, so that we have the diagram

$$
\begin{array}{ccc}
W = \mathrm{Bl}_Z X & \xleftarrow{\ \ j\ \ } & E \\
{\scriptstyle \pi}\downarrow & & \downarrow{\scriptstyle \pi_E} \\
X & \xleftarrow{\ \ i\ \ } & Z
\end{array}
$$

We write $\mathcal{N} = \mathcal{N}_{Z/X}$ for the normal bundle of Z in X.

Recall from Hartshorne [1977] or Eisenbud and Harris [2000, Theorem IV-23] that if $\mathcal{I} = \mathcal{I}_Z \subset \mathcal{O}_X$ denotes the ideal sheaf of Z then the blow-up W is Proj of the *Rees algebra*

$$\mathcal{A} = \mathcal{O}_X \oplus \mathcal{I} \oplus \mathcal{I}^2 \oplus \cdots.$$

The preimage E of $Z = V(\mathcal{I}) \subset X$ is then

$$E = \mathrm{Proj}(\mathcal{A} \otimes \mathcal{O}_X/\mathcal{I}) = \mathrm{Proj}(\mathcal{O}_X/\mathcal{I} \oplus \mathcal{I}/\mathcal{I}^2 \oplus \mathcal{I}^2/\mathcal{I}^3 \oplus \cdots).$$

Since $Z \subset X$ is smooth it is locally a complete intersection, so the conormal bundle $\mathcal{N}^*_{Z/X} = \mathcal{I}/\mathcal{I}^2$ is locally free and

$$\mathcal{I}^k/\mathcal{I}^{k+1} = \operatorname{Sym}^k \mathcal{N}^*_{Z/X}$$

(see Eisenbud [1995, Exercise 17.14]). Thus we may make the identification

$$E = \operatorname{Proj}(\operatorname{Sym} \mathcal{N}^*_{Z/X}) = \mathbb{P}\mathcal{N}_{Z/X}.$$

We write $\zeta \in A^1(E)$ for the first Chern class of the line bundle $\mathcal{O}_{\mathbb{P}\mathcal{N}}(1)$.

13.6.1 The normal bundle of the exceptional divisor

Proposition 13.11. *The normal bundle of* $E = \mathbb{P}\mathcal{N}_{Z/X}$ *in* W *is*

$$\mathcal{N}_{E/W} = \mathcal{O}_{\mathbb{P}\mathcal{N}_{Z/X}}(-1),$$

the tautological subbundle on $\mathbb{P}\mathcal{N}_{Z/Y}$.

Proof: With notation as above, $\mathcal{I}_E/\mathcal{I}_E^2$ is the line bundle associated to the module

$$\mathcal{I}/\mathcal{I}^2 \oplus \mathcal{I}^2/\mathcal{I}^3 \oplus \cdots = \mathcal{O}_{\mathbb{P}\mathcal{N}_{Z/X}}(1),$$

so that

$$\mathcal{N}_{E/W} = \mathcal{H}om_W(\mathcal{I}_E/\mathcal{I}_E^2, \mathcal{O}_W) \cong \mathcal{O}_{\mathbb{P}\mathcal{N}_{Z/X}}(-1),$$

as stated. $\square$

The proof of Proposition 13.11 just given works for any locally complete intersection subscheme of any scheme (and in complete generality, if we replace $\mathcal{N}_{Z/X}$ by the normal cone). In the case where Z and X are smooth, we can give a geometric proof and show that the isomorphism is induced by the differential $d\pi$ of the projection $\pi : W \to X$.

To this end, we first observe that, since the restriction of π to E is the projection from $\mathbb{P}\mathcal{N}$ to Z, the differential $d\pi$ induces a surjection from the tangent space $T_q E$ at a point q of E to $T_p Z$, where $p = \pi(q)$. Thus $d\pi$ induces a map

$$\overline{d\pi} : \mathcal{N}_{E/W} = \mathcal{T}_W/\mathcal{T}_E \to \pi^*(\mathcal{T}_X/\mathcal{T}_Z) = \pi^*\mathcal{N}_{Z/X}.$$

It now suffices to see that $\overline{d\pi}$ carries the one-dimensional vector space that is the fiber of $\mathcal{N}_{E/W}$ at a point q to the one-dimensional subspace of $\mathcal{N}_{Z/X,\pi(q)} = \pi^*(\mathcal{N}_{Z/X})_q$ that corresponds to the point q regarded as an element of $\mathbb{P}\mathcal{N}$. Indeed, if $C \subset X$ is the germ of a smooth curve passing through p with tangent space $L \subset T_p X$ corresponding to q, then the proper transform $\widetilde{C}$ of C passes through the point $q \in E$, and $d\pi$ carries the tangent space of $\widetilde{C}$ at q to L, as required.

13.6.2 Generators of the Chow ring

We maintain the notation introduced at the beginning of Section 13.6. Using the identification $E \cong \mathbb{P}\mathcal{N}$, we let $\zeta \in A^1(E)$ be the first Chern class of the line bundle $\mathcal{O}_{\mathbb{P}\mathcal{N}}(1)$. In these terms Proposition 13.11 implies that the first Chern class of the normal bundle of E in W is $-\zeta$.

Proposition 13.12. *The Chow ring $A(W)$ is generated by $\pi^*A(X)$ and $j_*A(E)$, that is, classes pulled back from X and classes supported on E. The rules for multiplication are*

$$\pi^*\alpha \cdot \pi^*\beta = \pi^*(\alpha\beta) \qquad \text{for } \alpha, \beta \in A(X),$$
$$\pi^*\alpha \cdot j_*\gamma = j_*(\gamma \cdot \pi_E^*i^*\alpha) \quad \text{for } \alpha \in A(X), \ \gamma \in A(E),$$
$$j_*\gamma \cdot j_*\delta = -j_*(\gamma \cdot \delta \cdot \zeta) \quad \text{for } \gamma, \delta \in A(E).$$

Proof: The first formula is the statement that pullback $\pi^* : A(X) \to A(W)$ is a ring homomorphism. For the second, we note that $\pi^*\alpha \cdot j_*\gamma = j_*(j^*\pi^*\alpha \cdot \gamma)$ by the push-pull formula, while $j^*\pi^*\alpha = \pi_E^*i^*\alpha$ by functoriality (both part of Theorem 1.23). Since the normal bundle of E in W has first Chern class $-\zeta$, the third formula is a special case of Proposition 13.11.

To conclude we must show that $A(W)$ is generated by $\pi^*A(X)$ and $j_*A(E)$. Let $U = W \setminus E \cong X \setminus Z$. Let $k : U \to W$ be the inclusion. Suppose that $A \subset W$ is a subvariety. If $A \subset E$ we are done; else π maps A generically isomorphically onto $\pi(A)$, so $\pi_*[A] = [\pi(A)]$.

By the moving lemma there is a cycle B on X that is rationally equivalent to $\pi(A)$ and generically transverse to π (Definition 1.22), so that, by Theorem 1.23, $\pi^*[B] = [\pi^{-1}(B)]$. It is enough to show that $[A] - \pi^*[B]$ is in the image of j_*.

From the right exact sequence

$$A(E) \longrightarrow A(W) \overset{k^*}{\longrightarrow} A(U) \longrightarrow 0$$

of Proposition 1.14, we see that it suffices to show that $k^*\pi^*([B]) = k^*[A]$. Since $[\pi(A)] = [B]$, we have

$$k^*([A]) = [A \cap U] = \pi_U^*([\pi(A)]) = \pi_U^*[B] = k^*\pi^*[B]$$

by functoriality, completing the argument. $\qquad\square$

13.6.3 Example: the blow-up of $\mathbb{P}^3$ along a curve

The first nontrivial case, the blow-up of a smooth curve in $\mathbb{P}^3$, is already interesting. We first establish some notation:

Let $C \subset \mathbb{P}^3$ be a smooth curve of degree d and genus g, and let $\pi : W \to \mathbb{P}^3$ be the blow-up of $\mathbb{P}^3$ along C. We write E, ζ, i and j as in the general discussion above, so that we have the diagram

$$
\begin{array}{ccc}
W = \mathrm{Bl}_C \, \mathbb{P}^3 & \xleftarrow{\quad j \quad} & E \cong \mathbb{P}\mathcal{N} \\
\pi \downarrow & & \downarrow \pi_E \\
\mathbb{P}^3 & \xleftarrow{\quad i \quad} & C
\end{array}
$$

In addition, we write $h \in A^1(\mathbb{P}^3)$ for the class of a plane, $e = [E]$, and $\tilde{h}$ for the pullback $\pi^* h \in A^1(W)$. Finally, for $D \in Z^1(C)$ any divisor, we will denote by $F_D = \pi_E^* D \in Z^1(E)$ the corresponding linear combination of fibers of $E \to C$, and similarly for divisor classes.

Proposition 13.13. *If W is the blow-up of $\mathbb{P}^3$ along a smooth curve C, then, with notation as above,*

$$
\begin{aligned}
A^0(W) &= \mathbb{Z}, & \textit{generated by the class of } W; \\
A^1(W) &= \mathbb{Z}^2, & \textit{generated by } e, \tilde{h}; \\
A^2(W) & & \textit{is generated by } e^2 = -j_*(\zeta), F_D \textit{ for } D \in A^1(C) \textit{ and } \tilde{h}^2; \\
A^3(W) &= \mathbb{Z}, & \textit{generated by the class of a point.}
\end{aligned}
$$

Other products among these classes are

$$
\begin{aligned}
\deg(e \cdot F_D) &= \deg D, & \tilde{h} \cdot F_D &= 0, & \deg(\tilde{h}^3) &= 1, \\
\deg(\tilde{h}^2 e) &= 0, & \deg(\tilde{h} e^2) &= -d, & \deg(e^3) &= -4d - 2g + 2.
\end{aligned}
$$

Proof: As for any variety, $A^0(W) \cong \mathbb{Z}$, generated by the fundamental class $[W]$ of W. Since W is rational, $A^3(W) \cong \mathbb{Z}$ is generated by the class of a point.

Proposition 13.12 shows that $A^1(W)$ is generated by the classes $\tilde{h}$ and $e := [E]$. The map π_* sends $\tilde{h}$ to h and sends $[E]$ to 0, so we have an exact sequence

$$
A^0(E) \xrightarrow{\;j_*\;} A^1(W) \xrightarrow{\;\pi_*\;} \mathbb{Z} \cdot h \cong \mathbb{Z} \longrightarrow 0.
$$

Also, by Proposition 13.12, $j^*(e) = j^* j_*[E] = -\zeta$. By Theorem 9.6, ζ generates a subgroup of $A^1(E)$ that is isomorphic to $\mathbb{Z}$, and thus the map

$$
\mathbb{Z} = \mathbb{Z} \cdot [E] = A^0(E) \to A^1(W)
$$

is a monomorphism. It follows that $A^1(W) \cong \mathbb{Z}^2$, freely generated by $\tilde{h}$ and e.

It will be convenient to introduce the notation $\tilde{l} = \pi^* l \in A^2(W)$ for the pullback of a line. Since $A^2(\mathbb{P}^3)$ is freely generated by l, Proposition 13.12 and Theorem 9.6 together show that $A^2(W)$ is generated by $\tilde{l}$, $j_* \zeta$ and $j_* F_D$ for $D \in A^1(C)$.

If we represent h as the class of a general hyperplane H, then we can see by considering cycles that

$$\tilde{h}^2 = \pi^*[H]^2 = \pi^*l = \tilde{l} \quad \text{and} \quad \tilde{h} \cdot e = j_* F_{(C \cap H)},$$

while

$$e^2 = -j_*\zeta$$

by Proposition 13.12.

Similarly, we see by considering cycles that

$$\deg(\tilde{h} \cdot l) = 1 \quad \text{and} \quad \deg(\tilde{h} \cdot j_* f_D) = 0$$

for any divisor class D on C, while

$$\deg(\tilde{h} \cdot j_*\zeta) = d,$$

which follows from the push-pull formula and the equality $\pi_*(j_*\zeta) = [C] = dl \in A^2(\mathbb{P}^3)$. This determines the pairing between $A^1(W) \times A^2(W) \to A^3(W) = \mathbb{Z}$.

Likewise,

$$\deg(e \cdot l) = 0,$$

$$\deg(e \cdot j_* f_D) = -\deg D \qquad \text{for any divisor class } D \text{ on } C,$$

$$\deg(e \cdot j_*\zeta) = -\deg c_1(\mathcal{N}_{C/\mathbb{P}^3}) = -4d - 2g + 2. \qquad \square$$

Three surfaces in $\mathbb{P}^3$ revisited

As a first application of this description, we revisit the first keynote question of this chapter, or rather its generalization to Proposition 13.2. Again, let $S_1, S_2, S_3 \subset \mathbb{P}^3$ be surfaces of degrees s_1, s_2 and s_3 whose scheme-theoretic intersection consists of the disjoint union of a smooth curve C of degree d and genus g and a zero-dimensional scheme Γ.

We can get rid of the component C of the intersection by pulling back the problem to the blow-up W of $\mathbb{P}^3$ along C. If we let $\tilde{S}_i$ be the proper transform of S_i, then the fact that $\bigcap S_i = C$ scheme-theoretically implies that the intersection $\bigcap \tilde{S}_i$ does not meet E. Thus

$$\bigcap \tilde{S}_i = \pi^{-1}(\Gamma);$$

in particular,

$$\deg(\Gamma) = \deg\left(\prod[\tilde{S}_i]\right).$$

In terms of the generators of $A(W)$ given in Proposition 13.13,

$$[\tilde{S}_i] = s_i\tilde{h} - e,$$

and so our answer is

$$\deg(\Gamma) = \deg \prod(s_i\tilde{h} - e).$$

By Proposition 13.13, then

$$\deg(\Gamma) = \prod s_i - d\left(\sum s_i\right) + 4d + 2g - 2,$$

as before.

Tangencies along a curve

We can also use Proposition 13.13 to answer Keynote Question (e): If $C \subset \mathbb{P}^3$ is a smooth curve of degree d and genus g, and $S, T \subset \mathbb{P}^3$ smooth surfaces of degrees s and t containing C, at how many points of C are S and T tangent?

We will count these points with the multiplicity defined by the intersection multiplicity of the sections of the normal bundle of C determined by S and T. Equivalently, these are the intersection multiplicities of the proper transforms $\tilde{S}$ and $\tilde{T}$ with the exceptional divisor of the blow-up W of $\mathbb{P}^3$ along C. With notation above, Proposition 13.13 yields

$$\begin{aligned}
\deg([\tilde{S}] \cdot [\tilde{T}] \cdot [E]) &= \deg((s\tilde{h} - e)(t\tilde{h} - e)e) \\
&= -(s + t)\deg(\tilde{h}e^2) + \deg(e^3) \\
&= (s + t)d - 4d - 2g + 2.
\end{aligned}$$

Thus, for example, two planes meeting along a line are nowhere tangent, but two quadrics Q_1, Q_2 containing a twisted cubic curve C will be tangent twice along C — as we can see directly, since the intersection $Q_1 \cap Q_2$ will consist of the union of C and a line meeting C twice.

One can check that the intersection multiplicity of the intersection of $\tilde{S}$ and $\tilde{T}$ and E at a point q with image $p \in C$ is 1 when $S \cap T$ is in a neighborhood of p the union of C with a curve D meeting C transversely in S — in other words, when $S \cap T$ has a node at p.

13.6.4 Relations on the Chow ring of a blow-up

There is one item missing in the description of the Chow ring of a blow-up provided by Proposition 13.12: a criterion for deciding whether a given class is zero. This is provided by the following:

Theorem 13.14. *Let $i : Z \to X$ be the inclusion of a smooth codimension-m subvariety in a smooth variety X, $\pi : W \to X$ the blow-up of X along Z and E the exceptional divisor, with inclusion $j : E \to W$. If $\mathcal{Q}$ is the universal quotient bundle on $E \cong \mathbb{P}\mathcal{N}_{Z/X}$, there is a split exact sequence of additive groups, preserving the grading by dimension:*

$$0 \longrightarrow A(Z) \xrightarrow{(i_* \; h)} A(X) \oplus A(E) \xrightarrow{(\pi^* \; j_*)} A(W) \longrightarrow 0,$$

where $h : A(Z) \to A(E)$ is defined by $h(\alpha) = -c_{m-1}(\mathcal{Q})\pi_E^*(\alpha)$.

Proof: We adopt the notation from the beginning of Section 13.6.

Note that $\mathcal{Q} = \pi_E^* \mathcal{N}/\mathcal{O}_{\mathbb{P}\mathcal{N}}(-1)$, a bundle of rank $m - 1$ on E. Theorem 13.9 shows that

$$\pi^* i_*(\alpha) = j_*(c_{m-1}(\mathcal{Q})\pi_E^*(\alpha));$$

that is, the two maps in the exact sequence compose to zero. As for the surjectivity of the right map, this was part of our preliminary description of $A(W)$ in Proposition 13.12.

Finally, to prove that the left map is a split monomorphism, it is enough to prove that h is a split monomorphism. We will show that $\pi_E \circ h = 1$ on $A(Z)$. To this end, we compute the Chern class of $\mathcal{Q}$ in terms of ζ and the Chern class of $\mathcal{N}$:

$$c(\mathcal{Q}) = \frac{c(\mathcal{N})}{1 - \zeta}, \quad \text{so} \quad c_{m-1}(\mathcal{Q}) = \zeta^{m-1} + c_1(\mathcal{N})\zeta^{m-2} + \cdots + c_{m-1}(\mathcal{N}),$$

where we are regarding the $c_i(\mathcal{N})$ as elements of $A(E)$ via the ring homomorphism π_E^*. Since $c_{m-1}(\mathcal{Q})$ is monic in ζ, Lemma 9.7 shows that

$$\pi_{E*}(h(\alpha)) = \alpha,$$

as required. $\square$

13.7 Exercises

Exercise 13.15. Show that the formula of Proposition 13.2 applies more generally if we replace $\mathbb{P}^3$ by an arbitrary smooth projective threefold X — that is, under the hypotheses of the proposition, we have

$$\deg(\Gamma) =$$
$$\deg([S] \cdot [T] \cdot [U]) - \deg(\mathcal{N}_{S/X}|_L) - \deg(\mathcal{N}_{T/X}|_L) - \deg(\mathcal{N}_{U/X}|_L) + \deg(\mathcal{N}_{L/X}).$$

Exercise 13.16. (a) Show that a smooth quintic curve $C \subset \mathbb{P}^3$ of genus 2 is the scheme-theoretic intersection of three surfaces in $\mathbb{P}^3$.

(b) Show that a smooth rational quintic curve $C \subset \mathbb{P}^3$ is the scheme-theoretic intersection of three surfaces in $\mathbb{P}^3$ if and only if it lies on a quadric surface; conclude that some rational quintics are expressible as such intersections and some are not.

Exercise 13.17. Let $S, T, U, V \subset \mathbb{P}^4$ be smooth hypersurfaces of degrees d, e, f and g respectively, and suppose that

$$S \cap T \cap U \cap V = C \cup \Gamma,$$

with C a smooth curve of degree a and genus g and Γ a zero-dimensional scheme disjoint from C. What is the degree of Γ?

Exercise 13.18. Let $S, T, U, V, W \subset \mathbb{P}^5$ be smooth hypersurfaces of degree d, and suppose that

$$S \cap T \cap U \cap V \cap W = \Lambda \cup \Gamma$$

with Λ a 2-plane and Γ a zero-dimensional scheme disjoint from Λ. What is the degree of Γ?

Exercise 13.19. Verify the answer to Keynote Question (b), given in Section 13.3.3, in the following cases:

(a) S is a smooth surface of degree d in a hyperplane $\mathbb{P}^3 \subset \mathbb{P}^4$ containing a line L, and T is a general 2-plane in $\mathbb{P}^4$ containing L.

(b) S and T are smooth quadric surfaces.

Exercise 13.20. Let $S = S(1, 2) \subset \mathbb{P}^4$ be a cubic scroll, as in Section 9.1.1. Show directly that a general 2-plane $T \subset \mathbb{P}^4$ containing a line of the ruling of S meets S in one more point, but a general 2-plane containing the directrix of S (that is, the line of S transverse to the ruling) does not meet S anywhere else.

Exercise 13.21. Let $\mathbb{P}^8$ be the space of 3×3 matrices and $\mathbb{P}^5 \subset \mathbb{P}^8$ the subspace of symmetric matrices. Show that the Veronese surface in $\mathbb{P}^5$ is the intersection of $\mathbb{P}^5$ with the Segre variety $\mathbb{P}^2 \times \mathbb{P}^2 \subset \mathbb{P}^8$, and verify the excess intersection formula in this case.

Exercise 13.22. Let $X \subset \mathbb{P}^5$ be a smooth hypersurface of degree d and $\Lambda \cong \mathbb{P}^2 \subset X$ a 2-plane contained in X. Use Theorem 13.7 to determine the degree of the self-intersection $[\Lambda]^2 \in A^4(X)$ of Λ in X. Check this in the cases $d = 1$ and 2.

Exercise 13.23. Let $X = 2C \subset \mathbb{P}^2$ be a double conic, that is, a subscheme defined by the square of a quadratic polynomial whose zero locus is a smooth conic curve $C \subset \mathbb{P}^2$. Show that the dualizing sheaf ω_X is isomorphic to $\mathcal{O}_X(1)$, and applying Riemann–Roch deduce that X is not hyperelliptic; that is, it does not admit a degree-2 map to $\mathbb{P}^1$. Conclude that as asserted in Section 13.4 no analytic neighborhood of C in $\mathbb{P}^2$ is biholomorphic to an analytic neighborhood of the zero section in the normal bundle $\mathcal{N}_{C/\mathbb{P}^2}$. (See Bayer and Eisenbud [1995] for more about this.)

In Exercises 13.24–13.26 we adopt the notation of Section 13.6.3; in particular, we let $C \subset \mathbb{P}^3$ be a smooth curve of degree d and genus g, $W = \mathrm{Bl}_C \mathbb{P}^3$ the blow-up of $\mathbb{P}^3$ along C and $E \subset W$ the exceptional divisor.

Exercise 13.24. Let $q \in \mathbb{P}^3$ be any point not on the tangential surface of C, and let $\Gamma \subset E \subset W$ be the curve of intersections with E of the proper transforms of lines $\overline{p, q}$ for $p \in C$. Find the class of Γ in $A(W)$.

Exercise 13.25. Let $B \subset \mathbb{P}^3$ be another curve, of degree m, and suppose that B meets C in the points of a divisor D on C. Show that the class of the proper transform $\tilde{B} \subset W$ of B is

$$[\tilde{B}] = ml - F_D.$$

Exercise 13.26. Let $S \subset \mathbb{P}^3$ be a smooth surface of degree e containing C, $\tilde{S} \subset W$ its proper transform and $\Sigma_S = \tilde{S} \cap E$ the curve on E consisting of normal vectors to C contained in the tangent space to S. Find the class $[\Sigma_S] \in A(W)$ of Σ_S in the blow-up

(a) by applying Proposition 9.13; and
(b) by multiplying the class $[\tilde{S}]$ by the class $[E]$ in $A(W)$.

Exercise 13.27. In Section 9.3.2, we observed that the blow-up $X = \mathrm{Bl}_{\mathbb{P}^k} \mathbb{P}^n$ of $\mathbb{P}^n$ along a k-plane was a $\mathbb{P}^{k+1}$-bundle over $\mathbb{P}^{n-k-1}$, and used this to describe the Chow ring of X. We now have another description of the Chow ring of X. Compare the two, and in particular:

(a) Express the generators of $A(X)$ given in this chapter in terms of the generators given in Section 9.3.2.
(b) Verify the relations among the generators given here.

Exercise 13.28. Redo Exercise 13.17 by the method of Section 13.6.3, that is, by blowing up the positive-dimensional component of the intersection.

Exercise 13.29. Let $Q \subset \mathbb{P}^3$ be a quadric cone with vertex p and $\sigma : X = \mathrm{Bl}_p Q \to Q$ its blow-up at p. Let $\pi = \pi_q : Q \to \mathbb{P}^2$ be the projection from a general point $q \in \mathbb{P}^3$ and $f = \pi \circ \sigma : X \to \mathbb{P}^2$ the composition of the blow-up map with the projection. Find the degree of f by looking at the fiber over $\pi(p)$.

Exercise 13.30. Prove that if $Z \subset A$ is a locally complete intersection subscheme inside the projective variety A then $s(Z, A) = s(\mathcal{N}_{Z/A})$.

Exercise 13.31. Let $\mathbb{P}^N$ be the space of plane curves of degree d and $X \subset \mathbb{P}^N$ the locus of d-fold lines dL. Let $C \subset \mathbb{P}^2$ be a smooth curve of degree m, and let $\Sigma \subset \mathbb{P}^N$ be the locus of curves tangent to C (that is, intersecting C in fewer than dm distinct points).

(a) Let $\mathcal{D} \subset \mathbb{P}^N$ be a general line. Show that every curve $D \in \mathcal{D}$ is either transverse to C or meets C in exactly $dm - 1$ points, and use Riemann–Hurwitz to conclude that

$$\deg(\Sigma) = 2md + m(m - 3).$$

(b) Now suppose that $\mathcal{D} \subset \mathbb{P}^N$ is a general line meeting the locus X of d-fold lines. Show that $\mathcal{D}$ meets Σ in $2md + m(m-3) - m(d-1)$ other points, and conclude that

$$\mathrm{mult}_X(\Sigma) = m(d - 1).$$

Exercise 13.32. Recalling the definition of a circle from Section 2.3.1, we say that a *sphere* $Q \subset \mathbb{P}^3$ is a quadric containing the "circle at infinity" $W = X^2 + Y^2 + Z^2 = 0$. Let $Q_1, \ldots, Q_4 \subset \mathbb{P}^3$ be four general spheres, and let $S_i \subset \mathbb{G}(1, 3)$ be the locus of lines tangent to Q_i. Using Theorem 13.3 applied to the intersection $\bigcap S_i$, find the number of lines tangent to all four.

Exercise 13.33. Let $S \subset \mathbb{P}^5$ be the Veronese surface. Find the Chow ring $A(X)$, where $X = \mathrm{Bl}_S(\mathbb{P}^5)$ is the blow-up of $\mathbb{P}^5$ along S.

Exercise 13.34. Use the result of the preceding exercise to re-derive the number of conics tangent to five conics, as suggested in Section 8.1

Exercise 13.35. In Section 3.5.5 we saw how to determine the number of common chords of two general twisted cubics $C, C' \subset \mathbb{P}^3$ by specializing C and C' to curves of types $(1, 2)$ and $(2, 1)$ on a smooth quadric surface $Q \subset \mathbb{P}^3$. As noted there, if we specialized both curves to curves of type $(1, 2)$ on Q, there would be a positive-dimensional family of common chords. Use Theorem 13.3 to analyze this case, and to show that the intersection number $\deg([\Psi_2(C)][\Psi_2(C')])$ of the cycles $\Psi_2(C), \Psi_2(C') \subset \mathbb{G}(1, 3)$ is 10.

Chapter 14

The Grothendieck Riemann–Roch theorem

The goal of Riemann–Roch theorems is to relate the dimension of the space of solutions of an analytic or algebraic problem — typically realized as the space $H^0(\mathcal{F})$ of global sections of a coherent sheaf $\mathcal{F}$ on a compact analytic or projective algebraic variety X — to topological invariants, expressed in terms of polynomials in the Chern classes of the sheaf and of the tangent bundle of X. In practice, the formulas deal not with $h^0(\mathcal{F})$ but with the Euler characteristic $\chi(\mathcal{F}) = \sum(-1)^i h^i(\mathcal{F})$ of $\mathcal{F}$, so the strength and importance of Riemann–Roch theorems, which are very great in the case of curves and surfaces, decline as the dimension of X, and with it the number of potentially nonzero cohomology groups, grows. Nevertheless, Riemann–Roch theorems have played an important role in the history of algebraic geometry.

Our goal in this chapter is to state, explain and apply a version of the Riemann–Roch theorem proved by Grothendieck that deals not just with a sheaf on a variety X but with families of such sheaves. To clarify its context, we start this chapter with older versions of the theorem. Although some of these were first proven in an analytic context, we will stick with the category of projective algebraic varieties. Good references for the simplest forms of these theorems are Sections IV.1 and V.1 of Hartshorne [1977].

Convention: To simplify notation in this chapter, we sometimes identify a class in $A_0(X)$ with its degree when X is a projective algebraic variety.

14.1 The Riemann–Roch formula for curves and surfaces

14.1.1 Curves

The original Riemann–Roch formula deals with a smooth projective curve C over $\mathbb{C}$. It says in particular that the dimension $h^0(K_C)$ of the space of regular 1-forms on C, an algebraic/analytic invariant, is equal to the topological genus $g(C) = 1 - \chi_{top}(C)/2$. To express this in modern language and suggest the generalizations to come, we invoke Serre duality, which says that

$$h^0(K_C) = h^1(\mathcal{O}_C),$$

and the Hopf index theorem for the topological Euler characteristic, which says that $\chi_{top}(C) = c_1(\mathcal{T}_C)$. In these terms, we can state the Riemann–Roch theorem as the formula

$$\chi(\mathcal{O}_C) := h^0(\mathcal{O}_C) - h^1(\mathcal{O}_C) = \frac{c_1(\mathcal{T}_C)}{2}.$$

From this formula, and the additivity of the Euler characteristic, it is easy to prove the Riemann–Roch formula for any line bundle on C, as we will see now.

To start, recall that for a coherent sheaf $\mathcal{F}$ on an arbitrary projective variety the Euler characteristic is defined to be

$$\chi(\mathcal{F}) = \sum(-1)^i h^i(\mathcal{F}).$$

This formula is *additive on exact sequences*: if

$$0 \longrightarrow \mathcal{F}' \longrightarrow \mathcal{F} \longrightarrow \mathcal{F}'' \longrightarrow 0$$

is an exact sequence of coherent sheaves on a projective variety, then the resulting long exact sequence in cohomology

$$\cdots \longrightarrow H^{i+1}(\mathcal{F}'') \longrightarrow H^i(\mathcal{F}') \longrightarrow H^i(\mathcal{F}) \longrightarrow H^i(\mathcal{F}'') \longrightarrow H^{i-1}(\mathcal{F}') \longrightarrow \cdots$$

yields $\chi(\mathcal{F}) = \chi(\mathcal{F}') + \chi(\mathcal{F}'')$.

Returning to the case of a smooth curve, suppose that $\mathcal{L} = \mathcal{O}_C(D)$ for an effective divisor D of degree $c_1(\mathcal{L}) = d$. From the sequence

$$0 \longrightarrow \mathcal{O}_C \longrightarrow \mathcal{L} \longrightarrow \mathcal{L}|_D \longrightarrow 0,$$

it follows that

$$\chi(\mathcal{L}) = \chi(\mathcal{O}_C) + \chi(\mathcal{L}|_D) = c_1(\mathcal{L}) + \frac{c_1(\mathcal{T}_C)}{2} F = d + 1 - g,$$

and a similar sequence extends this formula to line bundles of the form $\mathcal{O}_C(D - E)$, that is, arbitrary line bundles. (See Appendix D for a fuller discussion of this theorem and its consequences.)

It is not hard to go from this to the version for an arbitrary coherent sheaf $\mathcal{F}$ on C, valid for a smooth curve over any field (see Section 14.2.1 for the definition of the Chern classes of coherent sheaves):

Theorem 14.1 (Riemann–Roch for curves). *If $\mathcal{F}$ is a coherent sheaf on a smooth curve C, then*

$$\chi(\mathcal{F}) = c_1(\mathcal{F}) + \text{rank}(\mathcal{F}) \frac{c_1(\mathcal{T}_C)}{2}.$$

14.1.2 Surfaces

To state a Riemann–Roch theorem for a smooth projective surface S, we start again from a special case,

$$\chi(\mathcal{O}_S) = \frac{c_1(T_S)^2 + c_2(T_S)}{12},$$

usually referred to as *Noether's formula* (see Bădescu [2001, Chapter 5] for references). From this, the prior Riemann–Roch for curves, and sequences of the form

$$0 \longrightarrow \mathcal{L} \longrightarrow \mathcal{L}(D) \longrightarrow \mathcal{L}(D)|_D \longrightarrow 0$$

for smooth effective divisors $D \subset S$, we can deduce the version for line bundles:

$$\chi(\mathcal{L}) = \frac{c_1(\mathcal{L})^2 + c_1(\mathcal{L})c_1(T_C)}{2} + \frac{c_1(T_S)^2 + c_2(T_S)}{12}.$$

For example, to prove the formula for $\mathcal{L} = \mathcal{O}_X(D)$ when D is a smooth curve on S, we use the sequence

$$0 \longrightarrow \mathcal{O}_S \longrightarrow \mathcal{O}_S(D) \longrightarrow \mathcal{O}_S(D)|_D \longrightarrow 0.$$

From the additivity of the Euler characteristic, we get

$$\chi(\mathcal{O}_S(D)) = \chi(\mathcal{O}_S) + \chi(\mathcal{O}_S(D)|_D)$$
$$= \frac{c_1(T_S)^2 + c_2(T_S)}{12} + \chi(\mathcal{O}_S(D)|_D).$$

To evaluate the last term, observe that $\mathcal{O}_S(D)|_D$ is a line bundle of degree $D \cdot D$ on the curve D, which by the adjunction formula has genus

$$g(D) = \frac{D \cdot D + D \cdot K_S}{2} + 1.$$

Using Riemann–Roch for curves we obtain

$$\chi(\mathcal{O}_S(D)|_D) = D \cdot D - \frac{D \cdot D + D \cdot K_S}{2}$$
$$= \frac{D \cdot D + D \cdot c_1(T_S)}{2},$$

and the Riemann–Roch formula above follows for $\mathcal{L} = \mathcal{O}_S(D)$.

As in the previous case of curves, this can be extended to apply to arbitrary coherent sheaves on S:

Theorem 14.2. *If $\mathcal{F}$ is a coherent sheaf on a smooth projective surface S, then*

$$\chi(\mathcal{F}) = \frac{c_1(\mathcal{F})^2 - 2c_2(\mathcal{F}) + c_1(\mathcal{F})c_1(T_C)}{2} + \mathrm{rank}(\mathcal{F})\frac{c_1(T_S)^2 + c_2(T_S)}{12}.$$

14.2 Arbitrary dimension

Much of the content of the formulas in Section 14.1 above was known to 19th century algebraic geometers, although the formulas were expressed without cohomology, and only for line bundles (represented by divisors). In the 20th century these formulas were extended to sheaves on varieties of arbitrary dimension by Hirzebruch. One key to this extension was the introduction of cohomology groups in general, and the recognition that the left-hand side of all the classical formulations of Riemann–Roch represented Euler characteristics of sheaves.

Equally important was understanding how to express the polynomials in the Chern classes that appear in the right-hand side of these formulas in a way that generalized to arbitrary dimensions. We digress to introduce the two useful power series in the Chern classes that are needed, the *Chern character* and the *Todd class*.

14.2.1 The Chern character

We have seen many uses of the Whitney formula (Theorem 5.3); we have also used special cases of the formula for the Chern classes of the tensor product of vector bundles, which we deemed too complicated to write down in closed form in general. But there is a better way to make sense of these two formulas, discovered by Hirzebruch [1966]; together, they say that a certain power series in the Chern classes, the *Chern character*, defines a ring homomorphism. To explain this useful fact, we first recall the definition of the Grothendieck ring of vector bundles:

The set of isomorphism classes $[\mathcal{A}]$ of vector bundles $\mathcal{A}$ of finite rank on a variety X forms a semigroup under direct sum that we will call $\mathrm{Bun}(X)$. The Euler characteristic χ defines a homomorphism of semigroups $\mathrm{Bun}(X) \to \mathbb{Z}$ and, as we have already remarked, this map is also additive on exact sequences of bundles $0 \to \mathcal{A} \to \mathcal{B} \to \mathcal{C} \to 0$. It is interesting to ask what other such maps there may be, and a natural step in investigating this is to form the *Grothendieck group* $K(X)$ of vector bundles on a variety X. This is defined as the free abelian group on the set of isomorphism classes $[\mathcal{A}]$ of vector bundles $\mathcal{A}$ on X, modulo relations $[\mathcal{A}] + [\mathcal{C}] = [\mathcal{B}]$ for every short exact sequence of vector bundles $0 \to \mathcal{A} \to \mathcal{B} \to \mathcal{C} \to 0$. The natural map of semigroups $\mathrm{Bun}(X) \to K(X)$ is *universal*, in the sense that any map from $\mathrm{Bun}(X)$ to a group that is additive on short exact sequences factors uniquely through the map to $K(X)$. A bonus of this construction is that, since tensoring with a vector bundle preserves exact sequences, $K(X)$ has a natural ring structure, where the product is given by tensor product.

The *Chern character* is a way of combining the Chern classes to produce a ring homomorphism

$$\mathrm{Ch} : K(X) \to A(X) \otimes \mathbb{Q}.$$

To see how such a combination could be defined (and how the rational coefficients arise), consider first the case of line bundles. If $\mathcal{L}$ and $\mathcal{M}$ are two line bundles, then the first Chern class of the tensor product $c(\mathcal{L} \otimes \mathcal{M})$ is $c_1(\mathcal{L}) + c_1(\mathcal{M})$, so if we set

$$\mathrm{Ch}([\mathcal{L}]) = e^{c_1(\mathcal{L})} = 1 + c_1(\mathcal{L}) + \frac{c_1(\mathcal{L})^2}{2} + \frac{c_1(\mathcal{L})^3}{6} + \cdots,$$

then

$$\mathrm{Ch}(\mathcal{L} \otimes \mathcal{M}) = e^{c_1(\mathcal{L}) + c_1(\mathcal{M})}$$

$$= e^{c_1(\mathcal{L})} e^{c_1(\mathcal{M})}$$

$$= \mathrm{Ch}([\mathcal{L}]) \, \mathrm{Ch}([\mathcal{M}]).$$

Note that the apparently infinite sums are actually finite, since $A^i(X)$ vanishes for $i > \dim X$.

If now $\mathcal{E} = \bigoplus \mathcal{L}_i$ is a direct sum of line bundles, then for Ch to preserve sums we must extend the definition above by setting

$$\mathrm{Ch}(\mathcal{E}) = \sum e^{c_1(\mathcal{L}_i)}.$$

The coefficients of this power series are symmetric in the "Chern roots" $c_1(\mathcal{L}_i)$, and thus can be expressed in terms of the elementary symmetric functions of these quantities — that is, in terms of the Chern classes of $\mathcal{E}$.

We define $\mathrm{Ch}(\mathcal{E})$ in general by using these expressions: If $\mathcal{E}$ is any vector bundle, we write $c(\mathcal{E}) = \prod(1 + \alpha_i)$, and then define

$$\mathrm{Ch}(\mathcal{E}) = \sum e^{\alpha_i};$$

in other words, the k-th graded piece $\mathrm{Ch}_k(\mathcal{E})$ of the Chern character is

$$\mathrm{Ch}_k(\mathcal{E}) = \sum \frac{\alpha_i^k}{k!},$$

expressed as a polynomial in the elementary symmetric functions of the α_i and applied to the Chern classes $c_i(\mathcal{E})$. The first few cases are

$$\mathrm{Ch}_0(\mathcal{E}) = \mathrm{rank}(\mathcal{E}),$$

$$\mathrm{Ch}_1(\mathcal{E}) = c_1(\mathcal{E}),$$

$$\mathrm{Ch}_2(\mathcal{E}) = \frac{c_1(\mathcal{E})^2 - 2c_2(\mathcal{E})}{2}.$$

The splitting principle implies that this formula does indeed give a ring homomorphism: First, Whitney's formula shows that if

$$0 \longrightarrow \mathcal{E}' \longrightarrow \mathcal{E} \longrightarrow \mathcal{E}'' \longrightarrow 0$$

is an exact sequence of vector bundles, then the Chern roots of $\mathcal{E}$ are the Chern roots of $\mathcal{E}'$ together with the Chern roots of $\mathcal{E}''$, so the definition above yields $\mathrm{Ch}(\mathcal{E}) = \mathrm{Ch}(\mathcal{E}') + \mathrm{Ch}(\mathcal{E}'')$. Further, the Chern roots of $\mathcal{E}' \otimes \mathcal{E}''$ are the pairwise sums of the Chern roots of $\mathcal{E}'$ and those of $\mathcal{E}''$, so the definition in terms of Chern roots again immediately yields the product formula $\mathrm{Ch}(\mathcal{E}' \otimes \mathcal{E}'') = \mathrm{Ch}(\mathcal{E}') \, \mathrm{Ch}(\mathcal{E}'')$, as required.

Since the Chern character is equivalent data to the rational Chern class, this yields a formula for the rational Chern class of a tensor product. The result is quite convenient for machine computation, but the conversion of polynomials in the power sums to polynomials in the elementary symmetric polynomials is complicated enough that it is not so useful for computation by hand; see Exercise 14.11 for an example.

Coherent sheaves

Let X be a smooth projective variety and $\mathcal{F}$ a coherent sheaf of X. By the Hilbert syzygy theorem, we can resolve $\mathcal{F}$ by locally free sheaves; that is, we can find an exact sequence

$$0 \longrightarrow \mathcal{E}_n \longrightarrow \mathcal{E}_{n-1} \longrightarrow \cdots \longrightarrow \mathcal{E}_1 \longrightarrow \mathcal{E}_0 \longrightarrow \mathcal{F} \longrightarrow 0$$

in which all the sheaves $\mathcal{E}_i$ are locally free. We can use this to extend the definitions of Chern classes and Chern characters to all coherent sheaves, in the only way possible that makes the Whitney formula and the product formula hold in general: We define the Chern polynomial and Chern character by

$$c(\mathcal{F}) = \prod_{i=0}^{n} c(\mathcal{E}_i)^{(-1)^i} \quad \text{and} \quad \mathrm{Ch}(\mathcal{F}) = \sum_{i=0}^{n} (-1)^i \, \mathrm{Ch}(\mathcal{E}_i).$$

Of course for these definitions to make sense we need to know that they are independent of the resolution chosen; the verification is left as Exercise 14.12.

Caution: If $Y \subset X$ is a Cartier divisor, then from the sequence

$$0 \longrightarrow \mathcal{O}_X(-Y) \longrightarrow \mathcal{O}_X \longrightarrow \mathcal{O}_Y \longrightarrow 0$$

we see the Chern class of the sheaf $\mathcal{O}_Y$, viewed as a coherent sheaf on X, is simply

$$c(\mathcal{O}_Y) = \frac{c(\mathcal{O}_X)}{c(\mathcal{O}_X(-Y))} = \frac{1}{1 - [Y]} = 1 + [Y].$$

The equality $c(\mathcal{O}_Y) = 1 + [Y]$ is emphatically *not* true for subvarieties $Y \subset X$ of codimension $c > 1$: In general, even when X and Y are smooth the Chern class of $\mathcal{O}_Y$ may have components of codimensions greater than c, and even the component in $A^c(X)$ differs from $[Y]$ by a factor of $(-1)^{c-1}(c-1)!$. This is a consequence of the Grothendieck Riemann–Roch theorem below; for examples, see Exercises 14.13–14.14.

The information in the Chern classes

Up to torsion, giving the Chern classes of a bundle $\mathcal{E}$ is equivalent to giving the class of $\mathcal{E}$ in $K(X)$:

Theorem 14.3 (Grothendieck). *If X is a smooth projective variety, then the map*

$$\mathrm{Ch} : K(X) \otimes \mathbb{Q} \to A(X) \otimes \mathbb{Q}$$

is an isomorphism of rings.

For more information, see Fulton [1984, Example 15.2.16b].

Strikingly, there is an analogous statement in the category of differentiable manifolds: If we define the topological K-group of a manifold M to be the group of formal linear combinations of C^∞ vector bundles modulo relations coming from exact sequences, with ring structure given as above by tensor products, then for a suitable filtration of the K-group the Chern character gives an isomorphism

$$\mathrm{gr}\, K(M) \otimes \mathbb{Q} \cong H^{2*}(X, \mathbb{Q}),$$

where the term on the right is the ring of even-degree rational cohomology classes (see Griffiths and Adams [1974]).

14.2.2 The Todd class

In the case of curves and surfaces we derived the Riemann–Roch formula by starting with an expression for the Euler characteristic $\chi(\mathcal{O}_X)$ of our variety X in terms of the Chern classes of the tangent bundle $\mathcal{T}_X$ of X. The *Todd class* of a vector bundle gives us a way of doing this in general: It is a polynomial in the Chern classes, with rational coefficients, such that for any smooth variety X we have

$$\chi(\mathcal{O}_X) = \{\mathrm{Td}(\mathcal{T}_X)\}_0;$$

that is, if X is n-dimensional, $\chi(\mathcal{O}_X) = \mathrm{Td}_n(\mathcal{T}_X)$, the degree of the n-th graded piece of the Todd class.

This is the property of these classes that led Todd to their definition (see Exercise 14.15). In fact the Todd class may be characterized as the only formula in the Chern classes that, when evaluated on the Chern classes of the tangent bundle $\mathcal{T}_\mathbb{P}$ of any product $\mathbb{P}$ of projective spaces, has zero-dimensional component $\{\mathrm{Td}(\mathcal{T}_\mathbb{P})\}_0 = 1$ (Hirzebruch [1966, p. 3]).

The naturality of this characterization does not imply that the formula will look simple! Suppose $\mathcal{E}$ is a vector bundle/locally free sheaf of rank n on a smooth variety X, and formally factor its Chern class:

$$c(\mathcal{E}) = \prod_{i=1}^{n}(1 + \alpha_i).$$

We define the *Todd class* of $\mathcal{E}$ to be

$$\mathrm{Td}(\mathcal{E}) = \prod_{i=1}^{n} \frac{\alpha_i}{1 - e^{-\alpha_i}},$$

written as a power series in the elementary symmetric polynomials $c_i(\mathcal{E})$ of the α_i.

This definition, together with Whitney's formula and the splitting principle, immediately implies a multiplicative property: If

$$0 \longrightarrow \mathcal{E}' \longrightarrow \mathcal{E} \longrightarrow \mathcal{E}'' \longrightarrow 0$$

is an exact sequence of vector bundles, then, as before, the Chern roots of $\mathcal{E}$ are the Chern roots of $\mathcal{E}'$ together with the Chern roots of $\mathcal{E}''$, so the definition above at once yields $\mathrm{Td}(\mathcal{E}) = \mathrm{Td}(\mathcal{E}') \, \mathrm{Td}(\mathcal{E}'')$.

To calculate the first few terms of the Todd class, write

$$1 - e^{-\alpha} = \alpha - \frac{\alpha^2}{2} + \frac{\alpha^3}{6} - \frac{\alpha^4}{24} + \cdots,$$

so

$$\frac{1 - e^{-\alpha}}{\alpha} = 1 - \frac{\alpha}{2} + \frac{\alpha^2}{6} - \frac{\alpha^3}{24} + \frac{\alpha^4}{120} - \cdots;$$

inverting this, we get

$$\frac{\alpha}{1 - e^{-\alpha}} = 1 + \frac{\alpha}{2} + \frac{\alpha^2}{12} - \frac{\alpha^4}{720} + \cdots,$$

so

$$\mathrm{Td}(\mathcal{E}) = \prod_{i=1}^{n} \left(1 + \frac{\alpha_i}{2} + \frac{\alpha_i^2}{12} - \frac{\alpha_i^4}{720} + \cdots \right).$$

Rewriting the first few of these in terms of the symmetric polynomials of the α_i — that is, the Chern classes of $\mathcal{E}$ — we get formulas for the first few terms of the Todd class:

$$\mathrm{Td}_0(\mathcal{E}) = 1,$$

$$\mathrm{Td}_1(\mathcal{E}) = \sum \frac{\alpha_i}{2} = \frac{c_1(\mathcal{E})}{2},$$

$$\mathrm{Td}_2(\mathcal{E}) = \frac{1}{12} \sum \alpha_i^2 + \frac{1}{4} \sum_{i<j} \alpha_i \alpha_j = \frac{c_1^2(\mathcal{E}) + c_2(\mathcal{E})}{12},$$

$$\mathrm{Td}_3(\mathcal{E}) = \frac{1}{24} \sum_{i \neq j} \alpha_i \alpha_j^2 = \frac{c_1(\mathcal{E}) c_2(\mathcal{E})}{24}.$$

14.2.3 Hirzebruch Riemann–Roch

We now have the language necessary to express the Hirzebruch Riemann–Roch theorem, one formula that specializes to all the Riemann–Roch theorems stated above and their higher-dimensional analogs. In fact, given the definitions above, it is remarkably simple:

Theorem 14.4 (Hirzebruch's Riemann–Roch formula). *If X is a smooth projective variety of dimension n and $\mathcal{F}$ a coherent sheaf on X, then*

$$\chi(\mathcal{F}) = \{\mathrm{Ch}(\mathcal{F}) \, \mathrm{Td}(\mathcal{T}_X)\}_n.$$

This formula was first stated and proved in the setting of algebraic varieties over $\mathbb{C}$ by Hirzebruch [1966]; it was generalized to differentiable manifolds and elliptic differential operators by Atiyah and Singer; see Palais [1965]. The generalization to varieties over arbitrary fields is a special case of the work of Grothendieck, which we will describe next.

14.3 Families of bundles

Grothendieck's version of the Riemann–Roch theorem introduces a fundamental new idea into the mix.

Briefly, suppose we have a family $\{X_b\}_{b\in B}$ of schemes, and a family of sheaves $\mathcal{F}_b$ on X_b — in other words, a morphism $\pi : X \to B$ and a sheaf $\mathcal{F}$ on X. As we see in Appendix B, the vector spaces $H^0(\mathcal{F}_b)$ form, at least for b in an open subset $U \subset B$, the fibers of a sheaf on B; this is the direct image $\pi_*\mathcal{F}$. We can think of the "classical" Hirzebruch Riemann–Roch theorem, applied to the sheaf $\mathcal{F}_b$ on the general fiber X_b, as a formula for the dimension $h^0(\mathcal{F}_b)$ (that is, the rank of the sheaf $\pi_*\mathcal{F}$), with "error terms" coming from the dimensions $h^i(\mathcal{F}_b)$ of the higher cohomology groups (i.e., the ranks of the higher direct images $R^i\pi_*\mathcal{F}$). The Grothendieck version of the Riemann–Roch theorem describes the way in which the spaces $H^0(\mathcal{F}_b)$ fit together as b varies as measured by the Chern classes of $\pi_*\mathcal{F}$.

14.3.1 Grothendieck Riemann–Roch

Theorem 14.5 (Grothendieck's Riemann–Roch formula). *If $\pi : X \to B$ is a projective morphism with X and B smooth, and $\mathcal{F}$ is a coherent sheaf on X, then*

$$\sum_{i=0}^{n}(-1)^i \operatorname{Ch}(R^i\pi_*\mathcal{F}) = \pi_*\left[\frac{\operatorname{Ch}(\mathcal{F})\cdot\operatorname{Td}(\mathcal{T}_X)}{\pi^*\operatorname{Td}(\mathcal{T}_B)}\right] \in A(B)\otimes\mathbb{Q}.$$

This was first stated and proved in Borel and Serre [1958]; a shorter and more natural argument may be found in Fulton [1984, Section 15.2]

Hirzebruch's Riemann–Roch is Grothendieck's Riemann–Roch in the special case when B is a single point.

There are equivalent formulations of the Grothendieck Riemann–Roch theorem. For example, using the push-pull formula we can rewrite it in the form

$$\left(\sum_{i=0}^{n}(-1)^i \operatorname{Ch}(R^i\pi_*\mathcal{F})\right)\operatorname{Td}(\mathcal{T}_B) = \pi_*[\operatorname{Ch}(\mathcal{F})\cdot\operatorname{Td}(\mathcal{T}_X)].$$

Also, in case the map π is a submersion — that is, the differential $d\pi$ is surjective everywhere — then from the short exact sequence

$$0 \longrightarrow \mathcal{T}^v_{X/B} \longrightarrow \mathcal{T}_X \longrightarrow \pi^* \mathcal{T}_B \longrightarrow 0$$

for the relative tangent bundle of π and the multiplicativity of the Todd class we have:

Corollary 14.6. *If $\pi : X \to B$ is a projective morphism and a submersion, and $\mathcal{F}$ is a coherent sheaf on X, then*

$$\sum_{i=0}^n (-1)^i \operatorname{Ch}(R^i \pi_* \mathcal{F}) = \pi_*[\operatorname{Ch}(\mathcal{F}) \cdot \operatorname{Td}(\mathcal{T}^v_{X/B})].$$

When applying these formulas it is crucial to know when the sheaves $R^i \pi_* \mathcal{F}$ have fibers at an arbitrary point b equal to $H^i(\mathcal{F}|_{\pi^{-1}(b)})$. The theorem on cohomology and base change (Theorem B.5 in Appendix B) gives conditions under which this happens. Given this, it is possible to use the Grothendieck Riemann–Roch formula to calculate the Chern classes of many of the bundles whose Chern classes we calculated by ad hoc methods, e.g., filtrations and the splitting principle, earlier in this book. Indeed, virtually all of the bundles we have introduced and analyzed in the preceding chapters can be defined as direct images, and the Grothendieck Riemann–Roch formula gives an expression of their Chern classes. In the next section we will work this out in a simple case. Other examples are suggested in Exercises 14.17–14.19.

Following this example, the remainder of this chapter will be concerned with two applications of the Grothendieck Riemann–Roch formula in situations where we do not have alternative ways of calculating the Chern classes of the bundles in question. In Section 14.4, we will describe an application of the formula to the geometry of vector bundles on projective space, and in Section 14.5 an application to the geometry of families of curves.

14.3.2 Example: Chern classes of $\operatorname{Sym}^3 \mathcal{S}^*$ on $\mathbb{G}(1,3)$

Recall that to compute the number of lines on a smooth cubic surface $S \subset \mathbb{P}^3$ in Chapter 6 we introduced a vector bundle $\mathcal{E}$ on the Grassmannian $G = \mathbb{G}(1,3)$ of lines in $\mathbb{P}^3$. Informally, we described $\mathcal{E}$ by saying that, for each line $L \subset \mathbb{P}^3$, the fiber of $\mathcal{E}$ at the point $[L] \in G$ was the vector space of homogeneous cubic polynomials on L; that is,

$$\mathcal{E}_{[L]} = H^0(\mathcal{O}_L(3)).$$

We explained that the number of lines, computed with multiplicity, was equal to $c_4(\mathcal{E})$. (The result that every smooth cubic surface actually has 27 distinct lines required further argument.)

Back then, we showed how $\mathcal{E}$ may be realized as a direct image, but did not use that construction to calculate its Chern class. (Rather, we calculated the Chern classes of $\mathcal{E}$ by realizing that $\mathcal{E} = \mathrm{Sym}^3 \, S^*$, the third symmetric power of the dual of the universal subbundle on G, and using the splitting principle.) We now have a tool that will allow us to calculate the Chern classes of $\mathcal{E}$ directly from its construction; as an illustration of the general technique, we will carry this out explicitly here.

To set this up, recall that the universal family of lines over G is the incidence correspondence

$$\Phi = \{(L, p) \in G \times \mathbb{P}^3 \mid p \in L\};$$

we will let $\alpha : \Phi \to G$ and $\beta : \Phi \to \mathbb{P}^3$ be the projection maps. The bundle $\mathcal{E}$ is the direct image of $\mathcal{L} = \beta^* \mathcal{O}_{\mathbb{P}^3}(3)$.

To compute the Chern classes of $\mathcal{E}$ we first observe that the restriction of $\mathcal{O}_L(3)$ of $\mathcal{L}$ to each fiber $\Phi_{[L]} = \alpha^{-1}([L]) = L \cong \mathbb{P}^1$ is $\mathcal{O}_{\mathbb{P}^1}(3)$, which has no higher cohomology. From the theorem on cohomology and base change (Theorem B.5), it follows that the direct image

$$\mathcal{E} = \alpha_* \mathcal{L} = \alpha_*(\beta^* \mathcal{O}_{\mathbb{P}^3}(3))$$

is locally free, with fiber $H^0(\mathcal{O}_L(3))$ at $[L]$.

Because of the vanishing of the higher cohomology of $\mathcal{L}$ on the fiber of α, the higher direct images $R^i \alpha_*(\mathcal{L})$ are 0 for $i > 0$, so the Grothendieck Riemann–Roch theorem becomes a formula for the Chern character of $\mathcal{E}$:

$$\mathrm{Ch}(\mathcal{E}) = \alpha_*(\mathrm{Ch}(\mathcal{L}) \cdot \mathrm{Td}(T^v_{\Phi/G})).$$

To evaluate this explicitly requires the following steps:

(a) Describe the Chow ring $A(\Phi)$.
(b) Describe the direct image map $\alpha_* : A(\Phi) \to A(G)$.
(c) Calculate the Chern character of $\mathcal{L}$ and the Todd class of the relative tangent bundle $T^v_{\Phi/G}$.
(d) Take the direct image of their product, to arrive at $\mathrm{Ch}(\mathcal{E})$.
(e) Finally, convert this back into the Chern classes of $\mathcal{E}$.

The necessary result for step (a) is Proposition 9.10: $\Phi = \mathbb{P}S$ is the projectivization of the universal subbundle on G, so that we have

$$A(\Phi) = A(G)[\zeta]/(\zeta^2 - \sigma_1 \zeta + \sigma_{1,1}),$$

where $\zeta = c_1(\mathcal{O}_{\mathbb{P}S}(1))$.

As for step (b), we have

$$\alpha_*(1 + \zeta + \zeta^2 + \cdots) = s(S) = \frac{1}{c(S)} = \frac{1}{1 - \sigma_1 + \sigma_{1,1}} = 1 + \sigma_1 + \sigma_2;$$

in other words,

$$\alpha_*\zeta = 1, \quad \alpha_*(\zeta^2) = \sigma_1, \quad \alpha_*(\zeta^3) = \sigma_2 \quad \text{and} \quad \alpha_*(\zeta^4) = 0.$$

(The last equation can also be seen directly: $\zeta^4 = 0$ in $A(\Phi)$, since ζ is the pullback of the hyperplane class on $\mathbb{P}^3$.) By the push-pull formula, this allows us to evaluate the pushforward of any product of a power of ζ with the pullback of a class from G.

To compute the Chern character of $\mathcal{L}$ for step (c), we first observe that, since the fiber of the line bundle $\mathcal{O}_{\mathbb{P}S}(1)$ at a point $(L, p) \in \Phi$ is the dual of the one-dimensional vector subspace of $\mathbb{C}^4$ corresponding to p, we have

$$\zeta = c_1(\mathcal{O}_{\mathbb{P}S}(1)) = \beta^* c_1(\mathcal{O}_{\mathbb{P}^3}(1)).$$

In particular, it follows that

$$c_1(\mathcal{L}) = 3\zeta,$$

and so

$$\text{Ch}(\mathcal{L}) = 1 + 3\zeta + \tfrac{9}{2}\zeta^2 + \tfrac{27}{6}\zeta^3,$$

since higher powers of ζ vanish.

In Section 11.1.2 we saw how to find the Chern classes of the relative tangent bundle of Φ over G: If we denote by $\mathcal{U}$ the tautological subbundle on $\Phi = \mathbb{P}S$, and by $\mathcal{Q}$ the tautological quotient bundle, we have

$$T^v_{\Phi/G} = \mathcal{U}^* \otimes \mathcal{Q}.$$

From the exact sequence

$$0 \longrightarrow \mathcal{U} \longrightarrow \alpha^*S \longrightarrow \mathcal{Q} \longrightarrow 0$$

we see that

$$c_1(\mathcal{Q}) = c_1(\alpha^*S) - c_1(\mathcal{U}) = -\sigma_1 + \zeta,$$

and hence

$$c_1(T^v_{\Phi/G}) = c_1(\mathcal{U}^* \otimes \mathcal{Q}) = \zeta + c_1(\mathcal{Q}) = -\sigma_1 + 2\zeta.$$

Plugging this into the formula for the Todd class, we have

$$\text{Td}(T^v_{\Phi/G}) = 1 + \frac{2\zeta - \sigma_1}{2} + \frac{(2\zeta - \sigma_1)^2}{12} - \frac{(2\zeta - \sigma_1)^4}{720}.$$

For step (d) we take the product

$$\begin{aligned}
\text{Ch}(\mathcal{L})\,\text{Td}(T^v_{\Phi/G}) = 1 &+ \tfrac{1}{2}(8\zeta - \sigma_1) + \tfrac{1}{12}(94\zeta^2 - 22\sigma_1\zeta + \sigma_1^2) \\
&+ \tfrac{1}{12}(120\zeta^3 - 39\sigma_1\zeta^2 + 3\sigma_1^2\zeta) \\
&+ \tfrac{1}{720}(-2668\sigma_1\zeta^3 + 246\sigma_1^2\zeta^2 + 8\sigma_1^3\zeta - \sigma_1^4) \\
&+ \tfrac{1}{720}(198\sigma_1^2\zeta^3 + 24\sigma_1^3\zeta^2 - 3\sigma_1^4\zeta).
\end{aligned}$$

Applying the direct image map, we find that by Grothendieck Riemann–Roch

$$\text{Ch}(\mathcal{E}) = 4 + 6\sigma_1 + (7\sigma_2 - 3\sigma_{1,1}) - 3\sigma_{2,1} + \tfrac{1}{3}\sigma_{2,2}.$$

It remains to convert this to the Chern class of $\mathcal{E}$ (step (e)). We have

$$c_1(\mathcal{E}) = \text{Ch}_1(\mathcal{E}) = 6\sigma_1$$

and

$$
\begin{aligned}
c_2(\mathcal{E}) &= \tfrac{1}{2}\text{Ch}_1(\mathcal{E})^2 - \text{Ch}_2(\mathcal{E}) \\
&= 18\sigma_1^2 - (7\sigma_2 - 3\sigma_{1,1}) \\
&= 11\sigma_2 + 21\sigma_{1,1}.
\end{aligned}
$$

Similarly,

$$
\begin{aligned}
c_3(\mathcal{E}) &= \tfrac{1}{6}\text{Ch}_1(\mathcal{E})^3 - \text{Ch}_1(\mathcal{E})\text{Ch}_2(\mathcal{E}) + 2\text{Ch}_3(\mathcal{E}) \\
&= 36\sigma_1^3 - 6\sigma_1(7\sigma_2 - 3\sigma_{1,1}) - 6\sigma_{2,1} \\
&= 72\sigma_{2,1} - 24\sigma_{2,1} - 6\sigma_{2,1} \\
&= 42\sigma_{2,1},
\end{aligned}
$$

and, finally, the payoff!

$$
\begin{aligned}
c_4(\mathcal{E}) &= \tfrac{1}{24}\text{Ch}_1(\mathcal{E})^4 - \tfrac{1}{2}\text{Ch}_1(\mathcal{E})^2\text{Ch}_2(\mathcal{E}) + \tfrac{1}{2}\text{Ch}_2(\mathcal{E})^2 + 2\text{Ch}_1(\mathcal{E})\text{Ch}_3(\mathcal{E}) - 6\text{Ch}_4(\mathcal{E}) \\
&= 54\sigma_1^4 - 18\sigma_1^2(7\sigma_2 - 3\sigma_{1,1}) + \tfrac{1}{2}(7\sigma_2 - 3\sigma_{1,1})^2 - 36\sigma_1\sigma_{2,1} - 2\sigma_{2,2} \\
&= (108 - 72 + 29 - 36 - 2)\sigma_{2,2} \\
&= 27\sigma_{2,2};
\end{aligned}
$$

we have calculated again the number of lines on a cubic surface, counted with multiplicities, under the assumption that the number is finite.

We have also illustrated a fact well-known to practicing algebraic geometers: One should almost never use Grothendieck Riemann–Roch to calculate the Chern classes of a bundle if there is an alternative!

14.4 Application: jumping lines

In this section, we will describe the notion of *jumping lines*, an invariant used to study the geometry of vector bundles on projective space. To keep the notation simple, we will deal just with the case of vector bundles of rank 2 on $\mathbb{P}^2$. The generalization to rank-2 bundles on $\mathbb{P}^n$ is indicated in Exercises 14.20–14.22.

We start by recalling that, though a vector bundle $\mathcal{E}$ of rank 2 on $\mathbb{P}^2$ may be indecomposable (see for example Exercise 5.41), its restriction to any line $L \cong \mathbb{P}^1 \subset \mathbb{P}^2$ can be expressed as a direct sum

$$\mathcal{E}|_L = \mathcal{O}_{\mathbb{P}^1}(a_1) \oplus \mathcal{O}_{\mathbb{P}^1}(a_2)$$

of line bundles, and of course

$$c_1(\mathcal{E}|_L) = a_1 + a_2.$$

This *splitting type* (a_1, a_2) of $\mathcal{E}$ on L provides a useful way to analyze vector bundles: For every pair $a = (a_1, a_2)$ with $a_1 < a_2$ and $a_1 + a_2 = c_1(\mathcal{E}|_L)$, we define a subset $\Gamma_a \subset \mathbb{G}(1,2) = \mathbb{P}^{2*}$ by

$$\Gamma_a = \{L \in \mathbb{P}^{2*} \mid \mathcal{E}|_L \cong \mathcal{O}_{\mathbb{P}^1}(a_1) \oplus \mathcal{O}_{\mathbb{P}^1}(a_2)\}.$$

As we will see in a moment, these will be locally closed subsets of $\mathbb{P}^{2*}$. In particular, the decomposition of $\mathcal{E}|_L$ will be constant for L in an open dense subset of $\mathbb{P}^{2*}$; the lines outside this open set are called *jumping lines*. Together the loci Γ_a give a stratification of $\mathbb{P}^{2*}$ whose geometry is an important invariant of $\mathcal{E}$; the closures of the strata Γ_a are called *loci of jumping lines*. It follows at once that the loci of jumping lines will not change (except for indexing) if we replace $\mathcal{E}$ with a bundle of the form $\mathcal{E}(n)$.

14.4.1 Loci of bundles on $\mathbb{P}^1$ with given splitting type

To continue the analysis we need some basic facts about how vector bundles on $\mathbb{P}^1$ behave in families. Let B be a connected scheme. By a *family of vector bundles on $\mathbb{P}^1$ with base B* we mean a vector bundle $\mathcal{E}$ on the product $B \times \mathbb{P}^1$. The rank of the bundle $\mathcal{E}_t$ on $\mathbb{P}^1$ that is the restriction of $\mathcal{E}$ to the fiber $\{t\} \times \mathbb{P}^1$ is locally constant in t since $\mathcal{E}$ is locally trivial, and by the theorem on cohomology and base change (Theorem B.5) the Euler characteristic, and thus the degree, is too. Since we have assumed that B is connected, both these are constant. Denote the common rank by r and the common degree by d.

The actual decomposition $\mathcal{E}_t \cong \bigoplus \mathcal{O}_{\mathbb{P}^1}(a_i(t))$, however, may vary with t; we want to describe the possible variation. After rearranging the summands, the decomposition of $\mathcal{E}_t$ is given by an increasing sequence of integers with sum d:

$$a = (a_1, \ldots, a_r) \quad \text{with} \quad a_1 \le a_2 \le \cdots \le a_r \in \mathbb{Z} \quad \text{and} \quad \sum a_i = d.$$

We partially order such sequences by initial subsums: we say

$$a \le b \quad \text{if} \quad \sum_{i=1}^{j} a_i \le \sum_{i=1}^{j} b_i \quad \text{for all } j = 1, \ldots, r;$$

and we say that $a < b$ if $a \le b$ and $a \ne b$. For example, if $a_{i+1} > a_i + 1$ for any $i < r$, then we can form a larger sequence by the replacements

$$a_i \leftarrow a_i + 1, \qquad a_{i+1} \leftarrow a_{i+1} - 1.$$

Thus there is a unique maximal sequence for a given r and d, determined by the condition $|a_i - a_j| \leq 1$ for all i, j. Equivalently, this largest sequence is the "most balanced:" it is the unique sequence of the form $(a, \ldots, a, a+1, \ldots, a+1)$. On the other hand, there is no minimal sequence; for example, with $r = 2, d = 0$ we have $(0, 0) > (-1, 1) > (-2, 2) > \cdots$. But if we impose an upper bound on a_r, say $a_r \leq e$, then the partially ordered set is finite. If $e \gg d$ the set has unique minimal element $(-(r-1)e + d, e, \ldots, e)$.

As a measure of the deviation of a given sequence a from the most balanced, we will set

$$u(a) = \sum_{i < j} \max\{a_j - a_i - 1, 0\}.$$

The quantity $u(a)$ should be thought of as the "expected codimension of the locus of bundles of splitting type a," as explained by the following result:

Theorem 14.7. *Let $\mathcal{E}$ be as above a vector bundle on $B \times \mathbb{P}^1$. If for each sequence a we set*

$$\Gamma_a = \{t \in B \mid \mathcal{E}_t \cong \bigoplus \mathcal{O}_{\mathbb{P}^1}(a_i)\},$$

then:

(a) For each a, the locus

$$\Psi_a = \bigcup_{a' \leq a} \Gamma_{a'}$$

is closed in B (so that in particular Γ_a is a locally closed subset of B).
(b) The codimension of Γ_a in B is at most $u(a)$.

We will prove (a) completely, but we prove (b) only for $r = 2$, the case we will use.

Proof: (a) Consider the function on B

$$\mu(t) = \max\{m \mid h^0(\mathcal{E}_t(-m)) > 0\}.$$

Since $h^0(\mathcal{E}_t(-m))$ is upper-semicontinuous in t, the function μ is as well; this shows that the degree $a_r(t)$ of the largest summand of $\mathcal{E}_t$ is upper-semicontinuous. Applying this to the bundle $\wedge^k(\mathcal{E}_t)$, we see that the function

$$\mu_k(t) = \max\{m \mid h^0((\wedge^k \mathcal{E}_t)(-m)) > 0\} = a_r(t) + \cdots + a_{r-k+1}(t)$$

is likewise upper-semicontinuous, and correspondingly

$$d - \mu_{r-k}(t) = a_1(t) + \cdots + a_k(t)$$

is lower-semicontinuous, establishing part (a).

(b) ($r = 2$) Suppose that $\mathcal{E}$ is a bundle of rank 2 on $B \times \mathbb{P}^1$ and let $b \in B$ be a point. The function $u(a)$ is invariant under the addition of a fixed quantity to all the a_i, so we can twist the bundle $\mathcal{E}$ by the pullback of a line bundle on $\mathbb{P}^1$ without affecting the truth of the statement; after such a twist we can assume that the fiber at b has the form $\mathcal{E}_b \cong \mathcal{O}_{\mathbb{P}^1} \oplus \mathcal{O}_{\mathbb{P}^1}(n)$ for some $n \geq 0$. If $n = 0$ there is nothing to prove, so we may assume $n \geq 1$, and we must show that the locus $\Gamma_{(0,n)}$ has codimension at most $u(0, n) = n - 1$ near b.

Since $h^1(\mathcal{E}_b) = 0$, the section $(1, 0)$ of $\mathcal{E}_b$ extends to a nowhere-zero section of $\mathcal{E}$ in a neighborhood of the fiber $\{b\} \times \mathbb{P}^1$. Replacing B by a suitably small open neighborhood of $b \in B$, then, we have a sequence

$$0 \longrightarrow \mathcal{O}_{B \times \mathbb{P}^1} \longrightarrow \mathcal{E} \longrightarrow \mathcal{O}_{B \times \mathbb{P}^1}(n) \longrightarrow 0,$$

where we write $\mathcal{O}_{B \times \mathbb{P}^1}(n)$ for the pullback of $\mathcal{O}_{\mathbb{P}^1}(n)$ to $B \times \mathbb{P}^1$.

Now, a sequence

$$0 \longrightarrow \mathcal{O}_{\mathbb{P}^1} \longrightarrow \mathcal{E} \overset{\alpha}{\longrightarrow} \mathcal{O}_{\mathbb{P}^1}(n) \longrightarrow 0$$

splits if and only if there exists a bundle map $\varphi : \mathcal{O}_{\mathbb{P}^1}(n) \to \mathcal{E}$ such that $\alpha \circ \varphi$ is the identity on $\mathcal{O}_{\mathbb{P}^1}(n)$. Accordingly, consider the exact sequence of bundles on $B \times \mathbb{P}^1$

$$0 \to \mathcal{H}om(\mathcal{O}_{B \times \mathbb{P}^1}(n), \mathcal{O}_{B \times \mathbb{P}^1}) \to \mathcal{H}om(\mathcal{O}_{B \times \mathbb{P}^1}(n), \mathcal{E})$$
$$\to \mathcal{H}om(\mathcal{O}_{B \times \mathbb{P}^1}(n), \mathcal{O}_{B \times \mathbb{P}^1}(n)) \to 0$$

and the coboundary map

$$\pi_*\big(\mathcal{H}om(\mathcal{O}_{B \times \mathbb{P}^1}(n), \mathcal{O}_{B \times \mathbb{P}^1}(n))\big) \overset{\delta}{\longrightarrow} R^1\pi_*(\mathcal{H}om(\mathcal{O}_{B \times \mathbb{P}^1}(n), \mathcal{O}_{B \times \mathbb{P}^1}))$$

appearing in the associated long exact sequence of sheaves on B. If we let $\sigma = \delta(\mathrm{id})$ be the image in

$$R^1\pi_*(\mathcal{H}om(\mathcal{O}_{B \times \mathbb{P}^1}(n), \mathcal{O}_{B \times \mathbb{P}^1}))$$

of the identity section of $\pi_*\big(\mathcal{H}om(\mathcal{O}_{B \times \mathbb{P}^1}(n), \mathcal{O}_{B \times \mathbb{P}^1}(n))\big)$, then the zero locus $(\sigma) \subset B$ of σ will be contained in the stratum $\Gamma_{(0,n)}$; since

$$R^1\pi_*(\mathcal{H}om(\mathcal{O}_{B \times \mathbb{P}^1}(n), \mathcal{O}_{B \times \mathbb{P}^1}))$$

is locally free of rank $n - 1$, it follows that the codimension of $\Gamma_{(0,n)}$ in B is at most $n - 1$, as required. $\qquad\square$

One can also realize the Γ_a as pullbacks of strata in a family of vector bundles directly: After suitable twisting, every member $\mathcal{E}_b$ of the family will be an extension of the form

$$0 \longrightarrow \mathcal{O}_{\mathbb{P}^1}^{r-1} \longrightarrow \mathcal{E}_b \longrightarrow \mathcal{O}_{\mathbb{P}^1}(d) \longrightarrow 0,$$

and the family over B will locally be a pullback of a family defined over

$$B' := \operatorname{Ext}^1_{\mathbb{P}^1}(\mathcal{O}_{\mathbb{P}^1}(d), \mathcal{O}^{r-1}_{\mathbb{P}^1}) \cong \mathbb{A}^{(r-1)d}.$$

The codimension of the locus Γ_a in B' is exactly $u(a)$. For a general study, including a conjecture on the equations of the strata Γ_a in B', see Eisenbud and Schreyer [2008, Example 5.2].

14.4.2 Jumping lines of bundles of rank 2 on $\mathbb{P}^2$

Let $\mathcal{E}$ be a vector bundle of rank 2 on $\mathbb{P}^2$. The nature of the locus of jumping lines depends on whether $c_1(\mathcal{E}) \in A^1(\mathbb{P}^2) \cong \mathbb{Z}$ is even or odd. In the even case the expected dimension of the locus in 1, and we will compute the degree of the curve of jumping lines; for a complete treatment see Barth [1977]. In the odd case the expected dimension is 0, and we will compute the degree of this finite scheme; for a complete treatment see Hulek [1979].

Vector bundles with even first Chern class

Let $c_1(\mathcal{E}) = 2k\zeta$, where $\zeta \in A^1(\mathbb{P}^2)$ is the class of a line. We want to think of the restrictions of $\mathcal{E}$ to each line in $\mathbb{P}^2$ in turn as a family of vector bundles on $\mathbb{P}^1$, parametrized by the Grassmannian $\mathbb{G}(1, 2) \cong \mathbb{P}^{2*}$. Let

$$\Phi = \{(L, p) \in \mathbb{P}^{2*} \times \mathbb{P}^2 \mid p \in L\}$$

be the universal line, and let $\pi_1 : \Phi \to \mathbb{P}^{2*}$ and $\pi_2 : \Phi \to \mathbb{P}^2$ be the projections. We can view $\pi_1 : \Phi \to \mathbb{P}^{2*}$ as the projectivization of the universal rank-2 subbundle $\mathcal{S}$ on $\mathbb{P}^{2*} = \mathbb{G}(1, 2)$. We let

$$\zeta = c_1(\mathcal{O}_{\mathbb{P}\mathcal{S}}(1)) \in A^1(\Phi)$$

be the tautological class; note that this is also the pullback to Φ of the hyperplane class $\zeta \in A^1(\mathbb{P}^2)$ on $\mathbb{P}^2$. Similarly, we denote by α both the hyperplane class on $\mathbb{P}^{2*}$ and its pullback to Φ. The Chow ring of Φ is given by

$$A(\Phi) = \mathbb{Z}[\alpha, \zeta]/(\alpha^3, \zeta^2 - \alpha\zeta + \alpha^2).$$

Note that $\zeta^3 = 0$; this follows either from the relations above or from the fact that ζ is the pullback of a class from $A^1(\mathbb{P}^{2*})$. We see also that the pushforward map $\pi_{1*} : A(\Phi) \to A(\mathbb{P}^{2*})$ is given by

$$\alpha^2 \mapsto 0, \quad \alpha\zeta \mapsto \alpha, \quad \zeta^2 \mapsto \alpha.$$

To realize the restrictions of $\mathcal{E}$ to the lines in $\mathbb{P}^2$ as a family, we consider the pullback bundle

$$\mathcal{F} = \pi_2^*(\mathcal{E})$$

on Φ.

We now assume that the "expected" codimensions of the loci Γ_a of Theorem 14.7 are actually attained; that is, that for an open dense subset U of $L \in \mathbb{P}^{2*}$, the restriction of $\mathcal{F}$ to the fiber over L splits as

$$\mathcal{E}|_L = \mathcal{F}|_{\pi_2^{-1}(L)} \cong \mathcal{O}_L(k) \oplus \mathcal{O}_L(k),$$

that there is a codimension-1 locus — a curve — $C \subset \mathbb{P}^{2*}$ of lines L such that

$$\mathcal{E}|_L = \mathcal{F}|_{\pi_2^{-1}(L)} \cong \mathcal{O}_L(k+1) \oplus \mathcal{O}_L(k-1),$$

and that no other splitting types occur.

To get some information about C we replace $\mathcal{E}$ by $\mathcal{E}' = \mathcal{E} \otimes \mathcal{O}_{\mathbb{P}^2}(-k-1)$ and consider $\mathcal{F}' = \pi_2^* \mathcal{E}'$. The restriction of $\mathcal{E}'$ to L will split as

$$\mathcal{E}'|_L \cong \begin{cases} \mathcal{O}_L(-1)^{\oplus 2} & \text{if } L \in U, \\ \mathcal{O}_L \oplus \mathcal{O}_L(-2) & \text{if } L \in C, \end{cases}$$

and thus

$$h^0(\mathcal{E}'|_L) = h^1(\mathcal{E}'|_L) = \begin{cases} 0 & \text{if } L \in U, \\ 1 & \text{if } L \in C. \end{cases}$$

By the theorem on cohomology and base change (Theorem B.5), the direct image $\pi_{1*}(\mathcal{F}')$ is 0: over the open subset $U \subset \mathbb{P}^{2*}$ there are no sections, and, since $\pi_{1*}(\mathcal{F}')$ is torsion-free, it follows that $\pi_{1*}(\mathcal{F}') = 0$. However, the jump in the cohomology groups $H^i(\mathcal{E}'|_L)$ is reflected in the higher direct image $R^1\pi_{1*}(\mathcal{F}')$; this will be a sheaf supported on the curve C. It follows that

$$c_1(R^1\pi_{1*}(\mathcal{F}'))$$

will give us the degree of this curve (counting each component with some positive multiplicity). The class $c_1(R^1\pi_{1*}(\mathcal{F}'))$ is something we can calculate from Grothendieck Riemann–Roch.

To make the computation, we first need the Todd class of the relative tangent bundle of $\pi_1 : \Phi \to \mathbb{P}^{2*}$. From Theorem 11.4,

$$c_1(\mathcal{T}^v_{\Phi/\mathbb{P}^{2*}}) = -\alpha + 2\zeta,$$

and correspondingly

$$\mathrm{Td}(\mathcal{T}^v_{\Phi/\mathbb{P}^{2*}}) = 1 + \tfrac{1}{2}(-\alpha + 2\zeta) + \tfrac{1}{12}(-\alpha + 2\zeta)^2$$

(note that since $\mathcal{T}^v_{\Phi/\mathbb{P}^{2*}}$ is a line bundle, there is no cubic term). If we write the Chern class of the bundle $\mathcal{F}'$ as

$$c(\mathcal{F}') = -2\zeta + e\zeta^2,$$

then we have

$$\begin{aligned} \mathrm{Ch}(\mathcal{F}') &= \mathrm{rank}(\mathcal{F}') + c_1(\mathcal{F}') + \tfrac{1}{2}(c_1^2(\mathcal{F}') - 2c_2(\mathcal{F}')) \\ &= 2 - 2\zeta + (2 - e)\zeta^2. \end{aligned}$$

Now the Grothendieck Riemann–Roch formula tells us that

$$
\begin{aligned}
[C] &= \mathrm{Ch}_1(R^1\pi_{1*}(\mathcal{F}')) \\
&= -\pi_{1*}\{(2 - 2\zeta + (2-e)\zeta^2)(1 + \tfrac{1}{2}(-\alpha + 2\zeta) + \tfrac{1}{12}(-\alpha + 2\zeta)^2)\}_2 \\
&= -\pi_{1*}(\tfrac{1}{6}(-\alpha + 2\zeta)^2 - \zeta(-\alpha + 2\zeta) + (2-e)\zeta^2) \\
&= (e - 1)\zeta;
\end{aligned}
$$

or, in other words, the degree of the curve of jumping lines, counted with multiplicities, is $e - 1$. Given that $\mathcal{E}' = \mathcal{E} \otimes \mathcal{O}_{\mathbb{P}^2}(-k - 1)$, we have

$$
e = c_2(\mathcal{E}') = c_2(\mathcal{E}) - (k^2 - 1)
$$

and so for our original bundle $\mathcal{E}$ we have:

Proposition 14.8. *If $\mathcal{E}$ is a vector bundle of rank 2 on $\mathbb{P}^2$ with even first Chern class $2k\zeta$, and the restriction of $\mathcal{E}$ to a general line $L \subset \mathbb{P}^2$ is balanced (that is, $\mathcal{E}|_L \cong \mathcal{O}_L(k) \oplus \mathcal{O}_L(k)$), then the degree of the curve of jumping lines, counted with multiplicities as above, is $c_2(\mathcal{E}) - k^2 = c_2(\mathcal{E}) - c_1^2(\mathcal{E})/4$.*

In fact there is a natural scheme structure on the curve C, defined by a Fitting invariant of $R^1\pi_{1*}(\mathcal{F}')$, that determines the multiplicities. See Maruyama [1983] for a treatment and further references. Although we have not described the multiplicities or given conditions under which they are equal to 1, Proposition 14.8 has nontrivial content: For example, we may deduce that the degree of C is at most $c_2(\mathcal{E}) - c_1^2(\mathcal{E})/4$, and if $c_2(\mathcal{E}) - c_1^2(\mathcal{E})/4 \neq 0$ we may conclude that the curve of jumping lines is nonempty. We give an explicit example below.

Vector bundles with odd first Chern class

We now assume that $c_1(\mathcal{E}) = (2k + 1)\zeta$ and, as before, that the "expected" codimensions of the loci Γ_a of Theorem 14.7 are attained. Thus, over an open dense subset U of $L \in \mathbb{P}^{2*}$, the restriction of $\mathcal{F}$ to the fiber over L splits as

$$
\mathcal{E}|_L = \mathcal{F}|_{\pi_2^{-1}(L)} \cong \mathcal{O}_L(k) \oplus \mathcal{O}_L(k + 1).
$$

But now our assumption means that the locus of lines L such that

$$
\mathcal{E}|_L = \mathcal{F}|_{\pi_2^{-1}(L)} \cong \mathcal{O}_L(k - 1) \oplus \mathcal{O}_L(k + 2)
$$

is a finite set Γ. Again, no other splitting types occur.

To get some information about Γ, we again twist the vector bundle $\mathcal{E}$ so that the jump in splitting type is reflected in the ranks of the cohomology groups of its restriction to lines. Specifically, we replace $\mathcal{E}$ by $\mathcal{E}' = \mathcal{E} \otimes \mathcal{O}_{\mathbb{P}^2}(-k - 1)$, so that the restriction of the bundle $\mathcal{E}'$ to L will split as

$$
\mathcal{E}'|_L \cong \begin{cases} \mathcal{O}_L \oplus \mathcal{O}_L(-1) & \text{if } L \in U, \\ \mathcal{O}_L(1) \oplus \mathcal{O}_L(-2) & \text{if } L \in \Gamma. \end{cases}
$$

We now have

$$h^0(\mathcal{E}'|_L) = \begin{cases} 1 & \text{if } L \in U, \\ 2 & \text{if } L \in \Gamma, \end{cases}$$

and correspondingly

$$h^1(\mathcal{E}'|_L) = \begin{cases} 0 & \text{if } L \in U, \\ 1 & \text{if } L \in \Gamma. \end{cases}$$

As before, this means that the sheaf $R^1\pi_{1*}(\mathcal{F}')$ is supported on the exceptional locus, in this case Γ, and thus the length of this sheaf counts the number of points in Γ, with some nonzero multiplicities.

An important difference between this case and that of even first Chern class is that here the direct image $\pi_{1*}(\mathcal{F}')$ is nonzero. By the theorem on cohomology and base change it is locally free of rank 1 away from Γ; in fact, it is locally free everywhere, as we will now show.

Let $L \in U$ be a point in an open subset of $\mathbb{P}^{2*}$ such that the splitting of $\mathcal{F}'|_L$ is $\mathcal{F}'|_L = \mathcal{O}_{\mathbb{P}^1}(-2) \oplus \mathcal{O}_{\mathbb{P}^1}(1)$ and the splitting of $\mathcal{F}'|_{L'}$ for $L' \in U$ other than L is $\mathcal{F}'|_{L'} = \mathcal{O}_{\mathbb{P}^1}(-1) \oplus \mathcal{O}_{\mathbb{P}^1}$. There is a short exact sequence

$$0 \longrightarrow \mathcal{O}_{\Phi_U}(-2)^2 \longrightarrow \mathcal{O}_{\Phi_U}(-2) \oplus \mathcal{O}_{\Phi_U}(-1)^3 \longrightarrow \mathcal{F}' \longrightarrow 0$$

(see Exercise 14.26). Applying $R\pi_{1*}$, we get an exact sequence of sheaves on U of the form

$$0 \longrightarrow \pi_{1*}\mathcal{F}' \longrightarrow R^1\pi_{1*}\mathcal{O}_{\Phi_U}(-2)^2 \longrightarrow R^1\pi_{1*}\mathcal{O}_{\Phi_U}(-2) \longrightarrow R^1\pi_{1*}\mathcal{F}' \longrightarrow 0.$$

This can be written as

$$0 \longrightarrow \pi_{1*}\mathcal{F}' \longrightarrow \mathcal{O}_U^2 \xrightarrow{(f,g)} \mathcal{O}_U \longrightarrow R^1\pi_{1*}\mathcal{F}' \longrightarrow 0,$$

where the common zero locus of f and g is the locus of jumping lines. Given that the locus of jumping lines has codimension 2, it follows that $\pi_{1*}\mathcal{F}'$ is locally free of rank 1, and thus

$$\pi_{1*}(\mathcal{F}') \cong \mathcal{O}_{\mathbb{P}^{2*}}(m)$$

for some m; the value of m will also emerge in the calculation.

Note that in a neighborhood of a point $p \in \Gamma \subset \mathbb{P}^{2*}$ corresponding to a line $L \subset \mathbb{P}^2$ the comparison map

$$\pi_{1*}(\mathcal{F}')|_p \to H_0(\mathcal{F}'|_L)$$

will be zero: none of the sections of $\mathcal{F}'|_L$ extend to a neighborhood of L.

We now apply Corollary 14.6. If we write

$$c(\mathcal{E}') = 1 - \zeta + e\zeta^2,$$

then the Chern character of $\mathcal{E}'$, and of its pullback $\mathcal{F}'$ to Φ, is

$$\mathrm{Ch}(\mathcal{F}') = 2 - \zeta + \left(\tfrac{1}{2} - e\right)\zeta^2.$$

By Corollary 14.6,

$$\mathrm{Ch}(\pi_{1*}\mathcal{F}') - \mathrm{Ch}(R^1\pi_{1*}\mathcal{F}')$$
$$= \pi_{1*}\left[\left(2 - \zeta + \left(\tfrac{1}{2} - e\right)\zeta^2\right)\left(1 + \tfrac{1}{2}(-\alpha + 2\zeta) + \tfrac{1}{12}(-\alpha + 2\zeta)^2\right)\right]$$
$$= 1 - e\alpha + \frac{e}{2}\alpha^2.$$

From the isomorphism $\pi_{1*}(\mathcal{F}') \cong \mathcal{I}_\Gamma(m)$, we get

$$\mathrm{Ch}(\pi_{1*}\mathcal{F}') = 1 + m\alpha + \frac{m^2}{2}\alpha^2.$$

Since

$$\mathrm{Ch}(R^1\pi_{1*}\mathcal{F}') = \mathrm{Ch}(\mathcal{O}_\Gamma) = [\Gamma] = \gamma\alpha^2,$$

we have

$$1 - e\alpha + \frac{e}{2}\alpha^2 = 1 + m\alpha + \left(\frac{m^2}{2} - 2\gamma\right)\alpha^2.$$

From the degree-1 terms, we see that $m = -e$, and then equating the degree-2 terms we find that the degree of Γ is

$$\gamma = \frac{e^2 - e}{2}.$$

To express this in terms of the Chern classes of our original bundle $\mathcal{E}$, we observe that, by the splitting principle, for an arbitrary bundle $\mathcal{E}$ of rank 2 on $\mathbb{P}^2$ with first Chern class $c_1(\mathcal{E}) = (2k + 1)\zeta$, the degree of the second Chern class of the twist $\mathcal{E}' = \mathcal{E}(-k - 1)$ is

$$e = c_2(\mathcal{E}) - (k + 1)c_1(\mathcal{E})\zeta + (k + 1)^2\zeta^2$$
$$= c_2(\mathcal{E}) - k(k + 1).$$

Proposition 14.9. *Let $\mathcal{E}$ be a vector bundle of rank 2 on $\mathbb{P}^2$ with odd first Chern class $(2k + 1)\zeta$. Under the hypotheses introduced above, the locus $\Gamma \subset \mathbb{P}^{2*}$ of jumping lines is a set of $(e^2 - e)/2$ points counted with multiplicities, where $e = c_2(\mathcal{E}) - k(k + 1)$.*

As in the case of Proposition 14.8, this statement implicitly invokes a scheme structure on Γ that we have not described; again, however, even in the absence of this we can deduce that the number of jumping lines is at most $(e^2 - e)/2$, and if $e^2 - e \neq 0$ we may conclude that the locus of jumping lines is nonempty.

14.4.3 Examples

Suppose that F_0, F_1 and F_2 are general homogeneous polynomials of degree d on $\mathbb{P}^2$, and let $\mathcal{E}$ be the kernel of the bundle map

$$\mathcal{O}_{\mathbb{P}^2}(-d)^{\oplus 3} \xrightarrow{(F_0, F_1, F_2)} \mathcal{O}_{\mathbb{P}^2}.$$

Note that since F_0, F_1 and F_2 have no common zeros this is a surjection, so that $\mathcal{E}$ is locally free of rank 2.

By Whitney's formula we have

$$c(\mathcal{E}) = \frac{1}{(1 - d\zeta)^3} = 1 + 3d\zeta + 6d^2\zeta^2 \in A(\mathbb{P}^2).$$

For odd degree $d = 2k + 1$, Proposition 14.9 suggests that the number of jumping lines should be the binomial coefficient $\binom{3k^2+3k+1}{2}$.

For example, when $d = 1$ (hence $k = 0$) this number is 0. It is easy to see that this is correct: If $L = \mathbb{P}^1$ is a line then (after applying a suitable element of GL_3 to $\mathcal{O}_{\mathbb{P}^2}(-d)^{\oplus 3}$, that is, replacing F_0, F_1 and F_2 with independent linear combinations) we may assume that L is given by the equation $F_2 = 0$; then F_0, F_1 necessarily restrict to independent linear forms $\overline{F}_1$, $\overline{F}_2$ on L. Thus

$$\mathcal{E}|_L = \mathcal{O}_{\mathbb{P}^1}(-1)^{\oplus 3} \xrightarrow{(\overline{F}_0, \overline{F}_1, 0)} \mathcal{O}_{\mathbb{P}^1} = \mathcal{O}_{\mathbb{P}^1} \oplus \mathcal{O}_{\mathbb{P}^1}(-1).$$

Put differently, the bundle $\mathcal{E}$ is invariant under the transitive group $PGL(3)$ (it is the cotangent bundle of $\mathbb{P}^2$), and this group must preserve the locus of jumping lines.

When $d = 3$, Proposition 14.9 suggests that the number of jumping lines should be 21; we will see how to verify this count in Exercise 14.24.

For even degree $d = 2k$, we would expect from Proposition 14.8 that the curve of jumping lines, with multiplicity, will have degree $3k^2 = \frac{3}{4}d^2$. In case $d = 2$ this number is 3; in Exercise 14.23 we explain how to verify that it is actually a nonsingular cubic.

14.5 Application: invariants of families of curves

We close this chapter by describing one of the most important applications of the Grothendieck Riemann–Roch formula, a result that is central to the study of the moduli space of curves.

Consider one-parameter families of curves of genus g; that is, flat morphisms $\alpha : X \to B$ from a surface X to a curve B, with fibers curves of genus g. How can we measure how much the isomorphism class of X_b is varying with $b \in B$?

In order to address this meaningfully, we have to make some restrictions on the type of curves that appear as fibers of α (as the footnote below will illustrate). For simplicity, in the present discussion we will assume that X and B are smooth, and that the fibers X_b of α are irreducible with at worst nodes as singularities. These conditions could be relaxed: we could drop the requirement that X and B be smooth, and we could weaken the hypothesis that X_b be irreducible to require only that X_b be *stable* (see for example Harris and Morrison [1998]) — but our goal here is just to acquaint the reader with the basic ideas.

A parenthetical note: This discussion touches on a topic that, had we the energy and expertise and you the willingness to countenance an 800-page book, we would take up: intersection theory on *algebraic stacks*. Briefly, the category of stacks is an enlargement of the category of schemes — schemes form a full subcategory of stacks — in which we can find parameter spaces that may not exist in the category of schemes. In the present setting, the correct framework for the calculations we are about to undertake is to view δ, λ and κ as classes in $A^1(\overline{\mathcal{M}}_g)$, where $\overline{\mathcal{M}}_g$ is the *moduli stack of stable curves*. A discussion of these ideas can be found in Harris and Morrison [1998, Chapters 2 and 3].

We consider three natural ways of quantifying the variation in the family:

(a) *Number of nodes*: If the family α were trivial, that is, if α were the projection of a product $C \times B$ to B, there would be no singular fibers. Thus the number of singular fibers is a measure of nontriviality.[1] It turns out (given that the total space X is smooth) to be natural to count a singular fiber with multiplicity equal to the number of nodes that appear in it. Thus we let $\delta(\alpha)$ be the total number of nodes appearing in the fibers of α. (Again, we are being unnecessarily restrictive here; it is possible to assign multiplicities without the assumption that X is smooth, but it requires additional complication.)

(b) *Degree of the Hodge bundle*: Recall that any projective curve C has a *dualizing sheaf* ω_C, which is the cotangent bundle if the curve is smooth and is defined in general to be $\omega_C := \mathrm{Ext}^{n-1}_{\mathbb{P}^n}(\mathcal{O}_C, \omega_{\mathbb{P}^n})$ for any embedding $C \subset \mathbb{P}^n$. Moreover, the dimension of $H^0(\omega_C)$ is equal to the arithmetic genus $p_a(C)$ of C.

By the theorem on cohomology and base change (Theorem B.5), the quantity

$$\chi(\mathcal{O}_{X_b}) = h^0(\mathcal{O}_{X_b}) - h^1(\mathcal{O}_{X_b}) = 1 - p_a(X_b)$$

is independent of the point $b \in B$, so $h^0(\omega_{X_b})$ is independent of b. The *relative dualizing sheaf*

$$\omega_{X/B} = \omega_X \otimes \alpha^* \omega_B^{-1}$$

of the family restricts on each fiber X_b to ω_{X_b}. Thus

$$\mathcal{E} := \alpha_*(\omega_{X/B})$$

is a vector bundle on B of rank g. It is called the *Hodge bundle* of the family.

Set

$$\lambda(\alpha) = c_1(\mathcal{E}),$$

the *degree of the Hodge bundle*. If the family were trivial then $\mathcal{E}$ would be the trivial bundle, so $\lambda(\alpha)$ would be 0.

---

[1] This would not be true if we did not assume that the fibers were irreducible: You could take a trivial family $B \times C \to B$ and blow-up a point in $B \times C$ to arrive at a family of curves of constant modulus with one singular fiber. The logic would be valid if we assumed all fibers to be *stable curves*; in the present context we avoid getting into a discussion of this notion by assuming all fibers to be irreducible.

(c) *Self-intersection of the relative canonical divisor*: Set

$$\kappa(\alpha) = c_1(\omega_{X/B})^2.$$

If the family were trivial then the relative cotangent bundle $\omega_{X/B}$ would be the pullback of ω_C via projection on the second factor, so its self-intersection number $\kappa(\alpha)$ would be 0.

The *Mumford relation* is a linear relation among these three numerical invariants:

Theorem 14.10. *If* $\alpha : X \to B$ *is a morphism from a smooth projective surface X to a smooth projective curve B whose fibers are irreducible curves g having at most nodes as singularities, then*

$$\lambda(\alpha) = \frac{\kappa(\alpha) + \delta(\alpha)}{12}.$$

We will give the proof in Section 14.5.2; see Harris and Morrison [1998, Section 3E] for a proof in greater generality and Mumford [1983] for many extensions. First, however, we show how to compute an example; further examples are in the exercises, and together these show that the Mumford relation is the only linear relation satisfied by δ, κ and λ.

Though there are no other linear relations, these invariants satisfy some subtle inequalities. For example, if $X \to B$ is a one-parameter family of irreducible curves of genus g having at most nodes as singularities and not all singular, then

$$\delta \le \left(8 + \frac{4}{g}\right)\lambda;$$

this is sharp, as shown by the example of Exercise 14.31. For a discussion of these questions, with references to the literature, see Harris and Morrison [1998].

14.5.1 Example: pencils of quartics in the plane

We will compute the invariants δ, κ, λ for a general pencil

$$\{C_t = V(t_0 F + t_1 G) \subset \mathbb{P}^2\}$$

of plane quartic curves. Set

$$X = \{(t, p) \in \mathbb{P}^1 \times \mathbb{P}^2 \mid p \in C_t\}.$$

We observe that X is smooth—the projection $\beta : X \to \mathbb{P}^2$ on the second factor expresses X as the blow-up of the plane at the 16 base points $F = G = 0$ of the pencil. We showed in Section 7.1 that every fiber C_t of $\alpha : X \to \mathbb{P}^1$ is either smooth or is irreducible with a single node; thus the family $\alpha : X \to \mathbb{P}^1$ satisfies the conditions of our discussion.

In Chapter 7 we computed that the number of nodes in the fibers of α is

$$\delta(\alpha) = 27.$$

To compute $\kappa(\alpha)$, let L be the preimage under β of a line in $\mathbb{P}^2$ and let E be the sum of the 16 exceptional divisors of β, the preimages of the base points in $\mathbb{P}^2$ of the family α. Let $l, e \in A^1(X)$ denote their classes. In these terms, the class of the fibers C_t of the projection α — that is, the pullback under α of the class η of a point in $\mathbb{P}^1$ — is given by

$$[C_t] = \alpha^*(\eta) = 4l - e.$$

In particular, this implies that

$$\alpha^* K_{\mathbb{P}^1} = -2\alpha^*(\eta) = -8l + 2e.$$

On the other hand, from the blow-up map $\beta : X \to \mathbb{P}^2$ we see that

$$K_X = \beta^*(K_{\mathbb{P}^2}) + E = -3L + E.$$

Thus, the first Chern class of the relative dualizing sheaf $\omega_{X/\mathbb{P}^1}$ is

$$c_1(\omega_{X/\mathbb{P}^1}) = c_1(K_X) - \alpha^* c_1(K_{\mathbb{P}^1}) = 5l - e$$

and the invariant $\kappa(\alpha)$ is given by

$$\kappa(\alpha) = c_1(\omega_{X/\mathbb{P}^1})^2 = 25 - 16 = 9.$$

To compute λ we describe the Hodge bundle: Since $5L - E = (4L - E) + L$, we can write the dualizing sheaf as

$$\omega_{X/\mathbb{P}^1} = \mathcal{O}_X(5L - E) = \alpha^* \mathcal{O}_{\mathbb{P}^1}(1) \otimes \beta^* \mathcal{O}_{\mathbb{P}^2}(1).$$

Now, the pushforward $\alpha_*(\beta^* \mathcal{O}_{\mathbb{P}^2}(1))$ is simply the trivial bundle on $\mathbb{P}^1$ with fiber $H^0(\mathcal{O}_{\mathbb{P}^2}(1))$; by Proposition B.7 we have then

$$\begin{aligned}
\alpha_*(\omega_{X/\mathbb{P}^1}) &= \alpha_*(\alpha^* \mathcal{O}_{\mathbb{P}^1}(1) \otimes \beta^* \mathcal{O}_{\mathbb{P}^2}(1)) \\
&= \mathcal{O}_{\mathbb{P}^1}(1) \otimes H^0(\mathcal{O}_{\mathbb{P}^2}(1)) \\
&\cong \mathcal{O}_{\mathbb{P}^1}(1)^3,
\end{aligned}$$

and correspondingly

$$\lambda(\alpha) = c_1(\alpha_*(\omega_{X/\mathbb{P}^1})) = 3.$$

Another, more concrete way to arrive at this is suggested in Exercise 14.27.

To summarize, we have

$$\delta(\alpha) = 27, \quad \kappa(\alpha) = 9 \quad \text{and} \quad \lambda(\alpha) = 3,$$

verifying Mumford's relation in this case.

Further examples of one-parameter families of curves for which we can verify the Mumford relation are given in Exercises 14.28–14.30.

14.5.2 Proof of the Mumford relation

Proof of Theorem 14.10: The Grothendieck Riemann–Roch theorem tells us that

$$\sum_i (-1)^i \operatorname{Ch}(R^i \alpha_* \omega_{X/B}) = \alpha_* [\operatorname{Ch}(\omega_{X/B}) \operatorname{Td}(\mathcal{T}^v_{X/B})],$$

and we recall that $\operatorname{Td}(\mathcal{T}^v_{X/B}) = \operatorname{Td}(\mathcal{T}_X)/\alpha^*(\operatorname{Td}(\mathcal{T}_B))$. Thus

$$
\begin{aligned}
c_1(\alpha_* \omega_{X/B}) &= \operatorname{Ch}_1(R^0 \alpha_* \omega_{X/B}) \\
&= \operatorname{Ch}_1(R^1 \alpha_* \omega_{X/B}) + \{\alpha_*(\operatorname{Ch}(\omega_{X/B}) \operatorname{Td}(\mathcal{T}^v_{X/B}))\}_0 \\
&= c_1(R^1 \alpha_* \omega_{X/B}) + \alpha_*\{\operatorname{Ch}(\omega_{X/B}) \operatorname{Td}(\mathcal{T}^v_{X/B})\}_0.
\end{aligned}
$$

We will compute the terms on the right.

We first claim that $R^1 \alpha_*(\omega_{X/B}) = \mathcal{O}_B$, and thus $c_1(R^1 \alpha_*(\omega_{X/B})) = 0$. This follows from a generalization of Serre duality due to Grothendieck, which we now describe:

Recall that Serre duality says that for any invertible sheaf $\mathcal{F}$ on a smooth curve C, we have a natural identification

$$H^1(\mathcal{F}) = \operatorname{Hom}(\mathcal{F}, \omega_C)^* = H^0(\mathcal{H}om(\mathcal{F}, \omega_C))^*.$$

Although this is sometimes stated just for nonsingular curves, it holds for any purely one-dimensional scheme C. Moreover, the identification is so natural that, properly formulated, it applies to families of such schemes: If $\alpha : X \to B$ is a family of projective curves of genus g and $\mathcal{F}$ is an invertible sheaf on X whose cohomology groups have constant rank on the fibers of α, then there is a natural isomorphism

$$R^1 \alpha_*(\mathcal{F}) \cong \alpha_*(\mathcal{H}om(\mathcal{F}, \omega_{X/B}))^*;$$

see Barth et al. [2004, Section III.12]. If we apply this to the case of $\mathcal{F} = \omega_{X/B}$, we get

$$R^1 \alpha_*(\omega_{X/B}) = \alpha_*(\mathcal{O}_X) = \mathcal{O}_B.$$

We next compute the necessary Todd classes. Recall the *Hopf index theorem* (Theorem 5.20): The degree of the top Chern class $c_2(\mathcal{T}_X)$ of the tangent bundle of the smooth projective surface X is the topological Euler characteristic of X. Also, the topological Hurwitz formula (Section 7.7) says that the topological Euler characteristic of X is the product of the Euler characteristics $2 - 2g$ of the general fiber of α and the Euler characteristic of the base curve B, plus the total number δ of nodes in the fibers of α. Combining these, we see that

$$
\begin{aligned}
c_2(\mathcal{T}_X) &= (2 - 2g)c_1(\mathcal{T}_B) + \delta \\
&= -c_1(\omega_{X/B}) \cdot \alpha^* c_1(\mathcal{T}_B) + \delta \\
&= (c_1(\mathcal{T}_X) - c_1 \alpha^* \mathcal{T}_B) \cdot \alpha^* c_1(\mathcal{T}_B) + \delta.
\end{aligned}
$$

Abbreviating the expression $\omega_{X/B}$ for the relative dualizing sheaf to simply ω, we can thus express the ratio $c(\mathcal{T}_X)/\alpha^* c(\mathcal{T}_B)$ as

$$\frac{c(\mathcal{T}_X)}{\alpha^* c(\mathcal{T}_B)} = (1 + c_1(\mathcal{T}_X) + c_2(\mathcal{T}_X))(1 - \alpha^* c_1(\mathcal{T}_B))$$

$$= 1 - c_1(\omega) + \delta,$$

where we have used the equality $c_1(\alpha^*(\mathcal{T}_B))^2 = \alpha^*(c_1(\mathcal{T}_B)^2) = 0$. Substituting these classes into the formulas for the Todd classes of degrees 0, 1, and 2, we get

$$\frac{\mathrm{Td}(\mathcal{T}_X)}{\alpha^* \mathrm{Td}(\mathcal{T}_B)} = 1 - \frac{c_1(\omega)}{2} + \frac{c_1(\omega)^2 + \delta}{12}.$$

We now have everything we need to apply Corollary 14.6, and we conclude that

$$\lambda(\alpha) = c_1(\alpha_* \omega)$$

$$= \left\{ \alpha_* \left(\left(1 + c_1(\omega) + \frac{c_1(\omega)^2}{2} \right) \left(1 - \frac{c_1(\omega)}{2} + \frac{c_1(\omega)^2 + \delta}{12} \right) \right) \right\}_0$$

$$= \alpha_* \left(\frac{c_1(\omega)^2}{2} - \frac{c_1(\omega)^2}{2} + \frac{c_1(\omega)^2 + \delta}{12} \right)$$

$$= \frac{\kappa(\alpha) + \delta(\alpha)}{12}. \qquad \square$$

14.6 Exercises

Exercise 14.11. (a) Find the Chern characters of the universal bundles $\mathcal{S}$ and $\mathcal{Q}$ on $\mathbb{G} = \mathbb{G}(1, 3)$.

(b) Use this to find the Chern character of the tangent bundle $\mathcal{T}_{\mathbb{G}} = \mathcal{S}^* \otimes \mathcal{Q}$.

(c) Use this in turn to find the Chern class of $\mathcal{T}_{\mathbb{G}}$.

Exercise 14.12. Verify that the definition of the Chern class of a coherent sheaf given in Section 14.2.1 is well-defined; that is, $c(\mathcal{F})$ does not depend on the choice of resolution.

Exercise 14.13. Let $p \in \mathbb{P}^n$ be a point. Using the Koszul complex, show that the Chern class of the structure sheaf $\mathcal{O}_p$, viewed as a coherent sheaf on $\mathbb{P}^n$, is

$$c(\mathcal{O}_p) = 1 + (-1)^{n-1}(n-1)! \, [p].$$

Exercise 14.14. Let $C \subset \mathbb{P}^3$ be a smooth curve. Find the Chern class of the structure sheaf $\mathcal{O}_C$, viewed as a coherent sheaf on $\mathbb{P}^3$, when:

(a) C is a twisted cubic.

(b) C is an elliptic quartic curve.

(c) C is a rational quartic curve.

Exercise 14.15. Consider the three varieties $X_1 = \mathbb{P}^3$, $X_2 = \mathbb{P}^1 \times \mathbb{P}^2$ and $X_3 = \mathbb{P}^1 \times \mathbb{P}^1 \times \mathbb{P}^1$.

(a) In each case, calculate the degrees of the classes $c_3(\mathcal{T}_{X_i})$, $c_1(\mathcal{T}_{X_i})c_2(\mathcal{T}_{X_i})$ and $c_1(\mathcal{T}_{X_i})^3$.

(b) Show that the resulting 3×3 matrix is nonsingular.

(c) Show that the Euler characteristic $\chi(\mathcal{O}_{X_i})$ is 1 for each i.

(d) Given that the Euler characteristic of the structure sheaf of a smooth projective threefold X is expressible as a polynomial of degree 3 in the Chern classes of its tangent bundle, show from the above examples that the polynomial must be

$$\mathrm{Td}_3(c_1, c_2, c_3) = \frac{c_1 c_2}{24}.$$

Exercise 14.16. Verify that the formula of Theorem 14.4 gives the classical Riemann–Roch formula for $n = 1$ or 2, and write down the analogous formula for $n = 3$.

Exercise 14.17. In Section 7.2 we introduced a vector bundle $\mathcal{E}$ on $\mathbb{P}^2$ whose fiber at a point $p \in \mathbb{P}^2$ is the space $H^0(\mathcal{O}_{\mathbb{P}^2}(d)/\mathcal{I}_p^2(d))$ of homogeneous polynomials of degree d, modulo those vanishing to order 2 or more at p; we also described $\mathcal{E}$ as a direct image. Use this and the Grothendieck Riemann–Roch formula to calculate the Chern classes of $\mathcal{E}$.

Exercise 14.18. In Chapter 11 we introduced a vector bundle $\mathcal{E}$ on the universal line

$$\Phi = \{(L, p) \in \mathbb{G}(1, 3) \times \mathbb{P}^3 \mid p \in L\}$$

whose fiber at a point $(L, p) \in \Phi$ is the space $H^0(\mathcal{O}_L(d)/\mathcal{I}_p^5(d))$ of homogeneous polynomials of degree d on L, modulo those vanishing to order 5 or more at p; we also described $\mathcal{E}$ as a direct image. Use this and the Grothendieck Riemann–Roch formula to calculate the Chern classes of $\mathcal{E}$.

Exercise 14.19. Let $C \subset \mathbb{P}^2$ be an irreducible plane curve of degree d. Construct a bundle $\mathcal{E}$ on the dual plane $\mathbb{P}^{2*}$ whose fiber at a point L is the space of sections of the structure sheaf $\mathcal{O}_\Gamma$ of the intersection $\Gamma = C \cap L$, and use the Grothendieck Riemann–Roch formula to calculate the Chern classes of $\mathcal{E}$.

Exercise 14.20. Let $\mathcal{E}$ be an indecomposable vector bundle of rank 2 on $\mathbb{P}^3$ with first Chern class 0, and let

$$\Gamma_1 = \{L \in \mathbb{G}(1, 3) \mid \mathcal{E}|_L \cong \mathcal{O}_L(k) \oplus \mathcal{O}_L(-k) \text{ with } k \geq 1\}.$$

Find the class of the divisor $\Gamma_1 \subset \mathbb{G}(1, 3)$.

Exercise 14.21. With $\mathcal{E}$ as in the preceding problem, let

$$\Gamma_2 = \{L \in \mathbb{G}(1, 3) \mid \mathcal{E}|_L \cong \mathcal{O}_L(k) \oplus \mathcal{O}_L(-k) \text{ with } k \geq 2\}.$$

Find the class of the locus $\Gamma_2 \subset \mathbb{G}(1, 3)$, assuming it has the expected dimension 1.

Exercise 14.22. Now let $\mathcal{E}$ be a vector bundle of rank 2 on $\mathbb{P}^3$ with first Chern class $c_1(\mathcal{E}) = \zeta$ (the hyperplane class), and let

$$\Phi_i = \{L \in \mathbb{G}(1,3) \mid \mathcal{E}|_L \cong \mathcal{O}_L(k+1) \oplus \mathcal{O}_L(-k) \text{ with } k \geq i\}$$

for $i = 1, 2$. Find the classes of the loci $\Phi_i \subset \mathbb{G}(1,3)$, assuming they have the expected codimension $2i$.

Exercise 14.23. Let F_0, F_1, F_2 be three general quadratic forms in three variables, and define a bundle $\mathcal{E}$ on $\mathbb{P}^2$ as the kernel of the surjection

$$\mathcal{O}_{\mathbb{P}^2}(-2)^{\oplus 3} \xrightarrow{(F_0, F_1, F_2)} \mathcal{O}_{\mathbb{P}^2}.$$

Prove the locus C of jumping lines of $\mathcal{E}$ is a nonsingular cubic curve in $\mathbb{P}^2$ as follows:

(a) Show that C is also the locus of jumping lines of

$$\mathcal{E}^* = \mathrm{coker}\left(\mathcal{O}_{\mathbb{P}^2} \xrightarrow{(F_0, F_1, F_2)} \mathcal{O}_{\mathbb{P}^2}(2)^{\oplus 3}\right).$$

(b) Show that $\mathcal{E}^*|_L$ contains a copy of $\mathcal{O}_L(2)$ as a summand if and only if some linear combination of F_0, F_1, F_2 vanishes identically on L.

(c) We may represent each F_i as a general symmetric 3×3 matrix of scalars Q_i. Introducing new variables z_0, z_1, z_2, we see from the previous item that the curve of jumping lines is a double cover of the smooth cubic curve defined in coordinates z_0, z_1, z_2 by the equation $\det(\sum_i z_i Q_i) = 0$.

Exercise 14.24. Let $\mathcal{E}$ be defined as in Exercise 14.23, but now take the F_i to be general cubics (and replace the occurrences of $\mathcal{O}_{\mathbb{P}^2}(-2)$ with $\mathcal{O}_{\mathbb{P}^2}(-3)$). Show that the jumping lines of 0 are exactly the lines in $\mathbb{P}^2$ contained in some curve of the form $\sum a_i F_i = 0$. Show that there are exactly 21 of these by observing that this is the degree of the seven-dimensional variety of reducible plane cubics in the $\mathbb{P}^9$ of all plane cubics.

Exercise 14.25. As an example, let $q : \Bbbk^4 \times \Bbbk^4 \to \Bbbk$ be a nondegenerate skew-symmetric bilinear form. We can define a bundle of rank 2 on $\mathbb{P}^3$ by setting

$$0_p = \langle p \rangle^{\perp} / \langle p \rangle.$$

Describe the locus of jumping lines for such a bundle.

Exercise 14.26. As in Section 14.4.2, let $L \in U$ be a point in an open subset of $\mathbb{P}^{2*}$ such that the splitting of $\mathcal{F}'|_L$ is $\mathcal{F}'|_L = \mathcal{O}_{\mathbb{P}^1}(-2) \oplus \mathcal{O}_{\mathbb{P}^1}(1)$ and the splitting of $\mathcal{F}'|_{L'}$ for $L' \in U$ other than L is $\mathcal{F}'|_{L'} = \mathcal{O}_{\mathbb{P}^1}(-1) \oplus \mathcal{O}_{\mathbb{P}^1}$. Show that there is a short exact sequence

$$0 \longrightarrow \mathcal{O}_{\Phi_U}(-2)^2 \longrightarrow \mathcal{F}' \longrightarrow \mathcal{O}_{\Phi_U}(1) \longrightarrow 0$$

and a resolution

$$0 \longrightarrow \mathcal{O}_{\Phi_U}(-2)^2 \longrightarrow \mathcal{O}_{\Phi_U}(-1)^3 \longrightarrow \mathcal{O}_{\Phi_U}(1) \longrightarrow 0,$$

and use these to deduce the presentation

$$0 \longrightarrow \mathcal{O}_{\Phi_U}(-2)^2 \longrightarrow \mathcal{O}_{\Phi_U}(-2) \oplus \mathcal{O}_{\Phi_U}(-1)^3 \longrightarrow \mathcal{F}' \longrightarrow 0.$$

Exercise 14.27. We will show how to arrive at the result $\lambda(\alpha) = 3$ of Section 14.5.1 more concretely. Choose affine coordinates t on $\mathbb{A}^1 \subset \mathbb{P}^1$ and (x, y) on $\mathbb{A}^2 \subset \mathbb{P}^2$, and write the equation of C_t as $f_t(x, y)$, where f_t is a quartic polynomial in x and y whose coefficients are linear in t.

(a) Show that the differential

$$\varphi_t = \frac{dx}{(\partial/\partial y) f_t(x, y)}$$

is regular on C_t, as are $x\varphi_t$ and $y\varphi_t$.

(b) Show that these differentials give rise to sections φ, $x\varphi$ and $y\varphi$ of the Hodge bundle $\mathcal{E} = \alpha_*(\omega_{X/B})$ that are everywhere linearly independent for $t \neq \infty$, and that in a neighborhood of $t = \infty$ the sections $t\varphi$, $tx\varphi$ and $ty\varphi$ are everywhere regular and linearly independent.

(c) Deduce that the Hodge bundle $\mathcal{E} = \alpha_*(\omega_{X/B})$ is isomorphic to $\mathcal{O}_{\mathbb{P}^1}(1)^{\oplus 3}$, and in particular that $\lambda(\alpha) = 3$.

Exercise 14.28. Let $\{C_t \subset \mathbb{P}^2\}_{t \in \mathbb{P}^1}$ be a general pencil of plane curves of degree d. Calculate the numerical invariants δ, κ and λ for the family, and verify the Mumford relation.

Exercise 14.29. Let $\{C_t \subset S\}_{t \in \mathbb{P}^1}$ be a general pencil of plane sections of a smooth surface $S \subset \mathbb{P}^3$ of degree d. Calculate the numerical invariants δ, κ and λ for the family, and verify the Mumford relation.

Exercise 14.30. Let $\{C_t \subset \mathbb{P}^1 \times \mathbb{P}^1\}_{t \in \mathbb{P}^1}$ be a general pencil of curves of bidegree (a, b) in $\mathbb{P}^1 \times \mathbb{P}^1$. Calculate the numerical invariants δ, κ and λ for the family, and verify the Mumford relation.

Exercise 14.31. Let $C \subset \mathbb{P}^1 \times \mathbb{P}^1$ be a general curve of bidegree $(2, 2g + 2)$, and let $X \to \mathbb{P}^1 \times \mathbb{P}^1$ be the double cover of $\mathbb{P}^1 \times \mathbb{P}^1$ branched over C. Viewing $X \to \mathbb{P}^1$ as a family of hyperelliptic curves of genus g via projection on the first factor, calculate the invariants δ, κ and λ for the family; verify the Mumford relation, and also show that the inequality

$$\delta \leq \left(8 + \frac{4}{g}\right)\lambda$$

stated in Section 14.5 is sharp.

Appendix A

The moving lemma

For many years, the development of intersection theory was based on a result known as the "moving lemma." This came in two flavors, the *basic moving lemma* and the *strong moving lemma*.

Lemma A.1 (Basic moving lemma). *Let X be a smooth, quasi-projective variety.*

(a) *Given cycles A, B on X, there exists a cycle A' rationally equivalent to A and generically transverse to B.*

(b) *The resulting class $[A' \cap B] \in A(X)$ is independent of the choice of such an A'.*

Given the validity of these assertions, the intersection product may be defined by the formula $[A][B] = [A' \cap B]$.

Lemma A.2 (Strong moving lemma). *Let $f : Y \to X$ be a morphism of smooth, quasi-projective varieties.*

(a) *Given a cycle A on X, there exists a cycle $A' = \sum n_i A_i$ rationally equivalent to A and generically transverse to f; that is, such that the preimage $f^{-1}(A_i) \subset Y$ is generically reduced of the same codimension as A_i.*

(b) *The class $\left[\sum n_i f^{-1}(A') \right] \in A(Y)$ is independent of the choice of such an A'.*

Given this, we can define a pullback map $f^* : A(X) \to A(Y)$, making the Chow ring into a contravariant functor. Note that the strong moving lemma is not just a generalization of the basic one — the basic moving lemma is just the strong one in case the map f is an inclusion — but also a strengthening: even in case f is an inclusion $Y \hookrightarrow X$, it produces a class in $A(Y)$ whose pushforward to X is the product $[A][Y]$, rather than simply a class in $A(X)$. This is called a *semi-refined* intersection product (the prefix "semi-" is because this is not the full degree of refinement possible; in Fulton's theory, subject to mild hypotheses it is possible to associate to a pair of subvarieties $A, B \subset X$ a class supported on the actual intersection $A \cap B$).

A method of proving part (a) of the basic moving lemma was put forward in Chow [1956] and Samuel [1956], following ideas of Severi [1933], and we will give the details in the first section of this appendix. (For other treatments see Samuel [1971], Roberts [1972a] and Hoyt [1971], as well as the discussion in Fulton [1984, Chapter 11].) In addition, part (a) of Lemma A.2 may be deduced from part (a) of Lemma A.1; we will do this in Section A.2 below.

It is also possible to use the same argument to prove part (b) of the basic version, by moving the second cycle B to a cycle B' generically transverse not only to all the components of A' but also to all subvarieties appearing in the rational equivalence between A and A'; this is carried out in the Stacks Project [2015, Tag 0AZ6] of de Jong and others. None of these approaches, however, suffice to prove part (b) of the strong moving lemma.

The Fulton–MacPherson approach to the definition of the intersection product, extended and detailed in Fulton [1984], has made the moving lemma unnecessary, and gives a technically superior and more general approach to intersection products. Though the direct proofs of part (b) (in either version) have remained controversial, the Fulton–MacPherson theory implies that the statements are correct. (If one is willing to work with rational coefficients, there is an alternative approach via K-theory as well.)

In our view, part (a) of the moving lemma, even though superseded (and rendered unnecessary) by the Fulton–MacPherson approach, still has heuristic importance, hence this appendix.

A.1 Generic transversality to a cycle

All existing proofs of part (a) of Lemma A.1 are based on an approach proposed by Severi, called the *cone construction*. The idea is this: We are given cycles A and B in a smooth variety $X \subset \mathbb{P}^N$, and want to find a cycle A', rationally equivalent to A and generically transverse to B. We will do this by expressing the cycle A as a difference of two cycles $A = E - A^1$, where E is the generically transverse intersection of X with another subvariety $\Phi \subset \mathbb{P}^N$ (so that E can be moved, by applying a linear transformation g of $\mathbb{P}^N$, to a cycle $E^1 = g\Phi \cap X$ generically transverse to B), and A^1 is better situated with respect to B than A. "Better situated" here means two things: if the intersection $A \cap B$ was not dimensionally transverse, then $A^1 \cap B$ will have strictly smaller dimension than $A \cap B$, and if $A \cap B$ is dimensionally transverse then $A^1 \cap B$ will actually be generically transverse. (It is called the cone construction because the variety Φ used is a cone over A, with vertex a general linear space $\Gamma \subset \mathbb{P}^N$.) If we carry out this process repeatedly, we will arrive at the desired cycle A'.

We remark that most of the salient points of the proof are already present in the very simplest case, where $X \subset \mathbb{P}^3$ is a smooth surface and $A = B \subset X$ a (possibly) singular curve; see Figure A.1.

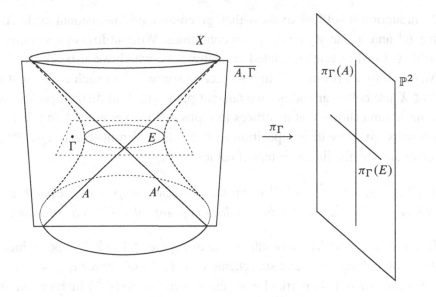

Figure A.1 $\overline{A,\Gamma} \cap X = A + A'$ and $\langle A \rangle \sim \langle E \rangle - \langle A' \rangle$.

Proof of part (a) of Lemma A.1: Set $n = \dim X$ and $a = \operatorname{codim}_X A$. We may assume that B is equidimensional, and we set $b = \operatorname{codim}_X B$.

We will construct sequences of equidimensional cycles of codimension a on X

$$A = A^0,\ A^1,\ A^2,\ \dots \quad \text{and} \quad E^1,\ E^2,\ \dots$$

satisfying the following conditions:

(a) A^i is rationally equivalent to $E^{i+1} - A^{i+1}$ on X.

(b) E^i is generically transverse to B.

(c) If C is a component of $A^{i+1} \cap B$ and $C \subset A^i \cap B$, then C is properly contained in a component of $A^i \cap B$; in particular, $\operatorname{codim} C > \operatorname{codim} A^i \cap B$.

(d) A^{i+1} is generically transverse to $B \setminus (A^i \cap B)$.

In what follows, the word "component" refers to an irreducible, nonembedded component. By Theorem 0.2, every component of $A \cap B$ has codimension $\leq a + b$. Thus part (c) shows that for $m > a + b - \operatorname{codim}(A \cap B)$ there are no components of $A^m \cap B$ that are contained in A^{m-1}, and part (d) then shows that A^m is generically transverse to B. By part (a) we have

$$A \sim \sum_{i=1}^{m} (-1)^{i+1} E^i + (-1)^m A^m,$$

and by part (b) all the E^i are generically transverse to B, so this will establish the theorem.

By induction it suffices to show that, given an equidimensional cycle A, we can produce E^1 and A^1 satisfying the given conditions. Without loss of generality we may assume that A is the cycle associated to a subvariety, which we also call A.

We begin by embedding X in a projective space $\mathbb{P}^N$ in such a way that no three points of X are collinear and no two tangent planes to X at distinct points meet. The following lemma shows that it suffices to replace whatever embedding $X \subset \mathbb{P}^{N'}$ we are originally given by its composition with the third Veronese map $v_3 : \mathbb{P}^{N'} \to \mathbb{P}^N$ (the embedding by the linear system of cubics):

Lemma A.3. *Let $v_3 : \mathbb{P}^N \to \mathbb{P}^M$ be the third Veronese map. No three points of $v_3(\mathbb{P}^N)$ are collinear, and the tangent planes at distinct points of $v_3(\mathbb{P}^N)$ are disjoint.*

Proof: By Proposition 7.10, any subscheme of degree ≤ 4 in $\mathbb{P}^N$ imposes independent conditions on cubics. Thus any subscheme of $v_3(\mathbb{P}^N)$ of degree $d \leq 4$ spans a plane of dimension $\geq d - 1$. In particular, no three points of $v_3(\mathbb{P}^N)$ lie on a line (this also follows from the fact that $v_3(\mathbb{P}^N)$ is cut out by quadrics and contains no lines). If the tangent planes to $v_3(\mathbb{P}^N)$ at points p, q met in some point r, then the lines $L_1 = \overline{p, r}$ and $L_2 = \overline{q, r}$ would be contained in the 2-plane $\overline{p, q, r}$, and this plane would contain the subscheme $(L_1 \cap X) \cup (L_2 \cap X)$, which has length at least 4, a contradiction. $\square$

Thus we may assume from the outset that no three points of X in $\mathbb{P}^N$ are collinear, and that any two tangent planes to X at distinct points are disjoint.

Lemma A.4. *Let X and A be as above and $\pi_\Gamma : X \to \mathbb{P}^n$ be the linear projection from a general $N - n - 1$ plane $\Gamma \subset \mathbb{P}^N$. We may write*

$$\pi_\Gamma^{-1}(\pi_\Gamma(A)) = A \cup A_\Gamma'$$

as schemes, where A_Γ' is a generically reduced scheme of pure codimension a that does not contain A.

Write $\mathbb{G} := \mathbb{G}(N - n - 1, N)$ for the Grassmannian of $(N - n - 1)$-dimensional planes in $\mathbb{P}^N$; the statement above means that the conclusion holds for all planes in an open dense subset of $\mathbb{G}$.

Proof: To simplify the notation, we set $\widetilde{A}_\Gamma := \pi_\Gamma^{-1}(\pi(A)) = X \cap \overline{A, \Gamma}$ (see Figure A.1).

By Theorem 0.2, the components of $\widetilde{A}_\Gamma$ have codimension in X at most the codimension of the irreducible variety $\pi_\Gamma(A)$ in $\mathbb{P}^n$. Since $\pi_\Gamma : X \to \mathbb{P}^n$ is finite, $\pi_\Gamma(A)$ has dimension $n - a$. Thus every component of $\pi_\Gamma^{-1}(\pi_\Gamma(A))$ is of pure codimension a and maps surjectively to $\pi_\Gamma(A)$. In particular, A itself is a component of $\widetilde{A}_\Gamma$.

Thus it suffices to prove that $\widetilde{A}_\Gamma$ is generically reduced. Since every component of $\widetilde{A}_\Gamma$ surjects onto $\pi_\Gamma(A)$, it even suffices to show that the general fiber of π_Γ is reduced. The fibers of π_Γ are the intersections of X with the $(N - n)$-dimensional planes containing Γ, and a general such plane Σ is a general $(N - n)$-plane containing a general $(N - n - 1)$-plane Γ. Thus Σ is general in the space of all $(N - n)$-planes, so reducedness follows from Bertini's theorem.　　　　　　　　　　　　　　　　□

With notation as in Lemma A.4, we set $A_\Gamma^1 := A_\Gamma'$. The situation is illustrated in Figure A.1. We can now establish conditions (a) and (b) of the proof of Lemma A.1:

Lemma A.5. *With notation as above, if $\Gamma \in \mathbb{G}$ is general then*

$$[A] = [E^1] - [A_\Gamma^1] \in A(X),$$

where E^1 is generically transverse to B.

Proof: We have

$$A = \widetilde{A}_\Gamma - A_\Gamma^1$$

where $\widetilde{A}_\Gamma$ is the generically transverse intersection of X with the cone $\overline{A, \Gamma}$. Let $g \in$ PGL_{N+1} be a general automorphism of $\mathbb{P}^N$. By the argument of part (c) of the Kleiman transversality theorem (Theorem 1.7),

$$\widetilde{A}_\Gamma \sim E' := X \cap g(\overline{A, \Gamma}),$$

and by part (a) of the same theorem $g(\overline{A, \Gamma})$, and hence E', will be generically transverse to B.　　　　　　　　　　　　　　　　　　　　　　　　　　　　　□

Completion of the proof of Lemma A.1: With notation and hypotheses as above, it suffices to show that, for general Γ, the cycle corresponding to the scheme A_Γ^1 satisfies (c) and (d). Since an intersection of open dense subsets of $\mathbb{G}$ is again open and dense, it suffices to do this for one component of B at a time, so we may assume that B is (the cycle associated to) a subvariety, which we also call B.

Condition (c): Consider a component C of $A_\Gamma^1 \cap B$ that is contained in A, so that in fact

$$C \subset A \cap A_\Gamma^1 \cap B \subset A \cap B.$$

We must show that C is not a component of $A \cap B$.

Since Γ is general, every component of $A \cap B$ contains points p such that Γ does not meet the tangent plane to X at p. The map $\pi_\Gamma : X \to \mathbb{P}^n$ is nonsingular at such points. Since $\pi_\Gamma^{-1}(\pi_\Gamma(A)) = A' \cup A$, such points cannot lie in $A \cap A'$. Consequently C must be properly contained in some component of $A \cap B$, as required.

Condition (d): Finally, we wish to show that for general Γ the intersection of A_Γ^1 with $B^* = B \setminus A \cap B$ is generically transverse, or equivalently that $\widetilde{A}_\Gamma$ and B^* are generically transverse.

We first prove the weaker statement that $\widetilde{A}_\Gamma$ and B^* are dimensionally transverse. Consider the incidence correspondence Ψ defined by

$$\Psi := \{(\Gamma, p, q) \in \mathbb{G} \times A \times B^* \mid \Gamma \cap \overline{p, q} \neq \varnothing\}.$$

The fiber of Ψ over any point $(p, q) \in A \times B^*$ is isomorphic to the set $\Sigma(\overline{p, q})$ of $(N - n - 1)$-planes $\Gamma \in \mathbb{G}^*$ meeting the line $\overline{p, q}$. By Theorem 4.1 this is an irreducible variety of codimension n in $\mathbb{G}^*$. Since the projection $\Psi \to A \times B^*$ is proper, it follows that Ψ is irreducible of dimension

$$\dim \Psi = \dim \mathbb{G}^* + \dim A + \dim B - n.$$

This implies that for general Γ the fiber of Ψ over Γ has dimension $\dim A + \dim B - n$. Since this fiber surjects onto $A_\Gamma^1 \cap B^*$, we see that $\dim A_\Gamma^1 \cap B^* \leq \dim A + \dim B - n$, so by Theorem 0.2 we have equality. That is, for general Γ the sets A_Γ and B^* are dimensionally transverse; every component of their intersection has dimension $\dim A + \dim B - n$.

Let $\mathbb{G}^* \subset \mathbb{G}$ be the open dense set consisting of the planes Γ disjoint from X such that $\pi_\Gamma : A \to \pi_\Gamma(A)$ is birational and $\widetilde{A}_\Gamma$ is generically reduced.

To prove the generic transversality of $\widetilde{A}_\Gamma$ and B^* for general Γ, we next consider $\Psi_0 \subset \Psi$, where

$$\Psi_0 := \{(\Gamma, p, q) \in \mathbb{G}^* \times A \times B^* \mid \pi_\Gamma(p) = \pi_\Gamma(q), \widetilde{A}_\Gamma \text{ is not transverse to } B^* \text{ at } q\}.$$

If $\widetilde{A}_\Gamma$ and B^* were not generically transverse for generic Γ, then for an open set of Γ in $\mathbb{G}$ the fiber of Ψ_0 over Γ would surject to at least one component of $A_\Gamma^1 \cap B^*$, and thus would have dimension $\geq \dim A + \dim B - n$. This would imply that $\dim \Psi_0 \geq \dim \mathbb{G} + \dim A + \dim B - n = \dim \Psi$. Thus it suffices to show that $\dim \Psi_0 < \dim \Psi$.

To do this, we will write Ψ_0 as the union of five subsets

$$\Psi_0 = \Psi_1 \cup \Psi_2 \cup \Psi_3 \cup \Psi_4 \cup \Psi_5,$$

defined in terms of the reasons why A_Γ might not be transverse to B^* at q. To start, the intersection of A_Γ^1 and B at q will be nontransverse if:

(1) q is a singular point of B.

The intersection will also be nontransverse if q is a singular point of $\widetilde{A}_\Gamma$. Since $q \in X$, we have $q \notin \Gamma$, so q is singular on $\widetilde{A}_\Gamma$ if and only if $\pi_\Gamma(q) = \pi_\Gamma(p)$ is singular on $\pi_\Gamma(A)$. This can occur only if one of the following occurs:

(2) $q \in \overline{\Gamma, p}$ with p a singular point of A.
(3) $q \in \overline{\Gamma, p}$ and $q \in \overline{\Gamma, p'}$ for two distinct points $p, p' \in A$.
(4) $q \in \overline{\Gamma, p}$ and $\Gamma \cap \mathbb{T}_p A \neq \emptyset$.

Accordingly, we set

$$\Psi_1 := \{(\Gamma, p, q) \in \Psi_0 \mid q \in B_{\text{sing}}\} \subset \Psi,$$

$$\Psi_2 := \{(\Gamma, p, q) \in \Psi_0 \mid p \in A_{\text{sing}}\} \subset \Psi,$$

$$\Psi_3 := \{(\Gamma, p, q) \in \Psi_0 \mid \text{there exists } p' \neq p \in A \text{ with } \Gamma \cap \overline{p', q} \neq \emptyset\} \subset \Psi,$$

$$\Psi_4 := \{(\Gamma, p, q) \in \Psi_0 \mid p \in A_{\text{sm}} \text{ and } \Gamma \cap \mathbb{T}_p A \neq \emptyset\} \subset \Psi,$$

where $A_{\text{sing}} \subset A$ denotes the singular locus, $A_{\text{sm}} = A \setminus A_{\text{sing}}$, and similarly for B.

Now suppose that $(p, q) \in \Psi_0$ is not in $\bigcup_1^4 \Psi_i$, so that $\widetilde{A}_\Gamma$ and B^* are both smooth at q. The intersection of $\widetilde{A}_\Gamma$ and B^* will be nontransverse if and only if the tangent spaces of these two varieties at q fail to be transverse. The tangent plane to the cone $\overline{A, \Gamma}$ at q is the span of Γ and the tangent space $\mathbb{T}_p A$. This span fails to intersect B transversely at q only if the three linear spaces Γ, $\mathbb{T}_p A$ and $\mathbb{T}_q B$ fail to span all of $\mathbb{P}^N$. From our hypothesis on the embedding of X, it follows that $\mathbb{T}_p A$ and $\mathbb{T}_q B$ are disjoint. Thus a necessary condition for nontransversality at q in this case is that:

(5) Γ is not transverse to $\overline{\mathbb{T}_p A, \mathbb{T}_q B}$.

Since $\dim \overline{\mathbb{T}_p A, \mathbb{T}_q B} = 2n - a - b + 1$ and $\dim \Gamma = N - n$, the relevant set is

$$\Psi_5 := \{(\Gamma, p, q) \in \Psi_0 \mid p \in A_{\text{sm}}, \ q \in B_{\text{sm}}, \ \dim(\Gamma \cap \overline{\mathbb{T}_p A, \mathbb{T}_q B}) > n - a - b\}.$$

We can compute the dimensions of Ψ_1 and Ψ_2 just as we computed the dimension of Ψ itself. Since A and B are reduced, the sets A_{sing} and B^*_{sing} have strictly smaller dimension than A and B^*. Noting again that the fibers of the projection $\Psi \to A \times B^*$ all have codimension n in $\mathbb{G}$, we see that Ψ_1 and Ψ_2 have strictly smaller dimensions than Ψ.

The set Ψ_4 dominates $A_{\text{sm}} \times B^*$, but with strictly smaller fibers than Ψ: By our hypothesis on the embedding of X in $\mathbb{P}^N$, we have $q \notin \mathbb{T}_p A$. Also $p \notin \Gamma$. If $\Gamma \cap \mathbb{T}_p A \neq \varnothing$ then, in addition to meeting the line $\overline{p, q}$ in at least a point, Γ must intersect the $(a + 1)$-plane $\overline{q, \mathbb{T}_p A}$ in at least a line. This is a proper subvariety of the Schubert cycle $\Sigma_n(\overline{p, q})$, so Ψ_4 has smaller dimension than Ψ.

Similarly, the fiber of Ψ_5 over any point $(p, q) \in A_{\text{sm}} \times B^*_{\text{sm}}$ is a proper subvariety of $\Sigma_n(\overline{p, q})$, and again we conclude that $\dim \Psi_5 < \dim \Psi$.

To compute the dimension of Ψ_3 we introduce one more incidence correspondence. Set

$$\widetilde{\Psi}_3 := \{(\Gamma, p, p', q) \in \mathbb{G}^* \times A \times A \times B^* \mid p \neq p', \ \Gamma \cap \overline{p, q} \neq \varnothing \text{ and } \Gamma \cap \overline{p', q} \neq \varnothing\}.$$

Since Ψ_3 is the image of $\widetilde{\Psi}_3$ under a projection to $\mathbb{G}^* \times A \times B^*$, it suffices to show that $\dim \widetilde{\Psi}_3 < \dim \Psi$.

To estimate the dimension of $\widetilde{\Psi}_3$, consider the projection to $A \times A \times B^*$. By our hypothesis on the embedding of X in $\mathbb{P}^N$, any three points of X span a 2-plane. Also $q \in X$, so $q \notin \Gamma$. Thus the conditions $\Gamma \cap \overline{p, q} \neq \varnothing$ and $\Gamma \cap \overline{p', q} \neq \varnothing$ amount to saying that $\dim(\Gamma \cap \overline{p, p', q}) \geq 1$. This describes the Schubert cycle $\sigma_{n,n}(\overline{p, p', q})$, which has codimension $2n$ in $\mathbb{G}^*$ by Theorem 4.1. We thus have

$$\dim \widetilde{\Psi}_3 = \dim \mathbb{G} + 2a + b - 2n < \dim \Psi,$$

as required. Putting this all together, we get that $\dim \Psi_0 < \dim \Psi$, completing the proof of Lemma A.1. □

A.2 Generic transversality to a morphism

Let $f : Y \to X$ be a morphism of smooth varieties. Recall that a subvariety $A \subset X$ is said to be *dimensionally transverse to* f if the codimension of $f^{-1}(A)$ in Y is the same as the codimension of A in X, and *generically transverse to* f if in addition $f^{-1}(A)$ is generically reduced (Definition 1.22). In this section we will show that there is a finite collection of subvarieties of X, depending on f, such that if A is generically transverse to each of these subvarieties then A is generically transverse to f. (This would not be true without our standing hypothesis of characteristic 0, or at least a hypothesis that f is generically separable; if f is not separable, then $f^{-1}(A)$ is necessarily nonreduced, so A is never generically transverse to f.) For each k set

$$\Phi_k^\circ(f) := \{x \in f(Y) \mid \dim f^{-1}(x) \geq k\},$$

and write

$$\Psi^\circ(f) := \{x \in X \mid \text{for some } y \in f^{-1}(x),$$
$$\operatorname{rank} df_y : T_y Y \to T_x X \text{ is } < \min(\dim X, \dim Y)\}$$

for the image of the singular locus of f; let $\Psi(f)$ and $\Phi_k(f)$ be the closures of $\Psi^\circ(f)$ and $\Phi_k^\circ(f)$.

Theorem A.6. *Suppose that* $f : Y \to X$ *is a morphism of varieties. If a subvariety* $A \subset X$ *is dimensionally transverse to each* $\Phi_k(f)$ *then* A *is dimensionally transverse to* f. *If in addition* A *is generically transverse to* Ψ *then* A *is generically transverse to* f.

If f is not separable, then $f^{-1}(A)$ is necessarily nonreduced, so A is never generically transverse to f.

Proof: First suppose that A is dimensionally transverse to each Φ_k. The dimension of Φ_k is $\leq \dim Y - k$, with strict inequality for $k > \dim Y - \dim f(Y)$ since Y is irreducible.

Let $k_0 = \dim Y - \dim f(Y)$, so that $\Psi_{k_0} = f(Y)$. For $k > k_0$, transversality to Ψ_k yields

$$\dim(A \cap \Phi_k) \leq \dim A - \operatorname{codim} f(Y) - k + k_0 - 1,$$

from which it follows that

$$\dim(f^{-1}(A \cap \Phi_k)) \leq \dim A - \operatorname{codim} f(Y) + k_0 - 1.$$

By Theorem 0.2 every component of $f^{-1}(A)$ has dimension $\geq \dim A - \operatorname{codim} f(Y) + k_0$. It follows that no component of $f^{-1}(A)$ is contained in $f^{-1}(\Phi_k)$ for $k > k_0$, and hence

$$\dim f^{-1}(A) = k_0 + \dim(A \cap f(Y))$$
$$= k_0 + \dim f(Y) - \operatorname{codim} A$$
$$= \dim Y - \operatorname{codim} A,$$

as required.

Because the characteristic of the ground field is 0, the branch locus Ψ is strictly contained in $f(Y)$ (this is the algebraic version of Sard's theorem; see Milnor [1965, Theorem 6.1]). Thus $A \cap f(Y)$ is not contained in Ψ. It follows that a general point of each component of $f^{-1}(A)$ is smooth, so that $f^{-1}(A)$ is generically reduced. $\qquad\square$

Appendix B

Direct images, cohomology and base change

B.1 Can you define a bundle by its fibers?

To study the lines on a cubic surface back in Section 5.1, we needed to construct "the" vector bundle on the Grassmannian of lines in $\mathbb{P}^3$ whose fiber at the point representing a line L is the space of cubic forms on L. This specification is at best incomplete: the condition determines only the rank of the bundle. In the example we needed an additional property: we wanted the restriction map from the space of cubic forms on $\mathbb{P}^3$ to the line L to be induced by a map of bundles on the Grassmannian; that is, we needed the construction to be functorial in some reasonable sense.

To make things precise in this and many similar cases, we constructed the desired sheaves as *direct images*, and used the *theorem on cohomology and base change* to justify their properties. In this appendix we will give a gentle treatment of these important ideas. Much of the material is derived from Mumford [2008]; see also Arbarello et al. [1985, Chapter 4].

To state the problem more generally, suppose that we are given a family of varieties X_b with sheaves $\mathcal{F}_b$ on them, parametrized by the points b of a base variety B. As usual, by a family of varieties we mean a map $\pi : X \to B$, the "members" of the family being the fibers $X_b := \pi^{-1}(b)$. Similarly, by a family of sheaves we mean a sheaf $\mathcal{F}$ on X, with the members of the family being the sheaves $\mathcal{F}|_{X_b}$. We can expect nice results only if the members of the family "belong" together in some reasonable sense, which we generally take to be the condition that $\mathcal{F}$ is flat over B. Given such data, we ask whether there is a functorial construction of a sheaf $\mathcal{G}$ on B whose fiber $\mathcal{G}_b$ at a point b is the space of global sections of $\mathcal{F}_b$.

Such a sheaf $\mathcal{G}$ may or may not exist, as we shall soon see. Nevertheless, under very general circumstances we can define a sheaf $\pi_*\mathcal{F}$ on B, called the *direct image of $\mathcal{F}$ under π*, that is functorial in $\mathcal{F}$ and comes equipped with canonical maps φ_b : $(\pi_*\mathcal{F})_b \to H^0(\mathcal{F}|_{X_b})$ for $b \in B$. The theorem on cohomology and base change gives conditions under which all the φ_b are isomorphisms, in which case $\mathcal{G} := \pi_*\mathcal{F}$ will have the property we wish.

We present three versions of the theorem on cohomology and base change. The first — and the most often applied! — is Theorem B.5. A useful extension is the version given in Theorem B.9. The most general version, Theorem B.11, is paradoxically also the simplest, and easily implies the others. We prove these results, after various preliminaries, in Section B.5.

Before describing the results, we pause to explain an example that we will follow throughout this appendix:

Example B.1 (Two and three points in $\mathbb{P}^2$). Let $\{p, q\} \subset \mathbb{P}^2$ be a set of two distinct points. Since $h^0(\mathcal{O}_{\{p,q\}}(d)) = 2$ and $h^1(\mathcal{O}_{\{p,q\}}(d)) = 0$ for all $d \in \mathbb{Z}$, the long exact sequence in cohomology coming from the short exact sequence $0 \to \mathcal{I}_{\{p,q\}} \to \mathcal{O}_{\mathbb{P}^2} \to \mathcal{O}_{\{p,q\}} \to 0$ immediately yields

$$h^0(\mathcal{I}_{\{p,q\}}(d)) = \begin{cases} \binom{2+d}{2} - 2 & \text{if } d \geq 1, \\ 0 & \text{if } d \leq 0, \end{cases}$$

$$h^1(\mathcal{I}_{\{p,q\}}(d)) = \begin{cases} 0 & \text{if } d \geq 1, \\ 1 & \text{if } d = 0, \\ 2 & \text{if } d < 0. \end{cases}$$

Ideal sheaves of sets of three distinct points $\{p, q, r\}$ can be analyzed similarly, using the sequence $0 \to \mathcal{I}_{\{p,q,r\}} \to \mathcal{O}_{\mathbb{P}^2} \to \mathcal{O}_{\{p,q,r\}} \to 0$, but there is a difference: When $d = 1$ there are two cases, depending on whether some linear form vanishes on all three points; we have

$$h^0(\mathcal{I}_{\{p,q,r\}}(1)) = h^1(\mathcal{I}_{\{p,q,r\}}(1)) = \begin{cases} 1 & \text{if } r \text{ lies on the line } \overline{p,q}, \\ 0 & \text{otherwise.} \end{cases}$$

It is thus interesting to consider a family of ideal sheaves of triples $\{p, q, r\}$ of distinct points as one of the points crosses the line joining the other two.

We prefer to work with a projective family, so we fix points p and q in $\mathbb{P}^2$ and let r move along a line $B \subset \mathbb{P}^2$ containing neither p nor q. To set this up, we consider in $B \times \mathbb{P}^2$ three families of points: the constant families

$$\Gamma_p = B \times \{p\} \quad \text{and} \quad \Gamma_q = B \times \{q\},$$

contained in $B \times \mathbb{P}^2$, and a family of points moving along B, that is, the diagonal

$$\Delta = \{(r, r) \in B \times \mathbb{P}^2 \mid r \in B\} \subset B \times B.$$

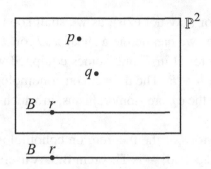

Figure B.1 The fibers of $\Gamma' \subset \Gamma$ over $r \in B$ are the sets $\{p, q\} \subset \{p, q, r\} \subset \mathbb{P}^2$.

We take

$$\Gamma = \Gamma_p \cup \Gamma_q \cup \Delta \subset B \times \mathbb{P}^2,$$

which we regard (via the projection map $\pi : \Gamma \to B$) as a family over B, with fiber over $r \in B$ the triple $\{p, q, r\} \subset \mathbb{P}^2$. Let Γ' be the trivial subfamily

$$\Gamma' = \Gamma_p \cup \Gamma_q \subset B \times \mathbb{P}^2$$

whose fiber over each point of B is the pair of fixed points $\{p, q\}$. See Figure B.1.

We now ask: Given an integer d, are there are sheaves $\mathcal{G}, \mathcal{G}'$ on B whose fibers at a point $r \in B$ are the spaces of forms of degree d vanishing on Γ and Γ', respectively, and a map $\mathcal{G} \to \mathcal{G}'$ inducing the obvious inclusion of spaces of forms?

To see that this question fits into the former context, let $\mathcal{F} \subset \mathcal{F}'$ be the ideal sheaves of Γ and Γ' in $B \times \mathbb{P}^2$. Abusing notation slightly, we write $\mathcal{O}_{\mathbb{P}^2}(d)$ for the pullback to $B \times \mathbb{P}^2$ of $\mathcal{O}_{\mathbb{P}^2}(d)$ on $\mathbb{P}^2$ and $\mathcal{F}(d)$ for $\mathcal{F} \otimes \mathcal{O}_{\mathbb{P}^2}(d)$. Thus

$$\mathcal{F}(d)|_{\pi^{-1}(r)} = \mathcal{I}_{\{p,q,r\}}(d),$$

the d-th twist of the ideal sheaf of $\{p, q, r\}$ in $\mathbb{P}^2$, and similarly for $\mathcal{F}'(d)$.

The answer is that such sheaves exist when $d \neq 1$ (and can, by Theorem B.5, be taken to be $\pi_* \mathcal{F}$ and $\pi_* \mathcal{F}'$, defined below). But no such $\mathcal{G} \to \mathcal{G}'$ exists when $d = 1$!

By our computation of $H^0(\mathcal{I}_{\{p,q,r\}}(1))$ above, the sheaf $\mathcal{G}$ would be a skyscraper sheaf concentrated at the unique point r_0 on the intersection of the line B with the line $\overline{p,q}$. Furthermore, the map $H^0(\mathcal{I}_{\{p,q,r_0\}}(1)) \to H^0(\mathcal{I}_{\{p,q\}}(1))$ is an isomorphism, so functoriality would imply that the map on fibers $\mathcal{G}_{r_0} \to \mathcal{G}'_{r_0}$ would be an isomorphism.

On the other hand, the fibers of $\mathcal{G}'$ would all be equal to the one-dimensional vector space $H^0(\mathcal{I}_{\{p,q\}}(1))$, so $\mathcal{G}'$ would be a line bundle. Since the only map from a skyscraper sheaf to a line bundle is 0, this is a contradiction, showing that the desired functorial construction is impossible! As we shall see in Example B.10, $\pi_* \mathcal{F} = 0$, and thus the fiber $(\pi_* \mathcal{F})_{r_0}$ is not equal to $H^0(\mathcal{F}|_{\pi^{-1}(r_0)})$.

B.2 Direct images

We remind the reader of our global convention that "sheaf" means coherent sheaf unless otherwise stated.

It becomes clear how to define $\pi_*\mathcal{F}$ if we add another natural condition to our desiderata: in cases where the fibers of $\pi_*\mathcal{F}$ are the spaces $H^0(\mathcal{F}|_{X_b})$, we would like algebraic families of elements of $H^0(\mathcal{F}|_{X_b})$ to give rise to sections of $\mathcal{F}$ in a way that is compatible with the identification of $\mathcal{G}_b$ with $H^0(\mathcal{F}|_{X_b})$. Here we interpret the phrase "algebraic family of elements" to mean "section of $\mathcal{G}$ defined over the preimage of an open set of B."

Definition B.2. Given a morphism of schemes $\pi : X \to B$ and a sheaf $\mathcal{F}$ on X, we define the *direct image* $\pi_*\mathcal{F}$ of $\mathcal{F}$ to be the quasi-coherent sheaf on B that assigns to each open subset $U \subset B$ the space of sections of $\mathcal{F}$ on the open set $\pi^{-1}(U)$, that is,

$$(\pi_*\mathcal{F})(U) = \mathcal{F}(\pi^{-1}(U)).$$

It is immediate that $\pi_*(\mathcal{F})$ is a sheaf if $\mathcal{F}$ is, and that the construction is functorial in $\mathcal{F}$.

This definition is particularly natural if we think of a presheaf $\mathcal{F}$ as a contravariant functor

$$\mathcal{U} \to (\text{Sets}), \quad U \mapsto \mathcal{F}(U),$$

from the category $\mathcal{U}$ whose objects are open sets of X and whose morphisms are inclusions to the category of sets. Since π induces a covariant functor $\pi^{-1} : V \mapsto \pi^{-1}(V)$ from the category of open sets of B to that of X, we may simply compose to get the presheaf

$$\pi_*\mathcal{F} = \mathcal{F} \circ \pi^{-1}.$$

Note that by definition $H^0(\pi_*\mathcal{F}) = (\pi_*\mathcal{F})(B) = \mathcal{F}(X) = H^0(\mathcal{F})$.

When X and B are affine varieties, or more generally when the morphism π is affine (for example when π is a finite map), the sheaf $\pi_*\mathcal{F}$ is easy to understand: Giving $\mathcal{F}$ is equivalent to giving an $\mathcal{O}_X$-module, and it follows at once from the definitions that $\pi_*\mathcal{F}$ corresponds to the *same* module, viewed as a $\mathcal{O}_B$-module via the map of (sheaves of) rings $\pi^* : \mathcal{O}_B \to \mathcal{O}_X$. In particular we see that even when $\mathcal{F}$ is coherent $\pi_*\mathcal{F}$ may be only quasi-coherent.

The situation is very different when π is a projective morphism. A fundamental result of Serre (Theorem B.8) shows that when $\mathcal{F}$ is a coherent sheaf and π is a projective morphism $\pi_*\mathcal{F}$ is coherent. (Recall that a projective morphism is one that factors as the inclusion of X as a closed subset of some $B \times \mathbb{P}^n$ and the projection to B; since any morphism can be factored through its graph, any morphism of projective varieties is a projective morphism.) With a little more effort, the results can all be extended to proper morphisms; see Grothendieck [1963, Theorem 3.2.2.1].

The theorems on cohomology and base change give information not only about fibers of the direct image sheaf, but about more general pullbacks ("base changes") as well; we pause to put things into this more general context.

We may think about the fiber X_b of π over a point b as coming from a pullback, or *base change* diagram

$$
\begin{array}{ccc}
X_b & \xrightarrow{\;\rho'\;} & X \\
\pi' \downarrow & & \downarrow \pi \\
\{b\} & \xrightarrow{\;\rho\;} & B
\end{array}
$$

The restriction of $\mathcal{F}$ to X_b can also be thought of as $\rho'^*(\mathcal{F})$, and from the definition of π'_* we see that $\pi'_*(\mathcal{F}|_{X_b}) = H^0(\mathcal{F}|_{X_b})$. Thus the theorem on cohomology and base change is about the comparison of $\pi'_*(\rho'^*\mathcal{F})$ and $\rho^*(\pi_*\mathcal{F})$.

More generally, for any map $\rho : B' \to B$ we can consider the pullback of the family $X \to B$ to B':

$$
\begin{array}{ccc}
X' = X \times_B B' & \xrightarrow{\;\rho'\;} & X \\
\pi' \downarrow & & \downarrow \pi \\
B' & \xrightarrow{\;\rho\;} & B
\end{array}
$$

We say that the map $\pi' : X' \to B'$ and sheaf $\mathcal{F}' = \rho'^*\mathcal{F}$ are obtained from the map $\pi : X \to B$ and sheaf $\mathcal{F}$ by *base change*. In this situation, there is a natural map

$$
\varphi_{B'} : \rho^*(\pi_*\mathcal{F}) \to \pi'_*(\rho'^*\mathcal{F}),
$$

constructed as follows:

Applying the definitions, we see that the module of sections of $\rho^*(\pi_*\mathcal{F})$ over an open set $U' \subset B'$ is the direct limit over all open sets $U \subset B$ such that $\rho^{-1}(U) \supset U'$ of the $\mathcal{O}_{B'}$-modules

$$
A_U := \mathcal{O}_{U'} \otimes \mathcal{O}_B \big(\mathcal{F}(\pi^{-1}(U)) \big).
$$

On the other hand, the module of sections of $\pi'_*(\rho'^*\mathcal{F})$ over U' is

$$
\rho'^*(\mathcal{F})(\pi'^{-1}(U')),
$$

which is the limit over all open subsets $V \subset X$ such that $\rho^{-1}(V) \supset \pi'^{-1}(U')$ of the $\mathcal{O}_{X'}$-modules

$$
B_V := \mathcal{O}_{\pi^{-1}(U')} \otimes_{\mathcal{O}_X} \mathcal{F}(V).
$$

Each open set U entering into the former limit gives rise to a $V = \pi^{-1}(U)$ that enters into the latter limit, and since $\mathcal{O}_{\pi^{-1}(U')}$ is a module over $\mathcal{O}_{U'}$ there is an induced map $A_U \to B_{\pi^{-1}(U)}$. The map

$$
\varphi_{B'} : \rho^*(\pi_*\mathcal{F}) \to \pi'_*(\rho'^*\mathcal{F})
$$

is the natural map induced between the limits.

When $B' = \{b\}$ is a closed point of B we write φ_b instead of $\varphi_{\{b\}}$. In this case, $\rho^*(\pi_*\mathcal{F})$ is the fiber of $\pi_*\mathcal{F}$ at b and $\pi'_*(\rho'^*\mathcal{F})$ is the set of global sections of $\mathcal{F}$ on the fiber X_b, and we see that *the image of φ_b is simply the set of global sections of $\mathcal{F}|_{X_b}$ that extend to an open neighborhood of X_b in X.* On the other hand, the kernel consists of sections defined in some small neighborhood V of $\pi^{-1}(b)$ that vanish on $\pi^{-1}(b)$, but cannot be expressed in terms of functions pulled back from any small neighborhood of b and vanishing at b.

In general, we ask when the natural maps $\varphi_{B'}$ are isomorphisms; in other words, *when does the formation of the direct image of $\mathcal{F}$ commute with base change?*

There are two cases that are easy:

(1) If π is affine, then the maps $\varphi_{B'}$ are isomorphisms for any quasi-coherent sheaf $\mathcal{F}$. (This follows at once from the description of π_* for affine morphisms given above.)

(2) If $\rho : B' \to B$ is a flat map, then the maps $\varphi_{B'}$ are isomorphisms. (This is an immediate consequence of Theorem B.11. See Hartshorne [1977, Proposition III.9.3] for a direct proof.) Since the inclusion of a point is not generally flat, this case is not, in practice, very useful.

The simplest case of a projective morphism is that of a finite morphism. The following proposition summarizes the situation in that case:

Proposition B.3. *If $\pi : X \to B$ is a finite morphism of quasi-projective varieties and $\mathcal{F}$ is a coherent sheaf on X, then $\pi_*\mathcal{F}$ is a coherent sheaf on B and the maps $\varphi : (\pi_*\mathcal{F})_b \to H^0(\mathcal{F}|_{X_b})$ are isomorphisms for all closed points $b \in B$. Moreover, the following are equivalent:*

(a) *$\pi_*\mathcal{F}$ is a vector bundle on B.*

(b) *$\mathcal{F}$ is flat over B.*

(c) *The dimension of $H^0(\mathcal{F}|_{X_b})$ as a vector space over the residue class field $\kappa(b)$ is independent of the closed point $b \in B$.*

Proof: Since sheaves are defined locally and the preimage of an affine subset under a finite map is again affine, we may suppose from the outset that both X and B are affine. Let $X = \operatorname{Spec} R$ and $B = \operatorname{Spec} S$, and let $M = H^0(\mathcal{F})$ be the R-module corresponding to $\mathcal{F}$. Because the varieties are affine, $\pi_*\mathcal{F}$ is represented by M regarded as a module over S via the map $\pi^* : S \to R$. The ring R is by hypothesis a finitely generated S-module, so M is a finitely generated S-module as well. The maps φ_b are isomorphisms because, writing $\mathfrak{m}_b \subset S$ for the maximal ideal corresponding to b, both $(\pi_*\mathcal{F})|_b$ and $H^0(\mathcal{F}|_{X_b})$ may be identified canonically with $M/\mathfrak{m}_b M$. (The same proof shows that the map $\varphi_{B'}$ is an isomorphism for any closed set B'.) The equivalence of parts (a) and (c) is proven in Proposition B.15 below. The equivalence of (a) and (b) is Eisenbud [1995, Exercise 6.2]. $\qquad\square$

Example B.4 (Direct images of $\mathcal{O}_{\Gamma'}(d)$ and $\mathcal{O}_{\Gamma}(d)$). Returning to the situation of Example B.1, note that the families Γ' and Γ are finite over B. For every p, q, r and every d, the space of global sections of $\mathcal{O}_{\{p,q\}}(d)$ is two-dimensional and that of $\mathcal{O}_{\{p,q,r\}}(d)$ is three-dimensional. From Proposition B.3 we see that $\pi_*\mathcal{O}_{\Gamma'}(d)$ and $\pi_*\mathcal{O}_{\Gamma}(d)$ are vector bundles of ranks 2 and 3, respectively. The inclusion $\Gamma' \subset \Gamma$ induces an inclusion of these bundles. Since we can choose a fixed basis of functions on the fibers of Γ', and the pullback of $\mathcal{O}_B(d)$ to Γ' is the trivial bundle, the bundle $\pi_*\mathcal{O}_{\Gamma'}(d)$ is the trivial bundle $\mathcal{O}_B^2$.

We have $\mathcal{O}_{\Gamma}(d) = \mathcal{O}_{\Gamma'}(d) \oplus \mathcal{O}_{\Delta}(d)$, where $\Delta = \{(r,r) \in B \times \mathbb{P}^2\}$, as before, and projects isomorphically to each factor. Thus $\mathcal{O}_{\Delta}(d)$, the pullback of $\mathcal{O}_B(d)$ from the first factor, pushes forward to $\mathcal{O}_B(d)$ on the second factor for every d, and $\pi_*\mathcal{O}_{\Gamma}(d) \cong \mathcal{O}_B^2 \oplus \mathcal{O}_B(d)$ is a nontrivial bundle.

For more general projective morphisms neither condition (b) nor condition (c) of Proposition B.3 alone will imply that $\pi_*\mathcal{F}$ is a vector bundle or that the maps φ_b are isomorphisms. But conditions (b) and (c) together do imply both of these conclusions. This result is the most often used special case of the theorem on cohomology and base change (Theorem B.9):

Theorem B.5 (Cohomology and base change, version 1). *Let $\pi : X \to B$ be a projective morphism of varieties and let $\mathcal{F}$ be a coherent sheaf on X that is flat over B. If the dimension of $H^0(\mathcal{F}|_{\pi^{-1}(b)})$ is independent of the closed point $b \in B$, then $\pi_*\mathcal{F}$ is a vector bundle of rank equal to $h^0(\mathcal{F}|_{\pi^{-1}(b)})$, and the comparison map*

$$\varphi_b : (\pi_*\mathcal{F})_b \to H^0(\mathcal{F}|_{\pi^{-1}(b)})$$

is an isomorphism for every closed point $b \in B$. More generally, if B' is any scheme, $\rho : B' \to B$ is a morphism, and

$$
\begin{array}{ccc}
X' = X \times_B B' & \xrightarrow{\ \rho'\ } & X \\
{\scriptstyle \pi'}\downarrow & & \downarrow{\scriptstyle \pi} \\
B' & \xrightarrow{\ \ \rho\ \ } & B
\end{array}
$$

is a pullback diagram, then the natural map

$$\varphi_{B'} : \rho^*\pi_*\mathcal{F} \to \pi'_*\rho'^*\mathcal{F}$$

is an isomorphism.

Theorem B.5 is subsumed by Theorem B.9, which will be proven in Section B.5.

We note that although we allow B' to be an arbitrary scheme it is necessary for the formulation above to assume that B is a variety (being reduced would be enough).

We will make use of the following natural "adjunction" maps, defined for any morphism $\pi : X \to B$ of schemes as follows:

(a) If $\mathcal{F}$ is a quasi-coherent sheaf on X, there is a natural map

$$\epsilon_{\mathcal{F}} : \pi^* \pi_* \mathcal{F} \to \mathcal{F},$$

defined by the condition that, on the preimage of an open set $U \subset B$, $\epsilon_{\mathcal{F}}$ is the map

$$\mathcal{O}_X(\pi^{-1}(U)) \otimes_{\mathcal{O}_Y(U)} \mathcal{F}(\pi^{-1}(U)) \to \mathcal{F}(\pi^{-1}(U))$$

sending each section $1 \otimes \sigma$ to σ. This is well-defined because $\mathcal{F}(\pi^{-1}(U))$ is an $\mathcal{O}_X(\pi^{-1}(U))$-module.

(b) Given a quasi-coherent sheaf $\mathcal{G}$ on B, there is a natural map

$$\eta_{\mathcal{G}} : \mathcal{G} \to \pi_* \pi^* \mathcal{G},$$

defined by the condition that, on any open set $U \subset B$, $\eta_{\mathcal{G}}$ is the map

$$\mathcal{G}(U) = \mathcal{O}_B(U) \otimes_{\mathcal{O}_B(U)} \mathcal{G}(U) \xrightarrow{\pi^* \otimes 1} \mathcal{O}_X(\pi^{-1}(U)) \otimes_{\mathcal{O}_B(U)} \mathcal{G}(U)$$
$$= \pi^*(\mathcal{G})(\pi^{-1}(U))$$
$$= (\pi_* \pi^* \mathcal{G})(U).$$

For example, $\pi^* \mathcal{O}_B = \mathcal{O}_X$, so $\eta_{\mathcal{O}_B}$ is a map $\mathcal{O}_B \to \pi_* \mathcal{O}_X$.

As an application that does not mention base change, we derive a result about line bundles that we used in studying projective bundles (Chapter 9):

Corollary B.6. *Suppose $\pi : X \to B$ is a flat, projective morphism and that all the fibers of π are reduced and connected.*

(a) *$\eta_{\mathcal{O}_B} : \mathcal{O}_B \to \pi_* \mathcal{O}_X$ is an isomorphism.*

(b) *If $\mathcal{L}, \mathcal{L}'$ are line bundles on X, then $\mathcal{L}|_{\pi^{-1}(b)} \cong \mathcal{L}'|_{\pi^{-1}(b)}$ for all $b \in B$ if and only if $\mathcal{L} \cong (\pi^* \mathcal{M}) \otimes \mathcal{L}'$ for some line bundle $\mathcal{M}$ on B, that is, if $\mathcal{L}$ and $\mathcal{L}'$ differ by tensoring with a line bundle pulled back from B.*

Remark. The result fails without flatness, for example in the case when π is the embedding of a proper closed subscheme of B.

Proof: To say that X is flat over B means that $\mathcal{O}_X$ is flat over B. Since flatness is a local property, this implies that any line bundle on X is flat over B, so we may apply Theorem B.9 to line bundles on X.

(a) The map $\eta_{\mathcal{O}_B}$ takes the global section 1 of $\mathcal{O}_B$ to the global section 1 in $\pi_* \mathcal{O}_X$. Because X is flat, $1 \in \pi_* \mathcal{O}_X$ is not annihilated by any nonzero (local) section of $\mathcal{O}_B$, so $\eta_{\mathcal{O}_B}$ is injective.

Since $\pi^{-1}(b)$ is a reduced, connected projective variety for every $b \in B$, the vector space $H^0(\mathcal{O}_X|_{\pi^{-1}(b)}) = H^0(\mathcal{O}_{\pi^{-1}(b)})$ is one-dimensional. With this, Theorem B.5 shows that $\pi_* \mathcal{O}_X$ is a line bundle with fiber $H^0(\mathcal{O}_{\pi^{-1}(b)})$. It follows that the restriction of $\eta_{\mathcal{O}_B}$ to any fiber is surjective. By Nakayama's lemma, $\eta_{\mathcal{O}_B}$ itself is surjective.

(b) First, suppose that $\mathcal{L}' = \mathcal{L} \otimes \pi^*\mathcal{M}$ for some line bundle $\mathcal{M}$ on B. We have

$$\mathcal{L}'|_{\pi^{-1}(b)} = (\mathcal{L} \otimes \pi^*\mathcal{M})|_{\pi^{-1}(b)} = \mathcal{M}_b \otimes \mathcal{L}|_{\pi^{-1}(b)} \cong \mathcal{L}|_{\pi^{-1}(b)}.$$

For the converse, given $\mathcal{L}$ and $\mathcal{L}'$, we may multiply both by $\mathcal{L}'^{-1}$, and thus reduce to the case where $\mathcal{L}' = \mathcal{O}_X$, so that $\mathcal{L}|_{\pi^{-1}(b)}$ is trivial for each $b \in B$. Our hypothesis then implies that $H^0(\mathcal{L}|_{\pi^{-1}(b)})$ is one-dimensional for every $b \in B$, so by Theorem B.9 $\pi_*\mathcal{L}$ is a line bundle.

We will complete the proof by showing that $\epsilon_\mathcal{L} : \pi^*\pi_*\mathcal{L} \to \mathcal{L}$ is an isomorphism. Since both the source and target are line bundles, it suffices to show that $\epsilon_\mathcal{L}$ is surjective, and for this we may by Nakayama's lemma restrict to a fiber. By Theorem B.5, the fiber of $\pi_*\mathcal{L}$ at a point b is $H^0(\mathcal{L}|_{\pi^{-1}(b)})$, so $(\pi^*\pi_*\mathcal{L})|_{\pi^{-1}(b)}$ is the trivial line bundle of rank 1, generated by $1 \otimes \sigma$ for any nonzero global section σ of $\mathcal{L}|_{\pi^{-1}(b)}$. Since $\epsilon_\mathcal{L}$ sends $1 \otimes \sigma$ to σ, and $\mathcal{L}|_{\pi^{-1}(b)}$ is a trivial line bundle, we see that $\epsilon_\mathcal{L}$ restricts to an isomorphism on each fiber, as required. $\square$

B.3 Higher direct images

Let $\pi : X \to B$ be a morphism and let $\mathcal{F}$ be a sheaf on X. The direct image functor $\mathcal{F} \mapsto \pi_*\mathcal{F}$ is a generalization of the functor sending a sheaf to its vector space of global sections. The *higher direct image functors* $\mathcal{F} \mapsto R^i\pi_*\mathcal{F}$ have the same relation to the higher cohomology, and may be defined as right derived functors of π_* or as the sheafification of the presheaf $U \mapsto H^i(\mathcal{F}|_{\pi^{-1}(U)})$ (in the case $i = 0$ the sheafification is unnecessary). For this see Hartshorne [1977, Section III.8]. Here we will take a more concrete approach, defining the right derived functors via the Čech complex and for simplicity sticking to the case of a projective morphism.

Just as the higher cohomology of a sheaf can be used to derive information about global sections, the higher direct image sheaves shed light on the direct image itself. There are other applications as well. For example, in Chapter 14 we used higher direct images to study jumping lines.

We will deal with both the cohomology of sheaves and the homology of complexes. To limit confusion, we will use H^i to denote the i-th Čech cohomology functor applied to a sheaf, while H^i will denote the i-th homology of a complex.

Suppose that $\pi : X \to B$ is a projective morphism; that is, π factors as a closed immersion $X \subset B \times \mathbb{P}^n$ and the projection $B \times \mathbb{P}^n \to B$. If $\mathcal{F}$ is a sheaf on X, we may regard $\mathcal{F}$ as a sheaf on $\mathbb{P} := B \times \mathbb{P}^n$. We write $\mathbb{P} = \mathrm{Proj}(S)$, where S is the sheaf of graded algebras $\mathcal{O}_B[x_0, \ldots, x_n]$, and thus $\mathcal{F}$ is the sheafification of a sheaf of graded S-modules.

Let $U_i \subset \mathbb{P}$ be the open subscheme $x_i \neq 0$, and let

$$\mathcal{C}^\bullet : \quad \bigoplus_i \mathcal{O}_{\mathbb{P}}|_{U_i} \longrightarrow \bigoplus_{i,j} \mathcal{O}_{\mathbb{P}}|_{U_i \cap U_j} \longrightarrow \cdots$$

be the Čech complex on $\mathbb{P}$. Note that each term $\mathcal{C}^i$ is $\mathbb{Z}$-graded. If $\mathcal{F}$ is a quasi-coherent sheaf on X we define $R^i \pi_* \mathcal{F}$ to be the degree-0 part of the homology of the complex $\mathcal{F} \otimes_{\mathcal{O}_{\mathbb{P}}} \mathcal{C}^\bullet$ at the i-th term, that is,

$$R^i \pi_* \mathcal{F} := (\mathrm{H}^i (\mathcal{F} \otimes_{\mathcal{O}_{\mathbb{P}}} \mathcal{C}^\bullet))_0.$$

What makes this somewhat technical definition useful is that

$$(\mathcal{F} \otimes \mathcal{O}_{\mathbb{P}}|_{U_i})_0$$

is the sheaf of modules over

$$\mathcal{O}_B[x_0, \ldots, x_{i-1}, \hat{x}_i, x_{i+1}, \ldots, x_n]$$

corresponding to the restriction of the sheaf $\mathcal{F}$ to the open set $U_i = B \times \mathbb{A}^n$. This and the assumption that $\mathcal{F}$ is a sheaf show that $R^0 \pi_* \mathcal{F} = \pi_* \mathcal{F}$. Also, when $U \subset B$ is affine, so that each $U_i \cap \pi^{-1}(U)$ is affine, we get

$$R^i \pi_* \mathcal{F}|_{\pi^{-1}(U)} = H^i (\mathcal{F}|_{\pi^{-1}(U)}).$$

Since any sheaf is determined by its restriction to affine open subsets, this property (together with the restriction morphisms) characterizes $R^i \pi_* \mathcal{F}$ and shows that the definition is independent of the embedding $X \subset B \times \mathbb{P}^n$ that we chose. If $b \in B$ is a point or a subvariety, then

$$H^i (\mathcal{F}_b) = \mathrm{H}^i (\kappa(b) \otimes \mathcal{F} \otimes \mathcal{C}^\bullet).$$

This is generally not equal to the fiber

$$(R^i \pi_* \mathcal{F})_b = \kappa(b) \otimes \mathrm{H}^i (\mathcal{F} \otimes \mathcal{C}^\bullet)$$

of the higher direct image. However, if z is a cycle or boundary in $\mathcal{F} \otimes \mathcal{C}^\bullet$, then $1 \otimes z$ is a cycle or boundary in $\kappa(b) \otimes \mathcal{F} \otimes \mathcal{C}^\bullet$, so we get maps $R^i \pi_* \mathcal{F} \to H^i (\mathcal{F}|_{X_b})$ that in turn induce comparison maps

$$\varphi_b^i : (R^i \pi_* \mathcal{F})_b \to H^i (\mathcal{F}|_{X_b}).$$

In the previous section we asked when the groups $H^i (\mathcal{F}|_{X_b})$ are the fibers of a sheaf, and when this sheaf is a vector bundle. Again, we will give sufficient conditions for these things to be the case by giving conditions for the maps φ_b^i to be isomorphisms and for $R^i \pi_* \mathcal{F}$ to be a vector bundle.

We start with some properties of the sheaves $R^i \pi_* \mathcal{F}$:

Proposition B.7. *Let $\pi : X \to B$ be a projective morphism.*

(a) *(Restriction to open sets) Let $U \subset B$ be an open subset, and let $\pi' : \pi^{-1}(U) \to U$ be the restriction of π. If $\mathcal{F}$ is any quasi-coherent sheaf on X, then $(R^i \pi_* \mathcal{F})|_U = R^i \pi'_*(\mathcal{F}|_{\pi^{-1}(U)})$.*

(b) *(Long exact sequence) The functor π_* is left exact, and the functors $R^i \pi_*$ are the right derived functors of π_*. In particular, if*

$$\epsilon : 0 \longrightarrow \mathcal{F} \xrightarrow{\alpha} \mathcal{G} \xrightarrow{\beta} \mathcal{H} \longrightarrow 0$$

is a short exact sequence of quasi-coherent sheaves on X, then there are natural "connecting homomorphisms" η_i making the sequence

$$\cdots \longrightarrow R^i \pi_* \mathcal{F} \xrightarrow{R^i \pi_* \alpha} R^i \pi_* \mathcal{G} \xrightarrow{R^i \pi_* \beta} R^i \pi_* \mathcal{H} \xrightarrow{\eta_i} R^{i+1} \pi_* \mathcal{F} \longrightarrow \cdots$$

exact.

(c) *(Push-pull formula) If $\mathcal{E}$ is a vector bundle on B and $\mathcal{F}$ is a quasi-coherent sheaf on X, then*

$$R^i \pi_*(\pi^* \mathcal{E} \otimes \mathcal{F}) \cong \mathcal{E} \otimes R^i \pi_* \mathcal{F}.$$

Proof: (a) Since $\mathcal{O}_U$ is flat over $\mathcal{O}_B$, the restriction to U commutes with taking homology.

(b) The terms of the complex $C^\bullet$ are flat over $\mathcal{O}_X$, so when we tensor $C^\bullet$ with the short exact sequence ϵ we get a short exact sequence of complexes, and thus a long exact sequence of homology sheaves. Taking the degree-0 part preserves exactness. Since the long exact sequence begins with

$$0 \longrightarrow \pi_* \mathcal{F} \longrightarrow \pi_* \mathcal{G} \longrightarrow \cdots,$$

we see that π_* is left exact.

To show that $R^i \pi_*$ is the i-th right derived functor of π_*, it now suffices to show that $R^i \pi_* \mathcal{F} = 0$ when $\mathcal{F}$ is injective (Eisenbud [1995, Appendix A3.9]), or more generally flasque. It suffices to prove this after restricting to an affine subset $U \subset B$. Since the restriction of a flasque sheaf to an open subset is flasque, the result follows from the corresponding result for cohomology.

(c) The sheaf $\pi^* \mathcal{E}$ is also a vector bundle, and thus flat, so tensoring with $\pi^* \mathcal{E}$ commutes with taking homology:

$$H^i(\pi^* \mathcal{E} \otimes \mathcal{F} \otimes C^\bullet) = \pi^* \mathcal{E} \otimes H^i(\mathcal{F} \otimes C^\bullet).$$

Taking the degree-0 part yields the desired formula. $\qquad\square$

Theorem B.8 (Serre's coherence theorem). *If $\pi : X \to B$ is a projective morphism and $\mathcal{F}$ is a coherent sheaf on X, then $R^i \pi_* \mathcal{F}$ is coherent for each i.*

The proof involves some useful ideas from homological commutative algebra.

Proof: Since the formation of $R^i \pi_* \mathcal{F}$ commutes with the restriction to an open set in the base, it suffices to treat the case where $B = \operatorname{Spec} A$ is affine. The Čech complex $\mathcal{C}^\bullet$ is the direct limit of the duals of the Koszul complexes that are S-free resolutions of the ideals $(x_0^m, \ldots, x_n^m) \subset S$, and direct limits commute with taking homology. Thus if M is any finitely generated graded $S = A[x_0, \ldots, x_n]$-module representing the sheaf $\mathcal{F}$, the homology of $\mathcal{F} \otimes \mathcal{C}^\bullet$ is

$$R^i \pi_* \mathcal{F} = \lim_m \operatorname{Ext}_S^i((x_0^m, \ldots, x_n^m), M)_0.$$

Write $\mathfrak{m}$ for the "irrelevant" ideal $(x_0, \ldots, x_n) \subset S$. For each m there is an integer $N(m)$ such that $\mathfrak{m}^{N(m)} \subset (x_0^m, \ldots, x_n^m) \subset \mathfrak{m}^m$. It follows that

$$\lim_m \operatorname{Ext}_S^i((x_0^m, \ldots, x_n^m), M) = \lim_m \operatorname{Ext}_S^i(\mathfrak{m}^m, M).$$

Each term $\operatorname{Ext}_S^i(\mathfrak{m}^m, M)$ of this limit is a finitely generated S-module, so its degree-0 part is a finitely generated A-module, and it suffices to show that the natural map

$$\operatorname{Ext}_S^i(\mathfrak{m}^m, M)_0 \to \operatorname{Ext}_S^i(\mathfrak{m}^{m+1}, M)_0$$

is an isomorphism for large m. From the long exact sequence in Ext_S, and the fact that $\mathfrak{m}^m / \mathfrak{m}^{m+1}$ is a direct sum of copies of $A(-m)$ (the free A-module of rank 1 with generator in degree m), we see that it is enough to prove that $\operatorname{Ext}_S^i(A(-m), M)_0 = 0$ when m is large. Disentangling the degree shifts, we see that

$$\operatorname{Ext}_S^i(A(-m), M)_0 = \operatorname{Ext}_S^i(A, M)(m)_0 = \operatorname{Ext}_S^i(A, M)_m.$$

However, A is annihilated (as an S-module) by $\mathfrak{m}$, so $\operatorname{Ext}_S^i(A, M)$ is annihilated by $\mathfrak{m}$. Since it is a finitely generated S-module, it can only be nonzero in finitely many degrees, whence, indeed, $\operatorname{Ext}_S^i(A, M)_m = 0$ when m is large. $\quad\square$

We remark that it is possible, using the notion of Castelnuovo–Mumford regularity, to bound the degree m for which $\operatorname{Ext}_S^i(A, M)_m = 0$ in terms of the data in a free resolution of M (similar to Smith [2000]), so the proof just given allows effective computation of the functors $R^i \pi_* \mathcal{F}$. For a different proof see Hartshorne [1977, Theorem III.8.8].

Theorem B.9 (Cohomology and base change, version 2). *Let $\pi : X \to B$ be a projective morphism of schemes, with B connected and quasi-projective, and let $\mathcal{F}$ be a coherent sheaf on X.*

(a) *If B is reduced then there is a dense open set $U \subset B$ such that $R^i \pi_* \mathcal{F}|_U$ is a vector bundle, and such that for all closed points $b \in U$ the fiber $(R^i \pi_* \mathcal{F}|_U)_b$ is equal to $H^i(\mathcal{F}|_{X_b})$.*

(b) *Suppose that $\mathcal{F}$ is flat over B. Let i be an integer. If $H^j(\mathcal{F}|_{X_b}) = 0$ for all $j > i$ and all closed points $b \in B$, then for every closed point $b \in B$ the comparison map*

$$\varphi_b^i : (R^i \pi_* \mathcal{F})_b \to H^i(\mathcal{F}|_{X_b})$$

is an isomorphism.

(c) *Suppose that $\mathcal{F}$ is flat over B and B is reduced. If for some i the function $b \mapsto$ $\dim_{\kappa(b)} H^i(\mathcal{F}|_{X_b})$ is constant, then $R^i \pi_* \mathcal{F}$ is a vector bundle of rank equal to $\dim_{\kappa(b)} H^i(\mathcal{F}|_{X_b})$, and for every closed point $b \in B$ the comparison maps*

$$\varphi_b^i : (R^i \pi_* \mathcal{F})_b \to H^i(\mathcal{F}|_{X_b}),$$
$$\varphi_b^{i-1} : (R^{i-1} \pi_* \mathcal{F})_b \to H^{i-1}(\mathcal{F}|_{X_b})$$

are isomorphisms. More generally, if B' is any scheme, $\rho : B' \to B$ is a morphism, and

$$
\begin{array}{ccc}
X' = X \times_B B' & \xrightarrow{\rho'} & X \\
{\scriptstyle \pi'} \downarrow & & \downarrow {\scriptstyle \pi} \\
B' & \xrightarrow{\rho} & B
\end{array}
$$

is a pullback diagram, then the natural map

$$\varphi_{B'}^i : \rho^* R^i \pi_* \mathcal{F} \to R^i \pi'_* \rho'^* \mathcal{F}$$

is an isomorphism.

Theorem B.9 will be proven in Section B.5.

Example B.10 (Continuation of Example B.1). Using Theorem B.9 we can easily compute the sheaves $R^i \pi_* \mathcal{I}_\Gamma(d)$ and $R^i \pi_* \mathcal{I}_{\Gamma'}(d)$ introduced in Example B.1. The sheaves $\mathcal{I}_\Gamma$ and $\mathcal{I}_{\Gamma'}$ are flat over $B \cong \mathbb{P}^1$ because they have no torsion (see for example Eisenbud and Harris [2000, Theorem II-29]). The functions $b \mapsto h^i(\mathcal{I}_{\Gamma'_b}(d))$ are constant for all d, since Γ'_b is itself constant. The computation discussed in Example B.1 shows that the same is true for $\mathcal{I}_{\Gamma_b}(d)$ as long as $d \geq 2$. It follows that for all d

$$R^0 \mathcal{I}_{\Gamma'}(d) = \mathcal{O}_B^{\binom{d+2}{2}-2},$$
$$R^1 \mathcal{I}_{\Gamma'}(d) = 0,$$

and for $d \geq 2$

$$R^0 \mathcal{I}_\Gamma(d) = \mathcal{O}_B^{\binom{d+2}{2}-3},$$
$$R^1 \mathcal{I}_\Gamma(d) = 0.$$

On the other hand, if $d = 1$, we will prove that

$$R^0 \mathcal{I}_\Gamma(d) = 0,$$
$$R^1 \mathcal{I}_\Gamma(d) = \mathcal{O}_{r_0}.$$

To this end we first apply Theorem B.9, which shows that $R^i \pi_* \mathcal{O}_{B \times \mathbb{P}^2}(1)$ is a vector bundle for each i and its fiber over $r \in B$ is $H^i \mathcal{O}_{\mathbb{P}^2}(1)$, while $R^0 \pi_* \mathcal{O}_\Gamma(1)$ is a vector bundle of rank 3 whose fiber over r is the set of functions on $\{p, q, r\}$. We now use Proposition B.7(b) to obtain the exact sequence

$$0 \longrightarrow R^0 \mathcal{I}_\Gamma(1) \longrightarrow R^0 \pi_* \mathcal{O}_{B \times \mathbb{P}^2}(1) \xrightarrow{e} R^0 \pi_* \mathcal{O}_\Gamma(1) \longrightarrow R^1 \mathcal{I}_\Gamma(1) \longrightarrow 0.$$

The map labeled e, restricted to the fiber over r, is (after choosing coordinates) the map $\Bbbk^3 \to \Bbbk^3$ sending each linear form on $\mathbb{P}^2$ to the vector of its values at the (homogeneous coordinates of the) three points p, q, r. When the points are non-collinear, this map is an isomorphism. A map of vector bundles that is generically an isomorphism is a monomorphism of sheaves, so $R^0 \mathcal{I}_\Gamma(1) = \operatorname{Ker} e = 0$. The unique fiber where the rank of e drops is r_0, and there the rank of e is 2. At the fiber over r_0 the image of the map is two-dimensional. It follows that $R^1 \mathcal{I}_\Gamma(1) = \operatorname{coker} e$ is the skyscraper sheaf of length 1 concentrated at the point r_0, as claimed.

B.4 The direct image complex

We now turn to the most general and simplest version of the theorem on base change and cohomology. To simplify the notation, we will identify quasi-coherent sheaves over an affine scheme $B = \operatorname{Spec} A$ with their modules of global sections.

Theorem B.11 (Cohomology and base change, version 3). *Let $\pi : X \to B$ be a projective morphism to an affine scheme $B = \operatorname{Spec} A$, and let $\mathcal{F}$ be a sheaf on X that is flat over B. Suppose that the maximum dimension of a fiber of π is n. There is a complex*

$$\mathcal{P}^\bullet : \cdots \longrightarrow P^0 \longrightarrow \cdots \longrightarrow P^n \longrightarrow 0$$

of finitely generated projective A-modules such that:

(a) $R^i \pi_*(\mathcal{F}) \cong H^i(\mathcal{P}^\bullet)$ for all i.
(b) For every $b \in B$ and $i \in \mathbb{Z}$ there is an isomorphism

$$H^i(\mathcal{F}|_{X_b}) \cong H^i(\kappa(b) \otimes_A \mathcal{P}^\bullet).$$

More generally, for every pullback diagram

$$\begin{array}{ccc} X' & \xrightarrow{\rho'} & X \\ {\scriptstyle \pi'}\downarrow & & \downarrow{\scriptstyle \pi} \\ B' & \xrightarrow{\rho} & B \end{array}$$

with $B' = \operatorname{Spec} A'$,

$$R^i \pi'_* \rho'^*(\mathcal{P}^\bullet) \cong H^i(A' \otimes_A \mathcal{P}^\bullet).$$

Moreover, if B is reduced, then we may choose $\mathcal{P}^\bullet$ so that $P^i = 0$ for $i < 0$.

Theorem B.11 is proven at the end of this section. A version where the base B is any quasi-projective scheme is given in Exercise B.18.

The second statement of (b) could be improved to say that $\rho^* R\pi_* \mathcal{F}$ is quasi-isomorphic (or, equivalently, isomorphic in the derived category) to $R\pi_* \rho'^* \mathcal{F}$.

The complex $\mathcal{P}^\bullet$ in Theorem B.11 is not unique up to isomorphism, but it is unique up to the equivalence relation, called *quasi-isomorphism*, generated by maps of complexes that induce isomorphisms on homology (for right-bounded projective complexes, as in our case, this comes down to homotopy equivalence). The class of $\mathcal{P}^\bullet$ modulo quasi-isomorphism is called the *direct image complex of $\mathcal{F}$*, written $R\pi_*\mathcal{F}$, which is usually treated as an element of the derived category of (right-bounded) complexes of coherent sheaves on B. We say that $\mathcal{P}^\bullet$ *represents $R\pi_*\mathcal{F}$*.

Abstract as Theorem B.11 may seem, the construction of $R\pi_*\mathcal{F}$ can be performed explicitly in examples of modest size, for instance by the computer algebra package *Macaulay2*; see Exercise B.20.

Theorem B.11 makes the proof of most other statements about base change easy, as we shall see in the next section. Here is a taste:

Corollary B.12. *Let $\pi : X \to B$ be a projective morphism of schemes and let $\mathcal{F}$ be a sheaf on X that is flat over B. For each i, the dimension function*

$$B \ni b \mapsto \dim_{\kappa(b)} H^i(\mathcal{F}|_{X_b})$$

is an upper-semicontinuous function (in particular, it takes its smallest value on an open set). Moreover, the Euler characteristic

$$\chi(\mathcal{F}|_{X_b}) := \sum (-1)^i \dim_{\kappa(b)} H^i(\mathcal{F}|_{X_b})$$

is constant on connected components of B.

Proof: It suffices to prove the result in the case where B is affine and connected, say $B = \operatorname{Spec} A$. Let $\mathcal{P}^\bullet$ be a complex of finitely generated projective A-modules with the properties given in Theorem B.11. Restricting to some possibly smaller open set of B, we may assume that $\mathcal{P}^\bullet$ is a complex of finitely generated free modules. For each $b \in B$ we get a complex of vector spaces by taking the fiber of $\mathcal{P}^\bullet$ at b. The maps $\varphi^i : P^i \to P^{i-1}$ in $\mathcal{P}^\bullet$ are given by matrices with entries in A, and thus the rank of $\varphi_b^i := \kappa(b) \otimes_A \varphi^i$ is a semicontinuous function of b. It follows that

$$\dim_{\kappa(b)} H^i(\mathcal{P}^\bullet|_b) = \dim_{\kappa(b)} P^i|_b - \operatorname{rank} \varphi_b^{i+1} - \operatorname{rank} \varphi_b^i$$

is a semicontinuous function of b. Further, the Euler characteristic

$$\begin{aligned}
\chi(\mathcal{F}|_{X_b}) &= \sum (-1)^i \dim_{\kappa(b)} H^i(\mathcal{P}^\bullet|_b) \\
&= \sum (-1)^i \dim_{\kappa(b)}(P^i|_b) \\
&= \sum (-1)^i \operatorname{rank} P^i
\end{aligned}$$

is a constant function of b. $\qquad\square$

Example B.13 (Further continuation of Example B.1)**.** The tools above can be converted into algorithms for computing the direct image complex of a coherent sheaf — see Exercise B.20 — but sometimes one can understand the result without computation.

The derived category $D^b(\mathbb{P}^1)$ of bounded complexes of coherent sheaves on $\mathbb{P}^1$ (or on any smooth curve) is *formal*, in the sense that every bounded complex of locally free sheaves is quasi-isomorphic to the direct sum of its homology sheaves, or equivalently to the direct sum of locally free resolutions of its homology sheaves (see Exercise B.17).

Thus, if $\mathcal{F}$ is a family of sheaves on the family of varieties $\pi : X \to C$, where C is a smooth curve and $\mathcal{F}$ is flat over C with support of relative dimension n, the direct image complex may be taken simply to be the direct sum of locally free resolutions of the coherent sheaves $R^0\pi_*\mathcal{F}, R^1\pi_*\mathcal{F}, \ldots, R^n\pi_*\mathcal{F}$, each in its appropriate homological degrees.

Returning to Example B.1, we see for example that the direct image complex $R\pi_*\mathcal{I}_{\Gamma'}(1)$ is the quasi-isomorphism class of

$$0 \longrightarrow \mathcal{O}_B \longrightarrow 0,$$

with nonzero term in cohomological degree 0, while $R\pi_*\mathcal{I}_\Gamma(1)$ is the quasi-isomorphism class of

$$0 \longrightarrow \mathcal{O}_B(-1) \xrightarrow{\sigma} \mathcal{O}_B \longrightarrow 0,$$

where $\mathcal{O}_B$ is in cohomological degree 1 and the differential is multiplication by a section (unique up to scalars) that vanishes at the point r_0. Further, the map $\mathcal{I}_\Gamma(1) \subset \mathcal{I}_{\Gamma'}(1)$ induces

$$
\begin{array}{ccccccc}
R\pi_*(\mathcal{I}_{\Gamma'}(1)) : & 0 & \longrightarrow & \mathcal{O}_B & \longrightarrow & 0 & \longrightarrow & 0 \\
& & & \uparrow & & \uparrow & & \\
R\pi_*(\mathcal{I}_\Gamma(1)) : & 0 & \longrightarrow & \mathcal{O}_B & \xrightarrow{\sigma} & \mathcal{O}_B(1) & \longrightarrow & 0
\end{array}
$$

The following example shows that the hypothesis of flatness in Theorem B.11 is essential:

Example B.14. Let $B = \mathbb{A}^2$ and let $\pi : X \to B$ be the blow-up of B at the origin. Let $\mathbb{P}^1 \cong E \subset X$ be the exceptional divisor and let $\mathcal{F}$ be the line bundle $\mathcal{O}(E)$. Note that $\mathcal{F}$ is not flat over B. We have $\chi(\mathcal{F}|_b) = 1$ for $b \neq 0$, but $\mathcal{F}|_0 = \mathcal{O}_{\mathbb{P}^1}(-1)$, so the dimension of $H^0(\mathcal{F}|_b)$ is not upper-semicontinuous and the Euler characteristic $\chi(\mathcal{F}|_b)$ is not constant.

B.5 Proofs of the theorems on cohomology and base change

We require two tools from commutative algebra. First, a fundamental method for proving that a sheaf is a vector bundle (that is, is locally free):

Proposition B.15. *A coherent sheaf $\mathcal{G}$ on a connected reduced scheme B is a vector bundle if and only the dimension of the $\kappa(b)$-vector space $\mathcal{G}_b$ is the same for all points $b \in B$; if B is quasi-projective, then it even suffices to check this for closed points. These conditions are satisfied, in particular, if $\mathcal{G}$ has a resolution*

$$\mathcal{P}^\bullet : \cdots \longrightarrow P^{-1} \longrightarrow P^0 \longrightarrow \mathcal{G} \longrightarrow 0$$

by vector bundles of finite rank that remains a resolution when tensored with $\kappa(b)$ for every $b \in B$.

Proof: If $\mathcal{G}$ is a vector bundle, then it has constant rank because B is connected, and this rank is the common dimension of $\mathcal{G}_b$ over $\kappa(b)$.

To prove the converse, and also the last statement of the proposition, we note that the problem is local, so we may assume that $B = \operatorname{Spec} A$, where A is a local ring with maximal ideal $\mathfrak{m}$ corresponding to the closed point $b_0 \in B$, and that $\mathcal{G}$ is the sheaf associated to a finitely generated A-module G. By Nakayama's lemma, a minimal set of generators of G corresponds to a map $f : F \to G$ from a free A-module F such that the induced map $(A/\mathfrak{m}) \otimes F \to (A/\mathfrak{m}) \otimes G$ is an isomorphism. In particular, the rank of the free module F is $\dim_{A/\mathfrak{m}}(A/(\mathfrak{m}) \otimes G) = \dim_{\kappa(b)} \mathcal{G}|_b$.

Let $K = \operatorname{Ker} \varphi$, and let P be a minimal prime of A. Since A is reduced, A_P is a field. Localizing at P, we get an exact sequence of finite dimensional vector spaces

$$0 \longrightarrow K_P \longrightarrow F_P \longrightarrow G_P \longrightarrow 0,$$

from which we have rank $F = \dim_{A_P} G_P + \dim_{A_P} K_P$. By hypothesis, $\dim_{A_P} G_P = \dim_{A/\mathfrak{m}} G/\mathfrak{m}G = \operatorname{rank} F$, so $K_P = 0$. Since A is reduced, the only associated primes of F are the minimal primes of A, and thus $K \subset F$ itself must be zero.

To prove that it suffices to assume the constancy of the dimension of $\mathcal{G}_b$ at closed points in the quasi-projective case, it suffices to show that

$$\dim_{\kappa(\eta)} \mathcal{G}_\eta = \min_b \dim_{\kappa(b)} \mathcal{G}_b,$$

where η is a generic point and the minimum is take over the closed points in the closure of η. The inequality $\leq$ follows at once from Nakayama's lemma, since $\mathcal{G}$ is generated locally at b — and thus at η — by $\dim_{\kappa(b)} \mathcal{G}_b$ elements.

For the opposite inequality, choose elements $\{x_i\} \subset \mathcal{G}$ that form a $\kappa(\eta)$-basis for $\mathcal{G}_\eta$ and a set of generators $\{y_i\} \subset \mathcal{G}$ for $\mathcal{G}$. We can express all the y_i as linear combinations of the x_i with rational coefficients, using finitely many denominators. By the Nullstellensatz, there is a closed point b in the closure of η such that the product of these denominators is invertible at b, and it follows that the elements x_i span $\mathcal{G}_b$, so $\dim_{\kappa(\eta)} \mathcal{G}_\eta \geq \dim_{\kappa(b)} \mathcal{G}_b$ as required.

For the last statement of the corollary, suppose that $\mathcal{G}$ has a resolution $\mathcal{P}^\bullet$ with the given property. Since B is the spectrum of the local ring A, we may identify $\mathcal{P}^\bullet$ with a free resolution of $\mathcal{G}$. Since a minimal free resolution is a summand of any resolution

(Eisenbud [1995, Theorem 20.2]) the minimal free resolution $\mathcal{P}'^\bullet$ of $\mathcal{G}$ has the same property. But after tensoring with the residue class field $\kappa(b_0)$ the differentials in $\mathcal{P}'^\bullet$ become zero. Since by hypothesis the resolution remains acyclic, we must have $P'^0 = \mathcal{G}$, so $\mathcal{G}$ is free. $\qquad\qquad\qquad\qquad\qquad\qquad\qquad\qquad\qquad\qquad\qquad\qquad\square$

The proof of Theorem B.11 requires one more tool, a way of approximating a complex by a complex of free modules with good properties.

Proposition B.16. *Let A be a Noetherian ring, and let*

$$\mathcal{K}^\bullet : \quad \cdots \xrightarrow{\ d\ } K^i \xrightarrow{\ d\ } K^{i+1} \xrightarrow{\ d\ } \cdots$$

be a complex of (not necessarily finitely generated) flat A-modules whose homology modules are finitely generated and such that $K^m = 0$ for $m \gg 0$. There is a complex of finitely generated free A-modules $\mathcal{P}^\bullet$ with $P^m = 0$ for $m \gg 0$ and a map of complexes $r : \mathcal{P}^\bullet \to \mathcal{K}^\bullet$ such that for every A-module M the map

$$r \otimes_A M : \mathcal{P}^\bullet \otimes_A M \longrightarrow \mathcal{K}^\bullet \otimes_A M$$

induces an isomorphism on homology.

Proof: We will construct a complex of finitely generated free modules $\mathcal{P}^\bullet$ with a map r to $\mathcal{K}^\bullet$ inducing an isomorphism on homology without the assumption of flatness, and then we will use the flatness hypothesis to show that any such $r : \mathcal{P}^\bullet \to \mathcal{K}^\bullet$ induces an isomorphism on homology after tensoring with the arbitrary module M.

We will construct the complex $\mathcal{P}^\bullet$ by downward induction on i, using the hypothesis that a map of complexes

$$
\begin{array}{ccccc}
K^{i+1} & \xrightarrow{\ d^{i+1}\ } & K^{i+2} & \xrightarrow{\ d^{i+2}\ } & \cdots \\[4pt]
{\scriptstyle r_{i+1}}\big\uparrow & & {\scriptstyle r_{i+2}}\big\uparrow & & \\[4pt]
P^{i+1} & \xrightarrow{\ e^{i+1}\ } & P^{i+2} & \xrightarrow{\ e^{i+2}\ } & \cdots
\end{array}
$$

inducing isomorphisms $H^j(\mathcal{P}^\bullet) \to H^j(\mathcal{K}^\bullet)$ for all $j \geq i+2$ has been constructed, with the additional property that the composite map $\operatorname{Ker} e^{i+1} \to \operatorname{Ker} d^{i+1} \to H^{i+1}(\mathcal{K}^\bullet)$ is surjective.

If i is sufficiently large that $K^m = 0$ for all $m \geq i+1$, then the choice $P^m = 0$ and $r_m = 0$ for $m \geq i+1$ satisfies these conditions, giving a base for the induction.

To make the inductive step from $i+1$ to i, we choose P^i to be the direct sum of two projective modules, $P^i = P_1^i \oplus P_2^i$, where P_1^i is chosen to map onto the kernel of the composite $\operatorname{Ker} e^{i+1} \to \operatorname{Ker} d^{i+1} \to H^{i+1}(\mathcal{K}^\bullet)$ and P_2^i is chosen to map onto $H^i(\mathcal{K}^\bullet)$. We define the differential e^i to be the given map on P_1^i and zero on P_2^i. Also, we define r_i on P_2^i by lifting the chosen map $P_2^i \to H^i(\mathcal{K}^\bullet)$ to a map $P_2^i \to \operatorname{Ker} d^i$ and composing with the inclusion $\operatorname{Ker} d^i \subset K^i$. On the other hand, since r_{i+1} carries the image of P_1^i to the kernel of the map $\operatorname{Ker} d^{i+1} \to H^{i+1}(\mathcal{K}^\bullet)$, which is by definition

$\operatorname{Im} d^i$, we may define r_i on P_1^i to be the lifting of this map $P_1^i \to \operatorname{Im} d^i$ to a map $P_1^i \to K^i$.

This gives a map of complexes

$$
\begin{array}{ccccccc}
K^i & \xrightarrow{\ d^i\ } & K^{i+1} & \xrightarrow{\ d^{i+1}\ } & K^{i+2} & \xrightarrow{\ d^{i+2}\ } & \cdots \\
{\scriptstyle r_i}\uparrow & & {\scriptstyle r_{i+1}}\uparrow & & {\scriptstyle r_{i+2}}\uparrow & & \\
P^i = P_1^i \oplus P_2^i & \xrightarrow{\ e^i\ } & P^{i+1} & \xrightarrow{\ e^{i+1}\ } & P^{i+2} & \xrightarrow{\ e^{i+2}\ } & \cdots
\end{array}
$$

It is clear from the construction that the r_i induce isomorphisms $\mathrm{H}^j(\mathcal{P}^\bullet) \to \mathrm{H}^j(\mathcal{K}^\bullet)$ for all $j \geq i+1$ and the composite map $\operatorname{Ker} e^{i+1} \to \operatorname{Ker} d^{i+1} \to \mathrm{H}^{i+1}(\mathcal{K}^\bullet)$ is surjective, so the induction is complete.

We now use the hypothesis that the K^i are flat, and suppose that we have proven that r_j induces an isomorphism $\mathrm{H}^j(\mathcal{P}^\bullet \otimes M) \to \mathrm{H}^j(\mathcal{K}^\bullet \otimes M)$ for every $j > i$ and for every module M. This is trivial in the range where K^j and P^j are both zero, so again we can do a downward induction.

Choose a surjection $F \to M$ from a free A-module, and let L be the kernel, so that

$$
0 \longrightarrow L \longrightarrow F \longrightarrow M \longrightarrow 0
$$

is a short exact sequence. Since all the K^i and the P^i are flat, we get short exact sequences of complexes by tensoring with $\mathcal{P}^\bullet$ and $\mathcal{K}^\bullet$, from which we get two long exact sequences by applying the higher direct image functors, and the comparison map $r : \mathcal{P}^\bullet \to \mathcal{K}^\bullet$ induces a commutative diagram

$$
\begin{array}{ccccccccc}
H^i(\mathcal{K}^\bullet \otimes L) & \longrightarrow & H^i(\mathcal{K}^\bullet \otimes F) & \longrightarrow & H^i(\mathcal{K}^\bullet \otimes M) & \longrightarrow & H^{i+1}(\mathcal{K}^\bullet \otimes L) & \longrightarrow & H^{i+1}(\mathcal{K}^\bullet \otimes F) \\
{\scriptstyle r_i \otimes L}\uparrow & & {\scriptstyle r_i \otimes F}\uparrow{\scriptstyle \cong} & & {\scriptstyle r_i \otimes M}\uparrow & & {\scriptstyle r_{i+1} \otimes L}\uparrow{\scriptstyle \cong} & & {\scriptstyle r_{i+1} \otimes F}\uparrow{\scriptstyle \cong} \\
H^i(\mathcal{P}^\bullet \otimes L) & \longrightarrow & H^i(\mathcal{P}^\bullet \otimes F) & \longrightarrow & H^i(\mathcal{P}^\bullet \otimes M) & \longrightarrow & H^{i+1}(\mathcal{P}^\bullet \otimes L) & \longrightarrow & H^{i+1}(\mathcal{P}^\bullet \otimes F)
\end{array}
$$

where, for any module N, we write $r_i \otimes N$ to denote the map $H^i(\mathcal{P}^\bullet \otimes N) \to H^i(\mathcal{K}^\bullet \otimes N)$ induced by r_i. The maps marked "$\cong$" are isomorphisms: $r_i \otimes F$ and $r_{i+1} \otimes F$ are isomorphisms because F is free, while $r_{i+1} \otimes L$ is an isomorphism by induction. A diagram chase (sometimes called the "five-lemma") immediately shows that the map $r_i \otimes M$ is a surjection. Since the module M was arbitrary, $r_i \otimes L$ is a surjection as well. Using this information, a second diagram chase shows that $r_i \otimes M$ is injective, completing the induction. $\qquad\square$

Proof of Theorem B.11: Since π is projective we may write $X \subset \mathbb{P} := \mathbb{P}_A^n$ for some n, and we let U_i, $i = 0, \ldots, n$, be the standard open covering of $\mathbb{P}$ as in Section B.3. Let $\mathcal{C}^\bullet$ be the Čech complex defined there. Since $\mathcal{F}$ is flat and $(\mathcal{F} \otimes \mathcal{O}_{U_i \cap U_j \cap \cdots})_0$ is the module corresponding to the restriction of $\mathcal{F}$ to the affine open set $U_i \cap U_j \cap \cdots$, the modules of the complex $(\mathcal{F} \otimes \mathcal{C}^\bullet)_0$ are flat. By Theorem B.8 the homology of this complex is finitely generated, so we may apply Proposition B.16 and obtain a complex $\mathcal{P}^\bullet$ whose

i-th homology is $R^i \pi_* \mathcal{F}$. Taking $M = \kappa(b)$ in the proposition, we see that $\mathcal{P}^\bullet$ has the second required property as well.

Finally, to show that we may choose $\mathcal{P}^\bullet$ with $P^i = 0$ for $i < 0$, note that for any choice of $\mathcal{P}^\bullet$ satisfying the proposition the homology $H^i(\mathcal{P}^\bullet)$ is zero for $i < 0$. The last statement of Proposition B.15 shows that $P'^0 := \mathrm{coker}(P^{-1} \to P^0)$ is projective. The map r_0 induces a map $P'^0 \to \mathrm{coker}(K^{-1} \to K^0)$, and since P'^0 is projective we may lift this to a new map $r_0' : P'^0 \to K^0$. It follows from the construction that

$$
\begin{array}{ccccccccc}
\cdots & \longrightarrow & K^{-1} & \longrightarrow & K^{-1} & \longrightarrow & K^0 & \longrightarrow & K^1 & \longrightarrow & \cdots \\
& & \uparrow{\scriptstyle 0} & & \uparrow{\scriptstyle 0} & & \uparrow{\scriptstyle r_0'} & & \uparrow & & \\
\cdots & \longrightarrow & 0 & \longrightarrow & 0 & \longrightarrow & P'^0 & \longrightarrow & P^1 & \longrightarrow & \cdots
\end{array}
$$

again induces an isomorphism on homology. □

Proof of Theorem B.9: The statements being local on B, we may assume from the outset that B is affine. In parts (b) and (c) we have assumed that $\mathcal{F}$ is flat, and by the generic flatness theorem (see Eisenbud [1995, Section 14.2]) there is in any case an open set $U_1 \subset B$ over which $\mathcal{F}$ is flat. Thus even for part (a) we may assume that $\mathcal{F}$ is flat over B. Let $\mathcal{P}^\bullet$ be a complex of projective modules representing $R\pi_* \mathcal{F}$, as in Theorem B.11.

(a) Removing the intersections of the components of X and then passing to a connected component, we may harmlessly assume that X is integral. Shrinking the open set U further, we may assume that the ranks of the maps in the restricted complex $(\mathcal{P}^\bullet)_b$ are constant for all $b \in U$. It follows from Proposition B.15 that all the homology modules of $\mathcal{P}^\bullet$ are vector bundles. Thus $\mathcal{P}^\bullet$ is locally split, and forming its homology commutes with any pullback.

(b) Since $H^i(\mathcal{P}^\bullet|_b) = H^i(\mathcal{F}|_{X_b})$, these spaces are zero for all $j > i$. If m is the greatest integer for which $\mathcal{P}^m \neq 0$, and $m > i$, then Nakayama's lemma implies that the map $\mathcal{P}^{m-1} \to \mathcal{P}^m$ is surjective, and thus split. Consequently we can build a quasi-isomorphic complex $\mathcal{P}'^\bullet$ by replacing $\mathcal{P}^{m-1}$ by $\mathcal{P}'^{m-1} := \mathrm{Ker}(\mathcal{P}^{m-1} \to \mathcal{P}^m)$ and truncating the complex there:

$$
\begin{array}{ccccccccc}
\mathcal{P}'^\bullet : & \cdots & \longrightarrow & \mathcal{P}^{m-2} & \longrightarrow & \mathcal{P}'^{m-1} & \longrightarrow & 0 \\
& & & \downarrow & & \downarrow & & \\
\mathcal{P}^\bullet : & \cdots & \longrightarrow & \mathcal{P}^{m-2} & \longrightarrow & \mathcal{P}^{m-1} & \longrightarrow & \mathcal{P}^m & \longrightarrow & 0
\end{array}
$$

Continuing in this way, we may assume that $\mathcal{P}^j = 0$ for all $j > i$. Since forming cokernels commutes with pullback,

$$
\begin{aligned}
H^i(\mathcal{F}|_{\pi^{-1}b}) &= \mathrm{coker}(\mathcal{P}_b^{i-1} \to \mathcal{P}_b^i) \\
&= (\mathrm{coker}(\mathcal{P}^{i-1} \to \mathcal{P}^i))_b \\
&= (R^i \pi_* \mathcal{F})_b.
\end{aligned}
$$

(c) Let

$$\cdots \longrightarrow \mathcal{P}^{i-1} \xrightarrow{d^{i-1}} \mathcal{P}^{i} \xrightarrow{d^{i}} \mathcal{P}^{i+1} \longrightarrow \cdots$$

be the differentials of $\mathcal{P}^{\bullet}$. Since the ranks of the differentials d_b^{i-1} and d_b^{i} are lower-semicontinuous, the constancy of $\dim_{\kappa(b)} H^{i}(\mathcal{F}_{\pi^{-1}(b)})$ implies the constancy of the ranks of d_b^{i-1} and d_b^{i}.

Focusing for a moment on d^{i-1}, we see that since taking fibers commutes with taking images the fibers of the module $\operatorname{Im} d^{i-1}$ have constant rank. By Proposition B.15, $\operatorname{Im} d^{i-1}$ is a projective module, so the short exact sequence

$$0 \longrightarrow \operatorname{Ker} d^{i-1} \longrightarrow \mathcal{P}^{i-1} \longrightarrow \operatorname{Im} d^{i-1} \longrightarrow 0$$

splits, and thus stays exact under pullback along $\rho : B' \to B$. It follows that

$$R^{i-1}\pi'_*(\rho'^* \mathcal{F}) = H^{i-1}(\rho^* \mathcal{P}^{\bullet}) \xrightarrow{\varphi_{B'}^{i-1}} \rho^* H^{i-1}(\mathcal{P}^{\bullet}) = \rho^* R^{i-1}\pi_*(\mathcal{F})$$

is an isomorphism.

Of course, the same considerations hold for $R^{i}\pi_*\mathcal{F}$. Furthermore, since the map $\operatorname{Im} d^{i-1} \to \operatorname{Ker} d^{i}$ has constant rank on restriction to each closed point b, so does the cokernel $R^{i}\pi_*\mathcal{F}$, so this sheaf is projective, proving part (c). $\qquad\square$

Remark. Suppose that $\pi : X \to B$ is a projective morphism and $\mathcal{F}$ is a coherent sheaf on X, flat over B, as in Theorem B.9, and suppose that $b \in B$ is a point at which $\dim_{\kappa(b)} H^{i}(\mathcal{F}_{X_b})$ "jumps" — i.e., is larger than it is for some points in any open neighborhood of b. From the constancy of the Euler characteristic $\chi(\mathcal{F}|_{X_b})$, it follows that some $\dim_{\kappa(b)} H^{j}(\mathcal{F}|_{X_b})$ with $j \neq i$ mod 2 must jump too. But the proof above gives a tiny bit more: Since the rank of d_b^{i} or of d_b^{i-1} must have gone down, either $\dim_{\kappa(b)} H^{i+1}(\mathcal{F}|_{X_b})$ or $\dim_{\kappa(b)} H^{i-1}(\mathcal{F}|_{X_b})$ must jump at b. Colloquially, "the jumps occur in adjacent pairs."

B.6 Exercises

Exercise B.17. (a) Suppose that R is a ring. Show that if

$$\mathcal{C} : \cdots \to C^{i-1} \to C^{i} \to C^{i+1} \to \cdots$$

is a complex of R-modules and $H := H^{i}\mathcal{C}$ has a projective resolution of length 1 $\mathcal{P}_i : 0 \to Q^{i-1} \to P^{i} \to H^{i} \to 0$ then there is a map $\mathcal{P} \to \mathcal{C}$ inducing the identity map on H. Conclude that if R is a Dedekind domain any complex is quasi-isomorphic to a direct sum of projective resolutions of its homology modules, and thus to a direct sum of its homology modules themselves. Note that there is generally no map from the direct sum of the homology modules to C; rather, there is a "roof" diagram with two maps that are quasi-isomorphisms, pointing in opposite directions:

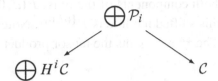

(b) Now suppose that C is a smooth curve. Given any coherent sheaves $\mathcal{H}, \mathcal{G}$ on C, show that there is a resolution $0 \to \mathcal{P} \to \mathcal{Q} \to \mathcal{H} \to 0$ by coherent sheaves such that $\mathrm{Ext}^1(\mathcal{P} \oplus \mathcal{Q}, \mathcal{G}) = 0$. Use this to imitate the argument of part (a), proving that every complex of coherent sheaves on C is quasi-isomorphic to its homology.

Exercise B.18. In Theorem B.11 we made the hypothesis that the base was affine in order to make use of projective resolutions, which do not generally exist over quasi-projective bases. Show that if B is quasi-projective, and $\mathcal{G}$ is a coherent sheaf on B, then one can resolve $\mathcal{G}$ by sums of line bundles, and that the twists of these bundles may be taken to be arbitrarily negative. Use this to give a version of Proposition B.16 that works for quasi-projective schemes. Show that this suffices to prove a version of Theorem B.11 where the object $R\pi_*\mathcal{F}$ is represented by a complex whose terms are sums of line bundles.

Exercise B.19. Let B be a curve of genus 1 over k and let $p \in B$ be a point. Let $X = B \times B$, and let $\pi : X \to B$ the projection onto the first factor. Let Δ be the diagonal in X, and consider the line bundle $\mathcal{F} = \mathcal{O}_X(\Delta - B \times p)$, so that $\mathcal{F}|_{b \times B} = \mathcal{O}_B(b - p)$. Show that $\pi_*\mathcal{F} = 0$, but the natural map $H^0(\mathcal{F}|_{p \times B}) \to \kappa(p) \otimes R^0\pi_*\mathcal{F}$ is not an isomorphism.

Show that $R^1\pi_*\mathcal{F}$ is a torsion sheaf, supported at p with fiber isomorphic to $H^1(\mathcal{F}|_{p \times B})$, and that the complex

$$0 \longrightarrow \mathcal{L}(-p) \longrightarrow \mathcal{L} \longrightarrow 0$$

represents $R\pi_*\mathcal{F}$.

Exercise B.20. Get a copy of *Macaulay2* from the website `http://www.math.uiuc.edu/Macaulay2/`, and compute some direct image complexes, using the following computation as a model. In the code below, we work over the field QQ of rational numbers (for a larger computation we would use a finite field such as $\mathbb{Z}/31003$ for efficiency). The projection map $\mathbb{P}^2 \times \mathbb{P}^1 \to \mathbb{P}^1$ is expressed by writing the homogeneous coordinate ring of $\mathbb{P}^2 \times \mathbb{P}^1$ as a polynomial ring P in three variables over the polynomial ring B in two variables representing the base. The ideals $I \subset I' \subset P$ are the bihomogeneous ideals of the families $\Gamma \supset \Gamma'$. Thus the computation is prepared as follows:

```
B = QQ[s,t];
P = B[x_0,x_1,x_2];
I' = intersect(ideal(x_0,x_1), ideal(x_0,x_2));
I = intersect(I', ideal(x_1-x_2, s*x_0-t*x_1));
```

We now compute the complex $R\pi_*(\mathcal{I}_{\Gamma'}(d))$ for d from 0 to 3. The computation uses code in the *Macaulay2* package *BGG*. *Macaulay2* does not abuse notation as we have in this chapter; both components of the twist $P(d, 0)$ must be made explicit: *Macaulay2* notation for this shifted module is $P^{\{\{d,0\}\}}$. Note that the ideal I' is made into a module explicitly. The ** represents the tensor product.

```
needsPackage "BGG";
for d from 0 to 3 do
<<directImageComplex(module I'**P^{{d,0}}) << endl<<endl
```

The output is something like the following (with the difference that in the actual output the complexes are indexed homologically instead of cohomologically):

$$
\begin{array}{cccc}
2 & 1 & 0 & -1 \\
0 \longleftarrow B \longleftarrow 0 \longleftarrow 0 \\
0 \longleftarrow 0 \longleftarrow B \longleftarrow 0 \\
0 \longleftarrow 0 \longleftarrow B^4 \longleftarrow 0 \\
0 \longleftarrow 0 \longleftarrow B^8 \longleftarrow 0
\end{array}
$$

We can do the same with $R\pi_*(\mathcal{I}_\Gamma(d))$

```
for d from 0 to 3 do
<<directImageComplex(module I**P^{{d,0}}) << endl<<endl
```

and obtain

$$
\begin{array}{cccc}
2 & 1 & 0 & -1 \\
0 \longleftarrow B^2 \longleftarrow 0 \longleftarrow 0 \\
0 \longleftarrow B \longleftarrow B \longleftarrow 0 \\
0 \longleftarrow 0 \longleftarrow B^3 \longleftarrow 0 \\
0 \longleftarrow 0 \longleftarrow B^8 \longleftarrow B
\end{array}
$$

where the map on the right in the last complex is a split inclusion, so that the last complex is quasi-isomorphic to

$$
0 \longleftarrow 0 \longleftarrow B^7 \longleftarrow 0.
$$

Appendix C

Topology of algebraic varieties

Throughout this appendix we work with projective varieties $X \subset \mathbb{P}^N_\mathbb{C}$ —that is, with complex projective varieties. We can also view such a variety as a complex analytic, or *holomorphic*, subvariety of $\mathbb{P}^N_\mathbb{C}$ —that is, a subset locally defined by the vanishing of analytic equations—or, if X is smooth, as a complex submanifold of $\mathbb{P}^N_\mathbb{C}$. The topology induced from the standard topology on $\mathbb{P}^N_\mathbb{C}$, referred to as the *classical*, or sometimes *analytic*, topology, is much finer than the Zariski topology with which we have dealt in this text. Using it, we can consider geometric invariants of X such as the singular homology and cohomology groups $H_*(X, \mathbb{Z})$ and $H^*(X, \mathbb{Z})$.

In this appendix, we explain a little of what is known about such invariants. Throughout, when we speak of topological properties of X, we refer to the classical, or analytic, topology.

C.1 GAGA theorems

One might think that there would be many more holomorphic subvarieties of $\mathbb{P}^N$ than algebraic subvarieties, or that in passing from a smooth projective variety X over $\mathbb{C}$ to its underlying complex manifold we would be losing information, since regular functions are holomorphic but not conversely. But this is not the case:

Theorem C.1 (Chow). *Every holomorphic subvariety of $\mathbb{P}^N_\mathbb{C}$ is algebraic.*

See for example Griffiths and Harris [1994, Section I.3] for a proof. Many further results in this direction were proven in Serre [1955/1956]. These are collectively known as the GAGA theorems, after the name of Serre's paper ("Géométrie algébrique et géométrie analytique").

It follows immediately from Chow's theorem that if X and Y are projective varieties over $\mathbb{C}$ then *any holomorphic map $f : X \to Y$ is algebraic* (Proof: Apply Theorem C.1 to the graph $\Gamma_f \subset X \times Y$). Not quite so immediate are the facts that any holomorphic vector bundle on a projective variety is algebraic and that if $\mathcal{E}$ is any such vector bundle on X then any global holomorphic section of $\mathcal{E}$ is algebraic. More generally, the Čech cohomology groups of $\mathcal{E}$ will be the same, whether computed for the sheaf of holomorphic sections of $\mathcal{E}$ in the analytic topology or the sheaf of regular sections in the Zariski topology. Thus, for example, the tangent space to the Picard group of X, which can be identified with the first Zariski cohomology $H^1(\mathcal{O}_X)$, may also be identified with the cohomology $H^1(\mathcal{O}_{X,\text{an}})$ in the analytic topology.

In sum, as it applies to projective varieties, we should think of the classical topology and the analytic tools it brings as a new approach to the study of the same projective algebraic varieties and phenomena. This is the point of view taken, for example, in Griffiths and Harris [1994], and we will give references to that book.

Once one goes beyond projective varieties, there are many complex manifolds that are not algebraic: while every one-dimensional compact complex manifold is an algebraic curve, there are many natural examples of compact complex surfaces that are not algebraic.

C.2 Fundamental classes and Hodge theory

C.2.1 Fundamental classes

Let $X \subset \mathbb{P}^N_{\mathbb{C}}$ be a smooth projective variety. Since $\mathbb{P}^N_{\mathbb{C}}$ is compact as a topological space and X is a closed subspace, X is compact. Because $\mathbb{C}$ has an orientation preserved by holomorphic functions, X is automatically orientable. In particular, if X has dimension n then it is a compact oriented real $2n$-dimensional manifold, and has a fundamental class $[X] \in H_{2n}(X, \mathbb{Z})$. (In this appendix, when we talk about the homology or cohomology groups of X or of a subvariety of X we mean the singular homology or cohomology.) By Poincaré duality, capping with this class induces an isomorphism

$$H^{2n-k}(X, \mathbb{Z}) \xrightarrow{\ \cap [X]\ } H_k(X, \mathbb{Z}).$$

More generally, let $Y \subset \mathbb{P}^r$ be any k-dimensional variety. By a theorem of Łojasiewicz [1964] (see Hironaka [1975]), Y admits a finite triangulation in which the singular locus is a subcomplex. Since the singularities of Y occur in real codimension ≥ 2, the sum of the simplices of dimension $2k$ in the triangulation is a cycle, called the *fundamental cycle* of Y; the class of this cycle is called the *fundamental class* of Y and is again denoted by $[Y] \in H_{2k}(Y, \mathbb{Z})$. If $\nu : \widetilde{Y} \to Y$ is a desingularization of Y, then the fundamental class of Y is equal to the pushforward $[Y] = \nu_*[\widetilde{Y}]$.

If X is an n-dimension projective variety (possibly singular), we can use this idea to define a homomorphism

$$Z_k(X) \to H_{2k}(X, \mathbb{Z})$$

from the group of k-cycles on X to its homology group, defined by associating to any k-dimensional subvariety $i : Y \hookrightarrow X$ the pushforward $i_*[Y] \in H_{2k}(X, \mathbb{Z})$ of the fundamental class $[Y]$ of Y. We say that two k-cycles A and A' are *homologically equivalent* if the pushforwards of their fundamental classes are equal. This is a coarser equivalence relation than rational equivalence; that is, this map factors through a map

$$A_k(X) \to H_{2k}(X, \mathbb{Z}).$$

If we suppose that X is smooth, then composing with Poincaré duality we get a homomorphism

$$\eta : A^{n-k}(X) = A_k(X) \to H^{2n-2k}(X, \mathbb{Z}),$$

which we call the *fundamental class map*; the image of the class of a subvariety $Y \subset X$ will be denoted by η_Y. (If one wants to avoid invoking the triangulability of complex varieties, the map η can be defined in de Rham cohomology by arguing that if $Y \subset X$ is any k-dimensional subvariety then integration over Y gives a well-defined linear functional on closed modulo exact $2k$-forms on X; see Griffiths and Harris [1994, p. 61].) For example, if $X = \mathbb{P}^n$ then η is an isomorphism — $A(X) = \mathbb{Z}[\zeta]/(\zeta^{n+1}) = H^*(X, \mathbb{Z})$ — so the information in the fundamental class $[Y]$ is the dimension and degree of Y.

A crucial fact is that intersection products in $A(X)$ corresponds to cup products in $H^*(X, \mathbb{Z})$:

Theorem C.2. *The map η is a ring homomorphism; that is, it takes the intersection product in $A(X)$ to the cup product on $H^*(X, \mathbb{Z})$:*

$$\eta([A][B]) = \eta[A] \cup \eta[B].$$

For a proof, see Griffiths and Harris [1994, Section 0.4].

The fundamental class map underlies much of the relevance of topology to geometry. The simplest invariant of a subvariety $A \subset X$ of a variety in general is its fundamental class in $H^*(X, Z)$, just as the simplest invariants of a subvariety of projective space are its dimension and degree.

This picture of the classes of subvarieties has a missing piece, encapsulated in one of the major open problems in the field: Which cohomology classes of a smooth variety are represented by linear combinations of classes of subvarieties? That is, what is the image of η? The *Hodge conjecture* (Section C.2.4) is an attempt to answer this question. Before explaining the statement, we provide some background.

C.2.2 The Hodge decomposition

Hodge noticed that if one tensors the real cotangent bundle of a complex manifold with the complex numbers it splits in a natural way. We can explain this phenomenon as follows:

Let $p \in X \subset \mathbb{P}_{\mathbb{C}}^N$ be a point on a smooth, n-dimensional projective algebraic variety. Viewing X as a real $\mathcal{C}^\infty$ manifold of dimension $2n$, we consider the $\mathcal{C}^\infty$ cotangent bundle $T^* X$. The fiber $T_p^* X$ has a natural complex structure coming from the fact that the differentials of real functions on a small open set are spanned by the differentials of the real and imaginary parts of complex analytic functions: for any such function $\varphi : X \to \mathbb{C}$ we have $i \cdot d(\text{Re}(\varphi)) := d(\text{Re}(i \cdot \varphi)) = -d(\text{Im}(\varphi))$.

Now, any n-dimensional complex vector space V can be regarded as a real vector space of dimension $2n$. Though there is no natural splitting into real and imaginary parts (that is, no splitting that is invariant under complex-linear transformations), the *complexification* $\mathbb{C} \otimes_{\mathbb{R}} V$ *does* split naturally as $\mathbb{C} \otimes_{\mathbb{R}} V \cong \mathbb{C}^n \oplus \mathbb{C}^n$, where the complex structure is defined by multiplication in the first factor and the two summands are the $+1$ and -1 eigenspaces under the real linear transformation that is multiplication by $i \otimes i$.

In particular, the complexification of the real cotangent space of a complex analytic manifold X at a point p is naturally a direct sum of the complex cotangent space (the complex vector space of differentials of complex analytic functions at p) and a space that may be identified as the space of differentials of anti-holomorphic functions. If $z_1 = x_1 + i y_1, \ldots, z_n = x_n + i y_n$ are complex analytic coordinates on an open set $U \subset X$ then the splitting may be written as

$$\mathbb{C} \otimes_{\mathbb{R}} T_p^* X \cong \mathbb{C}\langle dz_1, \ldots, dz_n \rangle \oplus \mathbb{C}\langle d\bar{z}_1, \ldots, d\bar{z}_n \rangle,$$

where $dz_\alpha = dx_\alpha + i\, dy_\alpha$ and $d\bar{z}_\alpha = dx_\alpha - i\, dy_\alpha$. A direct computation shows that $T_p^* X$, with the complex structure defined above, is mapped isomorphically to the fiber of the complex cotangent bundle under the composition of the inclusion into $\mathbb{C} \otimes T_p^* X$ and the projection modulo the space spanned by the differentials of the anti-holomorphic functions. Putting this together, we get:

Theorem C.3 (Hodge). *If X is a complex manifold, then the complexified cotangent bundle $T(X) \otimes \mathbb{C}$ splits as a direct sum*

$$\mathbb{C} \otimes T^* X = T^{*\prime} X \oplus T^{*\prime\prime} X, \tag{C.1}$$

where, with respect to any analytic local coordinates $\{z_\alpha\}$ at any point p, the fibers of $T^{*\prime} X$ and $T^{*\prime\prime} X$ are given by

$$T_p^{*\prime} X = \mathbb{C}\langle dz_1, \ldots, dz_n \rangle, \quad T_p^{*\prime\prime} X = \mathbb{C}\langle d\bar{z}_1, \ldots, d\bar{z}_n \rangle.$$

Moreover, $T^{*\prime} X$ is the underlying holomorphic vector bundle of Ω_X, the algebraic cotangent bundle of X.

The decomposition (C.1) immediately gives rise to a direct sum decomposition of the exterior powers of $T_p^* X \otimes \mathbb{C}$:

$$\wedge^k(\mathbb{C} \otimes T_p^* X) = \bigoplus_{p+q=k} \wedge^p T_p^{*\prime} X \otimes \wedge^q T_p^{*\prime\prime} X.$$

From this we get a decomposition of the sheaf $\mathcal{A}^k(X)$ of complex-valued $\mathcal{C}^\infty$ differential forms of degree k on X:

$$\mathcal{A}^k(X) = \bigoplus_{p+q=k} \mathcal{A}^{p,q}, \qquad (C.2)$$

where $\mathcal{A}^{p,q}$ is the sheaf of complex-valued $\mathcal{C}^\infty$ k-forms whose value at every point $p \in X$ lies in the summand $\wedge^p T_p^{*\prime} X \otimes \wedge^q T_p^{*\prime\prime} X$. A section of $\mathcal{A}^{p,q}$ is called a *form of type* (p, q), or a (p, q)-form. Note that while the bundle $\wedge^k \mathcal{T}^{*\prime} X$ has the structure of a holomorphic vector bundle, the sheaf $\mathcal{A}^{k,0}$ consists of all $\mathcal{C}^\infty$ sections of $\wedge^k \mathcal{T}^{*\prime} X$.

Hodge proved that when X is a smooth projective variety, and more generally when X is a Kähler variety, the decomposition (C.2) descends to the level of the de Rham cohomology of X: Any closed differential form φ on X can be written naturally as a sum

$$\varphi = \varphi^{(k,0)} + \varphi^{(k-1,1)} + \cdots + \varphi^{(0,k)},$$

where $\varphi^{(p,q)}$ is closed of type (p, q) and the de Rham cohomology class of $\varphi^{(p,q)}$ depends only on the class of φ.[1] Since the de Rham cohomology may be identified with the complexified singular cohomology, we get a decomposition of that space:

Theorem C.4 (Hodge decomposition). *If X is a smooth projective variety, then the singular cohomology of X with complex coefficients decomposes as*

$$H^k(X, \mathbb{C}) = H_{\mathrm{dR}}^k(X, \mathbb{C}) = \bigoplus_{p+q=k} H^{p,q}(X),$$

where $H^{p,q}(X)$ is the subspace of de Rham cohomology classes represented by forms of type (p, q). Complex conjugation interchanges $H^{p,q}(X)$ and $H^{q,p}(X)$.

Hodge also showed that the spaces $H^{p,q}(X)$, despite their apparently transcendental character, could be computed algebraically:

Theorem C.5. $H^{p,q}(X) = H^q(\Omega_X^p),$

the q-sheaf cohomology space of the p-th exterior power of the sheaf of differential forms on X; in particular, $H^{k,0}(X)$ is the space of global holomorphic k-forms on X.

---

[1] More precisely, the Hodge theorem asserts that with respect to any given Hermitian metric on X every de Rham cohomology class is uniquely represented by a harmonic form; the Kähler condition on the metric ensures that the Laplacian commutes with the decomposition of a form by type.

Proof sketch: This result follows from computing the homology of the *Dolbeault complex* which resolves Ω_X^p:

$$0 \to \Omega_X \to \mathcal{A}^{p,0} \xrightarrow{\bar{\partial}} \mathcal{A}^{p,1} \xrightarrow{\bar{\partial}} \cdots ;$$

since the sheaves $\mathcal{A}^{p,i}$ are flasque, we can compute the Zariski cohomology of Ω_X^p as the homology of the complex

$$H^0 \mathcal{A}^{p,0} \xrightarrow{\bar{\partial}} H^0 \mathcal{A}^{p,1} \xrightarrow{\bar{\partial}} \cdots$$

and this homology, at the q-th step of the resolution, is $H^{p,q}(X)$. $\qquad\square$

The data we have just described, consisting of a lattice (that is, finitely generated free $\mathbb{Z}$-module) $\Lambda = H^k(X, \mathbb{Z})/\text{tors}$, together with a decomposition of its complexification

$$\Lambda \otimes_{\mathbb{Z}} \mathbb{C} = \bigoplus_{p+q=k} H^{p,q} \quad \text{with } H^{p,q} = \overline{H^{q,p}},$$

where $\overline{H^{q,p}}$ denotes the complex conjugate of $H^{q,p}$, is called a *Hodge structure* of weight k. When dim $X = k$, so that $\Lambda = H^k(X, Z)$ is the middle-dimensional cohomology, the cup product is a unimodular inner product on the lattice Λ, and, since the wedge product of a (p,q)-form with a (p',q') form is a $(p+p', q+q')$-form, the subspaces $H^{p,q}$ and $H^{p',q'}$ are orthogonal, that is, $H^{p,q} \cup H^{p',q'} = 0$ unless $(p',q') = (q,p)$. Hodge structures with these additional properties are said to be *polarized*.

C.2.3 The Hodge diamond

The dimensions $h^{p,q} = h^{p,q}(X)$ of the Hodge groups $H^{p,q} = H^{p,q}(X)$ of a smooth projective variety are often represented in a diamond, called the *Hodge diamond*, with $h^{n,n}$ at the top. For example, the Hodge diamond of a smooth quartic surface $X \subset \mathbb{P}^3$ is

$$
\begin{array}{ccccc}
 & & 1 & & \\
 & 0 & & 0 & \\
1 & & 20 & & 1 \\
 & 0 & & 0 & \\
 & & 1 & &
\end{array}
$$

meaning, for example, that $h^{1,1}(X) = 20$. (Given the Lefschetz hyperplane theorem of Section C.4 below, the tools of Section 5.7.3 enable us to compute these numbers.)

The Hodge diamond is left-right symmetric through complex conjugation since $H^{p,q}(X) = \overline{H^{q,p}(X)}$, but it is also top-bottom symmetric: the cup product is defined on de Rham cohomology by multiplication of differential forms, and thus $H^{p,q} \cup H^{p',q'} = 0$ if either $p + p' > n$ or $q + q' > n$. Given this, Poincaré duality shows that $H^{n-p,n-q}(X)$ is dual to $H^{p,q}(X)$, and in particular they have the same dimension.

An immediate consequence of the symmetry is that the odd Betti numbers of a smooth projective variety are even; for example, $b_1(X) = h^{1,0} + h^{0,1} = 2h^{1,0}$. This tells us, for example, that the compact complex manifold X given as the quotient of $\mathbb{C}^2 \setminus \{(0,0)\}$ by the group of automorphisms generated by $\varphi : (z, w) \mapsto (2z, 2w)$, which is homeomorphic to $S^1 \times S^3$, cannot be a complex algebraic variety.

Holomorphic forms can be pulled back under birational transformations, and it follows that the Hodge numbers $h^{k,0}(X) = H^0(\Omega_X^k)$ are birational invariants. However, the other Hodge numbers $h^{p,q}(X)$ are not in general birational invariants.

The Hodge numbers are however deformation invariant, in the sense that they are constant in flat families X_λ of smooth projective varieties. Indeed, the Hodge numbers are upper-semicontinuous because they are the dimensions of the cohomology groups of coherent sheaves (Corollary B.12). On the other hand, their sum

$$\sum_{p+q=k} h^{p,q}(X_\lambda) = \dim H^k(X_\lambda)$$

is a topological invariant, and any two fibers in a flat family of smooth varieties are homeomorphic.

C.2.4 The Hodge conjecture

We now return to the question of which cohomology classes on a smooth projective variety X can be represented as linear combinations of the fundamental classes of algebraic varieties; that is, what is the image of $\eta : A(X) \to H^*(X)$?

Since the map on cohomology induced by a holomorphic map (such as the inclusion $Z \hookrightarrow X$, or the composition of the inclusion with a resolution $\widetilde{Z} \to Z$ of the singularities of Z) respects Hodge type, the fundamental class of a codimension-k subvariety $Z \subset X$ is of type (k, k). Thus the image of $\mathbb{C} \otimes_{\mathbb{Z}} \eta$ lies in $\bigoplus_k H^{k,k}(X) \subset H^*(X, \mathbb{C})$.

The Hodge conjecture asserts that modulo torsion the converse is true:

Conjecture C.6 (Hodge). *Every class in $H^{2k}(X, \mathbb{Q})$ whose image in $H^{2k}(X, \mathbb{C})$ lies in $H^{k,k}(X)$ is a rational linear combination of fundamental classes of algebraic subvarieties of X.*

It might seem more natural to make this conjecture without tensoring with $\mathbb{Q}$; but this statement, sometimes known as the "integral Hodge conjecture," is known to be false; see Atiyah and Hirzebruch [1961] and, for a recent survey, Totaro [2013].

An important special case of the Hodge conjecture is the following result of Lefschetz, which is the integral Hodge conjecture in the codimension-1 case:

Theorem C.7 (Lefschetz $(1, 1)$-theorem). *Let X be a smooth projective variety. If $\alpha \in H^2(X, \mathbb{Z})$ is any class whose image in $H^2(X, \mathbb{C})$ lies in $H^{1,1}(X)$, then α is the fundamental class of a divisor on X.*

In particular, *any* torsion class in $H^2(X, \mathbb{Z})$ is algebraic — that is, it is the class of a divisor on X.

Proof: We will give an outline of the proof; the explicit calculations may be found in Griffiths and Harris [1994, p. 163]. As above, let $\mathcal{O}_{X,\mathrm{an}}$ be the sheaf of holomorphic functions, defined as a sheaf in the classical topology. Let $\mathcal{O}^*_{X,\mathrm{an}}$ denote the sheaf of nowhere-zero holomorphic functions; this is a sheaf of abelian groups with respect to multiplication (in particular, it is *not* a coherent sheaf). There is an exact sequence

$$0 \longrightarrow \mathbb{Z} \longrightarrow \mathcal{O}_{X,\mathrm{an}} \xrightarrow{\ \exp\ } \mathcal{O}^*_{X,\mathrm{an}} \longrightarrow 0,$$

where $\mathbb{Z}$ denotes the sheaf of locally constant integer-valued functions and the exponential map $\exp$ sends a function f to e^f. (Note that this sequence is not exact in the Zariski topology, where a point $p \in X$ may not have any simply connected open neighborhoods.) The cohomology groups $H^i(\mathbb{Z})$ (in the classical topology) are naturally isomorphic to the singular homology groups $H^i(X, \mathbb{Z})$.

By the GAGA theorems

$$H^1(\mathcal{O}^*_{X,\mathrm{an}}) = \mathrm{Pic}\, X,$$

and the coboundary map $\delta : H^1(\mathcal{O}^*_{X,\mathrm{an}}) \to H^2(X, \mathbb{Z})$ in the associated long exact sequence is the composition of the Chern class map $\mathrm{Pic}\, X \to A^1(X)$ with the fundamental class map $\eta : A^1(X) \to H^2(X, \mathbb{Z})$. A calculation shows that the maps

$$H^i(X, \mathbb{Z}) \to H^i(\mathcal{O}_{X,\mathrm{an}}) = H^{0,i}(X)$$

in the long exact sequence are the compositions of the maps $H^i(X, \mathbb{Z}) \to H^i(X, \mathbb{C})$ with the projections $H^i(X, \mathbb{C}) \to H^{0,i}(X)$. It follows that a class $\alpha \in H^2(X, \mathbb{Z})$ that maps to $H^{1,1}(X)$ is in the kernel of the induced map $H^i(X, \mathbb{Z}) \to H^i(\mathcal{O}_{X,\mathrm{an}})$, and hence in the image of δ. $\qquad\square$

C.3 Comparison of rational equivalence with other cycle theories

In addition to the Chow ring $A(X)$ and the cohomology ring $H^*(X, \mathbb{Z})$ defined above, there are other cycle theories that we could have used in this text. In this section we will describe:

- The ring A_{alg} of algebraic cycles modulo algebraic equivalence.
- The ring $H_{\text{alg}}^*(X, \mathbb{Z})$ of algebraic cycles modulo homological equivalence.
- The ring $N^*(X)$ of algebraic cycles modulo numerical equivalence.

The reader will find more complete treatments in Fulton [1984, Chapter 19] and Fulton and MacPherson [1981]. A further way to treat algebraic cycles, which we will not discuss, is through K-theory (Fulton [1984, Section 20.5]). For interesting speculations about how this and other cycle theories might work even in the context of singular varieties, see Srinivas [2010].

For an example where the singular cohomology ring is useful (in a purely algebraic setting!), see Appendix D.

C.3.1 Algebraic equivalence

Rather than defining cycles to be equivalent only if there is a *rational* family of cycles interpolating between them — that is, a family parametrized by $\mathbb{P}^1$ — we can instead allow families parametrized by curves of any genus; that is, we say that two subvarieties $Z_1, Z_2 \subset X$ are *algebraically equivalent* if there is a reduced irreducible curve C and a subvariety $Y \subset C \times X$ such that the fibers of Y over two points $p, q \in C$ are Z_1 and Z_2. The resulting equivalence relation is called *algebraic equivalence*. The corresponding group $A_{\text{alg}}(X)$ comes with a natural map

$$A(X) \to A_{\text{alg}}(X).$$

All the items of the basic theory of Chow groups are true for the groups of cycles modulo algebraic equivalence; in particular, there is a decomposition by dimension $A_{\text{alg}}(X) = \bigoplus A_{\text{alg},d}(X)$, for a smooth variety X the groups $A_{\text{alg}}^k(X) := A_{\text{alg},\dim X-k}(X)$ form a graded ring, and there are pushforward and pullback maps as in the case of the Chow rings.

By Bertini's theorem there is an irreducible curve through any two points of a projective variety X, and thus unlike in the case of the Chow groups we have $A_{\text{alg},0}(X) \cong \mathbb{Z}$ via the degree map. The description of the group of cycles in codimension 1 is also simpler modulo algebraic equivalence than modulo rational equivalence: if X is smooth and projective over $\mathbb{C}$, then two divisors are algebraically equivalent if and only if they are homologically equivalent, and so $A_{\text{alg}}^1(X) \hookrightarrow H^2(X, \mathbb{Z})$; in particular, $A_{\text{alg}}^1(X)$ is finitely generated.

In other codimensions, however, the group of cycles modulo algebraic equivalence is as little understood as the Chow group. For example, if $X \subset \mathbb{P}^4$ is a smooth quintic threefold then we shall see in Section C.4.1 that two curves $C, C' \subset X$ are homologically equivalent if and only if they have the same degree. On the other hand, the 2875 lines on a general smooth quintic threefold X are linearly independent in $A_{\text{alg}}^2(X)$ by Ceresa and Collino [1983], and it is not known whether $A_{\text{alg}}^2(X)$ is even finitely generated.

C.3.2 Algebraic cycles modulo homological equivalence

The group of algebraic cycles modulo homological equivalence is the image of the map $\eta : A(X) \to H^*(X, \mathbb{Z})$ defined in Section C.2.4, and is denoted by $H^*_{\mathrm{alg}}(X, \mathbb{Z})$. Since it is a subgroup of a finitely generated group, it is finitely generated (again, unlike the Chow groups). When X is smooth, η is a ring homomorphism, so this is a subring of $H^*(X, \mathbb{Z})$.

However, it still has quirks: For example, by the Noether–Lefschetz theorem (see for example Griffiths and Harris [1985]), if $U \subset \mathbb{P}^{34}$ is the space of smooth quartic surfaces, the function $r : U \to \mathbb{Z}$ associating to each surface S the rank of $H^2_{\mathrm{alg}}(S, \mathbb{Z})$ is nowhere continuous.

C.3.3 Numerical equivalence

We say that a class $\alpha \in A^k(X)$ is *numerically equivalent to 0* if the degree of the product $\alpha\beta$ is 0 for all classes $\beta \in A^{n-k}(X)$ of complementary dimension. The group of Chow classes modulo numerical equivalence is denoted by $N^*(X)$; when X is smooth, the intersection product gives this group the structure of a ring, and the quotient map $A(X) \to N^*(X)$ is a ring homomorphism.

Numerical equivalence has some of the advantages of cohomology and is easily available in all characteristics. In characteristic 0, any class homologically equivalent to 0 is evidently numerically equivalent to 0, so numerical equivalence represents a further coarsening of the notion of homological equivalence for varieties over $\mathbb{C}$.

Conjecturally, it is not much coarser: If the Hodge conjecture is true, then by Poincaré duality and the hard Lefschetz theorem (Theorem C.13) the pairing

$$H^{2k}_{\mathrm{alg}}(X, \mathbb{Q}) \times H^{2n-2k}_{\mathrm{alg}}(X, \mathbb{Q}) \to \mathbb{Q}$$

given by the intersection/cup product would be nondegenerate. Given this, it would follow that any class $\alpha \in H^*_{\mathrm{alg}}(X)$ numerically equivalent to 0 is torsion in $H^*(X, \mathbb{Z})$, or, in other words,

$$N^k(X) = H^{2k}_{\mathrm{alg}}(X)/\mathrm{tors}.$$

Since the Hodge conjecture is known in codimension 1, this statement is known for $k = 1$ and $n - 1$. This makes the notion of numerical equivalence particularly suitable for applications of intersection theory that deal largely with divisors and curves, such as the minimal model program; most papers in that area work with numerical equivalence.

C.3.4 Comparing the theories

Summarizing the relationships of the theories we have defined, we have

$$A^*(X) \longrightarrow A_{\mathrm{alg}}^*(X) \longrightarrow H_{\mathrm{alg}}^*(X) \longrightarrow H^*(X, \mathbb{Z})$$

$$N^*(X)$$

with $N^*(X)$ conjecturally isomorphic to $H_{\mathrm{alg}}^*(X)/\mathrm{tors}$.

In fact, with small modifications to our arguments (and ignoring the necessary restriction to characteristic 0 when using cohomology), we could use any of the five cycle theories we have described for almost every intersection we compute in this book. The class of smooth projective varieties for which the theories coincide includes any variety with an affine stratification, and in particular all products of projective spaces and Grassmannians; it includes all homogeneous spaces for affine algebraic groups; it includes any projective bundle over a variety of this class, and all blow-ups of varieties in this class along subvarieties in this class. This class represents a tiny fraction of all varieties, but a large fraction of the set of varieties on which we can effectively carry out intersection theory.

C.4 The Lefschetz hyperplane theorem

The Lefschetz hyperplane theorem says that if $X \subset Y$ is a smooth, ample divisor on a smooth projective variety then in all but one dimension the cohomology of X is induced (in a sense we make precise below) from that of Y. The name comes from the use of the theorem to compare the topology of a projective variety with that of its general hyperplane section.

Theorem C.8 (Lefschetz hyperplane theorem)**.** *Let* $X \subset Y$ *be a smooth subvariety of codimension 1 on a smooth projective variety. If, as a divisor,* X *is ample, then the restriction map*

$$H^k(Y, \mathbb{Z}) \to H^k(X, \mathbb{Z})$$

is an isomorphism for $k < \dim(X)$ *and injective for* $k = \dim(X)$*. Similarly, the map*

$$H_k(X, \mathbb{Z}) \to H_k(Y, \mathbb{Z})$$

is an isomorphism for $k < \dim(X)$ *and surjective for* $k = \dim(X)$*.*

Despite the apparent symmetry of these two assertions they are rather different: the morphism

$$H^{2(\dim X)-k} \cong H_k(X, \mathbb{Z}) \to H_k(Y, \mathbb{Z}) \cong H^{2(\dim X+1)-k}$$

is induced by inclusion, whereas the homomorphism $H^k(Y, \mathbb{Z}) \to H^k(X, \mathbb{Z})$ may be defined by the intersection of X with a generic translate of a cycle in $H^k(Y, \mathbb{Z})$. We will see the importance of this difference in the examples of Section C.4.1.

By Poincaré duality, the groups $H^k(X, \mathbb{Z})$ for $k > \dim(X)$ are isomorphic to the groups $H_k(X, \mathbb{Z})$ with $k < \dim(X)$, so this describes all the cohomology groups of X in terms of those of Y, except for one: the middle-dimensional cohomology of X.

See Milnor [1963] for an elegant proof of Theorem C.8 following an argument of Andreotti and Frankel [1959]. Here is a sketch: The hypothesis that X is ample in Y implies that the line bundle $\mathcal{L} = \mathcal{O}_Y(X)$ has a Hermitian metric $\|\cdot\|$ with positive curvature. If σ is a global section of $\mathcal{L}$ vanishing on X, then $\|\sigma\|$ is a Morse function on Y with minimum 0 along X. We may apply Morse theory and use the curvature statement to bound the index of the critical points of $\|\sigma\|$. It follows that Y has the homotopy type of a space obtained by attaching cells of dimension $\geq \dim(X)$ to X, from which the Lefschetz theorem follows.

Lefschetz' original proof [1950] applied only to very ample divisors $X \subset Y$, that is, divisors that could be realized as hyperplane sections of Y under some embedding $Y \subset \mathbb{P}^N$ in projective space (hence the name of the theorem). In this setting, Lefschetz took a general pencil of divisors $\{X_\lambda\}_{\lambda \in \mathbb{P}^1}$ including X, and so gave a map

$$\pi : \widetilde{Y} = \mathrm{Bl}_\Gamma(Y) \to \mathbb{P}^1$$

from the blow-up of Y along the base locus $\Gamma = \bigcap X_\lambda$ of this pencil to $\mathbb{P}^1$. The map π is almost a fiber bundle: the fibers are all homeomorphic to X except over the finite number of values $\lambda \in \mathbb{P}^1$ with X_λ singular. By studying the local geometry of the map around these points, Lefschetz was able to relate the cohomology of the variety $\widetilde{Y}$, and hence that of Y, to that of X. Lefschetz' analysis in the end yields more than just the statement above — for example, it tells us about the monodromy action on the cohomology of the elements of the pencil — and is worth reading, despite the difficulties of translation (both linguistic and, more challengingly, mathematical!).

C.4.1 Applications to hypersurfaces and complete intersections

We can apply Theorem C.8 to smooth hypersurfaces in $\mathbb{P}^N$, but also, inductively using Bertini's theorem, to smooth complete intersections in projective space. In the following discussion we will write ζ for the class of a hyperplane in projective space, as well as its restriction to a projective variety X.

Corollary C.9. *If $X \subset \mathbb{P}^{n+c}$ is a smooth complete intersection of dimension n, then the map*

$$H_k(X, \mathbb{Z}) \to H_k(\mathbb{P}^{n+c}, \mathbb{Z}) = \begin{cases} \mathbb{Z} & \text{if } k \text{ is even,} \\ 0 & \text{if } k \text{ is odd,} \end{cases}$$

induced by the inclusion $X \subset \mathbb{P}^{n+1}$ is an isomorphism for $k < n$ and surjective for $k = n$, and the restriction map

$$H^k(\mathbb{P}^{n+c}, \mathbb{Z}) \to H^k(X, \mathbb{Z})$$

is an isomorphism for $k < n$ and injective for $k = n$.

Proof: We induct on the codimension c of the complete intersection X; the case $c = 1$ is the Lefschetz theorem itself. Now suppose that

$$X = V(f_1) \cap \cdots \cap V(f_c)$$

with $\deg f_1 \geq \cdots \geq \deg f_c$. To carry out the inductive step, it suffices to know that

$$X = V(f_1) \cap \cdots \cap V(f_{c-1})$$

is smooth. To achieve this, we may replace $f_1, \ldots, f_{c-1}$ by general forms $f_1', \ldots, f_{c-1}'$ of the same degrees in the ideal $(f_1, \ldots, f_c)$. By Bertini's theorem,

$$X' = V(f_1') \cap \cdots \cap V(f_{c-1}')$$

is smooth away from the base locus X. But since X is a complete intersection in X' any singular point on X' that lies on X would be a singular point on X as well.   □

As an application of Corollary C.9, observe that if $Z \subset X$ is any subvariety of codimension $k < n/2$ in X then Corollary C.9 tells us that the class $[Z] \in H^{2k}(X, \mathbb{Z})$ of Z is the restriction to X of an integral cohomology class on $\mathbb{P}^{n+c}$; in other words,

$$[Z] = (\alpha \cdot \zeta^k)|_X$$

for some $\alpha \in \mathbb{Z}$.

In Corollary 6.26 we showed that a smooth hypersurface in projective space cannot contain a linear space of more than half its dimension. Using Corollary C.9, we can say much more:

Corollary C.10. *Let $X \subset \mathbb{P}^{n+c}$ be a smooth complete intersection of dimension n and degree d. If $Z \subset X$ is any subvariety with $\dim(Z) > n/2$, then the degree of Z is divisible by d. In particular, if $Z \subset \mathbb{P}^{n+1}$ is any nondegenerate subvariety of dimension $> n/2$ and prime degree, then Z is contained in no smooth hypersurface.*

Proof: Let k be the codimension of Z in X. By Lefschetz, we have $[Z] = \alpha \zeta^k$ for some $\alpha \in \mathbb{Z}$, whence

$$
\begin{aligned}
\deg(Z) &= \deg([Z] \cdot \zeta^{n-k}) \\
&= \deg(\alpha \zeta^n) \\
&= \alpha d.
\end{aligned}
$$
$\square$

Thus, for example, a smooth nondegenerate surface of degree 3 in $\mathbb{P}^4$ (every such surface is a cubic scroll $S(1, 2)$) lies on no smooth threefold of degree $d > 3$ in $\mathbb{P}^4$. (In fact, it lies on no smooth hypersurfaces at all; the cases $d = 2$ and 3 can be handled by direct examination.)

There is a substantial strengthening in the case of codimension-1 subvarieties:

Corollary C.11. *If X is a smooth complete intersection of dimension $n \geq 3$ in projective space, any subvariety $Z \subset X$ of codimension 1 in X is the intersection of X with a hypersurface; equivalently, the homogeneous coordinate ring of X is factorial, so every unmixed codimension-1 subscheme of X is the intersection of X with a hypersurface.*

Proof: Since $n = \dim X > 1$, we have

$$
H^1(X, \mathbb{Z}) = 0.
$$

From the identification of Pic X with $H^1(\mathcal{O}_X^*)$ and the exponential sequence

$$
0 \longrightarrow \mathbb{Z} \longrightarrow \mathcal{O}_{X,\mathrm{an}} \xrightarrow{\exp} \mathcal{O}_{X,\mathrm{an}}^* \longrightarrow 0
$$

introduced in Section C.2.4, it follows that every line bundle on X is determined by its topological Chern class in $H^2(X, \mathbb{Z})$. Since $n > 2$, the map $H^2(\mathbb{P}^N, \mathbb{Z}) \to H^2(X, \mathbb{Z})$ is surjective, so every class of codimension 2 on X is the restriction of a class on projective space. Thus every line bundle on X has the form $\mathcal{O}_X(m)$ for some m.

Now suppose that X is a hypersurface of degree d in $\mathbb{P}^{n+1}$. From the sheaf sequence

$$
0 \longrightarrow \mathcal{O}_{\mathbb{P}^{n+1}}(m - d) \longrightarrow \mathcal{O}_{\mathbb{P}^{n+1}}(m) \longrightarrow \mathcal{O}_X(m) \longrightarrow 0
$$

and the fact that $H^1(\mathcal{O}_{\mathbb{P}^{n+1}}(m - d)) = 0$, we deduce that every global section of $\mathcal{O}_X(m)$ is the restriction to X of a global section of $\mathcal{O}_{\mathbb{P}^{n+1}}(m)$ — that is, a homogeneous polynomial of degree m. The same argument, applied inductively, yields the same statement for complete intersections.

Putting this together, we see that if $Z \subset X$ is a divisor then $\mathcal{O}_X(Z) = \mathcal{O}_X(m)$ for some m, and moreover the global section of $\mathcal{O}_X(Z)$ vanishing on Z is the restriction to X of a homogeneous polynomial of degree m. This proves the first statement of the corollary.

For the factoriality, it suffices to show that every codimension-1 prime ideal P in the homogeneous coordinate ring S_X is principal. We know that the subvariety S defined by P is the complete intersection of X and a hypersurface $V(f)$, and it follows that $(f) = P \cap Q$, where Q contains a power of the irrelevant maximal ideal. But since $S_X/(f)$ is a complete intersection, it is unmixed (see for example Eisenbud [1995, Theorem 11.5]), and we deduce that $(f) = P$.

The statement for unmixed codimension-1 subschemes follows at once.   □

Grothendieck eliminated topology and generalized this result substantially; see Call and Lyubeznik [1994] for the statement and a relatively simple proof.

C.4.2 Extensions and generalizations

There have been numerous extensions of the Lefschetz hyperplane theorem:

- One can replace the hypothesis "complete intersection of ample divisors" with the hypothesis "zero locus of a section of an ample vector bundle" (see for example Matsumura [2014]).

- More fundamentally, it became clear from Milnor's proof that over $\mathbb{C}$ the result should be strengthened to a result on homotopy in place of homology; see for example Barth [1975] for a development in this direction.

- Singularities on Y away from X seem to play a secondary role. For statements allowing such singularities, see for example Okonek [1987] and Hamm [1995].

- Another stream of activity has centered on subvarieties "of low codimension" in projective space that are *not* complete intersections. It turns out that these are very special (see for example the survey Hartshorne [1974]), so it is reasonable to hope that there might be analogous theorems for them. For example, we have:

Theorem C.12 (Larsen [1973]). *If $X \subset \mathbb{P}^{n+r}$ is a smooth codimension-r subvariety then the restriction map*

$$H^k(\mathbb{P}_{\mathbb{C}}^{n+r}, \mathbb{Z}) \to H^k(X, \mathbb{Z})$$

is an isomorphism for $k \le n - r$ and injective for $k = n - r + 1$.

Thus, for example, if $X \subset \mathbb{P}^5$ is a smooth threefold then $H^1(X, \mathbb{Z}) = 0$. See also the simplified proof in Barth [1975], which includes a homotopy-theoretic version.

C.5 The hard Lefschetz theorem and Hodge–Riemann bilinear relations

In this section we briefly describe two further results on the topology of varieties that are frequently used (though not in this text). A reference for both is Griffiths and Harris [1994, Section 0.7]. The first is called the "Hard Lefschetz theorem" in English; the more colorful French name is "Théorème de Lefschetz vache."

Theorem C.13 (Hard Lefschetz theorem). *Let $X \subset \mathbb{P}^r$ be a smooth projective variety of dimension n. If $\zeta \in H^2(X, \mathbb{C})$ is the class of a hyperplane section of X, then the map*

$$\bigcup \zeta^k : H^{n-k}(X, \mathbb{C}) \to H^{n+k}(X, \mathbb{C})$$

is an isomorphism for all $k = 1, \ldots, n$.

We define the *primitive cohomology groups* of X for $m \le n$ as

$$P^{n-k}(X) := \mathrm{Ker}\Big(\bigcup \zeta^{k+1} : H^{n-k}(X, \mathbb{C}) \to H^{n+k+2}(X, \mathbb{C})\Big),$$

and using Theorem C.13 we get the *Lefschetz decomposition*

$$H^m(X, \mathbb{C}) = \bigoplus \zeta^k \cdot P^{m-2k}(X)$$

for $m \le n$.

Since the class ζ is of type $(1, 1)$, the Lefschetz decomposition is compatible with the Hodge decomposition: If for $k = p + q$ we set

$$P^{p,q}(X) = P^k \cap H^{p,q}(X),$$

then we have

$$P^k(X) = \bigoplus_{p+q=k} P^{p,q}(X).$$

In these terms, we can state the *Hodge–Riemann bilinear relations*. First, we define a Hermitian inner product

$$Q : H^{n-k}(X, \mathbb{C}) \times H^{n-k}(X, \mathbb{C}) \to \mathbb{C}$$

on the cohomology of X in middle degree or lower by

$$Q(\alpha, \beta) = (\alpha \cup \overline{\beta} \cup \zeta^k)[X].$$

Theorem C.14 (Hodge–Riemann bilinear relations). *For $\alpha \in P^{p,q}(X)$,*

$$i^{p-q}(-1)^{\binom{n-k}{2}} Q(\alpha, \overline{\alpha}) > 0.$$

When n is even we can use this to express the index of the cup product on $H^n(X, \mathbb{R})$ in terms of the dimensions of the primitive cohomology groups of X. For example, when X is a surface, the Lefschetz decomposition of $H^2(X, \mathbb{C})$ has the form

$$H^2(X, \mathbb{C}) = P^2(X) \oplus \mathbb{C}\langle \zeta \rangle,$$

and we conclude the signature of the cup product on $H^2(X, \mathbb{R})$ is $(2h^{2,0} + 1, h^{1,1} - 1)$. As a consequence:

Corollary C.15. *Let $S \subset \mathbb{P}^N$ be a smooth surface of degree d. If $\gamma = [C]$ is the class of a curve C of degree e on S, then*

$$\deg(\gamma^2) \leq \frac{e^2}{d}.$$

Proof: If γ is a multiple of the class ζ of a hyperplane section this is immediate (and we have equality above); if not, this is just the statement that the intersection pairing on the group $\mathbb{Z}\langle \zeta, \gamma \rangle \subset A^1(S)$ generated by the classes ζ and γ has signature $(1, 1)$ (and we have a strict inequality). $\qquad\square$

Note that the statement of Corollary C.15 makes no reference to homology, and in fact it is true over any field; see Beauville [1996].

C.6 Chern classes in topology and differential geometry

If X is a smooth complex projective algebraic variety and $\mathcal{E}$ is an algebraic vector bundle on X, then applying the map $\eta : A(X) \to H^*(X, \mathbb{Z})$ to the Chern class of $\mathcal{E}$ we get a class in $H^*(X, \mathbb{Z})$, which we also denote by $c(\mathcal{E})$ and call the *topological Chern class* of $\mathcal{E}$.

It is possible to define $c(E) \in H^*(X, \mathbb{Z})$ much more generally, for any continuous complex vector bundle E on any reasonably nice topological space, say a simplicial complex. Here we think of E as a topological space, with a map $\pi : E \to X$ whose fibers are complex vector spaces $\mathbb{C}^n$, rather than as a locally free sheaf, and we denote by $E^\times$ the complement of the zero section in E.

The topological Chern classes are among the *characteristic classes* of vector bundles. Such classes were defined by Stiefel and Whitney (see for example Whitney [1941]) before they came into algebraic geometry. In this section we will sketch the original topological idea behind the construction.

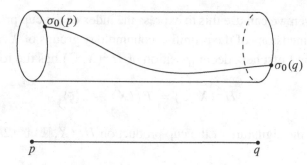

Figure C.1 Extending the section σ_0 over a 1-simplex.

C.6.1 Chern classes and obstructions

As we have seen, the top Chern class in algebraic geometry may be thought of as an obstruction to the existence of a nowhere-vanishing global section. In the category of manifolds (or more generally simplicial complexes) and continuous maps, we can make this precise, as follows:

Let E be a complex vector bundle on a topological space X that is the underlying space of a simplicial complex. We start by choosing an arbitrary section σ_0 of $E^\times$ over the 0-skeleton X_0 of X — that is, we select arbitrary nonzero vectors $\sigma_0(p) \in E_p^\times$ in the fibers of E over the vertices $p \in X$ of our complex.

Next we extend σ_0 to a section σ_1 of $E^\times$ defined over the 1-skeleton of X. We can always do this: Since a closed 1-cell $\Delta_1 \subset X$ is contractible, the restriction of E to Δ_1 is trivial, so that finding a nonzero section of E over Δ_1 with the assigned values $\sigma_0|_{\partial\Delta_1}$ on the boundary $\partial\Delta_1$ amounts to finding an arc $\gamma : [0, 1] \to \mathbb{C}^n \setminus \{0\}$ with given starting and ending points; since $\mathbb{C}^n \setminus \{0\}$ is connected, we can always do this (see Figure C.1).

We continue in this way, extending σ_1 to a section of $E^\times$ over successively larger skeleta of X: Given a section σ_{k-1} of $E^\times$ over the $(k-1)$-skeleton X_{k-1}, for each k-simplex $\Delta_k \subset X$ we trivialize the bundle on Δ_k, and view the problem of extending σ_{k-1} over Δ_k as that of giving a homotopy of $\sigma_{k-1}|_{\partial\Delta_k}$ with the constant map. We can certainly make the extension as long as $\pi_{k-1}(\mathbb{C}^n \setminus \{0\}) = 0$.

We first encounter difficulty when $k = 2n$: Since $\pi_{2n-1}(\mathbb{C}^n \setminus \{0\}) \cong \mathbb{Z}$, we may not be able to extend σ_{2n-1} over X_{2n}. As a measure of the obstruction, we define a simplicial $2n$-cochain $\alpha \in C^{2n}(X, \mathbb{Z})$ by

$$\alpha(\Delta_{2n}) = [\sigma_{2n-1}|_{\partial\Delta_k}] \in \pi_{2n-1}(\mathbb{C}^n \setminus \{0\}) \cong \mathbb{Z}.$$

One can show that the cochain α is a cocycle and that while α depends on the choices made its cohomology class does not.

In this context we define the *topological Chern class $c_n(E)$ of E* to be

$$c_n(E) = [\alpha] \in H^{2n}(X, \mathbb{Z}).$$

The nonvanishing of $c_n(E)$ is an *obstruction* to finding a nowhere-zero section of E in the sense that the nonvanishing of $[\alpha]$ implies that we cannot have a nowhere-zero section of E, but the converse is not true: the vanishing of $[\alpha]$ says only that we can find such a section over the $2n$-skeleton X_{2n}.

The other topological Chern classes are defined similarly. To measure the obstruction to finding k everywhere-independent sections of E, we introduce the bundle of k-frames in E: the space

$$F_k(E) = \{(p, v_1, \ldots, v_k) \mid p \in X \text{ and } v_1, \ldots, v_k \in E_p \text{ are independent}\}.$$

The fibers of $F_k(E)$ over X are *frame manifolds*

$$F_k = \{v_1, \ldots, v_k \in \mathbb{C}^n \mid v_1 \wedge \cdots \wedge v_k \neq 0\},$$

and a simple calculation with the long exact sequence in homotopy shows that we have

$$\pi_i(F_k) = \begin{cases} 0 & \text{if } i < 2n - 2k + 1, \\ \mathbb{Z} & \text{if } i = 2n - 2k + 1. \end{cases}$$

Thus we get a class in $H^{2n-2k+2}(X, \mathbb{Z})$, which is an obstruction to finding a section of $F_k(E)$. This class agrees with the Chern class $c_{n-k+1}(E)$ in the cases where the latter is defined.

One can show that when E is an algebraic vector bundle with enough sections this definition of the Chern classes agrees with their characterization as classes of degeneracy loci. For example, if σ is a global holomorphic section of E vanishing on a generically reduced subscheme $A \subset X$ of codimension n, we can choose the simplicial structure on X to be transverse to the zero locus of σ, meaning that $V(\sigma)$ is disjoint from the $(2n - 1)$-skeleton of X and intersects each $2n$-simplex of X transversely in a finite number of points. If we choose the section σ_{2n-1} of $E^\times$ over the $(2n - 1)$-skeleton on X to be the restriction of σ to X_{2n-1}, then the obstruction cocycle α will associate to each $2n$-simplex Δ_{2n} the number of its points of intersection with $V(\sigma)$, counted with appropriate sign. Unwinding the definitions, the cohomology class of this cocycle is exactly the Poincaré dual of the fundamental class of $V(\sigma)$.

C.6.2 Chern classes and curvature

Under slightly stronger hypotheses, Chern [1946] characterized the classes $c_k(E)$ in terms of curvature. Suppose that X is a differentiable manifold and E a differentiable complex vector bundle on X. Choose a Riemannian metric on the manifold X and a Hermitian metric on the vector bundle E — that is, a Hermitian inner product on each fiber of the bundle, varying differentiably with the point, so that in terms of a local trivialization of E the entries of the Hermitian matrix giving the inner product on E_p are C^∞ functions of p. The metric on the bundle E defines a notion of *parallel transport*:

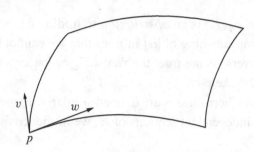

Figure C.2 A small geodesic quadrilateral with vertex at p.

given an arc γ in X, the metric determines a canonical way to identify the fibers of E over the points of the arc.

Now, suppose that $v, w \in T_p X$ are tangent vectors at a point $p \in X$. We can form a small geodesic square with a vertex at p and sides at p that are geodesics of length ϵ in the directions v and w; see Figure C.2.

Parallel transport around the perimeter of this square yields an automorphism $\varphi(v, w, \epsilon)$ of the fiber E_p; since $\varphi(v, w, \epsilon)$ goes to the identity as $\epsilon \to 0$, we arrive at an endomorphism $A(v, w)$ of E_p

$$A(v, w) = \lim_{\epsilon \to 0} \frac{\varphi(v, w, \epsilon) - \mathrm{Id}}{\epsilon}.$$

The endomorphism $A(v, w)$ is bilinear and skew-symmetric in v and w, so that we can think of A as an element of $\mathcal{H}om(\wedge^2 T_p X, \mathrm{End}(E_p))$; as p varies this gives us a global section Θ of the bundle

$$\mathcal{H}om(\wedge^2 T(X), \mathrm{End}(E)) = \Omega^2_X \otimes \mathrm{End}(E).$$

The section Θ is called the *curvature form* of the metric.

In terms of a local trivialization of E near p, we may think of Θ as an $n \times n$ matrix of differentiable 2-forms, called the *curvature matrix* of the metric. If we change the trivialization of E, the matrix Θ is replaced by its conjugate under the change-of-basis matrix. The coefficients of the characteristic polynomial of this matrix are thus well-defined global forms ω_{2k} on X for $k = 0, 1, \ldots, n$.

Chern showed that:

(a) The forms ω_{2k} are closed.

(b) The de Rham cohomology classes $[\omega_{2k}] \in H^{2k}_{\mathrm{dR}}(X, \mathbb{C})$ are independent of the choice of metric.

(c) The class $[\omega_{2k}]$ is the image of the class $c_k(E) \in H^{2k}(X, \mathbb{Z})$ under the natural map $H^{2k}(X, \mathbb{Z}) \to H^{2k}(X, \mathbb{C}) = H^{2k}_{\mathrm{dR}}(X, \mathbb{C})$.

See Chern [1946] for a proof.

The analogs of many of the theorems proved in Chapter 5 and elsewhere for the algebraic Chern classes $c(E) \in A(X)$ hold true in the topological setting as well: The Whitney product formula and the splitting principle, and their consequences, such as the formula for the Chern class of a tensor product with a line bundle (Proposition 5.17), hold more generally for topological vector bundles, and indeed can be proved along similar lines. In particular, the analog of the formula of Theorem 9.6 for the Chow ring of a projective bundle holds true for the cohomology ring of a projective bundle as well (see for example Bott and Tu [1982]), a fact we will use in Appendix D.

Appendix D

Maps from curves to projective space

Keynote Questions

(a) What is the smallest degree of a nonconstant map from a general curve of genus g to $\mathbb{P}^1$? (Answer on page 567.)

(b) In how many ways can a general curve C of genus 4 be expressed as a 3-sheeted cover of $\mathbb{P}^1$, up to automorphisms of $\mathbb{P}^1$? (Answer on page 577.)

(c) What is the smallest degree of a nondegenerate map from a general curve of genus g to $\mathbb{P}^2$? (Answer on page 567.)

(d) In how many ways can a general curve C of genus 6 be expressed as a curve of degree 6 in $\mathbb{P}^2$, up to automorphisms of $\mathbb{P}^2$? (Answer on page 578.)

(e) What is the smallest degree of an embedding of a general curve of genus g in $\mathbb{P}^3$? (Answer on page 567.)

(f) In how many ways can a general curve C of genus 8 be embedded as a curve of degree 9 in $\mathbb{P}^3$, up to automorphisms of $\mathbb{P}^3$? (Answer on page 578.)

In this book we have treated problems in enumerative geometry as interesting for the aspects of algebraic geometry that they illuminate, and for their own sake — there is a certain fascination with being able to enumerate solutions to a geometric problem, even when we cannot find those solutions explicitly. In this appendix we will see a striking example of another kind, where the methods of enumerative geometry are crucial in the analysis of a qualitative questions in geometry. We will describe and prove half of a foundational result in the theory of algebraic curves: the *Brill–Noether theorem*, which answers the question of when a general curve of given genus admits a map of given degree to projective space.

This material also illustrates the value of considering cycle theories other than the Chow ring of a variety. We will be working with Jacobians and symmetric powers of curves (see Section 10.3.1 for the definition and basic properties of symmetric powers in general, and Section D.2 below for a description of the Jacobian of a curve and its relation to the symmetric powers). These are spaces whose Chow rings remain opaque but whose cohomology rings are readily accessible (the Jacobian of a curve of genus g is homeomorphic to a product of $2g$ copies of S^1, so we can apply the Künneth formula to give a compact description of its cohomology ring). In particular, *Poincaré's formula* (Section D.5.1), which is crucial to our calculation, is readily verified in cohomology but difficult even to state in the Chow ring.

We will therefore assume for the duration of this appendix that we are working over the complex numbers, and use the results of Appendix C relating the Chow ring of a projective variety to its cohomology ring. The techniques developed can be extended to all characteristics using étale cohomology or numerical equivalence (Kleiman and Laksov [1972] and [1974], respectively).

D.1　What maps to projective space do curves have?

Until the 20th century, varieties were defined as subsets of projective space. In this respect, algebraic geometry was much like other fields in mathematics; for example, in the 19th century a group was by definition a subset of GL_n or S_n closed under the operations of composition and inversion; the modern definition of an abstract group did not appear until well into the 20th century. But in about 1860 Riemann's work introduced a way of talking about curves that crystallized, over the next hundred years, into our notion of an *abstract variety*—a geometric object defined independently of any particular embedding in projective space.

The basic problem of classifying all curves in projective space was thus broken down into two parts: the description of the family of abstract curves (the study of *moduli spaces* of curves), and the problem of describing all the ways in which a given curve C might be embedded in or, more generally, mapped to a projective space. To continue our analogy with group theory, the latter question is the analog of representation theory, that is, the study of the ways in which a given abstract group G can be mapped to GL_n.

Among the most basic questions we can pose along these lines is: "What maps to projective space do most curves of genus g have?" To focus on the objects of principal interest and avoid redundancies, we consider only *nondegenerate* maps $\varphi : C \to \mathbb{P}^r$, that is, maps whose image does not lie in any hyperplane. We define the *degree* of such a map to be the degree of the line bundle $\varphi^* \mathcal{O}_{\mathbb{P}^r}(1)$, or equivalently the cardinality of the preimage $\varphi^{-1}(H)$ of a general hyperplane $H \subset \mathbb{P}^r$.

There are really two parallel questions: First, "For which d, g and r is it the case that every curve of genus g admits a nondegenerate map of degree d or less to $\mathbb{P}^r$?" and second, "For which d, g and r is it the case that a *general* curve of genus g admits a nondegenerate map of degree d or less to $\mathbb{P}^r$?"

The second version of this question begs the further question of the meaning of the phrase "general curve of genus g." As explained in the introduction to this book, such a statement invokes the existence of a family of smooth projective curves. In this case we refer to the universal family of curves that exists over an open set of the *moduli space* parametrizing smooth, projective curves of genus g — a space whose points correspond naturally to isomorphism classes of such curves; this moduli space, denoted M_g, is irreducible. We will take this existence as given; details can be found in Harris and Morrison [1998].

Recall that maps $\varphi : C \to \mathbb{P}^r$ (modulo the group PGL_{r+1} of automorphisms of the target $\mathbb{P}^r$) correspond bijectively to pairs $(\mathcal{L}, V)$ with $\mathcal{L}$ a line bundle of degree d on C and $V \subset H^0(\mathcal{L})$ an $(r + 1)$-dimensional vector space of sections without common zeros (base locus). If we drop the requirement that the sections of V have no common zeros, such an object is called a *linear series of degree d and dimension r*; classically, it was referred to as a g_d^r. (Note that if a g_d^r has a nonempty base locus then we can subtract that locus and get a map $\varphi : C \to \mathbb{P}^r$ with degree smaller than d.) The linear series is called *complete* if $V = H^0(\mathcal{L})$.

It is easy to produce high-degree maps and high-degree embeddings: An application of the Riemann–Roch theorem shows that on a curve of genus g any line bundle of degree $\geq 2g$ defines a morphism, and any line bundle of degree $\geq 2g + 1$ defines an embedding (see Hartshorne [1977, Section IV.3]); a slightly more refined argument shows that on any curve of genus g a general line bundle of degree $g + 1$ defines a morphism and a general line bundle of degree $g + 3$ defines an embedding.

If we are interested in the simplest representation of the curve, our primary question becomes: How low a degree line bundle can we find that gives a morphism, or an embedding? For the case of a general curve, these are Keynote Questions (a) and (e). The correct answers were given by Brill and Noether [1874] soon after Riemann's work, but the first complete proofs followed only about 100 years later! The numerical function

$$\rho(g, r, d) := g - (r + 1)(g - d + r)$$

plays a central role. We will see how it arises in Section D.3.1. Here is the simplest version of the Brill–Noether theorem:

Theorem D.1. *(a) If $\rho(g, r, d) \geq 0$ then every curve of genus g admits a nondegenerate map of degree d or less to $\mathbb{P}^r$.*

(b) For a general curve this bound is sharp; that is, a general curve C of genus g admits a nondegenerate map of degree d or less to $\mathbb{P}^r$ if and only if $\rho(g, r, d) \geq 0$.

For example, we see that:

- All curves of genus g are expressible as covers of $\mathbb{P}^1$ of degree $\lceil (g + 2)/2 \rceil$ or less, and for general curves this is sharp; thus all curves of genus 1 or 2 are expressible as 2-sheeted covers of $\mathbb{P}^1$, but general curves of genus 3 are not.

- All curves of genus g admit maps to $\mathbb{P}^2$ of degree $\lceil (2g + 6)/3 \rceil$ or less, and for general curves this is sharp; for example, curves of genus 2 or 3 admit maps of degree 4 to $\mathbb{P}^2$, but general curves of genus 4 do not.

Part (a) of Theorem D.1, the existence, was first proved by Kleiman and Laksov [1972] and Kempf [1971], while part (b) was first proved by Griffiths and Harris [1980]. The two statements require quite different methods, part (a) using enumerative geometry and part (b) involving specialization techniques. We give successively stronger forms of part (a) in Theorems D.1, D.9 and D.17, and we will give a proof of the strongest form; in Section D.3.2 we sketch some of the steps needed for part (b).

It is sometimes interesting to ask what happens for the complete linear series corresponding to a general line bundle $\mathcal{L}$ of degree d with $h^0(\mathcal{L}) \geq r + 1$ on a specific curve C; again, this makes sense because the set $\mathrm{Pic}^d(C)$ of line bundles of degree d is a variety, isomorphic to the Jacobian variety of the curve, and the locus of $\mathcal{L} \in \mathrm{Pic}^d(C)$ with $h^0(\mathcal{L}) \geq r + 1$ is a closed subset (we will explain both of these assertions below). It turns out that general linear series on general curves are as well-behaved as possible. The following result, together with the Brill–Noether theorem, answers Keynote Questions (c) and (e).

Theorem D.2. *Let C be a general curve of genus g.*

- $r = 1$: *The map $\varphi : C \to \mathbb{P}^1$ given by a general g_d^1 is simply branched.*
- $r = 2$: *The map $\varphi : C \to \mathbb{P}^2$ given by a general g_d^2 is birational onto a plane curve $C_0 \subset \mathbb{P}^2$ having only nodes as singularities.*
- $r \geq 3$: *The map $\varphi : C \to \mathbb{P}^r$ given by a general g_d^r is an embedding.*

See Eisenbud and Harris [1983a, Theorems 1,2] for the proof of parts (a) and (c), and Zariski [1982] or Caporaso and Harris [1998] for part (b).

Thus, for example, a general curve of genus g is birational to a plane curve of degree $\lceil (2g + 6)/3 \rceil$, but no less, and is embeddable in projective space as a curve of degree $\lceil (3g + 12)/4 \rceil$ but no less.

Before stating the more refined versions of the Brill–Noether theorem, we pause to describe three classic and more elementary results that provide limitations on what g_d^r's and embeddings in projective space a curve can possess: the theorems of Riemann and Roch, Clifford, and Castelnuovo.

D.1.1 The Riemann–Roch theorem

The Riemann–Roch theorem for curves (Chapter 14) says that for a line bundle $\mathcal{L}$ on a smooth curve of genus g

$$h^0(\mathcal{L}) = d - g + 1 + h^0(\omega_C \otimes \mathcal{L}^{-1}),$$

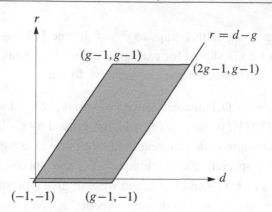

Figure D.1 Points in the shaded region correspond to degrees and dimension (d, r) of complete linear series whose existence is not excluded by the Riemann–Roch theorem.

where ω_C denotes the sheaf of differential forms on C. We will often exploit the equivalence between the notions of line bundles and divisors on a smooth curve, and write a divisor D in place of the line bundle $\mathcal{O}(D)$. Thus we allow ourselves to rewrite the Riemann–Roch theorem as

$$h^0(D) = d - g + 1 + h^0(K - D),$$

where K denotes a canonical divisor.

For line bundles $\mathcal{L}$ of degree $d > 2g - 2$ the last term is zero, and so the Riemann–Roch theorem tells us the dimension precisely:

$$h^0(\mathcal{L}) = d - g + 1.$$

For line bundles of degree close to $2g - 2$ it gives us approximate information: For example, if $d = 2g - 2$, the Riemann–Roch theorem says that

$$h^0(\mathcal{L}) = \begin{cases} g & \text{if } \mathcal{L} = \omega_C, \\ g - 1 & \text{otherwise,} \end{cases} \tag{D.1}$$

and if $g > 0$ and $d = 2g - 3$ it says that

$$h^0(\mathcal{L}) = \begin{cases} g - 1 & \text{if } \mathcal{L} = \omega_C(-p) \text{ for some point } p \in C, \\ g - 2 & \text{otherwise.} \end{cases} \tag{D.2}$$

It also tells us that for any line bundle

$$d + 1 \geq h^0(\mathcal{L}) \geq d - g + 1.$$

Beyond these values the Riemann–Roch theorem gives less precise information.

We summarize what the Riemann–Roch theorem tells us about the possible existence of g_d^r's on curves of genus g in Figure D.1, which shows the possible pairs (d, r) where $\mathcal{L}$ is a line bundle of degree d on C and $r = h^0(\mathcal{L}) - 1$.

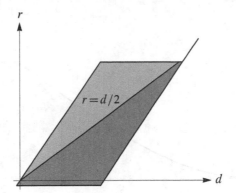

Figure D.2 Points in the shaded region correspond to degrees and dimension (d, r) of complete linear series whose existence is not excluded by Clifford's theorem. In fact, the result is sharp: every such degree and dimension occurs for some complete linear series on some curve.

D.1.2 Clifford's theorem

Clifford's theorem (Hartshorne [1977, Theorem IV.5.4]) says that if C is a curve of genus g and D is a divisor of degree $d \leq 2g - 2$ on C then

$$h^0(D) \leq d/2 + 1.$$

In the case when $h^0(K - D) = 0$, this inequality follows at once from the Riemann–Roch theorem, so its import is for effective divisors D such that $h^0(K - D) \neq 0$ — these are called *special divisors*. An extension of Clifford's theorem says that if, moreover, equality holds then either $\mathcal{L} = \mathcal{O}$, $\mathcal{L} = \omega_C$ or C is hyperelliptic and $\mathcal{L}$ is a multiple of the g_2^1 on C. If we exclude the cases of degree $d = 0$ and $d \geq 2g - 2$ (where after all Clifford tells us nothing new), we may state this as:

Theorem D.3 (Clifford). *If C is a curve of genus g and $\mathcal{L}$ a line bundle of degree d on C with $0 < d < 2g - 2$, then*

$$h^0(\mathcal{L}) \leq d/2 + 1,$$

with equality holding only if C is hyperelliptic and $\mathcal{L}$ a multiple of the g_2^1.

This cuts the above graph of allowed values of d and r essentially in half, as shown in Figure D.2.

Clifford's theorem is sharp: For every d, g and r allowed by Theorem D.3, there exist curves of genus g and g_d^r's on them. This does not represent a satisfactory answer to our basic problem, however, for two distinct reasons:

(a) Given our motivation for studying g_d^r's on curves — the classification of curves in projective space — you may say that our real object of interest is not g_d^r's in general but those whose associated maps give generically one-to-one morphisms

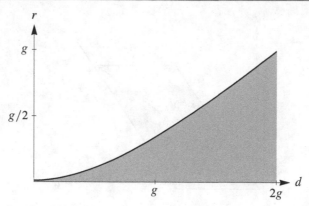

Figure D.3 Points in the shaded region correspond to degrees and dimension (d, r) of complete linear series defining birational maps whose existence is not excluded by Castelnuovo's theorem.

to $\mathbb{P}^r$. (When the morphism is generically one-to-one, or *birationally very ample* in more classical terminology, the source curve is the normalization of the target, so the image curves can be described as curves of geometric genus g — see Section 2.4.6.) The linear systems satisfying equality in Clifford's theorem — and, as we will see in a moment, those that are close to this — are *not* birationally very ample. Thus, we refine our original question and ask: "What birationally very ample linear series may exist on a curve of genus g?" — in other words, for which d, g and r do there exist irreducible, nondegenerate curves of degree d and geometric genus g in $\mathbb{P}^r$?

(b) As we have seen, interesting linear series that achieve equality in Clifford's theorem exist only on hyperelliptic curves, which are very special: they form a closed subset of codimension $g - 2$ in the space M_g of all smooth curves of genus g. Clifford's theorem thus leaves unanswered the question of what linear series exist on a *general curve* of genus g.

D.1.3 Castelnuovo's theorem

The issue of which linear series can embed a curve is dealt with in a theorem of Castelnuovo:

Theorem D.4 (Castelnuovo). *Let $C \subset \mathbb{P}^r$ be an irreducible, nondegenerate curve of degree d and geometric genus g. Then*

$$g \le \pi(d, r) := \binom{m}{2}(r - 1) + m\epsilon,$$

where $d = m(r - 1) + \epsilon + 1$ and $0 \le \epsilon \le r - 2$.

Castelnuovo showed that this bound is sharp, and using his analysis it is not hard to see that curves of all geometric genera between 0 and the bound do occur. Figure D.3 shows the values of d and r allowed by Castelnuovo's bound in the case $g = 30$. See Arbarello et al. [1985, Chapter 3] for a proof. It is worth pointing out, though, that a

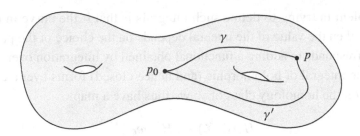

Figure D.4 The integral $\int_{p_0}^p \omega$ may depend on the choice of path.

slight variant of this question — "For which d, g and r do there exist *smooth*, irreducible and nondegenerate curves of degree d and genus g in $\mathbb{P}^r$?" remains open in general. It was solved for $r = 3$ in Gruson and Peskine [1982]; see Ciliberto [1987] and Harris and Eisenbud [1982, Chapter 3] for a discussion of some of the relevant issues in general.

It should be said that Castelnuovo's theorem answers question (a) in the preceding section (the question of what birationally very ample linear series a curve may have). It does not, however, address question (b); that is, it does not tell us what linear series are present on a general curve. To put it another way, inside the space M_g parametrizing smooth projective curves of genus g, the locus of "Castelnuovo curves" — that is, curves $C \subset \mathbb{P}^r$ of degree d with $g = \pi(d, r)$ — is contained in a subvariety of high codimension. Thus, the question remains of what linear series exist on all or most curves of genus g. This is the question addressed by the Brill–Noether theorem, of which Theorem D.1 is a weak version. To state it in a strong form and to prove it we will have to analyze the geometry of curves in greater depth.

D.2 Families of divisors

D.2.1 The Jacobian

A deeper study of linear series requires us to make sense of the set of linear systems as a variety. For this purpose we introduce the *Jacobian*.

One of the early motivations for studying algebraic curves came from calculus, specifically the desire to make sense of integrals of algebraic functions. In modern terms, this means integrals

$$\int_{p_0}^p \omega,$$

where ω is a holomorphic (or more generally meromorphic) differential on a smooth projective curve C over the complex numbers, p_0 and p are points of C and the integral is taken along a path from p_0 to p on C.

One problem in trying to define such integrals is that if the curve in question has positive genus then the value of the integral depends on the choice of the path. A change of the path corresponds to adding a functional obtained by integration over a closed loop on C. Now, the integral of holomorphic (and hence closed) forms over a closed loop γ depends only on the homology class of γ; we thus have a map

$$H_1(C, \mathbb{Z}) \to H^0(\omega_C)^*$$

which embeds $H_1(C, \mathbb{Z}) \cong \mathbb{Z}^{2g}$ as a discrete lattice in $H^1(\omega_C)^* \cong \mathbb{C}^g$ (see for example Griffiths and Harris [1994, p. 228]). Thus we may view the integral $\int_{p_0}^{p}$ as an element of the quotient

$$H^0(\omega_C)^*/H_1(C, \mathbb{Z}).$$

We define the *Jacobian* $\mathrm{Jac}(C)$ of the curve C to be this quotient. By our construction, $\mathrm{Jac}(C)$ is a complex torus, and in particular a compact complex manifold; in fact, it is a projective variety over $\mathbb{C}$, and the map $p \to \int_{p_0}^{p}$ is a map of projective varieties. We will use this often; see Griffiths and Harris [1994, §§2.2–2.3] for a treatment in the complex-analytic setting.

Having defined the Jacobian, we see that, after choosing a base point $p_0 \in C$, integration defines a map $C \to \mathrm{Jac}(C)$ and, more generally, maps

$$u = u_d : C_d \to \mathrm{Jac}(C)$$

from the symmetric powers $C_d = C^d/\mathfrak{S}_d$ of C to its Jacobian, defined by

$$D = \sum p_i \mapsto \sum \int_{p_0}^{p_i}.$$

These maps are called *Abel–Jacobi* maps.

D.2.2 Abel's theorem

Theorem D.5 (Abel). *Let $u : C_d \to \mathrm{Jac}(C)$ be the Abel–Jacobi map. Divisors D, E on C of the same degree are linearly equivalent if and only if $u(D) = u(E)$.*

One direction of Abel's theorem — the "if" half — is relatively easy to prove: If D and E are linearly equivalent, there is a pencil of divisors, parametrized by $\mathbb{P}^1$, interpolating between them. But if $f : \mathbb{P}^1 \to A$ is any map from $\mathbb{P}^1$ to a torus, the pullbacks $f^*\eta$ of holomorphic 1-forms on A vanish identically. Since these 1-forms generate the cotangent space at every point of A, it follows that the differential df is identically zero and hence that f is constant; thus $u(D) = u(E)$. The hard part (which was in fact proved by Clebsch) is the converse. See Griffiths and Harris [1994, p. 235] for a treatment.

The import of Abel's theorem is that we may, for each d, identify the set of linear equivalence classes of effective divisors of degree d on C with the Jacobian Jac(C). The identification is not canonical; it depends on the choice of a base point $p_0 \in C$. In Section D.2.3 we will use this correspondence to show that there exists a *fine moduli space* $\text{Pic}^d(C)$ for line bundles of degree d on C — that is, a space together with a universal family — and that $\text{Pic}^d(C)$ is isomorphic (again, non-canonically) to Jac(C).

Abel's theorem tells us that the fiber $u^{-1}(u(D))$ of u through a point $D \in C_d$ is the complete linear system $|D| = \{E \in C_d \mid E \sim D\}$ — set-theoretically, at least, a projective space. (We will see in Theorem D.6 that it is indeed isomorphic to $\mathbb{P}^{h^0(D)-1}$.) Beyond this, the behavior of the map u depends very much on d.

If $p_1, \ldots, p_d \in C$ are general points with $d \leq g$, then the conditions of vanishing at the p_i are independent linear conditions on differential forms. Writing $D = p_1 + \cdots + p_d$, we get $h^0(\omega - D) = g - d$. From the Riemann–Roch theorem, we see that $h^0(D) = 1$; that is, no other effective divisor of degree d is linearly equivalent to D. It follows from Abel's theorem that the map $u : C_d \to \text{Jac}(C)$ is birational onto its image, and in particular that the image $W_d := u(C_d) \subset \text{Jac}(C)$ is again d-dimensional.

In particular, the map $u : C_g \to \text{Jac}(C)$ is birational (this statement is called the *Jacobi inversion theorem*; see Exercise D.18 for a more classical version). Further, the image of $u : C_{g-1} \to \text{Jac}(C)$ is a divisor in Jac(C), called the *theta divisor* and written Θ.

When $d \geq g$ the same argument shows that the map $u : C_d \to \text{Jac}(C)$ is surjective. When $g \leq d \leq 2g - 2$ the dimensions of the fibers will vary, but when $d > 2g - 2$ the picture becomes regular: the fibers of $u : C_d \to \text{Jac}(C)$ are all of dimension $d - g$. We will see in this case that $u : C_d \to \text{Jac}(C)$ is in fact a projective bundle, and in Section D.5.2 we will identify the vector bundle $\mathcal{E}$ on Jac(C) such that $C_d \cong \mathbb{P}\mathcal{E}$.

Note that the subset of line bundles $\mathcal{L}$ of degree d such that $h^0(\mathcal{L}) \geq r + 1$ is Zariski closed in $\text{Pic}^d(C)$: it is the locus where the fiber dimension of u is r or greater. We define the *Brill–Noether locus* $W_d^r(C)$ to be the set

$$W_d^r(C) = \{\mathcal{L} \in \text{Pic}^d(C) \mid h^0(\mathcal{L}) \geq r + 1\}.$$

This Zariski closed subset of $\text{Pic}^d(C)$ has a natural scheme structure, which we will explain in Section D.4.2 below.

Here is the first step toward proving that for large d the Abel–Jacobi map is the projection from a projectivized vector bundle on Jac(C):

Theorem D.6. *For any d, the scheme-theoretic fiber of the Abel–Jacobi map $u : C_d \to \text{Jac}(C)$ through a point $D \in C_d$ is the projective space $|D| = \mathbb{P}H^0(\mathcal{O}_C(D))$. For $d > 2g - 2$, the map u is a submersion, that is, the differential $du : T_D C_d \to T_{u(D)} \text{Jac}(C)$ is surjective everywhere.*

Proof: From Abel's theorem above, we see that the projective space

$$|D| \cong \{(\sigma, D') \in \mathbb{P}H^0(\mathcal{O}_C(D)) \times C_d \,|\, \sigma \text{ vanishes on } D\}$$

projects onto the fiber $u^{-1}u(D) \subset C_n$, so it is enough to show that $u^{-1}u(D)$ is smooth and of dimension equal to $h^0(\mathcal{O}_C(D)) - 1$.

We start with the case $d > 2g - 2$, and consider a point $D \in C_d$ corresponding to a reduced divisor, that is, $D = p_1 + \cdots + p_d \in C_d$ with the points p_i distinct. From the definition of the Jacobian as $H^0(\omega_C)^*/H_1(C, \mathbb{Z})$ we see that the cotangent space at any point is

$$T_{u(D)}^* \operatorname{Jac}(C) = H^0(\omega_C),$$

the space of regular differentials on C. We may similarly identify the cotangent space to C_d at D: because the p_i are distinct we have

$$T_D^* C_d = \bigoplus T_{p_i}^* C = \bigoplus H^0((\omega_C)_p) = H^0(\omega_C/\omega_C(-D)).$$

Differentiating the Abel–Jacobi map

$$D = \sum p_i \mapsto \sum \int_{p_0}^{p_i}$$

with respect to the points p_i, we see that, in terms of these identifications, the differential du_D of u at a point $D \in C_d$ is given as the transpose of the evaluation map

$$H^0(\omega_C) \to \bigoplus T_{p_i}^* C,$$
$$\omega \mapsto (\omega(p_1), \ldots, \omega(p_d));$$

in particular, the cokernel of the differential du_D of u at a point $D \in C_d$ is the annihilator of the subspace $H^0(\omega_C(-D)) \subset H^0(\omega_C)$ of differentials vanishing along D. Since we are working in the range $d > 2g - 2 = \deg \omega_C$, there are no such differentials, and we are done.

In fact, the identification $T_D^* C_n = H^0(\omega_C/\omega_C(-D))$ extends to all divisors $D \in C_d$ and in these terms the differential is again the transpose of the evaluation map

$$H^0(\omega_C) \to H^0(\omega_C/\omega_C(-D))$$

(see for example Arbarello et al. [1985, §IV.1]); so the same logic applies.

Finally, in the case $d \le 2g - 2$ the Riemann–Roch theorem tells us that the dimension of the kernel of the differential du at any point D — that is, the dimension of the cokernel of the evaluation map $H^0(\omega_C) \to H^0(\omega_C/\omega_C(-D))$ — is exactly the dimension $r(D) = h^0(\mathcal{O}_C(D)) - 1$ of the fiber of u through D; thus the fibers of u are smooth in this case as well, even though they are not all of the same dimension.   □

D.2.3　Moduli spaces of divisors and line bundles

Abel's theorem tells us that when $d \geq g$ the fibers of the Abel–Jacobi map $C_d \to$ Jac are in one-to-one correspondence with the line bundles $\mathcal{L}$ of degree d on C, making the set of bundles $\mathrm{Pic}^d(C)$ into an analytic variety; the same follows for every d via the isomorphisms $\mathrm{Pic}^d(C) \cong \mathrm{Pic}^e(C)$. The defining isomorphism $\mathrm{Pic}^d(C) \cong \mathrm{Jac}(C)$ is not canonical: it depends on the choice of a base point p_0. If we chose a different point p_0' then the identifications would take $\mathcal{L}$ to $\mathcal{L}(d(p_0' - p_0))$. For clarity, it is usually best to think of the $\mathrm{Pic}^d(C)$ as distinct schemes.

The variety $\mathrm{Pic}^d(C)$ is actually a *fine moduli space*, in the sense that $\mathrm{Pic}^d(C) \times C$ carries a *universal line bundle* $\mathcal{P}$. The key property of $\mathcal{P}$ is that its restriction to each fiber $\{\mathcal{L}\} \times C$ is isomorphic to the corresponding line bundle $\mathcal{L}$. In fact, it satisfies a stronger functorial characterization:

Proposition D.7. *Let C be a smooth, projective curve of genus g, and d any integer. Let $p_0 \in C$ be a point. There exists a projective scheme $\mathrm{Pic}^d(C)$ and a line bundle $\mathcal{P}$ on $\mathrm{Pic}^d(C) \times C$ such that:*

(a) $\mathcal{P}$ is trivial on $\mathrm{Pic}^d(C) \times \{p_0\}$; and

(b) for any scheme B and any line bundle $\mathcal{M}$ on $B \times C$ of relative degree d that is trivial on $B \times \{p_0\}$, there exists a unique map $\varphi : B \to \mathrm{Pic}^d(C)$ such that

$$\mathcal{M} = (\varphi \times \mathrm{Id}_C)^* \mathcal{P}.$$

(The condition $\mathcal{P}|_{\{\mathcal{L}\} \times C} \cong \mathcal{L}$ is the special case of part (b) where B is a point.) The "universal line bundle" $\mathcal{P}$ on $\mathrm{Pic}^d(C) \times C$ is called the *Poincaré bundle* (with respect to p_0). We postpone its construction to Section D.4.1.

We have constructed $\mathrm{Pic}^d(C)$ as an analytic variety, but it has in fact the structure of an algebraic variety, and can be constructed for curves over any field. This was first done by André Weil. He observed that via the birational map $u : C_g \to \mathrm{Jac}(C)$ an open subset of $\mathrm{Jac}(C)$ was isomorphic to an open subset of C_g. Composing u with translations, we see that $\mathrm{Jac}(C)$ may be covered by such open sets, and these can be glued together to construct $\mathrm{Jac}(C)$. (Indeed, it was the desire to carry out this construction that led Weil to the definition of an abstract variety.) In even greater generality, Grothendieck applied his theory of *étale equivalence relations* to construct $\mathrm{Pic}^d(C)$ as the quotient of C_d by linear equivalence for large d; see Milne [2008] for a description.

The symmetric power C_d that is the source of the Abel–Jacobi map is also a fine moduli space, supporting a universal family of divisors; equivalently, we may regard C_d as the Hilbert scheme of subschemes of degree d on C:

Proposition D.8. *There exists a divisor $\mathcal{D} \subset C_d \times C$ such that, for any scheme B and any effective divisor $\Delta \subset B \times C$ finite over B of relative degree d, there exists a unique map $\varphi : B \to C_d$ such that*

$$\Delta = (\varphi \times \mathrm{Id}_C)^{-1} \mathcal{D}.$$

It is easy to make the "universal divisor" explicit: it is just the reduced divisor

$$\mathcal{D} = \{(D, p) \in C_d \times C \mid p \in D\},$$

which we will encounter repeatedly in what follows.

For proofs of Propositions D.7 and D.8, see, e.g., Arbarello et al. [1985, §IV.2].

D.3 The Brill–Noether theorem

Here is our second version of the Brill–Noether theorem. Recall that we have defined $\rho(g, r, d) := g - (r + 1)(g - d + r)$.

Theorem D.9 (Brill–Noether). *(a) For every curve C of genus g*

$$\dim W_d^r(C) \geq \rho(g, r, d).$$

(b) If C is a general curve of genus g then equality holds.

The appearance of ρ in this theorem can be understood as follows: g is the dimension of the Jacobian of C, which may be thought of as the space $\mathrm{Pic}^d(C)$ of all line bundles of degree d on C. Furthermore, if a line bundle $\mathcal{L} \in \mathrm{Pic}^d(C)$ has $h^0(\mathcal{L}) = r + 1$, so that $\mathcal{L} \in W_d^r(C) \setminus W_d^{r+1}(C)$, the Riemann–Roch theorem asserts that $h^1(\mathcal{L}) = g - d + r$. Thus the Brill–Noether theorem asserts that if C is a general curve of genus g then the codimension of $W_d^r \subset \mathrm{Pic}^d(C)$ is $h^0(\mathcal{L})h^1(\mathcal{L})$. Indeed, as we shall see, W_d^r can be thought of as the rank-k locus of a map between vector bundles of ranks $k + h^0(\mathcal{L})$ and $k + h^1(\mathcal{L})$ (for any large k!), so this is the "expected" codimension in the sense of Chapter 12.

In general, the formula of Theorem D.9 is far more restrictive than Castelnuovo's bound. For example, Figure D.5 shows the values of d and r allowed in the case $g = 100$; we see that only a tiny fraction of the complete linear series allowed by Castelnuovo actually occur on a general curve.

In the final section of this appendix we will give an enumerative proof of the existence half of Theorem D.9, namely:

Theorem D.10. *If $\rho = g - (r + 1)(g - d + r) \geq 0$, then every smooth curve C of genus g has*

$$\dim W_d^r(C) \geq \rho.$$

In particular, there exist linear systems on C of degree d and dimension r.

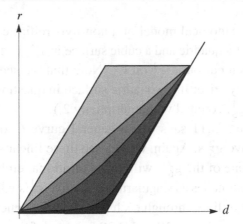

Figure D.5   The shaded regions correspond to points (d, r) such that there exists:

- a complete linear series whose existence is not contradicted by the Riemann–Roch theorem;
- a complete linear series that actually exists on some curve of genus 100;
- a complete linear series on some curve of genus 100, which defines a birational map; and
- a complete linear series on a general curve of genus 100 (this necessarily defines a birational map when $r > 1$).

The heart of our proof of Theorem D.10 is an enumerative formula for the class of $W_d^r(C)$, given in the stronger Theorem D.17. When $g = (r + 1)(g - d + r)$—that is, when $\rho = 0$—this formula becomes a number:

Corollary D.11. *If C is a smooth curve of genus $g = (r + 1)(g - d + r)$ and if W_d^r is finite, then C possesses*

$$g! \prod_{i=1}^{r} \frac{i!}{(g - d + r + i)!}$$

linear series of degree d and dimension r, counted with multiplicity. When $r = 1$ and $g = 2(g - d + 1) = 2k$, this number is the Catalan number

$$\frac{1}{k + 1} \binom{2k}{k}.$$

Proof: By Poincaré's formula (Proposition D.13 below), $\deg(\theta^g) = g!$; substituting in the formula of Theorem D.17 yields the corollary. Note that the multiplicity with which a given linear series occurs is equal to the multiplicity of the scheme W_d^r at the corresponding point. ☐

For example, in the first case that is not answered by the Riemann–Roch theorem, we can ask if a general curve of genus 4 is expressible as a 3-sheeted cover of $\mathbb{P}^1$, and if so in how many ways; this is the content of Keynote Question (b). Corollary D.11 gives an answer: It says that C will admit two such maps. Indeed, we can see directly that there are two: the canonical model of a non-hyperelliptic curve of genus 4 is the complete intersection of a quadric and a cubic surface in $\mathbb{P}^3$, and if the quadric is smooth its two rulings will each cut out a g_3^1 on C. (Note that we see in this example a case where multiplicities may arise: If the quadric surface in question is a cone, the curve C will possess only one g_3^1, counted with multiplicity 2.)

Similarly, Corollary D.11 says that a general curve C of genus 6 will possess five g_4^1's, and dually five g_6^2's. Again, we can see these linear series explicitly: By the extension below, any one of the g_6^2's will give a birational embedding of C as a plane sextic $C_0 \subset \mathbb{P}^2$ with four nodes as singularities. The five g_4^1's will then be the pencils cut out on C by the pencil of lines through each node and the pencil cut by conics passing through all four. (See Exercises D.19 and D.20 for a proof that these are all the g_4^1's on C, and Exercise D.21 for another example.)

There are various extensions of this theorem for general curves C of genus g: Fulton and Lazarsfeld [1981] showed that if $\rho > 0$ then $W_d^r(C)$ is irreducible; Gieseker proved that the singular locus of $W_d^r(C)$ is exactly $W_d^{r+1}(C)$, and hence (given Exercise D.23) that $W_d^r(C)$ is reduced (see Harris and Morrison [1998] for a discussion of this theorem and its proofs). In particular, we see that for a general curve there are in fact no multiplicities in the formula in Corollary D.11.

D.3.1 How to guess the Brill–Noether theorem and prove existence

Here is one way to describe the locus $W_d^r(C)$. Fix a divisor

$$D = p_1 + \cdots + p_m$$

consisting of m distinct points of C. For any line bundle $\mathcal{L}$ on C, there is an exact sequence

$$0 \longrightarrow \mathcal{L} \longrightarrow \mathcal{L}(D) \xrightarrow{b_\mathcal{L}} \bigoplus_{i=1}^{m} \mathcal{L}(D)_{p_i} \cong \bigoplus_{i=1}^{m} \mathcal{O}_{p_i} \longrightarrow 0,$$

and taking cohomology we see that

$$H^0(\mathcal{L}) = \mathrm{Ker}\left(H^0(\mathcal{L}(D)) \xrightarrow{h^0 b_\mathcal{L}} \bigoplus_{i=1}^{m} \mathbb{C} \right).$$

If the number m is large — say $m > 2g - 2 - d$ — then the Riemann–Roch formula tells us that $h^0(\mathcal{L}(D)) = m + d - g + 1$, independently of $\mathcal{L}$. Thus, as $\mathcal{L}$ varies over the set $\mathrm{Pic}^d(C)$ of line bundles of degree d, the locus $W_d^r(C) \subset \mathrm{Pic}^d(C)$ is the locus where the $m \times (m + d - g + 1)$ matrix $h^0 b_\mathcal{L}$ has rank at most $(m + d - g + 1) - (r + 1)$. The expected codimension of the locus where an $s \times t$ matrix with $s \leq t$ has rank u, in the sense of Chapter 12, is $(s - u)(t - u)$. Thus the "expected" codimension of W_d^r in

$\mathrm{Pic}^d(C)$ is $(r+1)(m-(m+d-g+1-r-1)) = (r+1)(g-d+r)$, exactly the codimension predicted for a general curve by the Brill–Noether theorem.

As we will see, the maps $h^0 b_{\mathcal{L}}$ vary algebraically with $\mathcal{L} \in \mathrm{Pic}^d(C)$. It follows that *if $W_d^r(C)$ is nonempty* for a given curve C then its dimension is at least $\rho(g,r,d)$, and, given the existence of one curve C_0 for which W_d^r is really nonempty and of dimension $\rho(g,r,d)$, it would follow that this is true for an open set of curves in any family containing C_0. Brill and Noether must have known many cases where these conditions were all satisfied, but they lacked the tools to give a proof of the theorem.

In Section D.4.2 we will identify the map $b_{\mathcal{L}}$ as the fiber of a map of vector bundles $b : \mathcal{F} \to \mathcal{G}$ over $\mathrm{Pic}^d(C)$, which is isomorphic to the Jacobian $\mathrm{Jac}(C)$ of C (this implies that the map $h^0 b_{\mathcal{L}}$ varies algebraically with $\mathcal{L}$). As remarked above, this implies that $W_d^r(C)$ has dimension at least $\rho(g,r,d)$ provided that it is nonempty.

To prove that $W_d^r(C)$ is nonempty when $\rho(g,r,d) \geq 0$, we will compute the Chern classes of the vector bundles $\mathcal{F}$ and $\mathcal{G}$. Porteous' formula allows us to compute the class $\alpha \in H^{2(r+1)(g-d+r)}(\mathrm{Jac}(C))$ that the locus W_d^r *would* have *if* it had dimension $\rho(g,r,d)$. We will show that this class is nonzero when $0 \leq (r+1)(g-d+r) \leq g$, and this suffices to prove the desired existence.

D.3.2 How the other half is proven

The proof of the other half of the Brill–Noether theorem — the statement that for a general curve C the dimension $\dim W_d^r(C)$ is at most $\rho(g,r,d)$, and in particular that C possesses no g_d^r's when $\rho < 0$ — requires very different ideas. One could prove it by exhibiting for each g, r and d a smooth curve C of genus g with $\dim W_d^r(C) = \rho(g,r,d)$ (or with $W_d^r(C) = \varnothing$ if $\rho < 0$), but no one has ever succeeded in doing this explicitly for large g. The known proofs fall into two families:

Degeneration to singular curves

One approach to this problem is to consider a one-parameter family of curves $\{C_t\}$ specializing from a smooth curve to a singular one, C_0. What needs to be done in this setting is first of all to describe the limit as $t \to 0$ of a g_d^r on C_t, and then to prove that such limits do not exist on C_0 when $\rho < 0$. This was done in the original proof, with C_0 a general g-nodal curve (that is, $\mathbb{P}^1$ with g pairs of general points identified); the possible limits of a g_d^r on C_t were identified in Kleiman [1976] and the proof that no such limit exists when $\rho < 0$ was given in Griffiths and Harris [1980]. Another proof (Eisenbud and Harris [1983a]) used a g-cuspidal curve as C_0, and in Eisenbud and Harris [1983b] the role of C_0 was played by a curve consisting of a copy of $\mathbb{P}^1$ with g elliptic tails attached. See Figure D.6 for examples of these curves. Much more recently, a proof was given using the methods of tropical geometry in Cools et al. [2012].

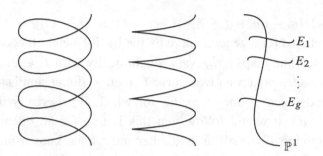

Figure D.6 Three singular curves used in specialization arguments for the nonexistence half of Brill–Noether.

Curves on a very general K3 surface

A completely different proof was given by Lazarsfeld [1986], who showed that a smooth curve C embeddable in a very general K3 surface — specifically, one whose Picard group was generated by the class of the curve C — necessarily satisfied the statement of the basic Brill–Noether theorem. The theorem was thus proved by specializing to a smooth curve, rather than a singular one (though the smooth curve in question could still not be explicitly given, inasmuch as we have no way to explicitly produce K3 surfaces with Picard number 1).

D.4 W_d^r as a degeneracy locus

In the remainder of this appendix we will deal with a fixed curve C. To simplify notation, we will write Jac for the Jacobian $\mathrm{Jac}(C)$ and Pic^d for the Picard variety $\mathrm{Pic}^d(C)$ parametrizing line bundles of degree d on C.

In this section we will explain how to construct the family of all line bundles of a given degree, and how to put the maps $b_{\mathcal{L}}$ of Section D.3.1 together into a map of bundles. To do this, we first need to construct the *Poincaré bundle*, a fundamental object in the theory.

D.4.1 The universal line bundle

Choose a base point $p_0 \in C$. The Poincaré bundle is a line bundle on the product $\mathrm{Pic}^d \times C$ whose restriction to the fiber $\{\mathcal{L}\} \times C$ over $\mathcal{L} \in \mathrm{Pic}^d$ is isomorphic to $\mathcal{L}$ and whose restriction to the cross-section $\mathrm{Pic}^d \times \{p_0\}$ is trivial.

Without the normalizing condition of triviality on $\mathrm{Pic}^d \times \{p_0\}$ the bundle $\mathcal{P}$ would not be determined uniquely: We could tensor with the pullback of any line bundle on $\mathrm{Pic}^d(C)$ and get another. But with the normalizing condition, Corollary B.6(b) shows that the Poincaré bundle is unique — if it exists.

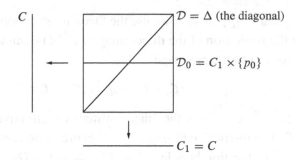

Figure D.7 The divisors $\mathcal{D}$ and $\mathcal{D}_0$ in the case $d = 1$.

We will construct the Poincaré bundle as the direct image of a line bundle $\mathcal{M}$ on $C_d \times C$ under the map

$$\eta = u \times \mathrm{Id} : C_d \times C \to \mathrm{Pic}^d \times C.$$

For $p \in C$, let $X_p \subset C_d$ be the divisor that is the image of $\{p\} \times C^{d-1}$ in C_d, that is, the set of divisors containing p. To describe $\mathcal{M}$, we write $\mathcal{D}_0$ for the divisor $X_{p_0} \times C \subset C_d \times C$, and let $\mathcal{D} \subset C_d \times C$ be the *universal divisor* of degree d as in Proposition D.8, that is,

$$\mathcal{D} = \{(D, p) \in C_d \times C \mid D - p \geq 0\}.$$

Thus the restriction of $\mathcal{D}$ to a fiber $\{D\} \times C$ of the projection to C_d is the divisor D, and the restriction of $\mathcal{D}$ to the fiber $C_d \times \{p\}$ of the projection to C is the divisor X_p. Finally, define

$$\mathcal{M} = \mathcal{O}_{C_d \times C}(\mathcal{D} - \mathcal{D}_0),$$

and set $\mathcal{P} = \eta_* \mathcal{M}$.

Proposition D.12. $\mathcal{P} = \eta_*(\mathcal{M})$ *is a Poincaré bundle on* $\mathrm{Pic}^d \times C$*; in particular,*

$$\mathcal{P}|_{\{\mathcal{L}\} \times C} \cong \mathcal{L}$$

for any point $\mathcal{L} \in \mathrm{Pic}^d$ *and*

$$\mathcal{P}_{\mathrm{Pic}^d \times \{p_0\}} \cong \mathcal{O}_{\mathrm{Pic}^d}.$$

Proof: Since the restriction of $\mathcal{M}$ to any fiber $\mathbb{P}^r$ of η is trivial (both divisors $\mathcal{D}$ and $\mathcal{D}_0$ intersect $\mathbb{P}^r$ in a hyperplane), the theorem on cohomology and base change (Theorem B.9) shows that the direct image $\eta_* \mathcal{M}$ is a line bundle and the formation of this direct image commutes with base change.

The proof that $\mathcal{P}_{\mathrm{Pic}^d \times \{p_0\}} \cong \mathcal{O}_{\mathrm{Pic}^d}$ is immediate: If we restrict to the preimage

$$\eta^{-1}(\mathrm{Pic}^d \times \{p_0\}),$$

the divisors $\mathcal{D}$ and $\mathcal{D}_0$ agree, so that $\mathcal{M}|_{\eta^{-1}(\mathrm{Pic}^d \times \{p_0\})}$ is trivial, and so is its direct image.

To prove that $\mathcal{P}|_{\{\mathcal{L}\}\times C} \cong \mathcal{L}$ we use the theorem on cohomology and base change. It implies that the formation of the direct image $\eta_*(\mathcal{M})$ commutes with base change, so we can first restrict to the preimage

$$|\mathcal{L}| \times C = \eta^{-1}(\{\mathcal{L}\} \times C),$$

where $|\mathcal{L}| \cong \mathbb{P}^r \subset C_d$ is the linear system of effective divisors D on C with $\mathcal{O}_C(D) \cong \mathcal{L}$. The restriction of η to $|\mathcal{L}| \times C$ is projection on the second factor.

As observed, the line bundle $\mathcal{M} = \mathcal{O}_{C_d \times C}(\mathcal{D} - \mathcal{D}_0)$ is trivial on each fiber of $\eta : |\mathcal{L}| \times C \to C$, so that the restriction $\mathcal{M}|_{|\mathcal{L}| \times C}$ must be a pullback of some line bundle on C; to prove that $\eta_*(\mathcal{M})|_{\{\mathcal{L}\}\times C} \cong \mathcal{L}$ amounts to showing that this line bundle is $\mathcal{L}$. Thus, it suffices to prove that

$$\mathcal{M}|_{|\mathcal{L}| \times C} = \eta^* \mathcal{L}. \tag{D.3}$$

For this it is enough to show that, for a general divisor D,

$$\mathcal{M}|_{\{D\} \times C} \cong \mathcal{L}.$$

This is immediate if D does not contain the point p_0: By definition, the divisor $\mathcal{D}_0 = X_{p_0} \times C \subset C_d \times C$ is disjoint from $\{D\} \times C$, while the divisor $\mathcal{D}$ intersects $\{D\} \times C$ in the divisor D. $\qquad\square$

D.4.2 The evaluation map

Fix a reduced divisor $D = p_1 + \cdots + p_m$ of degree $m \geq 2g - 1 - d$ on C, and set $n = m + d$. Choosing D gives us an identification of Pic^d with Pic^n. We will describe the locus $W_d^r + D \subset \text{Pic}^n$ as the degeneracy locus of an evaluation map.

On the product $\text{Pic}^n \times C$, consider the evaluation map

$$\mathcal{P} \to \mathcal{P}|_\Gamma,$$

where

$$\Gamma = \text{Pic}^n \times D = \bigcup_{i=1}^{m} (\text{Pic}^n \times \{p_i\})$$

is the union of the horizontal sections of $\text{Pic}^n \times C$ over Pic^n corresponding to the points p_i. Taking the direct image of this map under the projection $\pi : \text{Pic}^n \times C \to \text{Pic}^n$, we have a map of vector bundles

$$\rho : \mathcal{E} := \pi_*(\mathcal{P}) \to \mathcal{F} = \bigoplus_{i=1}^{m} \mathcal{L}_i,$$

where $\mathcal{L}_i$ is the restriction of $\mathcal{P}$ to the cross-section $\text{Pic}^n \times \{p_i\}$. For each point $\mathcal{L} \in \text{Pic}^n$, this is the map

$$\mathcal{E}_\mathcal{L} = H^0(\mathcal{L}) \to \bigoplus \mathcal{L}_{p_i}$$

obtained by evaluating sections of $\mathcal{L}$ at the points p_i. In particular, the kernel of this map is the vector space $H^0(\mathcal{L}(-D)) \subset H^0(\mathcal{L})$ of sections vanishing along D. We have now proven that *the locus* $W_d^r + D \subset \text{Pic}^n$ *is the locus where the map* ρ *has rank* $n-g-r$ *or less*.

In particular, the determinantal ideal defines a scheme structure on W_d^r. Though this structure appears to depend on the choice of the divisor D, in fact it does not: for any choice of D one can prove that the scheme W_d^r has a universal property independent of D that characterizes it. This is done explicitly, for example, in Arbarello et al. [1985, Chapter 4].

It remains to show that $W_d^r(C)$ is nonempty. To do so we will compute the Chern classes of the bundles $\mathcal{E}$ and $\mathcal{F}$ and apply Porteous' theorem. Before we do so, however, we must develop some basic information about the cohomology ring of the Jacobian, where these Chern classes live, and we must also identify the bundle $\mathcal{E}$ in a more useful way.

D.5 Natural classes in the cohomology ring of the Jacobian

Why are we working with the cohomology ring of the Jacobian rather than with the Chow ring? For one thing, it is computable: Since the Jacobian of a curve of genus g is topologically the product of $2g$ copies of the circle, its cohomology ring is an exterior algebra on $2g$ generators of degree 1, whereas the Chow groups of an abelian variety of dimension ≥ 2 are largely unknown. In particular, since any deformation of a cycle along a continuous path preserves the homology class of the cycle, the cohomology class of a cycle does not change under translation. The same is not true modulo rational equivalence, and this is very much an issue here: Most of the cycles whose classes we might hope to determine are in fact only defined after a choice of base point — in effect, only up to translation.

That said, our first goal will be to identify the classes in $H^*(\text{Jac}, \mathbb{Z})$ of certain basic cycles. To start, the Jacobian is the quotient of the contractible space $H^0(\omega_C)^* \cong \mathbb{C}^g$ by the subgroup $H_1(C, \mathbb{Z}) \cong \mathbb{Z}^{2g}$, so there is a natural identification

$$H_1(\text{Jac}, \mathbb{Z}) = H_1(C, \mathbb{Z}).$$

The first cohomology $H^1(\text{Jac}, \mathbb{Z})$ is similarly identified with $H^1(C, \mathbb{Z})$. These identifications are induced by the Abel–Jacobi map $u : C \to \text{Jac}$ because the integral takes a closed path $\gamma : [0, 1] \to \mathbb{C}$ to the path $\tilde{\gamma} : [0, 1] \to H^0(\omega_C)^*$ defined by

$$\tilde{\gamma}(t) = \int_{\gamma([0,t])} \in H^0(\omega_C)^*,$$

which joins the origin to the lattice point corresponding to the homology class of γ.

We choose a basis $\alpha_1, \ldots, \alpha_g, \beta_1, \ldots, \beta_g$ for $H^1(C, \mathbb{Z})$, normalized so that the cup product has the form

$$(\alpha_i \cup \beta_j)[C] = \begin{cases} 1 & \text{if } i = j, \\ 0 & \text{otherwise,} \end{cases}$$

and

$$\alpha_i \cup \alpha_j = \beta_i \cup \beta_j = 0 \quad \text{for all } i, j.$$

By abuse of notation, we will also use the symbols α_i and β_i to denote the corresponding cohomology classes in $H^1(\text{Jac}, \mathbb{Z})$.

Since the cohomology ring of the Jacobian is the exterior algebra generated by the α_i and β_j, we can compute arbitrary products of these elements. Given a multi-index $I = (i_1, \ldots, i_k)$ with $i_1, \ldots, i_k \in \{1, \ldots, g\}$, we will write α_I and β_I for the classes

$$\alpha_I = \alpha_{i_1} \cup \cdots \cup \alpha_{i_k} \quad \text{and} \quad \beta_I = \beta_{i_1} \cup \cdots \cup \beta_{i_k}$$

in $H^k(\text{Jac}, \mathbb{Z})$. The classes

$$\{\alpha_I \cup \beta_J \mid I, J \subset \{1, \ldots, g\}\}$$

form a basis for $H^*(\text{Jac}, \mathbb{Z})$. The cup product in complementary dimension is given (via the identification $H^{2g}(\text{Jac}, \mathbb{Z}) \cong \mathbb{Z}$) for $I, J, K, L \subset \{1, \ldots, g\}$ by

$$((\alpha_I \cup \beta_J) \cup (\alpha_K \cup \beta_L))[\text{Jac}] = \begin{cases} \pm 1 & \text{if } K = I' \text{ and } L = J', \\ 0 & \text{otherwise,} \end{cases} \tag{D.4}$$

where I' denotes the complement of I. To determine the signs, we can pull back to the direct product of g copies of C, and use (D.10) below to prove that

$$(\alpha_1 \cup \beta_1 \cup \alpha_2 \cup \beta_2 \cup \cdots \cup \alpha_g \cup b_g)[\text{Jac}] = +1;$$

the signs of the other expressions in (D.4) are determined by skew-symmetry.

Of special interest are the classes $\eta_i \in H^2(\text{Jac}, \mathbb{Z})$ defined as

$$\eta_i = \alpha_i \cup \beta_i \quad \text{for } i = 1, \ldots, g.$$

For a multi-index $I = (i_1, \ldots, i_k)$ with $i_1, \ldots, i_k \in \{1, \ldots, g\}$, we will write η_I for the class

$$\eta_I = \eta_{i_1} \cup \cdots \cup \eta_{i_k} \in H^{2k}(\text{Jac}, \mathbb{Z}).$$

For example, by what we have just said $\eta_{(1,\ldots,g)} = 1 \in H^{2g}(\text{Jac}, \mathbb{Z}) = \mathbb{Z}$. Rearranging the terms, we see that in general

$$\eta_I = (-1)^{\binom{k}{2}} \alpha_I \cup \beta_I.$$

The cup product in complementary dimension is easy to compute: For $I, J \subset \{1, \ldots, g\}$ we have

$$(\eta_I \cup \eta_J)[\text{Jac}] = \begin{cases} 1 & \text{if } J = I', \\ 0 & \text{otherwise.} \end{cases} \tag{D.5}$$

D.5.1 Poincaré's formula

The objects of primary interest to us are the classes of the subvarieties $W_d \subset \mathrm{Pic}^d$ parametrizing effective divisor classes of degree d, that is, the images of the maps $u = u_d : C_d \to \mathrm{Jac} \cong \mathrm{Pic}^d$. Like most of the objects in our treatment, the map u depends on the choice of base point $p_0 \in C$, so the subvarieties $W_d \subset \mathrm{Jac}$ are really only defined up to translation, though their classes in $H^*(\mathrm{Jac}, \mathbb{Z})$ are well-defined. Here is the basic result:

Proposition D.13 (Poincaré's formula).

$$[W_d] = \sum_{\substack{I \subset \{1,\dots,g\} \\ |I|=g-d}} \eta_I \in H^{2g-2d}(\mathrm{Jac}, \mathbb{Z}).$$

The divisor $W_{g-1} \subset \mathrm{Jac}$ occurs often, and is usually called the *theta divisor* on Jac and denoted by Θ; its class is denoted by $\theta \in H^2(\mathrm{Jac}, \mathbb{Z})$. By Proposition D.13 we have

$$\theta = \eta_1 + \cdots + \eta_g.$$

With this notation we can restate the proposition as

$$[W_d] = \frac{\theta^{g-d}}{(g-d)!}.$$

The formula $[W_d] = [W_{g-1}]^{g-d}/(g-d)!$ makes sense in the ring of cycles on Jac modulo numerical equivalence — we do not need to introduce the topological cohomology of Jac to state it — and indeed it was proven in this numerical form for curves and their Jacobians over arbitrary fields in Kleiman and Laksov [1974]. We do not know if there is an analogous formula in a finer cycle theory such as the group of cycles modulo rational or algebraic equivalence.

Proof of Proposition D.13: By Poincaré duality, it suffices to take the product of both sides of the formula with an arbitrary element $\alpha_I \cup \beta_J$ with $|I| = |J| = d$, evaluate on the fundamental class of Jac and show that they are the same. In view of (D.4) and (D.5) above, in the case $I \neq J$ we have $\alpha_I \cup \beta_J \cup \eta_K = 0$ for any η_K of complementary degree, while if $I = J$ we have $(\alpha_I \cup \beta_I \cup \eta_K)[\mathrm{Jac}] = (-1)^{\binom{g}{2}}$ if $K = I'$ and 0 otherwise. Equivalently, since the map $C_d \to W_d$ is generically one-to-one for $d \leq g$, Proposition D.13 will be proven if we show that

$$(u^*(\alpha_I \cup \beta_J))[C_d] = \begin{cases} (-1)^{\binom{d}{2}} & \text{if } I = J, \\ 0 & \text{otherwise.} \end{cases} \tag{D.6}$$

To evaluate the expression on the left, it is useful to pull back from the symmetric power C_d of the curve to C^d, the ordinary d-fold product. Let $\pi : C^d \to C_d$ be the quotient map, and let $\nu = u \circ \pi : C^d \to$ Jac be the composition. Since π is a $d!$-fold cover,

$$\pi_*[C^d] = d! \cdot [C_d],$$

so (D.6) is equivalent to

$$(\nu^*(\alpha_I \cup \beta_J))[C^d] = \begin{cases} (-1)^{\binom{d}{2}} d! & \text{if } I = J, \\ 0 & \text{otherwise;} \end{cases} \tag{D.7}$$

that is,

$$(\nu^*(\alpha_I \cup \beta_J))[C^d] = 0 \quad \text{if } I \neq J \tag{D.8}$$

and

$$(\nu^* \eta_I)[C^d] = d! \quad \text{for all } I. \tag{D.9}$$

Let $\rho_k : C^d \to C$ be projection on the k-th factor, and set $\alpha_i^k = \rho_k^* \alpha_i$ and $\beta_i^k = \rho_k^* \beta_i$. By the Künneth formula, $H^1(C^d) = \bigoplus_k \rho_k^*(H^1(C))$. Writing $\iota_k : C \to C^d$ for the inclusion sending C to

$$\{p_0\} \times \cdots \times \{p_0\} \times C \times \{p_0\} \times \cdots \times \{p_0\},$$

with C in the k-th position, we see that $\rho_k \iota_k : C \to C$ is the identity, while $\rho_j \iota_k$ is the constant map when $j \neq k$. It follows that if $\gamma \in H^1(C^d)$ then $\gamma = \sum_k \rho_k^*(\iota_k^*(\gamma))$. Applying this to $\nu^* \alpha_i$ and $\nu^* \beta_i$, we see that

$$\nu^* \alpha_i = \alpha_i^1 + \cdots + \alpha_i^d \quad \text{and} \quad \nu^* \beta_i = \beta_i^1 + \cdots + \beta_i^d. \tag{D.10}$$

By symmetry we may assume that $I = \{1, \ldots, d\}$. Applying Formula (D.10) to

$$\nu^* \eta_I = \nu^* \alpha_1 \cup \nu^* \beta_1 \cup \cdots \cup \nu^* \alpha_d \cup \nu^* \beta_d,$$

we get the sum of all products of the form

$$\alpha_1^{j_1} \cup \beta_1^{k_1} \cup \cdots \cup \alpha_d^{j_d} \cup \beta_d^{k_d}.$$

This product is zero unless each $j_i = k_i$ and the set $\{j_1 \ldots, j_d\}$ is equal to $\{1, \ldots, d\}$. For these $d!$ terms,

$$\alpha_1^{j_1} \cup \beta_1^{j_1} \cup \cdots \cup \alpha_d^{j_d} \cup \beta_d^{j_d}[C^d] = 1,$$

because $\alpha_i^{j_i} \cup \beta_i^{j_i}$ is the pullback of the class of a point under ρ_{j_i}. $\qquad\qquad\square$

D.5.2 Symmetric powers as projective bundles

We now return to the argument of Section D.4.2. Recall that we have chosen a base point p_0, and that $\mathcal{P}$ denotes the Poincaré bundle on $\mathrm{Pic}^n \times C$ for some $n > 2g - 2$, normalized so that $\mathcal{P}$ is trivial on $\mathrm{Pic}^n \times \{p_0\}$. To complete the argument of Theorem D.10 we need to compute the Chern classes of $\pi_*\mathcal{P}$, where $\pi : \mathrm{Pic}^n \times C \to \mathrm{Pic}^n$ is the projection. To do this we will identify $\pi_*\mathcal{P}$ in another way:

Theorem D.14. *With notation as above, the map* $u : C_n \to \mathrm{Pic}^n$ *is isomorphic to the projective bundle* $\mathbb{P}(\pi_*\mathcal{P}) \to \mathrm{Pic}^n$, *via an isomorphism* $\mathbb{P}(\pi_*\mathcal{P}) \to C_n$ *sending* $\mathcal{O}_{\mathbb{P}(\pi_*\mathcal{P})}(1)$ *to* $\mathcal{O}_{C_n}(X_{p_0})$.

Note that there is a natural identification of the fibers of the Abel–Jacobi map $u : C_n \to \mathrm{Pic}^n$ with the fibers of the projective bundle: By Theorem D.6, the scheme-theoretic fiber of u over a point $\mathcal{L} \in \mathrm{Pic}^n$ is the projective space $|\mathcal{L}|$. On the other hand, the restriction of $\mathcal{P}$ to the fiber $C \cong C \times \{\mathcal{L}\}$ is $\mathcal{L}$. Since the degree n of $\mathcal{L}$ is large, its higher cohomology vanishes, and the theorem on cohomology and base change (Theorem B.9) shows that the fiber of $\pi_*\mathcal{P}$ at a point $\mathcal{L} \in \mathrm{Pic}^n$ is the vector space $H^0(\mathcal{L})$ of sections of the line bundle $\mathcal{L}$. The projectivization of this space is again the projective space $|\mathcal{L}| = u^{-1}\{\mathcal{L}\}$. This fiber-by-fiber argument does not constitute a proof, but it suggests how we might go about giving one.

The divisor $X_{p_0} \subset C_n$, consisting of divisors containing p_0, cuts out the hyperplane section in each fiber of u, so, by Proposition 9.4, $C_n \to \mathrm{Pic}^n$ is a projective bundle in the Zariski topology. That is, u is the projection $\mathbb{P}\mathcal{G} \to \mathrm{Pic}^n$, where $\mathcal{G} := (u_*\mathcal{O}_{C_n}(X_{p_0}))^*$.

To prove Theorem D.14, accordingly, it suffices to show that the direct image $\pi_*\mathcal{P}$ is isomorphic to the dual of $u_*\mathcal{O}_{C_n}(X_{p_0})$. The key is to consider the direct image $\pi_*\mathcal{P}$ as the direct image of the line bundle $\mathcal{M} = \mathcal{O}_{C_n \times C}(\mathcal{D} - \mathcal{D}_0)$ in two ways: by definition $\pi_*\mathcal{P} = \pi_*(\eta_*\mathcal{M})$, and since $\pi \circ \eta = u \circ \pi_1$ we can also write it as $u_*(\pi_{1*}\mathcal{M})$. It may be helpful to have a diagram of the relevant objects:

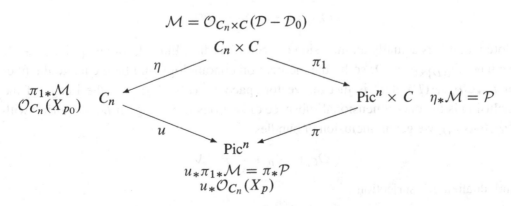

It will also be helpful to have the following lemma:

Lemma D.15. *With notation as above,*

$$\pi_{1*}\mathcal{M} \cong u^*(\pi_*\mathcal{P}).$$

Proof: There is a natural evaluation map

$$u^*u_*(\pi_{1*}\mathcal{M}) \to \pi_{1*}\mathcal{M};$$

since $u_*\pi_{1*}\mathcal{M} = \pi_*\mathcal{P}$, this gives a map

$$u^*(\pi_*\mathcal{P}) \to \pi_{1*}\mathcal{M}.$$

We claim this map is an isomorphism. This follows by looking at the map on fibers at a point $D \in C_n$. Since the sheaves involved have no higher cohomology of the fibers of the morphisms, the theorem on cohomology and base change allows us to identify the fibers of $u^*(\pi_*\mathcal{P})$ and $\pi_{1*}\mathcal{M}$ at D with the spaces

$$u^*(\pi_*\mathcal{P})_D = H^0(\mathcal{M}|_{|D|\times C}) \quad \text{and} \quad (\pi_{1*}\mathcal{M})_D = H^0(\mathcal{M}|_{\{D\}\times C}),$$

and in these terms the induced map $u^*(\pi_*\mathcal{P})_D \to (\pi_{1*}\mathcal{M})_D$ is the restriction map $H^0(\mathcal{M}|_{|D|\times C}) \to H^0(\mathcal{M}|_{\{D\}\times C})$. By equation (D.3), the bundle $\mathcal{M}|_{|D|\times C}$ on $|D|\times C$ is the pullback of a bundle on C, and so the restriction map is an isomorphism on global sections. $\qquad\square$

We now proceed with the proof of Theorem D.14:

Proof of Theorem D.14: Since $\mathcal{D}_0 = \pi_1^*(X_{p_0})$, we can write

$$\mathcal{M} = \mathcal{O}_{C_n\times C}(\mathcal{D} - \mathcal{D}_0) = \mathcal{O}_{C_n\times C}(\mathcal{D}) \otimes \pi_1^*\mathcal{O}_{C_n}(-X_{p_0}),$$

and so

$$\pi_{1*}\mathcal{M} = \pi_{1*}(\mathcal{O}_{C_n\times C}(\mathcal{D})) \otimes \mathcal{O}_{C_n}(-X_{p_0}).$$

We have an inclusion of sheaves $\mathcal{O}_{C_n\times C} \hookrightarrow \mathcal{O}_{C_n\times C}(\mathcal{D})$ coming from the effective divisor $\mathcal{D}$; taking the direct image under π_1 gives an inclusion

$$\mathcal{O}_{C_n} \hookrightarrow \pi_{1*}\mathcal{O}_{C_n\times C}(\mathcal{D}).$$

Note that this is actually an inclusion of vector bundles. Indeed, for any point $D \in C_n$ we have $\mathcal{D}|_{\{D\}\times C} = D$, so by the theorem on cohomology and base change the fiber of $\pi_{1*}\mathcal{O}_{C_n\times C}(\mathcal{D})$ at D is just the vector space $H^0(\mathcal{O}_C(D))$, and the image of the inclusion is the one-dimensional subspace of sections vanishing on D. Tensoring with $\mathcal{O}_{C_n}(-X_{p_0})$, we get an inclusion of bundles

$$\mathcal{O}_{C_n}(-X_{p_0}) \hookrightarrow \pi_{1*}\mathcal{M}$$

and, dualizing, a surjection

$$\rho : (\pi_{1*}\mathcal{M})^* \to \mathcal{O}_{C_n}(X_{p_0}).$$

To finish, we claim that the pushforward of ρ gives an isomorphism

$$u_*\rho : u_*((\pi_{1*}\mathcal{M})^*) \longrightarrow u_*\mathcal{O}_{C_n}(X_{p_0});$$

given the identification of Lemma D.15, this will complete the proof of Theorem D.14. Once more invoking the theorem on cohomology and base change, it is enough to prove that ρ induces an isomorphism on global sections on each fiber $|\mathcal{L}| = u^{-1}(\mathcal{L}) \subset C_n$ of u; so, let us consider the restriction of the sheaves in question to $|\mathcal{L}| = \mathbb{P}H^0(\mathcal{L})$. To begin with, by Lemma D.15 the restriction of the bundle $\pi_{1*}\mathcal{M} = u^*(\pi_*\mathcal{P})$ to $|\mathcal{L}|$ is the trivial vector bundle with fiber $V = H^0(\mathcal{L})$, and so

$$(\pi_{1*}\mathcal{M})^*|_{|\mathcal{L}|} = V^* \otimes \mathcal{O}_{\mathbb{P}V}.$$

Next, the restriction of $\mathcal{O}_{C_n}(X_{p_0})$ to $|\mathcal{L}|$ is given by

$$\mathcal{O}_{C_n}(X_{p_0})|_{|\mathcal{L}|} \cong \mathcal{O}_{\mathbb{P}V}(1),$$

and in these terms the map $\rho|_{|\mathcal{L}|}$ is just the quotient map

$$V^* \otimes \mathcal{O}_{\mathbb{P}V} \to \mathcal{O}_{\mathbb{P}V}(1)$$

on the projective space $\mathbb{P}V$. This induces an isomorphism on global sections, and we are done. $\square$

D.5.3 Chern classes from the symmetric power

We can now calculate the Chern class of $\mathcal{E} := \pi_*\mathcal{P}$:

Theorem D.16. *For $n \geq 2g - 1$, the pushforward $\mathcal{E}$ of the Poincaré bundle $\mathcal{P}$ from $\mathrm{Pic}^n \times C$ to Pic^n has Chern class $c(\mathcal{E}) = e^{-\theta}$; that is, $c_i(\mathcal{E}) = (-1)^i \theta^i / i!$ for each i.*

Proof: Computing the Chern class is equivalent to computing the Segre class $s(\mathcal{E}) = \sum s_i(\mathcal{E})$, since $c(\mathcal{E}) = 1/s(\mathcal{E})$ by Proposition 10.3. Recall that

$$s_k(\mathcal{E}) = u_*(\zeta^{k+n-g}),$$

where $\zeta = [X_p] \in H^2(C_n)$.

Since we are working in $H^*(C_n)$ rather than $A(C_n)$, the class ζ is also the class of the divisor $X_q = C_{n-1} + q \subset C_d$ for any point $q \in C$. To represent the class ζ^{k+n-g} we can just choose distinct points $p_1, \dots, p_{k+n-g} \in C$ and consider the intersection

$$\bigcap X_{p_i} = \{D \in C_n \mid D - p_i \geq 0 \text{ for all } i\} = C_{g-k} + E \subset C_d,$$

where $E = p_1 + \dots + p_{k+n-g}$. This intersection is generically transverse — it is visibly transverse at a point $E + D'$, where D' consists of $g - k$ distinct points distinct from $p_1, \dots, p_{k+n-g}$, and no component of the intersection is contained in the complement of this locus — and so we have

$$\zeta^{k+n-g} = [C_{g-k} + E] \in H^{2k+2n-2g}(C_d).$$

We have

$$s_k(\mathcal{E}) = u_*[C_{g-k} + E] = [W_{g-k}] \in H^{2k}(\text{Jac}).$$

Applying Poincaré's formula, this yields

$$s_k(\mathcal{E}) = \frac{\theta^k}{k!}.$$

We can express this compactly as

$$s(\mathcal{E}) = e^\theta,$$

from which the theorem follows. □

Theorem D.16 allows us to give a description of the cohomology ring of C_d; though we will not use it in what follows, we state it here. By the analog of Theorem 9.6 for topological cohomology, we have for $n \geq 2g - 1$ that

$$H^*(C_n) = H^*(\text{Jac})[\zeta] \left/ \left(\zeta^{n-g+1} - \theta\zeta^{n-g} + \frac{\theta^2}{2}\zeta^{n-g-1} - \cdots \right) \right.$$

Finally, we remark that there is an alternative way to derive the Chern classes of $\mathcal{E}$: Given that $\mathcal{E} = \eta_*\mathcal{M}$, we can apply Grothendieck–Riemann–Roch to the morphism η to arrive at Theorem D.16. This approach involves a larger initial investment — we have to have more knowledge of products in $H^*(C_d)$ than we currently do — but is also much more broadly applicable. This approach is carried out in Arbarello et al. [1985, Chapter 8], where many other applications are given.

D.5.4 The class of W_d^r

Here is our third and final version of the Brill–Noether theorem. It sharpens Theorem D.9 by giving the class of $W_d^r(C)$ in case this locus has the expected dimension. This enumerative statement is the key to the proof of the qualitative existence statement, which, surprisingly, does not depend on an a priori knowledge of the dimension.

Theorem D.17 (Enumerative Brill–Noether). *(a) For every curve C of genus g, the locus $W_d^r(C)$ is nonempty of dimension*

$$\dim W_d^r(C) \geq \rho(g, r, d).$$

(b) If C is a curve of genus g such that $\dim W_d^r(C) = \rho(g, r, d)$, then the class of $W_d^r(C)$ in the cohomology ring of the Jacobian $\text{Jac}(C)$ is

$$[W_d^r] = \prod_{i=1}^{r} \frac{i!}{(g - d + r + i)!} \theta^{(r+1)(g-d+r)}.$$

If $\rho \geq 0$ then this class is nonzero.

(c) If C is a general curve of genus g then $\dim W_d^r(C) = \rho(g, r, d)$.

Proof of parts (a) and (b): Part (a) follows from part (b) as in the discussion of the content of enumerative formulas generally (Section 3.1) because, first, a determinantal locus is either empty or of dimension at least the "expected dimension" (see Lemma 5.2), and second, if it were empty, then we could consider it to have been of the correct dimension, and thus it would have to have the nonzero homology class of part (b), a contradiction.

We will use Porteous' formula to calculate the class of W_d^r as a degeneracy locus of the map $\mathcal{E} \to \mathcal{F}$ of vector bundles on Pic^n obtained by pushing forward the evaluation map $\mathcal{P} \to \bigoplus \mathcal{L}_i$.

The necessary ingredients are the Chern class of $\mathcal{E}$, computed in the previous section, and the Chern class of $\mathcal{F}$. The line bundles $\mathcal{L}_i$ can all be continuously deformed to the bundle $\mathcal{P}_{p_0} = \mathcal{P}|_{\mathrm{Pic}^n \times \{p_0\}}$, which is trivial by our normalization of $\mathcal{P}$. The Chern classes $c_1(\mathcal{L}_i) \in H^2(\mathrm{Pic}^n)$ are thus all 0, so that

$$c(\mathcal{F}) = 1 \in H^*(\mathrm{Pic}^n).$$

We have

$$\frac{c(\mathcal{F})}{c(\mathcal{E})} = \frac{1}{e^{-\theta}} = e^\theta,$$

and so Porteous' formula tells us that if W_d^r has pure dimension $\rho = g-(r+1)(g-d+r)$ then its class is the determinant

$$[W_d^r] = \begin{vmatrix} \dfrac{\theta^{g-d+r}}{(g-d+r)!} & \dfrac{\theta^{g-d+r+1}}{(g-d+r+1)!} & \cdots & \dfrac{\theta^{g-d+2r}}{(g-d+2r)!} \\[2mm] \dfrac{\theta^{g-d+r-1}}{(g-d+r-1)!} & \dfrac{\theta^{g-d+r}}{(g-d+r)!} & \cdots & \dfrac{\theta^{g-d+2r-1}}{(g-d+2r-1)!} \\[2mm] \vdots & \vdots & \ddots & \vdots \\[2mm] \dfrac{\theta^{g-d}}{(g-d)!} & \dfrac{\theta^{g-d+1}}{(g-d+1)!} & \cdots & \dfrac{\theta^{g-d+r}}{(g-d+r)!} \end{vmatrix}.$$

In other words,

$$[W_d^r] = D_{a,r} \cdot \theta^{(r+1)(g-d+r)},$$

where $D_{a,r}$ is the $(r+1) \times (r+1)$ determinant

$$D_{a,r} = \begin{vmatrix} \dfrac{1}{a!} & \dfrac{1}{(a+1)!} & \cdots & \dfrac{1}{(a+r)!} \\[2mm] \dfrac{1}{(a-1)!} & \dfrac{1}{a!} & \cdots & \dfrac{1}{(a+r-1)!} \\[2mm] \vdots & \vdots & \ddots & \vdots \\[2mm] \dfrac{1}{(a-r)!} & \dfrac{1}{(a-r+1)!} & \cdots & \dfrac{1}{a!} \end{vmatrix}.$$

It remains to evaluate $D_{a,r}$. To do this, we clear denominators by multiplying the first column by $a!$, the second column by $(a+1)!$, and so on; we arrive at the expression

$$D_{a,r} = \prod_{i=0}^{r} \frac{1}{(a+i)!} \cdot M,$$

where M is the determinant

$$\begin{vmatrix} 1 & 1 & \cdots & 1 \\ a & a+1 & \cdots & a+r \\ a(a-1) & (a+1)a & \cdots & (a+r)(a+r-1) \\ \vdots & \vdots & \ddots & \vdots \\ a\cdots(a-r+1) & (a+1)\cdots(a-r+2) & \cdots & (a+r)\cdots(a+1) \end{vmatrix}.$$

Since the columns of M all consist of the same sequence of monic polynomials, applied to the arguments $a, \ldots, a+r$, the determinant is equivalent to the Vandermonde determinant, and thus has value

$$\prod_{0 \leq i < j \leq r} (j - i) = \prod_{i=0}^{r} i!.$$

Thus

$$D_{a,r} = \prod_{i=1}^{r} \frac{i!}{(a+i)!}.$$

It follows that if the dimension of $W_d^r(C)$ is ρ then it has the class given in the theorem. In particular, if it were empty then it would have this class, which is nonzero, a contradiction. Thus it must be nonempty. Since it is defined as a degeneracy locus, it must have dimension at least the "expected dimension" locally at each of its points. This completes the proof of both Theorems D.17 and D.10. $\qquad\square$

D.6 Exercises

Exercise D.18. Use the statements of Section D.2.2 to prove the original form of Jacobi inversion: Given two g-tuples of points $p_1, \ldots, p_g$ and $q_1, \ldots, q_g \in C$ on a smooth curve C of genus g, there exists a g-tuple of points $r_1, \ldots, r_g \in C$, whose coordinates are rational functions of the coordinates of the p_i and q_i, such that

$$\sum \int_{p_0}^{p_i} + \sum \int_{p_0}^{q_i} = \sum \int_{p_0}^{r_i}.$$

Exercise D.19. Let $C_0 \subset \mathbb{P}^2$ be a plane sextic with four nodes as singularities, whose normalization is a general curve of genus 6. Show that no three of the nodes are collinear. *Hint:* Use a dimension count.

Exercise D.20. Let $C_0 \subset \mathbb{P}^2$ be a plane sextic with four nodes as singularities, whose normalization is a general curve of genus 6. We have seen that there are five g_4^1's on C: the pencils cut out on C by the pencil of lines through each node and the pencil cut by conics passing through all four. Show that there are no others.

Exercise D.21. Let C be a curve of genus 8, embedded in $\mathbb{P}^3$ by one of the g_8^3's on C. Show that if C is general then the image curve $C_0 \subset \mathbb{P}^3$ does not lie on a cubic surface. In case it does, can you locate the 14 g_5^1's on C?

Exercise D.22. Let C be a general curve of genus 9. How many plane octic curves $C_0 \subset \mathbb{P}^2$ are birational to C?

Exercise D.23. Show that $W_d^r(C) \setminus W_d^{r+1}(C)$ is dense in $W_d^r(C)$.
Hint: For any point $\mathcal{L} \in W_d^{r+1}(C)$, consider the line bundle $\mathcal{L}(p-q)$ for general points $p, q \in C$.

References

Abhyankar [1984] S. Abhyankar, "Combinatoire des tableaux de Young, variétés déterminantielles et calcul de fonctions de Hilbert", *Rend. Sem. Mat. Univ. Politec. Torino* **42**:3 (1984), 65–88.

Alexander and Hirschowitz [1995] J. Alexander and A. Hirschowitz, "Polynomial interpolation in several variables", *J. Algebraic Geom.* **4**:2 (1995), 201–222.

Altman and Kleiman [1970] A. Altman and S. Kleiman, *Introduction to Grothendieck duality theory*, Lecture Notes in Math. **146**, Springer, Berlin-New York, 1970.

Aluffi [1990] P. Aluffi, "The enumerative geometry of plane cubics, I: Smooth cubics", *Trans. Amer. Math. Soc.* **317**:2 (1990), 501–539.

Aluffi [1991] P. Aluffi, "The enumerative geometry of plane cubics, II: Nodal and cuspidal cubics", *Math. Ann.* **289**:4 (1991), 543–572.

Andreotti and Frankel [1959] A. Andreotti and T. Frankel, "The Lefschetz theorem on hyperplane sections", *Ann. of Math.* (2) **69** (1959), 713–717.

Arbarello et al. [1985] E. Arbarello, M. Cornalba, P. A. Griffiths, and J. Harris, *Geometry of algebraic curves, I*, Grundlehren der Mathematischen Wissenschaften **267**, Springer, New York, 1985.

Artin [1982] M. Artin, "Brauer–Severi varieties", pp. 194–210 in *Brauer groups in ring theory and algebraic geometry* (Wilrijk, 1981), edited by A. Verschoren, Lecture Notes in Math. **917**, Springer, Berlin-New York, 1982.

Atiyah [1957] M. F. Atiyah, "Complex analytic connections in fibre bundles", *Trans. Amer. Math. Soc.* **85** (1957), 181–207.

Atiyah and Hirzebruch [1961] M. F. Atiyah and F. Hirzebruch, "Vector bundles and homogeneous spaces", pp. 7–38 in *Differential Geometry*, edited by C. B. Allendoerfer, Proc. Sympos. Pure Math. **3**, Amer. Math. Soc., Providence, RI, 1961.

Bădescu [2001] L. Bădescu, *Algebraic surfaces*, Springer, New York, 2001.

Barth [1975] W. Barth, "Larsen's theorem on the homotopy groups of projective manifolds of small embedding codimension", pp. 307–313 in *Algebraic geometry* (Arcata, CA, 1974), edited by R. Hartshorne, Proc. Sympos. Pure Math. **29**, Amer. Math. Soc., Providence, RI, 1975.

Barth [1977] W. Barth, "Moduli of vector bundles on the projective plane", *Invent. Math.* **42** (1977), 63–91.

Barth et al. [2004] W. Barth, K. Hulek, C. A. M. Peters, and A. Van de Ven, *Compact complex surfaces*, 2nd ed., Ergebnisse der Mathematik und ihrer Grenzgebiete (3) **4**, Springer, Berlin, 2004.

Bayer [1982] D. Bayer, *The division algorithm and the Hilbert scheme*, Ph.D. thesis, Harvard University, Ann Arbor, MI, 1982. Available at http://search.proquest.com/docview/303209159.

Bayer and Eisenbud [1995] D. Bayer and D. Eisenbud, "Ribbons and their canonical embeddings", *Trans. Amer. Math. Soc.* **347**:3 (1995), 719–756.

Beauville [1996] A. Beauville, *Complex algebraic surfaces*, 2nd ed., London Mathematical Society Student Texts **34**, Cambridge University Press, 1996.

Beheshti [2006] R. Beheshti, "Lines on projective hypersurfaces", *J. Reine Angew. Math.* **592** (2006), 1–21.

Beheshti and Mohan Kumar [2013] R. Beheshti and N. Mohan Kumar, "Spaces of rational curves on hypersurfaces", *J. Ramanujan Math. Soc.* **28A** (2013), 1–19.

Bifet et al. [1990] E. Bifet, C. De Concini, and C. Procesi, "Cohomology of regular embeddings", *Adv. Math.* **82**:1 (1990), 1–34.

Boissière and Sarti [2007] S. Boissière and A. Sarti, "Counting lines on surfaces", *Ann. Sc. Norm. Super. Pisa Cl. Sci.* (5) **6**:1 (2007), 39–52.

Borel [1991] A. Borel, *Linear algebraic groups*, 2nd ed., Graduate Texts in Mathematics **126**, Springer, New York, 1991.

Borel and Serre [1958] A. Borel and J.-P. Serre, "Le théorème de Riemann–Roch", *Bull. Soc. Math. France* **86** (1958), 97–136.

Bott and Tu [1982] R. Bott and L. W. Tu, *Differential forms in algebraic topology*, Graduate Texts in Mathematics **82**, Springer, New York-Berlin, 1982.

Brambilla and Ottaviani [2008] M. C. Brambilla and G. Ottaviani, "On the Alexander–Hirschowitz theorem", *J. Pure Appl. Algebra* **212**:5 (2008), 1229–1251.

Brieskorn and Knörrer [1986] E. Brieskorn and H. Knörrer, *Plane algebraic curves*, Birkhäuser, Basel, 1986.

Brill and Noether [1874] A. Brill and M. Noether, "Über die algebraischen Functionen und ihre Anwendung in der Geometrie", *Math. Ann.* **7** (1874), 269–310.

Buchsbaum and Eisenbud [1977] D. A. Buchsbaum and D. Eisenbud, "What annihilates a module?", *J. Algebra* **47**:2 (1977), 231–243.

Call and Lyubeznik [1994] F. Call and G. Lyubeznik, "A simple proof of Grothendieck's theorem on the parafactoriality of local rings", pp. 15–18 in *Commutative algebra: syzygies, multiplicities, and birational algebra* (South Hadley, MA, 1992), edited by W. J. Heinzer et al., Contemp. Math. **159**, Amer. Math. Soc., Providence, RI, 1994.

Caporaso and Harris [1998] L. Caporaso and J. Harris, "Counting plane curves of any genus", *Invent. Math.* **131**:2 (1998), 345–392.

Cavazzani [2016] F. Cavazzani, *A geometric invariant theory compactification of the space of twisted cubics*, Ph.D. thesis, Harvard University, 2016.

Ceresa and Collino [1983] G. Ceresa and A. Collino, "Some remarks on algebraic equivalence of cycles", *Pacific J. Math.* **105**:2 (1983), 285–290.

Chasles [1864] M. Chasles, "Construction des coniques qui satisfont à cinque conditions", *C. R. Acad. Sci. Paris* **58** (1864), 297–308.

Chern [1946] S.-S. Chern, "Characteristic classes of Hermitian manifolds", *Ann. of Math.* (2) **47** (1946), 85–121.

Chow [1956] W.-L. Chow, "On equivalence classes of cycles in an algebraic variety", *Ann. of Math.* (2) **64** (1956), 450–479.

Ciliberto [1987] C. Ciliberto, "Hilbert functions of finite sets of points and the genus of a curve in a projective space", pp. 24–73 in *Space curves* (Rocca di Papa, 1985), edited by F. Ghione et al., Lecture Notes in Math. **1266**, Springer, Berlin, 1987.

Clebsch [1861] A. Clebsch, "Zur Theorie der algebraischen Flächen", *J. Reine Angew. Math.* **58** (1861), 93–108.

Coşkun [2009] I. Coşkun, "A Littlewood–Richardson rule for two-step flag varieties", *Invent. Math.* **176**:2 (2009), 325–395.

Collino [1975] A. Collino, "The rational equivalence ring of symmetric products of curves", *Illinois J. Math.* **19**:4 (1975), 567–583.

Cools et al. [2012] F. Cools, J. Draisma, S. Payne, and E. Robeva, "A tropical proof of the Brill–Noether theorem", *Adv. Math.* **230**:2 (2012), 759–776.

Cox and Katz [1999] D. A. Cox and S. Katz, *Mirror symmetry and algebraic geometry*, Mathematical Surveys and Monographs **68**, Amer. Math. Soc., Providence, RI, 1999.

Dale [1985] M. Dale, "Severi's theorem on the Veronese-surface", *J. London Math. Soc.* (2) **32**:3 (1985), 419–425.

De Concini and Procesi [1983] C. De Concini and C. Procesi, "Complete symmetric varieties", pp. 1–44 in *Invariant theory* (Montecatini, 1982), edited by F. Gherardelli, Lecture Notes in Math. **996**, Springer, Berlin, 1983.

De Concini and Procesi [1985] C. De Concini and C. Procesi, "Complete symmetric varieties, II: Intersection theory", pp. 481–513 in *Algebraic groups and related topics* (Kyoto/Nagoya, 1983), edited by R. Hotta, Adv. Stud. Pure Math. **6**, North-Holland, Amsterdam, 1985.

De Concini et al. [1980] C. De Concini, D. Eisenbud, and C. Procesi, "Young diagrams and determinantal varieties", *Invent. Math.* **56**:2 (1980), 129–165.

De Concini et al. [1982] C. De Concini, D. Eisenbud, and C. Procesi, *Hodge algebras*, Astérisque **91**, Société Mathématique de France, Paris, 1982.

De Concini et al. [1988] C. De Concini, M. Goresky, R. MacPherson, and C. Procesi, "On the geometry of quadrics and their degenerations", *Comment. Math. Helv.* **63**:3 (1988), 337–413.

Decker et al. [2015] W. Decker, G.-M. Greuel, G. Pfister, and H. Schönemann, "SINGULAR 4-0-2 — A computer algebra system for polynomial computations", 2015. Available at http://singular. uni-kl.de.

Dieudonné [1969] J. Dieudonné, "Algebraic geometry", *Advances in Math.* **3** (1969), 233–321.

Donagi and Smith [1980] R. Donagi and R. Smith, "The degree of the Prym map onto the moduli space of five-dimensional abelian varieties", pp. 143–155 in *Journées de Géometrie Algébrique d'Angers* (Angers, 1979), edited by A. Beauville, Sijthoff & Noordhoff, Alphen aan den Rijn, 1980.

Doubilet et al. [1974] P. Doubilet, G.-C. Rota, and J. Stein, "On the foundations of combinatorial theory, IX: Combinatorial methods in invariant theory", *Studies in Appl. Math.* **53** (1974), 185–216.

Ein [1986] L. Ein, "Varieties with small dual varieties, I", *Invent. Math.* **86**:1 (1986), 63–74.

Eisenbud [1995] D. Eisenbud, *Commutative algebra with a view toward algebraic geometry*, Graduate Texts in Mathematics **150**, Springer, New York, 1995.

Eisenbud [2005] D. Eisenbud, *The geometry of syzygies*, Graduate Texts in Mathematics **229**, Springer, New York, 2005.

Eisenbud and Harris [1983a] D. Eisenbud and J. Harris, "Divisors on general curves and cuspidal rational curves", *Invent. Math.* **74**:3 (1983), 371–418.

Eisenbud and Harris [1983b] D. Eisenbud and J. Harris, "A simpler proof of the Gieseker–Petri theorem on special divisors", *Invent. Math.* **74**:2 (1983), 269–280.

Eisenbud and Harris [1987] D. Eisenbud and J. Harris, "On varieties of minimal degree (a centennial account)", pp. 3–13 in *Algebraic geometry* (Bowdoin, ME, 1985), edited by S. J. Bloch, Proc. Sympos. Pure Math. **46**, Amer. Math. Soc., Providence, RI, 1987.

Eisenbud and Harris [1992] D. Eisenbud and J. Harris, "Finite projective schemes in linearly general position", *J. Algebraic Geom.* **1**:1 (1992), 15–30.

Eisenbud and Harris [2000] D. Eisenbud and J. Harris, *The geometry of schemes*, Graduate Texts in Mathematics **197**, Springer, New York, 2000.

Eisenbud and Schreyer [2008] D. Eisenbud and F.-O. Schreyer, "Relative Beilinson monad and direct image for families of coherent sheaves", *Trans. Amer. Math. Soc.* **360**:10 (2008), 5367–5396.

Eisenbud et al. [1996] D. Eisenbud, M. Green, and J. Harris, "Cayley–Bacharach theorems and conjectures", *Bull. Amer. Math. Soc.* (*N.S.*) **33**:3 (1996), 295–324.

Eisenbud et al. [2003] D. Eisenbud, F.-O. Schreyer, and J. Weyman, "Resultants and Chow forms via exterior syzygies", *J. Amer. Math. Soc.* **16**:3 (2003), 537–579.

Fröberg [1985] R. Fröberg, "An inequality for Hilbert series of graded algebras", *Math. Scand.* **56**:2 (1985), 117–144.

Fulton [1984] W. Fulton, *Intersection theory*, Ergebnisse der Mathematik und ihrer Grenzgebiete (3) **2**, Springer, Berlin, 1984.

Fulton [1993] W. Fulton, *Introduction to toric varieties*, Annals of Mathematics Studies **131**, Princeton University Press, 1993.

Fulton [1997] W. Fulton, *Young tableaux*, London Mathematical Society Student Texts **35**, Cambridge University Press, 1997.

Fulton and Harris [1991] W. Fulton and J. Harris, *Representation theory*, Graduate Texts in Mathematics **129**, Springer, New York, 1991.

Fulton and Lazarsfeld [1981] W. Fulton and R. Lazarsfeld, "On the connectedness of degeneracy loci and special divisors", *Acta Math.* **146**:3-4 (1981), 271–283.

Fulton and MacPherson [1978] W. Fulton and R. MacPherson, "Defining algebraic intersections", pp. 1–30 in *Algebraic geometry* (Univ. Tromsø, 1977), edited by L. D. Olson, Lecture Notes in Math. **687**, Springer, Berlin, 1978.

Fulton and MacPherson [1981] W. Fulton and R. MacPherson, *Categorical framework for the study of singular spaces*, Mem. Amer. Math. Soc. **243**, Amer. Math. Soc., Providence, RI, 1981.

Fulton and Pandharipande [1997] W. Fulton and R. Pandharipande, "Notes on stable maps and quantum cohomology", pp. 45–96 in *Algebraic geometry* (Santa Cruz, 1995), edited by J. Kollár et al., Proc. Sympos. Pure Math. **62**, Amer. Math. Soc., Providence, RI, 1997.

Fulton et al. [1983] W. Fulton, S. Kleiman, and R. MacPherson, "About the enumeration of contacts", pp. 156–196 in *Algebraic geometry — open problems* (Ravello, 1982), edited by C. Ciliberto et al., Lecture Notes in Math. **997**, Springer, Berlin, 1983.

van Gastel [1990] L. van Gastel, "Excess intersections in projective space", pp. 109–124 in *Topics in algebra, II* (Warsaw, 1988), edited by S. Balcerzyk et al., Banach Center Publ. **26**, PWN, Warsaw, 1990.

Gelfand et al. [2008] I. M. Gelfand, M. M. Kapranov, and A. V. Zelevinsky, *Discriminants, resultants and multidimensional determinants*, Birkhäuser, Boston, 2008. Reprint of the 1994 edition.

Golubitsky and Guillemin [1973] M. Golubitsky and V. Guillemin, *Stable mappings and their singularities*, Graduate Texts in Mathematics **14**, Springer, New York-Heidelberg, 1973.

Grayson and Stillman [2015] D. Grayson and M. Stillman, "Macaulay2: a software system for research in algebraic geometry", 2015. Available at http://math.uiuc.edu/Macaulay2.

Grayson et al. [2012] D. Grayson, A. Seceleanu, and M. Stillman, "Computations in intersection rings of flag bundles", preprint, 2012. Available at http://arxiv.org/abs/1205.4190.

Green [1989] M. Green, "Restrictions of linear series to hyperplanes, and some results of Macaulay and Gotzmann", pp. 76–86 in *Algebraic curves and projective geometry* (Trento, 1988), edited by E. Ballico and C. Ciliberto, Lecture Notes in Math. **1389**, Springer, Berlin, 1989.

Greuel et al. [2007] G.-M. Greuel, C. Lossen, and E. Shustin, *Introduction to singularities and deformations*, Springer, Berlin, 2007.

Griffiths and Adams [1974] P. Griffiths and J. Adams, *Topics in algebraic and analytic geometry*, Mathematical Notes **13**, Princeton University Press and University of Tokyo Press, 1974.

Griffiths and Harris [1979] P. Griffiths and J. Harris, "Algebraic geometry and local differential geometry", *Ann. Sci. École Norm. Sup.* (4) **12**:3 (1979), 355–452.

Griffiths and Harris [1980] P. Griffiths and J. Harris, "On the variety of special linear systems on a general algebraic curve", *Duke Math. J.* **47**:1 (1980), 233–272.

Griffiths and Harris [1985] P. Griffiths and J. Harris, "On the Noether–Lefschetz theorem and some remarks on codimension-two cycles", *Math. Ann.* **271**:1 (1985), 31–51.

Griffiths and Harris [1994] P. Griffiths and J. Harris, *Principles of algebraic geometry*, Wiley, New York, 1994. Reprint of the 1978 original.

Grothendieck [1958] A. Grothendieck, "La théorie des classes de Chern", *Bull. Soc. Math. France* **86** (1958), 137–154.

Grothendieck [1963] A. Grothendieck, "Eléments de géométrie algébrique, III: Étude cohomologique des faisceaux cohérents, II", *Inst. Hautes Études Sci. Publ. Math.* **17** (1963), 5–91.

Grothendieck [1966a] A. Grothendieck, "Le groupe de Brauer, I: Algèbres d'Azumaya et interprétations diverses", in *Séminaire Bourbaki* 1964/1965 (Exposé 290), W. A. Benjamin, Amsterdam, 1966. Reprinted as pp. 199–219 in *Séminaire Bourbaki* **9**, Soc. Math. France, Paris, 1995.

Grothendieck [1966b] A. Grothendieck, "Techniques de construction et théorèmes d'existence en géométrie algébrique, IV: Les schémas de Hilbert", in *Séminaire Bourbaki* 1960/1961 (Exposé 221), W. A. Benjamin, Amsterdam, 1966. Reprinted as pp. 249–276 in *Séminaire Bourbaki* **6**, Soc. Math. France, Paris, 1995.

Gruson and Peskine [1982] L. Gruson and C. Peskine, "Genre des courbes de l'espace projectif, II", *Ann. Sci. École Norm. Sup.* (4) **15**:3 (1982), 401–418.

Hamm [1995] H. A. Hamm, "Affine varieties and Lefschetz theorems", pp. 248–262 in *Singularity theory* (Trieste, 1991), edited by D. T. Lê et al., World Sci. Publ., 1995.

Harris [1979] J. Harris, "Galois groups of enumerative problems", *Duke Math. J.* **46**:4 (1979), 685–724.

Harris [1982] J. Harris, "Theta-characteristics on algebraic curves", *Trans. Amer. Math. Soc.* **271**:2 (1982), 611–638.

Harris [1995] J. Harris, *Algebraic geometry*, Graduate Texts in Mathematics **133**, Springer, New York, 1995. Corrected reprint of the 1992 original.

Harris and Eisenbud [1982] J. Harris and D. Eisenbud, *Curves in projective space*, Séminaire de Mathématiques Supérieures **85**, Presses de l'Université de Montréal, Quebec, Canada, 1982.

Harris and Morrison [1998] J. Harris and I. Morrison, *Moduli of curves*, Graduate Texts in Mathematics **187**, Springer, New York, 1998.

Harris et al. [1998] J. Harris, B. Mazur, and R. Pandharipande, "Hypersurfaces of low degree", *Duke Math. J.* **95**:1 (1998), 125–160.

Hartshorne [1966] R. Hartshorne, "Connectedness of the Hilbert scheme", *Inst. Hautes Études Sci. Publ. Math.* **29** (1966), 5–48.

Hartshorne [1974] R. Hartshorne, "Varieties of small codimension in projective space", *Bull. Amer. Math. Soc.* **80** (1974), 1017–1032.

Hartshorne [1977] R. Hartshorne, *Algebraic geometry*, Graduate Texts in Mathematics **52**, Springer, New York-Heidelberg, 1977.

Hassett [2007] B. Hassett, *Introduction to algebraic geometry*, Cambridge University Press, 2007.

Herzog and Trung [1992] J. Herzog and N. V. Trung, "Gröbner bases and multiplicity of determinantal and Pfaffian ideals", *Adv. Math.* **96**:1 (1992), 1–37.

Hironaka [1975] H. Hironaka, "Triangulations of algebraic sets", pp. 165–185 in *Algebraic geometry* (Arcata, CA, 1974), edited by R. Hartshorne, Proc. Sympos. Pure Math. **29**, Amer. Math. Soc., Providence, RI, 1975.

Hirzebruch [1966] F. Hirzebruch, *Topological methods in algebraic geometry*, 3rd ed., Grundlehren der Mathematischen Wissenschaften **131**, Springer, New York, 1966.

Hochster [1973] M. Hochster, "Grassmannians and their Schubert subvarieties are arithmetically Cohen–Macaulay", *J. Algebra* **25** (1973), 40–57.

Hochster [1977] M. Hochster, "The Zariski–Lipman conjecture in the graded case", *J. Algebra* **47**:2 (1977), 411–424.

Hochster and Laksov [1987] M. Hochster and D. Laksov, "The linear syzygies of generic forms", *Comm. Algebra* **15**:1-2 (1987), 227–239.

Hodge [1943] W. V. D. Hodge, "Some enumerative results in the theory of forms", *Proc. Cambridge Philos. Soc.* **39** (1943), 22–30.

Hodge and Pedoe [1952] W. V. D. Hodge and D. Pedoe, *Methods of algebraic geometry, II*, Cambridge University Press, 1952.

Hoyt [1971] W. L. Hoyt, "On the moving lemma for rational equivalence", *J. Indian Math. Soc. (N.S.)* **35** (1971), 47–66.

Hulek [1979] K. Hulek, "Stable rank-2 vector bundles on $\mathbf{P}_2$ with c_1 odd", *Math. Ann.* **242**:3 (1979), 241–266.

Hulek [2012] K. Hulek, *Elementare algebraische Geometrie*, 2nd ed., Springer Spektrum, Wiesbaden, 2012.

Illusie [1972] L. Illusie, *Complexe cotangent et déformations, II*, Lecture Notes in Math. **283**, Springer, Berlin-New York, 1972.

Kempf [1971] G. Kempf, "Schubert methods with an application to algebraic curves", *Publ. Math Centrum* **6** (1971).

Kempf and Laksov [1974] G. Kempf and D. Laksov, "The determinantal formula of Schubert calculus", *Acta Math.* **132** (1974), 153–162.

Kleiman [1974] S. L. Kleiman, "The transversality of a general translate", *Compositio Math.* **28** (1974), 287–297.

Kleiman [1976] S. L. Kleiman, "r-special subschemes and an argument of Severi's", *Advances in Math.* **22**:1 (1976), 1–31.

Kleiman [1984] S. L. Kleiman, "About the conormal scheme", pp. 161–197 in *Complete intersections* (Acireale, 1983), edited by S. Greco and R. Strano, Lecture Notes in Math. **1092**, Springer, Berlin, 1984.

Kleiman [1986] S. L. Kleiman, "Tangency and duality", pp. 163–225 in *Proceedings of the 1984 Vancouver conference in algebraic geometry*, edited by J. Carrell et al., CMS Conf. Proc. **6**, Amer. Math. Soc., Providence, RI, 1986.

Kleiman and Laksov [1972] S. L. Kleiman and D. Laksov, "On the existence of special divisors", *Amer. J. Math.* **94** (1972), 431–436.

Kleiman and Laksov [1974] S. L. Kleiman and D. Laksov, "Another proof of the existence of special divisors", *Acta Math.* **132** (1974), 163–176.

Kleiman and Speiser [1991] S. L. Kleiman and R. Speiser, "Enumerative geometry of nonsingular plane cubics", pp. 85–113 in *Algebraic geometry* (Sundance 1988), edited by B. Harbourne and R. Speiser, Contemp. Math. **116**, Amer. Math. Soc., Providence, RI, 1991.

Laksov [1987] D. Laksov, "Completed quadrics and linear maps", pp. 371–387 in *Algebraic geometry* (Bowdoin, ME, 1985), edited by S. J. Bloch, Proc. Sympos. Pure Math. **46**, Amer. Math. Soc., Providence, RI, 1987.

Landsberg [2012] J. M. Landsberg, *Tensors: geometry and applications*, Graduate Studies in Mathematics **128**, Amer. Math. Soc., Providence, RI, 2012.

Larsen [1973] M. E. Larsen, "On the topology of complex projective manifolds", *Invent. Math.* **19** (1973), 251–260.

Lazarsfeld [1986] R. Lazarsfeld, "Brill–Noether–Petri without degenerations", *J. Differential Geom.* **23**:3 (1986), 299–307.

Lazarsfeld [1994] R. Lazarsfeld, "Lectures on Linear Series", lecture notes, 1994. Available at `http://arxiv.org/abs/alg-geom/9408011`.

Lefschetz [1950] S. Lefschetz, *L'analysis situs et la géométrie algébrique*, Gauthier-Villars, Paris, 1950.

Lipman [1965] J. Lipman, "Free derivation modules on algebraic varieties", *Amer. J. Math.* **87** (1965), 874–898.

Łojasiewicz [1964] S. Łojasiewicz, "Triangulation of semi-analytic sets", *Ann. Scuola Norm. Sup. Pisa* (3) **18** (1964), 449–474.

Manin [1986] Y. I. Manin, *Cubic forms*, 2nd ed., North-Holland Mathematical Library **4**, North-Holland, Amsterdam, 1986.

Maruyama [1983] M. Maruyama, "Singularities of the curve of jumping lines of a vector bundle of rank 2 on $\mathbf{P}^2$", pp. 370–411 in *Algebraic geometry* (Tokyo/Kyoto, 1982), edited by M. Raynaud and T. Shioda, Lecture Notes in Math. **1016**, Springer, Berlin, 1983.

Mather [1971] J. N. Mather, "Stability of C^∞ mappings, VI: The nice dimensions", pp. 207–253 in *Singularities — Symposium I* (Univ. Liverpool, 1969/70), edited by C. T. C. Wall, Lecture Notes in Math. **192**, Springer, Berlin, 1971.

Mather [1973] J. N. Mather, "Generic projections", *Ann. of Math.* (2) **98** (1973), 226–245.

Matsumura [2014] S.-I. Matsumura, "Weak Lefschetz theorems and the topology of zero loci of ample vector bundles", *Comm. Anal. Geom.* **22**:4 (2014), 595–616.

McCrory and Shifrin [1984] C. McCrory and T. Shifrin, "Cusps of the projective Gauss map", *J. Differential Geom.* **19**:1 (1984), 257–276.

Milne [2008] J. S. Milne, "Abelian varieties (v2.00)", notes, 2008. Available at `http://jmilne.org/math/CourseNotes/AV.pdf`.

Milnor [1963] J. Milnor, *Morse theory*, Annals of Mathematics Studies **51**, Princeton University Press, 1963.

Milnor [1965] J. Milnor, *Topology from the differentiable viewpoint*, The University Press of Virginia, Charlottesville, VA, 1965.

Milnor [1997] J. Milnor, *Topology from the differentiable viewpoint*, Princeton University Press, 1997. Revised reprint of the 1965 original.

Morrison [1993] D. R. Morrison, "Mirror symmetry and rational curves on quintic threefolds: a guide for mathematicians", *J. Amer. Math. Soc.* **6**:1 (1993), 223–247.

Mumford [1962] D. Mumford, "Further pathologies in algebraic geometry", *Amer. J. Math.* **84** (1962), 642–648.

Mumford [1966] D. Mumford, *Lectures on curves on an algebraic surface*, Annals of Mathematics Studies **59**, Princeton University Press, 1966.

Mumford [1976] D. Mumford, *Algebraic geometry, I: Complex projective varieties*, Grundlehren der Mathematischen Wissenschaften **221**, Springer, Berlin-New York, 1976.

Mumford [1983] D. Mumford, "Towards an enumerative geometry of the moduli space of curves", pp. 271–328 in *Arithmetic and geometry, II*, edited by M. Artin and J. Tate, Progr. Math. **36**, Birkhäuser, Boston, 1983.

Mumford [2008] D. Mumford, *Abelian varieties*, Tata Institute of Fundamental Research Studies in Mathematics **5**, Hindustan Book Agency, New Delhi, 2008. Corrected reprint of the 1974 edition.

Okonek [1987] C. Okonek, "Barth–Lefschetz theorems for singular spaces", *J. Reine Angew. Math.* **374** (1987), 24–38.

Palais [1965] R. S. Palais (editor), *Seminar on the Atiyah–Singer index theorem*, Annals of Mathematics Studies **57**, Princeton University Press, 1965.

Pardue [1996] K. Pardue, "Deformation classes of graded modules and maximal Betti numbers", *Illinois J. Math.* **40**:4 (1996), 564–585.

Peeva and Stillman [2005] I. Peeva and M. Stillman, "Connectedness of Hilbert schemes", *J. Algebraic Geom.* **14**:2 (2005), 193–211.

Perkinson [1996] D. Perkinson, "Principal parts of line bundles on toric varieties", *Compositio Math.* **104**:1 (1996), 27–39.

Peskine and Szpiro [1974] C. Peskine and L. Szpiro, "Liaison des variétés algébriques, I", *Invent. Math.* **26** (1974), 271–302.

Piene and Schlessinger [1985] R. Piene and M. Schlessinger, "On the Hilbert scheme compactification of the space of twisted cubics", *Amer. J. Math.* **107**:4 (1985), 761–774.

Porteous [1971] I. R. Porteous, "Simple singularities of maps", pp. 286–307 in *Singularities — Symposium I* (Univ. Liverpool, 1969/70), edited by C. T. C. Wall, Lecture Notes in Math. **192**, Springer, Berlin, 1971.

Ran [2005a] Z. Ran, "Geometry on nodal curves", *Compos. Math.* **141**:5 (2005), 1191–1212.

Ran [2005b] Z. Ran, "A note on Hilbert schemes of nodal curves", *J. Algebra* **292**:2 (2005), 429–446.

Re [2012] R. Re, "Principal parts bundles on projective spaces and quiver representations", *Rend. Circ. Mat. Palermo* (2) **61**:2 (2012), 179–198.

Reeves [1995] A. A. Reeves, "The radius of the Hilbert scheme", *J. Algebraic Geom.* **4**:4 (1995), 639–657.

Reid [1988] M. Reid, *Undergraduate algebraic geometry*, London Mathematical Society Student Texts **12**, Cambridge University Press, 1988.

Riedl and Yang [2014] E. Riedl and D. Yang, "Kontsevich spaces of rational curves on Fano hypersurfaces", preprint, 2014. Available at http://arxiv.org/abs/1409.3802.

Roberts [1972a] J. Roberts, "Chow's moving lemma", pp. 89–96 in *Algebraic geometry* (Fifth Nordic Summer School, Oslo, 1970), edited by F. Oort, Wolters-Noordhoff, Groningen, 1972. Appendix 2 to "Motives", by Steven L. Kleiman, pp. 53–82 of the same reference.

Roberts [1972b] J. Roberts, "The variation of singular cycles in an algebraic family of morphisms", *Trans. Amer. Math. Soc.* **168** (1972), 153–164.

Russell [2003] H. Russell, "Counting singular plane curves via Hilbert schemes", *Adv. Math.* **179**:1 (2003), 38–58.

Samuel [1956] P. Samuel, "Rational equivalence of arbitrary cycles", *Amer. J. Math.* **78** (1956), 383–400.

Samuel [1971] P. Samuel, "Séminaire sur l'équivalence rationnelle", pp. 1–17 in *Séminaire sur l'équivalence rationnelle* (Paris-Orsay, 1971), edited by M. Flexor and J.-J. Risler, Publ. Math. Orsay **425**, Dép. Math. Fac. Sci., Univ. Paris, Orsay, 1971.

Schubert [1979] H. Schubert, *Kalkül der abzählenden Geometrie*, Springer, Berlin-New York, 1979. Reprint of the 1879 original.

Segre [1943] B. Segre, "The maximum number of lines lying on a quartic surface", *Quart. J. Math., Oxford Ser.* **14** (1943), 86–96.

Serre [1955] J.-P. Serre, "Faisceaux algébriques cohérents", *Ann. of Math.* (2) **61** (1955), 197–278.

Serre [1955/1956] J.-P. Serre, "Géométrie algébrique et géométrie analytique", *Ann. Inst. Fourier, Grenoble* **6** (1955/1956), 1–42.

Serre [1979] J.-P. Serre, *Local fields*, Graduate Texts in Mathematics **67**, Springer, New York-Berlin, 1979.

Serre [2000] J.-P. Serre, *Local algebra*, Springer, Berlin, 2000.

Seshadri [2007] C. S. Seshadri, *Introduction to the theory of standard monomials*, Texts and Readings in Mathematics **46**, Hindustan Book Agency, New Delhi, 2007. Revised reprint of the 1985 original.

Severi [1933] F. Severi, "Über die grundlagen der algebraischen Geometrie", *Abh. Math. Sem. Univ. Hamburg* **9**:1 (1933), 335–364.

Shafarevich [1994] I. R. Shafarevich, *Basic algebraic geometry, I*, 2nd ed., Springer, Berlin, 1994.

Smith [2000] G. G. Smith, "Computing global extension modules", *J. Symbolic Comput.* **29**:4-5 (2000), 729–746.

Smith et al. [2000] K. E. Smith, L. Kahanpää, P. Kekäläinen, and W. Traves, *An invitation to algebraic geometry*, Springer, New York, 2000.

Srinivas [2010] V. Srinivas, "Algebraic cycles on singular varieties", pp. 603–623 in *Proceedings of the International Congress of Mathematicians, II*, edited by R. Bhatia et al., Hindustan Book Agency, New Delhi, 2010.

Stacks Project [2015] T. Stacks Project, "*Stacks Project*", 2015. Available at http://stacks.math.columbia.edu.

Stanley [1999] R. P. Stanley, *Enumerative combinatorics, II*, Cambridge Studies in Advanced Mathematics **62**, Cambridge University Press, 1999.

Steiner [1848] J. Steiner, "Elementare Lösung einer geometrischen Aufgabe, und über einige damit in Beziehung stehende Eigenschaften der Kegelschnitte", *J. Reine Angew. Math.* **37** (1848), 161–192.

Teissier [1977] B. Teissier, "The hunting of invariants in the geometry of discriminants", pp. 565–678 in *Real and complex singularities* (Ninth Nordic Summer School/NAVF Sympos. Math., Oslo, 1976), edited by P. Holm, Sijthoff & Noordhoff, Alphen aan den Rijn, 1977.

Terracini [1911] A. Terracini, "Sulle V_k per cui la varietà degli $S_h (h + 1)$-seganti ha dimensione minore dell'ordinario", *Rend. Circ. Mat. Palermo* **31** (1911), 392–396.

Totaro [2013] B. Totaro, "On the integral Hodge and Tate conjectures over a number field", *Forum Math. Sigma* **1** (2013), e4, 13 pp.

Totaro [2014] B. Totaro, "Chow groups, Chow cohomology, and linear varieties", *Forum Math. Sigma* **2** (2014), e17, 25 pp.

Vakil [2006a] R. Vakil, "A geometric Littlewood–Richardson rule", *Ann. of Math.* (2) **164**:2 (2006), 371–421.

Vakil [2006b] R. Vakil, "Murphy's law in algebraic geometry: badly-behaved deformation spaces", *Invent. Math.* **164**:3 (2006), 569–590.

Vogel [1984] W. Vogel, *Lectures on results on Bezout's theorem*, Tata Institute of Fundamental Research Lectures on Mathematics and Physics **74**, Springer, Berlin, 1984.

Voloch [2003] J. F. Voloch, "Surfaces in $\mathbf{P}^3$ over finite fields", pp. 219–226 in *Topics in algebraic and noncommutative geometry* (Luminy/Annapolis, MD, 2001), edited by C. G. Melles et al., Contemp. Math. **324**, Amer. Math. Soc., Providence, RI, 2003.

Whitney [1941] H. Whitney, "On the topology of differentiable manifolds", pp. 101–141 in *Lectures in Topology*, edited by R. Wilder and W. Ayres, University of Michigan Press, Ann Arbor, MI, 1941.

Zak [1991] F. L. Zak, "Some properties of dual varieties and their applications in projective geometry", pp. 273–280 in *Algebraic geometry* (Chicago, IL, 1989), edited by S. Bloch et al., Lecture Notes in Math. **1479**, Springer, Berlin, 1991.

Zariski [1982] O. Zariski, "Dimension-theoretic characterization of maximal irreducible algebraic systems of plane nodal curves of a given order n and with a given number d of nodes", *Amer. J. Math.* **104**:1 (1982), 209–226.

Index

Printed in the United States
by Baker & Taylor Publisher Services

Printed in the United States
by Baker & Taylor Publisher Services